# Molecular Stress Physiology of Plants

# Foreword

Department of Life Earth and Environmental Sciences
West Texas A&M University
Canyon, Texas 79015

Dr. Arun Ghosh Ph.D., MAAAAI
Professor of Biology
Department of Life, Earth and
Environmental Sciences, Canyon,
Texas 79015, USA

**Texas, USA**
July 30, 2012

I am glad to foreword *Molecular Stress Physiology of Plants*, a book edited by Dr. G. R. Rout, Professor, and Dr. A. B. Das, Associate Professor, Department of Agricultural Biotechnology, College of Agriculture, Orissa University of Agriculture and Technology, Bhubaneswar, India. This book is divided in various chapters focusing on the effect of abiotic and biotic stresses exerted on plants' growth and its mechanism. *Molecular Stress Physiology of Plants* covers abiotic stresses like light, temperature, salinity, drought, heavy metals, osmotic, and submergence. The effect of growth regulators on plants' growth and molecular mechanisms including photosynthetic machinery has also been widely discussed. This is a much-needed book in this area that covers the topics that are essential to understand the molecular mechanisms that controls the stress physiology of plants. With the declining mangrove populations in the world, the aspects of salt stress genes in mangroves and molecular mechanism of salt tolerance and measurement of chlorophyll fluorescence are the essential topics to evaluate the declining plant populations. Biochemical and physiological adaptations in some halophytes are well documented in these chapters. Measurement of drought and high temperature stress signal in crop plants and its application is well discussed and can be correlated with the other aspects of stress physiology among angiosperm populations all over the world. The use of isothermal calorimetry and Raman spectroscopy to study plant abiotic and biotic stress was unique and added a novel flavor in the reviewed chapter. Crop physiologists would be delighted to read the discussion on the physiology of reproductive stage and abiotic stress tolerance in cereals. Marker-assisted breeding for stress resistance in crops has depicted a novel strategy for crop improvement. The chapters also cover the approach with proteomics to understand the stress tolerance in plants and the role of calcium-mediated CBI-CIPK network in plants' mineral nutrition and abiotic stress that provides high impact on plant growth and metabolism. This book will help expand knowledge of stress physiology and improve understanding the mechanism involved. This book will be worth reading for students and researchers in plant physiology and plant biology.

N. Ghosh
Nabarun Ghosh Ph.D.
Office: ANS, Room #340
West Texas A&M University
PO Box 60808, Canyon
Texas 79016-0001 USA
Phones: (806) 651-2571 (Office)
(806) 651-3870 (Molecular Biology Lab)
Fax: (806) 651-2928
E-mail: nghosh@mail.wtamu.edu

# Preface

Progressive and sustainable rise in food grain production is essential for the country to maintain the level of nutrient supply. Growing population in both developing and developed countries already has alarmed to increased food grain production. Global climate changes exert multiple biotic and abiotic stresses which limit the crop production. The productivity of major staple food crops has reached to a plateau. There is very little scope to increase crop production area too. Abiotic stresses like drought, cold, salinity, and temperature are more emphasized with regard to crop productivity. Heavy metal stress creates the loss of crop production. Biotic stress also deals with the decrease in crop production. The soil reclamation is a costly affair, and it is temporary. Development of crop genotypes tolerant/resistant to the adverse conditions is the only alternative of such a problem. To develop tolerant/resistant genotypes, the plant breeder or plant biotechnologist should have keen knowledge regarding the injury and tolerance mechanisms in plant for specific stress and plant systems to identify the nature of abiotic stress, breeding methods, and modern biotechnological approaches. This book highlighted 17 invited chapters including various stresses like salt, drought, metal, osmotic, oxidative, submergence, temperature, chemical, hormonal, radiation, cold, and nutrient imbalance and its molecular mechanism, and stress mechanism in proteomic approaches. Emphases have been given to include latest development in the field of abiotic stresses with appropriate citations and application. Apart from this, the book also contains molecular mechanism of stress resistance of photosynthetic machinery, PS II fluorescence techniques for measurement of drought and high temperature stress signal in crop plants, isothermal calorimetry and Raman spectroscopy to study response of plants to abiotic and biotic stress, marker-assisted breeding for stress resistance in crop plants, physiology of reproductive abiotic stress tolerances in cereals, role of calcium-mediated CBL-CIPK network in plant mineral nutrition and abiotic stress, and DNA methylation-associated epigenetic changes in stress tolerance of plants. We hope that this book will help the students, researchers, teachers, and plant scientists in the field of basic and applied aspects of agriculture and botany.

We are extremely grateful to the contributors, specialist in the subject and also reviewers for their kind support in time. We are also thankful to all of our teachers for constant encouragement and support in promoting the development of this book on *Molecular Stress Physiology of Plants*. We are thankful to Prof. Prasanna Mohanty, Eminent Plant Physiologist and ex-Dean, School of Life Sciences, JNU and INSA, Sr. Scientist, for the constant encouragement.

Bhubaneswar

Gyana Ranjan Rout
Anath Bandhu Das

# About the Editors

Professor Gyana Ranjan Rout is the Head of Department of Agricultural Biotechnology, Orissa University of Agriculture and Technology, Bhubaneswar, India. Professor Rout has worked in various aspects of plant sciences for over 25 years. He has made significant contributions in the field of plant improvement and propagation of various plant species, heavy metal toxicity and its tolerant mechanism, reclamation of mine through phytoremediation, and DNA fingerprinting. Professor Rout was elected as a Fellow of National Academy of Sciences, India (FNASc), in 1999 in the field of plant biotechnology and molecular biology. He was awarded Samanta Chandra Sekhar Award in 2005 honored by Orissa Bigyan Academy, Govt. of Odisha, for contributions to Life Sciences. Professor Rout was recipient of British Council fellowship, UK; BOYSCAST fellowship by DST, Govt. of India, FAO/IAEA/BADA fellowship, Belgium and DBT Overseas fellowship by Govt. of India. Professor Rout has 25 years of research and teaching experience in the field of plant biotechnology and heavy metal stress mechanism. He has published 175 research papers and 16 review chapters published in national and international peer-reviewed journals and 20 book chapters in contributory volumes. He has also been a principal investigator of 10 major research projects funded by ICAR, DBT, and NMPB.

Dr. Anath Bandhu is an Associate Professor in the Department of Agricultural Biotechnology, Orissa University of Agriculture & Technology, Bhubaneswar, Orissa, India. Dr. Das has worked assiduously in various aspects of plant sciences for over 26 years. He has made significant contributions in the field of cytotaxonomy, cytometry, DNA fingerprinting, and molecular physiology in various groups of medicinal plants, mangroves, cacti, orchids, and sweet gourd. He has reported for the first time chromosome number, karyotype, and genome size in ~350 species of angiosperms, especially on Indian mangroves that underpinned mining of de novo genomic diversity in diploids and polyploids. Molecular phylogeny of mangroves using various DNA markers resolved many discrepancies in taxonomic classifications. He is also working in molecular basis of high salt adaptation of secretor and nonsecretor mangroves to find out salt stress-resistant gene. His work on salt stress on mangroves has generated interest to study these fascinating processes of molecular physiology in other laboratories as evidenced by extensive citation of his work. He has published more than 150 research papers in international journals, 2 books, 10 book chapters, and 10 review articles. Dr. Das is honored with Hira Lal Chakravarty Award of ISCA and Samanta Chandra Sekhar Award by Orissa Bigyan Academy and recipient of DBT Overseas and National Associateships, Govt. of India; RI-LAT Fellowship (UK); and MIF Fellowship (Japan).

# About the Book

## Book Title: Molecular Stress Physiology of Plants

Crop growth and production is dependent on various climatic factors. Both abiotic and biotic stresses have become an integral part of plant growth and development. There are several factors involved in plant stress mechanism. The information in the area of plant growth and molecular mechanism against abiotic and biotic stresses is scattered. The up-to-date information with cited references is provided in this book in an organized way. More emphasis has been given to elaborate the injury and tolerance mechanisms and growth behavior in plants against abiotic and biotic stresses. This book also deals with abiotic and biotic stress tolerance in plants, molecular mechanism of stress resistance of the photosynthetic machinery, stress tolerance in plants (special reference to salt stress – a biochemical and physiological adaptation of some Indian halophytes), PSII fluorescence techniques for measurement of drought and high temperature stress signal in crop plants (protocols and applications), salicylic acid (role in plant physiology and stress tolerance), salinity induced genes and molecular basis of salt tolerance mechanism in mangroves, reproductive stage abiotic stress tolerance in cereals, calorimetry and Raman spectrometry to study response of plant to biotic and abiotic stresses, molecular physiology of osmotic stress in plants and mechanisms, functions and toxicity of heavy metals stress in plants, submergence stress tolerance in plants and adoptive mechanism, Brassinosteroid modulated stress responses under temperature stress, stress tolerance in plants (a proteomics approach), Marker-assisted breeding for stress resistance in crop plants, DNA methylation associated epigenetic changes in stress tolerance of plants and role of calcium-mediated CBL-CIPK network in plant mineral nutrition and abiotic stress. Each chapter has been laid out with an introduction, up-to-date literature, possible stress mechanism and applications. Under abiotic stress, plants produce a large quantity of free radicals, which have been elaborated. We hope that this book will be of greater use for post-graduate students, researchers, physiologists and biotechnologists to sustain plant growth and development.

# Contents

# Contributors

**Suleyman I. Allakhverdiev** Institute of Basic Biological Problems, Russian Academy of Sciences, Pushchino, Moscow Region, Russia

Institute of Plant Physiology, Russian Academy of Sciences, Moscow, Russia

**Marian Brestic** Department of Plant Physiology, Slovak University of Agriculture, Nitra, Slovak Republic

**Nivedita Chakravarty** Bimr College of Professional Studies, Gwalior, Madhya Pradesh, India

**Anath Bandhu Das** Department of Agricultural Biotechnology, Orissa University of Agriculture and Technology, Bhubaneswar, Odisha, India

**Sauren Das** Agricultural and Ecological Research Unit, Indian Statistical Institute, Kolkata, India

**Nirjhar Dasgupta** Agricultural and Ecological Research Unit, Indian Statistical Institute, Kolkata, India

**Allah Ditta** Institute of Soil & Environmental Sciences, University of Agriculture, Faisalabad, Pakistan

**Rudy Dolferus** CSIRO Plant Industry, Canberra, ACT, Australia

**Jane Edlington** CSIRO, division of Plant Industry, Canberra, ACT, Australia

**Irina R. Fomina** Biosphere Systems International Foundation, Tucson, Arizona, USA

Institute of Basic Biological Problems, Russian Academy of Sciences, Pushchino, Moscow Region, Russia

**Xuemei JI** CSIRO, division of Plant Industry, Canberra, ACT, Australia

**Chitta Ranjan Kole** Institute of Nutraceutical Research, Clemson University, Clemson, SC, USA

**Vladimir D. Kreslavski** Institute of Basic Biological Problems, Russian Academy of Sciences, Pushchino, Moscow Region, Russia

**Dmitry A. Los** Institute of Plant Physiology, Russian Academy of Sciences, Moscow, Russia

**Ramya Ranjan Mishra** Plant Biotechnology Laboratory, School of Life Sciences, Sambalpur University, Sambalpur, Odisha, India

**Monalisa Mohanty** Laboratory of Environmental Physiology and Biochemistry, Post Graduate Department of Botany, Utkal University, Bhubaneswar, Odisha, India

**Paramita Nandy** Department of Botany, Barasat Government College, Kolkata, India

**Sandra Oliver** CSIRO, division of Plant Industry, Canberra, ACT, Australia

**Sanjib Kumar Panda** Plant Molecular Biotechnology Laboratory, Department of Life Science and Bioinformatics, Assam University, Silchar, India

**Amita Pandey** Department of Plant Molecular Biology, University of Delhi South Campus, New Delhi, India

**Girdhar K. Pandey** Department of Plant Molecular Biology, University of Delhi South Campus, New Delhi, India

**Jogeswar Panigrahi** Plant Biotechnology Laboratory, School of Life Sciences, Sambalpur University, Sambalpur, Odisha, India

**Nicola Powell** CSIRO, division of Plant Industry, Canberra, ACT, Australia

**Chinmay Pradhan** Laboratory of Plant Biotechnology, Post Graduate Department of Botany, Utkal University, Bhubaneswar, Odisha, India

**Sobha Chandra Rath** Plant Biotechnology Laboratory, School of Life Sciences, Sambalpur University, Sambalpur, Odisha, India

**Rudabe Ravash** Faculty of Agriculture, Shahrekord University, Shahrekord, Iran

**Susana Redondo-Gómez** Departamento de Biología Vegetal y Ecología, Facultad de Biología, Universidad de Sevilla, Sevilla, Spain

**Gyana Ranjan Rout** Department of Agricultural Biotechnology, Orissa University of Agriculture and Technology, Bhubaneswar, Odisha, India

**Lingaraj Sahoo** Department of Biotechnology, Indian Institute of Technology Guwahati, Gauhati, India

**Alok Ranjan Sahu** Plant Biotechnology Laboratory, School of Life Sciences, Sambalpur University, Sambalpur, Odisha, India

**Gopal K. Sahu** School of Life Sciences, MATS University, Raipur, India

**Sunil Kumar Senapati** Department of Biotechnology, Guru Ghasidas Viswa Vidhalaya, Bilaspur, Chatisgarh, India

**Jot Sharma** Bimr College of Professional Studies, Gwalior, Madhya Pradesh, India

**Behrouz Shiran** Faculty of Agriculture, Shahrekord University, Shahrekord, Iran

**Geetika Sirhindi** Department of Botany, Punjabi University, Patiala, Punjab, India

**Andrzej Skoczowski** Department of Plant Physiology, Institute of Biology, Pedagogical University of Cracow, Cracow, Poland

**Reto J. Strasser** Bioenergetics Laboratory, University of Geneva, JUSSY-Geneva, Switzerland

**Indu Tokas** Department of Plant Molecular Biology, University of Delhi South Campus, New Delhi, India

**Magdalena Troć** The Franciszek Górski Institute of Plant Physiology, Polish Academy of Sciences, Cracow, Poland

**Hrishikesh Upadhyaya** Department of Botany and Biotechnology, Karimganj College, Karimganj, Assam, India

**Joost Van Dongen** Max Planck Institute, division of Molecular Plant Physiology, Golm, Germany

**Mahmoud W. Yaish** Department of Biology, College of Science, Sultan Qaboos University, Muscat, Oman

**Marek Zivcak** Department of Plant Physiology, Slovak University of Agriculture, Nitra, Slovak Republic

**Anna A. Zorina** Institute of Plant Physiology, Russian Academy of Sciences, Moscow, Russia

# Abiotic and Biotic Stress Tolerance in Plants

# 1

Susana Redondo-Gómez

**Abstract**

Environmental stresses play crucial roles in the productivity, survival and reproductive biology of plants as well as crops. Plants are subjected to many forms of environmental stress, which can be included into two broad areas: abiotic (physical environment) and biotic (e.g. pathogen, herbivore). However, plants evolve different mechanisms of tolerance to cope with the stress effects. These mechanisms comprise physiological, biochemical, molecular and genetic changes. This chapter represents a general overview of the major mechanisms developed by plants to tolerate environmental stresses, both abiotic (drought, high temperature, chilling and freezing, UV-B radiation, salinity and heavy metals) and biotic (herbivory, pathogen and parasite and allelopathy). Since the length and complexity of the topic is so wide, the effects of the different stresses on plant physiology and biochemistry, as well as the synergies between types of stresses, are beyond the scope of this chapter.

## Introduction

Tolerance of plant refers to its capacity to survive and reproduce under environmental stresses (Simms 2000). Plants are subjected to many forms of environmental stress. Some are abiotic physicochemical, such as drought, cold, heat and high salinity. Other sources of stress are biotic, such as herbivory, disease and allelopathy (Leavitt 1980). Plants alter their physiologies, metabolic mechanisms, gene expressions and developmental activities to cope with the stress effects. Therefore, plants possess unique and sophisticated mechanisms to tolerate stresses (Madhava Rao 2006). However, the degree of tolerance varies from plant to plant, from low to high (Smith et al. 2001). Anyway, the knowledge on the physiology and molecular biology of stress tolerance are certainly helpful to facilitate the biotechnological improvement of crop productivity in the near future.

The common theme of stress is the formation of reactive oxygen species (ROS, see Table 1.1) at cellular and molecular level, strong oxidants that can do significant damage to membrane

S. Redondo-Gómez (✉)
Departamento de Biología Vegetal y Ecología, Facultad de Biología, Universidad de Sevilla, Apartado 1095, 41080 Sevilla, Spain
e-mail: susana@us.es

G.R. Rout and A.B. Das (eds.), *Molecular Stress Physiology of Plants*,
DOI 10.1007/978-81-322-0807-5_1, © Springer India 2013

**Table 1.1** Formation of ROS by different types of stress

| Stress | Reference |
|---|---|
| Drought | Smirnoff (1993) |
| Extremes of temperature | Rao and Dubey (1993) |
| UV radiation | Murphy and Huerta (1990) |
| Heavy metals | Cakmak and Marschner (1988) |
| Air pollutants | Mehlhorn (1990) |
| Mechanical and physical stresses | Legendre et al. (1993) |
| Pathogens | Sutherland (1991) |

**Table 1.2** Signal molecules in plants

| Molecule | Reference |
|---|---|
| Nitric oxide | Shi et al. (2012) |
| Salicylic acid | Catinot et al. (2008) |
| Systemin | Rocha-Granados et al. (2005) |
| Jasmonic acid | Creelman and Mullet (1995) |
| Abscisic acid | Fujita et al. (2006) |
| Ethylene | Ludwig et al. (2005) |

systems and DNA. ROS include superoxide, hydrogen peroxide, and superhydroxide (Scandalios 1993). Antioxidative systems, both enzymatic (superoxide dismutases, catalase, peroxidases, phenol oxidase, and ascorbic acid oxidase) and nonenzymatic systems (compounds that are strong reductants such as glutathione, phenols, flavonoids, and polyamines), play an important role in balancing and preventing oxidative damage (Foyer et al. 1994). Desikan et al. (2001) subjected *Arabidopsis* to oxidative stress and found an increase in genes involved in cell rescue and defence as well as other metabolic functions from the $H_2O_2$ treatment.

Additionally, ROS production, after exposure to biotic or abiotic stresses, has been described to be involved in signalling cascade. ROS are rapidly produced in plants as a defence response to pathogen attack (Bolwell et al. 2002). Furthermore, ROS signalling pathways are closely interwoven with hormone-signalling pathways in plant-insect interaction (Kerchev et al. 2012). Accumulation of the ROS, caused by salinity stress, seems to activate mitogen-activated protein kinase (MAPK) cascade (Kovtun et al. 2000), which is known as signalling module. Glutathione (GSH) and $H_2O_2$ act alone or in unison, in intracellular and systemic signalling systems, to achieve tolerance to abiotic and biotic stresses. Other signal molecules produced in response to both biotic and abiotic stresses appear in Table 1.2.

Stress response and adaptation in plants is complex; there is a whole cascade of genes involved in stress tolerance, starting from stress perception followed by the formation of gene products that are involved in cellular protection and repair (Mantri et al. 2012). One central process of tolerance responses in plants is the activation of defence-related genes in response to biotic and abiotic stresses in their living environment (Xiang 1999). In this way, the signal transduction pathways that detect stress play a crucial role in the induction of stress tolerance in plants (Smalle and Vierstra 2004). Some genes involved in the stress signalling pathway appear in Table 1.3.

Moreover, many drought-inducible genes are also induced by salt stress and cold, which suggests the existence of similar mechanisms of stress responses. Interestingly, there are signalling pathways that are shared during abiotic and biotic stress responses. In a recent study, abscisic acid-induced myb1 (*SlAIM1*) gene from tomato (*Solanum lycopersicum*) encoding an R2R3MYB transcription factor was induced by pathogens, plant hormones, salinity and oxidative stress (Abuqamar et al. 2009). ROS and phytohormone signalling are also essential components of the inter-pathway crosstalk that allows plants to respond to multiple environmental inputs (Fujita et al. 2006). The main objective of this chapter is to present the major mechanisms developed by plants to tolerate environmental stresses, both abiotic and biotic.

**Table 1.3** List of some genes in the stress signalling pathway

| Gene | Function | Reference |
|---|---|---|
| *ATHK1* (*Arabidopsis thaliana*) | Osmosensor | Urao et al. (1999) |
| *etr1*, *ein2* (*Arabidopsis thaliana*) | Ethylene signalling | Larkindale et al. (2005) |
| *AtCBF1* (*Populus* spp.) | Cold response | Benedict et al. (2006) |
| *UVR8* (*Arabidopsis thaliana*) | UV-B response | Brown and Jenkins (2008) |
| *OsCDPK7* (*Oryza sativa*) | Salinity response | Saijo et al. (2000) |
| *OsMAPK2* (*Oryza sativa*) | Cu response | Yeh et al. (2003) |
| *CYP79F1* (*Arabidopsis thaliana*) | Herbivory response | Mewis et al. (2006) |

## Abiotic Stressors

### Drought

The major environmental factor that constrains the productivity and stability of plants is water stress (Araus et al. 2002). Water deficit affects plants on several levels. Cell expansion and growth are among the first processes to decline under water stress. With progressive water deficit, photosynthesis is adversely affected. On the cellular level, membranes and proteins can be damaged by a reduction in hydration and an increase ROS (Artlip and Wisniewski 2001). To overcome this, plants are equipped with various mechanisms to tolerate drought:

- *Reduction in water loss*: Plant leaves close their stomata immediately on sensing an increase in leaf-to-air vapour pressure difference, even if the roots have sufficient water, thereby reducing water loss through transpiration (Assmann et al. 2000). This response is induced by abscisic acid (ABA). The key role of ABA as a plant hormone regulating metabolism and stomatal behaviour under conditions of water stress is well established (Voesenek and Van der Veen 1994). The ABA is synthesized from carotenoid by ABA-synthesizing enzymes induced in root tip cells or parenchyma cells of vascular bundles by drought stress (Koiwai et al. 2004). ABA synthesized in the roots enters the xylem vessels in a free form or as a conjugate with glucose and is transported from here to the leaves (Sauter et al. 2002). Also, expression of the gene encoding abscisic aldehyde oxidase has been revealed in the guard cells of dehydrated *Arabidopsis* leaves (Koiwai et al. 2004). Stomatal closure allows plants to preserve absorbed soil water, to improve water-use efficiency, to avoid damaging water deficits, or a combination of these. During long periods of water stress, stomatal closure is very important to maintain a favourable water balance and thus is an effective means of controlling cuticular water loss (Freitas 1997).
- *Protection of photosynthetic machinery*: As leaf water is lost, the turgor pressure of leaf tissues decreases and leaves begin to wilt. Wilting of the leaves works to protect photosynthetic machinery from direct rays of the sun (Larcher 1995). Moreover, stomatal closure under drought stress restricts the influx of $CO_2$ and hence photosynthesis, depriving plants of their largest consumer of solar energy. Different studies suggest that the cyclical electron flow around photosystem I (PSI) can lead to additional consumption of reducing equivalents and can thus act as an important electron sink for excessive excitation energy (Yokota et al. 2006), as well as an increase in thermal dissipation in the photosystem II (PSII) antennae (Teraza et al. 2003) and photorespiration (Parida and Das 2005). The thermal dissipation in the PSII is detected as non-photochemical quenching (NPQ) (Ma et al. 2003).
- *Osmotic adjustment*: Plant cells are required to maintain turgor pressure and synthesize and accumulate small molecule compounds and ions (fundamentally potassium). The ability of the cited organic molecules to balance ions sequestered in the vacuole and to

stabilize enzymes incubated with salt solutions has resulted in describing these compounds as compatible solutes. Compatible solutes include polyoles (e.g. sorbitol or mannitol), amino acids or amides (e.g. proline), quaternary ammonium compounds (e.g. betaine), and soluble carbohydrates (sugars) (Orcutt and Nilsen 2000). The compounds that accumulate most commonly are proline and glycine betaine (Hasegawa et al. 2000). High concentrations of compatible solutes can increase cellular osmotic pressure (Delauney and Verma 1993). Otherwise, their high hydrophilicity helps to maintain the turgor pressure and water content of cells which protect against water loss from leaves under drought (Yokota et al. 2006).

The plant response to drought is accompanied by the activation of genes involved in the perception of drought stress and in the transmission of the stress signal. There are genes that encode regulatory proteins that further regulate the transduction of the stress signal and modulate gene expression (Waseem et al. 2011). Other genes encode proteins that protect the cells from the effects of desiccation; these include those that govern the accumulation of compatible solutes, passive transport across membranes, energy-requiring water transport systems, and protection of cell structures from desiccation and damage by ROS. Low water status reduces the hydration of biomolecules such as proteins, which lead to denaturation and to disruption of membranes (Steponkus et al. 1993). Dehydrins have been proposed to ameliorate these effects by reducing hydrophobic aggregations or inappropriate interactions (Close 1996).

## High Temperature

Heat stress due to high ambient temperatures is a serious threat to crop production worldwide (Hall 2001). High temperature stress occurs when plants experience temperature above that to which they are adapted and that adaptation depends on the makeup of the proteins and membranes of plants since both are strongly affected by temperature (Sharkey and Schrader 2006). Moreover, thermotolerance refers to the ability of an organism to cope with excessively high temperatures, and the term basal thermotolerance describes the plant response to high temperature in the absence of any period of acclimatization (Penfield 2008). Basal thermotolerance is highly dependent on salicylic acid (SA) action (Clarke et al. 2004). As little as 15 min after exposure to high temperatures, plants begin to acclimatize and the maximum tolerated temperature increases (Kaplan et al. 2004).

Plants exhibit a complex response to extreme high temperatures, including long-term evolutionary phenological and morphological adaptations and short-term avoidance or acclimation mechanisms such as changing leaf orientation, transpirational cooling, or alteration of membrane lipid compositions (Wahid et al. 2007). The sum total of metabolic changes elicited when living cells are subjected to a sudden and transient increase in temperature is referred to as heat shock (HS) response (Singla et al. 1997). Some major tolerance mechanisms, including ion transporters, osmoprotectants, free-radical scavengers, late embryogenesis abundant proteins and factors involved in signalling cascades and transcriptional control are significant to counteract the stress effects (Wang et al. 2004). Immediately after exposure to high temperatures and perception of signals, changes occur at the molecular level altering the expression of genes and accumulation of transcript that leads to the synthesis of stress-related proteins as a stress tolerance strategy (Iba 2002). Expression of heat shock proteins (HSPs) is known to be an important adaptive strategy in this regard (Feder and Hoffman 1999). The tolerance conferred by HSPs results in improved physiological processes such as photosynthesis, assimilate partitioning, water and nutrient use efficiency, and membrane stability (Wahid et al. 2007). HSPs also appear to protect plants against oxidative stress. Otherwise, production of ROS in the organelles is of great significance for signalling as well as production of antioxidants. Key role for ROS has been proposed in acquired thermotolerance (Penfield 2008).

Different mechanisms of plant tolerance to high temperatures are given below (Wahid et al. 2007):

- *Anatomical changes*: These are similar to those under drought stress (see the section "Drought"); there is a general tendency of reduced cell size, closure of stomata and curtailed water loss (Añon et al. 2004).
- *Physiological changes*: Plant water status is the most important variable under high temperatures. In general, plants tend to minimize water loss and synthesize and accumulate compatible solutes and ions (see the section "Drought"). For instance, glycine betaine (GB) plays an important role as a compatible solute in plants under various stresses, such as salinity or high temperature (Sakamoto and Murata 2002). Similarly, accumulation of soluble sugars under heat stress has been found in sugarcane (Wahid and Close 2007). On the other hand, photosynthesis is considered as the physiological process most sensitive to high temperatures. The rate of photosynthesis in most species declines above 35°C (Sage and Reid 1994). Overall, the rate of photosynthesis decreases while dark- and photorespiration increase considerably under high temperatures (Wahid et al. 2007). Both pathways can act as an important electron sink for excessive excitation energy.

  The integrity and functions of biological membranes are sensitive to high temperatures, as heat stress alters the tertiary and quaternary structures of membrane proteins. In the same way, carotenoids of the xanthophyll family and some other terpenoids, such as isoprene, are synthesized by plants in order to stabilize and photoprotect the lipid phase of the thylakoid membranes (Velikova et al. 2005). The resulting interaction of the xanthophyll molecules and the membrane lipids brings about a decreased fluidity (thermostability) of membrane and a lowered susceptibility to lipid peroxidation under high temperatures (Havaux 1998).
- *Molecular changes*: Increased production of HSPs occurs when plants experience either abrupt or gradual increase in temperature (Nakamoto and Hiyama 1999). Immunolocalization studies have determined that HSPs normally associate with particular cellular structures, such as cell wall, chloroplasts, ribosomes and mitochondria (Yang et al. 2006). However, in tomato plants, HSPs aggregate into a granular structure in the cytoplasm under high temperature stress that possibly protects the protein biosynthesis machinery (Miroshnichenko et al. 2005). Other proteins or mRNAs also increase in abundance during elevated temperature but are not considered HSPs. They include several glycolytic enzymes, protein kinases, and ubiquitin (Burke et al. 1988; Lindquist and Craig 1988; Moisyadi and Harrington 1990). Veirling (1991) suggested that glycolytic enzymes and protein kinases are involved in metabolic readjustment. Ubiquitin is probably required to remove aberrant proteins resulting from damage to translational machinery or thermally denatured proteins (Artlip and Wisniewski 2001). Dehydrins are also synthesized in response to heat stress (Wahid and Close 2007).

## Chilling and Freezing

Plants experience chilling stress as a result of temperatures above 0°C and below some threshold temperature unique for each species. Freezing stress occurs at temperatures below 0°C or when radiative frosts occur with ice formation. Chill-sensitive plants comprise many major field crops, such as cotton, soybean, maize and rice. Plants face three major problems when exposed to low temperatures (Vézina et al. 1997) like (1) perturbation of membranes since a fall in temperature is accompanied by a decrease in membrane fluidity manifested by electrolyte leakage from tissues (Barták et al. 1998), (2) slow down of their chemical and biochemical reactions and (3) changes in water status and availability.

Intracellular ice crystals are immediately lethal, as they can pierce the plasma membrane. Plants tolerate only extracellular freezing, which is associated with cell dehydration and cell volume reduction. Osmotic potential of ice is lower than that of water; thus, cell water exits the cell towards the growing ice crystal in the apoplast (Rajashekar 2000). Desiccation via freezing is ameliorated by both biochemical and biophysical changes, particularly in woody plant species (Wisniewski and Arora 1993). Some freeze-tolerant plants can limit the growth of apoplastic ice crystals with proteins and polysaccharides, limiting the extent of protoplast dehydration from ice crystal growth (Breman 2006).

Tolerance to chilling is apparently a prerequisite for tolerance to freezing. Chilling tolerance is an inducible response, dependent on day length and temperature (Gray et al. 1997), and it is accompanied by an increase in the ABA content of cells. Low temperatures also induce numerous proteins or their mRNA, and evidence exists that some of these proteins are necessary for chilling tolerance (Artlip and Wisniewski 2001). For example, some of the heat shock proteins, or their transcripts, are cold inducible (Yacoob and Filion 1987; Guy and Li 1998). These proteins and others (including signalling molecules or transcription factors, metabolic enzymes and many hydrophobic or hydrophilic gene products) have shown the ability to inhibit ice propagation or recrystallization either in vitro or in vivo (Artlip and Wisniewski 2001).

Overall, the major mechanisms of plant tolerance to low temperatures are the following:

1. *Stabilization of membranes*: Perturbation of membranes could be due to phase transitions caused by the presence of minor lipid components in the membrane or, alternatively, failure to seal critical intrinsic membrane proteins into the cell membrane by non-bilayer-forming lipids (Williams 1990). Thus, changes in lipid composition have been proposed to augment the membrane stability against freezing stress (Rajashekar 2000). Yoshida and Uemura (1984) found that freezing tolerance was accompanied by an increase in phospholipids, especially phosphatidyl ethanolamine. Additionally, glutathione has been described to protect membrane protein, preventing protein denaturing. High levels of glutathione also have been found to be correlated with acclimated freeze hardiness in several plant species (Breman 2006).
2. *Cryoprotection*: Soluble sugars and other osmolytes have cryoprotective function. These compounds can protect cell membranes and organelles during freezing (Rajashekar 2000). The main sugars and sugar alcohols that increase are sucrose, glucose, fructose, sorbitol, mannitol, raffinose and stachyose. It is postulated that sugars replace water and decrease the degree of freeze-induced dehydration (Trischuk et al. 2006). Sugars also promote glass transitions that protect cells from desiccation injury (Wolkers et al. 1999). Furthermore, thylakoid membranes are protected from freezing inactivation by exogenous proline, argenine, threonine and lysine. Proline and glycine betaine are both postulated to act as cryoprotectants (Trischuk et al. 2006). Abscisic acid can activate the BADH gene, which encodes for one of the enzymes involved in the synthesis of glycine betaine (Ishitani et al. 1995).

## UV-B Radiation

Increased solar UV-B radiation (280–320 nm), as a consequence of reductions in stratospheric $O_3$, has been shown to cause significant reduction in growth and other physiological responses in many sensitive crops (Kulandaivelu et al. 1997). However, plants have the capacity to develop various mechanisms of protection from the deleterious effects of UV-B radiation:

- *DNA repair*: One of the most important and sensitive targets of UV-B radiation is DNA. UV radiation induces various lesions in DNA, and the best studied are cyclobutane-type pyrimidine dimmers (Stapleton 1992). The dimmers can be repaired via photoreactivation (photolyase), excision repair, or recombinatorial repair (Smith 1989). The former type of repair is the best known, and it has been reported in several species (Artlip and Wisniewski 2001).

- *Accumulation of secondary metabolites*: Flavonoids and/or anthocyanins are induced by UV-B exposure (Beggs et al. 1986). Flavonoids and anthocyanins absorb UV radiation, and they generally accumulate in the epidermis, where they could keep UV radiation from reaching photosynthetic tissues. UV photoreceptors are responsible for the initial perception, and additional photoreceptors are required for anthocyanin or flavonoid biosynthesis in parsley (Stapleton 1992). Flavonoids also possess free-radical scavenging activity (Rice-Evans et al. 1997), which might offer additional protection to cell accumulating these compounds. Polyamines, waxes and specific alkaloids have all been suggested to contribute to UV tolerance. In soybean, a correlation was found between levels of polyamines and tolerance to UV-B radiation (Kramer et al. 1992). In *Dudleya*, accumulation of glaucescence, a powdery wax, increases reflection of UV-B to a larger extent than that of photosynthetically active radiation (PAR) (Mulroy 1979). Levels of UV-absorbing tetrahydrocannabinol increase linearly with UV-B dose in *Cannabis* (Lyddon et al. 1987).
- *Morphogenic changes*: Leaf curling is a photomorphogenic response observable at low fluences of UV-B that helps diminish the leaf area exposed to UV. A protective function has also been hypothesized for leaf or epidermal thickening (Jansen et al. 1998). In pea, leaf thickening is accompanied by a redistribution of chlorophyll away from the adaxial surface (Day and Vogelmann 1995).
- *Photosynthetic machinery repair*: D1 and D2 proteins form the core of PSII, and a very sensitive UV-B response is the rapid light-driven degradation of these two proteins. In this way, Jansen et al. (1998) suggested UV-B driven D1–D2 turnover is also part of a repair cycle, preventing accumulation of UV-inactivated PSII.

## Salinity

Although salinity stress is related to water deficit by a decrease in water status, the presence of excess ions also appears to be detrimental to many plant processes. Thus, plants subjected to salinity stress appear to face two stresses at the same time. Based on general tolerance to salt stress, all plants can be roughly divided into two major groups: halophytes that can withstand even 20% of salts in the soil and non-halophytes or glycophytes that exhibit limited growth in the presence of sodium salts (usually higher than 0.01%). However, there are great differences in the level of salt tolerance within halophytes, which include eu-halophytes and facultative halophytes (Dajic 2006). Eu-halophytes show stimulation of productivity at moderate salinity (e.g. *Sarcocornia fruticosa*, *Arthrocnemum macrostachyum*; Redondo-Gómez et al. 2006; 2010a), while facultative halophytes show a slight growth enhancement at low salinity (e.g. *Plantago maritima*, *Aster tripolium*; Dajic 2006). Table 1.4 shows the salinity in which different halophytes have their optimal growth.

A few agricultural crops have moderate salt tolerance (e.g. barley, cotton, sugar beet, wheat, tomato, corn, rice, bean, beetle grass), and there can be a wide variation in salt tolerance among varieties or genetic lines of one crop species (Orcutt and Nilsen 2000).

**Table 1.4** Optimal salinities for different halophytes

| Species | Salinity (mM NaCl) | Reference |
|---|---|---|
| *Arthrocnemum macrostachyum* | 171–510 | Redondo-Gómez et al. (2010a) |
| *Atriplex portulacoides* | 200 | Redondo-Gómez et al. (2007) |
| *Sarcocornia fruticosa* | 510 | Redondo-Gómez et al. (2006) |
| *Suaeda fruticosa* | 200–600 | Khan et al. (2000) |
| *Suaeda salsa* | 200 | Lu et al. (2002) |
| *Suaeda splendens* | 200–400 | Redondo-Gómez et al. (2008) |

Overall, mechanisms of salt tolerance are of two main types: those minimizing the entry of salt into the plant (or at least their accumulation in photosynthetic tissues) and those minimizing the concentration of salt in the cytoplasm (Munns 2002). This corresponds with two major adaptive strategies of plants to tolerate high salinity: *stress avoidance*, related to different physical, physiological and/or metabolic barriers with which the negative effects of stress are ameliorated, and *stress tolerance*, the linkage of adaptive mechanisms which enable successful survival despite the effects of stress internally (Dajic 2006).

Mechanisms of salt tolerance:

- *Salt exclusion*: Plants can limit salt accumulation in its tissues by inhibition of root uptake. However, in most plants such a mechanism is not efficient. Therefore, strategies have evolved to restrict salt transport into sensitive organs or tissues (Munns et al. 2002). Salt tolerance in beans (Awada et al. 1995), wheat or barley (Gorham 1993) is associated with $Na^+$ exclusion. Additionally, the presence of potassium (and calcium) ions has been shown to decrease $Na^+$ influx into plant cells. Salt tolerance in bread wheat and *Triticum turgidum* is associated with low rates of sodium transport to the shoots and high $K^+/Na^+$ discrimination (Gorham 1990; Munns et al. 2000). It has recently been reported that the genus *Triticum* expresses a range of genetic variation related to $K^+/Na^+$ discrimination (Munns et al. 2002). Salt tolerance in barley has been also associated with the ability to retain $K^+$ at elevated salinity (Chen et al. 2007).

  Sodium exclusion is accomplished by $H^+$-ATPase pumps and $Na^+/H^+$ antiporters. The $H^+$-ATPase complex creates the membrane electrical potential and provides the energy base for $Na^+/H^+$ antiporters. In fact, it has been suggested that the response of the $H^+$-ATPase genes to salinity may be a good indication of salinity tolerance in plants (Perez-Prat et al. 1994). According to Munns et al. (2002), the ability of plants to regulate the uptake and transport of salts is dependent on the following mechanisms: selectivity of uptake by root cells; preferential loading of $K^+$ rather than $Na^+$ into the xylem by the cells of the stele; removal of salts from the xylem in the upper parts of roots, the stem and leaf sheaths, based upon exchange of $K^+$ and $Na^+$; and loading of the phloem.
- *Salt excretion*: Halophytes frequently have anatomical structures designed for eliminating of excess salt ions from the plant into its environment. Salt glands and salt bladders are the main salt-excluding structures identified in plants. Both structures derive from epidermal tissue and have similar physiological function. Salt glands are embedded in the surface of leaves and salt bladders are specialized trichomes. The former are characteristics for, for example, *Avicennia* spp. mangroves (Griffiths et al. 2008) and the latter for members of the family Chenopodiaceae (Hagemeyer 1997; Orcutt and Nilsen 2000).
- *Intracellular ion compartmentation*: Sequestration of salts into leaf and/or shoot vacuoles is typical attribute of dicotyledonous halophytes. This accumulation is dependent on vacuolar $H^+$-translocating enzymes and tonoplast $Na^+/H^+$ antiporters, which are induced by saline environment (Barkla and Pantoja 1996). An immediate effect of salt stress is vacuolar alkalization, linked with $Na^+/H^+$ antiporters activity of tonoplast vesicles (Hasegawa et al. 2000). In this case, potassium ions and compatible solutes (see the section "Drought") should be accumulated in the cytoplasm in order to prevent dehydration and maintain the osmotic and ionic balance between these two compartments (Munns 2002). The effective capacity of halophytes to accumulate and utilize ions for osmotic adjustment to maintain turgor might explain their enhanced growth and control of their water regime in saline conditions (Dajic 2006). Succulence results from increased water uptake of the tissues, which may help to dilute absorbed salt ions (Munns et al. 1983). It is associated with the ability of intracellular compartmentation, to provide a larger capacity (volume of vacuoles) for salt storage.

*Sarcocornia fruticosa* (Chenopodiaceae) showed different photosynthetic areas with salinity treatments, which was a growth response (an increase in diameter of photosynthetic portions) mediated by an increase in turgor pressure (Redondo-Gómez et al. 2006). Otherwise, the *SOS1* (salt overly sensitive) locus has been described to be essential for $Na^+$ and $K^+$ homeostasis in *Arabidopsis*, as well as for the control of the long-distance $Na^+$ transport and loading $Na^+$ into the xylem under severe and mild salt stress, respectively (Shi et al. 2002). The *SOS2* gene is required for intracellular ion homeostasis (Liu et al. 2000).

## Heavy Metals

Heavy metal pollution is gaining in importance day by day due to its obvious impact on human health through the food chain. Several vegetable, fruit and cereal crops are reported to accumulate heavy metals (Pieczonka and Rosopulo 1985; Mejuto-Marti et al. 1988; Prasad 1997). Heavy metal pollution is increasing in the environment due to industrial and agricultural activities such as mining and smelting of metalliferous ores, wastewater irrigation, and abuse of chemical fertilizers and pesticides (Redondo-Gómez et al. 2010a, b). Certain phosphate fertilizers have been found to contain high levels of cadmium and other trace metals (Roberts et al. 1994).

An excess of an essential or nonessential element can have detrimental effects on plant growth and development. The survival of plants growing on contaminated soils is considered to be the result of tolerance rather than avoidance since no plant has the ability to prevent metal uptake but can only restrict it (Baker 1981). However, plants differ among species with respect to the concentration of a specific metal they can tolerate. Plants growing on soils contaminated with high levels of metals are referred to as metallophytes, which have developed three basic strategies for growing in metalliferous soils (Baker 1981): (1) excluders, which prevent metal from entering their aerial parts over a broad range of metal concentrations in the soil; (2) indicators, which take up metals at a linear rate relative to the concentration of metal in the soil; and (3) accumulators, which allow the uptake of very high levels of ions to the extent of exceeding the levels in the soil. Plants with enhanced tolerance versus pollutants are a promising tool in efficient bioremediation of areas contaminated with heavy metals.

Overall, metal tolerance mechanisms in plants include (Tomsett and Thurman 1988; Prasad 1997) the following:

- *Compartmentation*: There is some relationship between tolerance and accumulation characteristics in higher plants (Kuboi et al. 1987). The sequestering of metals in tissues (cell walls of roots and leaves) or cellular compartments (vacuoles), which are less sensitive to metals, away from metabolically active compartments (cytosol, mitochondria or chloroplast), has been described as a tolerance mechanism (Weis and Weis 2004). Manganese was accumulated in the cell walls of epidermis, collenchyma, bundle sheath cells and in a vacuolar compartment in the petioles of *Acanthopanax sciadophylloides* and in the leaves of tea plant (*Thea sinensis*) (Memon et al. 1980, 1981). Furthermore, Memon and Yatazawa (1984) explained that Mn was chelated with oxalic acid in a vacuolar compartment. With excess of Ni, vacuolization was observed in leaf mesophyll cells of *Brassica oleracea* (Molas 1997). Also, cereals are reported to accumulate trace metals (Piotrowska and Dudka 1994; Rivai et al. 1990). The mechanism involved in this preferential accumulation is not known.
- *Metal excretion*: Excretion is one of the important mechanisms of heavy metal tolerance. It has been shown that metals can be excreted in salt crystals released through salt glands of some halophytes (Krauss 1988). Redondo-Gómez et al. (2011) suggested that salinity could increase metal excretion and favour the tolerance and recovery of the photosynthetic apparatus of *Spartina densiflora* to the toxic action of zinc. Tobacco plants actively exclude Cd by forming and excreting

Cd/Ca-containing crystals through the head cells of trichomes (Choi et al. 2001).

- *Chelation*: A major factor governing the toxicity of a metal in soil is its bioavailability. Thus, to avoid undesirable metal penetration, plants are able to extrude material that can chelate free metallic cations in the extracellular space. Toxic metals can also be trapped once they are inside the cells (Bertrand et al. 2001).
- *Extracellular metal sequestration*: Differences in Al tolerance between several bean species have been attributed to the capacity of roots to exude citric acid, a strong Al chelator. Similar results were found for monocotyledons (barley, wheat, maize) for which better resistance to Al toxicity is associated with root exudation of citric acid, succinic acid and other organic acids (Bertrand et al. 2001). Root secretion includes organic ligands (e.g. carbohydrates, organic acids, nucleic acids) and inorganic ligands (e.g. $Cl^-$, $SO_4^{2-}$, $NH_4^+$, $CO_3^{2-}$). These substances functions as ligands to be chelated with heavy metals ions (Dong et al. 2007). Malate, citrate and oxalate are carboxylates exuded in the rhizosphere and implicated in the complexation of metals (Hinsinger 2001). Rice plant secretes phytosiderophores (amino acids) that can form much more stable complexes than carboxylates with Cd, Cu, Fe and Zn (Xu et al. 2005; Dong et al. 2007). Cambrollé et al. (2008) found that heavy metals combined as complex with oxides of Fe and Mn was accumulated in *Spartina densiflora* and *S. maritima* rhizospheres much more than in non-rhizosphere soil.
- *Intracellular metal sequestration*: Metallothioneins (MTs) and phytochelatins (PCs) are two protein families capable of sequestering metals. They are both cysteine-rich polypeptides having the ability to form metal-thiolate clusters (Cobbett and Goldsbrough 2002). MTs have characterized as gene-encoded proteins, whereas PCs are smaller enzymatically synthesized polypeptides (molecular weights of 5,000–20,000 and 500–2,300 Da, respectively). Although the precise physiological function of MTs has not yet been fully elucidated, expression and regulation of *Arabidopsis* MT genes have revealed that MTs have distinct functions in metal homeostasis, especially for Cu (Guo et al. 2003; Gasic and Korban 2006).

PCs are enzymatically synthesized directly from glutathione in the presence of metal ions by the enzyme PC synthase (EC.2.3.2.15) (Grill et al. 1989), and they are considered to have an important role in the cellular metal homeostasis (Steffens 1990). PCs also protect plant enzymes from heavy metal toxicity, and metal-requiring apoenzymes have been reactivated by PCs (Kneer and Zenk 1992). PC synthase genes in wheat (*TaPCS1*) (Clemens et al. 1999) and in *Arabidopsis* (*AtPCS1*) (Lee and Korban 2002) are regulated at the transcriptional level. However, transcriptional regulation of *AtPCS1* in *Arabidopsis* disappears as plants grow older (Lee and Korban 2002). Otherwise, under heavy metal stress, a high cysteine biosynthesis rate is required for the synthesis of GSH and PCs. *O*-acetylserine(thiol)-lyase (OASTL) is a key enzyme of a plant sulphur metabolism that catalyses the formation of Cys which serves as a precursor for GSH (Gasic and Korban 2006). Arabidopsis OASTL gene (*Atcys-3A*) has been described to be involved in cadmium tolerance (Domingues-Solis et al. 2001).

## Biotic Stressors

### Herbivory

Most of what we know about the mechanisms of herbivory tolerance in plants derives from studies of plant responses to mammalian (McNaughton 1983; Belsky et al. 1993; Lennartsson et al. 1998; Tiffin 2000). Empirical evidence indicates that relatively low levels of damage can be completely compensated by plants in terms of fitness. Further increments in the intensity of damage result in a decreasing ability to maintain complete tolerance (Chapin and McNaughton 1989; Fornoni and Núñez-Farfán 2000; Fornoni et al. 2003).

Several plant traits that may buffer losses of fitness have been considered as components of tolerance (Strauss and Agrawal 1999; Fornoni et al. 2003):

- *Photosynthetic enhancement*: Partial defoliation may result in an increased supply of leaf cytokinins or root-derived cytokinins, which have been shown to increase net $CO_2$ fixation as a result of enhanced assimilate transport and nutrient uptake (Trumble et al. 1993). Furthermore, increases in production of the carboxylating enzyme ribulose bisphosphate carboxylase and chlorophyll contents in the remaining leaf tissue have been described, which might increase photosynthetic activity (Thorne and Koller 1974; Satoh et al. 1977).
- *High relative growth rates*: New leaves may become larger due to increase in cell size via mesophyll elongation or increased cell division (Satoh et al. 1977). Cell expansion may be influenced by chloroplast enlargement due to starch accumulation or via increased turgor pressure (Milthorpe and Moorby 1979).
- *Activation of dormant meristems*: When dominant apical meristems are removed, as the new growth tissue tends to be more succulent than older (lignified foliage), an increase in meristematic activity at nonapical locations can lead to increased branching (Trumble et al. 1993). Simulated herbivory in *Convolvulus chilensis* significantly affected plant architecture; there was an increase in number of stems/plant height (González-Teuber and Gianoli 2007).
- *Reallocation of stored resources*: Induced reallocation can occur rapidly while the herbivore is still present in order to safeguard resources, sequestering the primary metabolites from exposed tissues into storage organs (Orians et al. 2011). Also, resources can be mobilized from storage organs to growing tissues after the herbivory threat is over (Steinbrenner et al. 2011). Smith et al. (1990) found that amino acid concentrations in extrafloral nectaries increased following simulated herbivory on *Impatiens sultani*. Root feeding by the scarabaeid *Phyllopertha horticola* caused reallocation of resources to reproductive growth in an annual herb (Gange and Brown 1989).
- *Delay in senescence*: Increased supply of root-derived cytokinins as consequence of partial defoliation, which inhibit mRNA, suppress protein and enzyme degradation, increase stomatal opening, and maintain cell membrane integrity (Waring et al. 1968; Trumble et al. 1993). Studies suggest that plant architecture like branching ability and the activation of secondary meristems are the most important traits that allow plants to regrow and tolerate grazing (McNaughton 1983; Martínez Moreno et al. 1999). In contrast, plant usually respond biochemically and physiologically at the leaf or branch level when damaged by small herbivores or pathogens (Welter 1989; Marquis 1996). Traits that confer tolerance are controlled genetically and therefore are heritable traits under selection (Strauss and Agrawal 1999).

## Pathogens and Parasites

Host plants have evolved defence mechanisms (i.e. resistance and/or tolerance) against pathogen and parasites attacks. Tolerance is defined as the ability to compensate in part for fitness decrements caused by pathogens or parasites, while resistance refers to traits that prevent infection or limit its extent (Boots and Bowers 1999). This allows infected host to live longer, which increases the infectious period and therefore increases rather than decreases pathogen prevalence, leading to a positive feedback (Best et al. 2008).

Clarke (1986) proposed the partitioning of tolerance into three types, namely, (1) tolerance to the parasite, the ability of a plant to endure the effects of levels of parasite infection; (2) tolerance to disease, the ability of a plant to endure the effects of levels of disease (i.e. host physiological damage); and (3) overall tolerance, the ability of a plant to endure the levels of parasite infection and disease.

Strategies that limit the extent of disease in an infected host (i.e. barriers to infection, immune response and rapid cell death in the immediate region around the wound) are sometimes interpreted as helping the host tolerate infection, but these are normally termed resistance strategies (Clarke 1986) because they combat the pathogen by limiting its spread (Roy and Kirchner 2000).

Tolerance often involves some degree of compensation for disease damage, for example (Roy and Kirchner 2000), namely, (1) *Photosynthetic enhancement*: Plants can tolerate infection by increasing the chlorophyll concentration in leaves, which might increase photosynthetic activity. (2) *Growth enhancement*: Plants can increase the size of new leaves or the number of new branches in response to a pathogen or parasite attack. (3) *Advancing the timing of bud break*: The advancement of bud break can increase the available photosynthetic area of the plant, and thus the net $CO_2$ fixation. (4) *Delaying the senescence of infected tissue*: The onset of senescence may be delayed due to increased levels of cytokinins (Trumble et al. 1993). (5) *Increasing nutrient uptake*: An enhancement of phosphorous uptake by a pathogen may increase plant development (Wehner et al. 2010).

These mechanisms of plant tolerance are similar to those developed against herbivory (Marquis 1992; Rosenthal and Welter 1995; Strauss and Agrawal 1999). However, in contrast to pathogen attack, herbivore attack is frequently associated with wounding, and the recognition of herbivore attack frequently involves modifications of a plant's wound response (Walling 2000). Blouin et al. (2005) demonstrated that the presence of belowground invertebrate activities improved tolerance of rice to parasitic nematodes, increasing plant biomass and photosynthetic activity. This response was mediated by the expression of three stress-responsive genes, coding for lipoxygenase, phospholipase D and cysteine protease. The host plant response to parasite attack is also associated with the formation of pathogenesis-related proteins (PRs) throughout the plant. The PRs occurring in the inter- and intracellular spaces are quite soluble, highly reactive and induced by signalling compounds such as salicylic acid, jasmonic acid, ethylene, xylanase and polypeptides (Orcutt and Nilsen 2000). Furthermore, the PR genes were found to be negatively modulated by an abscisic acid-inducible mitogen-activated protein kinase (MAPK) gene (*OsMAPK5*) in rice (Xiong and Yang 2003).

## Allelopathy

*Allelopathy* refers to any direct or indirect harm of one plant (or microorganism) on the germination, growth or development of other plants through the production of chemicals (*allelochemicals*) deposited into the environment (Molisch 1937). It differs from competition wherein plants compete for a common resource. Allelopathic interactions between plants have been implicated in the pattering of vegetation and weed growth in agricultural systems and in inhibition of growth of several crops (Rice 1987; Liu and Lovette 1993; Devi et al. 1997).

The allelochemicals (see Table 1.5) concerned in higher plants interaction are typical secondary metabolites and appear to be mainly low-molecular-weight compounds of relatively simple structure. Most allelochemicals that have been positively identified are either volatile terpenes or phenolic compounds (Harborne 1993). The standard modes of release for allelochemicals are volatilization, residue decay, leaching or root exudation (Devi et al. 1997).

Tolerance of allelopathic compounds in plants could be due to a number of different processes like (1) *Exclusion*: The capacity to exclude allelochemicals at the root or leaf surface is due to morphological characteristics of the organ surface. Thus, the permeability of phenolic compounds into leaves varies depending on the lipid composition of the cuticle (Shafer and Schönherr 1985). (2) *Compartmentation*: Some tolerance to allelochemicals can be attributed to the ability to deposit these compounds in nonmetabolic compartments, such as vacuole or cell wall. Esculin and scopolin are accumulated in wheat vacuoles (Werner and Maitile 1985). (3) *Excretion*: The excretion of toxins that have been absorbed from

**Table 1.5** List of allelochemical compounds

| Allelochemical | Chemical structure | Reference |
|---|---|---|
| Glucosinolates | 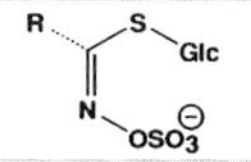 | Fahey et al. (2001) |
| Phenolic compounds | 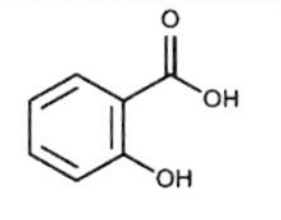 Salicylic acid | Einhellig (2004) |
| Terpenoids | 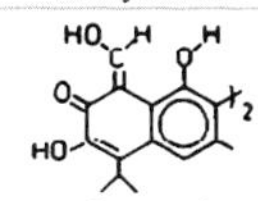 Gossypol | Fischer (1986) |
| Alkaloids | 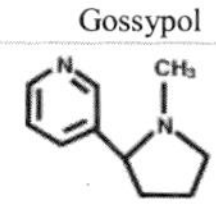 Nicotine | Wink (1998) |
| Hydroxamic acid | 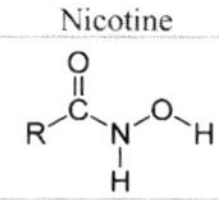 | Sicker and Schulz (2002) |
| Flavonoids | 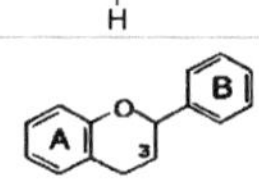 Isoflavone | Rice (1984) |
| Quinones | 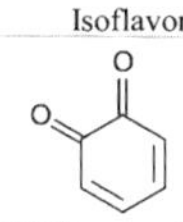 1,2-Benzoquinone | Thiboldeaux et al. (1994) |
| Polyacetylenes | 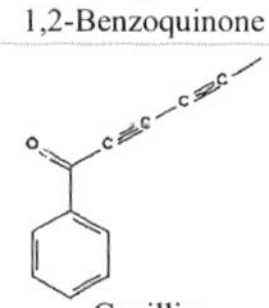 Capillin | Towers and Wat (1978) |

the environment is most likely to occur in plants that already excrete allelochemicals. *Urtica* species that exude toxins from glandular hairs are likely to be able to exclude toxins accumulated from extrinsic sources (Lei 2000). (4) *Detoxification*: Plants have acquired a variety of mechanisms that are involved in the detoxification of allelopathic compounds. These include three phases (Devi et al. 1997; Lei 2000):

- Phase I, a metabolic attack of the allelochemical. This usually involves increasing the polarity of the compound by hydroxylation and dealkylation.
- Phase II, conjugation with sugars, amino acids, or malonic acid, which increases the water solubility of the compound. The most important conjugation mechanisms in plants include glucosylation, which involves the conjugation of toxins with glucose mediated by an enzyme glucosyltransferase.
- Phase III, translocation of the compound to a storage location or excretion onto the plant surface.

For instance, two glucosyltransferases (BX8 and BX9) are involved in detoxification of benzoxazinoids in maize, which are stored in the vacuoles as glucosides (von Rad et al. 2001). Baerson et al. (2005) performed transcriptional profiling experiments with *Arabidopsis* seedlings exposed to Benzoxazolin-2(3*H*)-one (BOA) concentrations, an allelochemical most commonly associated with monocot species. One of the largest functional categories observed for BOA-responsive genes corresponded to protein families known to participate in cell rescue and defence, with the major of these genes potentially associated with chemical detoxification pathways.

## Future Perspectives

This chapter dealt with plant responses to biotic and abiotic stresses, which include a wide range of effects at the molecular, cellular, tissue and whole-plant level. However, the mechanisms for sensing environmental stimuli and transducing signals between organs and cells require further elucidation. Since the various mechanisms and adaptive responses of plant to biotic and abiotic stresses are multigenic traits, further efforts are necessary to comprehend the gene expression for groups of functionally related genes; it is important to gather information on what are the 'useful genes' responsible for better stress tolerance (Grover et al. 1998). Further insights are also expected from genetic engineering technologies in order to improve stress tolerance in crops.

**Acknowledgements** I am grateful to the Spanish Science and Technology Ministry, Govt. of Spain for its financial support (project CTM2008-04453).

## References

Abuqamar S, Luo H, Laluk K, Mickelbart MV, Mengiste T (2009) Crosstalk between biotic and abiotic stress responses in tomato is mediated by the AIM1 transcription factor. Plant J 58:347–360

Añon S, Fernandez JA, Franco JA, Torrecillas A, Alarcón JJ, Sánchez-Blanco MJ (2004) Effects of water stress and night temperature preconditioning on water relations and morphological and anatomical changes of *Lotus creticus* plants. Sci Hortic 101:333–342

Araus JL, Slafer GA, Reynolds MP, Royo C (2002) Plant breeding and relations in $C_3$ cereals: what to breed for? Ann Bot 89:925–940

Artlip TS, Wisniewski ME (2001) Induction of proteins in response to biotic and abiotic stresses. In: Pessarakli M (ed) Handbook of plant and crop physiology. Marcel Dekker, New York

Assmann SM, Snyder JA, Lee YRJ (2000) ABA-deficient (*aba1*) and ABA-insensitive (*abi1-1*, *abi2-1*) mutants of Arabidopsis have a wild-type stomatal response to humidity. Plant Cell Environ 23:387–395

Awada S, Campbell WF, Dudley M, Jurinak JJ (1995) Interactive effects of sodium chloride, sodium sulphate, calcium sulphate, and calcium chloride on snapbean growth, photosynthesis and ion uptake. J Plant Nutr 18:889–900

Baerson SR, Sánchez-Moreiras A, Pedrol-Bonjoch N, Schulz M, Kagan IA, Agarwal AK, Reigosa MJ, Duke SO (2005) Detoxification and transcriptome response in *Arabidopsis* seedlings exposed to the allelochemical benzoxazolin-2(3 *H*)-one. J Biol Chem 280:21867–21881

Baker AJM (1981) Accumulators and excluders – strategies in the response of plants to heavy metals. J Plant Nutr 3:643–654

Barkla BJ, Pantoja O (1996) Physiology of ion transport across the tonoplast of higher plants. Annu Rev Plant Physiol Mol Biol 47:159–184

Barták M, Nijs I, Impens I (1998) The susceptibility of PS II of *Lolium perenne* to a sudden fall in air temperature – response of plants grown in elevated $CO_2$ and/or increased air temperature. Environ Exp Bot 39:85–95

Beggs CJ, Schneider-Ziebert U, Wellmann E (1986) UV-B radiation and adaptive mechanisms in plants. In: Worrest RC, Caldwell MM (eds) Stratopheric ozone reduction, solar ultraviolet radiation and plant life, vol 8, NATO ASI series. Springer, Berlin

Belsky AJ, Carson WP, Jensen CL, Fox GA (1993) Over compensation by plants: herbivore optimization or red herring? Evol Ecol 7:109–121

Benedict C, Skinner JS, Meng R, Chang Y, Bhalerao R, Huner NP, Finn CE, Chen TH, Hurry V (2006) The CBF1-dependent low temperature signalling pathway, regulon and increase in freeze tolerance are conserved in *Populus* spp. Plant Cell Environ 29:1259–1272

Bertrand M, Guary JC, Schoefs B (2001) How plants adapt their physiology to an excess of metals. In: Pessarakli M (ed) Handbook of plant and crop physiology. Marcel Dekker, New York, pp 751–762

Best A, White A, Boots M (2008) Maintenance of host variation in tolerance to pathogens and parasites. PNAS 105:20786–20791

Blouin M, Zuily-Fodil Y, Pham-Thi AT, Laffray D, Reversat G, Pando A, Tondoh J, Lavelle P (2005) Belowground organism activities affect plant aboveground phenotype, inducing plant tolerance to parasites. Ecol Lett 8:202–208

Bolwell GP, Bindschedler LV, Blee KA, Butt VS, Davies DR, Gardner SL, Gerrish C, Minibayeva F (2002) The apoplastic oxidative burst in response to biotic stress in plants: a three-component system. J Exp Bot 53:1367–1376

Boots M, Bowers G (1999) Three mechanisms of host resistance to microparasites – avoidance, recovery and tolerance – show different evolutionary dynamics. J Theor Biol 201:13–23

Breman JW (2006) Leaf-tissue freeze-tolerance mechanisms in bahiagrass (*Paspalum notatum* Flüegge). University of Florida, Florida, pp 1–23

Brown BA, Jenkins GI (2008) UV-B signaling pathways with different fluence-rate response profiles are distinguished in mature *Arabidopsis* leaf tissue by requirement for UVR8, HY5, and HYH. Plant Physiol 146:576–588

Burke TJ, Callis J, Vierstra RD (1988) Characterisation of a polyubiquitin gene from *Arabidopsis thaliana*. Mol Gen Genet 213:435–443

Cakmak I, Marschner H (1988) Enhanced superoxide radical production in roots of zinc-deficient plants. J Exp Bot 39:1449–1460

Cambrollé J, Redondo-Gómez S, Mateos-Naranjo E, Figueroa ME (2008) Comparison of the role of two *Spartina* species in terms of phytostabilization and bioaccumulation of metals in the esturine sediment. Mar Pollut Bull 56:2037–2042

Catinot J, Buchala A, Abou-Mansour E, Métraux JP (2008) Salicylic acid production in response to biotic and abiotic stress depends on isochorismate in *Nicotiana benthamiana*. FEBS Lett 582:473–478

Chapin FS III, McNaughton SJ (1989) Lack of compensatory growth under phosphorus deficiency in grazing-adapted grasses from the Serengeti plains. Oecologia 79:551–557

Chen ZH, Zhou MX, Mewman IA, Mendham NJ, Zhang GP, Shabala S (2007) Potassium and sodium relations in salinised barley tissues as a basis of differential salt tolerance. Funct Plant Biol 34:150–162

Choi YE, Harada E, Wada M, Tsuboi H, Morita Y, Kusano T, Sano H (2001) Detoxification of cadmium in tobacco plants: formation and active excretion of crystals containing cadmium and calcium through trichomes. Planta 213:45–50

Clarke DD (1986) Tolerance of parasites and disease in plants and its significance in host-parasite interactions. In: Ingram DS, Williams PH (eds) Advances in plant pathology, vol 5. Academic, London, pp 161–198

Clarke SM, Mur LA, Wood JE, Scott IM (2004) Salicylic acid dependent signalling promotes basal thermotolerance but is not essential for acquired thermotolerance in *Arabidopsis thaliana*. Plant J 38:432–447

Clemens S, Kim EJ, Neumann D, Schroeder JJ (1999) Tolerance to toxic metals by a gene family of phytochelatin syntheses from plants and yeast. EMBO J 18:3325–3333

Close TJ (1996) Dehydrins: emergence of a biochemical role of a family of plant dehydration proteins. Physiol Plantarum 97:795–803

Cobbett CS, Goldsbrough P (2002) Phytochelatins and metallothioneins: roles in heavy metal detoxification and homeostasis. Annu Rev Plant Biol 53:159–182

Creelman RA, Mullet JE (1995) Jasmonic acid distribution and action in plants: regulation during development and response to biotic and abiotic stress. Proc Natl Acad Sci USA 92:4114–4119

Dajic Z (2006) Salt stress. In: Madhava Rao KV, Raghavendra AS, Janardhan Reddy K (eds) Physiology and molecular biology of stress tolerance in plants. Springer, Dordrecht, pp 41–100 (http://www.link.springer.com/book/10.1007/1-4020-4225-6/page/1)

Day TA, Vogelmann TC (1995) Alterations in photosynthesis and pigment distributions in pea leaves following UV-B exposure. Physiol Plant 94:433–440

Delauney AJ, Verma DPS (1993) Proline biosynthesis and osmoregulation in plants. Plant J 4:215–223

Desikan R, Mackerness SAH, Hancock JT, Neill SJ (2001) Regulation of the *Arabidopsis* transcriptome by oxidative stress. Plant Physiol 127:159–172

Devi SR, Pellissier F, Prasad MNV (1997) Allelochemicals. In: Prasad MNV (ed) Plant ecophysiology. Wiley, New York, pp 253–303

Domingues-Solis JR, Gutierrez-Alcala G, Romero LC, Gotor C (2001) The cytosolic *O*-acetyl-serine(thiol) lyase gene is regulated by heavy metals and can function in cadmium tolerance. J Biol Chem 276: 9297–9302

Dong J, Mao WH, Zhang GP, Wu FB, Cai Y (2007) Root excretion and plant tolerance to cadmium toxicity – a review. Plant Soil Environ 53:193–200

Einhellig FA (2004) Mode of allelochemical action of phenolic compounds. In: Macias FA, Galindo JCG, Molinillo JMG, Cutler HG (eds) Allelopathy: chemistry and mode of action of allelochemicals. CRC Press, Boca Raton, pp 217–238

Fahey JW, Zalcmann AT, Talalay P (2001) The chemical diversity and distribution of glucosinolates and isothiocyanates among plants. Phytochemistry 56:5–51

Feder ME, Hoffman GE (1999) Heat-shock proteins, molecular chaperones, and the stress response: evolutionary and ecological physiology. Annu Rev Physiol 61:243–282

Fischer NH (1986) The function of mono and sesquiterpenes as plant germination and growth regulators. In: Putnam AR, Tang CS (eds) The science of allelopathy. Wiley, New York, pp 203–218

Fornoni J, Núñez-Farfán J (2000) Evolutionary ecology of *Datura stramonium*: genetic variation and cost of tolerance to defoliation. Evolution 54:789–797

Fornoni J, Núñez-Farfán J, Valverde PL (2003) Evolutionary ecology of tolerance to herbivory: advances and perspectives. Comments Theor Biol 8:643–663

Foyer CH, Lelandais M, Kunert KJ (1994) Photooxidative stress in plants. Physiol Plant 92:696–717

Freitas HMO (1997) Drought. In: Prasad MNV (ed) Plant ecophysiology. Wiley, New York, pp 129–147

Fujita M, Fujita Y, Noutoshi Y, Takahashi F, Narusaka Y, Yamaguchi-Shinozaki K (2006) Crosstalk between abiotic and biotic stress responses: a current view from the point of convergence in the stress signalling network. Curr Opin Plant Biol 9:436–442

Gange AC, Brown VK (1989) Effects of root herbivory by an insect on a foliar-feeding species, mediated through changes in the host plant. Oecologia 81: 38–42

Gasic K, Korban SS (2006) Heavy metal stress. In: Madhava Rao KV, Raghavendra AS, Janardhan Reddy K (eds) Physiology and molecular biology of stress tolerance in plants. Springer, Dordrecht, pp 219–254 (http://www.link.springer.com/book/10.1007/1-4020-4225-6/page/1)

González-Teuber M, Gianoli E (2007) Tolerante to simulated herbivory in two populations of *Convolvulus chilensis* (Convolvulaceae). Acta Oecol 32: 119–123

Gorham J (1990) Salt tolerance in Triticeae: K/Na discrimination in synthetic hexaploid wheats. J Exp Bot 41:623–627

Gorham J (1993) Genetics and physiology of enhanced K/Na discrimination. In: Randall P (ed) Genetic aspects of plant mineral nutrition. Kluwer Academic, Dordrecht, pp 151–159 (http://www.link.springer.com/book/10.1007/1-4020-4225-6/page/1)

Gray GR, Chauvin LP, Sarhan F, Huner NPA (1997) Cold acclimation and freezing tolerance: a complex interaction of light and temperature. Plant Physiol 114:467–474

Griffiths ME, Rotjan RD, Ellmore GS (2008) Differential salt deposition and excretion on leaves of *Avicennia germinans* mangroves. Caribb J Sci 44:267–271

Grill E, Winnacker EL, Zenk MH (1989) Phytochelatins, a class of heavy metal binding peptides from plants are functionally analogous to metallothioneins. Proc Natl Acad Sci USA 84:439–443

Grover A, Pareek A, Singla SL, Minhas D, Katyar S, Ghawana S, Dubey H, Agarwal M, Rao GU, Rathee J, Grover A (1998) Engineering crops for tolerance against abiotic stresses through gene manipulation. Curr Sci 75:689–696

Guo WJ, Bundithya W, Goldsbrough PB (2003) Characterization of the *Arabidopsis* metallothionein gene family: tissue-specific expression and induction during senescence and in response to copper. New Phytol 159:369–381

Guy CL, Li QB (1998) The organization and the evolution of the spinach stress 70 molecular chaperone gene family. Plant Cell 10:539–556

Hagemeyer J (1997) Salt. In: Prasad MNV (ed) Plant ecophysiology. Wiley, New York, pp 173–206

Hall AE (2001) Crop responses to environment. CRC Press LLC, Boca Raton, pp 17–28

Harborne JB (1993) Biochemicals interaction between higher plants. In: Harborne JB (ed) Introduction to ecological biochemistry. Academic, London, pp 243–259

Hasegawa PM, Bressan RA, Zhu JK, Bohnert HJ (2000) Plant cellular and molecular responses to high salinity. Annu Rev Plant Physiol Mol Biol 51:463–499

Havaux M (1998) Carotenoids as membrane stabilizers in chloroplasts. Trends Plant Sci 3:147–151

Hinsinger P (2001) Bioavailability of soil inorganic P in the rhizosphere as affected by root-induced chemical changes: a review. Plant Soil 273:173–195

Iba K (2002) Acclimative response to temperature stress in higher plants: approaches of gene engineering for temperature tolerance. Annu Rev Plant Biol 53: 225–245

Ishitani M, Nakamura T, Han SY, Takabe T (1995) Expression of the betaine aldehyde dehydrogenase gene in barley in response to osmotic stress and abscisic acid. Plant Mol Biol 27:307–315

Jansen MAK, Gaba V, Greenberg BM (1998) Higher plants and UV-B radiation: balancing damage, repair and acclimation. Trends Plant Sci 3:131–135

Kaplan F, Kopka J, Haskell DW, Schiller C, Gatzke N, Sung DY, Guy CL (2004) Exploring the temperature-stress metabolome of *Arabidopsis*. Plant Physiol 136: 4159–4168

Kerchev PI, Fenton B, Foyer CH, Hancock RD (2012) Plant responses to insect herbivory: interactions between photosynthesis, reactive oxygen species and hormonal signalling pathways. Plant Cell Environ 35:441–453

Khan MA, Ungar IA, Showalter AM (2000) The effect of salinity on the growth, water status, and ion content of a leaf succulent perennial halophyte, *Suaeda fruticosa* (L.) Forssk. J Arid Environ 45:73–84

Kneer R, Zenk MH (1992) Phytochelatins protect plant enzymes from heavy metal poisoning. Phytochemistry 31:2663–2667

Koiwai H, Nakaminami K, Seo M, Mitsuhasi W, Toyomasu T, Koshiba T (2004) Tissue-specific localization of an abscisic acid biosynthetic enzyme, AAO3, in *Arabidopsis*. Plant Physiol 134:1697–1707

Kovtun Y, Chiu WL, Tena G, Sheen J (2000) Functional analysis of oxidative stress-activated mitogen-activated protein kinase cascade in plants. Proc Natl Acad Sci USA 97:2940–2945

Kramer GF, Krizek DT, Mirecki RM (1992) Influence of UV-B radiation and spectral quality on UV-B-induced polyamine accumulation in soybean. Phytochemistry 31:1119–1125

Krauss ML (1988) Accumulation and excretion of five heavy metals by the salt marsh grass *Spartina alterniflora*. Bull New Jersey Acad Sci 33:39–43

Kuboi T, Noguchi A, Yazaki J (1987) Relationship between tolerance and accumulation characteristics of cadmium in higher plants. Plant Soil 104: 275–280

Kulandaivelu G, Lingakumar K, Premkumar A (1997) UV-B radiation. In: Prasad MNV (ed) Plant ecophysiology. Wiley, New York, pp 41–60

Larcher W (1995) Physiological plant ecology, 3rd edn. Springer, Berlin, pp 48–55

Larkindale J, Hall JD, Knight MR, Vierling E (2005) Heat stress phenotypes of *Arabidopsis* mutants implicate multiple signaling pathways in the acquisition of thermotolerance. Plant Physiol 138: 882–897

Leavitt J (1980) Responses of plants to environmental stresses, vol II. Academic, New York

Lee S, Korban SS (2002) Transcriptional regulation of *Arabidopsis thaliana* phytochelatin synthase (*AtPCS1*) by cadmium during early stages of plant development. Planta 215:689–693

Legendre L, Yueh YG, Crain R, Haddock N, Heinstein PF, Low PS (1993) Phospholipase-C activation during elicitation of the oxidative burst in cultured plant cells. J Biol Chem 268:24559–24563

Lei TT (2000) Allelochemistry as a plant stress. In: Orcutt DM, Nilsen ET (eds) The physiology of plants under stress. Wiley, New York, pp 329–385

Lennartsson T, Tuomi J, Nilsson P (1998) Induction of overcompensation in the field gentian, *Gentianella campestris*. Ecology 79:1061–1072

Lindquist S, Craig EA (1988) The heat-shock proteins. Annu Rev Genet 22:631–677

Liu DL, Lovette JVE (1993) Biologically active secondary metabolites of barley II. Phyto-toxicity of barley allelochemicals. J Chem Ecol 19:2231–2244

Liu JP, Ishitani M, Halfter U, Kim CS, Zhu JK (2000) The *Arabidopsis thaliana* SOS2 gene encodes a protein-kinase that is required for salt tolerance. Proc Natl Acad Sci USA 97:3730–3734

Lu C, Qiu N, Lu Q, Wang B, Kuang T (2002) Does salt stress lead to increased susceptibility of photosystem II to photoinhibition and changes in photosynthetic pigment composition in halophyte *Suaeda salsa* grown outdoors? Plant Sci 163:1063–1068

Ludwig AA, Saitoh H, Felix G, Freymark G, Miersch O, Wasternack C, Boller T, Jones JDG, Romeis T (2005) Ethylene-mediated cross-talk between calcium-dependent protein kinase and MAPK signaling controls stress responses in plants. Proc Natl Acad Sci USA 102:10736–10741

Lyddon J, Teramura AH, Coffman CB (1987) UV-B radiation effects on photosynthesis, growth and cannabinoid production of two *Cannabis sativa* chemotypes. Photochem Photobiol 46:201–206

Ma YZ, Holt NE, Li XP, Niyogi KK, Fleming GR (2003) Evidence for direct carotenoid involvement in the regulation of photosynthetic light harvesting. Proc Natl Acad Sci USA 100:4377–4382

Madhava Rao KV (2006) Introduction. In: Madhava Rao KV, Raghavendra AS, Janardhan Reddy K (eds) Physiology and molecular biology of stress tolerance in plants. Springer, Dordrecht, pp 1–14

Mantri N, Patade V, Penna S, Ford R, Pang E (2012) Abiotic stress responses in plants: present and future. In: Ahmad P, Prasad MNV (eds) Abiotic stress responses in plants: metabolism, productivity and sustainability. Springer, New York

Marquis RJ (1992) Selective impact of herbivores. In: Fritz RS, Simms EL (eds) Plant resistance to herbivores and pathogens. University of Chicago Press, Chicago, pp 392–425

Marquis RJ (1996) Plant architecture, sectoriality and plant tolerance to herbivores. Vegetatio 127:85–97

Martínez Moreno D, Núñez-Farfán J, Terrazas T, Ruiz LM, Trinidad-Santos A, Trejo C, Larque-Saavedra A (1999) Plastic responses to clipping in two species of *Amaranthus* from the Sierra Norte the Puebla, Mexico. Gen Res Crop Evol 46:225–234

McNaughton SJ (1983) Compensatory plant growth as a response to herbivory. Oikos 40:329–336

Mehlhorn H (1990) Ethylene-promoted ascorbate peroxidase activity protects plants against hydrogen peroxide, ozone and paraquat. Plant Cell Environ 13:971–976

Mejuto-Marti M, Bollain-Rodriguez M, Herrero-Latorre C, Bermejo-Martinez F (1988) Selenium content of vegetables, fruits and cereals in Galicia (Northwest Spain). J Agric Food Chem 36:293–295

Memon AR, Yatazawa M (1984) Nature of manganese complexes in Mn accumulator plant *Acanthopanax sciadophylloides*. J Plant Nutr 7:961–974

Memon AR, Chino M, Takeoka Y, Hara K, Yatazawa M (1980) Distribution of manganese in leaf tissues of manganese accumulator: *Acanthopanax sciadophylloides* as revealed by electron probe X-ray microanalyzer. J Plant Nutr 2:457–477

Memon AR, Chino M, Hidaka H, Hara K, Yatazawa M (1981) Manganese toxicity in field grown tea plants and microdistribution of manganese in the leaf tissues as revealed by electron probe X-ray microanalyzer. Soil Sci Plant Nutr 27:317–328

Mewis I, Tokuhisa JG, Schultz JC, Appel HM, Ulrichs C, Gershenzon J (2006) Gene expression and glucosinolate accumulation in *Arabidopsis thaliana* in response to generalist and specialist herbivores of different feeding guilds and the role of defense signaling pathways. Phytochemistry 67:2450–2462

Milthorpe FL, Moorby J (1979) An introduction to crop physiology. Cambridge University Press, Cambridge, pp 148–166

Miroshnichenko S, Tripp J, Nieden U, Neumann D, Conrad U, Manteuffel R (2005) Immunomodulation of function of small heat shock proteins prevents their assembly into heat stress granules and results in cell death at sublethal temperatures. Plant J 41:269–281

Moisyadi S, Harrington HM (1990) Functional characterization of a low molecular weight heat shock protein in cultured sugarcane cells. Plant Physiol Suppl 93:88

Molas J (1997) Ultrastructural response of cabbage outer leaf mesophyll cells (*Brassica oleracea*) to excess of nickel. Acta Soc Bot Pol 66:307–317

Molisch H (1937) Der einfluss einer pflanzen auf die andere-allelopathie. Fischer, Jena, p 106

Mulroy TW (1979) Spectral properties of heavily glaucous and non-glaucous leaves of a succulent rosette-plant. Oecologia 38:349–357

Munns R (2002) Comparative physiology of salt and water stress. Plant Cell Environ 25:239–250

Munns R, Greenway H, Kirst GO (1983) Halotolerant eukaryotes. In: Lange OL, Nobel PS, Osmond CB, Ziegler H (eds) Encyclopedia of plant physiology, vol 12C, Physiological plant ecology III. Springer, New York, pp 59–135

Munns R, Hare RA, James RA, Rebetzke GJ (2000) Genetic variation for improving the salt tolerance of durum wheat. Aust J Agric Res 51:69–74

Munns R, Husain S, Rivelli AR, James RA, Condon AG, Linsay MP, Lagudah ES, Schachtman DP, Hare RA (2002) Avenues for increasing salt tolerance of crops, and the role of physiologically based selection traits. Plant Soil 247:93–105

Murphy TM, Huerta AJ (1990) Hydrogen peroxide formation in cultured rose cells in response to UV-C radiation. Physiol Plant 78:247–253

Nakamoto H, Hiyama T (1999) Heat-shock proteins and temperature stress. In: Pessarakli M (ed) Handbook of plant and crop stress. Marcel Dekker, New York, pp 399–416

Orcutt DM, Nilsen ET (2000) The physiology of plants under stress. Wiley, New York, p 680

Orians C, Thorn A, Gómez S (2011) Herbivore-induced resource sequestration in plants: why bother? Oecologia 167:1–9

Parida AK, Das AB (2005) Salt tolerance and salinity effects on plants: a review. Ecotoxicol Environ Saf 60:324–349

Penfield S (2008) Temperature perception and signal transduction in plants. New Phytol 179:615–628

Perez-Prat E, Narasimhan M, Niu X, Botella MA, Bressan R, Valpuesta V (1994) Growth cycle stage-dependent NaCl induction of plasma membrane $H^+$-ATPase mRNA accumulation in de-adapted tobacco cells. Plant Cell Environ 17:327–333

Pieczonka K, Rosopulo A (1985) Distribution of cadmium, copper, and zinc in the caryopsis of wheat (*Triticum aestivum* L.). Fresenius Z Anal Chem 322:697–699

Piotrowska M, Dudka S (1994) Estimation of maximum permissible levels of cadmium in a light soil by using cereal plants. Water Air Soil Pollut 73:179–188

Prasad MNV (1997) Trace metals. In: Prasad MNV (ed) Plant ecophysiology. Wiley, New York, pp 207–249

Rajashekar CB (2000) Cold response and freezing tolerance in plants. In: Wilkinson RE (ed) Plant environment interaction. Marcel Dekker, New York, pp 321–342

Rao MV, Dubey PS (1993) Response of hydrogen peroxide scavenging system in 2 soybean cultivars exposed to $SO_2$ – experimental evidence for the detoxification of $SO_2$ by enhanced $H_2O_2$ scavenging components. Environ Poll 82:99–105

Redondo-Gómez S, Wharmby C, Castillo JM, Mateos-Naranjo E, Luque CJ, de Cires A, Luque T, Davy AJ, Figueroa ME (2006) Growth and photosynthetic responses to salinity in an extreme halophyte, *Sarcocornia fruticosa*. Physiol Plantarum 128:116–124

Redondo-Gómez S, Mateos-Naranjo E, Davy AJ, Fernández-Muñoz F, Castellanos EM, Luque T, Figueroa ME (2007) Growth and photosynthetic responses to salinity of the salt-marsh shrub *Atriplex portulacoides*. Ann Bot 100:555–563

Redondo-Gómez S, Mateos-Naranjo E, Cambrollé J, Luque T, Figueroa ME, Davy AJ (2008) Carry-over of differential salt tolerance in plants grown from dimorphic seeds of *Suaeda splendens*. Ann Bot 102:103–112

Redondo-Gómez S, Mateos-Naranjo E, Figueroa ME, Davy AJ (2010a) salt stimulation of growth and photosynthesis in an extreme halophyte, *Arthrocnemum macrostachyum*. Plant Biol 12:79–87

Redondo-Gómez S, Mateos-Naranjo E, Andrades-Moreno L (2010b) Accumulation and tolerance characteristics of cadmium in a halophytic Cd-hyperaccumulator, *Arthrocnemum macrostachyum*. J Hazard Mater 184:299–307

Redondo-Gómez S, Andrades-Moreno L, Mateos-Naranjo E, Parra R, Valera-Burgos J, Aroca R (2011) Synergic effect of salinity and zinc stress on growth and photosynthetic responses of the cordgrass *Spartina densiflora*. J Exp Bot 62:5521–5530

Rice EL (1984) Allelopathy. Academic, London, p 422

Rice EL (1987) Role in agriculture and forestry. In: Waller GR (ed) Allelochemicals, vol 330. American Chemical Society, Washington, DC, pp 343–357

Rice-Evans CA, Miller NJ, Papaga G (1997) Antioxidant properties of phenolic compounds. Trends Plant Sci 2:152–159

Rivai IF, Koyama H, Suzuki S (1990) Cadmium content in rice and its daily intake in various countries. Bull Environ Contam Toxicol 44:910–916

Roberts AHC, Longhurst RD, Brown MW (1994) Cadmium status of soils, plants and grazing animals in New Zealand. N Z J Agric Res 37:119–129

Rocha-Granados MC, Sánchez-Hernández C, Sánchez-Hernández C, Martínez-Gallardo NA, Ochoa-Alejo N, Délano-Frier JP (2005) The expression of the hydroxyproline-rich glycopeptide systemin precursor A in response to (a)biotic stress and elicitors is indicative of its role in the regulation of the wound response in tobacco (*Nicotiana tabacum* L.). Planta 222:794–810

Rosenthal JP, Welter SC (1995) Tolerance to herbivory by a stemboring caterpillar in architecturally distinct maizes and wild relative. Oecologia 102:146–155

Roy BA, Kirchner JW (2000) Evolutionary dynamics of pathogen resistance and tolerance. Evolution 54:51–63

Sage RF, Reid CD (1994) Photosynthetic response mechanisms to environmental change in C3 plants. In: Wilkinson RE (ed) Plant environment interactions. Marcel Dekker, New York, pp 413–499

Saijo Y, Hata S, Kyozuka J, Shimamoto K, Izui K (2000) Over-expression of a single $Ca^{2+}$-dependent protein kinase confers both cold and salt/drought tolerance on rice plants. Plant J 23:319–327

Sakamoto A, Murata N (2002) The role of glycine betaine in the protection of plants from stress: clues from transgenic plants. Plant Cell Environ 25:163–171

Satoh M, Kriedeman PE, Loveys BR (1977) Changes in photosynthetic activity and related processes following decapitations in mulberry trees. Physiol Plant 41:203–210

Sauter A, Dietz KJ, Hartung W (2002) A possible stress physiological role of the abscisic acid conjugates in root-to-shoot signalling. Plant Cell Environ 25: 223–228

Scandalios JG (1993) Oxygen stress and superoxide dismutases. Plant Physiol 101:7–12

Shafer WE, Schönherr J (1985) Accumulation and transport of phenol, 2-nitrophenol, and 4-nitrophenol in plant cuticles. Ecotoxicol Environ Saf 10:239–252

Sharkey TD, Schrader SM (2006) High temperature stress. In: Madhava Rao KV, Raghavendra AS,

Janardhan Reddy K (eds) Physiology and molecular biology of stress tolerance in plants. Springer, Dordrecht, pp 101–130

Shi HZ, Quintero FJ, Pardo JM, Zhu JK (2002) The putative plasma membrane $Na^+/H^+$ antiporter SOS1 controls long distance $Na^+$ transport in plants. Plant Cell 14:465–477

Shi HT, Li RJ, Cai W, Liu W, Fu ZW, Lu YT (2012) *In vivo* role of nitric oxide in plant response to abiotic and biotic stress. Plant Signal Behav 7:438–440

Sicker D, Schulz M (2002) Benzoxazinones in plants: occurrence, synthetic access, and biological activity. In: Atta-ur-Rahman (ed) Bioactive natural products (part H), vol 27. Elsevier Science, Amsterdam, pp 185–232

Simms EL (2000) Defining tolerance as a norm of reaction. Evol Ecol 14:563–570

Singla SL, Pareek A, Grover A (1997) High temperature. In: Prasad MNV (ed) Plant ecophysiology. Wiley, New York, pp 101–127

Smalle J, Vierstra RD (2004) The ubiquitin 26S proteosome proteolytic pathway. Annu Rev Plant Biol 55:555–590

Smirnoff N (1993) The role of active oxygen in the response of plants to water deficit and desiccation. New Phytol 125:27–58

Smith KC (1989) The science of photobiology. Plenum Press, New York, pp 155–190

Smith LL, Lanza J, Smith GC (1990) Amino acid concentrations in extrafloral nectar of *Impatiens sultani* increase after simulated herbivory. Ecology 71:107–115

Smith BN, Harris LC, McCarlie VW, Stradling DL, Thygerson T, Walker J, Criddle RS, Hansen LD (2001) Time, plant growth, respiration, and temperature. In: Pessarakli M (ed) Handbook of plant and crop physiology. Marcel Dekker, New York, pp 1–12

Stapleton AE (1992) Ultraviolet radiation and plants: burning questions. Plant Cell 4:1353–1358

Steffens JC (1990) The heavy metal-binding peptides of plants. Annu Rev Plant Physiol Plant Mol Biol 41:553–575

Steinbrenner A, Gómez S, Osorio S, Fernie AR, Orians CM (2011) Herbivore-induced changes in tomato (*Solanum lycopersicum*) primary metabolism: a whole plant perspective. J Chem Ecol 37:1294–1303

Steponkus PL, Uemura M, Webb MS (1993) Membrane destabilization during freeze-induced dehydration. In: Close TJ, Bray EA (eds) Plant responses to cellular dehydration during environmental stress. American Society of Plant Physiologist, Rockville, pp 37–47

Strauss SY, Agrawal AA (1999) The ecology and evolution of plant tolerance to herbivory. Tree 14:179–185

Sutherland MW (1991) The generation of oxygen radicals during host plant responses to infection. Physiol Mol Plant Pathol 39:79–93

Teraza W, Martinez D, Rengifo E, Herrera A (2003) Photosynthetic responses of the tropical snipy shrub *Lycium nodosum* (*Solanaceae*) to drought, soil salinity and saline spray. Ann Bot 92:757–765

Thiboldeaux RL, Lindroth RL, Tracy JW (1994) Differential toxicity of juglone (5-hydroxy-1,4-naphthoquinone) and related naphthoquinones to saturniid moths. J Chem Ecol 20:1631–1641

Thorne JH, Koller HR (1974) Influence of assimilate demand on photosynthesis, diffusive resistances, translocation, and carbohydrate levels of soybean leaves. Plant Physiol 54:201–207

Tiffin P (2000) Mechanisms of tolerance to herbivore damage: what do we know? Evol Ecol 14:523–536

Tomsett AB, Thurman DA (1988) Molecular biology of metal tolerances of plants. Plant Cell Environ 11:383–394

Towers GHN, Wat CK (1978) Biological activity of polyacetylenes. Rev Latinoam Quim 9:162–170

Trischuk RG, Schilling BS, Wisniewski M, Gusta LV (2006) Freezing stress: systems biology to study cold tolerance. In: Madhava Rao KV, Raghavendra AS, Janardhan Reddy K (eds) Physiology and molecular biology of stress tolerance in plants. Springer, Dordrecht, pp 131–156

Trumble JT, Kolodny-Hirsch DM, Ting IP (1993) Plant compensation for arthropod herbivory. Annu Rev Entomol 38:93–119

Urao T, Yakubov B, Satoh R, Yamaguchi-Shinozaki K, Seki M, Hirayama T, Shinozaki K (1999) A transmembrane hybrid-type histidine kinase in *Arabidopsis* functions as an osmosensor. Plant Cell 11: 1743–1754

Veirling E (1991) The roles of heat shock proteins in plants. Annu Rev Plant Physiol Plant Mol Biol 42:579–620

Velikova V, Edreva A, Loreto F (2005) Endogenous isoprene protects *Phragmites australis* leaves against singlet oxygen. Plan Cell Environ 28:318–327

Vézina LP, Ferullo JM, Laliberté G, Laberge S, Willemot C (1997) Chilling and freezing. In: Prasad MNV (ed) Plant ecophysiology. Wiley, New York, pp 61–100

Voesenek LACJ, Van der Veen R (1994) The role of phytohormones in plant stress: too much or too little water. Acta Bot Neerl 43:91–127

von Rad U, Hüttl R, Lottspeich F, Gierl A, Frey M (2001) Two glucosyltransferases are involved in detoxification of benzoxazinoids in maize. Plan J 28:633–642

Wahid A, Close TJ (2007) Expression of dehydrins under heat stress and their relationship with water relations of sugarcane leaves. Biol Plant 51:104–109

Wahid A, Gelani S, Ashraf M, Foolad MR (2007) Heat tolerance in plants: an overview. Environ Exp Bot 61:199–223

Walling LL (2000) The myriad plant responses to herbivores. J Plant Growth Regul 19:195–216

Wang W, Vinocur B, Shoseyov O, Altman A (2004) Role of plant heat-shock proteins and molecular chaperones in the abiotic stress response. Trends Plant Sci 9: 244–252

Waring PF, Khalifa MM, Treharne KJ (1968) Rate-limiting processes in photosynthesis at saturating light intensities. Nature 220:453–458

Waseem M, Ali A, Tahir M, Nadeem MA, Ayub M, Tanveer A, Ahmad R, Hussain M (2011) Mechanisms of drought tolerance in plant and its management through different methods. Continental J Agric Sci 5:10–25

Wehner J, Antunes PM, Powell JR, Mazukatow J, Rilling MC (2010) Plant pathogen protection by arbuscular mycorrhizas: a role for fungal diversity? Pedobiologia 53:197–201

Weis JS, Weis P (2004) Metal uptake, transport and release by wetland plants: implications for phytoremediation and restoration. Environ Int 30:687–700

Welter SC (1989) Arthropod impact on plant gas exchange. In: Bernays EA (ed) Insect-plant interactions, vol I. CRC Press, Boca Raton, pp 135–215

Werner C, Maitile P (1985) Accumulation of coumaryl glycosides in vacuoles of barley mesophyll protoplasts. J Plant Physiol 118:237–249

Williams WP (1990) Cold-induced lipid phase transitions. Philos Trans R Soc Lond B Biol Sci 326:555–567

Wink M (1998) Chemical ecology of alkaloids. In: Roberts MF, Wink M (eds) Alkaloids: biochemistry, ecology, and medical applications. Plenum Press, New York, pp 265–300

Wisniewski M, Arora A (1993) Adaptation and response of fruit trees to freezing temperatures. In: Biggs AR (ed) Cytology, histology, and histochemistry of fruit tree diseases. CRC Press, Boca Raton, pp 299–320

Wolkers W, Tetteroo FAA, Alberde M, Hoekstra FA (1999) Changed properties of the cytoplasmic matrix associated with desiccation tolerance of dried carrot somatic embryos. An in situ Fourier transform infrared spectroscopic study. Plant Physiol 120:153–163

Xiang C (1999) The activation sequence-1 cognate promoter elements play critical roles in the activation of defense-related genes in higher plants. In: Pessarakli M (ed) Handbook of plant and crop stress. Marcel Dekker, New York, pp 527–538

Xiong L, Yang Y (2003) Disease resistance and abiotic stress tolerance in rice are inversely modulated by an abscisic acid-inducible mitogen-activated protein kinase. Plant Cell 15:745–759

Xu JK, Yang LX, Wang YL, Wang ZQ (2005) Advances in the study uptake and accumulation of heavy metal in rice (*Oryza sativa*) and its mechanisms. Chinese Bull Bot 22:614–622

Yacoob RK, Filion WG (1987) The effects of cold-temperature stress on gene expression in maize. Biochem Cell Biol 65:112–119

Yang KA, Lim CJ, Hong JK, Park CY, Cheong YH, Chung WS, Lee KO, Lee SY, Cho MJ, Lim CO (2006) Identification of cell wall genes modified by a permissive high temperature in Chinese cabbage. Plant Sci 171:175–182

Yeh CHM, Hung WCH, Huang HJ (2003) Copper treatment activates mitogen-activated protein kinase signalling in rice. Physiol Plant 119:392–399

Yokota A, Takahara K, Akashi K (2006) Water stress. In: Madhava Rao KV, Raghavendra AS, Janardhan Reddy K (eds) Physiology and molecular biology of stress tolerance in plants. Springer, Dordrecht, pp 15–40

Yoshida S, Uemura M (1984) Protein and lipid composition of isolated plasma membranes from orchard grass (*Dactylis glomerata* L.) and changes during cold acclimation. Plant Physiol 75:31–37

# Molecular Mechanisms of Stress Resistance of Photosynthetic Machinery

# 2

Vladimir D. Kreslavski, Anna A. Zorina, Dmitry A. Los, Irina R. Fomina, and Suleyman I. Allakhverdiev

**Abstract**

The mechanisms of action of stressors, such as high light intensity and heat stress, on the photosynthetic machinery, primarily on the photosystem II, are reviewed. First of all, stressors alter the chemical composition of thylakoid membranes and decrease the activity of photosynthesis. Photodamage is caused by the direct effect of light on oxygen-evolving complex, whereas accumulation of reactive oxygen species due to high light or high temperatures causes suppression of the *de novo* synthesis of the reaction center proteins and, ultimately, leads to the inhibition of the recovery of photosystem II. In addition to their destructive and inhibitory action, the reactive oxygen species and products of lipid peroxidation trigger protective processes that lead to acclimatization. Particular attention is paid to the mechanisms that protect photosynthetic machinery from injury and to the inhibitory effect of stressors in the light of varying intensity. The known stress sensory systems of cyanobacteria are also reviewed.

V.D. Kreslavski
Institute of Basic Biological Problems, Russian Academy of Sciences, Pushchino, Moscow Region 142290, Russia
e-mail: vkreslav@rambler.ru; vkreslav@mail.ru

A.A. Zorina • D.A. Los
Institute of Plant Physiology, Russian Academy of Sciences, Botanicheskaya street 35, Moscow 127276, Russia
e-mail: tarlonc@yandex.ru; losda@ippras.ru

I.R. Fomina
Institute of Basic Biological Problems, Russian Academy of Sciences, Pushchino, Moscow Region 142290, Russia

Biosphere Systems International Foundation, Tucson, AZ 85755, USA
e-mail: fomi@ibbp.psn.ru; irafomi@rambler.ru

S.I. Allakhverdiev (✉)
Institute of Basic Biological Problems, Russian Academy of Sciences, Pushchino, Moscow Region 142290, Russia

Institute of Plant Physiology, Russian Academy of Sciences, Botanicheskaya street 35, Moscow 127276, Russia
e-mail: suleyman.allakhverdiev@gmail.com

G.R. Rout and A.B. Das (eds.), *Molecular Stress Physiology of Plants*,
DOI 10.1007/978-81-322-0807-5_2, © Springer India 2013

## Introduction

Plants are often exposed to various abiotic and biotic stresses during their growth and development. The most common abiotic stresses for plants are drought or flood, high or low temperatures, high or low light, the presence of harmful organic substances or heavy metals, excess of salts in the soil, deficiency of oxygen (hypoxia), and ultraviolet radiation. In response to these factors, plants develop stress responses that lead to the suppression of many physiological functions.

In nature, plants and their photosynthetic machinery often experience simultaneous action of several stressors. The responses of plants to stress often involve similar protective and adaptive systems that operate through the same sensors. For example, cold stress may cause the reaction characteristic for osmotic stress (Thomashow 1999). It is also known that preexposure on plants to one stress can affect their resistance and adaptation to other stresses – the cross-tolerance phenomenon (Sabehat et al. 1998; Allen 1995; Pastori and Foyer 2002; Yang et al. 2007). For instance, the effect of short-term heat stress can lead to increased resistance to photoinhibition (Kreslavski et al. 2008). Photoinhibition and salinity, as well as low and high temperatures, alter the lipid composition of thylakoid membranes and their fluidity (Allakhverdiev and Murata 2004). The latter facilitates the development of various adaptive processes and ensures the increased stress resistance of photosynthetic apparatus and of whole plants (Wada et al. 1990; Gombos et al. 1994a, b; Allakhverdiev et al. 1999, 2001; Sharkey 2005; Los and Zinchenko 2009; Zorina et al. 2011). One of the intriguing observations in cyanobacteria was that the different environmental stresses can be perceived by one molecular sensor. In particular, the histidine kinase Hik33 of *Synechocystis*, that was first described as a component for resistance to a chemical inhibitor of photosynthesis (Zorina et al. 2011), appeared to be the multisensory protein, which perceives cold, salt, and oxidative stresses.

Light and oxygen are necessary for normal growth of plants and cyanobacteria. Oxygen participates in the process of formation of reactive oxygen species (ROS) (Nishiyama et al. 2001, 2004, 2005, 2006, 2011; Kreslavski et al. 2012b). The excess of ROS under unfavorable stress conditions causes a shift in the balance of oxidants/antioxidants toward oxidants, which leads to the intracellular oxidative stress (Nishiyama et al. 2001, 2004, 2005, 2006, 2011). The ROS pool is determined by the relative rates of formation and destruction of ROS and by the lifetime of their main species. The activity of antioxidant enzymes, superoxide dismutase (SOD), catalase, peroxidases, and several others, as well as the content of low molecular weight antioxidants, such as ascorbic acid, glutathione, tocopherols, carotenoids, and anthocyanins, plays a key role in regulation of the level of ROS and products of lipid peroxidation in cells (Apel and Hirt 2004; Pradedova et al. 2011). However, the mechanisms of neutralization and the distribution of ROS at the level of organelles, cells, and whole body mostly have not been clarified (Swanson and Gilroy 2010; Kreslavski et al. 2012b).

One of the basic processes that ensure stress resistance of plants is the energy-providing process of photosynthesis. Photosynthesis occurs in chloroplasts that are surrounded by two membranes: external and internal. The inner membrane surrounds the stroma, which contains the thylakoids – flattened discoid sacks. The internal cavities of thylakoids communicate with each other. They form the luminal space, which is separated from the stroma by the thylakoid membrane, which is impermeable to ions. Light reactions of photosynthesis – processes of energy conversion of sunlight into energy of chemical bonds – take place in the thylakoid membranes (Staehelin and van der Staay 1996). They contain four multienzyme complexes that are responsible for the charge separation and electron transfer through the membrane – photosystem I (PSI), photosystem II (PSII), cytochrome $b_6f$ complex, and ATP synthase.

Firstly, photosynthesis begins with absorption of light quantums with chlorophylls of reaction center (RC) of two photosystems associated with the membranes of thylakoids and next promotion of the photosynthetic electron transport generated by the integrated interaction of the photosystems. Therefore, the integrity of the thylakoid membranes is important for the efficiency of the photosystems, which, in turn, is crucial for plant resistance to stress.

Extreme environmental factors usually enhance formation of ROS (Nishiyama et al. 2001, 2004, 2005, 2006, 2011; Allakhverdiev and Murata 2004; Murata et al. 2007, 2012). In the process of photosynthesis, ROS are primarily produced by electron transport in chloroplasts (Foyer et al. 1994a, b, 1997; Foyer and Noctor 2005a, b; Asada 1999; Ivanov et al. 2007; Foyer and Shigeoka 2011) and in the process of $CO_2$ assimilation (Meloni et al. 2003). The accumulation of excessive amounts of ROS can lead to inhibition of the activity of several ROS-sensitive chloroplast enzymes, to a decrease in the rate of ATP formation, and to alterations in the structure of thylakoid membranes (Alscher et al. 1997; Foyer and Shigeoka 2011). Recently proposed new conception of mechanism of photoinhibition implies that excess of ROS ($H_2O_2$ or singlet oxygen) leads to inhibition of the *de novo* synthesis of protein D1 at the stage of its translational elongation (Nishiyama et al. 2001, 2004, 2006, 2011). Similarly, protein synthesis is inhibited under salt stress and upon high and low temperatures, and, accordingly, the restoration of photodamaged PSII is deteriorated (Allakhverdiev et al. 2002; Allakhverdiev and Murata 2004; Murata et al. 2007; Takahashi et al. 2009; Allakhverdiev 2011).

Besides inhibition of translation, ROS formed under stress are involved in signal transduction, triggering protective and adaptive mechanisms of the cell (Asada 1996; Demmig-Adams and Adams 2002; Dickinson and Chang 2011; Kreslavski et al. 2012b), in particular, at the level of photosynthetic reactions (El-Shitinawy et al. 2004; Ivanov et al. 2007). Other common messengers in the transduction of stress signal chain are lipids, signaling proteins, and second messengers, such as $Ca^{2+}$, $IP_3$, and cAMP.

Damage to the photosynthetic machinery under a moderate stress may be reversible. In such conditions, PSII of plants and cyanobacteria is often more susceptible to photodamage than the PSI (Berry and Björkman 1980; Aro et al. 1993; Andersson and Aro 2001; Thomas et al. 2001). However, under sufficiently hard stress, regardless of its nature, cells experience intense oxidative burst followed by lipid peroxidation and damage of proteins and of nucleic acids (Alscher et al. 1997; El-Shitinawy et al. 2004). All these result in the irreversible damage to the photosynthetic machinery, first of all – to thylakoid membranes.

Plants grow in the permanently changing environment. During the evolution, plants developed a variety of mechanisms for protection and acclimatization to stresses, including several regulatory pathways that help them to overcome the oxidative stress. These protective mechanisms imply diminishing of ROS generation with activation of ROS neutralization, acceleration of recovery of damaged cellular structures, increase in heat dissipation of the energy absorbed by plants, etc. All these protective pathways may be divided into fast and slow according to their molecular mechanisms. Fast mechanisms include rapid dissipation of light energy into heat due to nonphotochemical quenching (NPQ) and redistribution of absorbed energy between the photosystems (Choudhury and Behera 2001; Mohanty et al. 2002; Szabó et al. 2005; Shapiguzov et al. 2010). Two photosystems differ in light absorption characteristics and depend on the spectral quality of light. Regulation of the absorption properties of antenna complexes of PSI and PSII to optimize the linear flow of electrons is called the state transition (Allen 1992; Wollman 2001; Rochaix 2007). The transition from one state to another one modulates the cyclic electron flow and regulates the requirements of cellular ATP (Finazzi et al. 2002).

Slow mechanisms involve the synthesis of stress proteins (in particular, HSPs), antioxidant enzymes, membrane lipids (Yordanov 1993; Heckathorn et al. 1998a, b; Kreslavski et al. 2007; Allakhverdiev et al. 2008), and low molecular weight (LMW) compounds, such as proline and glycine betaine (Allakhverdiev et al. 2007). The precursor of glycine betaine is choline chloride, which can also increase stress resistance of the photosynthetic machinery (Kreslavski et al. 2001, 2011). One of the major stress defense mechanisms is the neutralization of ROS with various antioxidants that transform ROS into nontoxic products and eliminate the negative effects of stress (Asada 1996, 2006; Storz and Imlay 1999; Alscher et al. 2002; Foyer and Shigeoka 2011; Pradedova et al. 2011). The activity of antioxidant enzymes and the amounts of LMW antioxidants increase in response to higher levels of ROS (Allen 1995). ROS are mainly generated in chloroplasts (Foyer et al. 1997). Their accumulation is accompanied by an immediate increase in the activity of ascorbate-glutathione cycle and the activity of Fe- and Cu-Zn-containing superoxide dismutase (SOD) (Alscher et al. 2002; Foyer and Shigeoka 2011).

Another defense mechanism triggers changes in the lipid composition of membranes, especially in the ratio of unsaturated to saturated fatty acids (Gombos et al. 1994a, b; Vijayan and Browse 2002). It is assumed that the changes in fluidity of biological membranes are the primary signal in the perception of temperature and, possibly, osmotic and other stresses (Los and Zinchenko 2009; Los et al. 2010). However, the molecular mechanisms that control perception and transmission of these signals are still not fully understood. It remains unclear how do sensory transmembrane proteins recognize a change in membrane fluidity and which protein domains are involved in the perception of signals.

The specific sensor proteins perceive changes in environmental temperature, salinity, acidity, osmolarity, etc. They send a signal about these changes to another molecule, which acts as a signal transmitter. The latter may regulate the expression of stress response genes by itself or interacts with a downstream transcription factor that ultimately binds to the regulatory region of the responsive genes and regulates their transcription. Finally, protective proteins and/or metabolites are synthesized that allow cell's acclimation to new conditions (Los et al. 2010). Stress-induced changes may also directly alter chromosome packaging and cause changes in the transcription of many genes (Prakash et al. 2009).

Plants cannot escape from the unfavorable conditions, and their life is highly dependent on environment. They developed numerous sophisticated mechanisms to protect themselves from unfriendly surroundings. However, the organization of plants is very complex and difficult to study. For example, the routine generation of site-directed mutations in plant genome is either impossible or problematic. Thus, the studies of the molecular mechanisms of responses to stresses and identification of stress sensors in plants are still complicated. Various strains of cyanobacteria have been used to study plant-type stress responses at the molecular level. The advantages of cyanobacterial model system are as follows: (1) they quickly grow in batch cultures; (2) they are easily susceptible to any kind of genetic manipulations, including random and site-directed mutagenesis; (3) the complete genomes of many species of cyanobacteria have been already determined; (4) they perform oxygenic photosynthesis and carry the photosynthetic machinery similar to that of higher plants; (5) lipid and fatty acid composition of their thylakoid membranes resemble that of chloroplasts; and (6) they may acclimate to various stresses due to plasticity of their metabolism (Quist et al. 1995; Glatz et al. 1999; Thomas et al. 2001; Zorina et al. 2011). At present, the cyanobacteria *Synechocystis* sp. PCC 6803, *Synechococcus* sp. PCC 7942 or 6301, and *Anabaena variabilis*, along with *Escherichia coli* or *Bacillus subtilis*, became the most studied living organisms on Earth. In 1996, the complete nucleotide sequence of the genome of *Synechocystis* sp. PCC 6803 has been determined (Kanesaki et al. 2010), and this new era in studies of cyanobacteria has been initiated. In 1999, the company "Takara Bio" (Japan) started the production of DNA microarrays that carry 3,079 (97%) of the 3,264 individual chromosomal genes of *Synechocystis*. The use of DNA microarrays allows simultaneous monitoring of changes in expression of each individual gene in the whole genome of the organism. This technology allows determination of every single gene, whose transcription was altered or unchanged under stress conditions. This approach combined with directed mutagenesis of series of potential regulatory genes allows screening of genes that code for potential sensors and transducers, which are responsible for the perception of changes in temperature, light, salinity, lack or excess of various ions, etc. (Zorina et al. 2011).

All original results of gene expression can be found at the KEGG expression database (http://www.Genome.jp/kegg/expression/).

To assess the state of the photosynthetic machinery under stress, the methods of variable and delayed fluorescence are often applied that allow determination of parameters from fluorescence induction curves such as $F_v$, $F_m$, and $F_0$ reflecting the following: (a) the photoinduced changes in the fluorescence, (b) maximum yield of fluorescence after dark incubation of cells, and (c) yield of constant fluorescence, independent of photochemical reactions. The ratio ($F_v/F_m$) reflects the quantum efficiency of PSII in leaves and photosynthesizing cells. In some cases, however, this ratio is not suitable for assessment of the stress state of the photosynthetic machinery. For example, Li et al. (2009) investigated the heterogeneity of PSII in the soybean leaves. Heterogeneity was changed, whereas the $F_v/F_m$ ratio remained identical for all stress (heat) treatments.

Another parameter, which may be calculated from the fluorescence measurements, is photochemical quantum yield of PSII, which characterizes the state of the photosynthetic machinery in leaves adapted to light. This parameter allows evaluation of the effectiveness of PSII in real physiological conditions and appears to be more sensitive to many types of stressors than the fluorescence ratio of $F_v/F_m$ (Strasser et al. 2000; Crafts-Brandner and Salvucci 2000; Li et al. 2009). The methods for assessment of photosynthetic activity and stress acclimation of photosynthetic machinery in cyanobacteria and symbiotic microalgae are described (Campbell et al. 1998; Biel et al. 2009). Lately, a measurement of the induction curves of the photoinduced increase in *Chl* fluorescence (OJIP) is becoming a popular tool in studies of photosynthesis and is very sensitive to environmental stresses such as heat (Srivastava et al. 1997; Strasser 1997; Chen et al. 2008; Yordanov et al. 2008).

## Stress Signaling

When plants are under stress, forming ROS leads not only to damaging effect on cell systems but also to switching the signal transduction network and to induction of protectory (Hung et al. 2005; Mubarakshina et al. 2010; Kreslavski et al. 2012a). Intracellular ROS are involved in the complex system of stress signaling, which at the molecular level can be described as follows (Fig. 2.1). The response of cells starts with the perception of a stress treatment by certain sensor (s) and follows the transmitting of the signal to the regulatory systems (Kanesaki et al. 2010). The latter may consist of MAP-kinase cascades, various TFs, components of a phosphoinositol cycle, $Ca^{2+}$, phytohormones, etc. The detailed studies of such systems have been conducted in cyanobacteria. As mentioned above, cyanobacteria serve as efficient models for studying the molecular mechanisms of stress responses. The genes of these cells can be easily knocked out or overexpressed. All these circumstances allowed to conduct the successful studies of the potential stress sensors and signal transducers in cyanobacteria (Los et al. 2010; Kanesaki et al. 2010; Zorina et al. 2011).

## Multifunctional Sensory Systems

The systems of perception and transduction of stress signals, the hormonal regulation systems (in animals and plants), and the cascades that regulate growth and development certainly operate in close, finely tuned, coordination. While eubacteria, including cyanobacteria, actively use the two-component systems of signal perception and transduction, eukaryotes prefer serine-threonine (STPK) and tyrosine protein kinases, in combination with a broad range of phosphatases.

Two-component regulatory systems composed of a sensory histidine kinase (Hik) and a response regulator (Rre) form the central core of the phosphate signaling system in cyanobacteria (Los et al. 2010). Sensory histidine kinase perceives changes in the environment through the sensory domain. Changing in its conformation often leads to autophosphorylation of the conservative histidine residue in a Hik with a donor of ATP. Phosphate is then transferred to the conserved aspartate in a receiver domain of the response regulator protein, Rre. After phosphorylation, Rre also changes its conformation and

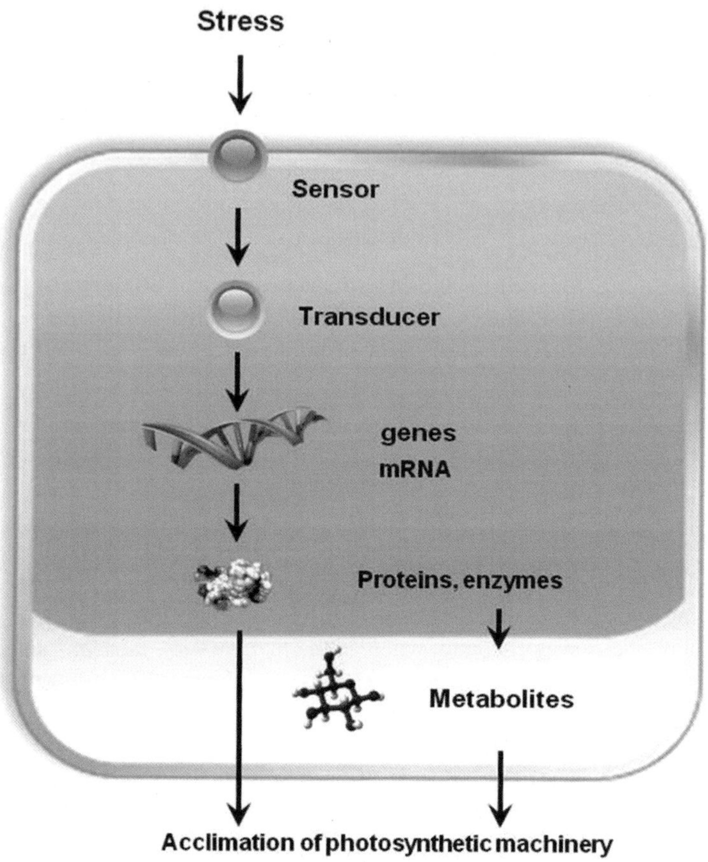


**Fig. 2.1** A scheme for perception, transduction of stress signals, and acclimatization of cells to stress conditions (Adopted from Zorina et al. 2011)

gains (positive regulation) or looses (negative regulation) the ability to binding DNA. Rre usually binds the promoter region (s) of genes for proteins that are involved in stress signal network or are linked to stress protection pathways.

Hik33 of *Synechocystis*, as mentioned above, is the multisensory protein, which perceives cold, salt, and oxidative stresses. The mechanisms by which Hik33 recognizes the stresses are still not entirely clear. The molecular mechanisms of activation of the Hik33 by autophosphorylation are not experimentally elucidated. These can be changes in the physical mobility of membrane lipids and changes in surface charge on the membrane, associated with changing mobility. Activation may also be caused by depolarization of the cytoplasmic membrane upon cold stress or due to changes in charge density of the membrane surface under stress (Nazarenko et al. 2003).

Sensory histidine kinases are also important for the functioning of genes involved in photosynthesis and/or regulated by high light intensity. The experiments with the *Synechocystis* mutant deficient in Hik33 (another name – DspA) revealed that low or moderate light intensity causes retardation in growth and decrease in photosynthetic oxygen evolution in mutant cells, compared to wild-type cells, under photoautotrophic conditions. The addition of glucose neutralized the difference. However, mutant cells were more sensitive to light intensity and quickly died under strong light (Hsiao et al. 2004). Multifunction sensor systems are also found in plants. Low temperature, drought, and salinity cause an increased concentration of $Ca^{2+}$ in the cytoplasm of plant cells. In this case, calcium channels may serve as multifunctional sensors that perceive stress-induced changes in the physical properties of cell membranes. The discovery of multisensory systems is important for understanding of a question – how perception and transmission of stress signals may operate? Apparently, changes in membrane fluidity, regardless of the nature of the stress effects, are a signal that is perceived by sensory histidine kinases or ion channels localized in the membranes.

## Signaling Role of Reactive Oxygen Species

It is known that ROS are produced in all cell compartments and their formation is necessary for the functioning of photosynthetic organisms (Suzuki and Mittler 2006). Currently, some ROS are considered as signaling molecules and regulators of expression of some chloroplast and nuclear genes (Minibayeva et al. 1998; Minibayeva and Gordon 2003; Desikan et al. 2001, 2003; Hung et al. 2005; Galvez-Valdivieso and Mullineaux 2010; Mubarakshina et al. 2010; Dickinson and Chang 2011). A new view on the effects of ROS as signaling molecules first appeared in the study of hormone signaling and the regulation of expression of genes involved in plant protection from pathogen infection (Chen et al. 1993), when interactions of ROS with salicylic acid and nitric oxide play the crucial role in regulation of the response to infection (Vallad and Goodman 2004).

One of the key points in understanding of the effect of ROS on photosynthesis was the discovery of the formation of superoxide anion and hydrogen peroxide in the pseudo-cyclic electron transport (Mehler reaction), which leads not to the reduction of $NADP^+$ but to the absorption of $O_2$ (Asada 1999). In addition, it was shown that the activation of plasma membrane redox systems and the increased formation of ROS in the apoplast are one of the universal reactions of plant cells to stress (Minibayeva et al. 1998, 2009; Minibayeva and Gordon 2003; Dickinson and Chang 2011).

It was found that the main generators of ROS in the apoplast of root cells are the cell wall peroxidases (Minibayeva et al. 2001, 2009; Minibayeva and Gordon 2003). Apparently, the release of ROS from cells followed by a switch of peroxidase/oxidase modes of extracellular peroxidases forms the basis for the fast response of plant cells to stress. In addition to ROS, the stress signaling functions may be attributed to some metabolites, whose formation is initiated by ROS, for example, the products of lipid peroxidation. The primary subjects for peroxidation in living cells are unsaturated fatty acids that constitute major components of phospho- and glycolipids of biological membranes. Signaling role of ROS and lipid peroxidation products is implemented through the regulation of calcium status of the cells by direct or indirect influence on the flow of $Ca^{2+}$ in the cytosol (Pei et al. 2000; Mori and Schroeder 2004), hormone signaling (Pei et al. 2000), MAP-kinase activity (Jaspers and Kangasjärvi 2010), as well as redox signaling systems (Pfannschmidt et al. 2009; Foyer and Noctor 2009; Foyer and Shigeoka 2011) and transcription factors (Pfannschmidt et al. 2009). ROS regulate the processes of polar growth, activity of stomata, light-induced movement of chloroplasts, and plant responses to the action of biotic and abiotic environmental factors (Pitzschke and Hirt 2006; Miller et al. 2007; Swanson and Gilroy 2010).

Signaling by ROS may be realized through changes in potential of the redox-sensitive cell systems and through phosphorylation/dephosphorylation cycles of signaling proteins (transcription factors, response regulators, etc.). The accumulation of redox-active compounds such as ROS within the chloroplast is associated with the rate of photosynthetic electron transport. Redox-sensitive thioredoxin or PQ pool may act as sensors of changes in redox properties under stress conditions (Fig. 2.2). Signals generated from modulation in the activity of electron transport chain (ETC) may also lead to changes in gene expression (Vallad and Goodman 2004).

There are still many gaps in understanding the mechanisms of action of ROS as signal molecules. The sensor(s) of $H_2O_2$ in higher plants remains largely unknown (Galvez-Valdivieso and Mullineaux 2010; Mubarakshina et al. 2010; Kreslavski et al. 2012b). There is no information about specific proteins that convert a signal about an increase in the intracellular ROS

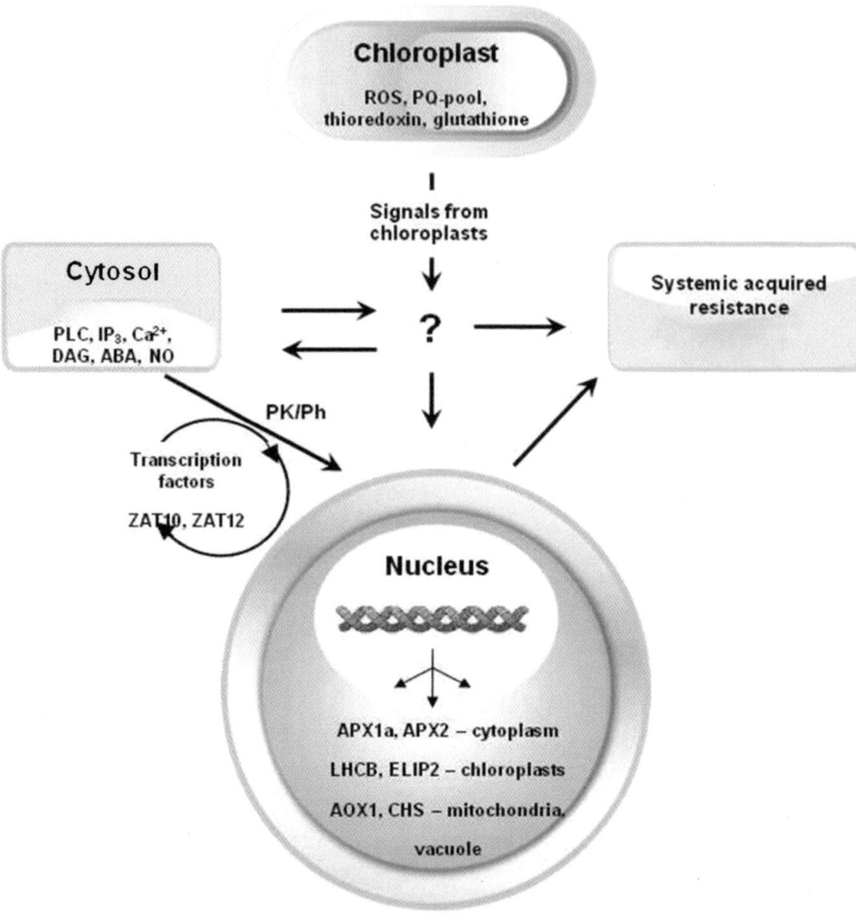


**Fig. 2.2** General scheme of photosynthesis-dependent regulation of nuclear gene expression. Accumulation of redox-active compounds, such as ROS, inside the chloroplast is related to the rate of photosynthetic electron transport. Redox-active systems, like thioredoxin and PQ, could function via definite pathways of signal transduction, or known second messengers appeared in the cytosol in response to abiotic stressors or interact with them. Signals arising during ETC functioning could also change expression of nuclear genes by generation of systemic signals in the process known as systemic acquired resistance. *PK* – protein kinases, *Ph* – phosphatases, *DAG* – diacylglycerol; *PhLC* phospholipase C, *IP3* inositol 1,4,5-triphosphate, *ZAT10* and *ZAT12* genes encoding transcription factors, *LHCB* gene encoding a protein of LHG in PSII, *ELIP2* gene encoding early light-inducible protein, *CHS* gene encoding chalcone synthase, *AOX1* gene encoding mitochondrial alternative oxidase. Known second messengers are marked. The question sign marks a hypothetic sensor of the cell redox state. Scheme is built on the basis of data from Pogson et al. (2008)

levels to a biochemical response in the cells. It is not known exactly which particular ROS play a signaling role in the chloroplast and other cellular compartments and how different signaling pathways respond to an increase in the level of different types of ROS? Knowledge of the mechanisms of regulation of these signaling pathways may help to construct biochemical pathways and to produce genetically engineered plants with enhanced stress resistance.

## Low Molecular Weight Protective Compounds

The ability of plants to limit the damage and/or repair the systems damaged by stress is critical for the survival of plants and for the maintenance of their productivity. Endogenous LMW compounds, such as hormones, antioxidants, and osmotic protectors, including proline, putrescine, and betaines, may serve as natural defense systems in plants (Raskin 1992; Leung and Giraudat 1998; Lankindale and Knight 2002; Kreslavski et al. 2007). LMW protective compounds also include growth-inhibiting substances (retardants), which can greatly enhance the stability of the photosynthetic machinery to high and low temperatures, ozone, and UVB radiation (Mackay et al. 1987; Sheng et al. 2006; Kreslavski et al. 2010, 2012a). Treatment of wheat plants with exogenous triazolic retardant S-3307 resulted in an increase of the stability of PSII to ozone (Mackay et al. 1987). Treatment of bean plants by exogenous choline chloride or chlorocholine chloride enhanced the resistance of the photosynthetic machinery to heat stress and UVB radiation (Kreslavski et al. 2001, 2010, 2011).

Nevertheless, the action mechanism(s) of the retardants on the stability of the photosynthetic machinery and its ability to recover from damage due to oxidative stress are still largely unknown. Sheng et al. (2006) investigated the effect of pretreatment with the choline chloride on cold resistance of the photosynthetic machinery of cucumber plants. Choline chloride promoted stress resistance of PSII and reduced the loss of chlorophyll under stress. The treatment also prevented the loss in the activities of guaiacol-dependent peroxidase, ascorbate peroxidase, and catalase under stress, as compared to untreated controls. In addition, pretreatment of plants with choline chloride increased the NPQ. Treatment of wheat plants by triazolic retardants S-3307 stimulated the accumulation of lipophilic antioxidants in microsomal membranes (Mackay et al. 1987). This diminished the damage to the membranes under stress induced by ozone fumigation. Choline chloride and chlorocholine chloride upregulated activities of antioxidant enzymes, SOD and ascorbate peroxidase, and caused the increase in the level of carotenoids and UVB-absorbing pigments (Kreslavski et al. 2010, 2012a). The effects of increased resistance to UV radiation and the activation of antioxidant enzymes are also evident in preparations of thylakoid membranes isolated from the leaves of treated plants. The action of these compounds was observed only after a few hours after their injection. Apparently, cholines induce transcription of genes involved in the synthesis of antioxidants and other stress-protective compounds, such as flavonoids and carotenoids, the content of which increased after treatment of plants with these retardants (Kreslavski et al. 2001, 2012a).

It can also be assumed that choline-contained compounds (CCC) may interact with membrane lipids and alter their composition (in particular in thylakoid membranes), thus providing the conditions for the increased stress resistance of photosynthetic machinery (Novitskaya et al. 2004). It is expected that CCC may stimulate synthesis of phospholipids and/or inhibit their decay, and maintain the proper ratio of unsaturated to saturated fatty acids (Novitskaya et al. 2004; Sheng et al. 2006).

Glycine betaine (GB) is a quaternary amine, known as the osmotic protector, which is accumulated in certain eukaryotic and bacterial cells under stress conditions (Caldas et al. 1999; Takabe et al. 1998). Its accumulation leads to the enhanced resistance of plants and the photosynthetic apparatus to various unfavorable conditions including salt and cold stress (Park et al. 2004; Vinocur and Altman 2005). Similarly to other osmotically active substances, it can stabilize the protein molecules and enzymes, probably by stabilizing the hydrophobic regions in these molecules (Takabe et al. 1998). Similar to a chaperone-like protein, trimethylamine-based compounds – choline and GB - protected citrate synthase *in vitro* from thermodenaturation and stimulated its renaturation with urea (Caldas et al. 1999). These effects, however, have been observed at relatively high concentrations of exogenous trimethylamine – more than 50 mM.

The influence of GB on the activity of some antioxidant enzymes was demonstrated (Alscher et al. 1997). GB reactivated the key enzyme of Calvin-Benson cycle, ribulose bisphosphate carboxylase (RuBisCo), denatured under salt stress. GB stabilized the system of $O_2$ production and ATP synthesis (Mamedov et al. 1993). It reduced the effect of high temperature inhibition of PSII (Allakhverdiev et al. 1996) and protected PSII from heat stress. In addition, GB stabilizes the electron-transfer reactions associated with the RC of PSII, including the restoration of pheophytin and oxidation of $P_{680}$ (Allakhverdiev et al. 1996). Glycine betaine also protects D1/D2/Cytb$_{559}$ complex of PSII from photo- and thermoinactivation (Allakhverdiev et al. 2003a). The suggested mechanism of protective action of GB is stabilization of the structure of membrane proteins, manganese cluster, and peripheral proteins of PSII – 33, 23, and 17 kDa (Allakhverdiev et al. 1996).

Organisms capable of synthesizing GB are characterized by high resistance of PSII to thermal- and photoinactivation. This was confirmed by studies of the recombinant cells of the cyanobacterium *Synechococcus* transformed with *codA* gene for bacterial choline oxidase, which is responsible for the synthesis of GB (Allakhverdiev et al. 2007). Accumulation of GB to a concentration of 60–80 mM in cyanobacterial cells enhanced their resistance to low temperatures through the restoration of the photosynthetic complex, which was photoinhibited (Hayashi and Murata 1998). Understanding the mechanisms of action of LMW compounds that improve plant resistance to the stressors (drought, salinity, UVB, heat, etc.) will help to improve the stability of a number of crop species under unfavorable environmental conditions (Chen and Murata 2011).

## The Role of Phytochromes in Increasing Stress Resistance of Photosynthetic Machinery

An evidence for a role of phytochromes in plant stress tolerance is explored and reviewed (Carvalho et al. 2011). However, there is a little information about a role of state and content of photoreceptors in the resistance of photosynthesis and PSII to abiotic stress factors. The transgenic plants of potato (*Solanum tuberosum*) that express the *PHYB* gene of *Arabidopsis* showed higher stomatal conductance and photosynthesis rates than the wild type (Thiele et al. 1999; Schittenhelm et al. 2004). The authors suggested that increased photosynthetic rates may be due to increased *Chl* (*a*+*b*) content in leaves of transgenic plants. Boccalandro et al. (2009) demonstrated that high, compared to low, R/FR ratios perceived by *phyB* increase stomata density, stomata index, and amphistomy in the leaves of *Arabidopsis*. This behavior results in an enhanced photosynthetic rate at high PAR at the expense of reduced water-use efficiency.

Irradiation of plants with red light (RL) may lead to an increase in the proportion of the active form of phytochrome B in its total pool that can be achieved by irradiating plants at the end of the dark period (Kreslavski et al. 2009). Recent studies (Kreslavski et al. 2004; Kreslavskii et al. 2012) demonstrated that irradiation of detached leaves of spinach with UVB and UVA caused a decrease in the activity of PSII and degradation of photosynthetic pigments after 12-h dark period. However, pre-irradiation with RL of low intensity (10 min; 620–660 nm) enhanced the resistance of PSII and photosynthetic machinery in whole to UVB and UVA. Also, a decrease in degradation of photosynthetic pigments had been detected.

Pre-irradiation of plants with RL resulted in the increased peroxidase activity and in the content of UV-absorbing pigments in leaves. The reduced pool of $H_2O_2$ was measured compared to that in leaves irradiated only with UVA. These RL effects were partially reversible by following FR light (10 min, 740 nm). On this basis, it is assumed that RL pre-irradiation induces resistance of the photosynthetic machinery to UV irradiation by increasing the peroxidase activity and the amount of the UV-absorbing pigments (UFPP) and different LMW compounds (Kreslavskii et al. 2012). In this case the active form of phytochrome B might be involved in the formation of the protective mechanisms upon the action of UV radiation. Phytochrome may also induce the synthesis of various protective compounds, including antioxidant enzymes and protective proteins (Fig. 2.3). Future studies should be directed toward the better understanding of the interrelations between phytochrome and stress signaling.

## Photoinhibition

Light intensities higher than intensity, which is necessary for the saturation of photosynthesis, may cause inhibition of the photosynthetic

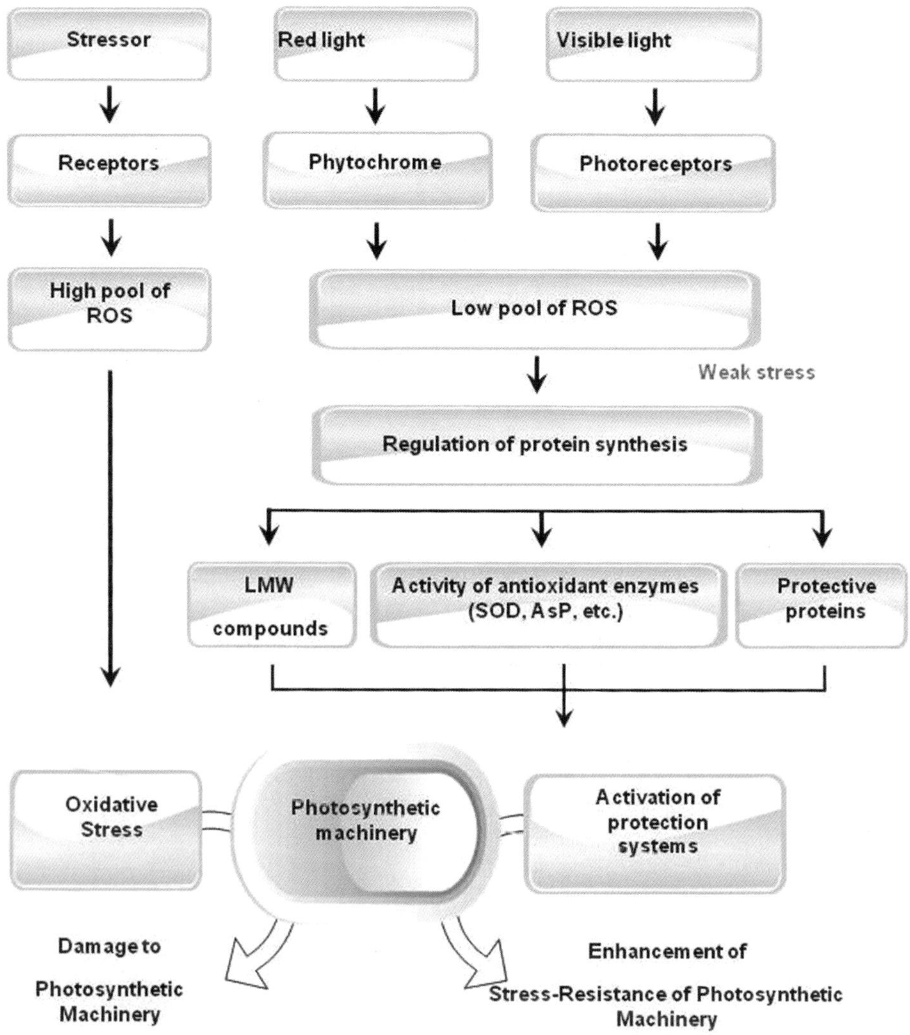


**Fig. 2.3** A general scheme illustrating the regulation of stress resistance of the photosynthetic machinery and whole plant to abiotic stressor as a result of pre-irradiation of plants with low-intensity red light and visible light. The arrows show the pathway of light signal transduction. The possible intermediates of light signaling – ROS, $Ca^{2+}$, TFs, etc. The scheme is based on data from Qi et al. (2000, 2002), Kreslavski et al. (2004, 2009, 2012a), Hung et al. (2005), Karu (2008), and Romanov (2009). *LMW* low molecular weight, *SOD* superoxide dismutase, *AsP* ascorbate peroxidase, *ROS* reactive oxygen species

machinery, especially of PSII, due to photodamage or photoinhibition under low light even under nonstress conditions (Kok 1956; Powles 1984; Allakhverdiev et al. 1987; Setlik et al. 1990; Adir et al. 2003). PSII is the main target of photoinhibition, whereas PSI is also inhibited, though to lesser extent (Barber and Andersson 1992). Although oxidative damage is evident even under light of low intensity, an irreversible damage to the PSII in intact leaves is only noticeable under irradiation with high light intensity, when the rate of recovery processes is lower than the rate of oxidative damage (Aro et al. 1993, 2005). Photoinhibition plays an important protective role since it reduces the rate of electron transfer under light of sufficiently high intensity, and it is characterized by the formation of photochemically inactive reaction centers (RCs) of PSII, in which light energy is transformed into heat (Powles 1984; Barber and Andersson 1992).

## Photoinhibitory Damage and the Repair of Damaged PSII

Photoinhibition occurs when the rate of photodamage of PSII exceeds the rate of its recovery (Aro et al. 1993; Melis 1999; Andersson and Aro 2001; Tyystjärvi 2008; Vass and Aro 2008; Murata et al. 2007, 2012; Takahashi and Murata 2008). Recently, it was shown in plants and cyanobacteria that the photodamage occurs by a two-step mechanism. The first step is the inactivation of manganese-containing oxygen-evolving complex (OEC) of PSII. Only after this event, light absorbed by photosynthetic pigments induces the photochemical inactivation of RC (Hakala et al. 2005; Ohnishi et al. 2005). Photoinhibition may be manifested under all visible light intensities and under ultraviolet radiation; however, UV radiation at comparable intensities is more destructive than visible light (Renger et al. 1989; Jung and Kim 1990). Since the initial absorption of light occurs with the participation of Mn-containing complex, rather than photosynthetic pigments, the rate of photodamage is proportional to the intensity of incident light, and it is not related to the rate of photosynthetic electron transport and appropriate formation of ROS (Allakhverdiev and Murata 2004; Ohnishi et al. 2005; Nishiyama et al. 2006, 2011; Murata et al. 2007, 2012; Takahashi and Murata 2008).

It is known that high light intensity leads to a damage of PSII (Allakhverdiev et al. 1987; Klimov et al. 1990; Setlik et al. 1990; Vass et al. 1992; Nishiyama et al. 2004, 2005, 2006, 2011; Hakala et al. 2005). Experiments with cyanobacteria demonstrated that this damage is mainly due to the action of ROS on the recovery of PSII than due to direct damage of the components of PSII (Allakhverdiev and Murata 2004; Nishiyama et al. 2001, 2004, 2011). It was also shown that the damage induced by light is determined mainly by the rate of degradation and resynthesis of one of the stress-sensitive proteins of PSII – D1 (Zhang et al. 2000; Andersson and Aro 2001; Thomas et al. 2001; Aro et al. 2005). Other proteins, D2 and PsbH, may also be the targets for damaging (Aro et al. 2005), though D1 protein is characterized by the maximum rate of turnover (Mattoo et al. 1989).

After photoinhibition, damaged RCs of PSII are not accumulated in the thylakoid membranes, and they are rapidly restored by *de novo* protein synthesis (Leitsch et al. 1994). Recovery involves the phosphorylation of several protein subunits of PSII core complex, monomerization and migration from the granal region to a lamellar area, partial proteolysis of damaged proteins and their removal from the complex, and replacement of the damaged proteins with their newly synthesized copies. Finally, restoration is accomplished with the formation of dimmers and their photoactivation (Aro et al. 2005). At high light intensities, when the rate of damage is fast, facilitated rate of repair is required to maintain PSII activity at a proper level.

There are a number of hypotheses generated in the last two decades that explain the mechanism of photoinactivation. Here, we mention only those that are supported by the solid experimental evidences (Sarvikas 2010; Takahashi and Badger 2011; Nishiyama et al. 2011). Photosynthesis and respiration produce sufficient amounts of different ROS. Photosynthetic electron transfer generates superoxide anion radical ($O_2^{\cdot -}$), hydrogen peroxide ($H_2O_2$), and hydroxyl radical ($OH^{\cdot}$), whereas singlet oxygen is formed during energy transfer from the triplet chlorophyll to triplet oxygen (Asada 1999).

The formation of ROS is the inevitable price that cells pay for the maintenance of their vital functions. ROS are produced in the chloroplast (in the light), in the mitochondria, and in the peroxisomes. The main species generated during the photosynthetic electron transport in chloroplasts are presented by singlet oxygen, superoxide radical ($O_2^{\cdot -}$), $H_2O_2$, and OH radical (Asada 1996, 1999; Kreslavski et al. 2012b). Excess of light is not used in photosynthesis, and the rate of formation of ROS may significantly increase. Numerous data indicate that under certain conditions ROS may be involved in the photodamage of PSII (Aro et al. 1993, 2005). For example, in model systems such as thylakoid membranes or partially purified particles of PSII, degradation of D1 protein was stimulated by the formation of OH radical (Miyao et al. 1995) or $^{1}O_2$ (Hideg et al. 1994; Okada et al. 1996).

The high level of ROS is observed during photoinhibition. This led to the assumption that amounts of ROS, in particular $^{1}O_2$, may be the primary reason of damage to the PSII (Krieger-Liszkay et al. 2008; Vass et al. 2007; Vass and Aro 2008; Vass and Cser 2009). The damaging effect of $^{1}O_2$ was protruded in a so-called *hypothesis of the acceptor side and charge recombination* (Setlik et al. 1990; Vass et al. 1992; Keren et al. 1997; Hideg et al. 2007). It was suggested that the stabilization of $Q_A$ by protonation or the formation of double-reduced primary acceptor leads to an increase in the lifetime of the ion-radical pair in the RC of the PSII (Setlik et al. 1990) and to an increase of the probability of formation of triplet chlorophyll molecules and, consequently, to the formation of singlet oxygen (Vass et al. 1992). However, according to some authors, the formation of $^{1}O_2$ is possible for unstabilized $Q_A^-$, implying that all the electron carriers (in particular the molecules of PQ pool) strongly recovered at high light intensity (Vass and Aro 2008; Vass and Cser 2009). According to the *hypothesis of the donor side*, strong light induces violations in the donor side of PSII. It could be, for example, due to acidification of the lumen due to proton migration through the thylakoid membrane. This may lead to a leak of calcium ions from the OEC of PSII and to its inactivation (Callahan et al. 1986; Theg et al. 1986; Klimov et al. 1990).

Singlet oxygen may be produced by the interaction of molecular oxygen with the $^{3}$P680 of the RC of PSII, which is formed under certain conditions in a process of recombination of ion-radical pair P680$^+$–Phe$^-$ (formation of $^{3}$[P680$^+$ Phe$^-$]) (Vass and Cser 2009). The redox potential of Phe and $Q_A$ might play a crucial role in this process. Some authors prefer *in vitro* studies of the damage to photosystems using isolated thylakoid membranes, which lack or are characterized by rather weak recovery together with weak activity of PSII. Moreover, sometimes the extreme (nonphysiological) conditions are applied to such samples, for example, strong light saturation. Such results often contradict to the data obtained *in vivo*, which show that the direct photodamage of PSII depends primarily on light intensity, but neither on the presence of molecular oxygen and/or ROS nor on the rate of photosynthetic electron transport (Murata et al. 2007; Nishiyama et al. 2011). A definite correlation between the initial rate of photodamage and light intensity exists, which cannot be changed either by withdrawal of oxygen or by inhibition of electron transport. Studies of the action spectrum of photodamage revealed bands in the UV and blue region. All this implies the possible existence of some blue-light photoreceptor in the PSII (Hideg et al. 1994; Ohnishi et al. 2005; Hakala et al. 2005).

In intact cells, photodamage of PSII is always accompanied by the recovery of damaged photosystem. Thus, photoinhibition *in vivo* is a result of the imbalance between photodamage and recovery of the damaged PSII (Nishiyama et al. 2001, 2011). Studies of cyanobacteria *in vivo* allowed *time and space* separation of the events of direct damage of PSII and its recovery. Photodamage can be detected in the presence of suitable inhibitors of protein synthesis – chloramphenicol and lincomycin - which block de novo protein synthesis and, consequently, the recovery of PSII. Recovery of PSII can be detected after transfer of the photosynthetic organism from strong to weak light.

Such methodology led to the development of a new concept, which explains the photoinhibition of PSII (Nishiyama et al. 2001, 2005, 2006, 2011; Allakhverdiev et al. 2005a, b). It was found that ROS are not involved in direct damage to PSII, but they inhibit recovery. Such inhibition results in a decrease in photochemical activity of the PSII (Allakhverdiev and Murata 2004; Nishiyama et al. 2001, 2005, 2011). Similar results have been also obtained in the experiments with higher plants (Moon et al. 1995; Takahashi and Murata 2005). Cellular mechanisms of resistance to stress factors had emerged at early stages of evolution. Evidently, the protective systems of higher plants and cyanobacteria have reasonable similarities. According to a new paradigm, the primary target of photodamage is the oxygen-evolving complex (OEC), primarily manganese-containing cluster. The secondary target of photodamage is the photochemical reaction center. The latter is not the key target for ROS action, as it was considered before. ROS inhibit the recovery of PSII (Nishiyama et al. 2001, 2005, 2006, 2011; Murata et al. 2007; Hakala et al. 2005; Ohnishi et al. 2005). ROS, for example, $H_2O_2$ inhibit the recovery of PSII by suppression of protein synthesis *de novo* - in particular the D1 protein (Nishiyama et al. 2001, 2004). ROS primarily inhibit translation of the *psbA* gene for the precursor of D1 (Fig. 2.4).

The double mutant of *Synechocystis* defective in *katG* and *tpx* genes for catalase and thioredoxin peroxidase, respectively, which are responsible for the neutralization of $H_2O_2$, displayed an elevated susceptibility to photodamage of PSII (Nishiyama et al. 2001). However, in the presence of the inhibitor of protein synthesis, chloramphenicol, photodamage was not amplified. This confirmed the statement that ROS suppress recovery of the damaged PSII at the level of *de novo* protein synthesis (Nishiyama et al. 2001).

It should be also noted that inhibition of protein synthesis and inhibition of PSII recovery by ROS are reversible under conditions of a moderate oxidative stress (Nishiyama et al. 2001; Kojima et al. 2007). The reason of this reversibility is explained by the molecular mechanism of translational inhibition by ROS. It is known that elongation factor G (EF-G) (Fig. 2.5), which catalyzes the translocation of peptidyl-tRNA, is sensitive to oxidation with $H_2O_2$ (Cabiscol et al. 2000). In *Synechocystis*, the disulfide bonds of EF-G have been oxidized by ROS and reduced by thioredoxin (Kojima et al. 2009). Moreover, EF-G of *Synechocystis* (Lindahl and Florencio 2003) and of spinach chloroplasts (Balmer et al. 2003) may be easily purified by affinity chromatography on thioredoxin. This indicates high affinity of thioredoxin to EF-G, suggesting that EF-G may serve as a potential target for thioredoxin.

The nature of the relationship between the processes of translation, changes in redox potential, and photosynthesis suggests that the photosynthetic electron transport activates the protein synthesis in

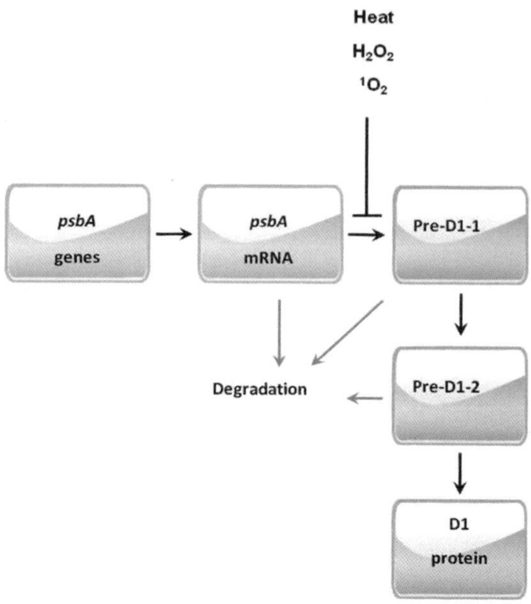


**Fig. 2.4** A scheme of *psbA* gene expression and protein D1 synthesis. Hypothetical targets for inhibiting effects of heat and photoinduced stress adopted from Allakhverdiev et al. (2002). Pre-D1-1 and pre-D1-2, protein D1 precursors. The D1 protein is encoded by *psbA* genes

redox-dependent manner (Nishiyama et al. 2011). ROS-dependent inhibition of translation is the result of violations of the redox regulation of this process (Fig. 2.5). In this scheme, translation elongation factor EF-G works as the molecular switch operating in the simple but elegant redox-dependent manner. EF-G ensures elongation of translation in the reduced form, but it does not support translation in its oxidized form in the presence of ROS.

It is obvious that the recovery of PSII requires balanced and cooperative action of different factors, such as light quality, synthesis of ATP, and presence of specific chaperones and proteases. Many details of this process still require detailed investigations. Similarly to $H_2O_2$, singlet oxygen also inhibits the recovery of PSII (Nishiyama et al. 2004). Singlet oxygen is generated during photosynthesis as a result of formation of the Chl triplet state in the RC of PSII. It is a very strong oxidant, and it can damage many components of the photosystems (Triantaphylides et al. 2008). Production of singlet oxygen, induced by rose bengal or ethyl eosin, stimulated visible damage to PSII. However, chloramphenicol completely prevented such effect of the inducers (Nishiyama

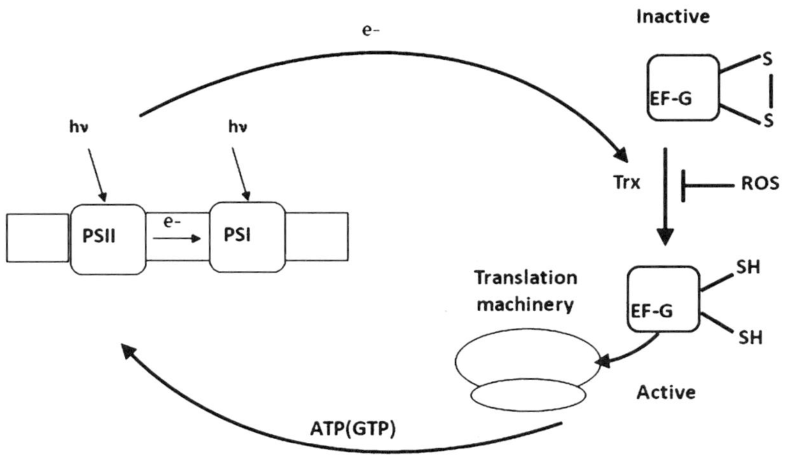


**Fig. 2.5** A schematic representation of the interactive regulation of photosynthesis and protein synthesis. In light, reducing equivalents that are generated by the photosynthetic transport of electrons (e−) are transmitted to EF-G via thioredoxin-dependent (Trx-dependent) redox pathways. The resultant reduction of EF-G activates the translational machinery, leading to induction of the synthesis of the D1 protein and other proteins. The light-dependent synthesis of the D1 protein and other proteins also requires ATP and GTP as sources of energy and is tightly coupled with the co-translational insertion of D1 into the PSII complex. The repair of PSII, in turn, requires the synthesis *de novo* of proteins and, in particular, that of the D1 protein. High levels of ROS, produced by the photosynthetic machinery, interrupt the redox signals by maintaining EF-G in an oxidized state, and, as a consequence, they suppress protein synthesis that is required for the repair of PSII (Adopted from Nishiyama et al. 2011)

et al. 2001). Moreover, disruption of a gene for the biosynthesis of α-tocopherol, which is known as the effective neutralizer of $^1O_2$, hampered the recovery of PSII with no effect on its photodamage (Inoue et al. 2011). All these experiments show that singlet oxygen also inhibits the recovery of damaged PSII.

Singlet oxygen did not affect the accumulation of mRNA for *psbA* gene encoding the precursor of D1 protein. Pulse labeling of the de novo synthesized proteins with [$^{35}$S]-methionine demonstrated that $^1O_2$ inhibited the synthesis of proteins of photosynthetic machinery, including the D1 protein (Fig. 2.4). It is very possible that singlet oxygen, similarly to $H_2O_2$, acts at the stage of translation elongation (Nishiyama et al. 2004). Overexpression of EF-G in *Synechocystis* supported *de novo* protein synthesis, including D1 protein, even under conditions of oxidative stress. This important data demonstrate the crucial role of EF-G for regulation of cellular responses by different ROS (Kojima et al. 2007). This mechanism is valid for non-photosynthesizing organisms as well (Ayala et al. 1996; Dukan and Nystrom 1999). As already mentioned above, the primary reaction in the photodamage may be the absorption of light quanta by manganese cluster in the OEC, followed by the dissociation of excited manganese ions (Tyystjärvi et al. 2002; Nishiyama et al. 2004, 2005; Allakhverdiev and Murata 2004; Ohnishi et al. 2005; Hakala et al. 2005) and the formation of long-lived $P680^+$, which is a strong oxidant (1.12 V) (Klimov et al. 1978; Allakhverdiev et al. 2010, 2011) and can damage protein D1 (Andersson and Aro 2001).

Irreversible photoinhibition of PSII takes place when cells of *Synechocystis* are exposed to a sufficiently strong light for extended period of time (Allakhverdiev et al. 2005a). Illumination of wild-type cells at 20°C (low temperature; optimum growth temperature for this organism lies in a range of 30–34°C) for 2 h with high light (2,500 μM photons $m^{-2}$ $s^{-1}$) completely and irreversibly inhibited $O_2$ evolution, whereas inactivation of the photochemical RCs of PSII was reversible.

Irreversible photoinhibition was intensified in wild-type cells at low temperatures as well as in the mutant *desA*$^-$/*desD*$^-$ cells at optimal temperatures. These mutant cells are defective in two genes for fatty acid desaturases, and they are characterized by the decreased fluidity of their cytoplasmic and thylakoid membranes. Under conditions of irreversible photoinhibition, cells synthesized the precursor of the D1 protein, but they could not produce the mature and active form of D1. Thus, irreversible photoinhibition of oxygen-evolving activity of PSII with strong light may be due to inability of cells to process the precursor of the D1. The D1 processing, therefore, may be affected by temperature and by the fluidity of the membranes (Allakhverdiev et al. 2005a).

Maintenance of the protein synthesis, in particular D1, requires sufficient intensity of the electron transport and ATP (Aro et al. 1993; Andersson and Aro 2001). Therefore, these parameters may be important for the recovery of the damaged PSII, but not so important in a process of the direct photodamage (Nishiyama et al. 2001). It was demonstrated that the rate of photodamage is proportional only to light intensity and it does not depend on the inhibition of electron transport in PSII, or on acceleration of electron transport in PSI, or on inhibition of ATP synthesis (Allakhverdiev et al. 2005b).

## Mechanisms that Protect Cells from Photoinhibition

During evolution plants have acquired a number of mechanisms to protect their photosynthetic apparatus from damaging by light. A detailed analysis of such mechanisms allows us to emphasize several basic pathways to protect the photosynthetic machinery (Sarvikas 2010; Kreslavski et al. 2007):

1. Activation of nonphotochemical quenching (NPQ) allows a downthrow of a significant portion of the excitation energy. In this way cells reduce accumulation of ROS (Müller et al. 2001) and prevent photoinhibition. Mutants with reduced efficiency of NPQ show increased sensitivity to light and, thus, demonstrate the important role of this process for photoprotection (Havaux and Niyogi 1999; Sarvikas et al. 2006). Light-induced electron transfer in photosynthetic ETC leads to the ejection of protons from stroma into lumen and to the formation of ΔpH, which is required for the synthesis of ATP from ADP and phosphate by the $CF_o$-$CF_1$ ATP synthase. This pH gradient is a key element in the initiation of NPQ. It is believed that light-regulated NPQ involves three components: energy-dependent quenching (qE), photoinhibitory quenching (qI), and the quenching associated with the redistribution of light energy from PSI to PSII (qT). The constitutive light-independent NPQ in leaves is also described (Kramer et al. 2004). The latter happens in the antenna chlorophyll or during a charge recombination in RCs of photosystems.

Nonphotochemical dissipation of the excess photons into heat is one of the most effective mechanisms for protection against excess light energy. In principle, the dissipation can occur in RCs of photosystems and in antenna (Schreiber and Kriger 1996). Anyhow, dissipation of the energy absorbed by chlorophyll into heat reduces the rate of formation of singlet oxygen and superoxide anion radical – the components of the signaling pathway, which involves lipids and redox compounds and which can trigger apoptosis (Demmig-Adams and Adams 2002; Kreslavski et al. 2012b).

It is known that chlorophyll-binding protein PsbS of PSII, zeaxanthin, and other xanthophylls participate in heat dissipation (Bugos and Yamamoto 1996; Niyogi et al. 2005). The ΔpH can be set up by cyclic electron flow around PSI as well as by photorespiration, which is accompanied by the reduction of $O_2$ into the $O_2^{\cdot -}$ followed by the formation of $H_2O_2$ in the reaction of disproportionation. The latter can be turned into $H_2O$ with photoreduced compounds as electron donors (water-water cycle) (Asada 1994, 1999). All these events are associated with

NPQ of chlorophyll fluorescence through nonphotochemical processes.

2. Accumulation of low molecular weight antioxidants such as tocopherols, carotenoids, and vitamin $B_6$ can be amplified by high light and protect the leaf and the photosynthetic machinery from light-induced photodamage (Havaux et al. 2005, 2009; Trebst et al. 2002; Krieger-Liszkay et al. 2008). In particular, singlet oxygen is efficiently quenched by two β-carotenes associated with core complex of PSII.
3. Recovery and synthesis of proteins, like D1 protein, damaged by oxidative stress.
4. Light-induced redistribution of chloroplasts in leaves with the change in the rotation of a long axis toward light direction.
5. The accumulation of compounds that absorb an excess of light, such as anthocyanins and carotenoids (Steyn et al. 2002; Solovchenko and Merzlyakov 2008).
6. Changes in fatty acid unsaturation in the lipids of thylakoid membranes. Thylakoid membranes are characterized by the presence of highly unsaturated fatty acids, 18:3 and 16:3, which constitute two thirds of the total amount of fatty acids in thylakoids. Tasaka et al. (1996) studied the effect of fatty acid (FA) unsaturation on the activity of PSII in *Synechocystis*. The importance of FA unsaturation for the recovery of PSII after photodamage has been postulated (Tasaka et al. 1996; Murata and Nishiyama 1998).

   Similar findings were obtained in studies of higher plants (Vijayan and Browse 2002). *Arabidopsis* mutants with reduced levels of unsaturated fatty acids (*fad5*, *fad6* and *fad3-2/fad7-2/fad8*) were characterized by higher sensitivity to photoinhibition than the wild-type plants. A detailed studies of the most light-sensitive triple mutant showed that the photoinactivation of PSII happens at the same rate in the mutant and in wild-type plants. As expected, the reduced amounts of polyunsaturated FAs in the membranes of the mutant plants affected the recovery of the photoinactivated PSII, which was slower in the mutant at temperatures below 27°C. This demonstrates the important role of unsaturation and, especially, of trienoic fatty acids, in the repair of the photodamaged PSII.
7. Change the heterogeneity of the antenna and the acceptor side of PSII during the development of responses of the photosynthetic machinery to stress (Mehta et al. 2011). Heterogeneity of the antenna is determined by its size and by energetic association between the subunits of PSII. Three types of antenna of PSII have been defined, which differ in size, location, and efficiency of energy transfer (Hsu and Lee 1991). The biggest size has a component of the antenna of PS 2 of α-type, localized in the granal region and considered as the most effective.

It is assumed that protective mechanisms against photoinhibition include cyclic electron transport in PSII (Allakhverdiev et al. 1997; Nishiyama et al. 2011) and PSI (Bukhov and Carpentier 2000; Miyake et al. 2002), aggregation of thylakoid proteins (Roberts et al. 1991), and photorespiration (Takeba and Kozaki 1998). Electron transport initiated in PSI, probably, aims to protect PSII against photoinhibition (Bukhov and Carpentier 2000; Thomas et al. 2001).

Redistribution of excitation energy between the photosystems associated with so-called states 1 and 2 is considered as one of the fastest stress-protective mechanisms. In state 1, the antenna is connected to PSII located in granal area with stacked thylakoid membranes. State 1 prevails when plants are irradiated with far-red light. State 2 prevails when they are irradiated with blue or red light. In state 2, the antenna migrates and binds to PSI, located in the lamellar region with concomitant unstacked thylakoid membranes (Allen 1992; Wollman 2001; Rochaix 2007; Tikkanen et al. 2008, 2010; Chuartzman et al. 2008). These transitions are regulated by protein kinase (STN7 in *Arabidopsis* or Stt7 in *Chlamydomonas*), which is involved in phosphorylation of LHCII proteins (Depege et al. 2003; Bellafiore et al. 2005; Bonardi et al. 2005). The activity of this kinase is controlled by the redox state of PQ pool and also depends on a complex $b_6f$ (Zito et al. 1999; Vener et al. 1997). Increase in the activity of PSII leads to reduction of the PQ pool or $b_6f$ (Qo site), activation of STN7, and phosphorylation of the light-harvesting complex (LHC) proteins with concomitant transition to state 2. Dephosphorylation of core complex of PSII and of LHC leads to transition to state 1 (Tikkanen et al. 2006).

Loss of Qo in the $b_6f$ complex due to mutation in the chloroplast gene *petD* in *Chlamydomonas* caused inability of cells to activate the protein kinase Stt7, which phosphorylates LHCII (Zito et al. 1999). There should be a correlation between the intensity and quality of light, phosphorylation of proteins of PSII, and conversion from state 1 to state 2 (and vice versa), as it was recently demonstrated in the *Arabidopsis* (Tikkanen et al. 2010). STN7-dependent phosphorylation of proteins of PSII leads to a balanced distribution of energy between PSII and PSI. Loss of activity of the STN7 kinase in the mutant *stn7* of *Arabidopsis* resulted in overexcitation of PSII even at low light. Thus, phosphorylation of thylakoid proteins in chloroplasts optimizes electron transfer under fluctuating light (Tikkanen et al. 2010).

It is known that nonphotochemical dissipation of light energy in both photosystems increased with increasing light intensity, but the mechanisms of dissipation in PSII and PSI are different (Bukhov et al. 2001). While thermal energy dissipation in PSI occurs in the antennae and, probably, through the oxidation of $P700^+$ (Nuijs et al. 1986), dissipation in the PSII is observed both in the RCs and in the antenna. It is believed that nonphotochemical quenching (NPQ) involves two processes activated by acidification of the lumen – transformation of violaxanthin to zeaxanthin and protonation of several protein components of the LHCII associated with PSII. This process is induced by strong light at decreasing pH values inside the thylakoid (Choudhury and Behera 2001). Low pH causes activation of violaxanthin de-epoxidase – an enzyme that converts violaxanthin to zeaxanthin – and protonation of PSII-associated protein PsbS (Li et al. 2002, 2004). That epoxidase switches the mechanism that transforms the energy of the absorbed light into heat (Demmig-Adams and Adams 2002).

Xanthophylls can take energy from the $^1$Chl, transforming it into heat, and neutralize Chl$^3$ and singlet oxygen. This means that they act as the antioxidants. X-ray analysis of the core complexes of PSII showed that two molecules of β-carotene associated with the protein D2 can also serve as the quenchers of singlet oxygen (Tefler 2005). These molecules protect protein complex against photodamage. Their protective effect is though more pronounced toward D2 than D1.

Both types of fluorescence quenching of Chl – photochemical and NPQ – can be easily measured by the Chl *a* fluorescence of intact leaves. The value of NPQ in leaves is a reliable indicator of the concentration of photosynthetic complexes, in which light energy is deactivated (Gilmore et al. 1995), and of the ability to "dump" the excess of light energy, which is not required for $CO_2$ assimilation. The leaves of plants grown in the shade have lower values of NPQ than the leaves of plants grown under strong light. Therefore, such leaves, which are not acclimated to high light, experience severe damage of photosynthetic machinery under strong light and complete loss of pigments (Burritt and Mackenzie 2003).

Preincubation of plants under moderate stress (*hardening*) may strengthen the stability of the photosynthetic machinery. Heat pretreatment of wheat plants at 40°C (but not at 42°C!) led to an increase in the rate of recovery of PSII after a second repetition of heat stress and also after exposure of plants to strong light (Fig. 2.6; Kreslavski et al. 2008). Such hardening resulted in higher rates of dark respiration and cyclic photophosphorylation as well as in higher activity of antioxidant enzymes (Kreslavski et al. 2008, 2009).

In addition, any stress factors (high-intensity light, UVB, heat, salt, hyperosmotic stress) lead to induction of genes encoding proteins of the *general stress response* (GSRPs) that have been previously considered as heat shock proteins (HSPs) (Los et al. 2010). The presence of these

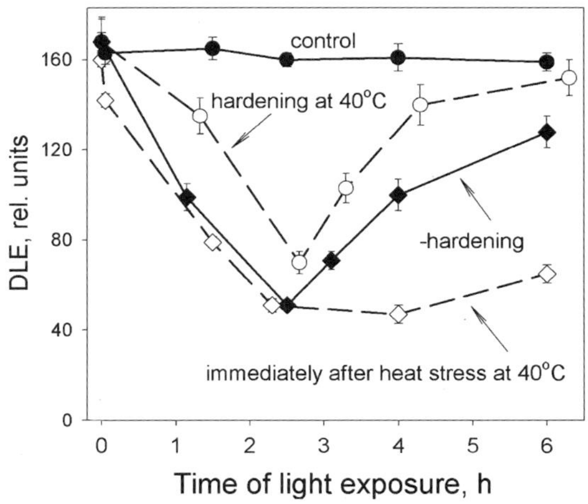


**Fig. 2.6** Effect of heat hardening at 40°C on photoinhibition. Some of the seedlings were pretreated (○) or non-pretreated (◇,◆) by heating, grown for 48 h at 120 μmol $m^{-2}$ $s^{-1}$ and then exposed to strong light ($I$ = 900 μmol $m^{-2}$ $s^{-1}$) either immediately (○,◆) or after heat treatment at 40°C for 20 min (◇). Other non-pretreated seedlings (control) were grown for 48 h at 120 μmol $m^{-2}$ $s^{-1}$ and then exposed to light at $I$ = 120 μmol $m^{-2}$ $s^{-1}$ for 6 h (●). Means ± SE ($n$ = 4) (Kreslavski et al. 2008)

proteins in the cells under subsequent stress conditions contributes to their stress resistance (Zorina et al. 2011). The following conclusions may be drawn from the above discussion:

1. Numerous data obtained *in vivo* suggest that both Chl-dependent and Chl-independent pathways are required to explain the process of photoinhibition *in vivo*.
2. Photosynthesis and protein synthesis interact with each other and are regulated cooperatively during photoinhibition.
3. Photoinhibition is regulated with such factors as redox systems of photosynthetic machinery, ATP content, and ratio of oxidants/antioxidants.
4. Repair of PSII is most sensitive to the inactivation by ROS.
5. Translation elongation factor EF-G works as a molecular switch, which, in its reduced form, allows translation and repair of PSII and which, oxidized by ROS, blocks translation and repair of PSII.

## Heat Stress

### The Influence of High Temperatures on the Physiological Processes in Plants

It is known that heat stress may inhibit growth and development of plants or reduce their productivity (Hall 2001). This may be due to heat stress itself or due to formation of ROS and other oxidants induced by heat stress. High temperatures affect the diffusion rates and, consequently, the rate of chemical reactions. They also change the structure of proteins, up to complete denaturation, and alter the activity of many enzymes. Disturbances in membranes structure, in their permeability and fluidity, may cause partial or total disintegration of cells. This is naturally accompanied by damage of the photosynthetic and mitochondrial ETC.

### Effect of High Temperature on Photosynthesis

Photosynthesis is one of the most sensitive physiological processes to heat (Berry and Björkman 1980; Sharkey 2005). PSII was always regarded as the most vulnerable part of the photosynthetic machinery (Aro et al. 1993; Andersson and Aro 2001; Allakhverdiev et al. 2008). However, even if the rate of photosynthesis is greatly reduced under moderate heat stress, the damage to PSII (irreversible decrease of photochemical activity) may be negligible. The real damage to the PSII happens only at sufficiently high temperatures that usually exceed 45°C (Yamane et al. 1998; Sharkey 2005). Under moderate heat stress and relatively low light, the activity of PSII is eventually restored (Enami et al. 1994; Kreslavski and Khristin 2003; Kreslavski et al. 2009). Among various components of PSII, the oxygen-evolving complex (OEC) is most sensitive to heat (Katoh and San Pietro 1967; Berry and Björkman 1980; Mamedov et al. 1993; Allakhverdiev et al. 1994, 1996; Havaux and

Tardy 1996; Carpentier 1999). One reason for such extreme sensitivity is heat-induced loss of two of four manganese ions and some proteins associated with Mn-containing OEC (Nash et al. 1985; Enami et al. 1994, 1998).

In addition to OEC, the temperature-sensitive targets in photosynthetic cells are ribulose-1,5-bisphosphate carboxylase (RuBisCo, a key enzyme of Calvin-Benson cycle) and the systems of protein translation and protein phosphorylation (Sharkey 2005). $CO_2$ fixation machine can be markedly inhibited even at moderate stress (Feller et al. 1998; Sharkey 2005; Kreslavski et al. 2008). In 1981, it was already known that the activity of RuBisCo drops sharply with increasing temperature (Weis 1981). This reduced activity of the RuBisCo correlated with inhibition of photosynthesis by heat (Law and Crafts-Brandner 1999). The efficiency of NPQ and activity of RuBisCo were more sensitive to heat stress than the quantum yield of PSII estimated as the $F_v/F_m$ ratio (Law and Crafts-Brandner 1999; Salvucci and Crafts-Brandner 2004). The value of the minimum Chl fluorescence ($F_0$ level) often indicates a critical temperature for the PSII inactivation, and it increases under heat stress. It appeared, however, that the value of $F_0$ may differ from species to species. In some plants (potato, tobacco, etc.), the heat-induced $F_0$ does not reflect inactivation of the photochemical reactions of PSII, but rather the reduction of $Q_A$ by PQ in the dark (Sazanov et al. 1998; Yamane et al. 2000; Bukhov and Carpentier 2000).

Furthermore, PSII remained active at temperatures that inhibit the activase of RuBisCo, the enzyme required for $CO_2$ assimilation (Salvucci and Crafts-Brandner 2004; Tang et al. 2007). The reaction of energy transfer from the LHC of the PSII to its core complex is also sensitive to heat. Even at 35°C, the migration of the LHC from PSII to PSI may occur (Pastenes and Horton 1996). It is also assumed that the main targets (or even sensors) of heat are associated with thylakoid membranes that change their physical properties under elevated temperatures (Horvath et al. 1998; Sharkey 2005). These changes may be provoked by the formation of ROS or products of lipid peroxidation (Allakhverdiev et al. 2008; Kreslavski et al. 2012b). One of the factors contributing to changes in thermal sensitivity of PSII is the change of fluidity of thylakoid membranes (Aminaka et al. 2006). However, the existence of thylakoid-associated thermo sensor is still under question (Horvath et al. 1998; Los and Murata 2004).

Plants use several mechanisms to protect the photosynthetic machinery from heat stress, which may be conventionally divided into slow and fast mechanisms. Slow mechanism implies the synthesis of proteins of general stress response (including HSPs), antioxidants, and membrane lipids (Kreslavski et al. 2007, 2012a). This increases the activity of antioxidant enzymes, and content of low molecular weight compounds serves as protectors of photosynthetic machinery against free radical oxidation (Sairam et al. 2000; Logan 2006). Fast processes are induced with phosphorylation and proceed with migration of LHC from PSII to PSI (Allen 1992; Mohanty et al. 2002; Shapiguzov et al. 2010). Fast processes also include the stabilization of PSII and thylakoid membranes due to conversion of violaxanthin to zeaxanthin, which enhances the resistance of PSII to heat (Havaux et al. 1996, 2004; Szabó et al. 2005). Redistribution of photoactive molecules in the inactive pool of PQ could also be regarded as a protective mechanism to limit the inactivation of PSII at the acceptor side (Pshybytko et al. 2008). Moderate or mild heat stress can also induce the increase in the rate of cyclic electron flow around PSI in the light (Bukhov and Carpentier 2000; Bukhov et al. 2002), which can diminish the formation of ROS that damage the PSII (Heber 2002). The activation of cyclic electron flow around PSI may be due to heat-induced reduction of the PQ pool by stromal electron donors in the dark (Sharkey 2005; Havaux 1996). Even at 36°C flow of electrons from the stroma to the PQ pool was indicated (Yamane et al. 2000).

The attempts to determine the compounds that may stabilize PSII under heat stress led to identification of carotenoids (Tefler and Barber 1995; Tefler 2005), isoprene (Sharkey 2005), several HSPs (Vierling 1991; Barua et al. 2003), and external PSII proteins –

PsbO, PsbP, PsbQ, PsbR, PsbU, and PsbV (Kimura et al. 2002; Roose et al. 2007). There is evidence that externally provided purified HSPs may protect PSII from damage induced by high temperature (Heckathorn et al. 1998b, 2002; Schroda et al. 1999; Neta-Sharir et al. 2005). The external proteins of PSII may be also important to stabilize photosynthetic machinery. The extrinsic protein of the PSII PsbU stabilized the PSII against heat-induced inactivation (Nishiyama et al. 1999). The inactivation of the *psbU* gene by site-directed knockout significantly diminished the thermal stability of OEC. Changes in FAs and lipid composition can affect photosynthesis by changing the bulk physical properties of the membranes or through the specific interactions of lipids with proteins (Vigh et al. 2007).

In *Chlamydomonas*, saturation of membrane lipids enhanced the thermal stability of PSII (Sato et al. 1996). In cyanobacteria, however, such effect was not so pronounced (Gombos et al. 1992, 1994b; Wada et al. 1994). Gombos et al. (1994b) monitored the changes in heat tolerance after the elimination of dienoic and trienoic fatty acids from membrane lipids of *Synechocystis* (the above mentioned *desA*$^-$/*desD*$^-$ mutant). Heat resistance of photosynthetic oxygen evolution was not associated with fatty acid unsaturation. Another set of data, however, suggests that the changes in the lipid desaturation levels of plasma and thylakoid membranes are among major factors responsible for the change in the thermosensitivity of PSII in *Synechocystis* (Aminaka et al. 2006). Research should be focused in several questions concerning susceptibility of photosynthetic machinery to heat, and its protection toward heat stress remains unanswered.

1. How heat-induced ROS regulate the activity of genes for HSPs?
2. How thylakoid membranes regulate the redox potential of components of photosynthetic ETC, particularly of the PQ pool, under heat stress?
3. How exactly heat stress inhibits the recovery of PS2? What is the individual impact of heat and ROS to this process?
4. Is there a specific heat sensor in plant-type cells, and what can one say on its nature and its location?
5. What are the molecular mechanisms that control of thermal stability of the photosynthetic machinery, and how we can enhance this stability?

## General Conclusion

It was assumed that enzymes that scavenge ROS may provide protection against different stressors (Allen 1995). Superproducers of glutathione increased antioxidant activity and resistance to photoinhibition in cotton (Payton et al. 2001). Transgenic tobacco that overproduces cytosolic ascorbate peroxidase, chloroplast SOD, or ascorbate peroxidase showed increased resistance to methyl viologen and photooxidation (Allen et al. 1997; Kwon et al. 2002). At the same time, overexpression of ascorbate peroxidase in tobacco chloroplasts could not protect photosynthetic machinery against ozone treatment (Torsethaugen et al. 1997).

In recent years, mutants and genetically modified plants have been constructed that can overproduce antioxidant enzymes, LMW protectors, and various defense proteins (Fig. 2.7). However, the results were not always unambiguous and straightforward because of the diversity of the investigated physiological targets, the strength of stressor, and the physiological state of plants (Allen 1995; Allen et al. 1997; Torsethaugen et al. 1997; Payton et al. 2001).

Our understanding of the mechanisms of stress resistance of the photosynthetic machinery is necessary for estimation of the possible impact of the global changes of climate on vital activity of humankind in the future. This knowledge is also important for the design of new biochemical pathways or rearrangement of the existing signaling and biochemical networks that would help to create stress-resistant plant species and varieties.

**Acknowledgement** This work was partially supported by grants from the Russian Foundation for Basic Research (Nos: 11-04-01389, 11-04-92690, 12-04-92101, 12-04-01035, and 12-04-00473), by a grant from the Molecular and Cell Biology Program of the Russian Academy of Sciences, by a grant from Russian Ministry of Education and Science (№ 16.740.11.0176), and by BMBF, Bilateral Cooperation between Germany and Russia (no. 8125).

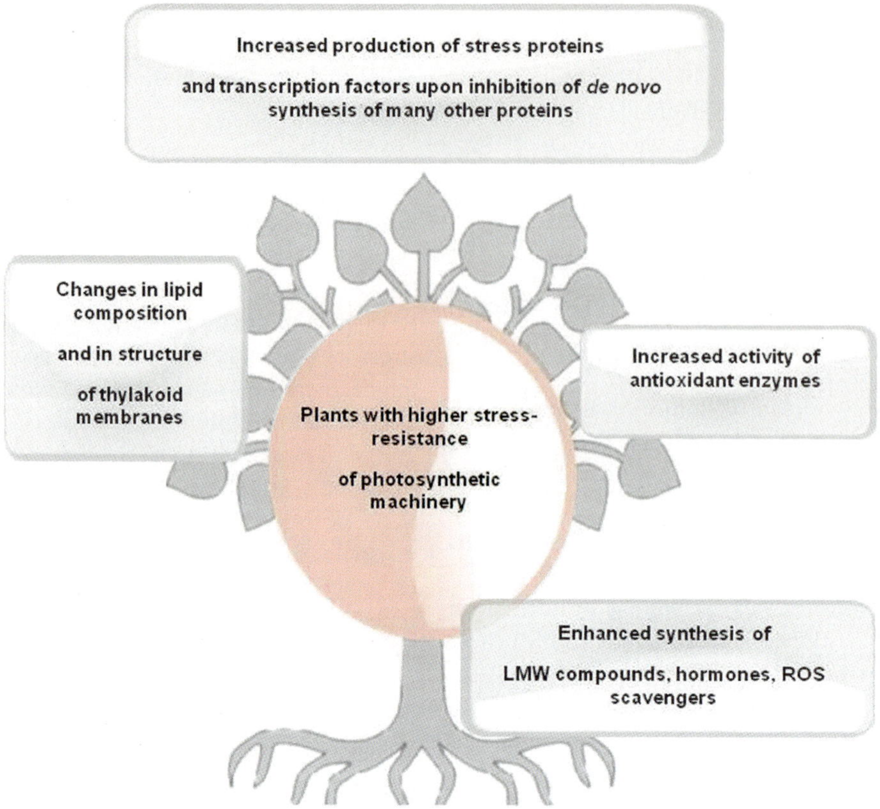


**Fig. 2.7** A hypothetical scheme of pathways to increasing the resistance of photosynthetic machinery through construction of stress-resistant transgenic plants. Adopted from Grover et al. (2000), Chinnusamy et al. (2005), and Kreslavski et al. (2007)

## References


Adir N, Zer H, Shochat S, Ohad I (2003) Photoinhibition – a historical perspective. Photosynth Res 76:343–370

Allakhverdiev SI (2011) Recent progress in the studies of structure and function of Photosystem II. J Photochem Photobiol B Biol 104:1–8

Allakhverdiev SI, Murata N (2004) Environmental stress inhibits the synthesis *de novo* of proteins involved in the photodamage – repair cycle of photosystem II in *Synechocystis* sp. PCC. 6803. Biochim Biophys Acta 1657:23–32

Allakhverdiev SI, Setlikova E, Klimov VV, Setlik I (1987) In photoinhibited photosystem II particles pheophytin photoreduction remains unimpaired. FEBS Lett 226:186–190

Allakhverdiev SI, Klimov VV, Carpentier R (1994) Variable thermal emission and chlorophyll fluorescence in photosystem II particles. Proc Natl Acad Sci USA 91:281–285

Allakhverdiev SI, Feyziev YM, Ahmed A, Hayashi H, Aliev JA, Klimov VV, Murata N, Carpentier R (1996) Stabilization of oxygen evolution and primary electron transport reactions in photosystem II against heat stress with glycinebetaine and sucrose. J Photochem Photobiol 34:149–157

Allakhverdiev SI, Klimov VV, Carpentier R (1997) Evidence for the involvement of cyclic electron transport in the protection of photosystem II against photoinactivation: influence of a new phenolic compound. Biochemistry (USA) 36:4149–4154

Allakhverdiev SI, Nishiyama Y, Suzuki I, Tasaka Y, Murata N (1999) Genetic engineering of the unsaturation of fatty acids in membrane lipids alters the tolerance of *Synechocystis* to salt stress. Proc Natl Acad Sci USA 96:5862–5867

Allakhverdiev SI, Kinoshita M, Inaba M, Suzuki I, Murata N (2001) Unsaturated fatty acids in membrane lipids protect the photosynthetic machinery against salt-induced damage in *Synechococcus*. Plant Physiol 125:1842–1853

Allakhverdiev SI, Nishiyama Y, Miyairi S, Yamamoto H, Inagaki N, Kanesaki Y, Murata N (2002) Salt stress inhibits the repair of photodamaged photosystem II by suppressing the transcription and translation of *psbA* genes in *Synechocystis*. Plant Physiol 130:1443–1453

Allakhverdiev SI, Hayashi H, Nishiyama Y, Ivanov AG, Aliev JA, Klimov VV, Murata N, Carpentier R (2003a) Glycinebetaine protects the D1/D2/Cytb559 complex of photosystem II against photo-induced and heat-induced inactivation. J Plant Physiol 160:41–49

Allakhverdiev SI, Mohanty P, Murata N (2003b) Dissection of photodamage at low temperature and repair in darkness suggests the existence of an intermediate form of photodamaged photosystem II. Biochemistry (USA) 42:14277–14283

Allakhverdiev SI, Tsvetkova N, Mohanty P, Szalontai B, Moon BY, Debreczeny M, Murata N (2005a) Irreversible photoinhibition of photosystem II is caused by exposure of *Synechocystis* cells to strong light for a prolonged period. Biochim Biophys Acta 1708: 342–351

Allakhverdiev SI, Nishiyama Y, Takahashi S, Miyairi S, Suzuki I, Murata N (2005b) Systematic analysis of the relation of electron transport and ATP synthesis to the photodamage and repair of photosystem II in *Synechocystis*. Plant Physiol 137:263–273

Allakhverdiev SI, Los DA, Mohanty P, Nishiyama Y, Murata N (2007) Glycinebetaine alleviates the inhibitory effect of moderate heat stress on the repair of photosystem II during photoinhibition. Biochim Biophys Acta 1767:1363–1371

Allakhverdiev SI, Kreslavski VD, Klimov VV, Los DA, Carpentier R, Mohanty P (2008) Heat stress: an overview of molecular responses in photosynthesis. Photosynth Res 98:541–550

Allakhverdiev SI, Tomo T, Shimada Y, Kindo H, Nagao R, Klimov VV, Mimuro M (2010) Redox potential of pheophytin a in photosystem II of two cyanobacteria having the different special pair chlorophylls. Proc Natl Acad Sci USA 107(8):3924–3929

Allakhverdiev SI, Tsuchiya T, Watabe K, Kojima A, Los DA, Tomo T, Klimov VV, Mimuro M (2011) Redox potentials of primary electron acceptor quinone molecule (QA)- and conserved energetics of photosystem II in cyanobacteria with chlorophyll a and chlorophyll d. Proc Natl Acad Sci USA 108(19):8054–8058

Allen JF (1992) Protein phosphorylation in regulation of photosynthesis. Biochim Biophys Acta 1098: 275–335

Allen R (1995) Dissection of oxidative stress tolerance using transgenic plants. Plant Physiol 107:1049–1054

Allen RD, Webb RP, Schake SA (1997) Use of transgenic plants to study antioxidant defenses. Free Radical Biol Med 23:473–479

Alscher RG, Donahue JL, Cramer CL (1997) Reactive oxygen species and antioxidants: relationships in green cells. Physiol Plant 100:224–233

Alscher RG, Erturk N, Heath LS (2002) Role of superoxide dismutases (SODs) in controlling oxidative stress in plants. J Exp Bot 53:1331–1341

Aminaka R, Taira Y, Kashino Y, Koike H, Satoh K (2006) Acclimation to the growth temperature and thermosensitivity of photosystem II in a mesophilic cyanobacterium, *Synechocystis* sp. PCC 6803. Plant Cell Physiol 47:1612–1621

Andersson B, Aro E-M (2001) Photodamage and D1 protein turnover in photosystem II. In: Aro E-M, Andersson B (eds) Regulation of photosynthesis. Kluwer Academic, Dordrecht

Apel K, Hirt H (2004) Reactive oxygen species: metabolism, oxidative stress, and signal transduction. Annu Rev Plant Biol 55:373–399

Aro E-M, Virgin I, Andersson B (1993) Photoinhibition of photosystem II: inactivation, protein damage and turnover. Biochim Biophys Acta 1143:113–134

Aro E-M, Suorsa M, Rokka A, Allahverdiyeva Y, Paakkarinen V, Saleem A, Battchikova N, Rintamäki E (2005) Dynamics of photosystem II: a proteomic approach to thylakoid protein complexes. J Exp Bot 56:347–356

Asada K (1994) Production and action of active oxygen in photosynthetic tissue. In: Foyer CH, Mullineaux PM (eds) Causes of photooxidative stress and amelioration of defense systems in plants. CRC Press, Boca Raton

Asada K (1996) Radical production and scavenging in the chloroplasts. In: Baker NR (ed) Photosynthesis and the environment. Kluwer Academic, Dordrecht

Asada K (1999) The water–water cycle in chloroplasts: scavenging of active oxygens and dissipation of excess photons. Annu Rev Plant Physiol Plant Mol Biol 50:601–639

Asada K (2006) Production and scavenging of reactive oxygen species in chloroplasts and their functions. Plant Physiol 141:391–396

Ayala A, Parrado J, Bougria M, Machado A (1996) Effect of oxidative stress, produced by cumene hydroperoxide, on the various steps of protein synthesis. Modifications of elongation factor-2. J Biol Chem 271:23105–23110

Balmer Y, Koller A, del Val G, Manieri W, Schurmann P, Buchanan BB (2003) Proteomics gives insight into the regulatory function of chloroplast thioredoxins. Proc Natl Acad Sci USA 100:370–375

Barber J, Andersson B (1992) Too much of a good thing: light can be bad for photosynthesis. Trends Biochem Sci 17:61–66

Barua D, Downs CA, Heckathorn SA (2003) Variation in chloroplast small heat-shock protein function is a major determinant of variation in thermotolerance of photosynthetic electron transport among ecotypes of *Chenopodium album*. Funct Plant Biol 30:1071–1079

Bellafiore S, Barneche F, Peltier G, Rochaix JD (2005) State transitions and light adaptation require chloroplast thylakoid protein kinase STN7. Nature 433: 892–895

Berry J, Björkman O (1980) Photosynthetic response and adaptation to temperature in higher plants. Annu Rev Plant Physiol 31:491–543

Biel K, Fomina IR, Kreslavski VD, Allakhverdiev SI (2009) Methods for assessment of activity and stress acclimation of photosynthetic machinery in cyanobacteria and symbiotic microalgae. In: Kliner W, Nath Bogchi S, Mohanty P (eds) Protocols on algal and cyanobacterial research, chapter 13. Narosa Publishing House, New Delhi

Boccalandro HE, Rugnone ML, Moreno JE, Ploschuk EL, Serna L, Yanovsky MJ, Casal JJ (2009) Phytochrome B enhances photosynthesis at the expense of water-use efficiency in *Arabidopsis*. Plant Physiol 150:1083–1092

Bonardi V, Pesaresi P, Becker T, Schleiff E, Wagner R, Pfannschmidt T, Jahns P, Leister D (2005) Photosystem II core phosphorylation and photosynthetic acclimation require two different protein kinases. Nature 437:1179–1182

Bugos RC, Yamamoto HY (1996) Molecular cloning of violaxanthin de-epoxidase from romaine lettuce and expression in *Escherichia coli*. Proc Natl Acad Sci USA 93:6320–6325

Bukhov NG, Carpentier R (2000) Heterogeneity of photosystem II reaction centers as influenced by heat treatment of barley leaves. Physiol Plant 110:279–285

Bukhov NG, Heber U, Wiese C, Shuvalov VA (2001) Energy dissipation in photosynthesis: does the quenching of chlorophyll fluorescence originate from antenna complexes of photosystem II or from the reaction center? Planta 212:729–758

Bukhov N, Egorova E, Carpentier R (2002) Electron flow to photosystem I from stromal reductants *in vivo*: the size of the pool of stromal reductants controls the rate of electron donation to both rapidly and slowly reducing photosystem I units. Planta 215:812–820

Burritt DJ, Mackenzie S (2003) Antioxidant metabolism during acclimation of *Begonia x erythrophylla* to high light levels. Ann Bot 91:783–794

Cabiscol E, Piulats E, Echave P, Herrero E, Ros J (2000) Oxidative stress promotes specific protein damage in *Saccharomyces cerevisiae*. J Biol Chem 275: 27393–27398

Caldas T, Demont-Caulet N, Ghasi A, Richarme G (1999) Thermoprotection by glycine betaine and choline. Microbiology 145:2543–2548

Callahan FE, Becker DW, Cheniae GM (1986) Studies on the photoactivation of the water-oxidizing enzyme: II. Characterization of weak light photoinhibition of PSII and its light-induced recovery. Plant Physiol 82:261–269

Campbell D, Hurry V, Clarke AK, Gustafsson P, Öquist G (1998) Chlorophyll fluorescence analysis of cyanobacterial photosynthesis and acclimation. Microbiol Mol Biol Rev 62(3):667–683

Carpentier R (1999) Effect of high-temperature stress on the photosynthetic apparatus. In: Pessarakli M (ed) Handbook of plant and crop stress. Marcel Dekker Inc, New York

Carvalho RF, Campos ML, Azevedo RA (2011) The role of phytochrome in stress tolerance. J Int Plant Biol 53:920–929

Chen TH, Murata N (2011) Glycinebetaine protects plants against abiotic stress: mechanisms and biotechnological applications. Plant Cell Environ 34:1–20

Chen Z, Silva H, Klessig DF (1993) Active oxygen species in the induction of plant systemic acquired resistance by salicylic acid. Science 262:1883–1886

Chen LS, Li P, Cheng L (2008) Effects of high temperature coupled with high light on the balance between photooxidation and photoprotection in the sun-exposed peel of apple. Planta 228:745–756

Chinnusamy V, Jagendorf A, Zhu J-K (2005) Understanding and improving salt tolerance in plants. Crop Sci 45:437–448

Choudhury NK, Behera RK (2001) Photoinhibition of photosynthesis: role of carotenoids in photoprotection of chloroplast constituents. Photosynthetica 39:481–488

Chuartzman SG, Nevo R, Shimoni E, Charuvi D, Kiss V, Ohad I, Brumfeld V, Reich Z (2008) Thylakoid membrane remodeling during state transitions in *Arabidopsis*. Plant Cell 20:1029–1039

Crafts-Brandner SJ, Salvucci ME (2000) Rubisco activase constrains the photosynthetic potential of leaves at high temperature and $CO_2$. Proc Natl Acad Sci USA 97:13430–13435

Demmig-Adams B, Adams WW (2002) Antioxidants in photosynthesis and human nutrition. Science 298: 149–2153

Depege N, Bellafiore S, Rochaix JD (2003) Role of chloroplast protein kinase Stt7 in LHCII phosphorylation and state transition in *Chlamydomonas*. Science 299:1572–1575

Desikan R, Mackerness SA-H, Hancock JT, Neill SJ (2001) Regulation of the *Arabidopsis* transcriptome by oxidative stress. Plant Physiol 127:159–172

Desikan R, Hancock JT, Neill SJ (2003) Oxidative stress signaling. In: Hirt H, Shinozaki K (eds) Plant responses to abiotic stress: topic in current genetics. Springer, Berlin/Heidelberg/New York

Dickinson BC, Chang CJ (2011) Chemistry and biology of reactive oxygen species in signaling or stress responses. Nat Chem Biol 7:504–511

Dukan S, Nystrom T (1999) Oxidative stress defense and deterioration of growth-arrested Escherichia coli cells. J Biol Chem 274:26027–26032

El-Shitinawy F, Ebrahim MKH, Sewelam N, El-Shourbagy MN (2004) Activity of photosystem 2, lipid peroxidation, and the enzymatic antioxidant protective system in heat shocked barley seedlings. Photosynthetica 42:15–21

Enami I, Kitamura M, Tomo T, Isokawa Y, Ohta H, Katoh S (1994) Is the primary cause of thermal inactivation of oxygen evolution in spinach PS II membranes release of the extrinsic 33 kDa protein or of Mn? Biochim Biophys Acta 1186:52–58

Enami I, Kikuchi S, Fukuda T, Ohta H, Shen J-R (1998) Binding and functional properties of four extrinsic proteins of photosystem II from a red alga, *Cyanidium caldarium*, as studied by release-reconstitution experiments. Biochemistry 37:2787–2793

Feller U, Grafts-Brander SJ, Salvucci ME (1998) Moderately high temperatures inhibit ribulose-1,5-biphosphate carboxylase/oxygenase (Rubisco) activase-mediated activation of Rubisco. Plant Physiol 116:539–546

Finazzi G, Rappaport F, Furia A, Fleischmann M, Rochaix JD, Zito F, Forti G (2002) Involvement of state transitions in the switch between linear and cyclic electron flow in *Chlamydomonas reinhardtii*. EMBO Rep 3:280–285

Foyer CH, Noctor G (2005a) Redox homeostasis and antioxidant signaling: a metabolic interface between stress perception and physiological responses. Plant Cell 17:1866–1875

Foyer C, Noctor G (2005b) Oxidant and antioxidant signaling in plants: a reevaluation of the concept of oxidative stress in a physiological context. Plant Cell Environ 28:1056–1071

Foyer CH, Noctor GD (2009) Redox regulation in photosynthetic organisms: signalling, acclimation, and practical implications. Antiox Redox Signal 11: 861–905

Foyer CH, Shigeoka S (2011) Understanding oxidative stress and antioxidant functions to enhance photosynthesis. Plant Physiol 155:93–100

Foyer CH, Descourvieres P, Kunert KJ (1994a) Protection against oxygen radicals: an important defense mechanism studied in transgenic plants. Plant Cell Environ 17:507–523

Foyer CH, Lelandais M, Kunert KJ (1994b) Photooxidative stress in plants. Physiol Plant 92:696–717

Foyer CH, Lopez-Delgado H, Dat JF, Scott IM (1997) Hydrogen peroxide- and glutathione-associated mechanisms of acclimatory stress tolerance and signaling. Physiol Plant 100:241–254

Galvez-Valdivieso G, Mullineaux PM (2010) The role of reactive oxygen species in signalling from chloroplasts to the nucleus. Physiol Plant 138:430–439

Gilmore AM, Hazlett TL, Govindjee (1995) Xanthophyll cycle-dependent quenching of photosystem II chlorophyll a fluorescence: formation of a quenching complex with a short fluorescence lifetime. Proc Natl Acad Sci USA 92:2273–2277

Glatz A, Vass I, Los DA, Vigh L (1999) The *Synechocystis* model of stress: from molecular chaperones to membranes. Plant Physiol Biochem 37:1–12

Gombos Z, Wada H, Murata N (1992) Unsaturation of fatty acids in membrane lipids enhances tolerance of the cyanobacterium *Synechocystis* PCC6803 to low-temperature photoinhibition. Proc Natl Acad Sci USA 89:9959–9963

Gombos Z, Wada H, Murata N (1994a) The recovery of photosynthesis from low-temperature photoinhibition is accelerated by the unsaturation of membrane lipids: a mechanism of chilling tolerance. Proc Natl Acad Sci USA 91:8787–8791

Gombos Z, Wada H, Hideg E, Murata N (1994b) The unsaturation of membrane lipids stabilizes photosynthesis against heat stress. Plant Physiol 104:563–567

Grover A, Agarwal M, Katiyar-Agarwal S, Sahi C, Agarwal S (2000) Production of high temperature tolerant transgenic plants through manipulation of membrane lipids. Curr Sci 79:557–559

Hakala M, Tuominen I, Keränen M, Tyystjärvi T, Tyystjärvi E (2005) Evidence for the role of oxygen-evolving manganese complex in photoinhibition of photosystem II. Biochim Biophys Acta 1706:68–80

Hall AE (2001) Crop responses to environment. CRS Press LLC, Boca Raton

Havaux M (1996) Short-term responses of photosystem I to heat stress – induction of a PS II-independent electron transport through PS I fed by stromal components. Photosynth Res 47:88–97

Havaux M, Niyogi KK (1999) The violaxanthin cycle protects plants from photooxidative damage by more than one mechanism. Proc Natl Acad Sci USA 96:8762–8767

Havaux M, Tardy F (1996) Temperature-dependent adjustment of the thermal stability of photosystem II *in vivo*: possible involvement of xanthophyll-cycle pigments. Planta 198:324–333

Havaux M, Tardy F, Ravenel J, Chanu D, Parot P (1996) Thylakoid membrane stability to heat stress studied by flash spectroscopic measurements of the electrochromic shift in intact potato leaves: influence of the xanthophylls content. Plant Cell Environ 19: 1359–1368

Havaux M, Dall' OL, Cuine S, Giuliano G, Bassi R (2004) The effect of zeaxanthin on structure and function of photosynthetic apparatus in *Arabidopsis thaliana*. J Biol Chem 279:13878–13888

Havaux M, Eymery F, Porfirova S, Rey P, Do P (2005) Vitamin E protects against photoinhibition and photooxidative stress in *Arabidopsis thaliana*. Plant Cell 17:3451–3469

Havaux M, Ksas B, Szewczyk A, Rumeau D, Franck F, Caffarri S, Triantaphylides C (2009) Vitamin B6 deficient plants display increased sensitivity to high light and photo-oxidative stress. BMC Plant Biol 9:130

Hayashi H, Murata N (1998) Genetically engineered enhancement of salinity tolerance in higher plants. In: Satoh K, Murata N (eds) Stress responses of photosynthetic organisms. Elsevier, Amsterdam/Tokyo

Heber U (2002) Irrungen, wirrungen? The Mehler reaction in relation to cyclic electron transport in C3 plants. Photosynth Res 73:223–231

Heckathorn SA, Bryan JK, Coleman JS (1998a) The methionine-rich low-molecular-weight chloroplast heat-shock protein: evolutionary conservation and accumulation in relation to thermotolerance. Am J Bot 85:175–183

Heckathorn SA, Downs SA, Sharkey TD, Soleman JS (1998b) The small, methionine-rich chloroplast heat-shock protein protects photosystem II electron transport during heat stress. Plant Physiol 116:439–444

Heckathorn SA, Ryan SL, Baylis JA, Wang D, Hamilton IIIEW, Cundiff L, Luthe DS (2002) *In vivo* evidence from an *Agrostis stolonifera* selection genotype that chloroplast small heat-shock proteins can protect photosystem II during heat stress. Funct Plant Biol 29:933–944

Hideg E, Spetea C, Vass I (1994) Singlet oxygen and free radical production during acceptor- and donor-side-induced photoinhibition. Studies with spin trapping EPR spectroscopy. Biochim Biophys Acta 1186: 143–152

Hideg EP, Kos B, Vass I (2007) Photosystem II damage induced by chemically generated singlet oxygen in tobacco leaves. Physiol Plant 131:33–40

Horvath I, Glatz A, Varvasovszki V, Török Z, Pali T, Balogh G, Kovacs E, Nadasdi L, Benko S, Joo F, Vigh L (1998) Membrane physical state controls the signaling mechanism of the heat shock response in *Synechocystis* PCC 6803: identification of *hsp17* as a fluidity gene. Proc Natl Acad Sci USA 95: 3513–3518

Hsiao H-Y, He Q, van Waasbergen LG, Grossman AR (2004) Control of photosynthetic and high-light-responsive genes by the histidine kinase DspA: negative and positive regulation and interactions between signal transduction pathways. J Bacteriol 186:3882–3888

Hsu BD, Lee YS (1991) A study on the fluorescence induction curve from DCMU poisoned chloroplast. Biochim Biophys Acta 1056:285–292

Hung S-H, Yu C-W, Lin CH (2005) Hydrogen peroxide functions as a stress signal in plants. Bot Bull Acad Sin 46:1–10

Inoue S, Ejima K, Iwai E, Hayashi H, Appel J, Tyystjärvi E, Murata N, Nishiyama Y (2011) Protection by α-tocopherol of the repair of photosystem II during photoinhibition in *Synechocystis* sp. PCC 6803. Biochim Biophys Acta 1807:236–241

Ivanov BN, Mubarakshina MM, Khorobrykh SA (2007) Kinetics of the plastoquinone pool oxidation following illumination. Oxygen incorporation into photosynthetic electron transport chain. FEBS Lett 581: 1342–1346

Jaspers P, Kangasjärvi J (2010) Reactive oxygen species in abiotic stress signaling. Physiol Plant 138: 405–413

Jung J, Kim H-S (1990) The chromatophores as endogenous sensitizers involved in the photogeneration of singlet oxygen in spinach thylakoids. Photochem Photobiol 52:1003–1009

Kanesaki Y, Los' DA, Suzuki I, Murata N (2010) Sensors and signal transducers of environmental stress in cyanobacteria. In: Pareek A, Sopory SK, Bohnert HJ, Govindjee (eds) Abiotic stress adaptation in plants: physiological, molecular and genomic foundation, chapter 2. Springer, Dordrecht

Karu TI (2008) Mitochondrial signaling in mammalian cells activated by red and near-IR radiation. Photochem Photobiol 84:1091–1099

Katoh S, San Pietro A (1967) Photooxidation and reduction of cytochrome-552 and NADP photoreduction by *Euglena* chloroplast. Arch Biochem Biophys 121:211–219

Keren N, Berg A, van Kan PJM, Levanon H, Ohad I (1997) Mechanism of photosystem II photoinactivation and D1 protein degradation at low light: the role of back electron flow. Proc Natl Acad Sci USA 94:1579–1584

Kimura A, Eaton-Rye JJ, Morita EH, Nishiyama Y, Hayashi H (2002) Protection of the oxygen-evolving machinery by the extrinsic proteins of photosystem II is also essential for development of cellular thermotolerance in *Synechocystis* sp. PCC 6803. Plant Cell Physiol 43:932–938

Klimov VV, Allakhverdiev SI, Demetr S, Krasnovsky AA (1978) Photorecovery of pheophytin in the photosystem 2 of chloroplasts as a function of the redox potential. Doklady AN SSSR (Russian) 249:2227–2230

Klimov VV, Schafiev MA, Allakhverdiev SI (1990) Photoinactivation of the reactivation capacity of photosystem II in pea subchloroplast particles after a complete removal of manganese. Photosynth Res 23: 59–65

Kojima K, Oshita M, Nanjo Y, Kasai K, Tozawa Y, Hayashi H, Nishiyama Y (2007) Oxidation of elongation factor G inhibits the synthesis of the D1 protein of photosystem II. Mol Microbiol 65:936–947

Kojima K, Motohashi K, Morota T, Oshita M, Hisabori T, Hayashi H, Nishiyama Y (2009) Regulation of translation by the redox state of elongation factor G in the cyanobacterium *Synechocystis sp.* PCC 6803. J Biol Chem 284:18685–18691

Kok B (1956) On inhibition of photosynthesis by intense light. Biochim Biophys Acta 21:305–322

Kramer DM, Avenson TJ, Kanazawa A, Cruz JA (2004) The relationship between photosynthetic electron transfer and its regulation. In: Papageorgiou C, Govindjee (eds) Chlorophyll a fluorescence: a signature of photosynthesis. Springer, Dordrecht

Kreslavski VD, Khristin MS (2003) Effect of heat shock on fluorescence induction and low-temperature fluorescence spectra in wheat leaves. Biofizika (Russian) 48:865–872

Kreslavski VD, Balakhnina TI, Khristin MS, Bukhov NG (2001) Pretreatment of bean seedlings by choline compounds increases the resistance of photosynthetic apparatus to UV radiation and elevated temperatures. Photosynthetica 39:353–358

Kreslavski VD, Ivanov AA, Kosobryukhov AA (2004) Low energy light in the 620–660 nm range reduces the UV-B-induced damage to photosystem II in spinach leaves. Biofizika (Russian) 49:767–771

Kreslavski VD, Carpentier R, Klimov VV, Murata N, Allakhverdiev CI (2007) Molecular mechanism of stress resistance of photosynthetic apparatus. Biochem (Moscow) Suppl Ser A Membr Cell Biol 1:185–205

Kreslavski V, Tatarinzev N, Shabnova N, Semenova G, Kosobryukhov A (2008) Characterization of the nature of photosynthetic recovery of wheat seedlings from short-time dark heat exposures and analysis of the mode of acclimation to different light intensities. J Plant Physiol 165:1592–1600

Kreslavski VD, Lyubimov VY, Shabnova NI, Balakhnina TI, Kosobryukhov AA (2009) Heat-induced impairments and recovery of photosynthetic machinery in wheat seedlings. Role of light and prooxidant-antioxidant balance. Physiol Mol Biol Plants 15: 115–122

Kreslavski VD, Balakhnina TI, Zharmukhamedov SK, Shabnova NI, Khristin MS, Lyubimov VY (2010) Mechanism of enhancing photosystem 2 thermoresistance in wheat plants by chlorocholine chloride. Russ Agric Sci 36:156–159

Kreslavski VD, Lyubimov VY, Kotova LM, Kotov AA (2011) Effect of common bean seedling pretreatment with chlorocholine chloride on photosystem II tolerance to UV B radiation, phytohormone content, and hydrogen peroxide content. Russ J Plant Physiol 58:324–329

Kreslavski VD, Los DA, Allakhverdiev SI, Kuznetsov VV (2012a) Signaling role of reactive oxygen species in plants under stress. Russ J Plant Physiol 59: 141–154

Kreslavski VD, Fomina IR, Los' DA, Allakhverdiev SI (2012b) Heat induced impairments and recovery of photosynthetic machinery. In: Josipovic S, Ludvig E (eds) Heat stress: causes, treatment and prevention, chapter 4. Nova Science Publishers, Inc, New York

Kreslavskii VD, Khristin MS, Shabnova NI, Lyubimov VY (2012) Preillumination of excised spinach leaves with red light increases resistance of photosynthetic apparatus to UV radiation. Russ J Plant Physiol 59(6):717–723

Krieger-Liszkay A, Fufezan C, Trebst A (2008) Singlet oxygen production in photosystem II and related protection mechanism. Photosynth Res 98:551–564

Kwon SY, Jeong YJ, Lee HS, Kim JS, Cho KY, Allen RD, Kwak SS (2002) Enhanced tolerances of transgenic tobacco plants expressing both superoxide dismutase and ascorbate peroxidase in chloroplasts against methyl viologen-mediated oxidative stress. Plant Cell Environ 25:873–882

Larkindale J, Knight MR (2002) Protection against heat stress-induced oxidative damage in *Arabidopsis* involves calcium, abscisic acid, ethylene, and salicylic acid. Plant Physiol 128:682–695

Law RD, Crafts-Brandner SJ (1999) Inhibition and acclimation of photosynthesis to heat stress is closely correlated with activation of ribulose-1,5-bisphosphate carboxylase/oxygenase. Plant Physiol 120:173–182

Leitsch J, Schnettger B, Critchley C, Krause CH (1994) Two mechanisms of recovery from photoinhibition *in vivo*: reactivation of photosystem II related and unrelated to Dl- protein turnover. Planta 194:15–21

Leung J, Giraudat J (1998) Abscisic acid signal transduction. Mol Biol 49:199–222

Li X-P, Müller-Moulé P, Gilmore AM, Niyogi KK (2002) PsbS-dependent enhancement of feedback de-excitation protects photosystem II from photoinhibition. Proc Natl Acad Sci USA 99:15222–15227

Li XP, Gilmore AM, Caffarri S, Bassi R, Golan T, Kramer D, Niyogi KK (2004) Regulation of photosynthetic light harvesting involves intrathylakoid lumen pH sensing by the PsbS protein. J Biol Chem 279: 22866–22874

Li Z, Ahn TK, Avenson TJ, Ballotari M, Cruz JA, Kramer DM, Bassi R, Fleming GR, Keasling JD, Niyogi KK (2009) Lutein accumulation in the absence of zeaxanthin restores nonphotochemical quenching in the *Arabidopsis thaliana npq1* mutant. Plant Cell 21: 1798–1812

Lindahl M, Florencio FJ (2003) Thioredoxin-linked processes in cyanobacteria are as numerous as in chloroplasts, but targets are different. Proc Natl Acad Sci USA 100:16107–16112

Logan BA (2006) Oxygen metabolism and stress physiology. In: Wise RR, Hoober JK (eds) The structure and function of plastids. Kluwer Academic, Dordrecht

Los DA, Murata N (2004) Membrane fluidity and its roles in the perception of environmental signals. Biochim Biophys Acta 1666:142–157

Los DA, Zinchenko VV (2009) Regulatory role of membrane fluidity in gene expression. Lipids in photosynthesis. In: Wada H, Murata N (eds) Essential and regulatory functions. Springer, Dordrecht

Los DA, Zorina A, Sinetova M, Kryazhov S, Mironov K, Zinchenko VV (2010) Stress sensors and signal transducers in cyanobacteria. Sensors 10:2386–2415

Mackay CE, Senaratna T, McKersie BD, Fletcher RA (1987) Ozone induced injury to cellular membranes in *Triticum aestivum* L. and protection by the Triasole S-3307. Plant Cell Physiol 128:1271–1278

Mamedov MD, Hayashi H, Murata N (1993) Effects of glycinebetaine and unsaturation of membrane lipids on heat stability of photosynthetic electron transport and phosphorylation reactions in *Synechocystis* PCC 6803. Biochim Biophys Acta 1142:1–5

Mattoo AK, Marder JB, Edelman M (1989) Dynamics of photosystem two reaction center. Cell 56:241–246

Mehta P, Kreslavski V, Bharti S, Allakhverdiev SI, Jajoo A (2011) Analysis of salt stress induced changes in Photosystem II heterogeneity by prompt fluorescence and delayed fluorescence in wheat (*Triticum aestivum*) leaves. J Photochem Photobiol B Biol 104:308–313

Melis A (1999) Photosystem II damage and repair cycle in chloroplasts: what modulates the rate of photodamage *in vivo*? Trends Plant Sci 4:130–135

Meloni DA, Oliva MA, Martinez CA, Cambraia J (2003) Photosynthesis and activity of superoxide dismutase, peroxidase and glutathione reductase in cotton under salt stress. Environ Exp Bot 49:69–76

Miller G, Nobuhiro S, Rizhsky L, Hegie A, Koussevitzky S, Mittler R (2007) Double mutants deficient in cytosolic and thylakoid ascorbate peroxidase reveal a complex mode of interaction between reactive oxygen species, plant development, and response to abiotic stresses. Plant Physiol 144:1777–1785

Minibayeva FV, Gordon LK (2003) Superoxide production and the activity of extracellular peroxidase in plant tissues under stress conditions. Russ J Plant Physiol 50:411–416

Minibayeva F, Kolesnikov OP, Gordon LK (1998) Contribution of a plasma membrane redox system to the superoxide production by wheat root cells. Protoplasma 205:101–106

Minibayeva FV, Gordon LK, Kolesnikov OP, Chasov AV (2001) Role of extracellular peroxidase in the superoxide production by wheat root cells. Protoplasma 217:125–128

Minibayeva F, Kolesnikov OP, Chasov A, Beckett RP, Lüthje S, Vylegzxanina NF, Buck F, Böttger M (2009) Wound-induced apoplastic peroxidase activities: their roles in the production and detoxification of reactive oxygen species. Plant Cell Environ 32:497–508

Miyake C, Yonekura K, Kobayashi Y, Yokota A (2002) Cyclic flow of electron within PSII functions in intact chloroplasts from spinach leaves. Plant Cell Physiol 43:951–957

Miyao M, Ikeuchi M, Yamamoto N, Ono T (1995) Specific degradation of the D1 protein of photosystem II by treatment with hydrogen peroxide in darkness: implication for the mechanism of degradation of the D1 protein under illumination. Biochemistry 34: 10019–10026

Mohanty P, Vani B, Prakash S (2002) Elevated temperature treatment induced alteration in thylakoid membrane organization and energy distribution between the two photosystems in *Pisum sativum*. Z Naturforsch C 57:836–842

Moon BY, Higashi S-I, Gombos Z, Murata N (1995) Unsaturation of the membrane lipids of chloroplasts stabilizes the photosynthetic machinery against low-temperature photoinhibition in transgenic tobacco plants. Proc Natl Acad Sci USA 92:6219–6223

Mori IC, Schroeder JI (2004) Reactive oxygen species activation of plant Ca2+ channels: a signaling mechanism in polar growth, hormone transduction, stress signaling, and hypothetically mechanotransduction. Plant Physiol 135:702–708

Mubarakshina MM, Ivanov BN, Naidov IA, Hillier W, Badger MR, Krieger-Liszkay A (2010) Production and diffusion of chloroplastic $H_2O_2$ and its implication to signaling. J Exp Bot 61:3577–3587

Müller P, Li X-P, Niyogi KK (2001) Non-photochemical quenching: a response to excess light energy. Plant Physiol 125:1558–1566

Murata N, Nishiyama Y (1998) Molecular mechanisms of the low-temperature tolerance of the photosynthetic machinery. In: Satoh K, Murata N (eds) Stress responses of photosynthetic organisms. Elsevier, Amsterdam/Tokyo

Murata N, Takahashi S, Nishiyama Y, Allakhverdiev SI (2007) Photoinhibition of photosystem II under environmental stress. Biochim Biophys Acta 1767: 414–421

Murata N, Allakhverdiev SI, Nishiyama Y (2012) The mechanism of photoinhibition *in vivo*: re-evaluation of the roles of catalase, α-tocopherol, non-photochemical quenching, and electron transport. Biochim Biophys Acta 1817:1127–1133

Nash D, Miyao M, Murata N (1985) Heat inactivation of oxygen evolution in photosystem II particles and its acceleration by chloride depletion and exogenous manganese. Biochim Biophys Acta 807:127–133

Nazarenko LV, Andreev IM, Lyukevich AA, Pisareva TV, Los DA (2003) Calcium release from *Synechocystis* cells induced by depolarization of the plasma membrane: MscL as an outward $Ca^{2+}$ channel. Microbiology 149:1147–1153

Neta-Sharir I, Isaacson T, Lurie S, Weiss D (2005) Dual role for tomato heat shock protein 21: protecting photosystem II from oxidative stress and promoting color changes during fruit maturation. Plant Cell 17: 1829–1838

Nishiyama Y, Los' DA, Murata N (1999) PsbU, a protein associated with photosystem II, is required for the acquisition of cellular thermotolerance in *Synechococcus* species PCC 7002. Plant Physiol 120:301–308

Nishiyama Y, Yamamoto H, Allakhverdiev SI, Inaba M, Yokota A, Murata N (2001) Oxidative stress inhibits the repair of photodamage to the photosynthetic machinery. EMBO J 20:5587–5594

Nishiyama Y, Allakhverdiev SI, Yamamoto H, Hayashi H, Murata N (2004) Singlet oxygen inhibits the repair of photosystem II by suppressing the transport elongation of the D1 protein in *Synechocystis* sp. PCC 6803. Biochemistry (USA) 43:11321–11330

Nishiyama Y, Allakhverdiev SI, Murata N (2005) Inhibition of the repair of photosystem II by oxidative stress in cyanobacteria. Photosynth Res 84:1–7

Nishiyama Y, Allakhverdiev SI, Murata N (2006) A new paradigm for the action of reactive oxygen species in the photoinhibition of photosystem II. Biochim Biophys Acta 1757:742–749

Nishiyama Y, Allakhverdiev SI, Murata N (2011) Protein synthesis is the primary target of reactive oxygen species in the photoinhibition of photosystem II. Physiol Plant 142:35–46

Niyogi KK, Li XP, Rosenberg V, Jung HS (2005) Is PsbS the site of non-photochemical quenching in photosynthesis? J Exp Bot 56:375–382

Novitskaya GV, Kocheshkova TK, Feofilaktova TV, Novitskii YN (2004) Effect of choline chloride on the lipid content and composition in the leaves of principal magnetically-oriented radish types. Russ J Plant Physiol 51:361–371

Nuijs AM, Shuvalov VA, van Gorkom HJ, Plijter JJ, Duysens LNM (1986) Picosecond absorbance difference spectroscopy on the primary reactions and the antenna-excited states in photosystem I particles. Biochim Biophys Acta 850:310–318

Ohnishi N, Allakhverdiev SI, Takanashi S, Higachi S, Watanabe M, Nishiyama Y, Murata N (2005) Two-step mechanism of photodamage to photosystem II: step 1 occurs at the oxygen–evolving complex and step 2 occurs at the photochemical reaction center. Biochemistry (USA) 44:9494–9499

Okada K, Ikeuchi M, Yamamoto N, Ono T, Miyao M (1996) Selective and specific cleavage of the D1 and

D2 proteins of photosystem II by exposure to singlet oxygen: factors responsible for the susceptibility to cleavage of the proteins. Biochim Biophys Acta 1274:73–79

Park EJ, Jeknic Z, Sakamoto A, Denoma J, Yuwansiri R, Murata N, Chen TH (2004) Genetic engineering of glycinebetaine synthesis in tomato protects seeds, plants, and flowers from chilling damage. Plant J 40:474–487

Pastenes C, Horton R (1996) Effect of high temperature on photosynthesis in beans. Plant Physiol 112: 1245–1251

Pastori GM, Foyer CH (2002) Common components, networks and pathways of cross-tolerance to stress. The central role of 'redox' and abscisic-acid-mediated controls. Plant Physiol 129:460–468

Payton P, Webb R, Kornyeyev D, Allen R, Holiday A (2001) Protecting cotton photosynthesis during moderate chilling at high light intensity by increasing chloroplastic antioxidant enzyme activity. J Exp Bot 52:2345–2354

Pei ZM, Murata Y, Benning G, Thomine S, Klusener B, Allen GJ, Grill E, Schroeder JI (2000) Calcium channels activated by hydrogen peroxide mediate abscisic acid signalling in guard cells. Nature 406:731–734

Pfannschmidt T, Brautigam K, Wagner R, Dietzel L, Schroter Y, Steiner S, Nykytenko A (2009) Potential regulation of gene expression in photosynthetic cells by redox and energy state: approaches towards better understanding. Ann Bot 103:599–607

Pitzschke P, Hirt H (2006) Mitogen activated protein kinases and reactive oxygen species signaling in plants. Plant Physiol 141:351–356

Pogson BJ, Woo NS, Forster B, Small ID (2008) Plastid signalling to the nucleus and beyond. Trends Plant Sci 13:602–609

Powles SB (1984) Photoinhibition of photosynthesis induced by visible light. Annu Rev Plant Physiol 35:15–44

Pradedova EV, Isheeva OD, Salyaev RK (2011) Classification of the antioxidant defense system as the ground for reasonable organization of experimental studies of the oxidative stress in plants. Russ J Plant Physiol 58:210–217

Prakash JS, Sinetova M, Zorina A, Kupriyanov E, Suzuki I, Murata N, Los DA (2009) DNA supercoiling regulates the stress-inducible expression of genes in the cyanobacterium *Synechocystis*. Mol Biosyst 5:1904–1912

Pshybytko NL, Kruk J, Kabashnikova LF, Strzalka K (2008) Function of plastoquinone in heat stress reactions of plants. Biochim Biophys Acta 1777: 1393–1470

Qi Z, Yue M, Wang XL (2000) Laser pretreatment protects cells of broad bean from UV-B radiation damage. J Photochem Photobiol B Biol 59:33–37

Qi Z, Yue M, Han R, Wang XL (2002) The damage repair role of He-Ne laser on plants exposed to different intensities of ultraviolet-B radiation. Photochem Photobiol 75(6):680–686

Quist G, Cambell D, Clarke A, Gustafsson P (1995) The cyanobacterium *Synechococcus* modulates photomodulates photosystem II function in response to excitation stress through D1 exchange. Photosynth Res 46:151–158

Raskin I (1992) Role of salicylic acid in plants. Annu Rev Plant Physiol Plant Mol Biol 43:439–463

Renger G, Volker M, Eckert HJ, Fromme R, Hohm-Veit S, Graber P (1989) On the mechanism of photosystem II deterioration by UV-B irradiation. Photochem Photobiol 49:97–105

Roberts DR, Kristie DN, Thompson JE, Dumbroff EB, Gepstein S (1991) *In vitro* evidence for the involvement of activated oxygen in light activated aggregation of thylakoid proteins. Physiol Plant 82: 389–396

Rochaix JD (2007) Role of thylakoid protein kinases in photosynthetic acclimation. FEBS Lett 581: 2768–2775

Romanov GA (2009) How do cytokinins affect the cell? Russ J Plant Physiol 56:268–290

Roose JL, Wegener KM, Pakrasi HB (2007) The extrinsic proteins of photosystem II. Photosynth Res 92:369–387

Sabehat A, Weiss D, Lurie S (1998) Heat shock proteins and cross-tolerance in plants. Physiol Plant 103 :437–441

Sairam RK, Svastava GC, Saxena DC (2000) Increased antioxidant activity under elevated temperatures: a mechanism of heat stress tolerance in wheat genotypes. Biol Plant 43:245–251

Salvucci ME, Crafts-Brandner SJ (2004) Inhibition of photosynthesis by heat stress: the activation state of Rubisco as a limiting factor in photosynthesis. Physiol Plant 120:179–186

Sarvikas P (2010) Photoinhibition of photosystem II. Kinetics, protection and mechanism. Turun Yliopisto, Turku

Sarvikas P, Hakala M, Pätsikkä E, Tyystjärvi T, Tyystjärvi E (2006) Action spectrum of photoinhibition in leaves of wild type and npq1-2 and npq4-1 mutants of *Arabidopsis thaliana*. Plant Cell Physiol 47:391–400

Sato N, Sonoike K, Kawaguchi A, Tsuzuki M (1996) Contribution of lowered unsaturation levels of chloroplast lipids to high temperature tolerance of photosynthesis in *Chlamydomonas reinhardtii*. J Photochem Photobiol B: Biol 36:333–337

Sazanov LA, Borrows PA, Nixon PJ (1998) The chloroplast Ndh complex mediates the dark reduction of the plastoquinone pool in response to heat stress in tobacco leaves. FEBS Lett 429:115–118

Schittenhelm S, Menge-Hartmann U, Oldenburg E (2004) Photosynthesis, carbohydrate metabolism, and yield of phytochrome-B-overexpressing potatoes under different light regimes. Crop Sci 44: 131–143

Schreiber U, Kriger A (1996) Two fundamentally different types of variable chlorophyll fluorescence *in vivo*. FEBS Lett 397:131–135

Schroda M, Vallon O, Wollman F-A, Beck CF (1999) A chloroplast-targeted heat shock protein 70 (HSP70) contributes to the photoprotection and repair of photosystem II during and after photoinhibition. Plant Cell 11:1165–1178

Setlik I, Allakhverdiev SI, Nedbal L, Setlikova E, Klimov VV (1990) Three types of photosystem 2 photoinactivation. I. Damaging processes on the acceptor side. Photosynth Res 23:39–48

Shapiguzov A, Ingelsson B, Samola I, Andres C, Kessler F, Rochaix J-D, Vener AV, Goldschmidt-Clermont M (2010) The PPH1 phosphatase is specifically involved in LHCII dephosphorylation and state transitions in *Arabidopsis*. Proc Natl Acad Sci USA 107:4782–4787

Sharkey TD (2005) Effects of moderate heat stress on photosynthesis: importance of thylakoid reactions, rubisco deactivation, reactive oxygen species, and thermotolerance provided by isoprene. Plant Cell Environ 28:269–277

Sheng RY, Li PM, Xue GX, Zhao XX, Gao HY (2006) Choline chloride protects cell membrane and the photosynthetic apparatus in cucumber seedling leaves at low temperature and weak light. J Plant Physiol Mol Biol 32:87–93

Solovchenko AE, Merzlyakov MN (2008) Screening of visible and UV radiation as a photoprotective mechanism in plants. Russ J Plant Physiol 55:719–737

Srivastava A, Guissre B, Greppin H, Strasser RJ (1997) Regulation of antenna structure and electron transport in photosystem II of *Pisum sativum* under elevated temperature probed by the fast polyphasic chlorophyll *a* fluorescence transient: OKJIP. Biochim Biophys Acta 1320:95–106

Staehelin LA, van der Staay GWM (1996) Structure, composition, functional organization and dynamic properties of thylakoid membranes. In: Ort DR, Yocum CF (eds) Oxygenic photosynthesis. The light reactions. Kluwer Academic Publishers, Dordrecht

Steyn WJ, Wand SJE, Holcroft DM, Jacobs G (2002) Anthocyanins in vegetative tissues: a proposed unified function in photoprotection. New Phytol 155:349–361

Storz G, Imlay JA (1999) Oxidative stress. Curr Opin Microbiol 2:188–194

Strasser BJ (1997) Donor side capacity of PS II probed by chlorophyll a fluorescence transients. Photosynth Res 52:147–155

Strasser RJ, Srivastava A, Tsimilli-Michael M (2000) The fluorescence transient as a tool to characterize and screen photosynthetic samples. In: Yunus M, Pathre U, Mohanty P (eds) Probing photosynthesis: mechanisms, regulation and adaptation. Taylor & Francis, London

Suzuki N, Mittler R (2006) Reactive oxygen species and temperature stresses: a delicate balance between signaling and destruction. Physiol Plant 126:45–51

Swanson S, Gilroy S (2010) ROS in plant development. Physiol Plant 138:384–392

Szabó I, Bergantino E, Giacometti GM (2005) Light and oxygenic photosynthesis: energy dissipation as a protection mechanism against photooxidation. EMBO Rep 6:629–634

Takabe T, Nakamura T, Nomura M, Hayashi Y, Ishitani M, Muramoto Y, Tanaka A, Takabe T (1998) Glycinebetaine and the genetic engineering of salinity tolerance in plants. In: Satoh K, Murata N (eds) Stress responses of photosynthetic organisms. Elsevier, Amsterdam/Tokyo

Takahashi S, Badger MR (2011) Photoprotection in plants: a new light on photosystem II damage. Trends Plant Sci 16:53–60

Takahashi S, Murata N (2005) Interruption of the Calvin cycle inhibits the repair of photosystem II from photodamage. Biochim Biophys Acta 1708:352–361

Takahashi S, Murata N (2008) How do environmental stresses accelerate photoinhibition? Trends Plant Sci 13:178–182

Takahashi S, Milward SE, Fan DY, Chow WS, Badger MR (2009) How does cyclic electron flow alleviate photoinhibition in *Arabidopsis*? Plant Physiol 149:1560–1567

Takeba G, Kozaki A (1998) Photorespiration is an essential mechanism for the protection of C3 plants from photooxidation. In: Satoh K, Murata N (eds) Stress responses of photosynthetic organisms: molecular mechanisms and molecular regulation. Elsevier, Amsterdam/Tokyo

Tang Y, Wen X, Lu O, Yang Z, Cheng Z, Lu C (2007) Heat stress induces an aggregation of the light-harvesting complex of photosystem II in spinach plants. Plant Physiol 143:629–638

Tasaka Y, Gombos Z, Nishiyama Y, Mohanty P, Ohba T, Ohki K, Murata N (1996) Targeted mutagenesis of acyl-lipid desaturases in *Synechocystis*: evidence for the important roles of polyunsaturated membrane lipids in growth, respiration and photosynthesis. EMBO J 15:6416–6425

Tefler A (2005) Too much light? How β–carotene protects the photosystem II reaction center. Photochem Photobiol Sci 4:950–956

Tefler A, Barber J (1995) Role of carotenoid bound to the photosystem II reaction center. In: Mathis P (ed) Photosynthesis: from light to the biosphere. Kluwer Academic, Dordrecht

Theg SM, Filar LJ, Dilley RA (1986) Photoinactivation of chloroplasts already inhibited on oxidizing side of photosystem II. Biochim Biophys Acta 849:104–111

Thiele A, Herold M, Lenk I, Quail PH, Gatz C (1999) Heterologous expression of *Arabidopsis* phytochrome B in transgenic potato influences photosynthetic performance and tuber development. Plant Physiol 120:73–82

Thomas DJ, Thomas J, Youderian PA, Herbert SK (2001) Photoinhibition and light-induced electron transport in *ndhB*⁻ and *psaE*⁻ mutants of *Synechocystis* sp. PCC 6803. Plant Cell Physiol 42:803–812

Thomashow MF (1999) Plant cold acclimation: freezing tolerance genes and regulatory mechanisms. Annu Rev Plant Physiol Plant Mol Biol 50:571–599

Tikkanen M, Piippo M, Suorsa M, Sirpio S, Mulo P, Vainonen J, Vener AV, Allahverdiyeva Y, Aro EM (2006) State transitions revisited: a buffering system for dynamic low light acclimation of *Arabidopsis*. Plant Mol Biol 62:779–793

Tikkanen M, Nurmi M, Suorsa M, Danielsson R, Mamedov F, Styring S, Aro EM (2008) Phosphorylation-dependent regulation of excitation energy distribution between the two photosystems in higher plants. Biochim Biophys Acta 1777:425–432

Tikkanen M, Grieco M, Kangasjärvi S, Aro EM (2010) Thylakoid protein phosphorylation in higher plant chloroplasts optimizes electron transfer under fluctuating light. Plant Physiol 152:723–735

Torsethaugen G, Pitcher LH, Zilinskas BA, Pell EJ (1997) Overproduction of ascorbate peroxidase in the tobacco chloroplast does not provide protection against ozone. Plant Physiol 114:529–537

Trebst A, Depka B, Hollander-Czytko H (2002) A specific role for tocopherol and for chemical singlet oxygen quenchers in the maintenance of photosystem II structure and function in *Chlamydomonas reinhardtii*. FEBS Lett 516:156–160

Triantaphylides C, Krischke M, Hoeberichts FA, Ksas B, Gresser G, Havaux M, van Breusegem F, Mueller MJ (2008) Singlet oxygen is the major reactive oxygen species involved in photooxidative damage to plants. Plant Physiol 148:960–968

Tyystjärvi E (2008) Photoinhibition of photosystem II and photodamage of the oxygen evolving manganese cluster. Coord Chem Rev 252:361–376

Tyystjärvi T, Tuominen I, Herranen M, Aro E-M, Tyystjärvi E (2002) Action spectrum of psbA gene transcription is similar to that of photoinhibition in *Synechocystis* sp. PCC 6803. FEBS Lett 516: 167–171

Vallad GE, Goodman RM (2004) Systemic acquired resistance and induced systemic resistance in conventional agriculture. Crop Sci 44:1920–1934

Vass I, Aro E-M (2008) Photoinhibition of photosynthetic electron transport. In: Renger G (ed) Primary processes of photosynthesis: basic principles and apparatus. Royal Society of Chemistry, Cambridge, UK

Vass I, Cser K (2009) Janus-faced charge recombinations in photosystem II photoinhibition. Trends Plant Sci 14:200–205

Vass I, Styring S, Hundal T, Koivuniemi A, Aro E-M, Andersson B (1992) The reversible and irreversible intermediates during photoinhibition of photosystem II – stable reduced $Q_A$ species promote chlorophyll triplet formation. Proc Natl Acad Sci USA 89: 1408–1412

Vass I, Cser K, Cheregi O (2007) Molecular mechanisms of light stress of photosynthesis. Ann N Y Acad Sci 1113:114–122

Vener AV, van Kan PJ, Rich PR, Ohad I, Andersson B (1997) Plastoquinol at the quinol oxidation site of reduced cytochrome bf mediates signal transduction between light and protein phosphorylation: Thylakoid protein kinase deactivation by a single turnover flash. Proc Natl Acad Sci USA 94:1585–1590

Vierling E (1991) The role of heat shock proteins in plants. Ann Rev Plant Physiol Plant Mol Biol 42:579–620

Vigh L, Horvath I, Maresca B, Harwood JL (2007) Can the stress protein response be controlled by 'membrane-lipid therapy'? Trends Biochem Sci 32:357–363

Vijayan P, Browse J (2002) Photoinhibition in mutants of *Arabidopsis* deficient in thylakoid unsaturation. Plant Physiol 129:876–885

Vinocur B, Altman A (2005) Recent advances in engineering plant tolerance to abiotic stress: achievements and limitations. Curr Opin Biotechnol 16:1–10

Wada H, Gombos Z, Murata N (1990) Enhancement of chilling tolerance of a cyanobacterium by genetic manipulation of fatty acid desaturation. Nature 347:200–203

Wada H, Gombos Z, Murata N (1994) Contribution of membrane lipids to the ability of the photosynthetic machinery to tolerate temperature stress. Proc Natl Acad Sci USA 91:4273–4277

Weis E (1981) Reversible heat inactivation of Calvin cycle: a possible mechanism of the temperature regulation of photosynthesis. Planta 151:33–39

Wollman FA (2001) State transitions reveal the dynamics and flexibility of the photosynthetic apparatus. EMBO J 20:3623–3630

Yamane Y, Kashino Y, Koike H, Satoh K (1998) Effects of high temperatures on the photosynthetic systems in spinach: oxygen-evolving activities, fluorescence characteristics and the denaturation process. Photosynth Res 57:51–59

Yamane Y, Shikanai T, Kashino Y, Koike H, Satoh K (2000) Reduction of $Q_A$ in the dark: another cause of fluorescence $F_o$ increases by high temperatures in higher plants. Photosynth Res 63:23–34

Yang X, Wen X, Gong H, Lu Q, Yang Z, Tang Y, Liang Z, Lu C (2007) Genetic engineering of the biosynthesis of glycinebetaine enhances thermotolerance of photosystem II in tobacco plants. Planta 225:719–733

Yordanov IT (1993) Response of photosynthetic apparatus to temperature stress and molecular mechanisms of its adaptation. Photosynthetica 26:517–531

Yordanov I, Goltsev V, Stefanov D, Chernev P, Zaharieva I, Kirova M, Gecheva V, Strasser RJ (2008) Preservation of photosynthetic electron transport from senescence-induced inactivation in primary leaves after decapitation and defoliation of bean plants. J Plant Physiol 165:1954–1963

Zhang L, Paakkarinen V, van Wijk KJ, Aro E-M (2000) Biogenesis of the chloroplast-encoded D1 protein: regulation of mechanisms of D1 protein translation elongation, insertion and assembly. Plant Cell 12:1769–1781

Zito F, Finazzi G, Delosme R, Nitschke W, Picot D, Wollman F-A (1999) The Qo site of cytochrome $b_6f$ complexes controls the activation of the LHCII kinase. EMBO J 18:2961–2969

Zorina AA, Mironov KS, Stepanchenko NS, Sinetova MA, Koroban NV, Zinchenko VV, Kupriyanova EV, Allakhverdiev SI, Los DA (2011) Regulation systems for stress responses in cyanobacteria. Russ J Plant Physiol 58:749–767

# Salinity-Induced Genes and Molecular Basis of Salt-Tolerant Strategies in Mangroves

# 3

Anath Bandhu Das and Reto J. Strasser

**Abstract**

Salinity is one of the major environmental stresses that affects the growth and productivity of plant by affecting photosynthesis and other metabolic process. Halophytes are capable of thriving and growing under high concentration of NaCl. However, soil salinity is among the leading environmental stresses affecting global agriculture, causing billions of dollars loss per annum due to crop damages. Regardless of its high salinity in the root zone, it severely impedes normal plant growth and development due to ion, osmotic toxicity, and water deficit, resulting in reduced of crop productivity. Salt tolerance is a genetically complex trait in plants, often modulated by multiple biosynthetic and signaling pathways. Cross-talks among various stress-controlling pathways have been observed under salt stress, many of which are regulated by transcription factors. Thus, a comprehensive knowledge of the up- and downregulating genes under salt stress is necessary, which would provide a better understanding of the interactions among pathways in response to salt stress. DNA microarray technology has been employed to study expression profiles in different plant species and at varying developmental stages in response to salt stress. As a result, large-scale gene expression profiles under salt stress are now available for many plant species. Examinations of such gene expression profiles will help to understand the complex regulatory pathways affecting plant salt tolerance and potentially functional characterization of unknown genes, which may be good candidate genes for

A.B. Das (✉)
Department of Agricultural Biotechnology,
Orissa University of Agriculture and Technology,
Bhubaneswar 751003, Orissa, India
e-mail: a_b_das@hotmail.com

R.J. Strasser
Bioenergetics Laboratory, University of Geneva,
10 Chemin des Embrouchis, CH 1254 JUSSY-Geneva,
Switzerland
e-mail: Reto.Strasser@bioveg.unige.ch

G.R. Rout and A.B. Das (eds.), *Molecular Stress Physiology of Plants*,
DOI 10.1007/978-81-322-0807-5_3, © Springer India 2013

developing salt tolerance in plants. In this chapter, current knowledge of plant salt tolerance is discussed for a better understanding of the genetic basis of plant salt tolerance. The understanding of salt-tolerant mechanism particularly in producing antioxidative enzymes and special salt stress marker proteins is discussed in adapting adversely high-salinity environment in tree mangroves. A detailed antioxidative defense system against the production of active oxygen species (ROS) – an early indicator for adverse condition in both mangroves and mangroves associates – is discussed. This aspect has been studied using a true mangrove *Bruguiera gymnorrhiza* to detect the changes in antioxidative enzyme system as well as leaf and thylakoid protein profile as a case study. Approaches for improving plant salt tolerance using various salt stress genes as well as tools of biotechnology have been discussed.

## Introduction

Abiotic stresses such as drought, salinity, waterlogging, extreme temperatures, mineral toxicities and deficiencies, and oxidative stresses cause severe crop losses worldwide (Jamil et al. 2005; Ashraf et al. 2008). Soil salinity is one of the most devastating environmental stresses, which causes major reductions in crop productivity and quality (Wang et al. 2009a). A saline soil is generally defined as one in which the electrical conductivity (EC) the root zone exceeds ~40 mM NaCl at 25°C and has an exchangeable sodium of 15%. Of the total 14 billion ha of land available on earth, about 1 billion ha are natural saline soils. Also, it is estimated that about 20% of total cultivated lands and 33% of irrigated agricultural lands are affected by high salinity globally (Flowers et al. 1986; Tanji 1990). Moreover, salinized areas are increasing at a rate of 10% annually due to low precipitation, high surface evaporation, weathering of native rocks, irrigation with saline water, and poor cultural practices (Tanji 1990; Szabolcs 1994). It is estimated that more than 50% of the arable land would be salinized by the year 2050 (Ashraf 2009). Meanwhile, human population is on the rise, and it is expected to reach 9.1 billion by 2050 (Stephenson et al. 2010). It is becoming more challenging to meet the demands of the growing population, with the available natural resources, in particular when additional lands are becoming unusable for agriculture due to various environmental factors and urban expansion. Thus, greater efforts must be devoted to increase crop productivity in stressful agricultural environments and bring marginal lands like saline soils under cultivation.

Salt stress inhibits plant growth and development and triggering programmed cell death, leading to reduced crop yield or total crop failure (Ueda et al. 2006). With the advent of modern molecular biology techniques, it is becoming possible to better understand plant response to salt stress and to improve plant salt tolerance via genetic means. However, knowledge of candidate salt-tolerant genes is a prerequisite to effective utilization transformation techniques for development of transgenic plants with improved salt tolerance. This chapter covers the application of new and emerging technologies, which potentially could be utilized for the purpose of improving plant salt tolerance. The current status of salt-tolerant and gene expression studies of different plant species under salt stress is also reviewed and discussed.

Mangroves grow in intertidal zones at the interface between land and sea along the tropical and subtropical coastlines (Tomlinson 1986). They are considered to be more salt adapted than any other halophytes, especially being capable of tolerating fluctuating salinities from hyposaline to hypersaline conditions (McMillan 1986). Mangroves at their native habitats get periodically inundated by low and high tides leading to permanent waterlogging at places and fluctuations in salinity at other sites (Parida et al. 2004a, b). These conditions become worse in tropical mangroves. Overheating and desiccation is high at low tide condition, and evaporation causes the remaining water to be highly saline. At high tide, the warmth of water depletes the soluble oxygen, so the mangrove plants face the anaerobic or anoxic conditions in their habitat. These situations repeat seasons after seasons. Mangroves do cope with these adverse environmental conditions by variety of adaptable features like salt secretion, production of pneumatophores, formation of prop roots, and viviparous germination of seeds or propagules. Besides these morphological adaptive features, the biochemical strategies include regulated ion uptake and transport of ions from and to different plant parts, selective accumulation and exclusion of ions, compartmentalization of ions, synthesis and accumulation of osmolytes, changes in metabolic pathways, alternations in membrane transport, and induction of hormone (see Das et al. 2006). The salt management strategies of mangroves are thus special group of salt-tolerant plants. Depending on their salt adaptation/-tolerant mechanisms, the mangroves are divided into two distinct groups: the "secretors," those that have salt glands like in *Avicennia marina* and the other "non-secretors," those that lack such morphological features for excretion of excess salt. Non-secretor mangroves like *Bruguiera* species are shown to exclude 99% of the salt in surrounding seawater by ultrafiltration (Scholander 1968). Recently, mangroves are being the focus plants for isolation of salt stress genes to mitigate the salt tolerance in crop plants. Hence, it is important to review the current situation on mangrove research on salt stress genes.

## Plant Response to Salt Stress

Plants respond to salinity stress in a complex fashion involving concerted functions and interactions of many genes, proteins, and metabolic and signaling pathways (Apse and Blumwald 2002; Ashraf 2009). Plants have the ability to sense environmental changes and cues (Vij and Tyagi 2007) and to respond and adapt to such conditions (Rensink et al. 2005; Matsui et al. 2008). Regulation of gene expression is one of the key phenomena in plants by which they respond and try to adapt to salt stress. One of the most intriguing examples of such salt-induced gene regulation in plants is the salt-induced activation of the phosphorylation/kinase cascade followed by activation of various transcription factors (Xiong et al. 2002). To reestablish cell homeostasis and normal functioning under salt stress, the activated transcription factors regulate the expression of genes, which encode proteins contributing to salt tolerance, or the activity of enzymes, which are involved in pathways leading to the protection and repair of cells under salt stress (Flowers 2004; Munns 2005). Salt uptake and transport through the cell is controlled by various membrane proteins. Entry of $Na^+$ into the cell is brought about either by nonselective cation channels (Demidchik et al. 2002) or through several low- or high-affinity $K^+$ carriers. Examples of the latter are the high-affinity $K^+$ transporter HKT1 mediating $Na^+$ uptake in *Saccharomyces cerevisiae* (Uozumi et al. 2000) and the low-affinity cation transporter LCT mediating $Na^+$ transport across plasma membrane in bread wheat (Schachtman et al. 1997; Amtmann et al. 2001). However, recently, the role of LCT in mediating Na+ influx across the membrane under saline conditions was questioned because of its considerable sensitivity to $Ca2^+$ (Zhang et al. 2010). A high concentration of Na+ is toxic to all plant cells, even in halophytes (Glenn et al. 1999), and maintenance of a high $K^+/Na^+$ ratio is essential for normal growth and development in most plant species. Plants maintain such a balance usually by any of the three mechanisms: (1) $Na^+$ efflux through either $Na^+$/H+ antiporters in plasma membranes like salt overly sensitive (SOS1) antiporter (Zhu 2001a), (2) vacuolar $Na^+/H^+$ antiporters such as NHX1 (Blumwald et al. 2000; Yokoi et al. 2002), and (3) $K^+$ retention in cells through differential functioning of high- and low-affinity K channels (Shabala and Cuin 2008; Coskun et al. 2010). Apart from antiporters, other regulatory pathways may be mediated by kinases and phosphatases (Diedhiou et al. 2008; Jung et al. 2008). The SOS pathway is regulated by binding of $Ca2^+$ to SOS3, a calcium-binding protein, which activates a serine/threonine-specific protein kinase SOS2, which in turn activates SOS1 (Zhu 2003). Salt stress also causes hyperosmolarity that activates the mitogen-activated protein kinase (MAPK) cascade (Kiegerl et al. 2000), which leads to the induction of transcription factors followed by increased synthesis of osmolytes, osmoprotectants, and detoxifying enzymes (Vinocur and Altman 2005).

Most plants, including halophytes, produce osmolytes and osmoprotectants under osmotic stress conditions to adjust to stress (Tal and Shannon 1983). Sugars, polyols, proline, and quaternary ammonium like glycine betaine and polyamines found suitable osmolytes to cope with saline environment (Rhodes et al. 2002). These compounds can be inherently produced in transgenic plants to contribute to salt tolerance. Furthermore, reactive oxygen species (ROS), generated in plants under salt stress, may be scavenged by osmolytes (Zhu 2001b; Ashraf 2009) or detoxified by enzymes induced by salt stress such as superoxide dismutase (SOD), catalase (CAT), and aldehyde dehydrogenase (Sunkar et al. 2003; Ashraf 2009). Furthermore, a number of free amino acids found very effective in regulating the NaCl-induced $K^+$ efflux (Cuin and Shabala 2007). Hence, free amino acids may play an effective role in plant adaptive responses to salt stress by regulating $K^+$ transport across the plasma membrane and thus maintaining an optimal $K^+/Na^+$ ratio. Therefore, there are alternative means by which plants respond to salt stress, and knowledge of the underlying mechanisms of such response may contribute to the development of plants with enhanced salt tolerance through genetic modifications.

## Salt Stress Imposes Oxidative Stresses

Most stresses cause oxidative stress; however, salt stress effect on plant is a complex phenomenon. It decreases the osmotic potential of soil solution creating a water stress in plants. This water deficit in plants accumulate abscisic acid which further triggers a signaling cascade in guard cells that ultimately results in stomatal closures. All plants' parts can produce reactive oxygen species (ROS), but it is mostly generated by mitochondria and chloroplasts. This stomatal closure limits $CO_2$ availability in leaf cells. As a whole, a disruption of coordination between carbon assimilation and chloroplast photo function in electron transport chains leads to the transfer of electrons in alternating electron acceptor like $O_2$ (Tanaka et al. 1999; Foyer and Noctor 2000; Tausz et al. 2004). A simplified path of ROS production is shown in Fig. 3.1. The excess electron reduces molecular oxygen to produce reactive oxygen species (ROS) such as singlet oxygen ($^1O_2$), superoxide radical ($O_2^-$), hydroxyl radicals ($OH^-$), and hydrogen peroxide ($H_2O_2$). Superoxide radicals are synthesized in the chloroplasts and mitochondria, and a little quantity is also reported to be produced in microbodies. Superoxide produces highly toxic hydroxyl radical ($OH^-$) in combination with $H_2O_2$. Hydroxyl radical can damage chlorophyll, protein, DNA, lipids, and

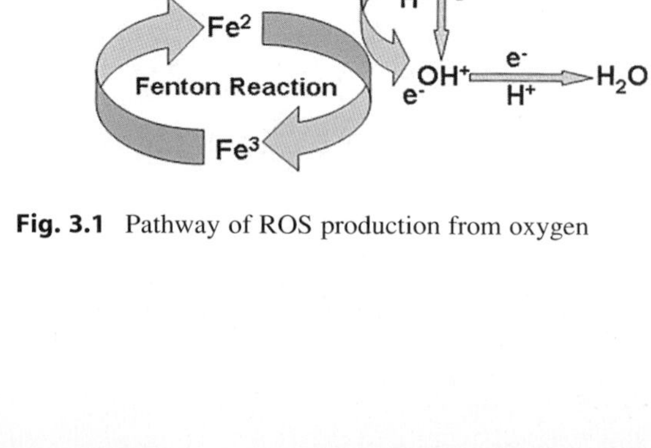


**Fig. 3.1** Pathway of ROS production from oxygen

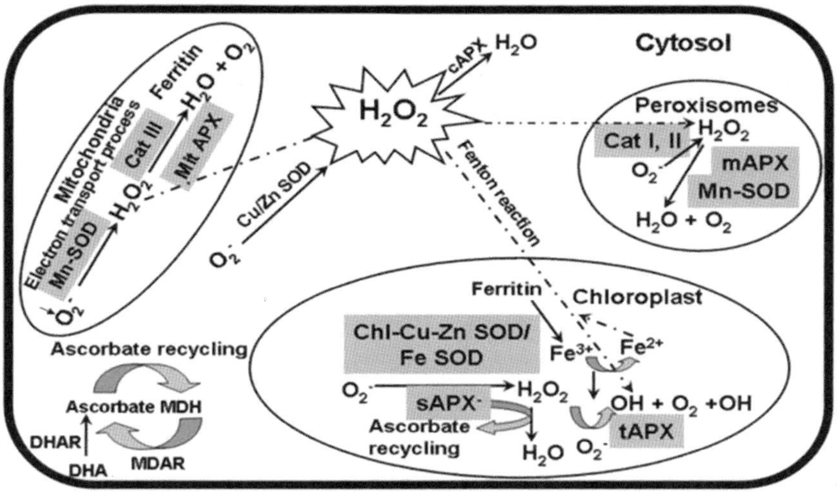


**Fig. 3.2** Schematic representation of antioxidant enzyme activity in different organelles of the plant cell

other important macromolecules. Thus, this likely affects plant metabolism and ultimately growth and yields (Figs. 3.1 and 3.2).

## Antioxidative Defense System: A True Scavenger of ROS

Major bioenergetic organelles like chloroplasts, mitochondria, and peroxisomes are the important intracellular generators of ROS. The formation of ROS is thus initiated by the univalent reduction of $O_2$ or by the transfer of excess excitation energy to $O_2$. Plant possesses both enzymatic and nonenzymatic mechanisms to scavenge these reactive oxygen species (ROS). The major enzymes induced in response to ROS are superoxide dismutase (SOD; EC1.15.1.1), catalase (CAT;EC1.11.1.6), ascorbate peroxidase (APX;EC1.11.1.11), glutathione reductase (GR;EC1.6.4.2), and guaiacol peroxidase (GPX;EC1.11.17). These enzymes are either upregulated or downregulated to minimize the concentration of ROS to maintained cellular homeostasis. In addition to antioxidative enzyme systems, nonenzymatic antioxidants like ascorbic acid, glutathione, α-tocopherol, and carotenoid as well as osmolytes like proline and glycine betaine also play a key role in scavenging free radicals in plants during salinity stress. Superoxide radical is regularly synthesized in the chloroplasts and mitochondria (Elstner 1991; Rich and Bonner 1987). Superoxide dismutase (SOD) scavenges these superoxide radicals which results in the production of $H_2O_2$. The produced $H_2O_2$ is removed by APX or catalase. The combination of $O^{\cdot}{}_2{}^{-}$ and $H_2O_2$ in presence of trace amount of $Fe^{2+}$ and $Fe^{3+}$ produced highly toxic hydroxyl radical (Figs. 3.1 and 3.2). Table 3.1 catalogues various antioxidant enzymes that have been reported in mangroves and their associates as a salt adaptive strategy. Earlier from our laboratory, Parida et al. (2004a, b) reported that salt stress enhances the activity of superoxide dismutase, ascorbate peroxidase, glutathione reductase, and guaiacol peroxidase, whereas the activity of catalase decreased in leaves of hydroponically maintained 250 mM NaCl-treated *Bruguiera parviflora* seedlings. It was found that the total activity of SOD enhanced 128% as compared to untreated control, whereas activity-staining data revealed that out of the five isoforms of SOD (one Mn-SOD, one Cu/Zn-SOD, and three Fe-SOD), the expression of Mn-SOD and Fe-SOD-2 was enhanced preferentially upon exposure to salt under hydroponic culture condition. A total of 35% decrease in catalase activity was found from spectrophotometry analysis, whereas activity-staining data indicated that out of the four

**Table 3.1** List of studies carried out on antioxidative enzymes during salt stress in mangroves and other halophytes

| Enzyme/ gene | Plant species | Method used | Enzyme expression | Reference |
|---|---|---|---|---|
| APX | *Bruguiera parviflora* | Activity staining on gel | Enhancement of APX activity in presence of NaCl stress in hydroponic culture | Parida et al. (2004a, b), Parida and Das (2005) |
| APX | *Thellungiella halophila* | EST analysis | Presence of three APX in cDNA library from salt-treated plants | Wang et al. (2004a) |
| APX | *Aegiceras corniculatum* | Spectrophotometric assay | Activity decrease 58% on 4th day in root | Mishra and Das (2003) |
| CAT | *Avicennia marina* | mRNA detection | Induction of AmCat I in salinity till 12 h and subsequent decreased | Jithesh et al. (2006a, b) |
| CAT | *Bruguiera parviflora* | Spectrophotometric assay and activity staining on gel | Loss of total catalase activity with NaCl stress, CAT-2, and CAT-3 isoforms decreased | Parida et al. (2004a, b) |
| CAT | *Bruguiera gymnorrhiza* | Spectrophotometric assay | Increase in total catalase up to sea water concentration | Takemura et al. (2000) |
| CAT | *Bruguiera gymnorrhiza* | mRNA detection | A particular catalase transcript level did not change with salinity stress | Takemura et al. (2002) |
| CAT | *Crithmum maritimum* | Spectrophotometric and isozyme analysis | Decrease in catalase activity with increase NaCl | Ben Amor et al. (2005) |
| Ferritin | *Mesembryanthemum crystallinum* | Electron microscopic studies | Ferritin deposits present with NaCl stress in chloroplasts | Paramonova et al. (2004) |
| Ferritin | *Avicennia marina* | mRNA analysis | Early induction of AmFer1 in salinity | Jithesh et al. (2006a, b) |
| CAT | *Aegiceras corniculatum* | Spectrophotometric assay | Activity decrease 72% in root | Mishra and Das (2003) |
| CAT | *Aegiceras corniculatum* | Activity staining on gel | Two isoforms CAT 1,2 decreased | Mishra and Das (2003) |
| CAT | *Suaeda nudiflora* | Spectrophotometric analysis | Decrease in catalase activity in callus cultures with salinity stress | Cherian and Reddy (2003) |
| POX | *Avicennia marina* | Spectrophotometric analysis | Enhanced peroxidase activity in root and shoot and no significant changes in leaf tissue | Cherian et al. (1999) |
| POX | *Crithmum maritimum* | Spectrophotometric analysis | Transient increase in peroxidase activity | Ben Amor et al. (2005) |
| POX | *Suaeda nudiflora* | Spectrophotometric analysis | Enhanced activity with NaCl stress | Cherian and Reddy (2003) |
| GPX | *Aegiceras corniculatum* | Spectrophotometric assay | Activity decrease 80% on 4th day | Mishra and Das (2003) |
| GPX | *Aegiceras corniculatum* | Activity staining on gel | Four isoforms of GPX-1,2,3,4 decreased upon salt treatment of 6d | Mishra and Das (2003) |
| GPX | *Bruguiera parviflora* | Spectrophotometric assay | 169% of GPX enhancement at 45d exposure of NaCl at 400 mM | Parida et al. (2004a, b), Parida and Das (2005) |
| GPX | *Bruguiera parviflora* | Activity staining on gel | GPX-6 shows a maximum of 73% increment in activity without any change in GPX-4,5, whereas GPX-1 increase by 61% and GPX-2 and 3 changed 34–38% | Parida et al. (2004a, b) |

(continued)

**Table 3.1** (continued)

| Enzyme/ gene | Plant species | Method used | Enzyme expression | Reference |
|---|---|---|---|---|
| SOD | *Avicennia marina* | Spectrophotometric analysis | Decrease in total activity in roots and shoots but increase in leaves with salinity stress | Cherian et al. (1999) |
| SOD | *Avicennia marina* | mRNA analysis | Constitutive expression of Cu/Zn-SOD in salinity | Jithesh et al. (2006a, b) |
| SOD | *Bruguiera parviflora* | Activity staining on gel | Increase in activity of Mn-SOD and Fe-SOD and no increase in activity of Cu/Zn-SOD isoform with salinity stress | Parida et al. (2004a, b), Parida and Das (2005) |
| SOD | *Bruguiera gymnorrhiza* | Spectrophotometric assay | Steep increase in SOD activity in salinity stress | Takemura et al. (2000) |
| SOD | *Bruguiera gymnorrhiza* | mRNA detection | Induction of Cu/Zn-SOD activity in salinity stress | Takemura et al. (2002) |
| SOD | *Crithmum maritimum* | Spectrophotometric assay and isozyme analysis | Increase in total SOD activity in shoots and decrease in roots | Ben Amor et al. (2005) |
| SOD | *Suaeda nudiflora* | Spectrophotometric assay | Decrease in total SOD activity in callus culture with salinity stress | Cherian and Reddy (2003) |
| SOD | *Suaeda salsa* | Isozyme assay and spectrophotometric assay | Increase in activity of thylakoid-bound SOD | Fang et al. (2005) |
| SOD | *Suaeda salsa* | Staining on gel | Increase in Mn-SOD and Fe-SOD isoforms, Cu/Zn-SOD I isoform activity decreased in presence of salinity stress, no change Cu/Zn-SOD II isoform | Wang et al. (2004a) |
| SOD | *Thellungiella halophila* | mRNA detection | Higher mRNA transcript levels of Fe-SOD | Taji et al. (2004a, b) |
| SOD | *Mesembryanthemum crystallinum* | mRNA detection | Increase in Fe-SOD and Cu/Zn-SOD during salt stress | Slesak et al. (2002), Hurst et al. (2004) |
| SOD | *Mesembryanthemum crystallinum* | mRNA detection | Induction of all three SOD isoforms in salt stress | Slesak et al. (2002) |
| Acid phosphatase | *Bruguiera parviflora* | Activity staining on gel | Out of the three isoforms of ACP, ACP-3 increased by 37% without much change in other isoforms | Parida et al. (2004a, b) |
| Alkaline phosphatase | *Bruguiera parviflora* | Spectrophotometric assay | Activity increased to 10%, 20%, and 39% dose dependent in 100, 200, and 400 mM NaCl | Parida et al. (2004a, b, 2005) |
| Nitrate reductase | *Bruguiera parviflora* | Spectrophotometric assay | Activity increased in 100 mM NaCl and gradually decline up to 63% in 400 mM NaCl | Parida et al. (2004a, b), Parida and Das (2005) |

isoforms (CAT-1, CAT-2, CAT-3, and CAT-4), CAT-2 isoform was decreased to 45% as compared to control. It was suggested that the inactivation of catalase was possibly due to enhancement of $H_2O_2$. The activity of APX and GR also enhanced in response to salinity in this species. Among the high molecular weight isoforms, GPX-1 was increased by 61%, whereas GPX-6 showed a ~70% increase in the concentration with regard to its control. It was further suggested that these two isoforms of GPX may have specific sensitivities to salinity.

When *Bruguiera gymnorrhiza* seedlings were treated with 500 mM NaCl for a period of 9 days, it was found that the catalase activity increased to about nearly 4.9 times as compared to untreated control plants (Takemura et al. 2000). This suggested that catalase retained its full activity at least up to seawater salt concentration up to 9 days and again lowered till the end of the experimental treatment time of 16 days. This type of opposite behavior in catalase activity in both species of *Bruguiera* is difficult to explain at present. SOD activity also increased about 8.1 times upon exposure to 500 mM NaCl for 9 days as compared to untreated control. Takemura et al. (2002) also reported that mRNA transcript of cytosolic Cu/Zn-SOD was increased in *B. gymnorrhiza* after the 5th and 6th day of salt treatment. However, there are only a few reports about the mechanisms of generation of superoxide radicals in the cytosol and role of cytosolic SOD in halophytes (Takemura et al. 2002). In *Avicennia marina*, it was found that the total SOD activity increased in leaf tissues after 6 days of salt treatment, whereas surprisingly, the total SOD activity decreased in root and shoot tissue (Cherian et al. 1999). The mRNA detection study by Jithesh et al. (2006a, b) reported an enhancement in Cu/Zn-SOD upon exposure to salinity. In another study, by the same group, a class II enzyme, CAT I from *Avicennia marina*, was induced upon salt and other oxidative stress, such as exposure to $H_2O_2$ and light, in leaves (Jithesh et al. 2006a, b). Increased POX activity was observed in root and shoot tissues in *Avicennia marina* under NaCl stress conditions, whereas no such significant changes in POX was found in leaf tissues (Cherian et al. 1999). However, in the roots of *Aegiceras corniculatum*, activity of antioxidative enzymes like CAT, APX, and GPX decreased upon salinity for a short period of 4 days (Mishra and Das 2003). A decrease in total SOD activity was also observed in NaCl-treated callus cultures of *Suaeda nudiflora* – a halophyte. This result is not in agreement with studies conducted on whole plants, where SOD activity was found to be increased under salt stress (Cherian et al. 1999). The catalase activity was found to decrease by 60%, which showed a declining capacity of callus tissues to scavenge $H_2O_2$ with increasing salt stress. An enhancement in POX activity in stalinized cell indicated that these cells had a higher capacity for decomposition of $H_2O_2$ and that POX performs a major role in this particular reaction (Cherian and Reddy 2003). Both spectrophotometric and in-gel assay of *Suaeda salsa* showed an increase in activity of thylakoid-bound SOD upon exposure to salinization (Fang et al. 2005). Wang et al. (2004a, b) reported an increase in Mn-SOD and Fe-SOD isoforms and a decrease in Cu/Zn-SOD I isoform, whereas no significant changes was found in Cu/Zn-SOD II isoforms in *S. salsa* on salinization. Again, the total SOD activity was shown to increase in shoot tissues of another halophyte, *Crithmum maritimum*, in the presence of 50 mM NaCl stress; however, at a concentration of 200 mM of NaCl, no significant difference in SOD activity was detected (Ben Amor et al. 2005). On the contrary, SOD activity in root tissues of *C. maritimum* decreased upon treatment with 50 mM as well as 200 mM NaCl (Ben Amor et al. 2005), whereas total catalase activity decreased with increasing NaCl concentration in *Crithmum maritimum* (Ben Amor et al. 2005). A transient increase of POX activity was also observed in this study. These data reflect that each mangrove plant has its own regulatory control of inducing specific antioxidative enzymes. Table 3.1 depicts different antioxidative enzyme systems in mangroves and their associates. However, the complexity of salt adaptive mechanisms necessitates the use of novel and advanced techniques for their elucidation and better understanding and in turn utilization of the knowledge in crop improvement for salt tolerance.

## Cross-Talks-Mediated Complexity of Salt Tolerance

A comprehensive knowledge of all signaling hierarchies involved in plant salt tolerance is essential in order to identify and introduce into plants different contributing components of salt tolerance and potentially produce transgenic plants that could successfully grow under saline field conditions. Such a task is obviously very difficult to accomplish due to the highly complex nature of the networks involved in stress signaling and their responses (Ma et al. 2006; Vij and Tyagi 2007). The polygenic nature of salt tolerance and co-occurrence of different stresses under field situations contribute to the complexity of this phenomenon. However, knowledge of molecular mechanisms of cross-talks among various stress-related pathways and transcription factors may contribute to understand better the underlying tolerance mechanisms. The machineries for various processes in the cell are intimately connected and functionally interlinked. Various signaling factors interact with each other, demonstrating existence of networks of different processes working in coordination to respond to multiple stress factors and conditions. Such communications among diverse pathways to converge into an integrated response, known as cross-talks, are now emerging as a common phenomenon in a variety of organisms including plants. In response to various environmental stresses, plants often initiate multiple perception and signaling pathways, which may cross-talk at different steps (Song and Matsuoka 2009; Hu et al. 2010; Urano et al. 2010). The MAPK cascade is a central point for various signaling pathways under both biotic and abiotic stresses. Plant responses to both cold and salt stresses exhibit cross-talks between cold and salt tolerance-related pathways (Mahajan and Tuteja 2005). Several auxin-responsive genes have been shown to cross-talk with abiotic stress signaling factors in rice (Jain and Khurana 2009). Protein kinases have also been shown to interact in different signaling pathways. At least 34 calcium-dependent protein kinases (CDPKs) have been identified in *Arabidopsis* which play central roles in $Ca_2^+$ signaling in multiple pathways including stress signaling (Ma and Wu 2007). Similarly, calcineurin B-like protein-interacting protein kinases (CIPKs) play important roles in stress and other signaling processes in plants (Xiang et al. 2007). A nuclear protein, ethylene insensitive 2 (EIN2), which interconnects several hormonal response pathways in plants, has been shown to regulate both osmotic and salt stress responses in *Arabidopsis* (Wang et al. 2007a). Cross-talks also exist between heat and non-heat-stress regulatory networks. Heat-shock transcription factors (Hsf) and heat-shock proteins (Hsp) were strongly induced under salt as well as heat, cold, and osmotic stresses (Swindell et al. 2007). Furthermore, cross-talk between nodule-specific genes of *Astragalus sinicus* (Chinese Milkvetch) and salt stress has been shown to facilitate the nitrogen fixation process (Chen et al. 2007). Similarly, cross-talks between salt stress and signaling pathways for potassium and iron deficiencies were found in the root of tomato via 14-3-3 proteins (TFT7; tomato 14-3-3 protein 7), which regulate a wide array of targets by protein-protein interactions (Xu and Shi 2006). A comprehensive study of interacting factors in response to different stresses is essential to unravel the functional links among such contributing factors. In this regard, analysis of the transcriptional profiles of plants in response to salt stress may provide a reference point for further understanding of stress mechanisms.

The complexity of various stress signaling pathways is exacerbated due to interactions and cross-talks among transcription factors at different steps (Kizis et al. 2001). ABA-dependent and ABA-independent osmotic stress signals activate tolerance genes through the inactivation of transcription factors (Zhu 2002); this is one of the best examples. At the same time, overexpression of transcription factors controlling multiple genes from various pathways may lead to attaining tolerance against multiple stresses (Sreenivasulu et al. 2007). Several recent reports emphasize the

**Table 3.2** Transcription factors with significant role in plant tolerance (Adapted from Jamil et al. (2011))

| Transcription factors | Source | Stress | References |
|---|---|---|---|
| TaSnRK2.4 | Wheat to Arabidopsis | Drought/salt/freezing | Mao et al. (2010) |
| ANAC092 | Arabidopsis | Salt | Balazadeh et al. (2010) |
| CaRAV1 | Pepper | Salt/osmotic | Lee et al. (2010) |
| GmDREB1 | Soybean to transgenic alfalfa | Salt | Jin et al. (2010) |
| Trihelix transcription factors GmGT-2A and GmGT-2B | Soybean to transgenic Arabidopsis | Salt/freezing/drought | Xie et al. (2009) |
| R2R3 MYB | Arabidopsis | Salt/drought | Ding et al. (2009) |
| NAC transcription factors NAM, ATAF, CUC | Rice | Salt/low temperature/drought | Fang et al. (2008) |
| Basic-leucine zipper (BZIP) factors GmbZIP132 | Soybean | Salt/drought | Liao et al. (2008) |
| WRKY-type factors | Soybean to transgenic | Salt/mannitol | Zhou et al. (2008) |
| GmWRKY13 | Arabidopsis | Cold | Zhou et al. (2008) |
| GmWRKY21 GmWRKY54 | | Salt drought | Zhou et al. (2008) |
| SNAC2 (NAC type) | Rice | Salt/drought/cold/wounding/ABA | Hu et al. (2008) |
| Homeodomain-leucine-zipper | Cotton | Salt | Ni et al. (2008) |
| (HD-Zip) ATMYB44 | Arabidopsis | Salt/dehydration/low temperature | Jung et al. (2008) |
| MtZpt2-1, MtZpt2-2 | Barrel Clover (Medicago truncatula) | Salt | de Lorenzo et al. (2007) |
| OsNAC6 (NAC type) | Rice | Salt/dehydration | Nakashima et al. (2007) |
| Zat12 (zinc finger) | Arabidopsis | Salt/cold/oxidative/osmotic/highlight/heat | Davletova et al. (2005) |
| 40 transcription factor genes | Arabidopsis | Salt/drought/cold/wounding/ABA | Seki et al. (2002) |

importance of transcription factors in mediating responses against abiotic stresses, as summarized in Table 3.2. Different families of transcription factors, such as WRKY-type and NAC (non-apical meristem), ATAF, and CUC (cup-shaped cotyledon) transcription factor families, have diverse roles in plant development and stress regulation (Nakashima et al. 2007; Hu et al. 2008; Zhou et al. 2008). Transcription factors from one pathway may beregulated by transcription factors from otherpathways under stress conditions. The activity of GmWRKY54 may be regulated through DREB2A (dehydration-responsive element-binding protein 2a) and STZ/Zat (salt-tolerant zinc finger/zinc finger *Arabidopsis thaliana*) transcription factors (Zhou et al. 2008). The leucine zipper-based transcription factors with well-established roles in plant development and biological processes have also been shown to induce stress tolerance in plants (Liao et al. 2008; Ni et al. 2008). Because transcription factors play important roles by regulating gene expression, functional characterization of various transcription factors under salt stress conditions may provide essential information. Overexpression of certain stress-responsive transcription factors in transgenic crops may lead to enhanced plant salt tolerance and improved crop productivity under saline conditions. The mutual interdependence and close relationships and interactions among regulatory cascades demand detailed studies of such phenomena so that the underlying mechanisms may be elucidated and subsequently, the knowledge is used to develop plants with improved salt tolerance. Due to the complex nature of such pathways and interactions among them, the techniques employed to define transcriptomes and/or proteomes involved under various stresses are of significant importance. Some of these techniques are reviewed and discussed below.

## Microarray for Global Gene Expression Profiling During Stress

Global gene expression profiling has been revolutionized with the development of the microarray technology as the entire gene complement of a genome could be studied in a single experiment (Vij and Tyagi 2007). The two major types of microarray analysis that are applied to studying gene expression profiles are cDNA and oligonucleotide microarrays. In cDNA microarray, expressed sequence tags (ESTs) are generated under specific conditions and immobilized on a solid support at a density of approximately 1,000 genes $cm^{-1}$ (Seki et al. 2002; Nguyen et al. 2006). The solid support may be a nylon or a chemically coated slide (Holtorf et al. 2002). The cDNA clones are sequenced in multiwell plates, amplified by PCR, and arrayed on the slides using microarray stamping machine. In oligonucleotide microarray, the target molecules are gene specific and are synthesized in situ on the surface of a chip by photolithography (Lipshutz et al. 1999). In both types of microarray, the transcripts from the samples to be compared are reverse transcribed and fluorescently labeled with different fluorophores (e.g., Cy5, Cy3) for both the sample and the control, followed by their simultaneous hybridization with the arrays (Baldwin et al. 1999). The gene expression pattern and relative abundance of each gene is determined with the help of a high-resolution laser of two different wavelengths (Lockhart and Winzeler 2000). The success of a microarray analysis depends on the quality control and normalization of the expression data between different experiments and on the reduction of noises as compared to the signal. The large-scale analysis of the transcripts depends mainly on the availability of relevant bioinformatics tools. Several programs and databases are utilized for this purpose. PipeOnline (Bohnert et al. 2001) and MEDIBA (Microarray Data Interface for Biological Annotation) (Law et al. 2008) are the two automated and highly interactive tools for the processing and analysis of large batches of DNA sequence data. The functional assignment algorithms in PipeOnline provide estimation of function utilizing the BLASTX homology analysis using MPW-based functional overview. MEDIBA provides a rapid analysis for the identification of the gene ontology terms relevant to each cluster and assists the researchers in understanding the meaning of co-expression of a cluster of genes. It helps determine the genomic localizations of the sequences, find putative transcriptional elements in the upstream regions, and visualize the metabolic pathways where the genes are implicated. The clustering analysis of genes is especially useful under abiotic stresses for understanding mechanistic details, as many genes are co-regulated under certain environmental conditions. Effective clustering algorithms and tools also are required for the identification of stress regulons (Sreenivasulu et al. 2007). "Genevestigator" is another web-based tool for the study of expression of small sets of genes from a variety of microarray experiments (Zimmermann et al. 2004). The difference in expression profile must be examined with at least two important considerations: the profile is developed from mRNA that may not necessarily correspond with changes in the relevant proteins, and certain variations may arise due to the adaptation of the plants to the changing environmental conditions. "Tiling-path" array is another type of microarray that has been used to study gene expression in plants (Vij and Tyagi 2007; Matsui et al. 2008). It provides unbiased and more accurate information about the transcriptomes as compared to the conventional arrays and provides information about transcriptional controls at the chromosomal level. Such advantages of the tiling arrays enhance its usefulness to acquire downloaded novel information about gene expression profiles under abiotic stresses.

Oligonucleotide microarrays have been applied in various plant species grown under different stresses to identify tolerance-related genes or transcription factors. Mohammadi et al. (2007) identified a novel transcription factor AP2/ERF by transcriptional profiling of wheat growing under drought stress. Affymetrix GeneChip provides oligomer microarrays for various plant species, and it has been used to study the transcriptome changes in different plant species such as *Arabidopsis* (El Ouakfaoui and Miki 2005; Kilian et al. 2007) and rice (Walia et al. 2005; Ray et al. 2007) under salt and other abiotic stresses. In most plant species, however, expression profiling is conducted mainly by using cDNA microarray format.

To use this technique, development of ESTs is an initial and necessary step, which needs careful consideration. An EST library is generally developed by using cDNA clones generated from differentially expressed mRNA isolated from different sources (e.g., leaves, stem, and seed) and at various developmental stages after initiating specific stress conditions. The cDNA clones are sequenced and arrayed on slides as discussed above. However, in using EST libraries for gene expression analysis, there are several issues of concern. First, transcripts with low-level expression may be underrepresented in EST libraries (Holtorf et al. 2002). Second, nonspecific cross-hybridization may occur between different PCR fragments due to gene duplications or presence of multigene families (Richmond and Somerville 2000). Both of these situations may affect the results of the microarray analysis. A primary requirement for studying complex abiotic stress responses is development of large-scale ESTs from different tissues at various developmental stages.

A large-scale collection of ESTs is also important as it would minimize highly repetitive DNA sequences normally found in genomic DNA, thus making the analysis rather simpler (Zhang et al. 2004). Detailed information on EST libraries generated for different plant species under various abiotic stress conditions has been presented in various websites and databases, as summarized in Table 3.3. For many plant species, there are numerous stress-related EST databases and resources, integration of which may lead to *in silico* inferences to genes related to responses against various abiotic stresses. Extensive EST collections are available for the model plants Arabidopsis and rice (Rensink and Buell 2005; Vij and Tyagi 2007). ESTs from salt-stressed rice have been generated and analyzed by many researchers (Bohnert et al. 2001; Walia et al. 2009; Senadheera et al. 2009). Large numbers of ESTs have also been generated for common wheat under different abiotic stresses (Houde et al. 2006). ESTs for corn, barley, and tobacco grown under salt stress have also been generated from roots and leaves at different developmental stages and varying time intervals (Bohnert et al. 2001; Zhang et al. 2004). ESTs from cDNA libraries derived from plants grown under stress conditions have higher representation of stress-inducible and stress-repressed tags compared to ESTs generated from non-stressed plants (Bohnert et al. 2001; Walia et al. 2009). Such comparisons would help identify stress-related genes. Analysis of ESTs from stress-tolerant species may also assist in the identification of biochemical pathways and genes involved in stress tolerance.

Although cDNA microarray analysis has been used extensively to investigate gene expression pattern under salt stress, further advancements in the oligonucleotide array and other technologies are needed to study the function and regulation of every known and predicted gene in the genome. Such advancement and the attained knowledge would guide directed manipulation of stress-responsive networks in plants toward improving crop productivity under stress conditions. EST analysis of the halophyte mangrove *Avicennia marina* resulted in isolation of monodehydroascorbate reductase gene, which conferred salt tolerance when transferred into tobacco (Kavitha et al. 2010).

**Table 3.3** Websites and databases for plant gene expression profiling

| Website/database | Information |
|---|---|
| http://www.ncbi.nlm.nih.gov/dbEST-/dbEST_summary.html | Database of EST by NCBI |
| www.tigr.org | The Institute for Genomics Research (TIGR) Gene Indices |
| http://mips.gsf.de/proj/sputnik | Sputnik database |
| http://stress-genomics.org/ | Stress Functional Genomics website |
| http://www.zmdb.iastate.edu | *Zea mays* database |
| http://mpss.udel.edu | MPSS site for some plants such as Arabidopsis, rice, grape |
| http://www.gsc.riken.go.jp/Plant/index.html | Arabidopsis genome, phenome, cDNA libraries, etc. |
| http://www.tigr.org/tdb/e2k1/osa1/data.download.html | Rice genome sequence |
| http://wheat.pw.usda.gov/ITMI/Repeats/index.shtml | Triticeae repeat element database |
| http://barleypop.vrac.iastate.edu/BarleyBase/content.php | GeneChip probe array search |
| http://bearlybase.org/ | Barley-related website |
| http://rgp.dna.affrc.go.jp/IRGSP/index.html | Monocot reference genome sequence form rice |
| http://www.mged.org/Annotations-wg/ | MIAME (minimum information about a microarray experiment) standards |
| http://web.unifrankfurt.de/fb15/botanik/mcb/AFGN/atgenex.htm | Profiling of transcriptome of Arabidopsis |
| http://www.Arabidopsis.org | |
| http://affymetrix.Arabidopsis.info/narrays/experimentbrowse.pl | |
| http://www.affymetrix.com/products/Arabidopsis_content.html | |
| http://atarrays.tigr.org/ | |
| http://www.genevestigator.ethz.ch | Genevestigator website |
| http://www.geneontogoly.org | Gene Ontology website |
| http://stress-genomics.org/stress.fls/tools/mutants.html | T-DNA insertion mutants lines |
| http://www.shigen.nig.ac.jp/wheat/komugi/top/top.jsp | Wheat EST database |
| http://www.tigr.org/tdb/Lgi/index.html | Tomat ESTs |
| http://macgrant.agron.iastate.edu/sobeanest.html | Soybean ESTs |
| http://microarray.rice.den.affrc.go.jp | Rice transcriptional profile |
| http://www.ebi.ac.uk/arrayexpress/ | ArrayExpress database |
| http://rgp.dna.affrc.go.jp/ | Rice Genome Project |

## Physiological Changes by Salt Stress in *Bruguiera gymnorrhiza*

*Bruguiera gymnorrhiza* popularly called as "Burma mangrove" is a prominent broad-leaved mangrove species in Indian mangrove forests. It belongs to the family Rhizophoraceae and is considered as a true non-salt-secreting tree mangrove (Fig. 3.3). This mangrove vegetation is mainly seen at the deltaic regions of Mahanadi river and Bhitarkanika in the east coastal region of India (Das et al. 1997). It is stated that *B. gymnorrhiza* is more salt tolerant than *B. parviflora*, but no detailed study has been made. As it is the dominant species for zonation, we have conducted a short-term experiment of high NaCl treatment on *B. gymnorrhiza* in hydroponic culture.

**Fig. 3.3** *Bruguiera gymnorrhiza* with vivipary seeds in natural condition at Bhitarkanika mangrove forests of Orissa

**Fig. 3.4** *Bruguiera gymnorrhiza* seedlings treated with NaCl in hydroponic culture under growth chamber condition at 22 ± 2°C temperature and 80% relative humidity

## Salt-Shock Experiments

In this section, we present a part of our findings in antioxidative enzyme response to high salt (500 mM NaCl) in leaf of hydroponically maintained mangrove, *Bruguiera gymnorrhiza* (Fig. 3.4). When 2-month-old seedlings of *B. gymnorrhiza* were exposed to 500 mM NaCl for a period of 9 days, specific changes in activity of selective antioxidative enzymes like CAT, SOD, APX, and GPX were observed in both spectrophotometry and activity-staining assays.

The total activity of SOD was increased twofold, whereas in gel study, a 2.5 time enhancement was noted in Fe-SOD and Cu/Zn-SOD upon salinization as compared to control (Fig. 3.5). The catalase activity increased 2.1 times in 9 days' salt stress leaf samples of *B.gymnorrhiza* as compared to untreated control samples. The activity-staining study revealed that out of four isoforms of catalase (CAT-1, CAT-2, CAT-3, CAT-4), CAT-1 and CAT-2 enhanced 1.2 and 1.4 times, respectively, upon 9 days salinization (Fig. 3.6). The ascorbate peroxidase activity

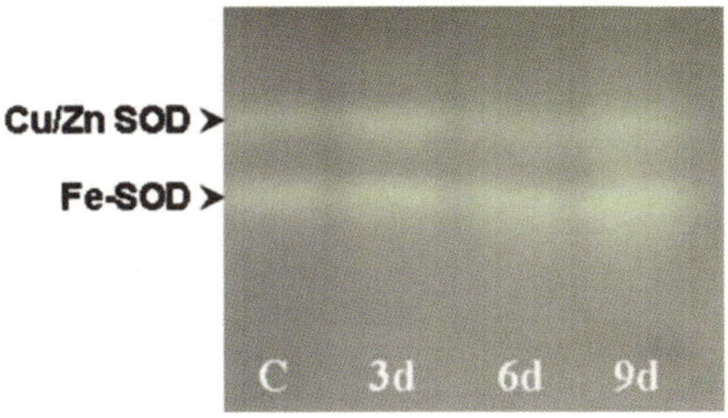


**Fig. 3.5** Activity staining of SOD in leaves of *B.gymnorrhiza* during 500 mM NaCl treatment

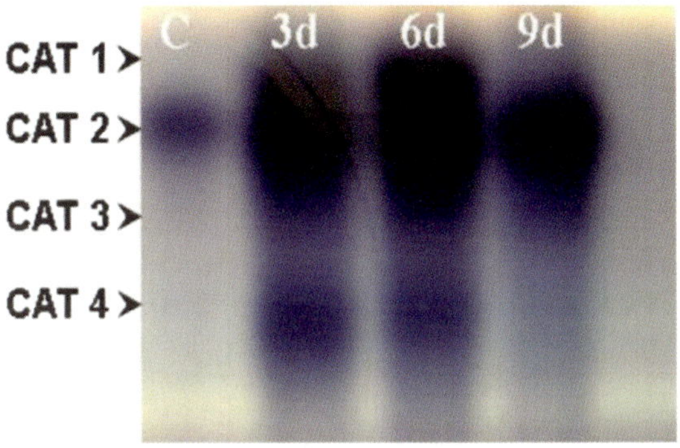


**Fig. 3.6** Activity staining of CAT in leaves of *B.gymnorrhiza* during 500 mM NaCl treatment

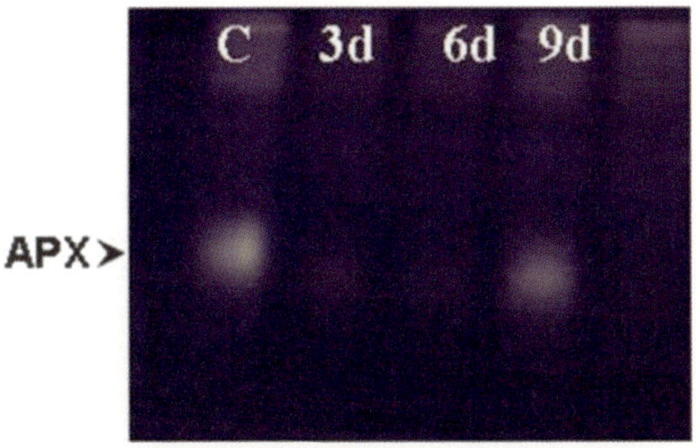


**Fig. 3.7** Activity staining of APX in leaves of *B.gymnorrhiza* during 500 mM NaCl treatment

increased 2.3 times as compared to untreated control (Fig. 3.7). The total activity of GPX increased 2.6 times as compared to control. Six isoforms of GPX (GPX-1, GPX-2, GPX-3, GPX-4, GPX-5, and GPX-6) were found from native electrophoresis and activity-staining data. Out of these six isoforms, GPX-1 and GPX-3 enhanced 2 and

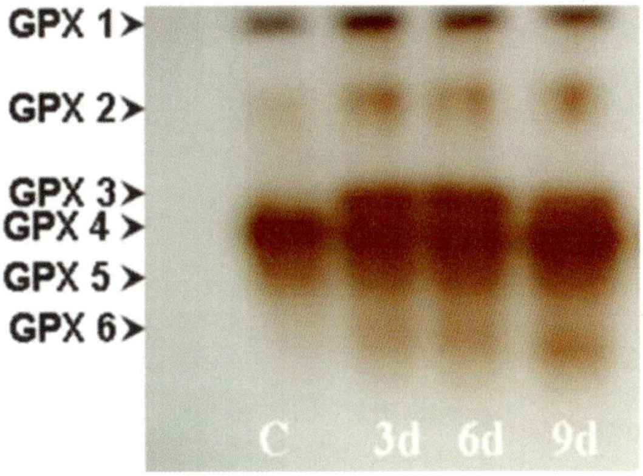


**Fig. 3.8** Activity staining of GPX in leaves of *B.gymnorrhiza* during 500 mM NaCl treatment

1.27 times, respectively, as compared to control, whereas GPX-2 increased 3.4 times in 3 days and 2.4 times in 9 days' salt treatment (Fig. 3.8).

## Changes of Soluble Leaf and Thylakoid Protein Profile

When 1-year-old seedlings of *B. gymnorrhiza* were exposed to 500 mM NaCl for a short period of 6 days under hydroponic culture condition, the total soluble leaf protein increased abruptly after 2 days of salt exposure and again decreased on 6 days as compared to control (Behera et al. 2009). From SDS-PAGE analysis of leaf protein, it was found that the intensity of several protein bands ranging from 10 to 86 kDa decreased on 6 days as compared to control (Fig. 3.9). The thylakoid membranes isolated from control and NaCl-treated leaves when analyzed by SDS-PAGE, and it was found that the intensity of many protein bands ranging from 10 to 86 kDa showed extensive decreased as compared to control (Fig. 3.10). The progressive decrease in protein bands due to increase in duration of salt exposure suggested that the target site of salt stress could be the degradation of specific thylakoid proteins. Parida et al. (2005) reported a novel protein of molecular mass 23 kDa in *B. parviflora* whose intensity gradually decreases with increasing concentration of salinity. In another study, several proteins were induced in salt-stressed *B. gymnorrhiza*, but the

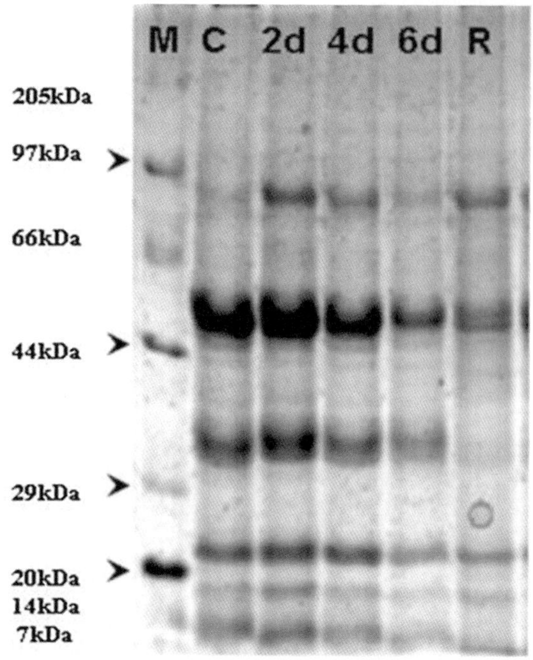


**Fig. 3.9** Changes in leaf total soluble protein profile in leaf of *B. gymnorrhiza* upon exposure to 500 mM NaCl treatment and post-recovery

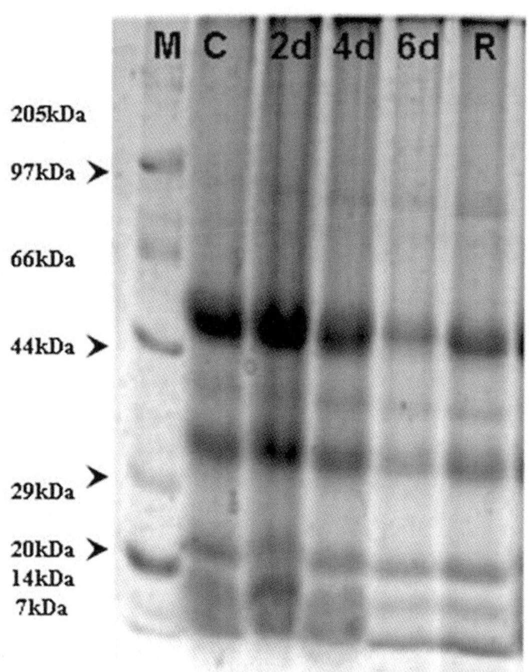


**Fig. 3.10** Changes in leaf total thylakoid protein profile in leaf of *B. gymnorrhiza* upon exposure to 500 mM NaCl treatment and post-recovery

induction of 33-kDa Mn-stabilizing protein is most important to understand the underlying mechanism of salt stress to maintain the capacity of PSII functioning (Sugihara et al. 2000). This 33-kDa protein is identified with oxygen evolving enhancer protein 1 (Sugihara et al. 2000).

## Salt Stress-Induced Changes in PSII Fluorescence in Mangroves

It is known that high light intensity leads to a damage of PSII (Klimov et al. 1990; Nishiyama et al. 2004, 2005; Hakala et al. 2005). Experiments with cyanobacteria demonstrated that this damage is mainly due to the action of ROS on the recovery of PSII rather than due to direct damage of the components of PSII (Allakhverdiev and Murata 2004; Nishiyama et al. 2001, 2004, 2011). A diagrammatic picture of PSII and its energy transfer is illustrated in Fig. 3.11. It was also shown that the damage induced by light is determined mainly by the rate of degradation and resynthesis of one of the stress-sensitive proteins of PSII – D1 (Zhang et al. 2000; Thomas et al. 2001; Aro 2006). Other proteins, D2 and PsbH, may also be the targets for damaging (Aro 2006), though D1 protein is characterized by the maximum rate of turnover (Mattoo et al. 1989). After photoinhibition damaged RC of PSII are not accumulated in the thylakoid membranes, and they are rapidly restored by de novo protein synthesis (Leitsch et al. 1994). Recovery involves the phosphorylation of several protein subunits of PSII core complex, monomerization and migration from the granal region to a lamellar area, partial proteolysis of damaged proteins and their removal from the complex, and the replacement of the damaged proteins with their newly synthesized copies. Finally, restoration is accomplished with the formation of dimers and their photoactivation (Aro 2006). At high light intensities, when the rate of damage is fast, facilitated rate of repair is required to maintain PSII activity at a proper level. Chlorophyll *a* fluorescence analysis is noninvasive, highly sensitive, and simple. It is highly versatile tools appropriate

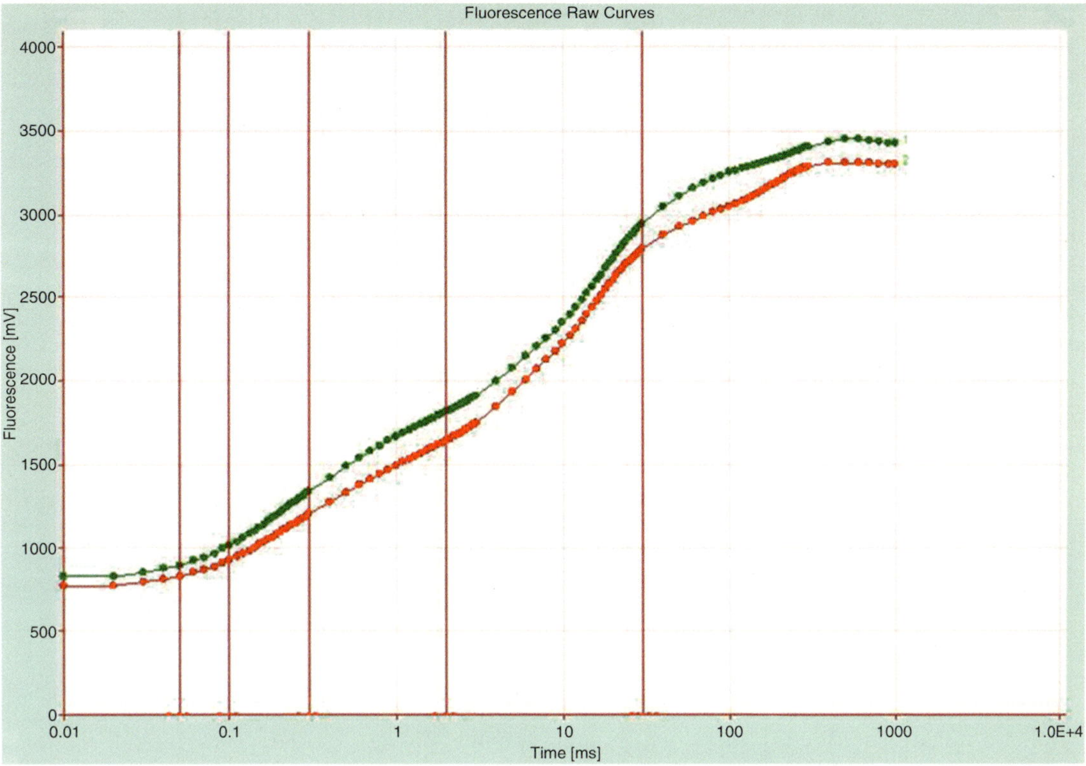


**Fig. 3.11** The OJIP curves presented as original measured data of chlorophyll fluorescence – $F_t$ (*left*), with normalized $F_0$ level (*middle*) plotted as $F_t/F_0$, and double normalized, i.e., shown in values of variable fluorescence $V_t$ (*right*). Raw averaged curves (green – control, red – salt-treated) measurements were realized on *B. gymnorrhiza* by fluorometer Handy PEA (Hansatech, GB)

for many applications, including plant physiology, biophysics, or biochemistry. The most useful chlorophyll fluorescence methods are based on variable chlorophyll *a* fluorescence measurements. However, there are several technical solutions enabling such a type of measurements, differing in the manner by which the photochemistry is saturated. The most frequently used is direct fluorescence measurement (in present mostly LED-based instruments) representing by fast chlorophyll *a* fluorescence measurement (Strasser and Govindjee 1991, 1992) or saturation pulse analysis method by pulse-amplitude modulation (PAM) fluorometry (Schreiber 2004). Fast chlorophyll fluorescence induction kinetics is now a widespread method used in photosynthesis research. This is because fluorescence induction is noninvasive and highly sensitive and fast and easily measured; it requires relatively inexpensive equipment and contains important information about photosynthetic apparatus (Lazar 1999). Illumination of dark-adapted photosynthetic samples leads to emission of the chlorophyll *a* fluorescence with a characteristic transient well known as the curve. Such a curve measured under continuous light has a fast (<1 s) exponential phase and a slow decay phase (duration of few min). The initial growing phase shows a typical polyphasic shape, which is well evident when the curve is plotted on the logarithmic time scale which is known as OJIP curve (Fig. 3.11).

The analysis of the OJIP curve taking the theoretical assumptions and probabilities derives different photosynthetic parameters for the dark-adapted state of the photosynthetic systems (Strasser et al. 2000, 2004; Stirbet and Govindjee 2011). The nomenclature for “OJIP” is O for origin or $F_O = F_0$ level measured at 50 μs (or less) after illumination, J and I represent intermediate states measured after 2 and 30 ms, and P is the peak or $F_P = F_m$ (maximal fluorescence). This is valid only if sufficient light intensity is used. In heat-stressed samples, another peak arises between $F_0$ and $F_J$ at app. 300 μs, which is usually called K-step (Guisse et al. 1995; Srivastava et al. 1997; Strasser et al. 2000); therefore, some authors call the fast chlorophyll fluorescence induction the OKJIP curve or transient. The OJIP curve from $F_0$ to $F_m$ is correlated with the primary photochemical reactions of PS II, and the fluorescence yield is controlled by a PS II acceptor quencher (the primary quinone acceptor, $Q_A$) (Van Gorkom 1986). Thus, the OJIP transient can be used for the estimation of the photochemical quantum yield of PS II photochemistry and the electron transport properties. The OJIP fluorescence curve analysis can be used to monitor the effect of various biotic and abiotic stresses and photosynthetic mutations affecting the structure and function of the photosynthetic apparatus (Strasser et al. 2004). The photosynthetic samples kept in darkness have the electron acceptor side of PS II in the oxidized state, as there is no electron flow in the photosynthetic electron transport chain and water oxidation by PS II. So the PS II reaction centers remain open, and the fluorescence intensity is minimum, i.e., equal to $F_0$ (="O" level in OJIP curve). Immediately after illumination with a strong intensity of light that can theoretically excite all the pigment molecules in the pigment protein complex of the thylakoid membrane, a fast electron transport process takes place and is recorded by a O–J transition or rise within 2 ms. This is followed by slow phases J–I and I–P (as evident on small upper plots in Fig. 3.5 plotted on regular time scale), which are known as thermal phases. The $F_m$ level ($F_P$) or $F_{t\max}$ is attained usually within interval 200 ms to 1 s, representing a closed PS II centers or complete reduction of all the primary electron acceptor in PS II, the $Q_A$ molecules, and saturating the electron flow on the acceptor side of PS II (Schansker et al. 2005).

The chlorophyll fluorescence emitted by higher plants upon illumination carries a lot of information about the structure and function of the photosynthetic apparatus (Strasser et al. 2010). In mangroves, chlorophyll fluorescence measurements data are lacking which can easily measure salt stress on photosystem II. We have stared small experiments on NaCl stress on two mangrove tree species *Bruguiera gymnorrhiza* and *B. parviflora* to check any salt stress signature on chlorophyll fluorescence for different salt sensitiveness of these two mangrove species. We found that there are so very high changes (Figs. 3.11, 3.12 and 3.13) of pattern of photosynthetic parameters (data not published).

## Halophytes as Potential Sources of Salt-Tolerant Genes

It is hypothesized that all plant species, including salt-tolerant halophytes and salt-sensitive glycophytes, possess similar genes for salt tolerance (Zhu 2000) and that mechanisms for stress response are active in both types of plants (Kant et al. 2006). Enhanced salt tolerance in halophytes, however, is presumed to be due to differences in the regulation and expression of the contributing genes (Cushman and Bohnert 2000; Jithesh et al. 2006a, b). This includes functional regulation of the activity of NHX vacuolar exchanger and the ability of halophytes to prevent backflow of $Na^+$ from the vacuole into the cytosol (e.g., via tonoplast channels' regulation). Therefore, it may be useful to generate tolerance-related ESTs from halophytic plants and used them to look for corresponding genes in glycophytes. There are many halophytic plant species that can tolerate high levels of salt stress (Table 3.4). In addition to its salt-tolerant behavior, *Thellungiella halophila* shares many advantages of *Arabidopsis*, including its small genome (actually considerably smaller than *A. thaliana*), short life cycle, abundant seed production, and easy transformation (Vinocur and Altman 2005). Due to the high sequence similarities at cDNA level between the two species, expression profiling of *T. halophila* would be possible

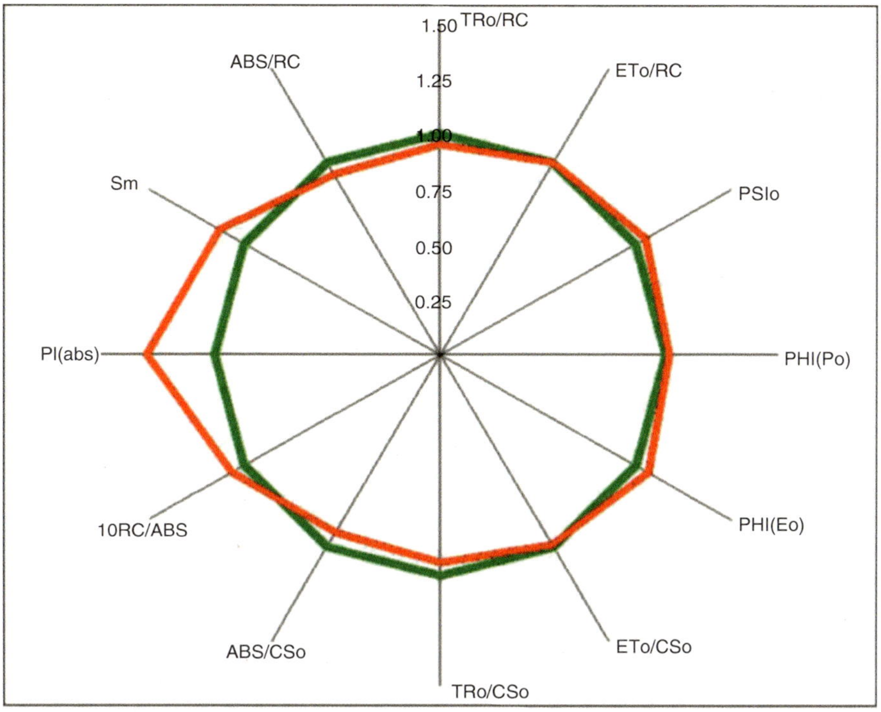


**Fig. 3.12** The "spider plot" of selected JIP-test parameters derived from fast chlorophyll fluorescence kinetics. The average values (shown relative to the corresponding value of the control, which thus is equal to 100%) measured in different salt treated (400 mM NaCl) and control *B. gymnorrhiza* (salt (*red*) versus control (*green*) as reference = 1 or 100% for each parameter)

using *Arabidopsis* cDNA microarrays. In one study, a comparison of microarray profiling under salt stress revealed that while only 6 genes were upregulated in *T. halophila*, 40 genes were upregulated in *A. thaliana* (Taji et al. 2004a, b). However, non-stressed *T. halophila* plants exhibited high-level expression of a large number of stress-inducible genes (Taji et al. 2004a, b), suggesting constitutive expression of stress-tolerant genes in this species. Higher expression of *AtP5CS* gene in *T. halophila* under normal conditions is accompanied by corresponding accumulation of proline at much higher levels. Interestingly, the genes being constitutively expressed in *T. halophila* have been determined to be expressed also in *Arabidopsis*, but only under salt stress, supporting the hypothesis that genes responsible for salt tolerance in halophytes do exist in glycophytes (Denby and Gehring 2005). In a different study, transcript profiling of *Thellungiella salsuginea*, another halophyte related to *A. thaliana*, revealed differential regulation of 154 genes under different stresses (Wong et al. 2006). Similar to the previous study, relatively only a few transcript changes were observed in *T. salsuginea* in response to salt stress. EST libraries from some other halophytes, including *Limonium sinense* (Chen et al. 2007), ice plant (*Mesembryanthemum crystallinum*), and green alga (*Dunaliella salina*) (Cushman and Bohnert 2000), which all are only distantly related to crop species, have been analyzed under high salinity. In ice plant, approximately additional 15% of functionally unknown genes were expressed under salt stress compared to the non-stress conditions. Transcript profiling of ice plant treated with 500 mm NaCl revealed 2–12-fold upregulation in some genes while up to fourfold downregulation in other genes (Koreeda et al. 2004). The gene expression pattern confirmed the switch from $C_3$ to CAM in the ice plant under salt stress, as CAM

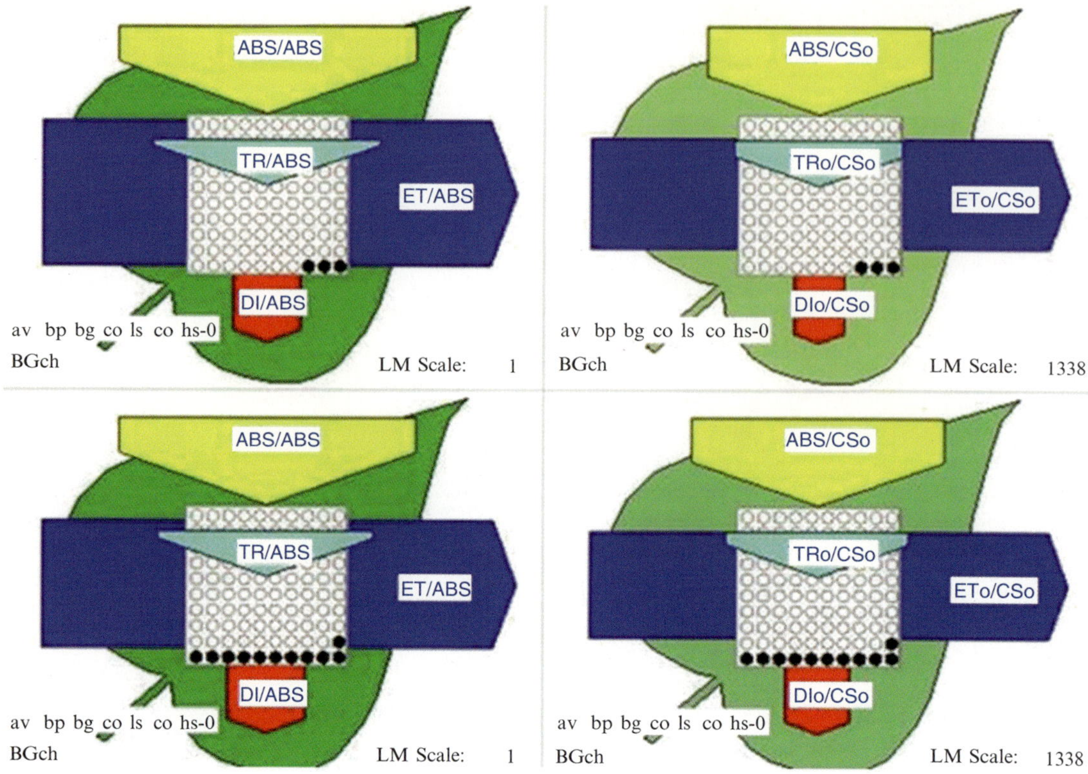


**Fig. 3.13** The use of phenomenological leaf model for comparison of differences among treated (400 mM NaCl) *B. gymnorrhiza* and control. Phenomenological leaf models are based on calculations of parameters per excited leaf cross section, based on equations derived by Strasser et al. (2000) for control, salt-stressed leaves. The thickness of each arrow represents the value of absorbance (ABS/CSm), trapping flux (TR/CSm), electron transport (ET/CSm), or heat dissipation of excess light (DI/CSm); all expressed per leaf cross section. The black points represent the fraction of inactive reaction centers. Measurements were done by fluorometer Handy PEA (Hansatech, England), and the models were generated using software Biolyzer 3.06 (Maldonado-Rodriguez, Laboratory of Bioenergetics, University of Geneva, Switzerland)

metabolism reduces the transpirational water loss and provides resistance to salt and drought stresses. In an oligonucleotide microarray analysis, a significant decrease in transcript abundance for photosynthetic-related genes was observed in ice plant (Cushman et al. 2008). A few other halophytic species such as *Cakile maritima* (Brassicaceae family) (Megdiche et al. 2009), and *Suaeda maritima* (Sahu and Shaw 2009) are known which could be suitable for transcript profiling and potentially identifying new and useful salt-tolerant genes. Furthermore, some members of the Triticeae family also display halophytic behavior, which could be exploited for the introduction of salt tolerance in wheat and other cereals (Colmer et al. 2006).

## Mangroves as Potential Source for Salt-Tolerant Genes

Transcript profiling in several other plant species has been conducted under salt or other abiotic stresses by microarray analysis. A list of such work is presented in Table 3.5. In Burma mangrove (*Bruguiera gymnorrhiza*), a high level of gene expression was observed 6 h post salt stress treatment; however, it was recovered 24 h after the treatment (Miyama and Tada 2008). Comparative analyses showed that the gene expression profile of plants under salt stress was distinctly different from the expression profile of plants under osmotic stress. The salt-responsive genes from Burma mangrove may be useful for

**Table 3.4** Salt-related genes reported in mangroves

| Mangrove species | Genes | Description | Function | Reference |
|---|---|---|---|---|
| *Aegiceras corniculatum* | P5CS | Delta 1-pyrroline-5-carboxylate synthase | A key enzyme of proline synthesis pathway; accumulation of transcript of this gene under salinity tended to accompany recruitment of proline in *A. corniculatum* | Fu et al. (2005) |
| | PIP1 | PIP1 aquaporin | This gene was upregulated by salt stress | Fu et al. (2005) |
| | PIP2 | PIP2 aquaporin | This gene was upregulated by salt stress | Fu et al. (2005) |
| | NHA | Na+/H + antiporter | This gene was upregulated by salt stress | Fu et al. (2005) |
| | CP1 | Cysteine proteinase inhibitor | Of this gene in transgenic Arabidopsis enhanced tolerance capacity of high saline medium | Fu (2006) |
| | BADH | Betaine-2-aldehyde dehydrogenase | High-salinity-induced increase of transcript level and such an increase was accompanied by accumulation of betaine. Although activity of this enzyme decreases with an increase in salinity, the extent of decrease is less than in homologues in *E. coli and spinach* | Hibino et al. (2001) |
| *Avicennia marina* | SodI | CU/Zn superoxide dismutase | High salinity did not lead to transcriptional change, but osmotic stress decreased transcript level of this gene. Under oxidative stress, its transcription was transiently upregulated | Prashanth et al. (2008) |
| | CatI | Catalase | It was upregulated by saline or oxidative stress but downregulated by osmotic stress | Jithesh et al. (2006a, b) |
| | FerI | Ferritin 1 | It was transcriptionally upregulated by saline or oxidative stress but did not change under osmotic stress | Jithesh et al. (2006a, b) |
| | AmT1; AmT2 | Betaine/proline transporter | Transgenic E. coil with such gene could accumulate betaine under salt stress in *A. marina, salt stress-induced transcription of such gene in root and leaf* | Waditee et al. (2002) |
| | AmT3 [partial] | Do | Do | (Do) |
| *Bruguiera gymnorrhiza* | OEE1 | OEE1 is one component of PSII | High-salinity-induced accumulation of its transcript and protein | Sugihara et al. (2000) |
| | DLDH | Dihydrolipoamide dehydrogenase | Upregulated when treated with 500 mmol/L NaCl for 1 day | Banzai et al. (2002c) |
| | LAS | Lipoic acid synthase | Being upregulated when treated with 500 mmol/L NaCl for 1 day | Banzai et al. (2002c) |
| | Unnamed gene | Fructose-6 phosphate, 2-kinase/fructose-2, 6-biphosphatase | Transcription of this gene increased after 6 h of salt stress. It was supposed to act in osmotic regulation process by controlling the content of Fru-2,6-P2 | Banzai et al. (2002c) |
| | Cytosolic Cu/Zn-SOD | Cytosolic Cu/Zn superoxide dismutase | High-salinity-, mannitol-, and ABA-induced accumulation of its transcripts in leaves; transcript was induced by high salinity in young and mature leaves rather than in old leaves | Takemura et al. (2002) |

(continued)

**Table 3.4** (continued)

| Mangrove species | Genes | Description | Function | Reference |
|---|---|---|---|---|
| | Cytosolic CAT [partial] | Catalase | No significant change occurred in the expression of this gene during the treatment with NaCl, mannitol, and ABA, but CEPA (2-chloroethylphosphonic acid) can increase its transcript level | Takemura et al. (2002) |
| *Bruguiera sexangula* | CCTα | α-subunit of CCT complex | Transgenic *E. coli* with one domain of this subunit displayed enhanced tolerance to high salinity | Yamada et al. (2002a) |
| | Mangrin | Partially homologous to gene-encoding Allene Oxidase Cyclase (AOC) | It was upregulated by high salinity and its overexpression enhanced salt tolerance of transgenic yeast and tobacco cells | Yamada et al. (2002a) |
| *Kandelia candel* | SIGKC1 & 2 | Cytosolic low molecular mass heat-shock proteins (sHSPs) | It acts as molecular chaperones to prevent thermal aggregation of protein by binding nonnative intermediates | Huang et al. (2003) |
| | SIGKC3 | ADP-ribosylation factor (AFR) | A ubiquitous, high conserved 21-kDa GTP-binding protein. The ARF proteins are through to function as regulators of membrane traffic | Huang et al. (2003) |
| | SIGKC4 & 5 | Unknown | – | Huang et al. (2003) |

**Table 3.5** Summary of transcript profiling of several salt-tolerant plants under salt stress

| Plant | Effect on gene regulation | References |
|---|---|---|
| Ice plant (*Mesembryanthemum crystallium*) | Upregulation of genes relevant to the proteins related to cell rescue, defense, cell death, aging, signal transduction, cellular organization, transport facilitation, protein destination | Bohnert et al. (2001) |
| | Upregulation of transcripts encoding CAM-related enzymes, pathogenesis-related, senescence-associated, cell death-related, and stress-related proteins such as heat shock proteins (HSPs), chaperones, early light-inducible proteins, ion homeostasis, antioxidative stress, detoxification, and biosynthetic enzymes for osmoprotectants. Downregulation of light-harvesting and photosystem complexes, C3 photosynthetic enzymes, ribulose bisphosphate carboxylase/oxygenase (RuBisCO) subunits | Koreeda et al. (2004) |
| | Upregulation of the genes active in CAM-related C4 acid carboxylation/decarboxylation, lycolysis/gluconeogenesis, polysaccharide, polyol, and starch biosynthesis/degradation, protein degradation, transcriptional activation, signalling, stress response, and transport facilitation. Downregulation of the genes encoding photosynthetic functions, protein synthesis, and cellular biogenesis functions | Cushman et al. (2008) |
| Salt cress (*Thellungiella halophila*) | Upregulation of the genes related to myoinositol-1-phosphate synthase, putative myoinositol 1-phosphate synthase, galactinol synthase, putative calcium-binding EF-hand protein, late embryogenesis abundant protein LEA-like, protein kinase family, etc. Upregulation of the genes encoding Fe-SOD, 9-cis-epoxycarotenoid dioxygenase, chitinase, plant defensin1.2, Δ1-pyrroline-5-carboxylate synthetase, SOS1, *β*-glucosidase, etc. | Taji et al. (2004) |
| Saltwater cress (*Thelungiella salsuginea*) | Induction of genes related to *β*-glucosidase (putative), lipid-associated family protein, lipid transfer protein, dehydrin RAB18-related protein, etc. Repression of genes related to chlorophyll A-B binding protein, oxygen-evolving enhancer protein, cysteine protease, RubisCO activase, carbonic anhydrase 1, zinc finger, calmodulin-6, etc. | Wong et al. (2006) |

introducing into crop plants via transgenic approaches to improve their salt tolerance. For example, it has been shown that expression of the genes for lipid transfer, zinc finger, and Ankyrin-repeat proteins from Burma mangrove in *Arabidopsis* and *Agrobacterium tumefaciens* resulted in improved salt tolerance in both of these organisms (Yamanaka et al. 2009). cDNA microarray analysis in the highly salt-tolerant species *Puccinellia tenuiflora* grown under saline-alkaline conditions led to the expression and detection of many genes, approximately 40% of which were novel or functionally unknown (Wang et al. 2007b). Cloning and expression of cycloartenol syntheses from *Rhizophora stylosa* and *Kandelia candel* have been studied by Basyuni et al. (2007, 2009, 2011). Salinity-tolerant genes from the mangrove plant *Bruguiera cylindrica* have been isolated using PCR-based technique suppression subtractive hybridization (SSH). Several genes in response to NaCl stress have been isolated by Takemura et al. (2002) and Zeng et al. (2006). Genes isolated from mangroves have been utilized for genetic transformation of crop plants (Hiei et al. 1994).

### Genes from *Avicennia marina*

The isolation of genes with the capability of improving stress tolerance (antistress genes) is an important factor for the breeding of stress-tolerant plants, and mangrove plants are attractive resources for anti-salt stress genes, since they grow under a condition of high salinity. In order to isolate antistress genes from mangrove plants, a cDNA library of *Avicennia marina* (Mehta et al. 2005) was constructed and screened for antistress genes by a functional expression screening with *Escherichia coli* cells. Several stress-related gene homologues, such as chaperonin-60, clpP protease of the clp/Hsp100 family of chaperones, ubiquitin, eEF1A, drought-induced AtDi19 gene of *Arabidopsis thaliana*, and secretory peroxidase, were successfully isolated.

### Genes from *Bruguiera gymnorrhiza*

About 7,029 gene expression patterns in Burmese mangrove under high-salinity stress were reported (Miyama and Hanagata 2007; Ezawa and Tada 2009). To identify genes of potential importance to salt tolerance in Burmese mangrove (*Bruguiera gymnorrhiza*), they analyzed the gene expression profiles in salt-stressed mangrove using cDNA microarray containing 7,029 clones based on 14,842 expressed sequence tags (ESTs) (Miyama et al. 2006). Combined results from all hybridization, they identified 287 genes with a greater than fivefold change of at least 1 time point after salt treatment (500 mM) compared with control water-treated plant; 228 genes were upregulated more than fivefold, and 61 genes were suppressed less than one-fifth. The major response observed by microarray analysis was the differential response to salt shock between leaf and roots. cDNA microarray analysis confirmed the stress-responsive expression of a number of previously reported stress-inducible genes, such as CDPK (Berberich and Kusano 1997), peroxidase, vacuolar ATPase, and several PR protein family genes (Hoffmann-Sommergruber 2002). These results indicate that there are similar molecular mechanisms of stress tolerance and responses between Burmese mangrove and model plants. Significant upregulation of unknown genes or mangrove-specific genes (Bg70) was also observed. Furthermore, several genes including BURP domain-containing protein (RD-22 homologue) showed completely opposite results (Banzai et al. 2002b). Although these results identify hundreds of potentially important transcriptome changes, the biochemical functions of many stress-regulated genes remain unknown. Computational analysis of these clones allows annotation of putative gene functions through similarity searches in nucleic acid and protein databases; over one-third (34.5%) of the possible coding sequences have no matches in current public databases and, hence, remain novel sequences with unknown functions. Determination of the biological functions of these genes is among the greatest challenge for post-genomic research. Further analysis using transgenic plants of these stress-inducible genes will identify salt-tolerant determinants from these results set and provide more information about the functions of the stress-inducible genes involved in stress tolerance. cDNA expression libraries were constructed from salt-treated roots and leaves using the host organism *Agrobacterium tumefaciens* to identify key genes in the regulation of salt tolerance in the mangrove plant *Bruguiera gymnorrhiza* by Ezawa and Tada (2009). Functional screening of the *Agrobacterium* libraries identified 44 putative salt-tolerant genes in *B. gymnorrhiza.* A cDNA clone which is homologous to an unknown cDNA from the mangrove plant *K. candel* and the *cyc02* gene from *Catharanthus roseus* conferred the highest level of salt tolerance at 450 mM NaCl to *A. tumefaciens*, indicating that it plays a major role in the regulation of salt tolerance in mangrove plants. Transgenic *Arabidopsis* plants expressing *Bg70* and *cyc02* homologue exhibited increased tolerance to NaCl. Although the generation and analysis of transgenic plants expressing each identified gene need to properly evaluate the gene function, at least two genes, *Bg70* and cyc02 homologue, successfully conferred increased salt tolerance to *Arabidopsis* plants.

### Genes from *Bruguiera sexangula* and *B. cylindrica*

Takemura et al. (2002) have been performed many studies on *Bruguiera* species to explore the potential of genes belonging to trees in this family in salinity tolerance. A $Na^+/H^+$ antiporter, catalyzing the exchange of $Na^+$ ion for $H^+$ ion across the vacuolar membrane and may be responsible for the ultrafiltration in roots, was reported from *Bruguiera sexangula* (Tanaka et al. 2000). Through the isolation and characterization of genes involved in salinity stress from *Bruguiera* species, detoxification of reactive oxygen species (ROS) by superoxide dismutase (Takemura et al. 2000), osmotic adjustments via sucrose biosynthesis (Banzai et al. 2002a), and increase in total amino acid pool especially proline (Parida et al. 2002) and polyphenol were suggested as some of the mechanisms that also contribute to salinity tolerance of these mangrove trees. Among the genes that were isolated were also genes involved in other salinity-tolerant adjustments, such as the BURP domain-containing proteins (Banzai et al. 2002a), allene-oxide cyclase (AOC) or "mangrin" (Yamada et al. 2002a, b), and cytosolic chaperonin-containing TCP-1a (CCTa) homologue (Yamada et al. 2002c). Isolation of salinity-tolerant genes from the mangrove plant, *Bruguiera cylindrica*, by using suppression subtractive hybridization (SSH) and bacterial functional screening was reported by Wong et al. (2005), and 126 salinity-tolerant cDNAs from the root were obtained by using suppression subtractive hybridization (SSH) and bacterial functional screening. Sequencing of 51 subtracted cDNA clones that were differentially expressed in the root of *B. cylindrica* exposed to 20 parts per thousand (ppt) NaCl water revealed 10 tentative unique genes (TUGs) with putative functions in protein synthesis, storage and destination, metabolism, intracellular trafficking, and other functions; and 9 unknown proteins. Meanwhile, the 75 cDNA sequences of *B. cylindrica* that conferred salinity tolerance to *Escherichia coli* consisted of 29 TUGs with putative functions in transportation, metabolism, and other functions; and 33 with unknown functions. Both approaches yielded 42 unique sequences that have not been reported elsewhere to be stress related and might provide further understanding of adaptations of this plant to salinity stress.

## Transcriptome and Transcriptional Analysis in Salt-Stressed Mangroves

Transcript profiling of plants with high levels of salt tolerance may be useful for revealing the mechanism(s) of salt tolerance involved. However, in microarray analyses of plants grown under salt stress, often a large number of up- or downregulated genes are found which could not be attributed to any known function. Functional analysis of such genes may lead to unraveling novel pathways for salt tolerance, which may facilitate development of crop cultivars with high salt tolerance. It is expected that expression profiling of more and more plants, especially those of economic significance, will be seen in near future. Comprehensive and across species analysis of expression profiles under salt stress conditions will provide additional candidate genes which may be useful for conferring salt tolerance in economically important crop species.

Transcriptional response of *Bruguiera gymnorrhiza* to high-salinity (salt stress; 500 mM NaCl) and hyperosmotic stress (osmotic stress; 1 M sorbitol) by microarray analysis was done by Miyama and Tada (2008). It revealed that 865 of 11,997 genes showed significant differential expression under salt and osmotic stress. Hierarchical clustering of the 865 genes showed that expression profiles under salt stress were distinctly different from those under osmotic stress. Comparison of gene ontology (GO) categories of differentially expressed genes under the stress conditions revealed that the adaptation of Burmese mangrove to salt stress was accompanied by the upregulation of genes categorized for "cell communication," "signal transduction," "lipid metabolic process," "photosynthesis," "multicellular organismal development," and "transport" and by downregulation of genes categorized for "catabolic process." Burmese mangrove maintained its leaf water potential and recovered from its photosynthesis rate that declined temporarily under salt stress, but not under osmotic stress. It was suggested that salt tolerance of *B. gymnorrhiza* might be attributed to their ability to accumulate high concentrations of $Na^+$ and $Cl^-$, even under non-stressed conditions; to uptake additional $Na^+$ and $Cl^-$ for use as osmolytes; and to maintain $K^+$ homeostasis under salt stress. The transcriptome of a highly salt-tolerant mangrove species, *Sonneratia alba,* was sequenced by Chen et al. (2011). Over 15 million 75-bp paired-end reads were assembled into 30,628 unique sequences with an average length of 581 bp. Of them, 2,358 SSRs were detected, with dinucleotide repeats (59.2%) and trinucleotide repeats (37.7%) being the most common. Analysis of codon usage bias based on 20,945 coding sequences indicated that genes of *S. alba* were less biased than those of some microorganisms and *Drosophila* and that codon usage variation in *S. alba*. Genome-wide gene ontology (GO) assignments showed that *S. alba* shared a similar GO slim classification with *Arabidopsis thaliana*. High percentages of sequences assigned to GO slim category "mitochondrion" and four KEGG pathways, such as carbohydrates and secondary metabolites, may contribute to salt adaptation of *S. alba*. In addition, 1,266 unique sequences matched to 273 known salt-responsive genes (gene families) in other species were screened as candidates for salt tolerance of *S. alba*, and some of these genes showed fairly high coverage depth. They identified four genes (AtRUB1 (related to ubiquitin), cyclin D3-2, LAG1 homolog 2, metallothionein-like protein 2) with signals of strong diversifying selection by comparing the transcriptome sequences of *S. alba* with 249 known ESTs from its congener *S. caseolaris*. Abundant SSR markers, salt-responsive genes, and four genes with signature of natural selection obtained from *S. alba* provide abundant sequence sources for future genetic diversity, salt adaptation, and speciation studies.

To identify key genes in the regulation of salt tolerance in the mangrove plant *Bruguiera gymnorrhiza,* transcriptome profiling in the lateral and main roots under conditions of salt stress was performed by Yamanaka et al. (2009). It revealed that 175 and 403 of 11,997 genes showed significantly increased high expression in the lateral and main roots, respectively. One hundred and sixty genes were upregulated in both types of roots in the early time period, 1–12 h after salt treatment. Expression vectors for 28 selected salt-responsive genes were constructed and transformed in *Agrobacterium tumefaciens* and then screened for salt tolerance. *A. tumefaciens* transformed with genes for lipid transfer, zinc finger, and ankyrin-repeat proteins showed enhanced salt tolerance. Transgenic *Arabidopsis* plants expressing these three genes also exhibited increased tolerance to NaCl. Analysis of transgenic plants expressing the genes identified in these types of study, and functional screening using plants as hosts, which should identify additional genes involved in salt tolerance in the mangrove plant. Yang et al. (2011) explored the expression profiles of *Hibiscus tiliaceus* under salt stress using a full-length cDNA microarray (Yang 2007). Four hundred and eighty-six salt-responsive uni-genes were identified in *H. tiliaceus,* 224 of which had high sequence similarity to *Arabidopsis*. Many genes identified are known to be salt stress responsive. The physiological performance of *H. tiliaceus* under salt stress suggests decrease in ratio of $K^+/Na^+$ and negative influence on photosynthesis of *H. tiliaceus*. It was evident that *H. tiliaceus* evolved its own mechanisms to regain both ionic and osmotic homeostasis through coordinated engagement of genes associated with gene transcription, signaling, and downstream cell transport and detoxification pathways. It revealed that to survive under high-salinity intertidal environments, *H. tiliaceus* evolved its own mechanisms to regain both ionic and osmotic homeostasis through coordinated engagement of genes associated with gene transcription, signaling, and downstream cell transport and detoxification pathways. The transcriptome of a highly salt-tolerant mangrove species, *Sonneratia alba*, was sequenced using the Illumina Genome Analyzer (Chen et al. 2011). Over 15 million 75-bp paired-end reads were assembled into 30,628 unique sequences with an average length of 581 bp. Of them, 2,358 SSRs were detected, with dinucleotide repeats (59.2%) and trinucleotide repeats (37.7%) being the most common. Genome-wide gene ontology (GO) assignments showed that *S. alba* shared a similar GO slim classification with *Arabidopsis thaliana*. High percentages of sequences assigned to GO slim category "mitochondrion" and four KEGG pathways, such as carbohydrates and secondary metabolites, may contribute to salt adaptation of *S. alba*. In addition, 1,266 unique sequences matched to 273 known salt-responsive genes (gene families) in other species were screened as candidates for salt tolerance of *S. alba*, and some of these genes showed fairly high coverage depth. This group identified four genes with signals of strong diversifying selection (Ka/KsN1) by comparing the transcriptome sequences of *S. alba* with 249 known ESTs from its congener *S. caseolaris*. Recently, the development of novel high-throughput DNA-sequencing methods has provided an opportunity to address this question by de novo assembly or mapping and quantification of transcriptomes (Wang et al. 2009b). As an early step in sequencing the transcriptomes of mangrove species, these high-throughput sequencing technologies were used to sequence the transcriptome of two mangrove species, Rhizophora mangle and *Heritiera littoralis* (Dassanayake et al. 2009). In this study, they sequenced the transcriptome of *S. alba* using the Illumina platform. After de novo assembly of the transcriptome, simple sequence repeats (SSRs) were identified, and codon usage bias was analyzed. Based on known salt-responsive genes, we searched their homologous sequences in *S. alba* and calculated their coverage depth. By comparing the transcriptome sequences of *S. alba* with the ESTs from *S. caseolaris*, genes under positive selection were identified, and we wish to gain insights into the role of natural selection in adaptation of mangroves to their environments.

## Conclusion and Future Perspectives

Salinity is one of the major environmental stresses that affect the growth and productivity of plant by affecting photosynthesis and other metabolic process. Halophytes are plants of salty environment capable of thriving and growing under high concentration of NaCl (Hellebust 1976; Flowers et al. 1986). So, the understanding of salt-tolerant mechanism would be useful in transferring salt-tolerant genes to crop plant. Among the halophytes, mangroves can be used as best model for salt stress research because they are able to adapt not only to stress condition ranging from high salinity at one extreme but also to a complete lack of salinity due to freshwater conditions. Mangroves possess specialized adoptable feature to cope with this adverse condition like high salinity. An early indicator of the adverse condition is production of reactive oxygen species (ROS) within different cellular compartment of the plant cell. The continuing rise in human population necessitates new approaches to increasing total crop production worldwide. Bringing marginal lands under cultivation, including saline and alkaline soils, is a major component of a multidisciplinary approach to increasing total food production. However, development of crop plants with enhanced salt tolerance is a complex task requiring involvement of many disciplines, including molecular biology and biotechnology. Genes contributing to different aspects of salt tolerance need to be identified, functionally characterized, and transferred to commercially acceptable genetic backgrounds to develop new cultivars with enhanced salt tolerance. During the past few decades, some progress has been made in the identification and characterization of genes, enzymes, or compounds with significant roles in plant salt tolerance, and some transgenic "salt-tolerant" plants have been developed. However, most of these plants have had limited success for commercial utilization under field saline conditions as they have not been equipped with all the necessary salt-tolerant genes or mechanisms required for successful growth under field saline conditions. A thorough understanding of all the essential physiological mechanisms and pathways leading to salt tolerance is needed, which in turn would provide great opportunities to identify new targets, including identification and characterization of novel genes with key roles in plant salt tolerance.

While there are many approaches to identifying genes with important roles in plant salt tolerance, investigating gene expression under various salt stress conditions and comparative analysis across salt-tolerant and salt-sensitive genotypes is potentially a useful approach to discovering new genes of importance. Microarray technology has emerged as a powerful technique to study gene expression patterns in various plant species growing under different conditions. Study of expression patterns of differentially regulated genes in plants grown under stress conditions could lead to the identification of useful genes for enhancing stress tolerance in crop species. Large-scale gene expression profiling has been conducted in various plant species grown under salt stress, which has revealed the involvement of very many genes. However, the function of most of such genes has not been determined. Functional characterization of the genes, using a combination of genetic, genomic, biochemical, and physiological approaches, may lead to the identification of genes with key roles in plant salt tolerance, which in turn may lead to the development of crop plants with improved salt tolerance. Moreover, under field conditions, often more than one stress factor are present at each time. For example, salinity stress is often accompanied by osmotic stress or sometimes temperature stress. Therefore, it would be useful to perform genomic and transcriptomic analyses of plants under simultaneous stresses, as such conditions may mimic field stress conditions. Microarray expression profiling of functionally known and unknown transcripts under such stress conditions would be helpful in the identification of genes with important contributions to general stress tolerance. Furthermore, comparative expression profiling of mutants, contrasting genotypes, and transgenic plants would expedite the identification of genes with important roles and thus the development of crop plants with stress tolerance. With the recent advances in the microarray technology, gene expression profiling of salt-stressed plants is becoming a convenient and promising approach to investigating the complex nature of salt tolerance and discerning the underpinning regulatory genes and mechanisms. This has led to the generation and deposition of a plethora of related data into public databases. Many such databases are being organized so that the information could be easily extracted and analyzed for testing hypotheses and designing future research. The real challenge, however, is to identify genes in various salt-tolerant plant species (halophytes or glycophytes) with essential roles in salt tolerance and try to either identify crop plants which may possess the same genes in silent forms or transform crop plants with such genes for breeding purposes. Nonetheless, it is expected that the volume of information that is being generated and their use will lead to the development of crop plants with improved salt tolerance and acceptable agronomic performance under field conditions.

## References


Allakhverdiev SI, Murata N (2004) Environmental stress inhibits the synthesis de novo of proteins involved in the photodamage – repair cycle of photosystem II in *Synechocystis* sp. PCC. 6803. Biochim Biophys Acta 1657:23–32

Amtmann A, Fischer M, Marsh EL, Stefanovic A, Sanders D, Schachtman DP (2001) The wheat cDNA LCT1 generates hypersensitivity to sodium in a salt-sensitive yeast strain. Plant Physiol 126:1061–1071

Apse MP, Blumwald E (2002) Engineering salt tolerance in plants. Curr Opin Biotechnol 13:146–150

Aro EM (2006) State transitions revisited: a buffering system for dynamic low light acclimation of *Arabidopsis*. Plant Mol Biol 62:779–793

Ashraf M (2009) Biotechnological approach of improving plant salt tolerance using antioxidants as markers. Biotechnol Adv 27:84–93

Ashraf M, Athar HR, Harris PJC, Kwon TR (2008) Some prospective strategies for improving crop salt tolerance. Adv Agron 97:45–110

Balazadeh S, Siddiqui H, Allu AD, Matallana-Ramirez LP, Caldana C, Mehrnia M, Zanor MI, Kohler B, Mueller-Roeber B (2010) A gene regulatory network controlled by the NAC transcription factor ANAC092/AtNAC2/ORE1 during salt-promoted senescence. Plant J 62:250–264

Baldwin D, Crane V, Rice D (1999) A comparison of gel-based, nylon filter and microarray techniques to detect differential RNA expression in plants. Curr Opin Plant Biol 2:96–103

Banzai T, Hershkovits G, Katcoff DJ, Hanagata N, Dubinsky Z, Karube I (2002a) Identification and characterization of mRNA transcripts differentially expressed in response to high salinity by means of differential display in the mangrove, *Bruguiera gymnorrhiza*. Plant Sci 162:499–505

Banzai T, Snmiya K, Hanagata N, Dubtnsky Z, Karube I (2002b) Molecular cloning and characterization of genes encoding BURP domain-containing protein in the mangrove, *Bruguiera gymnorrhiza*. Trees 16:87–93

Banzai T, Hershkovits G, Katcoff DJ, Hanagata N, Dubinsky Z, Karube I (2002c) Identification and characterization of mRNA transcripts differentially expressed in response to high salinity by means of differential display in the mangrove, *Bruguiera gymnorrhiza*. Plant Sci 162:499–505

Basyuni M, Oku H, Tsujimoto E, Kinjo K, Baba S, Takara T (2007) Triterpene synthases from the Okinawan mangrove tribe, Rhizophoraceae. FEBS J 274:5028–5042

Basyuni M, Baba S, Inafuku M, Iwasaki H, Kinjo K, Oku H (2009) Expression of terpenoid synthase mRNA terpenoid content in salt stressed mangrove. J Plant Physiol 166:1786–1800

Basyuni M, Kinjo Y, Baba S, Shinzato N, Iwasaki H, Siregar Edy BM, Oku H (2011) Isolation of salt stress tolerance genes from roots of mangrove plant, *Rhizophora stylosa* Griff., using PCR-based suppression subtractive hybridization. Plant Mol Biol Rep 29:533–543

Behera B, Das AB, Mohanty P (2009) Changes of soluble proteins in leaf and thylakoid exposed in high saline condition of mangrove taxa *Bruguiera gymnorrhiza*. Physiol Mol Biol Plants 15(1):55–59

Ben Amor N, Ben Hamed K, Dedez A, Grignon C, Abdelly C (2005) Physiological and antioxidative responses of the perennial halophyte *Crithmum maritimum* to salinity. Plant Sci 68:889–899

Berberich T, Kusano T (1997) Cycloheximide induces a subset of low temperature-inducible genes in maize. Mol Gen Genet 254:275–283

Blumwald E, Aharon GS, Apse MP (2000) Sodium transport in plant cells. Biochim Biophys Acta 1465: 140–151

Bohnert HJ, Ayoubi P, Borchert C, Bressan RA, Burnap RL, Cushman JC, Cushman MA, Deyholos M, Fischer R, Galbraith DW et al (2001) A genomics approach towards salt stress tolerance. Plant Physiol Biochem 39:295–311

Chen SH, Guo SL, Wang ZL, Zhao JQ, Zhao YX, Zhang H (2007) Expressed sequence tags from the halophyte Limonium sinense. DNA Seq 18:61–67

Chen S, Zhou R, Huang Y, Zhang M, Yang G, Zhong C, Shi S (2011) Transcriptome sequencing of a highly salt tolerant mangrove species *Sonneratia alba* using illumine platform. Mar Gen 4:129–136

Cherian S, Reddy MP (2003) Evaluation of NaCl tolerance in callus cultures of *Suaeda nudiflora*. Moq Biol Plant 46:193–198

Cherian S, Reddy MP, Pandya JB (1999) Studies on salt tolerance in *Avicennia marina* (Forstk.) Vierh: effect of NaCl salinity on growth, ion accumulation and enzyme activity. Indian J Plant Physiol 4:266–270

Colmer TD, Flowers TJ, Munns R (2006) Use of wild relatives to improve salt tolerance in wheat. J Exp Bot 57:1059–1078

Coskun D, Britto DT, Kronzucker HJ (2010) Regulation and mechanism of potassium release from barley roots: an in planta42K + analysis. New Phytol 188:1028–1038

Cuin TA, Shabala S (2007) Amino acids regulate salinity-induced potassium efflux in barley root epidermis. Planta 225:753–761

Cushman JC, Bohnert HJ (2000) Genomic approaches to plant stress tolerance. Curr Opin Plant Biol 3:117–124

Cushman JC, Tillett RL, Wood JA, Branco JM, Schlauch KA (2008) Large-scale mRNA expression profiling in the common ice plant, *Mesembryanthemum crystallinum*, performing C3 photosynthesis and Crassulacean acid metabolism (CAM). J Exp Bot 59:1875–1894

Das P, Basak UC, Das AB (1997) Restoration of mangrove vegetation in Mahanadi delta of Orissa, India. Mangroves & Salt Marshes 1:155–161

Das AB, Mishra S, Mohanty P (2006) Physiology of mangroves: strategies on stress adaptation with special emphasis on tolerance to high salinity. In:

Hemantarajan A (ed) Advances in plant physiology, vol 9. Scientific Publishers, Jodhpur, pp 101–154

Dassanayake M, Haas JS, Bohnert HJ, Cheeseman JM (2009) Shedding light on an extremophile lifestyle through transcriptomics. New Phytol 183:764–775

Davletova S, Schlauch K, Coutu J, Mittler R (2005) The zinc-finger protein Zat12 plays a central role in reactive oxygen and abiotic stress signaling in Arabidopsis. Plant Physiol 139:847–856

Demidchik V, Davenport RJ, Tester M (2002) Nonselective cation channels in plants. Annu Rev Plant Biol 53:67–107

Denby K, Gehring C (2005) Engineering drought and salinity tolerance in plants: lessons from genome-wide expression profiling in *Arabidopsis*. Trends Biotechnol 23:547–552

de Lorenzo L, Merchan F, Blanchet S, Megias M, Frugier F, Crespi M, Sousa C (2007) Differential expression of the TFIIIA regulatory pathway in response to salt stress between *Medicago truncatula* genotypes. Plant Physiol 145:1521–1532

Diedhiou CJ, Popova OV, Dietz KJ, Golldack D (2008) The SNF1- type serine-threonine protein kinase SAPK4 regulates stress-responsive gene expression in rice. BMC Plant Biol 8:49

Ding Z, Li S, An X, Liu X, Qin H, Wang D (2009) Transgenic expression of MYB15 confers enhanced sensitivity to abscisic acid and improved drought tolerance in *Arabidopsis thaliana*. J Genet Genomics 36:17–29

El Ouakfaoui S, Miki B (2005) The stability of the *Arabidopsis* transcriptome in transgenic plants expressing the marker genes nptII and uidA. Plant J 41:791–800

Elstner EF (1991) Mechanisms of oxygen activation in different compartments of plant cell. In: Pell EJ, Stefen KL (eds) Active oxygen/oxidative stress and plant metabolism. American Society of Plant Physiol, Rockville, pp 13–25

Ezawa S, Tada Y (2009) Identification of salt tolerance genes from the mangrove plant *Bruguiera gymnorrhiza* using *Agrobacterium* functional screening. Plant Sci 176:272–278

Fang ZQ, Yuan LY, Hong PC, Ming LC, Shan WB (2005) NaCl enhances thylakoid-bound SOD activity in the leaves of $C_3$ halophyte *Suaeda salsa* L. Plant Sci 168:423–430

Fang Y, You J, Xie K, Xie W, Xiong L (2008) Systematic sequence analysis and identification of tissue-specific or stress-responsive genes of NAC transcription factor family in rice. Mol Genet Genomics 280:547–563

Flowers TJ (2004) Improving crop salt tolerance. J Exp Bot 55:307–319

Flowers TJ, Hajibagheri MA, Clipson NCW (1986) *Halophytes*. Quart Rev Biol 61:313–337

Foyer HC, Noctor G (2000) Oxygen processing in photosynthesis: regulation and signaling. New Phytol 146:359–388

Fu XH (2006) Cloning and functional expression of salt tolerance related genes in *Aegiceras corniculatum* and their adaptive evolution analysis (in Chinese). PhD thesis. Sun Yat-sen University, Guangzhou

Fu X, Huang Y, Deng S, Zhou R, Yang G, Ni X, Li W, Shi S (2005) Construction of a SSH library of *Aegiceras corniculatum* under salt stress and expression analysis of four transcripts. Plant Sci 169:147–154

Glenn EP, Brown JJ, Blumwald E (1999) Salt tolerance and crop potential of halophytes. Crit Rev Plant Sci 18:227–255

Guisse B, Srivastava A, Strasser RJ (1995) The polyphasic rise of the chlorophyll *a* fluorescence (OKJIP) in heat stressed leaves. Arch Sci Geneve 48:147–160

Hakala M, Tuominen I, Keränen M, Tyystjärvi T, Tyystjärvi E (2005) Evidence for the role of oxygen-evolving manganese complex in photoinhibition of photosystem II. Biochim Biophys Acta 1706:68–80

Hellebust JA (1976) Osmoregulation. Annu Rev Plant Physiol 27:485–505

Hibino T, Meng YL, Kawamitsu Y, Uehara N, Matsuda N, Taiiaka V (2001) Molecular cloning and functional characterization of two kinds of betaine-aldehyde dehydrogenase in betaine-accumulating mangrove *Avicennia marina* (Forsk.) Vierh. Plant Mol Biol 45:353–363

Hiei Y, Ohta S, Komari T, Kumashiro T (1994) Efficient transformation of rice (*Oryza sativa* L.) mediated by *Agrobacterium* and sequence analysis of the boundaries of the T-DNA. Plant J 6:271–282

Hoffmann-Sommergruber K (2002) Pathogenesis-related (PR)-proteins identified as allergens. Biochem Soc Trans 30:930–935

Holtorf H, Guitton MC, Reski R (2002) Plant functional genomics. Naturwissenschaften 89:235–249

Houde M, Belcaid M, Ouellet F, Danyluk J, Monroy AF, Dryanova A, Gulick P, Bergeron A, Laroche A, Links MG, MacCarthy L, Crosby WL, Sarhan F (2006) Wheat EST resources for functional genomics of abiotic stress. BMC Genomics 7:149

Hu H, You J, Fang Y, Zhu X, Qi Z, Xiong L (2008) Characterization of transcription factor gene SNAC2 conferring cold and salt tolerance in rice. Plant Mol Biol 67:169–181

Hu X, Zhang Z, Xu P, Fu Z, Hu S, Song W (2010) Multifunctional genes: the cross-talk among the regulation networks of abiotic stress responses. Biol Plant 54:213–223

Huang W, Fang XD, Li GY, Lin QF, Zhao WM (2003) Cloning and expression analysis of salt responsive gene from *Kandelia candel*. Biol Plant 47:501–507

Hurst AC, Grams TEE, Ratajczak R (2004) Effects of salinity, high irradiance, ozone and ethylene on mode of photosynthesis, oxidative stress and oxidative damage in the $C_3$/CAM intermediate plant *Mesembryanthemum crystallinum* L. Plant Cell Environ 27:187–197

Jain M, Khurana JP (2009) Transcript profiling reveals diverse roles of auxin-responsive genes during reproductive development and abiotic stress in rice. FEBS J 276:3148–3162

Jamil A, Anwar F, Ashraf M (2005) Plant tolerance to biotic and abiotic stresses through modern genetic engineering techniques. In: Dris R (ed) Crops: growth, quality & biotechnology. WFL Publisher, Helsinki, pp 1276–1299

Jamil A, Riaz S, Ashraf M, Foolad MR (2011) Gene Expression Profiling of Plants under Salt Stress. Crit Rev Plant Sci 30(5):435–458

Jin T, Chang Q, Li W, Yin D, Li Z, Wang D, Liu B, Liu L (2010) Stress-inducible expression of GmDREB1 conferred salt tolerance in transgenic alfalfa. Plant Cell Tissue Organ Cult 100:219–227

Jithesh MN, Prashanth SR, Sivaprakash KR, Parida AK (2006a) Monitoring expression profiles of antioxidant genes to salinity, iron, oxidative, light and hyper osmotic stresses in the highly salt tolerant grey mangrove, *Avicennia marina* (Forsk.) Vierh. by mRNA analysis. Plant Cell Rep 25:865–876

Jithesh MN, Prashanth SR, Sivaprakash KR, Parida AK (2006b) Antioxidative response mechanisms in halophytes: their role in stress defense. J Genet 85: 237–254

Jung C, Seo JS, Han SW, Koo YJ, Kim CH, Song SI, Nahm BH, Choi YD, Cheong JJ (2008) Overexpression of AtMYB44 enhances stomatal closure to confer abiotic stress tolerance in transgenic Arabidopsis. Plant Physiol 146:623–635

Kant S, Kant P, Raveh E, Barak S (2006) Evidence that differential gene expression between the halophyte, *Thellungiella halophila*, and Arabidopsis thaliana is responsible for higher levels of the compatible osmolyte praline and tight control of $Na^+$ uptake in *T. halophila*. Plant Cell Environ 29:1220–1234

Kavitha K, George S, Venkataraman G, Parida A (2010) A salt-inducible chloroplastic monodehydroascorbate reductase from halophyte *Avicennia marina* confers salt stress tolerance on transgenic plants. Biochimie 92:1321–1329

Kiegerl S, Cardinale F, Siligan C, Gross A, Baudouin E, Liwosz A, Eklof S, Till S, Bogre L, Hirt H et al (2000) SIMKK, a mitogen-activated protein kinase (MAPK) kinase, is a specific activator of the salt stress-induced MAPK, SIMK. Plant Cell 12:2247–2258

Kilian J, Whitehead D, Horak J, Wanke D, Weinl S, Batistic O, D'Angelo C, Bornberg-Bauer E, Kudla J, Harter K (2007) The At-GenExpress global stress expression data set: protocols, evaluation and model data analysis of UV-B light, drought and cold stress responses. Plant J 50:347–363

Kizis D, Lumbreras V, Pages M (2001) Role of AP2/EREBP transcription factors in gene regulation during abiotic stress. FEBS Lett 498:187–189

Klimov VV, Schafiev MA, Allakhverdiev SI (1990) Photoinactivation of the reactivation capacity of photosystem II in pea subchloroplast particles after a complete removal of manganese. Photosynth Res 23:59–65

Koreeda S, Cushman MA, Akselrod I, Bufford D, Fredrickson M, Clark E, Cushman JC (2004) Transcript profiling of salinity stress responses by large-scale expressed sequence tag analysis in *Mesembryanthemum crystallinum*. Gene 341:83–92

Law PJ, Claudel-Renard C, Joubert F, Louw AI, Berger DK (2008) MADIBA: a web server toolkit for biological interpretation of Plasmodium and plant gene clusters. BMC Genomics 9:105

Lazar D (1999) Chlorophyll a fluorescence induction. Biochim Biophys Acta 1412:1–28

Lee SC, Choi DS, Hwang IS, Hwang BK (2010) The pepper oxidoreductase CaOXR1 interacts with the transcription factor CaRAV1 and is required for salt and osmotic stress tolerance. Plant Mol Biol 73:409–424

Leitsch J, Schnettger B, Critchley C, Krause CH (1994) Two mechanisms of recovery from photoinhibition *in vivo*: reactivation of photosystem II related and unrelated to Dl- protein turnover. Planta 194:15–21

Liao Y, Zhang JS, Chen SY, Zhang WK (2008) Role of soybean GmbZIP132 under abscisic acid and salt stresses. J Integr Plant Biol 50:221–230

Lipshutz RJ, Fodor SP, Gingeras TR, Lockhart DJ (1999) High density synthetic oligonucleotide arrays. Nat Genet 21:20–24

Lockhart DJ, Winzeler EA (2000) Genomics, gene expression and DNA arrays. Nature 405:827–836

Ma SY, Wu WH (2007) AtCPK23 functions in Arabidopsis responses to drought and salt stresses. Plant Mol Biol 65:511–518

Ma S, Gong Q, Bohnert HJ (2006) Dissecting salt stress pathways. J Exp Bot 57:1097–1107

Mao X, Zhang H, Tian S, Chang X, Jing R (2010) TaSnRK2.4, an SNF1-type serine/threonine protein kinase of wheat (*Triticum aestivum* L.), confers enhanced multistress tolerance in Arabidopsis. J Exp Bot 61:683–696

Mahajan S, Tuteja N (2005) Cold, salinity and drought stresses: an overview. Arch Biochem Biophys 444:139–158

Matsui A, Ishida J, Morosawa T, Mochizuki Y, Kaminuma E, Endo TA, Okamoto M, Nambara E, Nakajima M, Kawashima M et al (2008) Arabidopsis transcriptome analysis under drought, cold, high-salinity and ABA treatment conditions using a tiling array. Plant Cell Physiol 49:1135–1149

Mattoo AK, Marder JB, Edelman M (1989) Dynamics of photosystem two reaction center. Cell 56:241–246

McMillan C (1986) Isozyme patterns among populations of black mangrove *Avicennia germinans* L., from the gulf of Mexico-Caribbean and pacific panama. Contrib Mar Sci 29:130–162

Megdiche W, Passaquet C, Zourrig W, Zuily Fodil Y, Abdelly C (2009) Molecular cloning and characterization of novel cystatin gene in leaves Cakile maritima halophyte. J Plant Physiol 166:739–749

Mehta PA, Sivaprakash K, Parani M, Venkataraman G, Parida AK (2005) Generation and analysis of expressed sequence tags from the salt-tolerant mangrove species *Avicennia marina* (Forsk) Vierh. Theor Appl Genet 110:416–424

Mishra S, Das AB (2003) Effect of NaCl on leaf salt secretion and antioxidative enzyme level in roots of mangrove, *Aegiceras corniculatum*. Indian J Exp Biol 41:160–166

Miyama M, Hanagata N (2007) Microarray analysis of 7029 gene expression patterns in Burma mangrove under high-salinity stress. Plant Sci 172:948–957

Miyama M, Tada Y (2008) Transcriptional and physiological study of the response of Burma mangrove (*Bruguiera gymnorrhiza*) to salt and osmotic stress. Plant Mol Biol 68:119–129

Miyama M, Simizu H, Sugiyama M, Hanagata N (2006) Sequencing and analysis of 14,842 expressed sequence tags of Burma mangrove. Bruguiera gymnorrhiza Plant Sci 171:234–241

Mohammadi M, Kav NN, Deyholos MK (2007) Transcriptional profiling of hexaploid wheat (*Triticum aestivum* L.) roots identifies novel, dehydration-responsive genes. Plant Cell Environ 30:630–645

Munns R (2005) Genes and salt tolerance: bringing them together. New Phytologist 167:645–663

Nakashima K, Tran LS, Van Nguyen D, Fujita M, Maruyama K, Todaka D, Ito Y, Hayashi N, Shinozaki K, Yamaguchi-Shinozaki K (2007) Functional analysis of a NAC-type transcription factor OsNAC6 involved in abiotic and biotic stress-responsive gene expression in rice. Plant J 51:617–630

Nguyen PD, Ho CL, Harikrishna JA, Wong MCVL, Rahim RA (2006) Generation and analysis of expressed sequence tags from the mangrove plant, *Acanthus ebracteatus* Vahl. Tree Genet Genomes 2:196–201

Ni Y, Wang X, Li D, Wu Y, Xu W, Li X (2008) Novel cotton homeobox gene and its expression profiling in root development and in response to stresses and phytohormones. Acta Biochim Biophys Sin (Shanghai) 40:78–84

Nishiyama Y, Yamamoto H, Allakhverdiev SI, Inaba M, Yokota A, Murata N (2001) Oxidative stress inhibits the repair of photodamage to the photosynthetic machinery. EMBO J 20:5587–5594

Nishiyama Y, Allakhverdiev SI, Yamamoto H, Hayashi H, Murata N (2004) Singlet oxygen inhibits the repair of photosystem II by suppressing the transport elongation of the D1 protein in *Synechocystis* sp. PCC 6803. Biochemistry (USA) 43:11321–11330

Nishiyama Y, Allakhverdiev SI, Murata N (2005) Inhibition of the repair of photosystem II by oxidative stress in cyanobacteria. Photosynth Res 84:1–7

Nishiyama Y, Allakhverdiev SI, Murata N (2011) Protein synthesis is the primary target of reactive oxygen species in the photoinhibition of photosystem II. Physiol Plant 142:35–46

Paramonova NV, Shevyakova NI, Kuznetsov VV (2004) Ultrastructure of chloroplasts and their storage inclusions in the primary leaves of *Mesembryanthemum crystallinum* affected by putrescine and NaCl. Russ J Plant Physiol 1:86–96

Parida AK, Das AB (2005) Salt tolerance and salinity effects on plants: a review. Ecotoxicol Environ Saf 60:324–349

Parida A, Das AB, Das P (2002) NaCl stress on photosynthetic pigments, proteins and other metabolic changes in the leaves of *Bruguiera parviflora*: studies under hydroponic culture conditions. J Plant Biol 45:28–36

Parida AK, Das AB, Mohanty P (2004a) Investigation on the antioxidative defense responses to NaCl stress in a mangrove, *Bruguiera* parviflora: differential regulations of isoforms of some antioxidative enzymes. Plant Growth Regul 42(3):213–226

Parida AK, Das AB, Mohanty P (2004b) Defense potentials to NaCl in a mangrove, *Bruguiera parviflora*: differential changes of isoforms of some antioxidative enzymes. J Plant Physiol 161:531–542

Parida AK, Mittra B, Das AB, Das TK, Mohanty P (2005) High salinity reduces the content of a highly abundant 23 kDa protein of the mangrove, *Bruguiera parviflora*. Planta 221:135–140

Prashanth SR, Sadhasivam V, Panda A (2008) Over expression of cytosolic copper/zinc superoxide dismutase from a mangrove plant Avicennia marina in indica Rice var Pusa Basmati-1 confers abiotic stress tolerance. Transgenic Res 17:281–291

Ray S, Agarwal P, Arora R, Kapoor S, Tyagi AK (2007) Expression analysis of calcium-dependent protein kinase gene family during reproductive development and abiotic stress conditions in rice (*Oryza sativa* L. ssp. indica). Mol Genet Genomics 278:493–505

Rensink WA, Buell CR (2005) Microarray expression profiling resources for plant genomics. Trends Plant Sci 10:603–609

Rensink WA, Iobst S, Hart A, Stegalkina S, Liu J, Buell CR (2005) Gene expression profiling of potato responses to cold, heat, and salt stress. Funct Integr Genomics 5:201–207

Rhodes D, Nadolska-Orczyk A, Rich PJ (2002) Salinity, osmolytes and compatible solutes. In: Lauchli A, Liittge U (eds) Salinity: environment – plants- molecules. Kluwer, Dordrecht, pp 181–204

Rich PR, Bonner WD Jr (1987) The sites of superoxide anion generation in higher plant mitochondria. Arch Biochem Biophys 188:206–213

Richmond T, Somerville S (2000) Chasing the dream: plant EST microarrays. Curr Opin Plant Biol 3: 108–116

Sahu BB, Shaw BP (2009) Isolation, identification and expression analysis of salt-induced genes in Suaeda maritima, a natural halophyte, using PCR-based suppression subtractive hybridization. BMC Plant Biol 9:69

Schachtman DP, Kumar R, Schroeder JI, Marsh EL (1997) Molecular and functional characterization of a novel low-affinity cation transporter (LCT1) in higher plants. Proc Natl Acad Sci USA 94:11079–11084

Schansker G, Tóth SZ, Strasser RJ (2005) Methylviologen and dibromothymoquinone treatments of pea leaves reveal the role of photosystem I in the *Chl* a fluorescence rise OJIP. Biochim Biophys Acta 1706:250–261

Scholander PF (1968) How mangrove desalinate seawater. Physiol Plant 21:251–261

Schreiber U (2004) Pulse-amplitude-modulation (PAM) fluorometry and saturation pulse method: an overview. In: Papageorgiou GC, Govindjee (eds) Chlorophyll *a* fluorescence: a signature of photosynthesis, advances in photosynthesis and respiration, vol 19. Springer, Dordrecht, pp 279–319

Seki M, Narusaka M, Ishida J, Nanjo T, Fujita M, Oono Y, Kamiya A, Nakajima M, Enju A, Sakurai T et al

(2002) Monitoring the expression profiles of 7000 Arabidopsis genes under drought, cold and high-salinity stresses using a full-length cDNA microarray. Plant J 31:279–292

Senadheera P, Singh RK, Maathuis FJ (2009) Differentially expressed membrane transporters in rice roots may contribute to cultivar dependent salt tolerance. J Exp Bot 60:2553–2563

Shabala S, Cuin TA (2008) Potassium transport and plant salt tolerance. Physiol Plan 133:651–669

Slesak I, Miszalski Z, Karpinska B, Niewiadomska E, Ratajczak R, Karpinski S (2002) Redox control of oxidative stress responses in the $C_3$-CAM intermediate plant *Mesembryanthemum crystallinum*. Plant Physiol Biochem 40:669–677

Song X-J, Matsuoka M (2009) Bar the windows: an optimized strategy to survive drought and salt adversities. Genes Dev 23:1709–1713

Sreenivasulu N, Sopory SK, Kavi Kishor PB (2007) Deciphering the regulatory mechanisms of abiotic stress tolerance in plants by genomic approaches. Gene 388:1–13

Srivastava B, Guisse H, Greppin H, Strasser RJ (1997) Regulation of antenna structure and electron transport in photosystem II of *Pisum sativum* under elevated temperature probed by the fast polyphasic chlorophyll a fluorescence transient: OKJIP. Biochim Biophys Acta 1320:95–106

Stephenson J, Newman K, Mayhew S (2010) Population dynamics and climate change: what are the links? J Public Health (Oxf) 32:150–156

Stirbet A, Govindjee (2011) On the relation between the Kautsky effect (chlorophyll *a* fluorescence induction) and Photosystem II: basics and applications of the OJIP fluorescence transient. J Photochem Photobiol B 104:236–257

Strasser RJ, Govindjee (1991) The Fo and the OJIP fluorescence rise in higher plants and algae. In: Argyroudi-Akoyunoglou JH (ed) Regulation of chloroplast biogenesis. Plenum Press, New York, pp 423–426

Strasser RJ, Govindjee (1992) On the OJIP fluorescence transients in leaves and D1 mutants of *Chlamydomonas reinhardtii*. In: Murata N (ed) Research in photosynthesis, vol II. Kluwer Academic, Dordrecht, pp 29–32

Strasser RJ, Srivastava A, Tsimilli-Michael M (2000) The fluorescence transient as a tool to characterize and screen photosynthetic samples. In: Yunus M, Pathre U, Mohanty P (eds) Probing photosynthesis: mechanism, regulation and adaptation. Taylor & Francis, London, pp 443–480

Strasser RJ, Srivastava A, Tsimilli-Michael M (2004) Analysis of the chlorophyll a fluorescence transient, advances in photosynthesis and respiration. In: Papageorgiou GC, Govindjee E (eds) Chlorophyll fluorescence: a signature of photosynthesis. Kluwer Academic, Dordrecht, pp 321–362

Strasser RJ, Tsimili-Michael M, Qiang S, Goltsev V (2010) Simultaneous in vivo recording of prompt and delayed fluorescence and 820-nm reflection changes during drying and after rehydration of the resurrection plant *Haberlea rhodopensis*. Biochim Biophys Acta 1797:1313–1326

Sugihara K, Hanagata N, Dubinsky Z, Baba S, Karube J (2000) Molecular characterization of cDNA encoding oxygen evolving enhancer protein 1 increased by salt treatment in the mangrove, *Bruguiera gymnorrhiza*. Plant Cell Physiol 41:1279–1285

Sunkar R, Bartels D, Kirch HH (2003) Overexpression of a stress-inducible aldehyde dehydrogenase gene from *Arabidopsis thaliana* in transgenic plants improves stress tolerance. Plant J 35:452–464

Swindell WR, Huebner M, Weber AP (2007) Transcriptional profiling of Arabidopsis heat shock proteins and transcription factors reveals extensive overlap between heat and non-heat stress response pathways. BMC Genomics 8:125

Szabolcs I (1994) Prospects of soil salinity for the 21st century. 15th world congress of soil science, vol I,. Acapulco, pp 123–141

Taji T, Seki M, Satox, Sakurai T, Kobayashi M, Ishiyama K, Narusaka Y, Narusaka M, Zhu JK, Shinozaki K (2004a) Comparative genomics in salt tolerance between *Arabidopsis* and *Arabidopsis*-related halophyte salt stress using *Arabidopsis* microarray. Plant Physiol 135:1–13

Taji T, Seki M, Satou M, Sakurai T, Kobayashi M, Ishiyama K, Narusaka Y, Narusaka M, Zhu JK, Shinozaki K (2004b) Comparative genomics in salt tolerance between Arabidopsis and Arabidopsis-related halophyte salt cress using Arabidopsis microarray. Plant Physiol 135:1697–1709

Takemura T, Hanagata N, Sugihara K, Baba S, Karube I, Dubinsky Z (2000) Physiological and biochemical responses to salt stress in the mangrove, *Bruguiera gymnorrhiza*. Aquat Bot 68:15–28

Takemura T, Hanagata N, Dubinsky Z, Karube I (2002) Molecular characterization and response to salt stress of mRNAs encoding cytosolic Cu/Zn superoxide dismutase and catalase from *Bruguiera gymnorrhiza*. Trees 16:94–99

Tal M, Shannon MC (1983) Salt tolerance in the wild relatives of the cultivated tomato: responses of *Lycopersicon esculentum*, *Lycopersicon cheesmanii*, *Lycopersicon peruvianum*, *Solanum pennellii*, and F1 hybrids to high salinity. Aust J Plant Physiol 10:109–117

Tanaka Y, Hibino T, Hayashi Y, Tanaka A, Kishitani S, Takabe T et al (1999) Salt tolerance of transgenic rice overexpressing yeast mitochondrial Mn-SOD in chloroplasts. Plant Sci 148:131–138

Tanaka Y, Fukuda A, Nakamura A, Yamada A, Saito T (2000) Molecular cloning and characterization of mangrove Na+/H+ antiporter cDNA. Plant Cell Physiol 41 (suppl):27

Tanji KK (1990) Nature and extent of agricultural salinity. In: Tangi KK (ed) Agricultural salinity assessment and management. Am Soc Civil Engineers, New York, pp 1–13

Tausz M, Sircelj H, Grill D (2004) The glutathione system as a stress marker in plant ecophysiology: is a stress-response concept valid? J Exp Bot 55:1955–1962

Thomas DJ, Thomas J, Youderian PA, Herbert SK (2001) Photoinhibition and light-induced electron transport in *ndhB⁻* and *psaE⁻* mutants of *Synechocystis* sp. PCC 6803. Plant Cell Physiol 42:803–812

Tomlinson PB (1986) The botany of mangroves, 1st edn. Cambridge University Press, Cambridge, pp 413

Ueda A, Kathiresan A, Bennett J, Takabe T (2006) Comparative transcriptome analyses of barley and rice under salt stress. Theor Appl Genet 112:1286–1294

Uozumi N, Kim EJ, Rubio F, Yamaguchi T, Muto S, Tsuboi A, Bakker EP, Nakamura T, Schroeder JI (2000) The Arabidopsis HKT1 gene homolog mediates inward Na(+) currents in xenopus laevis oocytes and Na(+) uptake in Saccharomyces cerevisiae. Plant Physiol 122:1249–1259

Urano K, Kurihara Y, Seki M, Shinozaki K (2010) 'Omics' analyses of regulatory networks in plant abiotic stress responses. Curr Opin Plant Biol 13: 32–138

Van Gorkom HJ (1986) Fluorescence measurements in the study of photosystem II electrontransport. In: Amesz J, Fork DC, Govindjee (eds) Light emission by plants and bacteria. Academic, Ontario

Vij S, Tyagi AK (2007) Emerging trends in the functional genomics of the abiotic stress response in crop plants. Plant Biotechnol J 5:361–380

Vinocur B, Altman A (2005) Recent advances in engineering plant tolerance to abiotic stress: achievements and limitations. Curr Opin Biotechnol 16:123–132

Waditee R, Hibino T, Tanaka Y, Nakamura T, Incharoensakdi A, Hayakawa S, Suzuki S, Futsuhara Y, Kawamitsu Y, Takabe T, Takabe T (2002) Functional characterization of betaine/proline transporters in betaine-accumulating mangrove. J Biol Chem 277:18373–18382

Walia H, Wilson C, Condamine P, Liu X, Ismail AM, Zeng L, Wanamaker SI, Mandal J, Xu J, Cui X et al (2005) Comparative transcriptional profiling of two contrasting rice genotypes under salinity stress during the vegetative growth stage. Plant Physiol 139: 822–835

Walia H, Wilson C, Ismail AM, Close TJ, Cui X (2009) Comparing genomic expression patterns across plant species reveals highly diverged transcriptional dynamics in response to salt stress. BMC Genomics 10:398

Wang B, Luttge U, Ratajczak R (2004a) Specific regulation of SOD isoforms by NaCl and osmotic stress in leaves of the C3 halophyte *Suaeda salsa* L. J Plant Physiol 161:285–293

Wang ZL, Li PH, Fredricksen M, Gong ZZ, Kim CS, Zhang C, Bohnert HJ, Zhu JK, Bressan RA, Hasegawa PM, Zhao YX, Zhang H (2004b) Expressed sequence tags from *Thellungiella halophila*, a new model to study plant salt-tolerance. Plant Sci 166:609–616

Wang Y, Liu C, Li K, Sun F, Hu H, Li X, Zhao Y, Han C, Zhang W, Duan Y et al (2007a) Arabidopsis EIN2 modulates stress response through abscisic acid response pathway. Plant Mol Biol 64:633–644

Wang Y, Yang C, Liu G, Jiang J (2007b) Development of a cDNA microarray to identify gene expression of Puccinellia tenuiflora under saline-alkali stress. Plant Physiol Biochem 45:567–576

Wang X, Fan P, Song H, Chen X, Li X, Li Y (2009a) Comparative proteomic analysis of differentially expressed proteins in shoots of *Salicornia europaea* under different salinity. J Proteome Res 8:3331–3345

Wang Z, Gerstein M, Snyder M (2009b) RNA-Seq: a revolutionary tool for transcriptomics. Nat Rev Genet 10:57–63

Wong CE, Li Y, Whitty BR, Diaz-Camino C, Akhter SR, Brandle JE, Golding GB, Weretilnyk EA, Moffatt BA, Griffith M (2005) Expressed sequence tags from the Yukon ecotype of *Thellungiella* reveal that gene expression in response to cold, drought and salinity shows little overlap. Plant Mol Biol 58:561–574

Wong CE, Li Y, Labbe A, Guevara D, Nuin P, Whitty B, Diaz C, Golding GB, Gray GR, Weretilnyk EA, Griffith M, Moffatt BA (2006) Transcriptional profiling implicates novel interactions between abiotic stress and hormonal responses in *Thellungiella*, a close relative of Arabidopsis. Plant Physiol 140:1437–1450

Xiang Y, Huang Y, Xiong L (2007) Characterization of stress-responsive CIPK genes in rice for stress tolerance improvement. Plant Physiol 144:1416–1428

Xie ZM, Zou HF, Lei G, Wei W, Zhou QY, Niu CF, Liao Y, Tian AG, Ma B, Zhang WK, et al (2009) Soybean Trihelix transcription factors GmGT-2A and GmGT-2B improve plant tolerance to abiotic stresses in transgenic arabidopsis. PLoS One 4:e6898

Xiong L, Schumaker KS, Zhu JK (2002) Cell signaling during cold, drought, and salt stress. Plant Cell 14 (Suppl):S165–S183

Xu WF, Shi WM (2006) Expression profiling of the 14-3-3 gene family in response to salt stress and potassium and iron deficiencies in young tomato (*Solanum lycopersicum*) roots: analysis by real-time RT-PCR. Ann Bot 98:965–974

Yamada A, Saitoh T, Mimura T, Ozeki Y (2002a) Expression of mangrove allene cyclase enhances salt tolerance in *Escherichia coli*, yeast and tobacco cells. Plant Cell Physiol 43:903–910

Yamada A, Saitoh T, Mimura T, Ozeki Y (2002b) Detection of differences in mRNA expression regulated by salt-stress in mangrove cultured cells. Plant Biotech 19:145–148

Yamada A, Sekiguchi M, Mimura T, Ozeki Y (2002c) The role of plant CCTa in salt- and osmotic-stress tolerance. Plant Cell Physiol 43:1043–1048

Yamanaka T, Miyama M, Tada Y (2009) Transcriptome profiling of the mangrove plant Bruguiera gymnorrhiza and identification of salt tolerance genes by Agrobacterium functional screening. Biosci Biotechnol Biochem 73:304–310

Yang GL (2007) Comparative genomics in salt tolerance between two ecotypes of *Hibiscus tiliaceus* using cDNA microarray. PhD thesis. Sun Yat-sen University, Guangzhou

Yang G, Zhou R, Tang T, Chen X, Ouyang J, He L, Li W, Chen S, Guo M, Li X, Zhong C, Shi S (2011) Gene expression profiles in response to salt stress in *Hibiscus tiliaceus*. Plant Mol Biol Rep 29:609–617

Yokoi S, Quintero FJ, Cubero B, Ruiz MT, Bressan RA, Hasegawa PM, Pardo JM (2002) Differential expression and function of *Arabidopsis thaliana* NHX $Na^+/H^+$ antiporters in the salt stress response. Plant J 30:529–539

Zeng HC, Deng LH, Zhang CF (2006) Cloning of salt tolerance related cDNAs from the mangrove *plant Sesuvium portulacastrum* L. J Integr Plant Biol 48:952–957

Zhang L, Paakkarinen V, van Wijk KJ, Aro E-M (2000) Biogenesis of the chloroplast-encoded D1 protein: regulation of mechanisms of D1 protein translation elongation, insertion and assembly. Plant Cell 12:1769–1781

Zhang H, Sreenivasulu N, Weschke W, Stein N, Rudd S, Radchuk V, Potokina E, Scholz U, Schweizer P, Zierold U et al (2004) Large-scale analysis of the barley transcriptome based on expressed sequence tags. Plant J 40:276–290

Zhang JL, Flowers TJ, Wang SM (2010) Mechanisms of sodium uptake by roots of higher plants. 326:45–60

Zhou QY, Tian AG, Zou HF, Xie ZM, Lei G, Huang J, Wang CM, Wang HW, Zhang JS, Chen SY (2008) Soybean WRKY-type transcription factor genes, GmWRKY13, GmWRKY21, and GmWRKY54, confer differential tolerance to abiotic stresses in transgenic Arabidopsis plants. Plant Biotechnol J 6:486–503

Zhu JK (2000) Genetic analysis of plant salt tolerance using Arabidopsis. Plant Physiol 124:941–948

Zhu JK (2001a) Plant salt tolerance. Trends Plant Sci 6:66–71

Zhu JK (2001b) Cell signaling under salt, water and cold stresses. Curr Opin Plant Biol 4:401–406

Zhu JK (2002) Salt and drought stress signal transduction in plants. Annu Rev Plant Biol 53:247–273

Zhu JK (2003) Regulation of ion homeostasis under salt stress. Curr Opin Plant Biol 6:441–445

Zimmermann P, Hirsch-Hoffmann M, Hennig L, Gruissem W (2004) GENEVESTIGATOR. Arabidopsis microarray database and analysis toolbox. Plant Physiol 136:2621–2632

# PSII Fluorescence Techniques for Measurement of Drought and High Temperature Stress Signal in Crop Plants: Protocols and Applications

# 4

Marian Brestic and Marek Zivcak

**Abstract**

Field crops are frequently exposed to drought and high temperature in the field. As the stress tolerance is the major target of many research and breeding programmes, the efficient and reliable tools and methods useful in screening of the heat and drought stress effects are required. The techniques based on measurement of chlorophyll fluorescence induction belong recently to fundamentals of plant stress research; however, in most cases the very basic tools are used and its potential is not utilised sufficiently. This proposed chapter tries to summarise the knowledge, starting from basic theory through parameters and useful experimental protocols and results up to special kinds of application of chlorophyll fluorescence techniques. In addition to generally used pulse-amplitude-modulated (PAM) method with saturation pulse analysis, the fast fluorescence kinetics, the fluorescence imaging, as well as simultaneous measurements of chlorophyll fluorescence with other parameters and their potential application in drought and heat-stress research are discussed.

## Introduction

Climate change, climate variability and environmental stress play an important role in determining physiological performance, vulnerability and productivity of crop plants in their environment. Agricultural crops grown at midlatitudes may be less sensitive to climate variability, but crop production in some of these areas will become more environmentally limited and more risky under future climate as the frequency of high-temperature extremes and competition for water resources increases. During the twenty-first century it will be required to better understand interactions between plants and extreme environment. Curbing crop susceptibility to main environmental stresses could allow for higher yields during drought years in the agricultural areas of the world. The ability to produce high quality of agricultural products will depend on the resources available for agriculture, and maintaining yields at current levels often requires new cultivars and progressive management methods.

M. Brestic (✉) • M. Zivcak
Department of Plant Physiology, Slovak University of Agriculture, Tr. A. Hlinku 2, 949 76 Nitra, Slovak Republic
e-mail: marian.brestic@uniag.sk; marek.zivcak@uniag.sk

G.R. Rout and A.B. Das (eds.), *Molecular Stress Physiology of Plants*,
DOI 10.1007/978-81-322-0807-5_4, © Springer India 2013

It's clear that modern cropping systems will need to feed the several billion people living on our planet. Now there are increasing demands for new types of crop products that largely must be met by increased productivity by using the improved understanding of plant physiology and crop responses to changing environment that have been obtained in recent years. On the other hand, the production of high-quality food must increase with reduced inputs. This accomplishment will be particularly challenging in the face of global environmental change (Tester and Langridge 2010). The need to accelerate breeding for increased yield potential and better adaptation to drought and other abiotic stresses is an issue of increasing urgency. As the population continues to grow rapidly, the pressure on resources, mainly untouched land and water, is also increasing, and potential climate change poses further challenges (Araus et al. 2008). Crop ecophysiologists provide information on plant function and environment that could be used to determine traits that should be adaptive in specific environment. This information is useful for plant breeders who need to focus on traits with the greatest potential to increase yield. Hence, new technologies must be developed to accelerate breeding through improving genotyping and phenotyping methods and by increasing the available genetic diversity in breeding germplasm (Tester and Langridge 2010).

Many researchers have suggested chlorophyll fluorescence as a tool for understanding photosynthetic metabolism and thus identify plant performance in their environment or at least reactions of genotypes in relation to water deficit, high temperatures and other abiotic stress factors (Longenberger et al. 2009). The relationships between growth, leaf photosynthetic performance as well as productivity and chlorophyll fluorescence parameters are reviewed in the context of practical applications to potential screening programmes, which seek to identify improved plant performance. The value of fluorescence measurement lies in its relationship to photosynthesis since light absorbed by plants that does not drive the production of carbohydrates is dissipated as heat or reemitted as light in the form of fluorescence. Physiologists and plant breeders now seek to relate fluorescence measurements and genotype-specific responses to stress (Baker and Rosenqvist 2004; Longenberger et al. 2009).

This chapter gives a summary of key chlorophyll fluorescence parameters and protocols currently used in studies of plant stress physiology and ecophysiology. We also demonstrate how different fluorescence parameters can be applied to identify possible causes of changes in photosynthesis under drought and high-temperature conditions and to evaluate photosynthetic performance in vivo. We focus on practical aspects, protocols and applications, which allow studying effectively primary photosynthetic processes in crop plants. It is aimed at plant ecophysiologists, biologists and breeders who seek to use chlorophyll fluorescence as modern and progressive tool in their research.

## Acclimation and Limitations of Photosynthesis Under Drought Stress: Stomatal and Non-stomatal Effects

Drought is considered as the main environmental factor limiting plant growth and yield worldwide. Drought stress results in stomatal closure and reduced transpiration rates; a decrease in the water potential of plant tissues; growth inhibition and decrease in photosynthesis; accumulation of abscisic acid (ABA), sorbitol, mannitol and proline; and formation of radical scavenging compounds, e.g. ascorbate, glutathione and α-tocopherol (Yordanov et al. 2003). Besides these physiological responses, plants also undergo morphological changes (Vassileva et al. 2012). At the whole plant level, the effect of drought stress is usually perceived as a decrease in photosynthesis and growth and is associated with alteration in carbon and nitrogen metabolism. The plant response is complex because it reflects over time and space the integration of stress effects and responses at all underlying levels of organisation. Under field conditions these responses can be synergistically or antagonistically modified by the superimposition of other stresses (Blum 1996; Cornic and Massacci 1996; Yordanov et al. 2003).

Understanding the biochemical, biophysical and physiological bases for impairment of photosynthesis in plants which experience internal water deficits becomes of major interest in order to improve plant responses to environmental stresses. Drought stress effects on leaf photosynthetic parameters and consequences on plants productivity have been discussed over the last three decades (Genty et al. 1987; Sharkey and Seeman 1989; Cornic and Briantais 1991; Brestic et al. 1995; Lawlor and Tezara 2009). Results of several research groups have shown that inhibition of leaf photosynthesis caused by mild water stress is primarily the result of stomatal behaviour, and there is high degree of co-regulation of stomatal conductance and photosynthesis. It is clear that stomata close progressively as drought progresses, followed by parallel decreases of net photosynthesis (Medrano et al. 2002).

The leaf $CO_2$ assimilation rate is decreased at mild leaf drought stress or even before the water content in leaf change in soil water potential (Gollan et al. 1986; Davies and Zhang 1991) or in response to a drop in humidity of atmosphere (Bunce 1981). The proportion of photosynthetic stomatal effect depends on the severity of drought stress. Under mild water deficit, the stomatal closure is a first event, followed by changes of photosynthetic reactions (Cornic and Briantais 1991). Under field conditions, stomatal regulation of transpiration was shown as a primary event in plant response to water deficit leading to decrease of $CO_2$ uptake by the leaves (Chaves 1991; Cornic and Massacci 1996; Chaves et al. 2002). It was shown that stomatal close occurs in response either to a decrease in leaf turgor and/or water potential, or to low air humidity (Maroco et al. 1997). Stomatal responses are more closely linked to soil moisture content than to leaf water status. This suggests that stomata are responding to 'non-hydraulic' chemical signals (Yordanov et al. 2003). This chemical signal has been shown to be abscisic acid (ABA) synthesised in the roots in response to soil drying (Davies and Zhang 1991). Stomata often close in response to drought before any change in leaf water potential and/or leaf water content is detectable (Medrano et al. 2000). A significant time relationship in stomatal responsiveness to air humidity and water status suggests that a part of diurnal changes in stomatal function may result from metabolic processes with a circadian rhythm (Chaves et al. 2002). Lawlor (2002) has shown that variations in cell carbon metabolism are frequently occurring early in the dehydration process. The stomatal function in drought-tolerant species is controlled to allow some carbon fixation even in stress conditions; hence, the water use efficiency increases. Moreover, when water deficit is relieved, they open stomata more rapidly. Some studies have shown (Faver et al. 1996; Herppich and Peckmann 1997) that at severe drought stress, photosynthesis may be more controlled by the chloroplast's carbon fixation capacity than by increased resistance for $CO_2$ diffusion.

Non-stomatal responses of carbon fixation (PS II energy conversion, the dark reaction of RuBisCO carbon fixation) are referred as resistant to water deficits (Genty et al. 1987; Chaves 1991). The strong drought-induced reductions of RuBisCO activity was also reported (Maroco et al. 2002; Parry et al. 2002), but a lot of studies have observed negligible effect of drought (Lal et al. 1996). Flagella et al. (1998) have shown that quantum yield of PS II, as related to Calvin cycle metabolism, is reduced only under drastic water deficit. However, Lauer and Boyer (1992) identified a metabolic impairment of photosynthesis by evaluating intercellular $CO_2$ partial pressure measured directly in the leaves. Similarly, Tang et al. (2002) have shown that a combination of stomatal and non-stomatal effects on photosynthesis exists, depending on the extent of drought stress and even in plants well hydrated. Declines of the photosynthetic rate in drought-stressed plants can be caused by closed stomata (i.e. reduction of $CO_2$ availability) and/or impairments in photochemical processes (i.e. decrease in NADPH and ATP supply) and/or biochemical reactions (i.e. reduced RuBP regeneration and carboxylation efficiency). Low biochemical activity may cause, under drought stress, photochemical downregulation by decreasing the demand for photochemical products (Santos et al. 2009). Nogues and Baker (2000) concluded that in C3 plants when stomata close in response to drought and $CO_2$ assimilation is reduced, the photosynthetic reduction of $O_2$ via photorespiration increases and serves as a sink for excess excitation energy in the photosynthetic apparatus. However, increases in the rate of photorespiratory reduction of $O_2$ are not sufficient to dissipate the excess energy in PS II antennae, and consequently, increased thermal dissipation of this energy occurs in order to minimise photodamage to PS II reaction centres.

The drought stress inhibits photosynthesis through decrease of ribulose-1,5-bisphosphate (RuBP) supply imposed by low ATP synthesis. Water deficit can also increase the oxygenase activity of RuBisCO, reducing thus the carboxylation efficiency (Tezara et al. 1999). In contrary, Lawlor and Cornic (2002) concluded that the decrease of photosynthetic potential under low leaf water content is caused by impaired metabolism (shortage of ATP) limiting RuBP synthesis without inhibition or loss of photosynthetic carbon reduction cycle enzymes including RuBisCO. Decreased ATP content as well as imbalance in reductant status substantially affect cell metabolism.

The cycles of photosynthetic carbon reduction and oxidation represent the main electron sinks for PS II activity in conditions of mild drought (Cornic and Fresneau 2002). Functioning and regulation of PS II were not quantitatively changed during early phases of desiccation. The $CO_2$ content in the chloroplasts decreases after stomatal closure in drying leaves. Consequently, RuBP oxygenation in C3 plants increases and becomes the main sink for photosynthetic electrons. The $O_2$ through photorespiratory activity can entirely or partly replace $CO_2$ as an electron acceptor, depending on intensity of incident light. However, according to Girardi et al. (1996), long-term drought reduction in water content led to considerable depletion of pea PS II core. The remaining PS II complex appeared to be functional and reorganised with a unit size (LHCP/PS II core) twofold greater than that of well-irrigated plants and enhanced degradation of CP43 and Dl proteins. Yordanov et al. (2003) reported that drought stress increased PS II-inactive reaction centres in drought-sensitive more than in drought-tolerant bean cultivars.

Stress physiologists are particularly interested in photosynthesis because it is a very good sensitive indicator of the overall fitness of the plant. One of the first responses of plants to harmful environment is decrease in the rate of photosynthesis and inhibition of several molecular mechanisms. In this context, chlorophyll fluorescence analysis has become a good tool for estimating various photosynthetic parameters and for optimisation and control of crop photosynthesis in the field conditions. The in vivo effects of water deficit on chlorophyll fluorescence have been described for several species.

## Acclimation and Limitations of Photosynthesis Under High-Temperature Conditions

Description of general mechanisms for adjustment of photosynthesis to temperature encompassing higher plants is very difficult for three reasons: differential strategies in growth and development, inherent genetic diversity and because organisms respond to temperature changes rather than to absolute temperature. Acquired stress tolerances to temperature extremes are complex traits dependent on many attributes. Within defined narrow limits, the ability to survive a temperature stress that would be lethal can be conferred by exposure to a mild nonlethal temperature stress (Sung et al. 2003). Plants' responses to the external environmental changes may be separated in two principal components. Adaptations express stable genotypic responses to long-term changes, while acclimations cause phenotypic alteration over a single generation time without any corresponding genetic compositional change.

To examine the role of acclimation versus adaptation on the temperature responses of $CO_2$ assimilation, Silim et al. (2010) measured dark respiration and the $CO_2$ response of net photosynthesis in species *Populus* collected from warm and cool habitats and grown at warm and cool temperatures. Respiration and the rate of photosynthetic electron transport are significantly higher in plants grown at 19 versus 27°C, and respiration is not affected by the native thermal habitat. By contrast, both the maximum capacity of RuBisCO and photosynthesis are relatively insensitive to growth temperature, but both parameters are slightly higher in plants from cool habitats. Photosynthesis is limited by RuBisCO capacity from 17 to 37°C regardless of growth temperature, and there is little evidence for an electron transport limitation. Stomatal conductance is higher in warm-grown plants but declines with increasing measurement temperature from 17 to 37°C, regardless of growth temperature. The mesophyll conductance seems to be relatively temperature insensitive below 25°C but declines at 37°C in cool-grown plants.

Photosynthesis has long been recognised as one of the plant components most sensitive to high-temperature stress. Inhibition of photosynthesis has been observed after short exposure to moderately high temperature (from 35 to 40°C) in various plant species (Crafts-Brandner and Salvucci 2000; Sinsawat et al. 2004). Following heat stress, the loss of photosynthetic electron transport was attributed to the thermolability of photosystem II (PS II), while stomatal conductance appears to play only minor role in the temperature limitation of photosynthesis. Havaux (1996) has shown that PS II is damaged by severely high-temperature stress when temperature is normally higher than 45°C, while the significant decrease of $CO_2$ assimilation already occurs at moderate heat stress. Decrease of $CO_2$ assimilation is associated with the inhibition of RuBisCO activation via a direct effect on RuBisCO activase (Feller et al. 1998; Salvucci and Crafts-Brandner 2004). Photochemical reactions on the level of thylakoids and carbon metabolism in the stroma of chloroplast have been suggested as the primary sites of injury at high temperatures (Wise et al. 2004; Wahid et al. 2007).

PS II is highly susceptible to high temperature, and its activity is significantly impaired or even partially stopped under heat conditions (Bukhov et al. 1999; Camejo et al. 2005; Datko et al. 2008). It may be caused by the properties of thylakoid membranes (Sharkey and Zhang 2010). In addition to dissociation of manganese-stabilising protein at PS II reaction centre complex and the release of Mn atoms (Yamane et al. 1998), heat stress may also lead to damage of other components in the reaction centre, e.g. major protein units (D1, D2) or other proteins (De Las Rivas and Barber 1997). In barley, heat pulses led to damage of the PS II and to the decline of capacity of oxygen evolution and, hence, to a restricted electron transport (Toth et al. 2005). Sharkova (2001) has shown that high temperatures and excessive light led to damage of different sites of PS II. It was associated with reinforcing of different pathways for the recovery of its functional activity. The degradation of the impaired PS II units occurred in the light during this period of time. Following this, de novo synthesis of PS II units in the light led to a gradual increase of the observed PS II activities (Wahid et al. 2007).

Under high temperatures, photosystem I (PS I) chloroplast envelops and stromal enzymes are thermostable, and the PS I-driven cyclic electron flow, contributing to thylakoid proton gradient, is activated (Bukhov et al. 1999).

More recently, inhibition of the RuBisCO enzyme has been identified as one of the most heat-sensitive components of the photosynthetic apparatus (Feller et al. 1998; Law and Crafts-Brandner 1999; Crafts-Brandner and Salvucci 2002). Photosynthetic apparatus is sensitive to high temperature; anyway, the thermotolerance can be improved by exposure to moderately high temperature. Havaux (1993) observed that exposure of potato plants to 35°C for 20 min significantly increased the PS II thermostability. Similar effect was observed by Brestic et al. (2012) in field-grown wheat after air temperature exceeded 30°C. Such a rapid acclimation was probably due to the accumulation of the xanthophyll zeaxanthin (Havaux and Tardy 1996), which is able to stabilise the lipid phase of the thylakoid membrane.

It has long been reviewed that heat accelerates PS I activity. The increased rate of photosystem I reduction is associated with heat tolerance as it is connected with an increase in reduction status (Zhang and Sharkey 2009), which reflects its quantum yield (Harbinson et al. 1989; Datko et al. 2008). At heat conditions, such an increase is associated with an increased rate of PS I reduction in both dark-adapted and light-adapted leaves. Thus, the PS I electron transport rate is strongly accelerated by high temperature, and plants deficient in cyclic electron transport were more heat sensitive (Zhang and Sharkey 2009).

The accelerated PS I activity at high temperature is associated with increased cyclic electron flow around PS I. One or more pathways of cyclic electron transport can be stimulated and it can lead to an increase or maintaining the energy gradient across the thylakoid membranes without increase of NADPH production (Shikanai 2007; Rumeau et al. 2007). Heat tolerance of cyclic electron flow is long-time known as a mechanism of heat tolerance. If the thylakoid membrane becomes leaky as the result of heat stress, the cyclic electron transport keeps the proton gradient across the thylakoid membrane constant, and hence, the supply of ATP remains constant (Bukhov et al. 1999). The ATP levels do not decrease at heat stress, even if very large changes in thylakoid metabolism occur (Schrader et al. 2004). Recent papers brought some results (Shikanai 2007; Rumeau et al. 2007) and thanks to progresses in this field, the mechanistic linkages between heat and cyclic electron flow become possible to determine (Sharkey and Zhang 2010).

The cyclic electron flow is probably enhanced by the state 1 to state 2 transitions in the light-harvesting complex of PS II, LHCII, antennas moving from PS II to PS I (Bukhov et al. 1999). State transitions may increase the absorption cross section of PS I by as much as 25% (Ruban and Johnson 2009; Zhang and Sharkey 2009). Hence, the PS II fluorescence measurements underestimate electron transport rate at high temperature.

Although physiological mechanisms of heat tolerance are relatively well understood, further studies are essential to determine physiological basis of assimilate partitioning from source to sink, plant phenotypic flexibility which leads to heat tolerance and factors that modulate plant heat-stress response. Chlorophyll fluorescence has been used as an early, in vivo, indication of many types of plant stress including temperature stress. Good correlations have been found between chlorophyll fluorescence and crop growth and production potential, gas exchange, electrolyte leakage, visible leaf damage, leaf water potential and leaf temperature (Wahid et al. 2007). Yamada et al. (1996) has shown that chlorophyll fluorescence parameters, as the maximum quantum yield of PS II photochemistry ($F_v/F_m$) and the basal fluorescence ($F_0$), correlate with heat tolerance. Many researchers suggest that for monitoring heat stress, chlorophyll fluorescence may be a more reliable measurement of photosynthesis than $CO_2$ exchange, which can be influenced mostly by stomatal closure not induced primarily by heat. It has also been used to screen plants for drought and heat tolerance (Ierna 2007).

Stress physiologists are particularly interested in photosynthesis because it is a very good sensitive indicator of the overall fitness of the plant. One of the first responses of plants to unfavourable environment is decrease in the rate of photosynthesis and inhibition of numerous molecular mechanisms. In this context chlorophyll fluorescence analysis has become a good tool for estimating various photosynthetic parameters and for optimisation and control of crop photosynthesis in the field conditions. We can conclude that chlorophyll fluorescence provides useful information about crop photosynthetic performance under abiotic stresses.

## Methods of Chlorophyll Fluorescence Measurements Useful in Plant Stress Research

Chlorophyll *a* fluorescence techniques represent highly versatile tools appropriate for many applications, including plant physiology, biophysics or biochemistry. Chlorophyll fluorescence analysis is non-invasive, highly sensitive and simple. The several techniques and approaches of chlorophyll fluorescence were introduced with development of technical equipment. The most useful chlorophyll fluorescence methods are based on variable chlorophyll *a* fluorescence measurements. However, there are several technical solutions enabling such a type of measurements, differing in the manner by which the photochemistry is saturated. The most frequently used are direct fluorescence measurement (in present mostly LED-based instruments) represented by fast chlorophyll *a* fluorescence measurement (Strasser and Govindjee 1991, 1992) or saturation pulse analysis represented by pulse-amplitude modulation (PAM) fluorometry (Schreiber et al. 1986; Schreiber 2004). The several techniques bring similar results to PAM fluorometry and can be used in a similar way, like the pump and probe fluorometry (Mauzerall 1972; Falkowski et al. 1986), the fast repetition rate (FRR) fluorometry (Kolber et al. 1998) and the pump during probe (PDP) fluorometry (Olson et al. 1996); for simplicity, in the next chapters, we will describe saturation pulse method measured as a PAM fluorometry only, as it is the most common approach.

## Saturation Pulse Method

### The Principal and Basic Parameters

The saturation pulse method is recently the most frequently used and generally accepted chlorophyll fluorescence technique. It enables the so-called quenching analysis using modulated fluorescence and saturation pulses. In this type of measurement, measuring light is switched on and off (modulated pulse) at high frequency, and the detector measures the fluorescence emission only; it enables to measure the correct fluorescence yield (Bradbury and Baker 1981; Quick and Horton 1984; Schreiber et al. 1986; Schreiber 2004).

A leaf must be dark adapted for at least 15 min prior to the measurement. The basal fluorescence ($F_0$) in darkness is measured by a weak modulating light beam (ML). Then the application of a saturating pulse (SP) (high light intensity for a short period, e.g. 7,500 μmol photons $m^{-2}$ $s^{-1}$ for 0.7 s) raises the fluorescence to a maximum value, $F_m$. Subsequently, after this first light pulse, the actinic light (AL) is switched on (photosynthetic samples are illuminated). The fluorescence intensity ($F$) increases from the basal level and reaches the local maximum (plateau) at given actinic light intensity ($F_P$). After a few seconds, it starts decreasing as the photochemical and non-photochemical processes initialise. After several minutes, the fluorescence level decreases and reaches the stable level; it indicates the steady state ($F_s$). The SP can be turned on repeatedly in short time interval, enabling measurement of the $F_m'$ values (fluorescence maximum at light-adapted state). The $F_m'$ level is lower than $F_m$, as the acceptor side of PS II is partly reduced and the non-photochemical quenching occurs (Stirbet and Govindjee 2011; Misra et al. 2012). A typical fluorescence measurement using saturation pulse method indicating the most important values is shown in Fig. 4.1.

In general, five primary mutually independent chlorophyll fluorescence yields taken from a record of slow chlorophyll fluorescence kinetics ($F_m$, $F_0$, $F_m'$, $F_s'$, $F_0'$ in Fig. 4.1) are sufficient for definitions of most chlorophyll fluorescence parameters, which can be found in literature (for review, see Rohacek et al. 2008; Baker 2008, etc.). The calculation and short physiological interpretation of individual parameters is reviewed in Table 4.1.

*Maximum Quantum Yield of PS II Photochemistry, $F_v/F_m$* (in some cases called also $\Phi_{Po}$). It is the most frequently used parameter, applied often as the indicator of photoinhibition or other kind of injury caused to the PS II complexes (Rohacek et al. 2008). It quantifies the maximum photochemical efficiency (capacity) of open PS II reaction centres. It is almost constant for many different plant species when measured under non-stressed conditions and equals to 0.832 (Bjorkman and Demmig 1987). For stressed and/or damaged plants, $F_v/F_m$ is markedly reduced. Moreover, its value might also be lowered due to fluorescence emission from PS I contributing to the $F_0$ level (Pfündel 1998; Franck et al. 2002). At severe heat stress, it can be underestimated (Toth et al. 2007).

*Effective Quantum Yield of Photochemical Energy Conversion in PS II, $\Phi_{PSII}$ ($F_q'/F_m'$; $\Delta F'/F_m'$). It* can be interpreted as the effective quantum yield of the PS II photochemistry related to the actual fraction of photochemically active PS II reaction centres (*qP*). Assessment of $\Phi_{PSII}$ does not require knowledge of the $F_0'$ level,

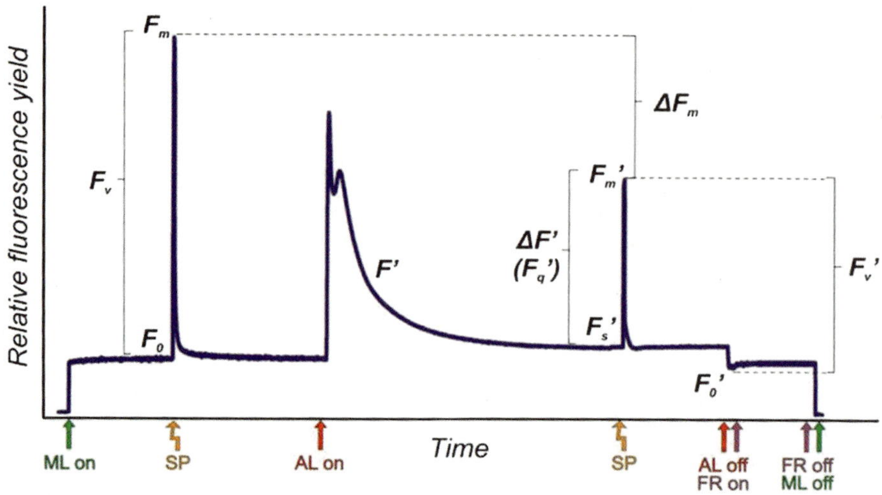


**Fig. 4.1** The principle of the fluorescence measurement using saturation pulse method with quenching analysis. *ML* measuring light, *SP* saturation pulse, *AL* actinic light, *FR* far-red light

**Table 4.1** Basic chlorophyll fluorescence parameters derived from slow fluorescence kinetics with saturation pulse analysis (Based on Baker 2008; Kramer et al. 2004; Rosenquist and van Kooten 2003; and Oxborough and Baker 1997)

| Parameter | Name and basic physiological interpretation | Calculation |
|---|---|---|
| *Basic parameters derived from fluorescence kinetics* | | |
| $F$, $F'$ | Fluorescence emission from dark- or light-adapted leaf, respectively | |
| $F_0$ | Minimal fluorescence from dark-adapted leaf, (PS II centres open) | |
| $F_0'$ | Minimal fluorescence from light-adapted leaf, (PS II centres open). Value can be directly measured or calculated (less accurate) | $F_0{\prime} = \frac{F_0}{(F_v/F_m) + (F_0/F_m{\prime})}$ |
| $F_m$, $F_m'$ | Maximum fluorescence from dark- or light-adapted leaf, respectively (PS II centres closed) | |
| $F_s'$ | Steady-state fluorescence at any light level | |
| $F_v$ | Variable fluorescence from dark-adapted leaf | $F_v = F_m - F_0$ |
| $F_q'$ (= $\Delta F'$) | Photochemical quenching of fluorescence by open PS II reaction centre | $F_q' = F_m' - F_s'$ |
| $F_v'$ | Variable fluorescence from light-adapted leaf | $F_v' = F_m' - F_0'$ |
| *Parameters derived from basic fluorescence parameters* | | |
| $F_v/F_m$ | Maximum quantum efficiency (yield) of PS II photochemistry | $F_v/F_m = 1 - \frac{F_0}{F_m}$ |
| $\Phi_{PSII}$ (= $F_q'/F_m'$) | Effective quantum yield (efficiency) of PS II photochemistry at given light intensity | $F_q{\prime}/F_m{\prime} = \frac{F_m{\prime} - F{\prime}}{F_m{\prime}}$ |
| $F_v'/F_m'$ | Maximum quantum efficiency of PS II photochemistry at given light intensity | $F_v{\prime}/F_m{\prime} = \frac{F_m{\prime} - F_0{\prime}}{F_m{\prime}}$ |
| $NPQ$ | Non-photochemical quenching | $NPQ = \frac{F_m - F_m{\prime}}{F_m{\prime}}$ |
| $qP$ | Coefficient of photochemical quenching based on the puddle model (unconnected PS II units) | $qP = \frac{F_m{\prime} - F_s{\prime}}{F_m{\prime} - F_0{\prime}}$ |
| $qL$ | Coefficient of photochemical quenching based on the lake model (connected PS II units) | $qL = \frac{F_m{\prime} - F_s{\prime}}{F_m{\prime} - F_0{\prime}} \cdot \frac{F_0}{F_s{\prime}}$ |
| $qN$ | Coefficient of non-photochemical quenching of variable fluorescence | $qN = \frac{F_m - F_m{\prime}}{F_m - F_0{\prime}}$ |
| $\Phi_{NO}$ | Quantum yield of nonregulated energy dissipation in PS II | $\Phi_{NO} = \frac{1}{NPQ + 1 + qL(F_m/F_0 - 1)}$ |
| $\Phi_{NPQ}$ | Quantum yield of regulated energy dissipation in PS II | $\Phi_{NPQ} = 1 - \Phi_{PSII} - \Phi_{NO}$ |

nor does it need previous dark adaptation of the sample. Therefore, it is often used for field investigations. It quantifies the efficiency of the (non-cyclic) electron transport, as well as a fraction of photons absorbed in PS II antennae and utilised in the PS II photochemistry. If the photochemical and biochemical processes of photosynthesis are equilibrated under non-stress conditions, $\Phi_{PSII}$ is often correlated with the quantum yield of $CO_2$ fixation or the rate of photorespiration (Genty et al. 1989; Rohacek et al. 2008).

*Quantum Yield of Regulated Energy Dissipation in PS II,* $\Phi_{NPQ}$, and *Quantum Yield of Non-regulated Energy Dissipation in PS II,* $\Phi_{NO}$. They are parameters introduced by Kramer et al. (2004). It is assumed that $\Phi_{PSII} + \Phi_{NPQ} + \Phi_{NO} = 1$. While $\Phi_{NPQ}$ reflects the quantum yield for dissipation by downregulation (active, organised process), $\Phi_{NO}$ is the yield of other non-photochemical losses.

*Photochemical Quenching of Variable Chlorophyll Fluorescence, qP or qL.* It indicates the actual photochemical capacity of PS II in light-adapted state, which is connected with the photochemical energy conversion by charge separation in reaction centres of PS II. It quantifies the actual fraction of PS II reaction centres being in the open state, i.e. with reoxidised $Q_A$ (Duysens and Sweers 1963; Kramer et al. 2004; Rohacek et al. 2008). The older parameter, *qP*, is based on the so-called 'puddle' model, ignoring the connectivity among PS II units. On the contrary, the parameter *qL* based on the 'lake' model considers the units to be fully connected. It was found that *qL* usually better reflects the reality; therefore, if an accurate assessment of the redox state of the $Q_A$ pool is required, then *qL*, and not *qP*, should be used (Kramer et al. 2004; Baker 2008).

*Non-photochemical Quenching of Variable Chlorophyll Fluorescence, qN.* It reflects the activation of several processes of non-photochemical nature during the light period and mostly leading to non-radiative dissipation of the excitation energy as heat (thermal dissipation). *qN* includes the pH-gradient build-up, state transitions, ATP-synthesis regulation, inactivation of reaction centres, conformational changes within thylakoid membranes, activation of the xanthophyll cycle, etc. (Schreiber et al. 1986; Krause and Weis 1991; Rohacek et al. 2008).

*Non-photochemical Quenching of Chlorophyll Fluorescence, NPQ.* It is often used as an indicator of the excess radiant energy dissipation to heat in the PS II antennae. The extent of *NPQ* is linearly correlated to xanthophyll deepoxidation through the xanthophyll cycle. *NPQ* reflects also the decrease of the light-harvesting antenna size, PS II inactivation, etc. (Bilger and Bjorkman 1990; Rohacek et al. 2008).

*Electron Transport Rate* (*ETR*). As a function of the quantum yield and illumination, it can be calculated as $ETR = I \cdot A_{leaf} \cdot \text{fraction}_{PSII} \cdot \Phi_{PSII}$, where $I$ is incident PAR on the leaf, $A_{leaf}$ is the proportion of the PAR absorbed by leaf (in most cases the value 0.84 is used) and $\text{fraction}_{PSII}$ is the fraction of absorbed PAR received by PS II (0.5 is mostly used assuming the equal distribution of absorbed light between PS II and PS I). Hence, the most frequently used formula is $ETR = 0.84 \cdot 0.5 \cdot I \cdot \Phi_{PSII}$. However, this calculation must be used carefully, especially in conditions of plant stress. If the chlorophyll content decrease occurs, the absorbance change must be considered (Baker 2008). It can be measured or estimated using several different methods.

## The Most Useful Protocols of Slow Fluorescence Kinetics

In the next paragraphs, we will try to introduce and discuss advantages and disadvantages of the most useful measuring protocols applying the saturation pulse method.

### Single Saturation Pulse on Dark-Adapted Sample ($F_v/F_m$ Measurement)

The $F_0$ and $F_m$ determination is a basic part of many protocols but can be measured even independently. The length of dark adaptation recommended is 15–20 min; however, it need not be sufficient for full relaxation if the illumination was too strong or if the sample is stressed. Therefore, it is crucial to keep the same time of dark adaptation (e.g. by using leaf clips) for all samples.

A pulsed measuring low light (ML) of low frequency is used to determine $F_0$ of dark-adapted leaves. The irradiance must be low enough not to induce a fluorescence induction kinetic. Then a saturation pulse (e.g. with intensity 6,000 μmol photons $m^{-2}$ $s^{-1}$ for 1 s) is applied, $F_m$ is determined (Fig. 4.1) and the variable fluorescence ($F_v$) as well as maximum quantum yield of PS II photochemistry ($F_v/F_m$) or other $F_0$, $F_m$ based ratios are calculated (Lichtenthaler et al. 2005a, b).

In many research papers, only the dark-adapted state measured parameters are used ($F_0$, $F_m$, $F_v$, $F_v/F_m$, $F_0/F_m$, $F_v/F_0$). This is because such an approach is very quick, and hence, a lot of samples can be measured within short time. The only limitation is the need of dark adaptation, which can be easily solved using leaf clips. However, it provides only very basic and in the most cases insufficient information about physiological status of photosynthetic apparatus (as discussed in subchapter Application of Chlorophyll Fluorescence in Drought Stress). In fact, the same rapid measuring procedure, but much more information, brings the technique of fast chlorophyll kinetics measurements (see subchapter Fast Chlorophyll Fluorescence Induction Kinetics), by which the same parameters can be measured, but provides also many other biophysical parameters that can be used in plant stress research.

### Simple Dark-to-Light Slow Fluorescence Induction

This measuring protocol consists of four steps: (1) dark adaptation, usually 15–20 min prior to measurements; (2) $F_0$ and $F_m$ determination by saturation pulse on dark-adapted sample; (3) the actinic light is switched on for the time needed to reach the steady state; and (4) next saturation pulse is applied to measure the $Fs'$ and $F_m'$. At the end, the $F_0'$ measurement using far-red light can be performed. This protocol is usually realised with fixed time between flashes, usually about 5 min. It is time sufficient for reaching initial steady state, but usually not for opening stomata. For this reason, the measured PS II quantum yield and the electron transport rate (*ETR*) usually do not reach their maximum values; however, they can serve for comparison of drought or heat-stress effects among samples, including assessment of non-photochemical processes. For obtaining data corresponding to photosynthetic potential in current physiological state, the sufficiently long induction curve, as described below, is necessary.

### Measurements of Sample Without Dark Adaptation ($\Phi_{PSII}$ Measurement)

The measurements composed by dark adaptation followed by several minutes of actinic light illumination, which is necessary to reach the steady state, are time-consuming. For some purposes it is possible to use the chlorophyll fluorescence measurements without previous dark adaptation (used e.g. by Kuckenberg et al. 2009; Morales et al. 2012). However, the parameters based on $F_0$ and $F_m$ values cannot be calculated. Anyway, it is possible to measure actual quantum yield of PS II ($\Phi_{PSII}$) and hence the electron transport rate (*ETR*) using the light intensity of actinic light provided by a device or measured value of the photosynthetic active radiation incident on leaf, if the fluorescence was measured with external (e.g. solar) irradiation. Such an approach enables to make much more measurements in a short time; however, the user have to get along without basic parameters providing information about the status of PS II and the photoprotective responses, like $F_v/F_m$, *qP*, *qN*, *NPQ* and many others. Kuckenberg et al. (2009) reported more heterogenous data measured without previous dark adaptation compared to dark-adapted samples, but the data from both approaches showed the same tendency.

### Slow Induction Curve and Recovery

In irradiated leaves, the PS II and PS I perform the photosynthetic light reactions and associated electron transport in a strict, highly coordinated cooperation. In dark-adapted green leaves, the cooperation of the photosystems is temporarily impaired. In the dark, the photosynthetic apparatus is in its non-functional 'state 1'. Upon irradiation of dark-adapted leaves, it takes a few minutes to induce again the cooperation of both photosystems and to bring about the joint photosynthetic electron transport reactions that lead to proper water splitting, oxygen evolution, as well

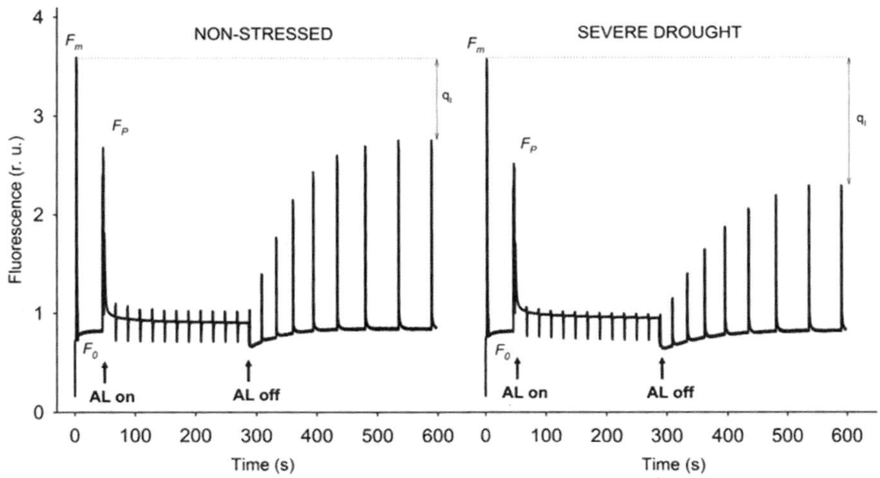


**Fig. 4.2** Example of chlorophyll fluorescence kinetics during slow induction curve followed by dark recovery period (red actinic light (AL) intensity app. 1,000 μmol photons $m^{-2}$ $s^{-1}$) in non-stressed wheat leaves and severely dehydrated wheat leaves (relative water content app. 60%). The curve starts with $F_0$ and $F_m$ determination; after that AL was switched on for 4 min followed by AL switch-off and measurement of dark recovery. Original data measured by the fluorometer DUAL-PAM (Walz, Germany)

as $NADP^+$ reduction and ATP formation needed for $CO_2$ assimilation in the Calvin–Benson cycle. This light-triggered induction period of the photosynthetic apparatus to its functional 'state 2' ('state 1' → 'state 2' transition), which is caused by several changes, e.g. also by a phosphorylation of the LHC2 (see below), can be detected and measured via the chlorophyll fluorescence induction kinetics (Lichtenthaler et al. 2005a, b).

Slow induction curve kinetics (Fig. 4.2) is regularly measured after dark adaptation period (usually 15–20 min) and $F_0$ and $F_m$ determination. After short delay enabling reoxidation after saturation pulse, the actinic light is switched on and the fluorescence yield quickly rises to a peak ($F_P$), which is followed by biphasic decline. This is called Kautsky effect (Kautsky and Hirsch 1931). Briefly after acinic light is on, the next saturation pulse is applied and it is applied regularly during whole induction curve. The $F_0'$ can be measured by switching off actinic light and switching on far-red red light for a few seconds after each saturation pulse. In some devices and applications, it is not possible to use far-red light (e.g. fluorescence imaging); in such cases the $F_0'$ is calculated by the formula of Oxborough and Baker (1997). After several minutes, the fluorescence ($F$) is becoming constant indicating the steady state of photosynthesis. Time needed for reaching the steady state can be very different depending on the type of sample, its physiological status, actinic light intensity, leaf temperature, etc.

After the steady state is documented, the dark relaxation measurements (Dark recovery) can be realised. It enables to recognise the major constituents of non-photochemical quenching: the energy quench $qE$ (related to the built-up pH gradient), the state transition quench $qT$ (related to state 1' to state 2' transitions of the photosynthetic apparatus including phosphorylation of the mobile light-harvesting protein LHC2 able to move dynamically from photosystem II to photosystem I and vice versa) and the photoinhibitory quench qI caused by photoinhibition of PS II units. For identifying these components, the first saturation pulse is applied immediately after turning off actinic light; one saturation pulse is applied after 1 min of darkness, and after that the saturation pulses are done every 2 min. The minimum length of dark period needed for estimate $qE$, $qT$ and $qI$ components is 20 min (Lichtenthaler et al. 2005a, b).

## Slow and Rapid Light Curves

The light response curves belong to basic protocols in photosynthetic research. They represent plots of fluorescence parameters related to increasing (or

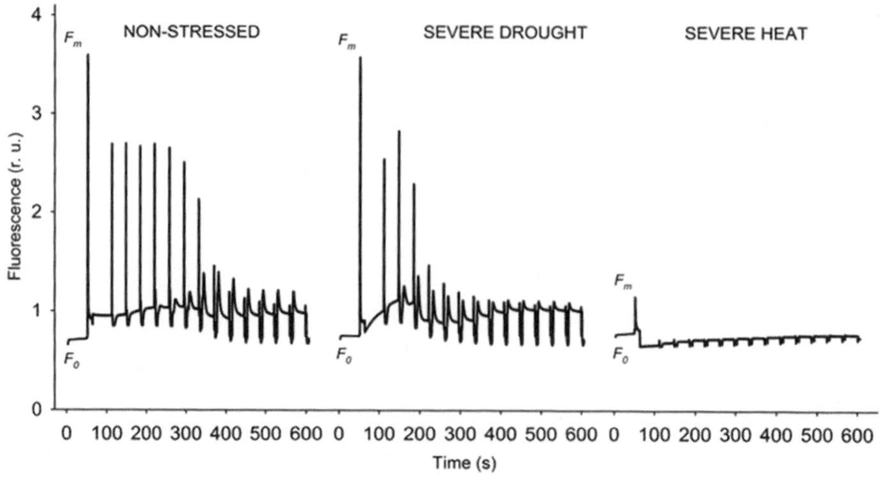


**Fig. 4.3** Example of chlorophyll fluorescence kinetics during a rapid light curve (gradually increased light intensities from 10 to 2,000 μmol photons $m^{-2}$ $s^{-1}$, 30 s at each light intensity) in non-stressed wheat leaves, severely dehydrated wheat leaves (relative water content app. 60%) and heat-exposed wheat leaves (at 45°C for 1 h in high light). The curve starts with $F_0$ and $F_m$ determination; after each saturation pulse, the measurement of $F_0'$ was performed using far-red light pulse. Original data measured by the fluorometer DUAL-PAM (Walz, Germany)

even decreasing) light intensities. In case of chlorophyll fluorescence measurements, the increasing light intensities are used, as the relaxation of non-photochemical processes needs more time than induction of photosynthesis. The time of steps with particular light intensities should be enough to get the steady state at each light level. Recently, the typical application of steady-state light curves is the simultaneous measurement of gas exchange ($CO_2$ uptake or $O_2$ release) with chlorophyll fluorescence measurements.

As this approach is very time-consuming, in plant stress research, the more efficient version called 'rapid light curves' is recently used more frequently. Guarini and Moritz (2009) reported that rapid light curves are used to study the physiological flexibility of plants' photosynthetic units to rapid changes in irradiation, similar to what occurs in natural environments (as referenced by Schreiber et al. 1997; White and Critchley 1999; Ralph and Gademann 2005). They provide detailed ecophysiological information on photosynthetic performances of plants as a function of their physiological condition (referenced by Wing and Patterson 1993; Kubler and Raven 1996; Hewson et al. 2001; Seddon and Cheshire 2001). An example of the fluorescence kinetics during rapid light curve is shown in Fig. 4.3.

The rapid light curves represent plots of electron transport rate (*ETR*) versus actinic irradiances applied for short time period (e.g. 10 s to 1 min). If the rapid light curve is measured after dark adaptation, even non-photochemical quenching (*NPQ*) and other parameters can be calculated. However, such measurements immediately after dark adaptation lead to very low *ETR* values, as the photosynthetic process is not fully induced (White and Critchley 1999). This can be avoided if the measurements are done without dark adaptation; in this case, we can calculate only effective quantum yield, but not *NPQ* and other parameters requiring correct $F_0$ and $F_m$ determination. Another solution is pre-illumination by moderate light following after dark adaptation. Usually period 4–10 min is required, depending on length of previous dark exposure. In this case, efficient way how to perform measurement is running induction curve at moderate light intensity after dark adaptation followed by the rapid light curve.

Rapid light curves were frequently used in many studies dealing with plant stress, like

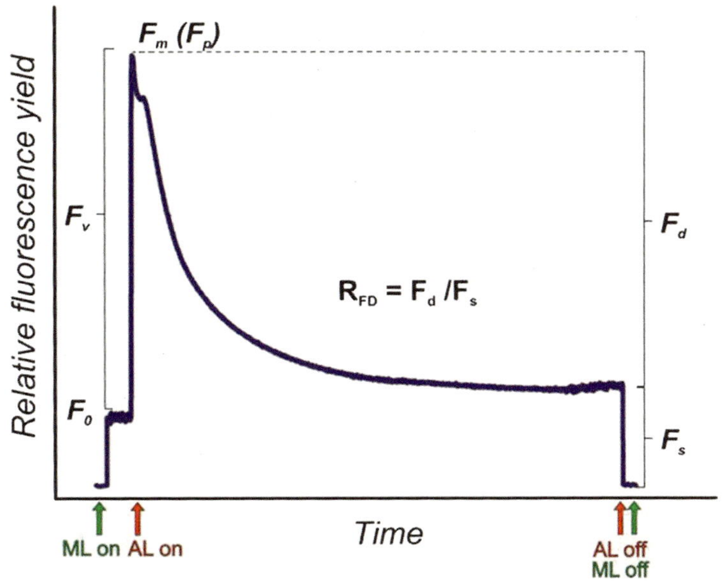


**Fig. 4.4** An example of chlorophyll fluorescence kinetics recorded for calculation of relative fluorescence decrease ratio ($R_{fd}$). *ML* modulated measuring light, *AL* actinic light (the PAR intensity used was 1,500 μmol photons $m^{-2}$ $s^{-1}$, total time of kinetics was app. 5 min)

drought stress (e.g. Li et al. 2008; Xu et al. 2009; Huang et al. 2011; Zhang et al. 2011b) or high temperature (Karim et al. 2003; Pérez et al. 2007; Datko et al. 2008). Their usefulness increase with development techniques of simultaneous measurements of chlorophyll fluorescence with other parameters as well as in fluorescence imaging measurements (more discussed below).

### Relative Fluorescence Decrease

A common approach similar to measurement of dark-to-light fluorescence kinetics can be used also for calculation of relative fluorescence decrease ratio, $R_{fd}$. For calculation of this parameter, however, the saturation light pulses are not needed, as the values of fluorescence (F, F′) measured at defined points during dark-to-light induction at strong actinic light are used. According to the methodic paper of Lichtenthaler et al. 2005a, b, this irradiance-induced chlorophyll fluorescence induction kinetic is characterised by a fast increase of chlorophyll fluorescence from the initial level $F_0$ (also termed base fluorescence) to a local maximum fluorescence level, plateau, $F_P$, within 100–200 ms as indicated in Fig. 4.4. With the saturating actinic light, one can obtain the $F_P = F_m$, but for the $R_{fd}$, the $F_P$ value is used in nomenclature. In the initial fluorescence rise, only PS II is involved. Thereafter, PS I starts to work and drains off electrons from PS II, then the cooperation of PS II and PS I as well as the photosynthetic electron transport goes into full operation, and the net $CO_2$ assimilation and oxygen evolution are triggered. This is seen in a gradual decline of the Chl fluorescence intensity from $F_m$ within 3–5 min to a much lower steady state $F$s (Fig. 4.4). The greater this Chl fluorescence decreases $F_d$, from $F_m$ to $F$s, the higher the net photosynthetic rate ($A_{co_2}$) of the leaf examined (Lichtenthaler and Rinderle 1988; Lichtenthaler and Miehé 1997; Lichtenthaler and Babani 2004; Lichtenthaler et al. 2005a, b). In fully or partially sun-exposed leaves of outdoor plants, this decrease is mainly caused by the photosynthetic quantum conversion process; however, it also includes some non-photochemical processes. This relationship can be quantified by the Chl fluorescence decrease ratio, $R_{Fd}$, defined as ratio $F_d/F$s or ($F_m/F$s) – 1. The $R_{Fd}$ is higher in sun leaves of trees (values of 3–5) than in shade leaves (values of 1.0–2.5) reflecting their higher photosynthetic capacity and $CO_2$ fixation rates (Lichtenthaler and Burkart 1999; Lichtenthaler and Babani 2004; Lichtenthaler et al. 2005a, b). In fact, the $R_{Fd}$ values being measured at the saturation irradiance of photosynthesis exhibit a highly significant linear correlation to $A_{co_2}$ as shown in Lichtenthaler et al. (2005a, b). This correlation particularly applies to outdoor plants that are much less affected by photoinhibition in comparison to leaves of greenhouse plants. In leaves from the extreme shade or from greenhouse plants, one should check by repetition of the induction kinetic measurement whether the applied irradiance for measuring the $R_{Fd}$ values might already cause a certain photoinhibition of the photosynthetic apparatus. If so, the excitation radiation can be reduced to ca. 1,000–1,200 μmol photons $m^{-2}$ $s^{-1}$ (Lichtenthaler et al. 2005a, b).

Thus, $R_{Fd}$ values permit a fast screening of the photosynthetic activity and vitality of plants also under stress, and such an approach was used by many researchers in studies of drought (e.g. Epron and Dreyer 1990; Georgieva et al. 2011) and heat stress (e.g. Georgieva and Lichtenthaler 1999 Dash and Mohanty 2001; Sarieva et al. 2010). In moderate heat stress, the first parameter influenced by 38°C in pea plants was $F$s, which started to increase at the first hours of heat treatment when the $F_P$ values remained close to those of the control. But this already caused some decline in $F_d$, which is the difference of $F_P$–$F$s. The longer heat treatment decreased $F_P$ and strongly increased $F$s, consequently leading to a strong reduction in $F_d$ and $R_{Fd}$ (Georgieva and Lichtenthaler 2006). Dash and Mohanty (2001) found $R_{fd}$ more sensitive to heat stress than PS II quantum yield ($\Phi_{PSII}$), and it correlated well with other physiological parameters measured in heat stress in wheat. The $R_{fd}$ ratio reflects the physiological changes associated with drought stress too (Georgieva et al. 2011).

## Fast Chlorophyll Fluorescence Induction Kinetics

Chlorophyll *a* fluorescence induction is now a widespread method used in photosynthesis research. This is because fluorescence induction is non-invasive and highly sensitive, fast and easily measured, it requires relatively inexpensive equipment and contains important information about photosynthetic apparatus (Lazár 1999). Illumination of dark-adapted photosynthetic samples leads to emission of the chlorophyll *a* fluorescence with a characteristic transient well known as the Kautsky curve (Kautsky and Hirsch 1931). Chlorophyll fluorescence induction represents a plot of measure fluorescence intensity as a function of time of continuous illumination. Such a curve measured under continuous light has a fast (less than 1 s) exponential phase and a slow decay phase (duration of few minutes). The initial growing phase shows a typical polyphasic shape, well evident when the curve is plotted on the logarithmic time scale (Fig. 4.5).

The analysis of the OJIP curve taking the theoretical assumptions and probabilities derives different photosynthetic parameters for the dark-adapted state of the photosynthetic systems (Strasser et al. 2000, 2004; reviewed in Stirbet and Govindjee 2011). The nomenclature for 'OJIP' is as follows: O is for origin or $F_O = F_0$ level measured at 50 μs (or less) after illumination; J and I represent intermediate states measured after 2 and 30 ms, respectively; and $P$ is the peak or $F_P = F_m$ (maximal fluorescence). This is valid only if sufficient light intensity is used. In heat-stressed samples, another peak arises between $F_0$ and $F_J$ at app. 300 μs, which is usually called K-step (Guisse et al. 1995; Srivastava et al. 1997; Strasser et al. 2000); therefore, some authors call the fast chlorophyll fluorescence induction the OKJIP curve or transient. The OJIP curve from $F_0$ to $F_m$ is correlated with the primary photochemical reactions of PS II (Duysens and Sweers 1963), and the fluorescence yield is controlled by a PS II acceptor quencher (the primary quinone acceptor, $Q_A$) (Van Gorkom 1986). Thus, the OJIP transient can be used for the estimation of the photochemical quantum yield of PS II photochemistry and the electron transport properties. The OJIP fluorescence curve analysis can be used to monitor the effect of various biotic and abiotic stresses and photosynthetic mutations affecting the structure and function of the photosynthetic apparatus (Strasser et al. 2004).

The photosynthetic samples kept in darkness have the electron acceptor side of PS II in the oxidised state, as there is no electron flow in the photosynthetic electron transport chain and water oxidation by PS II. So the PS II reaction centres remain open, and the fluorescence intensity is minimum, i.e. equal to $F_0$ (='O' level in OJIP curve).

Immediately after illumination with a strong intensity of light that can theoretically excite all

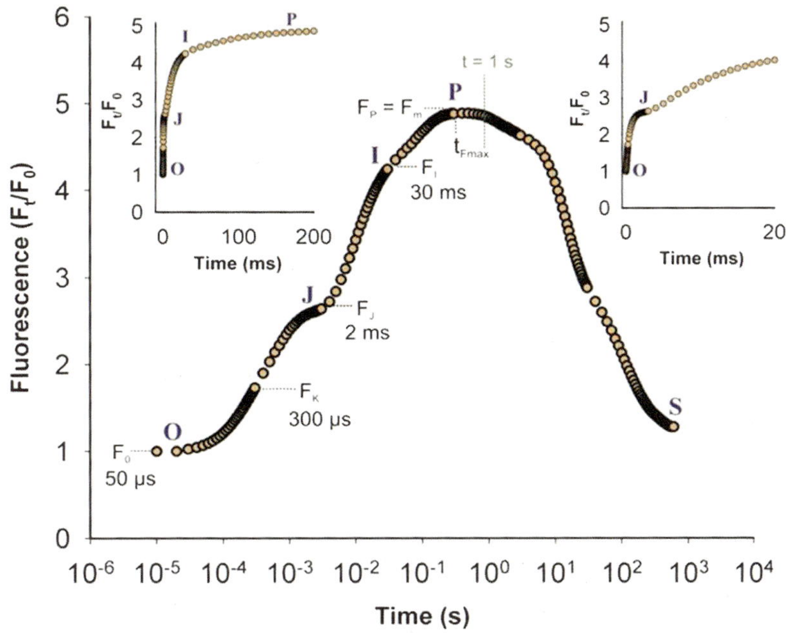


**Fig. 4.5** Typical OJIP transient of chlorophyll fluorescence (Kautsky curve) exhibited upon illumination of a dark-adapted leaf sample by saturating red light, plotted on logarithmic time scale (main graph) and on the regular time scale (minor graphs in upper part). Fluorescence values are expressed as $F_t/F_0$, where $F_t$ represents measured fluorescence intensity in each time interval and $F_0$ represents fluorescence intensity at 50 μs. The O–J–I–P–fluorescence rise is followed by decrease until the steady-state fluorescence level (S phase) is reached. The figure was created using data measured by the Handy PEA device (Hansatech, England) on wheat leaf and it was plotted according to scheme of Strasser et al. (2000)

the pigment molecules in the pigment protein complex of the thylakoid membrane, a fast electron transport process takes place and is recorded by an O–J transition or rise within 2 ms. This is followed by slow phases J–I and I–P (as evident on small upper plots in Fig. 4.5 plotted on regular time scale), which are known as thermal phases. The $F_m$ level ($F_P$) or $F_{tmax}$ is attained usually within interval 200 ms to 1 s, representing a closed PS II centre or complete reduction of all the primary electron acceptor in PS II, the $Q_A$ molecules, and saturating the electron flow on the acceptor side of PS II (Schansker et al. 2005).

## Parameters Derived from Fast Chlorophyll Fluorescence Kinetics

The chlorophyll fluorescence emitted by higher plants upon illumination carries a lot of information about the structure and function of the photosynthetic apparatus (Strasser et al. 2010). There are several groups of parameters derived from the fluorescence rise. The first group represents the data directly extracted from recorded fluorescence induction kinetics. If the PEA (plant efficiency analyser, Hansatech, GB) device is used, the data are recorded with maximum frequency 100 kHz; hence, the first point is $F_{10\mu s}$, but the reliable data starts from 30 μs. Another important data are time when the maximum fluorescence is reached ($t_{Fm}$) and the area parameter, which represents the area above the induction fluorescence curve between $F_0$ and $F_m$ and the $F_m$ asymptote; it expresses the size of the reduced plastoquinone pool (Malkin and Kok 1966).

The second group is represented by the basic parameters analogical to those used by saturation pulse method, as the $F_0$ and $F_m$ values are also determined by the direct method. Hence, the variable fluorescence $F_v$ can be calculated as well as maximum quantum yield of PS II photochemistry, $F_v/F_m$, and related parameters, like quantum yield of basal dissipation, $F_0/F_m$, and the ratio of maximum quantum yield of photochemistry and

**Table 4.2** Basic chlorophyll fluorescence parameters derived from OJIP test (Strasser et al. 1995, 2000, 2004, 2010; Stirbet and Govindjee 2011)

| Parameter | Name and basic physiological interpretation | Calculation |
|---|---|---|
| *Basic parameters derived from OJIP transient* | | |
| $F_t$ | Fluorescence intensity at time $t$ | |
| $t_{fmax}$ | Time of reaching maximum fluorescence | |
| $F_0$ | Minimum fluorescence, when all RC PS II are open (O-step of OJIP transient) | $F_0 = F_{50\mu s}$ |
| $F_m$ | Maximum fluorescence, when all RC PS II are closed (P-step of OJIP transient) | $F_m = F_P$ |
| $M_o$ | Initial slope of relative variable chlorophyll fluorescence; it express the rate of electron trapping | $Mo = 4 \cdot \frac{(F_{0.3ms} - F_0)}{(F_m - F_0)}$ |
| Area | Area above the OJIP curve; it express the size of the reduced plastoquinone pool | $\text{Area} = \int_{t=0}^{t=\text{tf max}} (F_m - F_t) \cdot t$ |
| $S_m$ | Normalised area; it is related to the number of electron carriers per electron transport chain | $Sm = \frac{\text{Area}}{F_m - F_0}$ |
| $V_J$ | Relative variable fluorescence at time 2 ms (J-step) after start of actinic light pulse | $V_J = \frac{(F_{2ms} - F_0)}{(F_m - F_0)}$ |
| $V_I$ | Relative variable fluorescence at time 30 ms (I-step) after start of actinic light pulse | $V_I = \frac{(F_{30ms} - F_0)}{(F_m - F_0)}$ |
| *Quantum yields* | | |
| $\varphi_{Po}$ | Maximum quantum yield of primary PSII photochemistry | $\varphi_{Po} = 1 - \frac{F_0}{F_m}$ |
| $\varphi_{Do}$ | Quantum yield of energy dissipation | $\varphi_{Do} = \frac{F_0}{F_m}$ |
| $\varphi_{ET2o}$ | Quantum yield of electron transport from $Q_A$ to $Q_B$ in PS II | $\varphi_{ET2o} = \varphi_{Po} \cdot (1 - V_J)$ |
| $\varphi_{RE1o}$ | Quantum yield of reduction of end electron acceptors at the PSI acceptor side | $\varphi_{RE1o} = \varphi_{Po} \cdot (1 - V_I)$ |
| $PI_{ABS}$ | Performance index for the photochemical activity (basic formula on absorption basis) | $PI_{ABS} = \frac{1-(F_0/F_m)}{(M_o/V_J)} \cdot \frac{(F_m - F_0)}{F_0} \cdot \frac{(1 - V_J)}{V_J}$ |
| $PI_{TOT}$ | Total performance index for the photochemical activity (including the flow beyond PS I) | $PI_{TOT} = PI_{ABS} \cdot \frac{(1-V_i)}{(1-V_j)}$ |
| $Df_{TOT}$ | Driving force of photochemical processes (based on $PI_{TOT}$) | $Df_{TOT} = \log(PI_{TOT})$ |

competitive non-photochemical processes in PS II in dark-adapted state, $F_v/F_0$ (Schreiber et al. 1986; Genty et al. 1989; Bilger and Bjorkman 1990).

Another group is represented by parameters derived from the JIP test, introduced by Strasser and Strasser (1995) and Strasser et al. (1995, 2000, 2004, 2010) and reviewed by Stirbet and Govindjee (2011). We can divide it into the fluorescence parameters derived from the data extracted from OJIP transient and the biophysical parameters calculated using previous group of fluorescence parameters (Strasser et al. 2010). Selected fluorescence and biophysical parameters are reviewed in Table 4.2.

Strasser et al. (2004) defined a group of biophysical parameters characterising photosynthetic sample, which can be used in screening physiological effects of different factors as well as the plant vitality. They can be divided into:

- Energy fluxes per absorbed photon flux = specific quantum yields ($TRo/ABS = \varphi_{Po}$; $DIo/ABS = \varphi_{Do}$; $ETo/ABS = \varphi_{ET2o}$; $REo/ABS = \varphi_{RE1o}$)
- Energy fluxes per reaction centre (*ABS/RC*; *TRo/RC*; *ETo/RC*, *DIo/RC*)
- Energy fluxes per excited cross section at fully open RC (*ABS/CSo*; *TRo/CSo*; *ETo/CSo*, *DIo/CSo*) or completely closed RC (*ABS/CSm*; *TRo/CSm*; *ETo/CSm*, *DIo/CSm*)
- Density of reaction centres (*RC/ABS*; *RC/CSo*; *RC/CSm*)
- Probabilities of electron transport ($\psi_{ET2}$; $\psi_{RE1}$, $\delta_{RE1}$)
- Performance indices and driving forces ($PI_{ABS}$, $PI_{TOT}$, $df_{TOT}$)

The listed parameters can serve for identifying the limitation within electron transport as well as

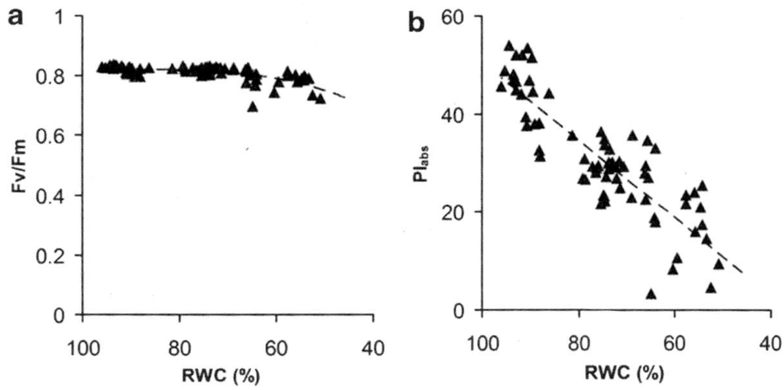


**Fig. 4.6** Correlation between values of performance index ($PI_{abs}$) or maximum quantum efficiency of PS II photochemistry ($F_v/F_m$) and relative water content (RWC) recorded in observed genotypes during drought period in natural climatic conditions. Lines show trend calculated for all observed genotypes (Data adapted from Zivcak et al. 2008a)

for comparison samples. Some examples of application are mentioned below.

As evident from previous paragraph and the table, there is high number of parameters derived from the fast fluorescence kinetics by the JIP test (only minor part of them was presented). In plant stress research, there are several possible ways of interpreting the data. A multiparametric approach is based on visualisation of data, e.g. by spider plots or pipeline models (described below). On the other hand, the model offers the integrative parameters enabling simple assessment of status and vitality of photosynthetic apparatus, which are sensitive and created mostly for possible practical applications in prescreening or selection in research and breeding programmes.

Plant vitality could be characterised by performance index $PI_{abs}$ (Strasser et al. 2000). This integrative parameter includes three independent parameters: (1) density of fully active reaction centres (RCs), (2) efficiency of electron movement by trapped exciton into the electron transport chain beyond the $Q_A$ and (3) the probability of that an absorbed photon will be trapped by RCs. $PI_{ABS}$ reflects the functionality of both photosystems II and I and gives us a quantitative information on the current state of plant performance under stress conditions (Strasser et al. 2004). Figure 4.6 demonstrates the effect of drought stress on values of performance index $PI_{ABS}$ in comparison with $F_v/F_m$.

Performance index $PI_{ABS}$ is found to be very sensitive parameter in different crops and for most of environmental stress situations (Strasser et al. 2000; Jiang et al. 2006a, b; Christen et al. 2007; Oukarroum et al. 2007; Zivcak et al. 2008a). Van Heerden et al. (2007) observed also a very good positive correlation between $CO_2$ assimilation capacity and $PI_{ABS}$ values under drought stress.

## Presentation of Rapid Fluorescence Kinetics Data

As the information brought by fast chlorophyll fluorescence kinetics is very complex, it is presented by several different ways. In this paragraph the most useful types of presentations are reviewed.

### Fluorescence Raw Curves and Normalised Plots

The fluorescence raw curve (fluorescent transient, OJIP curve) represents a plot of fluorescence data plotted on logarithmic time scale (Fig. 4.7 a, d).

The fluorescence raw curves contain all important information ($F_0$, $F_m$ values). As evident in Fig. 4.7, the severe drought stress (Fig. 4.7a) and severe heat stress (Fig. 4.7b) led to significant increase of $F_0$ and decrease of $F_m$. The difference in variable fluorescence is well visualised by the so-called $F_0$ normalisation (Fig. 4.7b, e), where the florescence data are divided by $F_0$ value ($F_t/F_0$) and hence $F_0$ equals 1. Anyway, besides the changes of $F_0$, $F_m$ and $F_v$, the shape of curves was dramatically changed in heat stress.

The typical polyphasic transient is very sensitive to high temperature, and the additional K-step was described within the O–J phase approximately 300 μs after light was switched on (Guisse et al. 1995); the K-step is clearly evident as an early peak in dark heated samples, becoming dominant at severe heat stress. As at moderate temperature, the K-step is hidden. The double normalisation (Fig. 4.7c, f), i.e. the plot of relative variable fluorescence values calculated for each point ($V_t$), enables to recognise proportional changes of fluorescence rise. The increase of variable fluorescence in 300 μs in heat-stressed samples is attributed to decrease of electron transport between the oxygen-evolving complex (OEC) and the RC PS II (Srivastava et al. 1995) as a result of loss of manganese cluster (Nash et al. 1985; Enami et al. 1994) associated with dissociation of a manganese-stabilising protein bound to the donor side of the RC PS II complex followed by the release of Mn atoms (Yamane et al. 1998). This was shown as the most heat susceptible site within the PS II photochemistry in observed wheat leaves (Brestic et al. 2012).

Another visualisation is also the plot of relative variable fluorescence in time 2 ms, $W_t$, presented later in this chapter (Fig. 4.17), showing the values of control samples as well as of drought-stressed samples running along the same line, whereas the values of heat-treated samples deviated upwards from the curve typical for control samples.

### Spider Plots

The relative values (relative to the corresponding value of the control, which thus become equal to unity) of selected expressions, such as the specific fluxes, can be plotted using a spider-plot presentation (Fig. 4.8). This is a multiparametric description of structure and function of each photosynthetic sample, presented by an octagonal line. This type of presentation provides a direct visualisation of the behaviour of a sample and thus facilitates the comparison of plant material as well as the classification of the effect of different environmental stressors on it in terms of the modifications it undergoes to adapt to new conditions (Strasser et al. 2004).

Figure 4.8 shows the typical spider plots of fluorescence parameters calculated from the data measured in drought-stressed and heat-stressed leaves of seven wheat genotypes. The spider plot enables to easily identify deviation (in positive or negative way) from a typical shape, as the spider plot represents some kind of 'fingerprint' of each stress or physiological status. It enables also to recognise the more sensitive or more resistant genotypes.

### Energy Pipeline Models

The derived parameters can also be visualised by means of an energy pipeline model of the photosynthetic apparatus (Strasser 1987; Strasser et al. 1996; Krüger et al. 1997). This is a dynamic model in which the value of each energy flux, either changing as a function of time or modified by the different imposed environmental conditions, is expressed by the appropriately adjusted width of the corresponding arrow. For each sample and/or state of the sample, two types of models can be presented; one refers to the reaction centre in the membrane and thus deals with the specific energy fluxes (per RC) and the other refers to the excited cross section of a leaf and thus deals with the phenomenological energy fluxes (per CS). The flux of dissipated excitation energy at time zero ($DI_0 = ABS - TR_0$) is also shown both per RC and per CS. The membrane model includes also a demonstration of the average 'antenna size', which follows the value of the ABS/RC. This value expresses the total absorption of PS II antenna chlorophylls divided by the number of active (in the sense of $Q_A$ reducing) reaction centres. Therefore, the antenna of inactivated reaction centres (non-$Q_A$ reducing, here due to the depletion of the oxygen-evolving side) is mathematically added to the antenna of the active reaction centres. In the leaf model the active reaction centres per cross section (RC/CS) are indicated by open circles, and those that have been inactivated are indicated by closed circles (Strasser et al. 2004). An example of leaf model for different levels of drought and

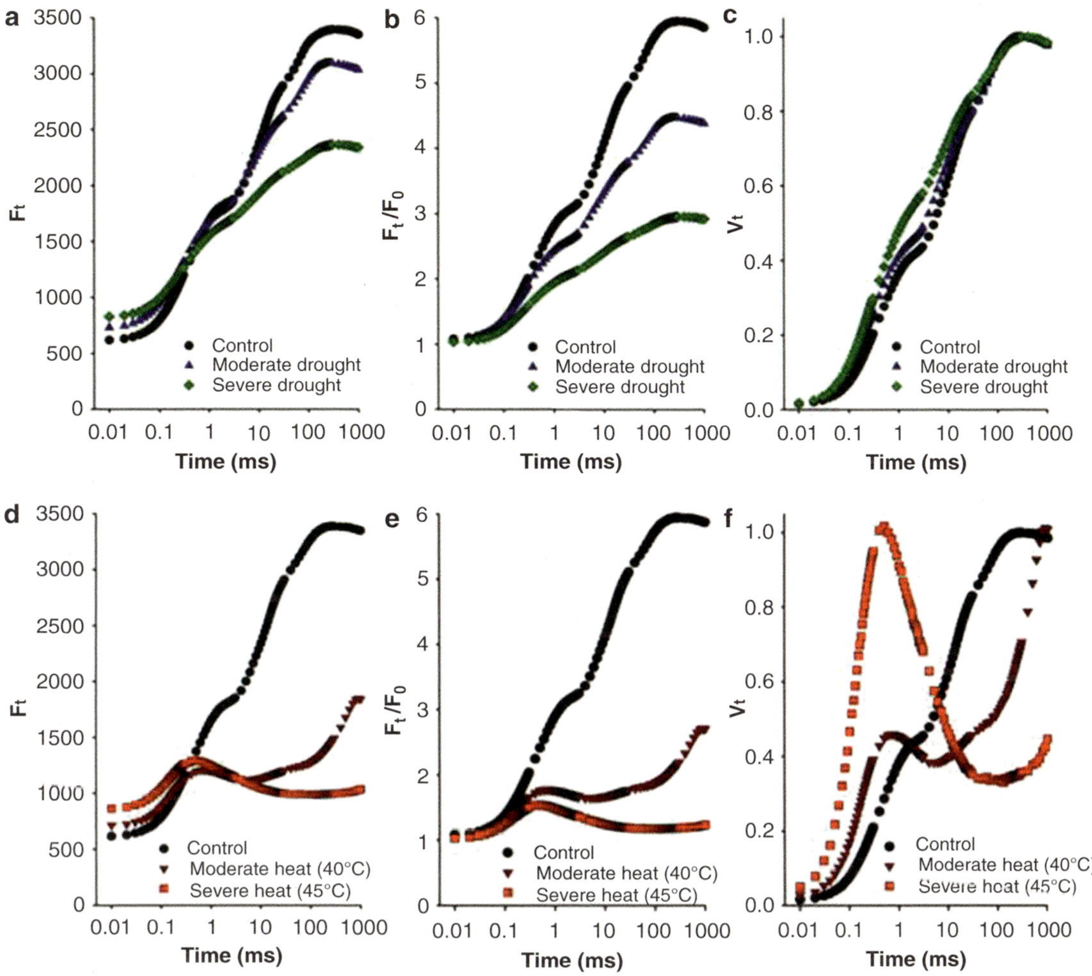


**Fig. 4.7** The OJIP curves presented as original measured data of chlorophyll fluorescence – $F_t$ (*left*), with normalised $F_0$ level (*middle*) plotted as $F_t/F_0$ and double normalised, i.e. shown in values of variable fluorescence $V_t$ (*right*). The *upper row* (**a**, **b**, **c**) shows the data measured during drought stress in well-hydrated plants (control), moderately dehydrated plants (about 75% RWC) and severely dehydrated plants (RWC below 60%). The *bottom row* (**d**, **e**, **f**) presents data measured after 1 h of exposure at high temperature in comparison with control plants. Measurements were realised on mature wheat leaves by fluorometer Handy PEA (Hansatech, GB). Data were published by Zivcak et al. 2008a, b, c

heat stress measured in wheat plants is shown as Fig. 4.9. The model well illustrated the fact that heat stress has much stronger effect on primary photochemical reactions than drought stress. While the drought stress moderately decreased absorption and electron transport rate as well as diminished partly the number of functional reaction centres (but 80% were still active even at very severe drought), heat stress acted mainly through the deactivation of reaction centres, and all components of electron transport chain (absorbance, trapping flux and electron transport) per excited leaf cross section were strongly impaired.

The model can serve also as a good educational tool for explanation of primary processes as well as for demonstrating results for people, who are not experts in the given area. The pipeline models are valuable tools for visualising the primary stages of photosynthesis. They give simple representation of the main energy fluctuations

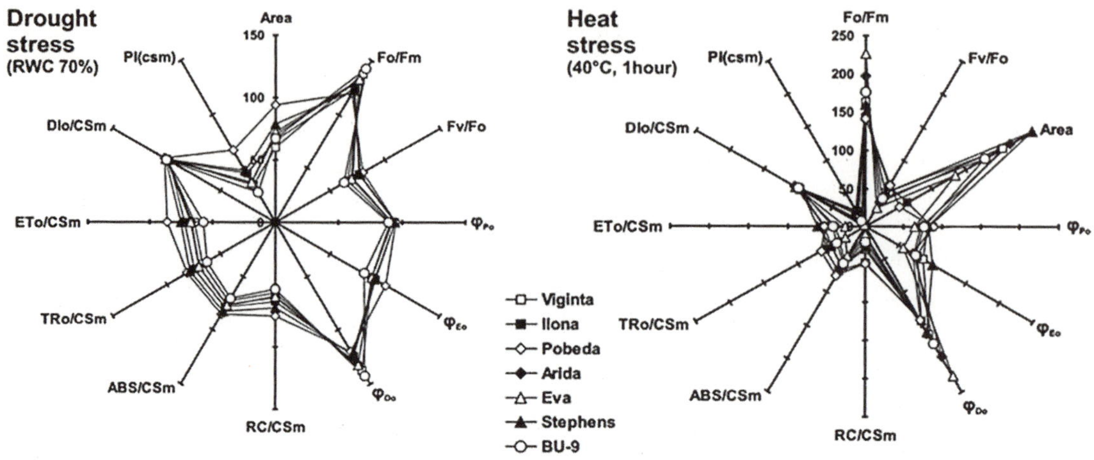


**Fig. 4.8** The 'spider plot' of selected JIP-test parameters derived from fast chlorophyll fluorescence kinetics. The average values (shown relative to the corresponding value of the control, which thus become equal to 100%) measured in different genotypes of winter wheat in conditions of water deficit with leaf relative water content (RWC) app. 70% (*left*) and after exposition at 40°C in dark for 1 h (*right*)

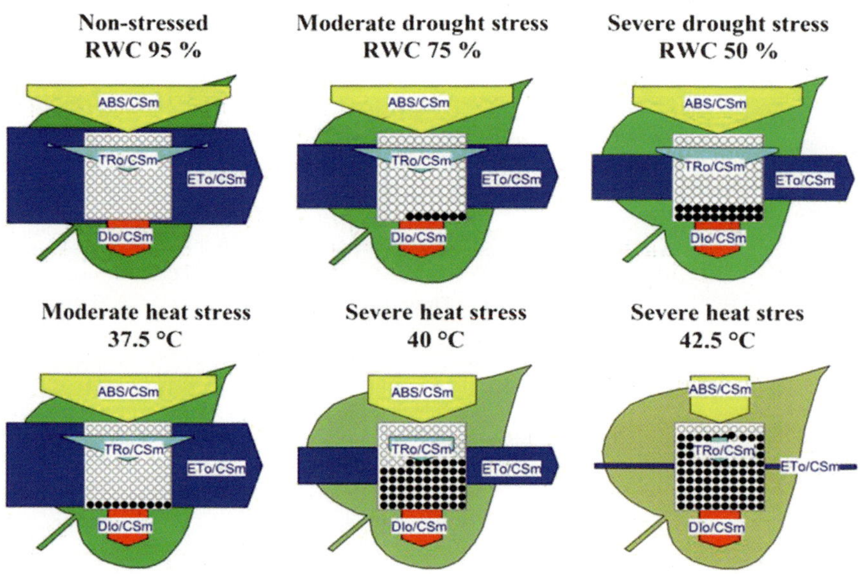


**Fig. 4.9** Phenomenological leaf models based on calculations of parameters per excited leaf cross section based on equations derived by Strasser et al. (2000) for control, drought-stressed and heat-stressed leaves. The thickness of each *arrow* represents the value of absorbance (ABS/CSm), trapping flux (TR/CSm), electron transport (ET/CSm) or heat dissipation of excess light (DI/CSm), all expressed per leaf cross section. The *black points* represent the fraction of inactive reaction centres. Measurements were done by fluorometer Handy PEA (Hansatech, England) and the models were generated using software Biolyzer 3.06 (Maldonado-Rodriguez, Laboratory of Bioenergetics, University of Geneva, Switzerland). Drought stress was induced by withholding of irrigation in plants cultivated in pots in natural climatic conditions. Heat stress was induced by exposure of leaf segments to high temperature for 1 h in dark (original data)

through the sample and it allows the visual comparison of assessed plants (Hermans et al. 2003). Another example (Fig. 4.10) compares the effect of drought and heat on photosynthetic apparatus in a set of wheat genotypes.

## Special Applications of Chlorophyll Fluorescence Measurements

### Chlorophyll Fluorescence Imaging

One of the most useful innovations of the chlorophyll fluorescence technique has been the development of chlorophyll fluorescence imaging, which involves advancements in the technology of light emission, imaging detectors and rapid data handling (Nedbal and Whitmarsh 2004; Gorbe and Calatayud 2012). Fluorescence imaging devices have been constructed for their use either at microscopic level (Oxborough and Baker 1997; Rolfe and Scholes 1995) or at plant, leaf and organ level (Omasa et al. 1987; Calatayud et al. 2006) or for remote sensing of chlorophyll fluorescence (Zarco-Tejada et al. 2009; Saito et al. 1999; Calatayud et al. 2006; Gorbe and Calatayud 2012). Fluorescence imaging represents an efficient tool for observing the fluorescence emission pattern of subcellular levels, cells, tissues, leaves or other plant organs or whole plants, providing precise visual information about plant stress (Calatayud et al. 2006). The fluorescence imaging method enables the observation of the temporal and spatial heterogeneities of photosynthetic processes over the relatively large observed area; such heterogeneities occur as a result of internal and/or environmental factors (Nedbal and Whitmarsh 2004). They can be just barely (or not at all) detected through the conventional point measurements by non-imaging chlorophyll fluorescence (Ellenson and Amundson 1982; Oxborough and Baker 1997; Omasa and Takayama 2003).

There are also special applications of chlorophyll fluorescence imaging. The dynamics and the heterogeneity of stomatal responses in the leaf (stomatal patchiness) were observed (Mott et al. 1993; Omasa and Takayama 2003; West et al. 2005). For the purpose of presenting the records of measurements, false colour palettes, with different colours encoding different numerical values of parameters, are usually used to show such heterogeneity in sufficient pixel resolution in images. It means that the presented images represent some kind of a topologic map showing the values of measured parameters on the sample. Hence, imaging technique avoids the imperfection of point chlorophyll fluorescence measurements that are responsible for many errors (Ehlert and Hincha 2008; Gorbe and Calatayud 2012).

The chlorophyll fluorescence imaging can be routinely used for the analysis of light induction kinetics, enabling the measurement of photochemical and non-photochemical yields and quenchings involved in the fluorescence induction in the presence of actinic light or in the dark. Usually, photosynthetic active radiation is used for excitation of chlorophyll fluorescence (Lichtenthaler et al. 2005a, b). Three types of light are used, i.e. weak pulse-modulated measuring light, continuous non-saturating actinic light and saturating light flashes, from which the five key fluorescence intensity levels are obtained, as with non-imaging PAM Chl fluorescence: $F_0$, $F_0'$, $F_m$, $F_m'$ and $Fs'$. Technical solution of many devices does not enable direct measurement of $F_0'$ by far-red light pulse; therefore, the calculated value of $F_0'$ (Oxborough and Baker 1997) is used for calculations where $F_0'$ is needed. Using these fluorescence values, other parameters can be calculated such as the effective quantum yield of energy conversion in PS II ($\Phi_{PSII}$), the non-photochemical quenching ($NPQ$), and the photochemical quenching ($qP$, $qL$).

Figure 4.11 shows the fluorescence images of well-hydrated leaves as well as leaves under moderate (RWC 75%) and severe (RWC 55%) drought stress. The figure in the left shows the map of values $F_v/F_m$ and the figure in the right represents values of effective quantum yield ($\Phi_{PSII}$).

The recently available fluorescence imaging devices represent the full operating fluorometers, which can be programmed for any protocol, like light response curve or induction curve. Figure 4.12 shows the results of measurement of dark relaxation kinetics followed after exposure

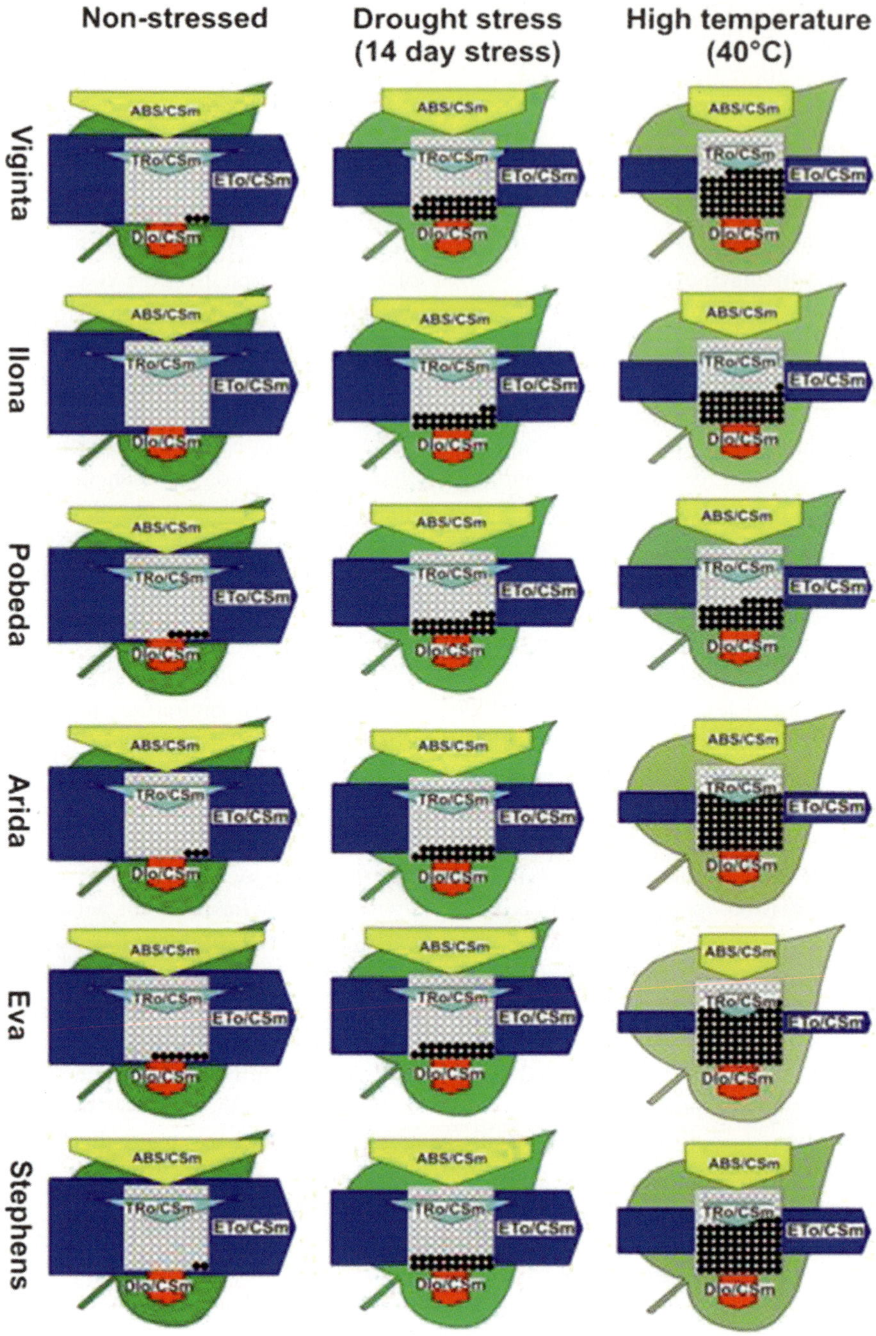


**Fig. 4.10** The use of phenomenological leaf model for comparison of differences among wheat genotypes in drought and heat-stress responses. Phenomenological leaf models are based on calculations of parameters per excited leaf cross section based on equations derived by Strasser et al. (2000) for control, drought-stressed and heat-stressed leaves. The thickness of each *arrow* represents the value of absorbance (ABS/CSm), trapping flux (TR/CSm), electron transport (ET/CSm) or heat dissipation of excess light (DI/CSm); all expressed per leaf cross section. The *black points* represent the fraction of inactive reaction centres. Measurements were done by fluorometer Handy PEA (Hansatech, England) and the models were generated using software Biolyzer 3.06 (Maldonado-Rodriguez, Laboratory of Bioenergetics, University of Geneva, Switzerland). Drought stress was induced by withholding of irrigation in plants cultivated in pots in natural climatic conditions. Heat stress was induced by exposure of leaf segments to high temperature for 1 h in dark (Zivcak 2006)

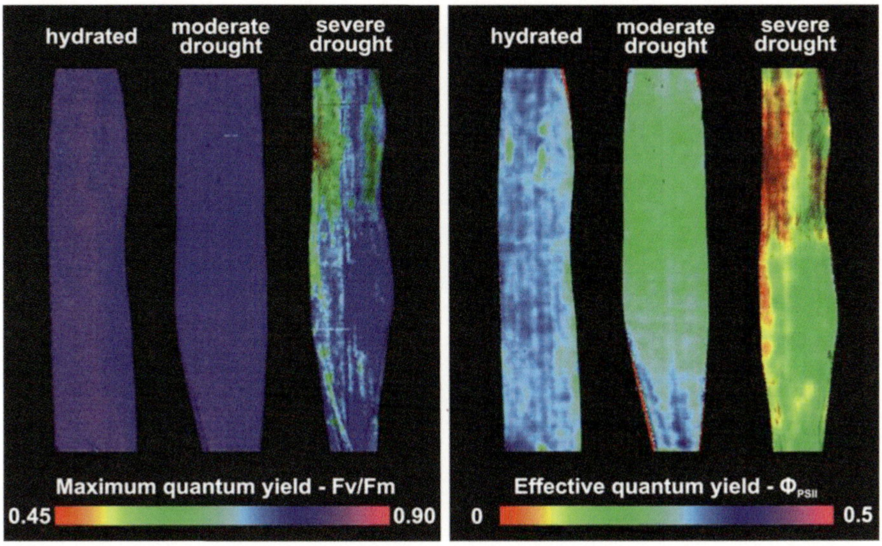


**Fig. 4.11** The chlorophyll fluorescence imaging screens of well-hydrated, moderately and severely drought-stressed wheat leaves. The figure in the *left* shows a two-dimensional distribution of recorded values of maximum quantum yield of PS II photochemistry ($F_v/F_m$); the figure in the *right* shows the effective quantum yield of PS II photochemistry ($\Phi_{PSII}$) measured at actinic light intensity 280 μmol photons $m^{-2}$ $s^{-1}$) within rapid light curve. Measurements were performed by Maxi-Imaging-PAM (Walz, Germany)

at high light intensity in barley as an example of application on the plant material differing in photosynthetic responses.

Several research papers have shown that the fluorescence imaging technique is sufficiently sensitive to record drought stress effects (Gorbe and Calatayud 2012). Water deficit resulted into heterogeneous distribution of chlorophyll fluorescence on the leaf surface. Similar to the presented measurement, Lang et al. (1996) reported a linear increase of the fluorescence ratios F440/F690 and F440/F740 as the response to decrease of the leaf water content in tobacco.

Lichtenthaler and Babani (2000) and Lichtenthaler et al. (2005a, b) measured gradients in photosynthetic capacity over different parts of bean leaves by chlorophyll fluorescence induction, showing values of $R_{fd}$ and $F_m/F_s$ ratio sensitive to decrease of water content in leaves. Calatayud et al. (2006) observed in rose plants exposed to decrease of leaf water content the spatial and temporal changes expressed by PS II fluorescence parameters, shown as the fluorescence images. The temporal changes were represented by non-photochemical processes, which increased within initial phase of drought stress; it was followed by the decrease of *NPQ* at severe water deficit. The spatial differences were represented by different distribution of parameters across the leaf. While $F_0$ and $F_m$ were higher next to the midrib throughout the entire drought period (9 days), $\Phi_{PSII}$ and *NPQ* showed spatial differences only in the middle of the drought period, when non-photochemical processes were stimulated at the expense of the remaining processes. Moreover, almost homogenous $F_v/F_m$, *qP* and *qL* were found over the whole leaf. Similarly, Massacci et al. (2008) also measured a quite homogenous distribution of $F_v/F_m$ but spatial variations of *NPQ* in cotton leaves under water stress.

One of probable future applications of chlorophyll fluorescence imaging is in plant breeding. Harbinson et al. 2012 stated that the objective is the high-throughput screening of genotypes tolerant to abiotic and biotic stress factors. Recently, the assessment of disease resistance or stress tolerance in breeding programmes relies mostly on visual scoring by experts; this can generate bias between different experts and

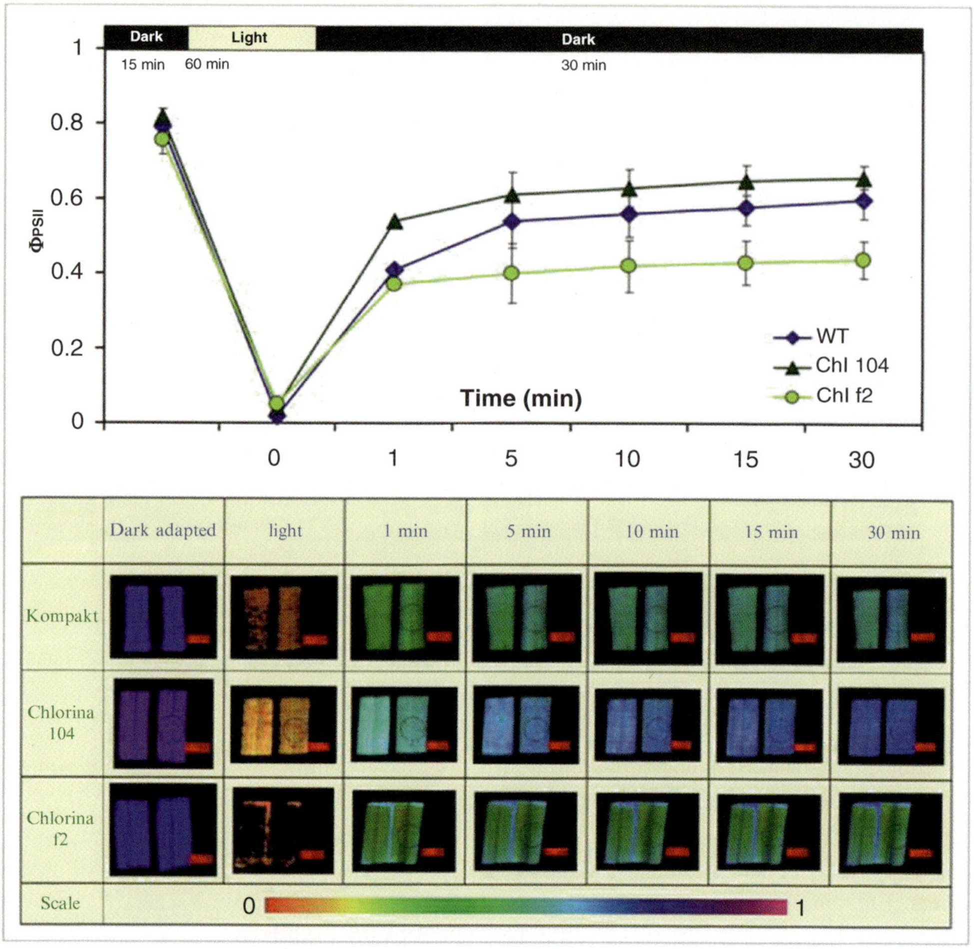


**Fig. 4.12** *Upper*: the light to dark relaxation kinetics of PS II quantum yield ($\Phi_{PSII}$) measured on well-hydrated leaf segments of wild-type barley (WT) and two antenna mutants of barley (Chlorina f2, Chlorina 104), with different susceptibility to photoinhibition. The points plotted on graph represent the average value of round areas of interest as shown in the figure below. *Bottom*: the chart of chlorophyll fluorescence imaging screens of leaf segments in dark-adapted state during strong illumination for 60 min (PAR intensity 1,250 μmol photons $m^{-2}$ $s^{-1}$) and during dark recovery. The imaging screens show a two-dimensional distribution of recorded values of the quantum yield of PS II photochemistry ($\Phi_{PSII}$). Measurements were performed by Maxi-Imaging-PAM (Walz, Germany). Data were published by Brestic et al. (2008)

experimental repeats; moreover, it is time-consuming. Therefore, high-throughput phenotyping tools are necessary to reduce the time invested and also to improve the objectivity of this procedure. In this regard, an important advantage of chlorophyll fluorescence imaging is that it can be used to screen a large number of small plants simultaneously. Moreover, fluorescence imaging can be integrated in robots for automatic measurements (Baker and Rosenqvist 2004; Chaerle et al. 2007; Gorbe and Calatayud 2012).

## Chlorophyll Fluorescence Measured Simultaneously with Photosynthetic Gas Exchange

Despite the fact that fluorescence emanates from only the top few layers of chlorenchyma,

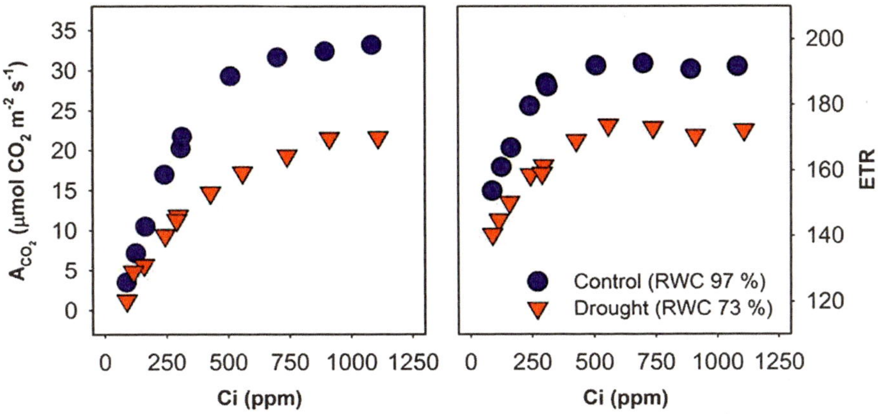


**Fig. 4.13** (a) An example of data obtained by simultaneous measurements of gas exchange and chlorophyll fluorescence. The measurements of A/Ci curves were done on leaves of well-hydrated (control) wheat plants and those under moderate drought stress (drought) as indicated by the values of relative water content (RWC). The first graph (*left*) shows the relationship between the content of $CO_2$ inside leaf (internal $CO_2$ content, Ci) and $CO_2$ uptake by leaf ($CO_2$ assimilation, $A_{CO2}$). The second graph (*right*) shows the values of apparent electron transport rate (ETR, calculated from the PS II quantum yield measured simultaneously with $CO_2$ uptake by chlorophyll fluorescence) related to Ci. The measurements were done by gasometric system LI-6400 of LI-COR with modulated fluorescence chamber head, and data were used for estimation of leaf mesophyll conductance (Gabris 2012)

whereas gas exchange is integrated across the thickness of the leaf, simultaneous measurements have emerged as a powerful tool for investigating the relationship between light use efficiency, $CO_2$ fixation and photoinhibition. A relatively simple technique involves exploring the empirical relationship between electron transport and $CO_2$ fixation in vivo (Maxwell and Johnson 2000).

Simultaneous measurement of the responses of leaf gas exchange and modulated chlorophyll fluorescence to light and $CO_2$ concentration now provide a means to determine a wide range of key biochemical and biophysical limitations on photosynthesis in vivo. Leaf $CO_2$ uptake (*A*) versus intercellular $CO_2$ concentration (*Ci*) curves may now be routinely obtained from commercial gas exchange systems. Precision in determining of metabolic limitations in intact leaves is improved by the simultaneous measurement of electron transport via modulated chlorophyll fluorescence. The *A*/*Ci* response also provides a simple practical method for quantifying the limitation that stomata impose on $CO_2$ assimilation. Again combining gas exchange and fluorescence provides a means to determine mesophyll conductance (Long and Bernacchi 2003). The *A*/*Ci* and *ETR*/*Ci* curves from gas exchange measurements combined with chlorophyll fluorescence are shown in Fig. 4.13.

It is evident that although the curves are partly similar, the relationship between $CO_2$ assimilation and *ETR* is not linear. At a very low *Ci*, the $CO_2$ assimilation was almost stopped, while the value of *ETR* decreased by app. 30% only.

Drought stress led to decrease of assimilation and electron transport more in high *Ci*. I suggest that in this case the non-stomatal effects (the decrease of carboxylation efficiency) were more important than the effect of stomata.

Simultaneous measurements of chlorophyll fluorescence were used also at measurement of $O_2$ evolution; such an approach brought an important knowledge that $O_2$ evolution and electron transport rate measured on the same leaves at saturating light and $CO_2$ concentration always remained identical to that of control plants, indicating dominant stomatal limitation of photosynthesis at the water deficit (Brestic et al. 1995).

## Chlorophyll Fluorescence Measured Simultaneously with Redox State of Photosystem I

The devices enabling simultaneous measurement of chlorophyll fluorescence and the PS I transmittance are able to detect imbalanced rates of

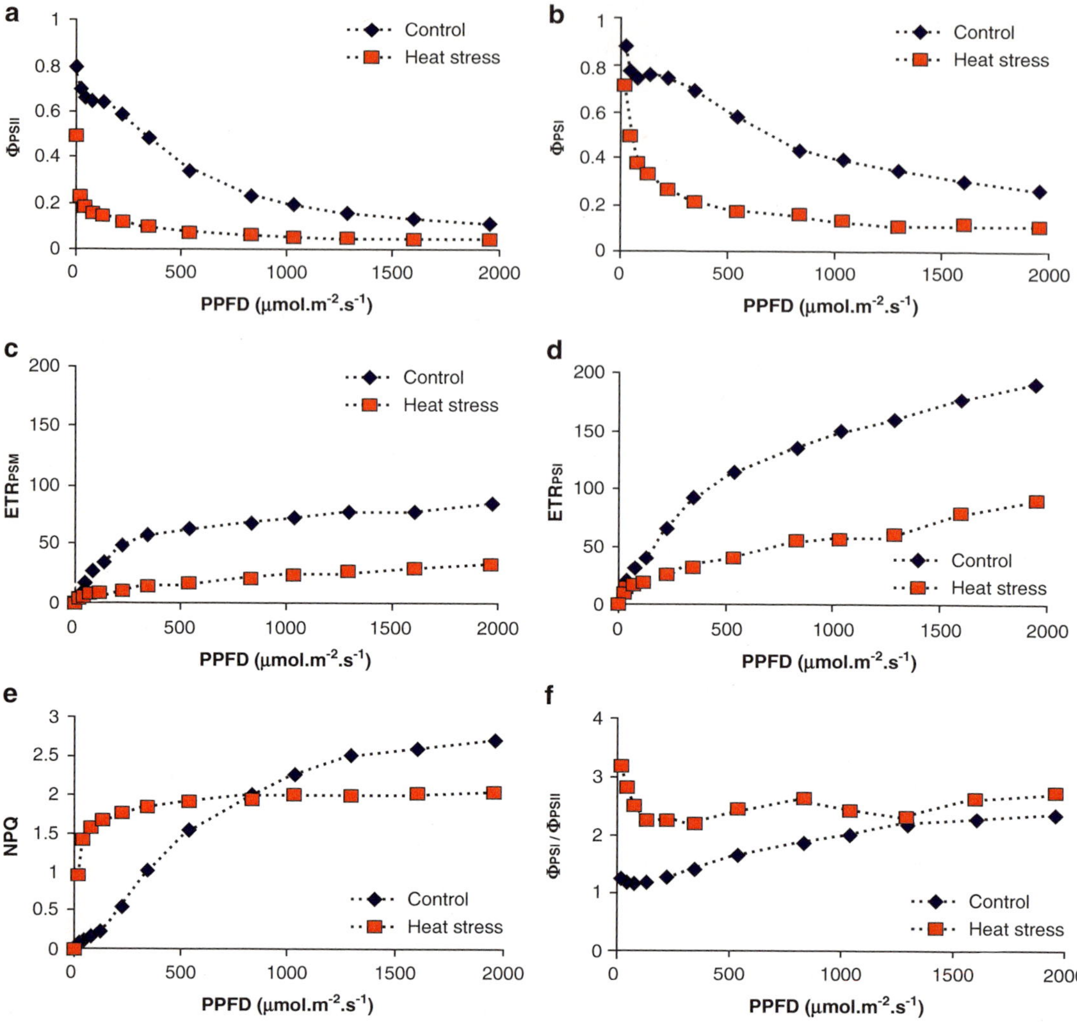


**Fig. 4.14** Photosynthetic parameters derived from light response curves in non-stressed and heat-treated leaves (42°C for 1 h in darkness) of barley leaves, recorded as a simultaneous measurement of chlorophyll fluorescence and PS I transmittance (redox state of photosystem I) by DUAL-PAM device (Walz, Germany): (**a**) effective quantum yield of PS II, (**b**) quantum yield of PS I, (**c**) apparent electron transport rate at PS II, (**d**) apparent electron transport rate at PSI, (**e**) non-photochemical quenching (NPQ) of PS II and (**f**) the PSI/PS II quantum yield ratio (Datko et al. 2008)

photochemistry of PS I and PS II with high accuracy by simultaneously recording quantum yields of both photosystems.

In general, the method is applicable to and capable of the characterisation of all types of factors affecting the interplay between photosystems, e.g. different types of stresses (Pfündel et al. 2008). Figure 4.14 presents the rapid light curves measured by this combined technique. It indicates that PSII is much more sensitive to heat than PSI. Hence, the cyclic electron transport is enhanced, as indicated by high $\Phi_{PSI}$ to $\Phi_{PSII}$ ratio.

## Application of Chlorophyll Fluorescence in Drought Stress

The effect of drought stress on chlorophyll fluorescence parameters depends on the degree of water deficit. It is generally accepted that mild

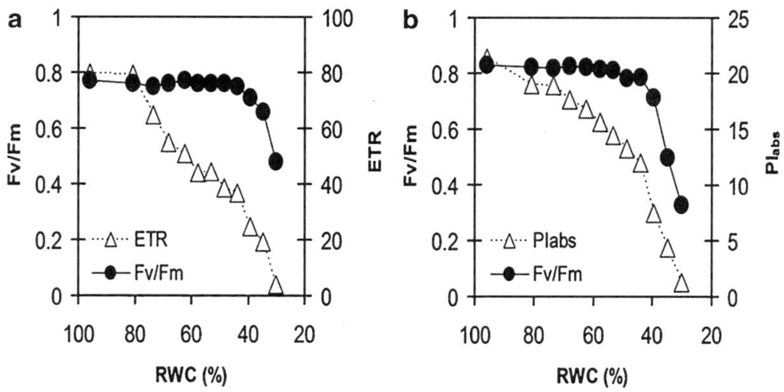


**Fig. 4.15** The comparison of values of commonly used parameters derived from measurements of either slow or rapid fluorescence kinetics related to relative water content in leaves. The rapid dehydration of wheat flag leaves was realised in laboratory conditions in low light within several hours after leaf detachment. Figure in the *left*: the effect of decrease of leaf relative water content (RWC) on values of maximum quantum yield of PS II photochemistry ($F_v/F_m$) and the electron transport rate (ETR) calculated using the values of effective quantum yield of PS II and the PAR intensity, both derived from measurements of slow fluorescence kinetics with analysis of saturation pulses at light intensity 1,000 μmol photons $m^{-2}$ $s^{-1}$ after 30 min of previous illumination. It is measured by the fluorometer DUAL-PAM (Walz, Germany). Figure in the *right*: calculated values of parameter performance index ($PI_{abs}$) during rapid dehydration. The measurements of fluorescence induction were performed by fluorometer Handy PEA (Hansatech, England)

to moderate drought stress decreases the photosynthetic rate mainly due to stomata closure, while the metabolic processes remain almost unaffected (Cornic and Massacci 1996). The critical leaf relative water content is app. 70%; below this value, the non-stomatal limitation of photosynthesis increases its importance. This phenomenon is reflected also into the measured values of chlorophyll fluorescence and calculated fluorescence parameters.

One of the most frequently used fluorescence parameters in plant physiological research, including drought stress research, is the maximum quantum yield of PS II photochemistry ($F_v/F_m$). It is mostly because this parameter is very easy to measure and it is generally a well-accepted measure of photosynthetic status. However, this parameter is highly insensitive to stomatal effects or to any other effects occurring in moderate drought stress; this is well documented in Fig. 4.15, recorded during quick dehydration of wheat leaf in darkness. It is obvious that $F_v/F_m$ values are extremely stable and they start to decrease at the level that can be entitled as the lethal level. If the drought stress runs in field conditions, the water deficit acts to the leaf much longer, and hence, the decrease of $F_v/F_m$ starts at higher values of relative water content; such a case is documented in Fig. 4.6. Even here, it is obvious that the $F_v/F_m$ values are kept high and start its decline below 70% of relative water content in leaf; anyway we did not observe the strong decrease even at 50% of relative water content. As reported in many other studies, we can conclude that any decrease of $F_v/F_m$ cannot be attributed to drought stress at the physiologically relevant level; however, the $F_v/F_m$ measurements during drought stress make sense, as they can draw attention to the effects of co-occurring stresses (heat stress, photoinhibition, etc.) or to the early phases of leaf senescence.

Anyway, both measurements of the slow and the fast chlorophyll fluorescence kinetics were shown to be sensitive to drought stress (Fracheboud and Leipner 2003; Oukarroum et al. 2007, 2009; Zivcak et al. 2008a, b, c); this is evident even from previously presented figures of slow kinetics within light curve (Fig. 4.3) or slow induction curve and recovery (Fig. 4.2), as well as from OJIP transients (Fig. 4.7). The decrease of effective PS II quantum yield ($\Phi_{PSII}$) and *ETR* in drought-stressed leaves compared to well-hydrated leaves is mainly due to lack of $CO_2$ inside the leaf (closed stomata). In C3 leaves, the decrease of $\Phi_{PSII}$ is not linearly correlated with net assimilation rate, as the increased photorespiration efficiently consumes part of electrons flowing within the linear electron transport chain; this is evident even from simultaneous measurements of chlorophyll fluorescence with $CO_2$ assimilation within A/Ci curve measurement (Fig. 4.13). This indicates that measurement of slow fluorescence kinetics and calculation of quantum yields and electron transport rate are useful for determination of the drought stress effects, reflecting both stomatal and non-stomatal effects; however, such measurements during drought stress cannot be directly related to $CO_2$ assimilation (Baker 2008). Anyway, Lichtenthaler et al. (2005a, b) reported the relative fluorescence decrease ratio ($R_{fd}$) being more sensitive and better correlating with photosynthetic assimilation compared to PS II quantum yield or *ETR*.

The gradual decrease of parameter performance index ($PI_{ABS}$), derived from fast chlorophyll fluorescence kinetics and the JIP test (Strasser et al. 2000) during dehydration (Fig. 4.15), indicates the drought-induced changes at the PS II electron acceptor side. Such a decrease was observed even in natural conditions during slowly advancing drought stress (Zivcak et al. 2008a, b, c). The decrease of performance index is associated with changes of chlorophyll fluorescence transients (Fig. 4.7). The mentioned results, as well as the high number of other scientific papers, describe the possibilities of chlorophyll fluorescence method in study of drought stress. The selected papers aimed at screening of crops using chlorophyll fluorescence are in Table 4.3.

Measurement of chlorophyll fluorescence has been used as a mean to evaluate the integrity of photosystem II (PS II) upon exposure to deficit (Shabala 2002). Chlorophyll fluorescence provides useful information about leaf photosynthetic performance of many plants under water deficit (Baker and Rosenqvist 2004). The in vivo effects of water deficit on chlorophyll fluorescence have been described for several crop species (Razavi et al. 2008). Interspecific differences for chlorophyll fluorescence parameters have been documented, e.g. in groundnut (Danièle et al. 2006)

## Application of Chlorophyll Fluorescence in Heat Stress

Photosynthesis is a process very sensitive to high temperature that can be partially or even completely inhibited by heat before other symptoms of the stress are detected. High temperature affects the photosynthetic functions of plants by its effect on the rate of chemical reactions and on structural organisation (Pastenes and Horton 1996).

According to Wahid et al. (2007), threshold temperature is the value of daily mean temperature at which a detectable reduction in growth begins. Upper and lower developmental threshold temperatures have been determined for many plant species through controlled laboratory and field experiments. A lower developmental threshold or a base temperature is one below which plant growth and development stop. Similarly, an upper developmental threshold is the temperature above which growth and development cease. Knowledge of lower threshold temperatures is important in physiological research as well as for crop production. Determining a consistent upper threshold temperature is difficult because the plant behaviour may differ depending on other environmental conditions. Upper threshold temperatures for some major crop species are displayed in Table 4.4.

High-temperature sensitivity is particularly important in tropical and subtropical climates as heat stress may become a major limiting factor for field crop production. There exists tremendous variation within and between species, providing opportunities to improve crop heat-stress tolerance through genetic means. Some attempts to develop heat-tolerant genotypes via conventional plant breeding protocols have been successful.

The extreme temperatures induce heat stress in exposed plant tissues. Exposure of plants to heat stress typically induces inactivation of photosynthesis regardless of whether the stress is applied to whole plants, intact tissues or isolated organelles (Berry and Björkman 1980). The primary targets of high-temperature effects in plants are the photosystem II (PS II), carbon fixation by RuBisCO

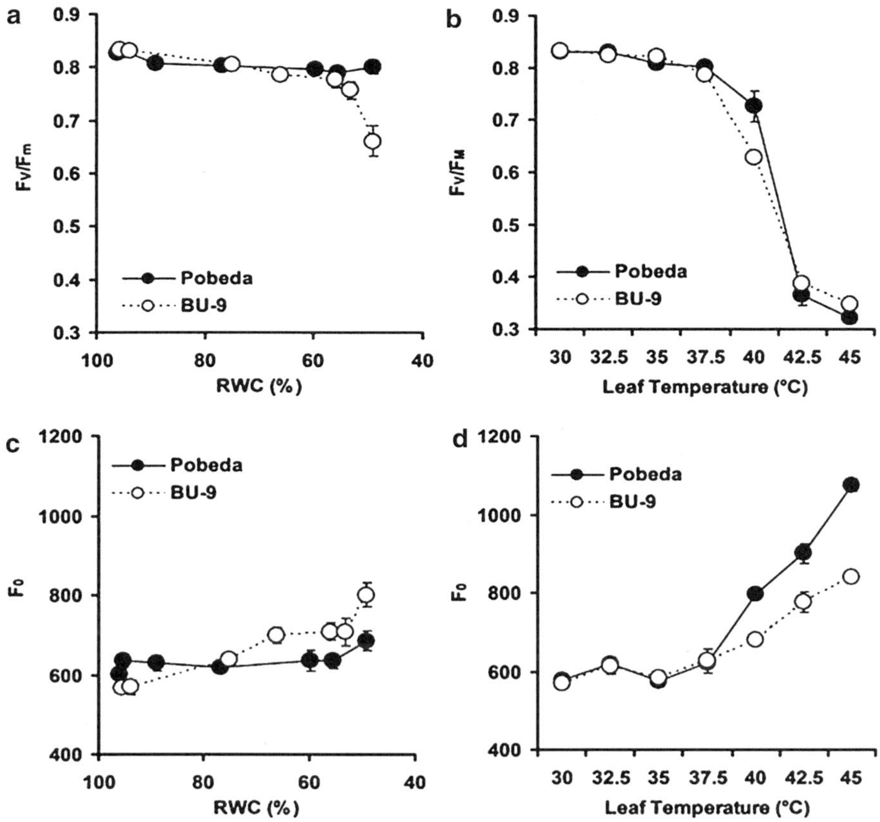


**Fig. 4.16** Relationship between values of maximum quantum efficiency of PS II photochemistry ($F_v/F_m$) and relative water content measured during dehydration. (**b**) Relationship between values of maximum quantum efficiency of PS II photochemistry ($F_v/F_m$) measured after 1 h exposition of leaf segments to high temperature and temperatures of corresponding expositions. (**c**) Relationship between values of initial fluorescence ($F_0$) and relative water content measured during dehydration. (**d**) Relationship between values of initial fluorescence ($F_0$) measured after 1 h exposition of leaf segments to high temperature and temperatures of corresponding expositions (Zivcak et al. 2009)

and the ATP-generating system (Salvucci and Crafts-Brandner 2004; Srivastava et al. 1997).

The assessment of heat-stress effects on photosynthetic apparatus, particularly on PS II status and linear electron transport rate, has been frequently performed through chlorophyll *a* fluorescence measurements (Krause and Weis 1991). The comparison of basic fluorescence parameters ($F_0$, $F_v/F_m$) measured in prolonged drought stress and at short (1 h) exposure at high temperature in two wheat genotypes is shown in Fig. 4.16.

The high-temperature effects on PS II photochemistry and heat tolerance at the PS II level have been frequently assessed even by the only $F_0$, $F_m$ and $F_v/F_m$ parameters. Basal and maximum fluorescence values are strongly dependent on sample, mainly on chlorophyll content. Hence, it causes high variability even in non-stressed samples and can become a source of mistake. The use of the $F_v/F_m$ ratio diminished this problem; however, it is based on the assumption that $F_0$ is measured in open reaction centres ($Q_A$ fully oxidised) and $F_m$ in closed reaction centres ($Q_A$ fully reduced). In heated samples, the $F_0$ can slightly grow up as the high temperature enhances the process of chlororespiration leading to partial reduction of $Q_A$ in the dark (Sazanov et al. 1998). At heat-stressed samples with highly impaired OEC, the electron transport

**Table 4.3** Chlorophyll fluorescence characteristics applied for acclimation potential study and screening of different crop genotypes under drought stress

| Crops | Parameters | References |
|---|---|---|
| Wheat | $F_v/F_m$ | Araus et al. (1998), Munns et al. (2010), Czyczyło-Mysza et al. (2011), Shefazadeh et al. (2012), Bogale et al. (2011), Sayar et al. (2008), Zhang et al. (2010) |
| | $F_v/F_m$, PI | Zivcak et al. (2008a), Zivcak et al. (2008c) |
| | $qP$, $qe$ | Flagella et al. (1994) |
| | $\Phi_{PSII}$, $NPQ$, $F_v/F_m$ | Wu and Bao (2011), Tari et al. (2008) |
| | Fo, $F_m$, $F_v/F_m$, $qP$, $NPQ$, $PI$ | Roostaei et al. (2011) |
| | Fo, $F_m$, $F_v/F_m$, $qP$. $qN$, $\Phi_{PSII}$ | Zlatev and Lidon (2012), Yang et al. (2006) |
| | $F_v/F_m$, $qP$. $qN$, $\Phi_{PSII}$, JIP test | Yang et al. (2007) |
| | $F_v/F_m$, $\Phi_{PSII}$ | Subrahmanyam et al. (2006), Yang et al. (2007) |
| | $F_t$, $F_m$, $ETR$, $\Phi_{PSII}$ | Slapakauskas and Ruzgas (2005) |
| Barley | $F_0$, $F_v$, $F_m$, $F_v/F_m$ | Kocheva et al. (2004), Mamnouie et al. (2006), Ashraf et al. (2007), Li et al. (2006) |
| | $F_v/F_m$ | Munns et al. (2010) |
| Maize | Fo, $F_v$, $F_v/F_m$ | Gholamin and Khayatnezhad (2011) |
| | $\Phi_{PSII}$, $ETR$ | O'Neill et al. (2006) |
| | $F_0$, $F_v$ | Smillie and Gibbons (1981) |
| | $qP$, $qN$, $ETR$, $F_v/F_m$ | De Carvalho et al. (2011) |
| | $F_v/F_m$, $PI$ | Lepedus et al. (2012), Zegada-Lizarazu et al. (2012) |
| Sesame | $F_v/F_m$, $PI$ | Boureima et al. (2012) |
| Amaranth | $F_v/F_m$, JIP | Slabbert and Krüger (2011) |
| Bean | $F_v/F_m$ | Brestic et al. (1995), Grzesiak et al. (1997) |
| | $R_{fd}$ | Lichtenthaler and Babani (2000) |
| | $F_v/F_m$, $\Phi_{PSII}$, $ETR$, $qP$, $qN$ | Santos et al. (2009), Terzi et al. (2010) |
| | $F_0$, $F_m$, $F_v/F_m$, $qN$ | Zlatev and Yordanov (2004) |
| Chickpea | $F_v/F_m$ | Rahbarian et al. (2011) |
| Soybean | $F_v/F_m$ | Rathod et al. (2011), De Ronde et al. (2004) |
| | $\Phi_{PSII}$ | Zhang et al. (2011b) |
| Canola | $F_v/F_m$ | Kauser et al. (2006) |
| Cotton | $\Phi_{PSII}$ | Zhang et al. (2011b) |
| Cassava | $F_v/F_m$ | Oyetunji et al. (2007) |
| Mango | $F_v/F_m$, $NPQ$ | Elsheery and Cao (2008) |
| Papaya | $F_v/F_m$ | Marler and Mickelbart (1998) |
| Strawberry | $qP$, $qN$, $F_v$/fm, $\Phi_{PSII}$ | Razavi et al. (2008) |
| Potato | Fo, $F_m$, $F_v/F_m$ | Van der Mescht et al. (1999) |
| | $F_m$/Fo | Schapendonk et al. (1989) |
| | $Fv'/Fm'$ | Jefferies (1994) |

leading to $Q_A$ reduction is low, and hence, it takes longer to reach the time of maximum fluorescence. In some cases, the time of saturation flash can be insufficient to reach $F_m$ value, and hence, the measured $F_m$ does not refer to fully reduced $Q_A$ state ($F_m$ underestimation) (Toth et al. 2007; Chen and Cheng 2009). Both the overestimation of $F_0$ and underestimation of $F_m$ lead to $F_v/F_m$ underestimation, showing the temperature effect to be more severe than it was.

**Table 4.4** Threshold temperatures reported for major field crops

| Crop | Threshold temperature (°C) | Sensitive phenophase (reference) |
|---|---|---|
| Wheat | 26 | Post-anthesis (Stone and Nicolas 1994) |
| Corn | 38 | Grain filling (Thompson 1988) |
| Cotton | 45 | Reproductive (Rehman et al. 2004) |
| Pearl millet | 30 | Emergence (Ashraf and Hafeez 2004) |
| Tomato | 29 | Flowering (Camejo et al. 2005) |
| Brassica | 25 | Flowering (Siddique et al. 1999) |
| Groundnut | 34 | Pollen production (Vara Prasad et al. 1999) |
| Cowpea | 41 | Flowering (Patel and Hall 1990) |
| Rice | 34 | Grain yield (Morita et al. 2004) |

Some studies have shown that the parameters derived from JIP test show a greater sensitivity to heat than the conventional fluorescence parameters such as $F_v/F_m$. It is caused by the fact that $F_v/F_m$ represents an average value of the efficiency for all the PS II units in the measured excited cross section but also the units with inactivated reaction centres.

General effects of heat stress on fast fluorescence kinetics are the decrease of maximum fluorescence $F_m$, increase of basal fluorescence $F_0$ (Fig. 4.16) and an appearance of new peak at approximately 300 μs (known as the K-step).

This is frequently followed by a dip in J–I phase and by the final fluorescence I–P increase (Fig. 4.7). The K-step is an important marker of decrease of oxygen-evolving complex capacity due to impairment of electron transport between OEC and RC PS II (Srivastava et al. 1997). As this effect is masked at moderate heat levels if the chlorophyll fluorescence transient in absolute units is plotted, mathematically double-normalised values of OJ phase are used in the analysis, i.e. the time plot of the values of relative variable fluorescence $W = V_t/V_J$ from 0.05 to 2 ms (Fig. 4.17a). If we focus at the level of this parameter at 300 μs, where the K-step can occur ($W_K = V_K/V_J$ value), very stable values of near 0.50 in nontreated samples are usually recorded. In heat-treated samples, the values increased gradually, eventually rising to several times higher values in comparison with non-stressed controls. This effect is caused only by high temperature, e.g. even sublethal level of drought stress does not induce the K-step (Fig. 4.17b).

The PS II thermostability, as a part of the overall leaf heat tolerance, has been determined frequently by the dependence of basal fluorescence and temperature ($F_0$-T curve), introduced by Schreiber and Berry (1977). The method is based on continuous increase of sample temperature and the permanent record of $F_0$ value. The critical temperature $T_C$ is the temperature at which the $F_0$ starts a steep increase. The measurements of critical temperature were performed in different species with the range of critical temperature from 42°C up to more than 50°C (Froux et al. 2004). Similarly, the method analogous to the continuous $F_0$ measurements can be used, which is the exposition of leaf samples at several temperature levels (graduated temperature approach) with the fresh sample used at each level; such measurements give more complex information about the heat effects on PS II photochemistry, but it also enables estimating the critical temperature for $F_0$ parameter (but even for any other measured parameter) similar to the previous technique. Figure 4.18a shows the slight increase of $F_0$ at moderate temperatures, which indicates that the primary PS II acceptor, $Q_A$, is not fully oxidised; it can be a result of chlororespiration (Sazanov et al. 1998) or by other processes causing partial re-reduction of fully oxidised $Q_A$ during dark heat treatment.

However, at temperature 42°C and more, we observed the start of exponential $F_0$ increase (discussed in Yamane et al. 1997); this observation

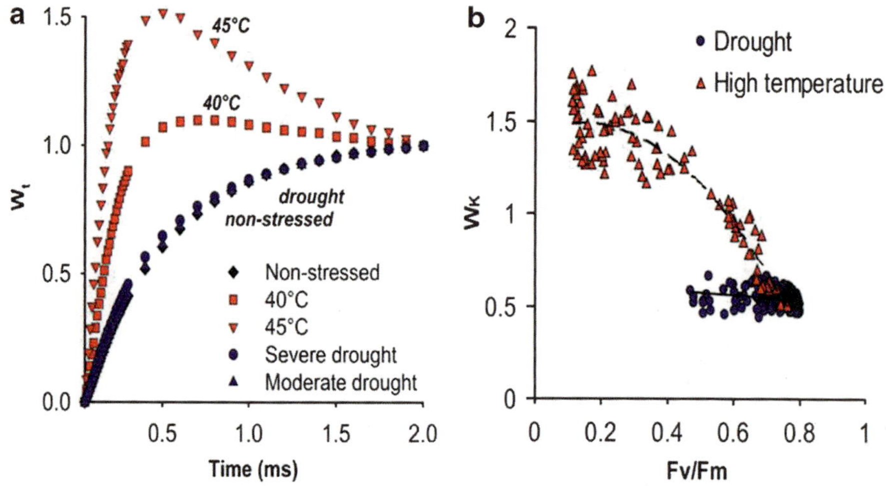


**Fig. 4.17** The K-step as a sole indicator of heat inhibition at the level of oxygen-evolving complex of photosystem II in wheat leaves. (**a**) The double-normalised initial O–J phase of OJIP fluorescence transient with typical peak (K-step) appeared in records of high temperature-treated leaves. (**b**) The relationship between maximum quantum yield of PS II photochemistry ($F_v/F_m$) and relative variable fluorescence in K-step ($W_K$) based on data observed at severe drought and heat stress

is consistent with critical temperatures published for annual species grown at moderate temperatures up to 25°C (Froux et al. 2004; Havaux et al. 1990). In moderate temperature-grown woody species as well as in tropical and desert species, the critical temperature ranged from 44 to 49°C (Dreyer et al. 2001; Robakowski et al. 2002; Froux et al. 2004).

The critical temperature can be stated also using any heat-sensitive parameter of fast fluorescence kinetics, if we have done measurements on graduated levels of high temperature (Fig. 4.18, right).

Critical temperature indeed represents the temperature at which the serious disorganisation of structure and loss of main functions occur. Hence, it is temperature mostly out of range of 'physiologically relevant' temperature, as in most cases, leaves in a field are not heated up to 50°C or more during their life. On the other hand, the occurrence of K-step even at temperature 4–5°C lower than a steep $F_0$ increase was found, and the $W_K$ values in nontreated plants were almost stable. Moreover, a higher capacity to increase thermostability on K-step level than on $F_0$ level was observed (Brestic et al. 2012).

The physiological meaning of the K-step increase is clear and it represents probably the first irreversible heat effect (and hence a key effect) inactivating the PS II reaction centres. For all these reasons the application of fast chlorophyll fluorescence measurements aimed at the K-step determination can be considered to be an efficient approach of testing PS II heat thermostability and plasticity in plants as shown also by other studies (Oukarroum et al. 2009). Moreover, it has with potential of practical use in testing, as this method is very reliable and expeditive. The record of fast chlorophyl fluorescence transient in dark-adapted samples represents an efficient tool for assessment of many external or intrinsic adverse effects on PS II photochemistry (Strasser et al. 2004).

Numerous research papers deal with application of chlorophyll fluorescence method in research of high-temperature stress on photosynthetic apparatus in different plant species. Part of the published reports on screening of crops for high-temperature tolerance using the chlorophyll fluorescence method is shown in Table 4.5. It is evident that, similarly to drought stress, the most frequently used approach is the measurement of

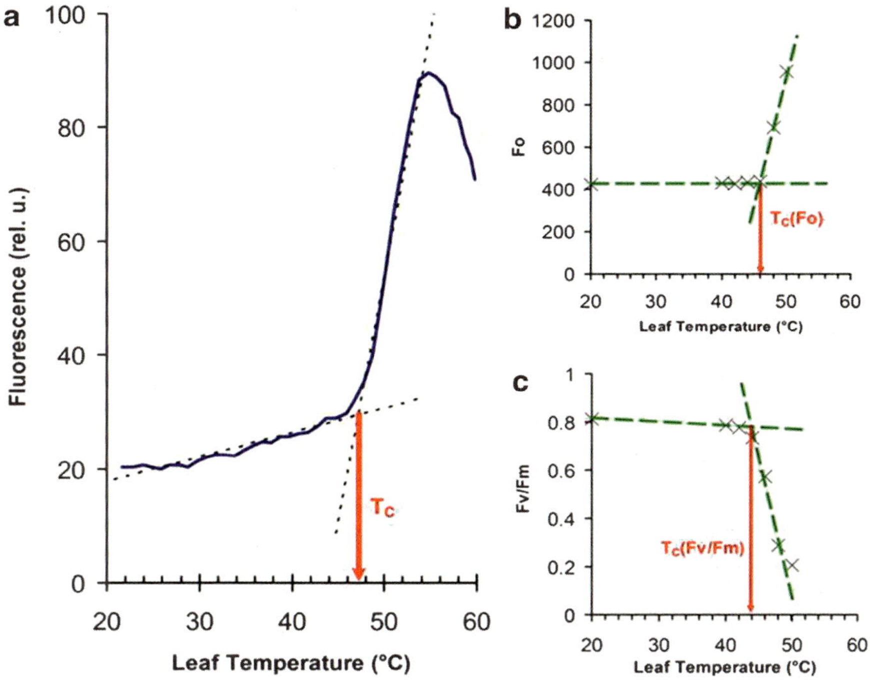


**Fig. 4.18** (**a**) The estimation of critical temperature ($T_C$) for PS II photochemistry using continuous measurement of basal fluorescence ($F_0$) on detached leaf segment during continuous heating with temperature increase rate app. 1°C per minute. The chlorophyll fluorescence measurement was realised by fluorometer PAM-2100 (Walz, Germany). The continuous heating of six samples in one time with monitoring temperature was realised using robotic system constructed at INRA Nancy (France). Measurements made on mature leaf of *Acacia*. (**b**) and (**c**) The critical temperature estimated according to measurements of rapid chlorophyll fluorescence kinetics by fluorometer Handy PEA (Hansatech, England) after heat treatment at graduated temperature levels (38–50°C by 2°C; the fresh sample for each level) for 30 min in the dark. The start of deep increase indicates the critical temperature; the value is determined graphically or calculated as an intersection of fitted lines. (**b**) The estimate of critical temperature for PS II photochemistry $T_C(F_0)$ based on recorded $F_0$ values at graduated temperature (analogical approach to those using continuous measurement of $F_0$ is shown above). (**c**) The estimate of critical temperature for function of oxygen-evolving complex in PS II reaction centre $T_C(F_v/F_m)$. The measurements in (**b**) and (**c**) were realised on leaf of barley; the data were published by Zivcak et al. (2013)

maximum quantum yield of PS II photochemistry ($F_v/F_m$) as well as application of parameters measured and calculated using saturation pulse method. However, in last the years, also approaches utilising the fast chlorophyll *a* fluorescence with the JIP-test analysis become more frequent.

In general, the last years brought promising data towards practical application of this technique in screening and breeding new genotypes characterised by higher heat and drought tolerance. The big challenges in this area come with new technical applications, like chlorophyll fluorescence imaging and its modifications, as well as simultaneous measurement of chlorophyll fluorescence (or even fluorescence imaging) with other physiological parameters, that can verify fluorescence measurements and bring more complex and useful information about the mechanisms contributing to plant stress tolerance.

## Conclusion

Chlorophyll fluorescence techniques represent unique and reliable non-invasive tools for screening effects of drought and high temperature in

**Table 4.5** Chlorophyll fluorescence characteristics applied for acclimation potential study and screening of different crop genotypes under heat stress

| Crops | Parameters | References |
|---|---|---|
| Cereals | $F_v/F_m$ | Sayed (2003) |
| C3, C4, CAM plants | $F_0$, $T_C$ | Weng and Lai (2005) |
| Wheat | $F_0$, $F_m$, $F_v/F_m$ | Xu et al. (2001), Efeoglu and Terzioglu (2009), Balouchi (2010) |
| | $F_v/F_m$, JIP-test parameters | Zivcak et al. (2008b), Sheikh et al. (2010), Brestic et al. (2010, 2012) |
| Barley | $F_v/F_m$, JIP-test parameters | Kalaji et al. (2011), Repkova et al. (2008) |
| Rice | $F_v/F_m$ | Han et al. (2009), Sikuku et al. (2010) |
| | $qP$, $qN$ | Zhang et al. (2011a) |
| Maize | $F_v/F_m$ | Crafts-Brandner and Salvucci (2002), Kebede et al. (2012) |
| | $F_v/F_m$, $qP$, $NPQ$, $\Phi_{PSII}$ | Sinsawat et al. (2004) |
| Sorghum | $F_v/F_m$, $qP$, $qe$ | Lu and Zhang (1998) |
| Soybean | $F_v/F_m$ | Srinivasan et al. (1996) |
| Bean | $F_0$, $F_m$, $F_v/F_m$ | Costa et al. (2002) |
| | $F_0$, $F_m$, $F_v/F_m$, $F_v$/Fo, JIP-test parameters | Petkova et al. (2007) |
| Mung bean | $\Delta F/F_m'$, $qP$, $ETR$, $NPQ$, $F_v/F_m$ | Karim et al. (2003) |
| Cabbage | $F_v/F_m$ | Nyarko et al. (2008) |
| Pineapple, peach and coconut | $F_v/F_m$ | Yamada et al. (1996) |
| Apple | $F_v/F_m$ | Brestic et al. (2011) |
| Grapevine | JIP test | Luo et al. (2011) |

plants. The method of fast fluorescence kinetics enables to quantify the stress effects within high number of samples in short time, thus fulfilling requirements of breeders for speed and sufficient sensitivity. On the other hand, the protocols of modulated chlorophyll fluorescence represent reliable and highly accepted tools, providing very complex information about photosynthetic processes and plant responses to environmental conditions. Moreover, applications of chlorophyll fluorescence measurements simultaneously with other photosynthetic parameters and the fluorescence imaging increase reliability of the results and provide new possibilities into plant research.

## Abbreviations

| | |
|---|---|
| A $A_{CO2}$ | Photosynthetic $CO_2$ assimilation |
| ABA | Abscisic acid |
| ABS | Absorbed photon flux |
| AL | Actinic light |
| $A_{leaf}$ | Absorbance of the light by leaf |
| area | Area above the OJIP curve |
| $C_i$ | Intercellular $CO_2$ concentration |
| $CS_m$ | Excited cross section (at $F_m$) |
| $CS_o$ | Excited cross section (at $F_0$) |
| D2 | Protein in reaction centre of PS II |
| $d_{fTOT}$ | Driving force (based on $PI_{TOT}$) |
| $DI_o$ | Dissipation from PS II (dark-adapted sample) |
| $ET_o$ | Electron transport beyond $Q_B$ (dark-adapted sample) |
| ETR | Electron transport rate |
| $F_0$ | Basal fluorescence |
| $F_d$ | Fluorescence decrease ($F_d = F_P - F_s$) |
| $F_m'$ $F_s'$, $F_0'$ | Maximum, steady state and minimum fluorescence on light |
| $F_P$ | Fluorescence maximum after actinic light is switched on |

| | |
|---|---|
| FR | Far-red light |
| FRR | Fast repetition rate |
| $F_s$ | Steady-state fluorescence |
| $F_{tmax}$ | Fluorescence value at time when $F_m$ reaches its maximum |
| $F_v/F_m$ | Maximum quantum yield of PS II photochemistry |
| $g_m$ | Mesophyll conductance |
| $g_s$ | Stomatal conductance |
| I | Irradiance |
| JIP JIP test | The mathematical model for calculating electron yields and fluxes, based on fast fluorescence kinetics |
| LED | Light-emitting diode |
| LHC2 | Light-harvesting complex of PS II |
| LHCP | Peripheral light-harvesting complex |
| ML | Measuring light |
| $M_o$ | Initial slope of relative variable chlorophyll fluorescence |
| *NPQ* | Non-photochemical quenching of maximum fluorescence |
| OEC | Oxygen-evolving complex |
| OJIP (OKJIP) | The fast chlorophyll fluorescence induction |
| PAM | Pulse-amplitude modulation |
| PDP | Pump during probe |
| PFD | Photon flux density |
| $PI_{ABS}$ | Performance index |
| $PI_{TOT}$ | Total performance index including the flow beyond PS I |
| PS I | Photosystem I |
| PS II | Photosystem II |
| $Q_A$ | Primary quinone electron acceptor in PS II |
| *qE* | The energy quenching |
| *qI* | The photoinhibitory quenching |
| *qL qP* | Photochemical quenching based on 'lake' and 'puddle' model, respectively |
| *qN* | Non-photochemical quenching of variable fluorescence |
| *qT* | The state transition quenching |

| | |
|---|---|
| $RE_o$ | Electron transport beyond PS I (dark-adapted sample) |
| $R_{fd}$ | Relative fluorescence decrease ratio |
| $R_n$ | Dark (night) respiration |
| RuBP | Ribulose bisphosphate |
| RWC | Relative water content |
| $S_m$ | Normalised area above OJIP curve |
| SP | Saturation pulse |
| TC | Critical temperature |
| $T_C(F_0)$ | Critical temperature based on $F_0$ increase |
| $T_C(F_v/F_m)$ | Critical temperature based on $F_v/F_m$ decrease |
| $TR_o$ | Trapping flux in PS II (dark-adapted sample) |
| $V_{cmax}$ | Maximum rate of carboxylation |
| $V_I$ | Relative variable fluorescence at time 30 ms (I-step) after start of actinic light pulse |
| $V_J$ | Relative variable fluorescence at time 2 ms (J-step) after start of actinic light pulse |
| $V_t$ | Relative variable fluorescence in time t |
| $W_K$ | Relative variable fluorescence at time 0.3 ms |
| WT | Wild type |
| $\delta_{RE1}$ | Probability of electron flow from $Q_B$ beyond the PS I |
| $\varphi_{Do}$ | Quantum yield of energy dissipation |
| $\varphi_{ET2o}$ | Quantum yield of electron transport |
| $\Phi_{NO}$ | Quantum yield of non-organised energy dissipation |
| $\Phi_{NPQ}$ | Quantum yield of energy-dependent non-photochemical dissipation |
| $\Phi_{Po}$ | Maximum quantum yield of PSII photochemistry ($F_v/F_m$); $\Phi_{PSII}$; $F_q'/F_m'$ |
| $\Delta F'/F_m'$ | Effective quantum yield of PS II photochemistry |

| | |
|---|---|
| $\varphi_{REIo}$ | Quantum yield of reduction of end electron acceptors at the PSI acceptor side |
| $\psi_{ET2}$ | Probability of electron flow from $Q_A$ beyond $Q_B$ |
| $\psi_{RE}$ | Probability of electron flow from $Q_A$ beyond the PS I. |

## References


Araus JL, Amaro T, Voltas J, Nakkoul H, Nachit MM (1998) Chlorophyll fluorescence as a selection criterion for grain yield in durum wheat under Mediterranean conditions. Field Crops Res 55:209–223

Araus JL, Slafer GA, Royo C, Serret MD (2008) Breeding for yield potential and stress. Adaptation in cereals. Crit Rev Plant Sci 27:377–412. doi:10.1080/07352680802467736

Ashraf M, Hafeez M (2004) Thermotolerance of pearl millet and maize at early growth stages: growth and nutrient relations. Biol Plant 48:81–86

Ashraf M, Nawazish S, Athar HUR (2007) Are chlorophyll fluorescence and capacity potential physiological determinants of drought tolerance inn maize (*Zea mays* L.). Pak J Bot 39:1123–1131

Baker NR (2008) Chlorophyll fluorescence: a probe of photosynthesis in vivo. Annu Rev Plant Biol 59:659–668

Baker NR, Rosenqvist E (2004) Applications of chlorophyll fluorescence can improve crop production strategies: an examination of future possibilities. J Exp Bot 403:1607–1621. doi:10.1093/jxb/erh196

Balouchi HR (2010) Screening wheat parents of mapping population for heat and drought tolerance, detection of wheat genetic variation. Int J Biol Life Sci 6:56–66

Berry JA, Björkman O (1980) Photosynthetic response and adaptation to temperature in higher plants. Annu Rev Plant Physiol 31:491–543

Bilger W, Bjorkman O (1990) Role of the xanthophyll cycle in photoprotection elucidated by measurements of light-induced absorbance changes, fluorescence and photosynthesis in leaves of *Hedera canariensis*. Photosynth Res 25:173–185

Bjorkman O, Demmig B (1987) Photon yield of $O_2$ evolution and chlorophyll fluorescence characteristics at 77 K among vascular plants of diverse origins. Planta 170:489–504

Blum A (1996) Crop responses to drought and the interpretation of adaptation. Plant Growth Regul 20:135–148

Bogale A, Tesfaye K, Geleto T (2011) Morphological and physiological attributes associated to drought tolerance of Ethiopian durum wheat genotypes under water deficit condition. J Biol Environ Sci 1:22–36

Boureima S, Oukarroum A, Diouf M, Cisse N, Van Damme P (2012) Screening for drought tolerance in mutant germplasm of sesame (*Sesamum indicum*) probing by chlorophyll a fluorescence. Environ Exp Bot 91:37–43. http://dx.doi.org/10.1016/j.envexpbot.2012.02.015

Bradbury M, Baker NR (1981) Analysis of the slow phases of the in vivo chlorophyll fluorescence induction curve. Changes in the redox state of photosystem II electron acceptors and fluorescence emission from photosystems I and II. Biochim Biophys Acta 635:542–555

Brestic M, Cornic G, Fryer MJ, Baker NR (1995) Does photorespiration protect the photosynthetic apparatus in French bean leaves from photoinhibition during drought stress? Planta 196:450–457

Brestic M, Zivcak M, Olsovska K, Repkova J (2008) Functional study of PS II and PSI energy use and dissipation mechanisms in barley wild type and chlorina mutants under high light conditions. In: Allen JF, Gantt E, Goldbeck JH, Osmond B (eds) Photosynthesis. Energy from the sun: 14th international congress on photosynthesis. Springer, Dordrecht

Brestic M, Olsovska K, Pivkova J (2010) Bioindication of thermotolerance of winter wheat (*Triticum aestivum*) photosynthetic apparatus. Acta fytotechnica et zootechnica 3:67–71

Brestic M, Shao HB, Ferus P, Malbeck J (2011) Peroxidases play important roles in abscisic acid (ABA)-simulating photosystem II (PSII) thermostability of apple tree rootstock leaves. Afr J Biotechnol 10:15891–15900. doi:10.5897/AJB11.1710

Brestic M, Zivcak M, Kalaji HM, Allakhverdiev SI, Carpentier R (2012) Photosystem II thermostability in situ: environmentally induced acclimation and genotype-specific reactions in *Triticum aestivum* L. Plant Physiol Biochem 57:93–105. doi:10.1016/j.plaphy.2012.05.012

Bukhov NG, Wiese C, Neimanis S, Heber U (1999) Heat sensitivity of chloroplasts and leaves: leakage of protons from thylakoids and reversible activation of cyclic electron transport. Photosynth Res 59:81–93

Bunce JA (1981) Comparative responses of leaf conductance to humidity in single attached leaves. J Exp Bot 32:629–634

Calatayud A, Roca D, Martínez PF (2006) Spatial-temporal variations in rose leaves under water stress conditions studied by chlorophyll fluorescence imaging. Plant Physiol Biochem 44:564–573

Camejo D, Rodriguez P, Morales MA, Dell'Amico JM, Torrecillas A, Alarcon JJ (2005) High temperature effects on photosynthetic activity of two tomato cultivars with different heat susceptibility. J Plant Physiol 162:281–289

Chaerle L, Leinonen I, Jones HG, Van Der Straeten D (2007) Display settings: monitoring and screening plant populations with combined thermal and chlorophyll fluorescence imaging. J Exp Bot 58:773–784. doi:10.1093/jxb/erl257

Chaves MM (1991) Effects of water deficits on carbon assimilation. J Exp Bot 42:1–16

Chaves MM, Pereira JS, Maroco J, Rodrigues ML, Ricardo CP, Osorio ML, Carvalho I, Faria T, Pinheiro C (2002) How plants cope with water stress in the field. Photosynthesis and growth. Ann Bot 89:907–916

Chen LS, Cheng L (2009) Photosystem 2 is more tolerant to high temperature in apple (*Malus domestica* Borkh) leaves than in fruit peel. Photosynthetica 47:112–120

Christen D, Schönmann S, Jermini M, Strasser RJ, Dèfago G (2007) Characterization and early detection of grapevine (*Vitis vinifera*) stress responses to esca disease by *in situ* chlorophyll fluorescence and comparison with drought stress. Environ Exp Bot 60:504–514

Cornic G, Briantais JM (1991) Partitioning of photosynthetic electron flow between $CO_2$ and $O_2$ reduction in a C3 leaf (*Phaseolus vulgaris* L.) at different $CO_2$ concentrations and during water stress. Planta 183:178–184

Cornic G, Fresneau C (2002) Photosynthetic carbon reduction and carbon oxidation cycles are the main electron sinks for photosystem 2 activity during a mild drought. Ann Bot 89:887–894

Cornic G, Massacci A (1996) Leaf photosynthesis under drought stress. In: Baker NR (ed) Photosynthesis and environment. Kluwer Academic, Dordrecht

Costa ES, Bressan-Smith R, Oliveira JG, Campostrini E, Pimentel C (2002) Photochemical efficiency in bean plants (*Phaseolus vulgaris* L. and *Vigna unguiculata* L. Walp) during recovery from high temperature stress. Braz J Plant Physiol 14:105–110

Crafts-Brandner SJ, Salvucci ME (2000) Rubisco activase constrains the photosynthetic potential of leaves at high temperature and $CO_2$. Proc Natl Acad Sci USA 97:13430–13435

Crafts-Brandner SJ, Salvucci ME (2002) Sensitivity of photosynthesis in a C4 plant, maize to heat stress. Plant Physiol 129:1773–1780. doi:http://dx.doi.org/10.1104/pp.002170

Czyczyło-Mysza I, Marcińska I, Skrzypek E, Chrupek M, Grzesiak S, Hura T, Stojałowski S, Myśków B, Milczarski P, Quarrie S (2011) Mapping QTLs for yield components and chlorophyll a fluorescence parameters in wheat under three levels of water availability. Plant Genet Res 9:291–295. doi:http://dx.doi.org/10.1017/S1479262111000207

Danièle C, Omar D, Jean L, Khalfaoui SB (2006) Genotypes variations in fluorescence parameters among closely related groundnut (*Arachis hypogaea* L.) lines and their potential for drought screening programs. Field Crops Res 96:296–306

Dash S, Mohanty N (2001) Evaluation of assays for the analysis of thermo-tolerance and recovery potentials of seedlings of wheat (*Triticum aestivum* L.) cultivars. J Plant Physiol 158:1153–1165

Datko M, Zivcak M, Brestic M (2008) Proteomic analysis of barley (Hordeum vulgare L.) leaves as affected by high temperature treatment. In: Allen JF, Gantt E, Goldbeck JH, Osmond B (eds) Photosynthesis. Energy from the sun: 14th international congress on photosynthesis. Springer, Dordrecht

Davies WJ, Zhang J (1991) Root signals and the regulation of growth and development of plant in drying soil. Annu Rev Plant Physiol Plant Mol Biol 42:55–76

De Carvalho RC, Cunha A, da Silva JM (2011) Photosynthesis by six Portuguese maize cultivars during drought stress and recovery. Acta Physiol Plant 33:359–374

De Las Rivas J, Barber J (1997) Structure and thermal stability of photosystem II reaction centers studied by infrared spectroscopy. Biochemistry 36:8897–8903

De Ronde JA, Cress WA, Kruger GHJ, Strasser RJ, Van Staden J (2004) Photosynthetic response of transgenic soybean plants, containing an *Arabidopsis* P5CR gen, during heat and drought stress. J Plant Physiol 161:1211–1224. doi:10.1016/j.jplph.2004.01.014

Dreyer E, Le Roux X, Montpied P, Daudet FA, Masson F (2001) Temperature response of leaf photosynthetic capacity in seedlings from seven temperate tree species. Tree Physiol 21:223–232

Duysens LMN, Sweers HE (1963) Mechanism of the two photochemical reactions in algae as studied by means of fluorescence. In: Japanese Society of Plant Physiologists (ed) Studies on microalgae and photosynthetic bacteria. University of Tokyo Press, Tokyo

Efeoglu B, Terzioglu S (2009) Photosynthetic responses of two wheat varieties to high temperature. EurAsia J BioSci 3:97–106. doi:10.5053/ejobios.2009.3.0.13

Ehlert B, Hincha DK (2008) Chlorophyll fluorescence imaging accurately quantifies freezing damage and cold acclimation responses in *Arabidopsis* leaves. Plant Methods 4:12. doi:10.1186/1746-4811-4-12

Ellenson JL, Amundson RG (1982) Delayed light imaging for the early detection of plant stress. Science 215:1104–1106

Elsheery NI, Cao CF (2008) Gas exchange, chlorophyll fluorescence, and osmotic adjustment in two mango cultivars under drought stress. Acta Physiol Plant 30:769–777

Enami I, Kitamura M, Tomo T, Isokawa Y, Ohta H, Katoh S (1994) Is the primary cause of thermal inactivation of oxygen evolution in spinach Ps-Ii membranes release of the extrinsic 33 kDa protein or of Mn? Biochim Biophys Acta, Bioenerg 1186:52–58

Epron D, Dreyer E (1990) Stomatal and non stomatal limitation of photosynthesis by leaf water deficits in three oak species: comparison of gas exchange and chlorophyll a fluorescence data. Ann Sci For 47:435–450

Falkowski PG, Wyman K, Ley AC, Mauzerall D (1986) Relationship of steady state photosynthesis to fluorescence in eucaryotic algae. Biochim Biophys Acta 849:183–192

Faver KL, Gerik TJ, Thaxton PM, El-Zik KM (1996) Late season water stress in cotton: II: leaf gas exchange and assimilation capacity. Crop Sci 36:922–928

Feller U, Crafts-Brandner SJ, Salvucci E (1998) Moderately high temperatures inhibit ribulose-1,5-bisphosphate carboxylase/oxygenase activase-mediated activation of Rubisco. Plant Physiol 116:539–546

Flagella Z, Pastore D, Campanile RG, Di Fonzo N (1994) Photochemical quenching of chlorophyll fluorescence and drought tolerance in different durum wheat (*Triticum durum*) cultivars. J Agric Sci 122:183–192

Flagella Z, Campanile RG, Stoppelli MC, De Caro A, Di Fonzo N (1998) Drought tolerance of photosynthetic electron transport under $CO_2$ enriched and normal air in cereal species. Physiol Plant 104:753–759

Fracheboud Y, Leipner J (2003) The application of chlorophyll fluorescence to study light, temperature, and drought stress. In: DeEll JR, Toivonen PMA (eds) Practical applications of chlorophyll fluorescence in plant biology. Kluwer Academic, Dordrecht

Franck F, Juneau P, Popovic R (2002) Resolution of Photosystem I and Photosystem II contributions to chlorophyll fluorescence of intact leaves at room temperature. Biochim Biophys Acta 162:239–246

Froux F, Ducrey M, Epron D, Dreyer E (2004) Seasonal variations and acclimation potential of the thermostability of photochemistry in four Mediterranean conifers. Ann For Sci 61:235–241

Gabris M (2012) The regulation mechanisms in plant water management and the crop productivity in dry conditions. Dissertation, Slovak University of Agriculture

Genty B, Briantais JM, Vieira Da Silva JB (1987) Effects of drought on primary photosynthetic processes of cotton leaves. Plant Physiol 83:360–364

Genty B, Briantais JM, Baker NR (1989) The relationship between the quantum yield of photosynthetic electron transport and quenching of chlorophyll fluorescence. Biochim Biophys Acta 990:87–92

Georgieva K, Lichtenthaler HK (1999) Photosynthetic activity and acclimation ability of pea plants to low and high temperature treatment as studied by means of chlorophyll fluorescence. J Plant Physiol 155:416–423

Georgieva K, Lichtenthaler HK (2006) Photosynthetic response of different pea cultivars to low and high temperature treatments. Photosynthetica 44:569–578

Georgieva K, Ivanova A, Doncheva S, Petkova S, Stefanov D, Péli E, Tuba Z (2011) Fatty acid content during reconstitution of the photosynthetic apparatus in the air-dried leaves of Xerophyta scabrida after rehydration. Biol Plant 55:581–585

Gholamin R, Khayatnezhad M (2011) The effect of end season drought stress on the chlorophyll content, chlorophyll fluorescence parameters and yield in maize cultivars. Sci Res Essays 25:5351–5357

Girardi MT, Cona B, Geiken B, Kucera T, Masojidek J, Matoo AK (1996) Long-term drought stress induces structural and functional reorganization of photosystem II. Planta 199:118–125

Gollan Y, Passioura JB, Munns R (1986) Soil water status affects the stomatal conductance of fully turgid wheat and sunflower leaves. Aust J Plant Physiol 48:575–579

Gorbe E, Calatayud A (2012) Applications of chlorophyll fluorescence imaging technique in horticultural research: a review. Sci Hortic 138:24–35

Grzesiak S, Iijima M, Kono Y, Yamauchi A (1997) Differences in drought tolerance between cultivars of field bean and field pea. A comparison of drought-resistance and drought-sensitive cultivars. Acta Physiol Plant 19:349–357. doi:10.1007/s11738-997-0012-y

Guarini JM, Moritz C (2009) Modelling the dynamics of the electron transport rate measured by PAM fluorimetry during Rapid Light Curve experiments. Photosynthetica 47:206–214

Guisse B, Srivastava A, Strasser RJ (1995) The polyphasic rise of the chlorophyll a fluorescence (OKJIP) in heat stressed leaves. Arch Sci Geneve 48:147–160

Han F, Chen H, Li X-J, Yang M-F, Liu G-S, Shen S-H (2009) A comparative proteomic analysis of rice seedlings under various high-temperature stresses. Biochim Biophys Acta Proteins Proteomics 1794:1625–1634. http://dx.doi.org/10.1016/j.bbapap.2009.07.013

Harbinson J, Genty B, Baker NR (1989) Relationship between the quantum efficiencies of Photosystems I and II in pea leaves. Plant Physiol 90:1029–1034

Harbinson J, Prinzenberg AE, Kruijer W, Aarts MG (2012) High throughput with chlorophyll fluorescence imaging and its use in crop improvement. Curr Opin Biotechnol 23:221–226

Havaux M (1993) Characterization of thermal damage to the photosynthetic electron transport system in potato leaves. Plant Sci 94:19–33

Havaux M (1996) Short-term responses of Photosystem I to heat stress. Induction of a PS II-independent electron transport through PS I fed by stromal components. Photosynth Res 47:85–97

Havaux M, Tardy F (1996) Temperature-dependent adjustment of the thermal stability of photosystem II in vivo: possible involvement of xanthophyll-cycle pigments. Planta 198:324–333. doi:10.1007/BF00620047

Havaux M, Strasser RJ, Greppin H (1990) In vivo photoregulation of photochemical and non-photochemical deactivation of photosystem II in intact plant leaves. Plant Physiol Biochem 28:735–746

Hermans C, Smeyers M, Rodriguez RM, Eyletters M, Strasser RJ, Delhaye JP (2003) Quality assessment of urban trees: a comparative study of physiological characterisation, airborne imaging and on site fluorescence monitoring by the OJIP-test. J Plant Physiol 160:81–90

Herppich WB, Peckmann K (1997) Responses of gas exchange, photosynthesis, nocturnal acid accumulation and water relations of *Aptenia cordifolia* to short-term drought and rewatering. J Plant Physiol 150:467–474

Hewson I, O'Neil JM, Dennison WC (2001) Virus-like particles associated with Lyngbya majuscula (*Cyanophyta*; *Oscillatoriacea*) bloom decline in Moreton Bay. Aquat Microb Ecol 25:207–213

Huang W, Yang SJ, Zhang SB, Zhang JL, Cao KF (2011) Cyclic electron flow plays an important role in photoprotection for the resurrection plant *Paraboea rufescens* under drought stress. Planta. doi:10.1007/s00425011115443

Ierna A (2007) Characterization of potato genotypes by chlorophyll fluorescence during plant aging in a Mediterranean environment. Photosynthetica 45:568–575

Jefferies RA (1994) Drought and chlorophyll fluorescence in field-grown potato (*Solanum tuberosum*). Physiol Plant 90:93–97

Jiang CD, Shi L, Gao HY, Schansker G, Tóth SZ, Strasser RJ (2006a) Development of photosystems 2 and 1 during leaf growth in grapevine seedlings probed by chlorophyll a fluorescence transient and 820 nm transmission *in vivo*. Photosynthetica 44:454–463

Jiang Q, Roche D, Monaco TA, Durham S (2006b) Gas exchange, chlorophyll fluorescence parameters and carbon isotope discrimination of 14 barley genetic lines in response to salinity. Field Crops Res 96:269–278. doi:10.1016/j.fcr.2005.07.010

Kalaji HM, Bosa K, Koscielniak J, Hossain Z (2011) Chlorophyll a fluorescence – a useful tool for the early detection of temperature stress in spring barley (*Hordeum vulgare* L.). OMICS 15:95–934. doi:10.1089/omi.2011.0070

Karim A, Fukamachi H, Hidaka T (2003) Photosynthetic performance of *Vigna radiata* L. leaves developed at different temperature and irradiance levels. Plant Sci 164:451–458

Kauser R, Athar HUR, Ashraf M (2006) Chlorophyll fluorescence: potential indicator for rapid assessment of water stress tolerance in canola (*Brassica napus* L.). Pak J Bot 38:1501–1509

Kautsky H, Hirsch A (1931) Neue Versuche zur Kohlensaure assimilation. Naturwissenschaften 19:964

Kebede H, Fisher DK, Young LD (2012) Determination of moisture deficit and heat stress tolerance in corn using physiological measurements and a low-cost microcontroller-based monitoring system. J Agron Crop Sci 198:118–129. doi:10.1111/j.1439-037X.2011.00493.x

Kocheva K, Lambrev P, Georgiev G, Goltsev V, Karabaliev M (2004) Evaluation of chlorophyll fluorescence and membrane injury in the leaves of barley cultivars under osmotic stress. Bioelectrochemistry 63:121–124

Kolber ZS, Prašil O, Falkowski PG (1998) Measurements of variable chlorophyll fluorescence using fast repetition rate techniques: defining methodology and experimental protocols. Biochim Biophys Acta 1367:88–110

Kramer DM, Johnson G, Kiirats O, Edwards GE (2004) New fluorescence parameters for the determination of $Q_A$ redox state and excitation energy fluxes. Photosynth Res 79:209–218

Krause GH, Weis E (1991) Chlorophyll fluorescence and photosynthesis: the basics. Annu Rev Plant Physiol Plant Mol Biol 42:313–349

Krüger GHJ, Tsimilli-Michael M, Strasser RJ (1997) Light stress provokes plastic and elastic modifications in structure and function of Photosystem II in Camellia leaves. Physiol Plant 101:265–277

Kubler JE, Raven J (1996) Nonequilibrium rates of photosynthesis and respiration under dynamic light supply. J Phycol 32:963–969

Kuckenberg J, Tartachnyk I, Noga G (2009) Temporal and spatial changes of chlorophyll fluorescence as a basis for early and precise detection of leaf rust and powdery mildew infections in wheat leaves. Precision Agric 10:34–44

Lal A, Ku MSB, Edwards GE (1996) Analysis of inhibition of photosynthesis due to water-stress in the C3 species *Hordeum vulgare* and *Vicia faba* – electron-transport, $CO_2$ fixation and carboxylation capacity. Photosynth Res 49:57–69

Lang M, Lichtenthaler HK, Sowinska M, Heisel F, Miehé JA (1996) Fluorescence imaging of water and temperature stress in plant leaves. J Plant Physiol 148:613–621.

Lauer MJ, Boyer JS (1992) Internal $CO_2$ measured directly in leaves. Plant Physiol 98:1310–1316

Law RD, Crafts-Brandner SJ (1999) Inhibition and acclimation of photosynthesis to heat stress is closely correlated with activation of ribulose-1,5-bisphosphate carboxylase/oxygenase. Plant Physiol 120:173–181

Lawlor DW (2002) Limitation to photosynthesis in water-stressed leaves: stomatal metabolism and the role of ATP. Ann Hort 89:871–885

Lawlor DW, Cornic G (2002) Photosynthetic carbon assimilation and associated metabolism in relation to water deficits in higher plants. Plant Cell Environ 25:275–294

Lawlor DW, Tezara W (2009) Causes of decreased photosynthetic rate and metabolic capacity in water-deficient leaf cells: a critical evaluation of mechanisms and integration processes. Ann Bot 103:561–579

Lazár D (1999) Chlorophyll a fluorescence induction. Biochim Biophys Acta 1412:1–28

Lepedus H, Brkic I, Cesar V, Jurkovic V, Antunovic J, Jambrovic A, Brkic J, Simic D (2012) Chlorophyll fluorescence analysis of photosynthetic performance in seven maize inbred lines under water-limited conditions. Period Biol 114:73–76

Li RH, Guo PG, Michael B, Stefania G, Cecceralli S (2006) Evaluation of chlorophyll content and fluorescence parameters as indicators of drought tolerance in barley. Agric Sci China 5(10):751–757

Li QM, Liu BB, Wu Y, Zou ZR (2008) Interactive effects of drought stresses and elevated $CO_2$ concentration on photochemistry efficiency of cucumber seedlings. J Integr Plant Biol 50:1307–1317

Lichtenthaler HK, Babani F (2000) Detection of photosynthetic activity and water stress by imaging the red chlorophyll fluorescence. Plant Physiol Biochem 38:889–895. doi:http://dx.doi.org/10.1016/S0981-9428(00)01199-2

Lichtenthaler HK, Babani F (2004) Light adaptation and senescence of the photosynthetic apparatus: changes in pigment composition, chlorophyll fluorescence parameters and photosynthetic activity. In: Papageorgiou GC, Govindjee (eds) Chlorophyll a fluorescence: a signature of photosynthesis, advances in photosynthesis and respiration, vol 19. Springer, Dordrecht

Lichtenthaler HK, Burkart S (1999) Photosynthesis and high light stress. Bulg J Plant Physiol 25:3–16
Lichtenthaler HK, Miehé JA (1997) Fluorescence imaging as a diagnostic tool for plant stress. Trends Plant Sci 2:316–320
Lichtenthaler HK, Rinderle U (1988) Role of chlorophyll fluorescence in the detection of stress conditions in plants. CRC Crit Rev Anal Chem 19(1):29–85
Lichtenthaler HK, Buschmann C, Knapp M (2005a) How to correctly determine the different chlorophyll fluorescence parameters and the chlorophyll fluorescence decrease ratio RFd of leaves with the PAM fluorometer. Photosynthetica 43:379–393
Lichtenthaler HK, Langsdorf G, Lenk S, Buschmann C (2005b) Chlorophyll fluorescence imaging of photosynthetic activity with the flash-lamp fluorescence imaging system. Photosynthetica 43:355–369
Long SP, Bernacchi CJ (2003) Gas exchange measurements, what can they tell us about the underlying limitations to photosynthesis? Procedures and sources of error. J Exp Bot 54:2393–2401
Longenberger PS, Smith CW, Duke SE, McMichael BL (2009) Evaluation of chlorophyll fluorescence as a tool for the identification of drought tolerance in upland cotton. Euphytica 166:25–33
Lu C, Zhang J (1998) Thermostability of photosystem II is increased in salt stressed sorghum. Aust J Plant Physiol 25:317–324. http://dx.doi.org/10.1071/PP97138
Luo HB, Ma L, Xi HF, Duan W, Li SH et al (2011) Photosynthetic responses to heat treatments at different temperatures and following recovery in grapevine (*Vitis amurensis* L.) leaves. PLoS One 6:e23033. doi:10.1371/journal.pone.0023033
Malkin S, Kok B (1966) Fluorescence induction studies in isolated chloroplasts. I. Number of components involved in the reaction and quantum yields. Biochim Biophys Acta 126:413–432
Mamnouie E, Fotouhi GR, Esfahany M, Nakhoda B (2006) The effects of water deficit on crop yield and the physiological characteristics of barley (*Hordeum vulgare* L.) varieties. J Agric Sci Technol 8:211–219
Marler TE, Mickelbart MV (1998) Drought, leaf gas exchange, and chlorophyll fluorescence of field-grown papaya. J Am Soc Hort Sci 123:714–718
Maroco JP, Pereira JS, Chaves MM (1997) Stomatal responses of leaf-to-air vapour pressure deficit in Sahelian species. Aust J Plant Physiol 24:381–387
Maroco JP, Rodriges ML, Lopes C, Chaves MM (2002) Limitation to leaf photosynthesis in grapevine under drought – metabolic and modeling approaches. Funct Plant Physiol 29:1–9
Massacci A, Nabiev SM, Pietrosanti L, Nematov SK, Chernikova TN, Thor K, Leipner J (2008) Response of the photosynthetic apparatus of cotton (Gossypium hirsutum) to the onset of drought stress under field conditions studied by gas-exchange analysis and chlorophyll fluorescence imaging. Plant Physiol Biochem 46:189–195
Mauzerall D (1972) Light-induced changes in Chlorella, and the primary photoreaction for the production of oxygen. Proc Natl Acad Sci 69:1358–1362
Maxwell K, Johnson GN (2000) Chlorophyll fluorescence-a practical guide. J Exp Bot 51:659–668
Medrano H, Escalona JM, Bota J, Gulias J, Flexas J (2002) Regulation of photosynthesis of C3 plants in response to progressive drought: stomatal conductance as a reference parameter. Ann Bot 89:895–905
Misra AN, Misra M, Singh R (2012) Chlorophyll fluorescence in plant biology. In Misra AN (ed) Biophysics. InTech, Shanghai
Morales SG, Trejo-Téllez LI, Merino FCG, Caldana C, Espinosa-Victoria D, Herrera Cabrera BE (2012) Growth, photosynthetic activity, and potassium and sodium concentration in rice plants under salt stress. Acta Sci 34:317–324
Morita S, Siratsuchi H, Takanashi J, Fujita K (2004) Effect of high temperature on ripening in rice plant. Analysis of the effect of high night and high day temperature applied to the panicle in other parts of the plant. Jpn J Crop Sci 73:77–83
Mott KA, Cardon ZG, Berry JA (1993) Asymmetric patchy stomatal closure for the two surfaces of *Xanthium strumarium* L. leaves at low humidity. Plant Cell Environ 20:1098–1107
Munns R, James RA, Sirault XRR, Furbank RT, Jones HG (2010) New phenotyping methods for screening wheat and barley for beneficial responses to water deficit. J Exp Bot 61:3499–3507. doi:10.1093/jxb/erq199
Nash D, Miyao M, Murata N (1985) Heat inactivation of oxygen evolution in photosystem II particles and its acceleration by chloride depletion and exogenous manganese. Biochim Biophys Acta 807:127–133
Nedbal L, Whitmarsh J (2004) Chlorophyll fluorescence imaging of leaves and fruits. In: Papageorgiou GC, Govindjee (eds) Chlorophyll a fluorescence: a signature of photosynthesis, advances in photosynthesis and respiration, vol 19. Springer, Dordrecht
Nogués S, Baker NR (2000) Effects of drought on photosynthesis in Mediterranean plants grown under enhanced UV-B radiation. J Exp Bot 51:1309–1317
Nyarko G, Alderson PG, Craigon E, Murchie E, Sparkes DL (2008) Comparison of cell membrane thermostability and chlorophyll fluorescence parameters for the determination of heat tolerance in ten cabbage lines. J Hort Sci Biotechnol 83:678–682
O'Neill PM, Shanahan JF, Schepers JS (2006) Use of chlorophyll fluorescence assessments to differentiate corn hybrid response to variable water conditions. Crop Sci 46:681–687
Olson RJ, Chekalyuk AM, Sosik HM (1996) Phytoplankton photosynthetic characteristics from fluorescence induction assays of individual cells. Limnol Oceanogr 41:1253–1263
Omasa K, Takayama K (2003) Simultaneous measurement of stomatal conductance, non-photochemical quenching, and photochemical yield of photosystem

II in intact leaves by thermal and chlorophyll fluorescence imaging. Plant Cell Physiol 44:1290–1300

Omasa K, Shimazaki KI, Aiga I, Larcher W, Onoe M (1987) Image analysis of chlorophyll fluorescence transients for diagnosing the photosynthetic system of attached leaves. Plant Physiol 84:748–752

Oukarroum A, El Madidi S, Schansker G, Strasser RJ (2007) Probing the responses of barley cultivars (*Hordeum vulgare* L.) by chlorophyll a fluorescence OLKJIP under drought stress and re-watering. Environ Exp Bot 60:438–446

Oukarroum A, Schansker G, Strasser RJ (2009) Drought stress effects on photosystem I content and photosystem II thermotolerance analyzed using Chl a fluorescence kinetics in barley varieties differing in their drought tolerance. Physiol Plant 137:188–199

Oxborough K, Baker NR (1997) Resolving chlorophyll a fluorescence images of photosynthetic efficiency into photochemical and non-photochemical components – calculation of $qP$ and $F_v'/F_m'$ without measuring $Fo'$. Photosynth Res 54:135–142

Oyetunji OJ, Ekanayake IJ, Osonubi O (2007) Chlorophyll fluorescence analysis for assessing water deficit and arbuscular mycorrhizal fungi (AMF) inoculation in cassava (*Manihot esculenta* Crantz). Adv Biol Res 1:108–117

Parry M, Andraloje PJ, Khan S, Lea PJ, Keys A (2002) Rubisco activity: effect of drought stress. Ann Bot 89:833–839

Pastenes C, Horton P (1996) Effect of high temperature on photosynthesis in beans. 1. Oxygen evolution and chlorophyll fluorescence. Plant Physiol 112:1245–1251

Patel PN, Hall AE (1990) Genotypic variation and classification of cowpea for reproductive responses to high temperatures under long photoperiods. Crop Sci 30:614–621

Pérez P, Zita G, Morcuende R, Martínez-Carrasco R (2007) Elevated $CO_2$ and temperature differentially affect photosynthesis and resource allocation in flag and penultimate leaves of wheat. Photosynthetica 45:9–17

Petkova V, Denev ID, Cholakov D, Porjazov I (2007) Field screening for heat tolerant common bean cultivars (Phaseolus vulgaris L.) by measuring of chlorophyll fluorescence induction parameters. Sci Hort 111:101–106. http://dx.doi.org/10.1016/j.scienta.2006.10.005

Pfündel E (1998) Estimating the contribution of photosystem I to total leaf chlorophyll fluorescence. Photosynth Res 56:185–195

Pfündel E, Klughammer C, Schreiber U (2008) Monitoring the effects of reduced PS II antenna size on quantum yields of photosystems I and II using the Dual-PAM-100 measuring system. PAM Appl. Notes 1:21–24

Quick WP, Horton P (1984) Studies on the induction of chlorophyll fluorescence in barley protoplasts. II. Resolution of fluorescence quenching by redox state and the trans-thylakoid pH gradient. Philos Trans R Soc Lond B Biol Sci 220:371–382

Rahbarian R, Khavari-Nejad R, Ganjeal A, Bagheri A, Najafi F (2011) Drought stress effects on photosynthesis, chlorophyll fluorescence and water relations in tolerant and susceptible chickpea (*Cicer arietinum* L.) genotypes. Acta Biol Cracov Ser Bot 53:47–56. doi:10.2478/v10182-011-0007-2

Ralph PJ, Gademann R (2005) Rapid light curve: a powerful tool to assess photosynthetic activity. Aquat Bot 82:222–237

Rathod DP, Brestic M, Shao HB (2011) Chlorophyll a fluorescence determines the drought resistance capabilities in two varieties of mycorrhized and non-mycorrhized *Glycine max* Linn. Afr J Microb Res 5:4197–4206. doi:10.5897/AJMR11.737

Razavi F, Pollet B, Steppe K, van Labeke MC (2008) Chlorophyll fluorescence as a tool for evaluation of drought stress in strawberry. Photosynthetica 46:631–633. doi:10.1007/s11099-008-0108-7

Rehman H, Malik SA, Saleem M (2004) Heat tolerance of upland cotton during the fruiting stage evaluated using cellular membrane thermostability. Field Crops Res 85:149–158

Repkova J, Brestic M, Zivcak M (2008) Bioindication of barley leaves vulnerability in conditions of water deficit. Cereal Res Commun 36:1747–1750

Robakowski P, Montpied P, Dreyer E (2002) Temperature response of photosynthesis of silver fir (*Abies alba* Mill.) seedlings. Ann For Sci 59:159–166

Rohacek K, Soukupova J, Bartak M (2008) Chlorophyll fluorescence: a wonderful tool to study plant physiology and plant stress. In: Schoefs B (ed) Plant cell compartments – selected topics. Research Signpost, Trivandrum

Rolfe SA, Scholes JD (1995) Quantitative imaging of chlorophyll fluorescence. New Phytol 131:69–79

Roostaei M, Mohammadi SA, Amri A, Majidi E, Nachit M, Haghparast R (2011) Chlorophyll fluorescence parameters and drought tolerance in a mapping population of winter bread wheat in the highlands of Iran. Russ J Plant Physiol 58:351–358. doi:http://dx.doi.org/10.1134/S102144371102018X

Rosenqvist E, van Kooten O (2003) Chlorophyll fluorescence: a general description and nomenclature. In: DeEll JR, Toivonen PMA (eds) Practical applications of chlorophyll fluorescence in plant biology. Kluwer Academic Publishers, Dordrecht, pp 31–78

Ruban AV, Johnson MP (2009) Dynamics of higher plant photosystem cross-section associated with state transitions. Photosynth Res 99:173–183

Rumeau D, Peltier G, Cournac L (2007) Chlororespiration and cyclic electron flow around PS I during photosynthesis and plant stress response. Plant Cell Environ 30:1041–1051

Saito Y, Saito R, Nomura E, Kawahara TD, Nomura A, Takaragaki S, Ida K, Takeda S (1999) Performance check of vegetation fluorescence imaging lidar through in vivo and remote estimation of chlorophyll concentration inside plant leaves. Optical Rev 6:155–159

Salvucci ME, Crafts-Brandner SJ (2004) Inhibition of photosynthesis by heat stress: the activation state of Rubisco as a limiting factor in photosynthesis. Physiol Plant 120:179–186

Santos MG, Ribeiro RV, Machado EC, Pimentel C (2009) Photosynthetic parameters and leaf water potential of five common bean genotypes under mild water deficit. Biol Plant 53:229–236

Sarieva GE, Kenzhebaeva SS, Lichtenthaler HK (2010) Adaptation potential of photosynthesis in wheat cultivars with a capability of leaf rolling under high temperature conditions. Russ J Plant Physiol 57:28–36

Sayar R, Khemira H, Kameli A, Mosbahi M (2008) Physiological tests as predictive appreciations for drought tolerance in durum wheat (*Triticum durum* Desf.). Agron Res 6:79–90

Sayed OH (2003) Chlorophyll fluorescence as a tool in cereal crop research. Photosynthetica 41:321–330

Sazanov LA, Burrows PA, Nixon PJ (1998) The chloroplast Ndh complex mediates the dark reduction of the plastoquinone pool in response to heat stress in tobacco leaves. FEBS Lett 429:115–118

Schansker G, Tóth SZ, Strasser RJ (2005) Methylviologen and dibromothymoquinone treatments of pea leaves reveal the role of photosystem I in the Chl a fluorescence rise OJIP. Biochim Biophys Acta 1706:250–261

Schapendonk AHC, Spitters CJT, Groot PJ (1989) Effects of water stress on photosynthesis and chlorophyll fluorescence of five potato cultivars. Potato Res 32:17–32

Schrader SM, Wise RR, Wacholtz WF, Ort DR, Sharkey TD (2004) Thylakoid membrane responses to moderately high leaf temperature in Pima cotton. Plant Cell Environ 27:725–735

Schreiber U (2004) Pulse-amplitude-modulation (PAM) fluorometry and saturation pulse method: an overview. In: Papageorgiou GC, Govindjee (eds) Chlorophyll a fluorescence: a signature of photosynthesis, advances in photosynthesis and respiration, vol 19. Springer, Dordrecht

Schreiber U, Berry JA (1977) Heat-induced changes of chlorophyll fluorescence in intact leaves correlated with damage of the photosynthetic apparatus. Planta 136:233–238

Schreiber U, Schliwa U, Bilger W (1986) Continuous recording of photochemical and non-photochemical chlorophyll fluorescence quenching with a new type of modulation fluorometer. Photosynth Res 10:51–62

Schreiber U, Gademann R, Ralph PJ, Larkum AWD (1997) Assessment of photosynthetic performance of Prochloron in Lissoclinum patella in hospite by chlorophyll fluorescence measurements. Plant Cell Physiol 38:945–951

Seddon S, Cheshire AC (2001) Photosynthetic response of Amphibolis antarctica and Posidonia australis to temperature and desiccation using chlorophyll fluorescence. Mar Ecol Progr Ser 220:119–130

Shabala SI (2002) Screening plants for environmental fitness: chlorophyll fluorescence as a "Holy Grail" for plant breeders. In: Hemantaranjan A (ed) Advances in plant physiology, vol 5. Scientific Publishers, Jodhpur

Sharkey TD, Seeman JR (1989) Mild water stress effects on carbon-reduction-cycle intermediates, ribulose bisphosphate carboxylase activity, and spatial homogeneity of photosynthesis in intact leaves. Plant Physiol 89:1060–1065

Sharkey TD, Zhang R (2010) High temperature effects on electron and proton circuits of photosynthesis. J Integr Plant Biol 52:712–722. doi:10.1111/j.1744-7909.2010.00975.x

Sharkova VE (2001) The effect of heat shock on the capacity of wheat plants to restore their photosynthetic electron transport after photoinhibition or repeated heating. Russ J Plant Physiol 48:793–797

Shefazadeh MK, Mohammadi M, Karimizadeh R (2012) Genotypic difference for heat tolerance traits under real field conditions. J Food Agric Environ 10:484–487

Sheikh S, Behl RK, Dhand SS, Munjal R (2010) Membrane thermostability and chlorophyll fluorescence as indices of high temperature stress tolerance and performance in wheat (*Triticum aestivum* L.). Cereal Res Commun 38:335–344. doi:10.1556/CRC.38.2010.3.4

Shikanai T (2007) Cyclic electron transport around photosystem I: genetic approaches. Annu Rev Plant Biol 58:199–217

Siddique KHM, Loss SP, Regan KL, Jettner RL (1999) Adaptation and seed yield of cool season grain legumes in Mediterranean environments of south-western Australia. Aust J Agric Res 50:375–387

Sikuku PA, Netondo GW, Onyango JC, Musyimi DM (2010) Chlorophyll fluorescence, protein and chlorophyll content of three NERICA rainfed rice varieties under varying irrigation regimes. ARPN J Agric Biol Sci 5:19–25

Silim SN, Ryan N, Kubien DS (2010) Temperature responses of photosynthesis and respiration in *Populus balsamifera* L.: acclimation versus adaptation. Photosynth Res 104:19–30. doi:10.1007/s11120-010-9527-y

Sinsawat V, Leipner J, Stamp P, Fracheboud Y (2004) Effect of heat stress on the photosynthetic apparatus in maize (Zea mays L.) grown at control or high temperature. Environ Exp Bot 52:123–129

Slabbert RM, Krüger GHJ (2011) Assessment of changes in photosystem II structure and function as affected by water deficit in *Amaranthus hypochondriacus* L. and *Amaranthus hybridus* L. Plant Physiol Biochem 49:978–984. doi:10.1016/j.plaphy.2011.05.001

Slapakauskas V, Ruzgas V (2005) Chlorophyll fluorescence characteristics of different winter wheat varieties (*Triticum aestivum* L.). Agron J 32:203–209

Smillie RM, Gibbons GC (1981) Heat tolerance and hardening in crop plants measured by chlorophyll fluorescence. Carlberg Res Commun 46:395–403. doi:10.1007/BF02907961

Srinivasan A, Takeda H, Senboku T (1996) Heat tolerance in food legumes as evaluated by cell membrane thermostability and chlorophyll fluorescence techniques. Euphytica 88:35–45. doi:10.1007/BF00029263

Srivastava A, Greppin H, Strasser RJ (1995) Acclimation of land plants to diurnal changes in temperature and light. In: Mathis P (ed) Photosynthesis: from light to biosphere. Kluwer Academic, Dordrecht

Srivastava B, Guisse H, Greppin H, Strasser RJ (1997) Regulation of antenna structure and electron transport in photosystem II of *Pisum sativum* under elevated temperature probed by the fast polyphasic chlorophyll a fluorescence transient: OKJIP. Biochim Biophys Acta 1320:95–106

Stirbet A, Govindjee (2011) On the relation between the Kautsky effect (chlorophyll a fluorescence induction) and Photosystem II: basics and applications of the OJIP fluorescence transient. J Photochem Photobiol B 104:236–257

Stone PJ, Nicolas ME (1994) Wheat cultivars vary widely in their responses of grain yield and quality to short periods of post-anthesis heat stress. Aust J Plant Physiol 21:887–900

Strasser, RJ (1987) Energy pipeline model of the photosynthetic apparatus. In: Biggins J (ed) Progress in Photosynthesis Research, 2:717–720. Martinus Nijhoff Publisher, Dordrecht

Strasser RJ, Eggenberg P, Strasser BJ (1996) How to work without stress but with fluorescence. Bull Royal Soc Liege 65:330–349

Strasser RJ, Govindjee (1991) The Fo and the OJIP fluorescence rise in higher plants and algae. In: Argyroudi-Akoyunoglou JH (ed) Regulation of chloroplast biogenesis. Plenum Press, New York

Strasser RJ, Govindjee (1992) On the OJIP fluorescence transients in leaves and D1 mutants of *Chlamydomonas reinhardtii*. In: Murata N (ed) Research in photosynthesis, vol II. Kluwer Academic, Dordrecht

Strasser BJ, Strasser RJ (1995) Measuring fast fluorescence transients to address environmental questions: the JIP-test. In: Mathis P (ed) Photosynthesis: from light to biosphere. Kluwer Academic, Dordrecht, pp 977–980

Strasser RJ, Srivastava A, Govindjee (1995) Polyphasic chlorophyll *a* fluorescence transient in plants and cyanobacteria. Photochem Photobiol 61:32–42

Strasser RJ, Srivastava A, Tsimilli-Michael M (2000) The fluorescence transient as a tool to characterize and screen photosynthetic samples. In: Yunus M, Pathre U, Mohanty P (eds) Probing photosynthesis: mechanism, regulation and adaptation. Taylor & Francis, London

Strasser RJ, Srivastava A, Tsimilli-Michael M (2004) Analysis of the chlorophyll a fluorescence transient, advances in photosynthesis and respiration. In: Papageorgiou GC, Govindjee (eds) Chlorophyll fluorescence: a signature of photosynthesis. Kluwer Academic, Dordrecht

Strasser RJ, Tsimilli-Michael M, Qiang S, Goltsev V (2010) Simultaneous in vivo recording of prompt and delayed fluorescence and 820-nm reflection changes during drying and after rehydration of the resurrection plant *Haberlea rhodopensis*. Biochim Biophys Acta 1797:1313–1326

Subrahmanyam D, Subash N, Haris A et al (2006) Influence of water stress on leaf photosynthetic characteristics in wheat cultivars differing in their susceptibility to drought. Photosynthetica 44:125–129. doi:10.1007/s11099-005-0167-y

Sung DY, Kaplan F, Lee KJ, Guy CHL (2003) Acquired tolerance to environmental extremes. Trends Plant Sci 8:179–187. doi:10.1016/S1360-1385(03)00047-5

Tang AC, Kawamitsu Y, Kanechi M, Boyer JS (2002) Photosynthetic oxygen evolution at low water potential in leaf discs lacking an epidermis. Ann Bot 89:861–870

Tari I, Camen D, Coradini G, Csiszar J, Fediuc E, Gemes K, Lazar A, Madosa E, Mihacea S, Poor P, Postelnicu S, Staicu M, Szepesi A, Nedelea G (2008) Changes in chlorophyll fluorescence parameters and oxidative stress responses of bush bean genotypes for selecting contrasting acclimation strategies under water stress. Acta Biol Hung 59:335–345

Terzi R, Saglam A, Kutlu N, Nar H, Kadioglu A (2010) Impact of soil drought stress on photochemical efficiency of photosystem II and antioxidant enzyme activities of *Phaseolus vulgaris* cultivars. Turk J Bot 34:1–10. doi:10.3906/bot-0905-20

Tester M, Langridge P (2010) Breeding technologies to increase crop production in a changing world. Science 12(327):818–822

Tezara W, Mitchell VJ, Driscoll SD, Lawlor DW (1999) Water stress inhibits plant photosynthesis by decreasing coupling factors and ATP. Nature 401:914–917

Thompson LM (1988) Effects of changes in climate and weather variability on the yields of corn and soybean. J Prod Agric 1:20–27

Toth SZ, Schansker G, Kissimon J, Kovacs L, Garab G, Strasser RJ (2005) Biophysical studies of photosystem II-related recovery processes after a heat pulse in barley seedlings (*Hordeum vulgare* L.). J Plant Physiol 162:181–194

Toth SZ, Schansker G, Strasser RJ (2007) A non-invasive assay of the plastoquinone pool redox state based on the OJIP-transient. Photosynth Res 93:193–203

Van der Meescht, de Ronde JA, Rossouw FT (1999) Chlorophyll fluorescence and chlorophyll content as a measure of drought tolerance in potato. S Afr J Sci 95:407–411

Van Gorkom HJ (1986) Fluorescence measurements in the study of photosystem II electron transport. In: Amesz J, Fork DC, Govindjee (eds) Light emission by plants and bacteria. Academic, Orlando

Van Heerden PDR, Swanepoel JW, Krüger GHJ (2007) Modulation of photosynthesis by drought in two desert scrub species exhibiting C3-mode $CO_2$ assimilation. Environ Exp Bot 61:124–136

Vara Prasad PV, Craufurd PQ, Summerfield RJ (1999) Fruit number in relation to pollen production and viability in groundnut exposed to short episodes of heat stress. Ann Bot 84:381–386

Vassileva V, Demirevska K, Simova-Stoilova L, Petrova T, Tsenov N, Feller U (2012) Long-term field drought affects leaf protein pattern and chloroplast ultrastructure of winter wheat in a cultivar-specific manner. J Agron Crop Sci 198:104–117

Wahid A, Gelani S, Ashraf M, Foolad MR (2007) Heat tolerance in plants: an overview. Environ Exp Bot 61:199–223. doi:10.1016/j.envexpbot.2007.05.011

Weng JH, Lai MF (2005) Estimating heat tolerance among plant species by two chlorophyll fluorescence parameters. Photosynthetica 43:439–444. doi:10.1007/s11099-005-0070-6

West JD, Peak D, Peterson JQ et al (2005) Dynamics of stomatal patches for a single surface of Xanthium strumarium L. leaves observed with fluorescence and thermal images. Plant Cell Environ 28:633–641

White AJ, Critchley C (1999) Rapid light curves: a new fluorescence method to assess the state of the photosynthetic apparatus. Photosynth Res 59:63–72

Wing SR, Patterson MR (1993) Effects of wave-induced light flecks in the intertidal zone on photosynthesis in the macroalgae Postelsia palmaeformis and *Hedophyllum sessile* (Phaeophyceae). Mar Biol 116:519–552

Wise RR, Olson AJ, Schrader SM, Sharkey TD (2004) Electron transport is the functional limitation of photosynthesis in field-grown Pima cotton plants at high temperature. Plant Cell Environ 27:717–724. doi:10.1111/j.1365-3040.2004.01171.x

Wu XL, Bao WK (2011) Effect of water deficit on growth and photosynthetic characteristics of 13 winter wheat. Afr J Biotechnol 10:11861–11869

Xu XL, Wang ZM, Zhang JP (2001) Effect of heat stress on photosynthetic characteristics of different green organs of winter wheat during grain-filling stage. Acta Bot Sin 43:571–577

Xu F, Guo WH, Wang RQ, Xu WH, Du N, Wang YF (2009) Leaf movement and photosynthetic plasticity of black locust (*Robinia pseudoacacia*) alleviate stress under different light and water conditions. Acta Physiol Plant 31:553–563

Yamada M, Hidaka T, Fukamachi H (1996) Heat tolerance in leaves of tropical fruit crops as measured by chlorophyll fluorescence. Sci Hortic 67:39–48

Yamane Y, Kashino Y, Koike H, Satoh K (1997) Increases in the fluorescence $F_0$ level and reversible inhibition of photosystem II reaction center by high temperature treatment in higher plants. Photosynth Res 52:57–64

Yamane Y, Kashino Y, Koike H, Satoh K (1998) Effects of high temperatures on the photosynthetic systems in spinach: oxygen-evolving activities, fluorescence characteristics and the denaturation process. Photosynth Res 57:51–59

Yang X, Chen X, Ge Q, Li B, Tong Y, Zhang A, Li Z, Kuang T, Lu C (2006) Tolerance of photosynthesis to photoinhibition, high temperature and drought stress in flag leaves of wheat: a comparison between a hybridization line and its parents grown under field conditions. Plant Sci 171:389–397

Yang X, Chen X, Qiaoying G, Li B, Tong Y, Li Z, Kuang T, Lu C (2007) Characterisation of photosynthesis of flag leaves in a wheat hybrid and its parents grown under field conditions. J Plant Physiol 164:318–326

Yordanov I, Velikova V, Tsonev T (2003) Plant responses to drought and stress tolerance. Bulg J Plant Physiol. Special Issue 187–206

Zarco-Tejada PJ, Berni JAJ, Suárez L, Sepulcre-Cantó G, Morales F, Miller JR (2009) Imaging chlorophyll fluorescence with an airborne narrow-band multispectral camera for vegetation stress detection. Remote Sens Environ 113:1262–1275

Zegada-Lizarazu W, Zatta A, Monti A (2012) Water uptake efficiency and above- and belowground biomass development of sweet sorghum and maize under different water regimes. Plant Soil 351:47–60. doi:10.1007/s11104-011-0928-2

Zhang R, Sharkey TD (2009) Photosynthetic electron transport and proton flux under moderate heat stress. Photosynth Res 100:29–43. doi:10.1007/s11120-009-9420-8

Zhang ZB, Xu P, Jia JZ, Zhou RH (2010) Quantitative trait loci for leaf chlorophyll fluorescence trait in wheat. Aust J Crop Sci 4:571–579

Zhang ST, Zhang GL, Chen LY, Xiao YH (2011a) Effects of high temperature stress on net photosynthetic rate and chlorophyll fluorescence parameters of flag leaf in rice. Chin J Rice Sci 25:335–338. doi:CNKI:SUN:ZGSK.0.2011-03-017

Zhang YL, Hu YY, Luo HH, Chow WS, Zhang WF (2011b) Two distinct strategies of cotton and soybean differing in leaf movement to perform photosynthesis under drought in the field. Funct Plant Biol 38:567–575. http://dx.doi.org/10.1071/FP11065

Zivcak M (2006) Application of physiological reaction diversity in screening of wheat genotypes for drought and high temperature tolerance. Dissertation, Slovak University of Agriculture

Zivcak M, Brestic M, Olsovska K, Slamka P (2008a) Performance index as a sensitive indicator of water stress in Triticum aestivum. Plant Soil Environ 54:133–139

Zivcak M, Brestic M, Olsovska K (2008b) Application of photosynthetic parameters in screening of wheat (Triticum aestivum L.) genotypes for improved drought and high temperature tolerance. In: Allen JF, Gantt E, Goldbeck JH, Osmond B (eds) Photosynthesis. Energy from the sun: 14th international congress on photosynthesis. Springer, Dordrecht

Zivcak M, Brestic M, Olsovska K (2008c) Physiological parameters useful in screening for improved tolerance to drought in winter wheat (*Triticum aestivum* L.). Cereal Res Commun 36:1943–1946

Zivcak M, Brestic M, Olsovska K (2009) Application of chlorophyll fluorescence for screening wheat (*Triticum aestivum* L.) genotype susceptibility to drought and high temperature. Vagos 82:82–87

Zivcak M, Olsovska K, Brestic M, Slabbert MM (2013) Critical temperature derived from the selected chlorophyll a fluorescence parameters of indigenous vegetable species of South Africa treated with high temperature. In: C. Lu (ed) Photosynthesis: Research for Food, Fuel and Future—15th International Conference on Photosynthesis, Zhejiang University Press, Springer-Verlag GmbH., in press.

Zlatev Z, Lidon FC (2012) An overview on drought induced changes in plant growth, water relations and photosynthesis. Emir J Food Agric 24:57–72

Zlatev ZS, Yordanov IT (2004) Effect of soil drought on photosynthesis and chlorophyll fluorescence in bean plants. Bulg J Plant Physiol 30:3–18

# Salt Tolerance in Cereals: Molecular Mechanisms and Applications

# 5

Allah Ditta

**Abstract**

Major abiotic stress that limits plant growth and agriculture productivity is the soil salinity. In order to minimize the detrimental effects of salinity, highly complex salt-responsive signaling and metabolic processes at the cellular, organ, and whole-plant levels have been evolved in the plants. Currently, it has become the need of the hour to understand the molecular basis of salt stress signaling and tolerance mechanisms in cereals for engineering and/or screening for more tolerance to salt stress. Valuable information will be provided through investigation of the physiological and molecular mechanisms of salt tolerance for effective engineering strategies. Current advancement in proteomics has helped us in studying the sophisticated molecular networks in plants. Reports of proteomics studies about plant salt response and tolerance mechanisms, especially that of cereals, have revealed the mechanisms that include changes in photosynthesis, scavenging system of reactive oxygen species (ROS), ion homeostasis, osmotic homeostasis, membrane transport, signaling transduction, transcription, protein synthesis/turnover, cytoskeleton dynamics, and cross talks with other stresses.

## Introduction

Salinity, one of the most significant abiotic stresses, not only limits the productivity and geographical distribution of plants but also causes ion imbalance, hyperosmotic stress, and oxidative damage in plants leading to molecular damage, growth and yield reduction, and even plant death (Wang et al. 2004). Infiltration and accumulation of NaCl (Tuteja 2007) is the major cause of salinization and can result in soil $Na^+$ concentration above 40 mM which can suppress the growth of most crops (Wong et al. 2006). Salt accumulation has been attributed to the natural phenomena and human activities like irrigation. About 1/5 of the earth's arable land and 50% of the irrigated one are under salinity (Mahajan and Tuteja 2005; Munns and Tester 2008). Plants have coped with this problem through various sophisticated mechanisms that include selective ion

A. Ditta (✉)
Institute of Soil & Environmental Sciences, University of Agriculture, Faisalabad 38040, Pakistan
e-mail: ad_abs@yahoo.com

G.R. Rout and A.B. Das (eds.), *Molecular Stress Physiology of Plants*,
DOI 10.1007/978-81-322-0807-5_5, © Springer India 2013

uptake/exclusion, compartmentalization of toxic ions, synthesis of compatible products, adjustment of photosynthetic and energy metabolism, accumulation of antioxidative enzymes, regulation of hormones, and modification of cell structure. Molecular and physiological aspects of plant salt stress tolerance have been revealed through physiological, molecular genetics, and functional genomics studies. A few important genes responsible for the above-stated mechanisms have been cloned, and their involvement in plants' response and adaptation to salinity has been confirmed (Tuteja 2007).

However, state-of-the-art transcriptomics studies have helped us in the collection of immense data on their expression at the mRNA level (Chand and Kumar 2006; Diedhiou et al. 2009; Jha et al. 2009; Wong et al. 2005, 2006; Zhang et al. 2001b, 2008; Zouari et al. 2007), and these data present a global vision about salt-responsive genes in different plants. But mRNA levels do not usually correlate with the expression levels of proteins due to posttranscriptional events and posttranslational modifications such as phosphorylation and glycosylation. These proteins are more directly related to signaling and metabolic processes under salt stress conditions. So the need of the hour is to study the salt tolerance mechanisms at the so-called protein level which is possible through the optimum utilization of proteomic technologies.

The plant taxa have an extensive genetic diversity for salt tolerance as it is distributed over numerous genera (Flowers and Colmer 2008) making them either glycophytes (salt-sensitive or hypersensitive plants) or halophytes (native flora of saline environments). In some halophytes, very special anatomical and morphological adaptations or avoidance mechanisms have been employed (Flowers and Colmer 2008), but on the basis of these, we are unable to introgress the responsible genes into crop plants. During the last decade, it has been established that most halophytes and glycophytes use analogous tactical processes rather than similar strategies to tolerate salinity (Hasegawa et al. 2000b). For example, cytotoxic ions ($Na^+$ and $Cl^-$) are compartmentalized into the vacuole and used as osmotic solutes under saline environments (Blumwald et al. 2000; Niu et al. 1995). It follows that there is similarity among many of the molecular entities that mediate ion homeostasis and salt stress signaling in all plants (Hasegawa et al. 2000b) as, for example, ion homeostasis that facilitates plant salt tolerance resembles that described for yeast (Bressan et al. 1998; Serrano et al. 1999). According to another finding of genomics, there is remarkable colinearity (gene synteny and homology) of gene sequences among different grass species, including cultivated crops (Tang et al. 2008; Zahn et al. 2008), and strongly suggests their resemblance in evolution and probably in functions. The facts stated above have made it feasible to use a model system for the dissection of the plant salt stress response (Bressan et al. 1998; Hasegawa et al. 2000a; Sanders et al. 1999; Serrano et al. 1999; Zhu 2000, 2001b). Since a salt-tolerant genetic model is required for complete delineation that if salt tolerance is affected most by form or function of genes or more by differences in the expression of common genes either due to transcriptional or posttranscriptional control (Zhu et al. 2007). In this way, our understanding of cellular salt tolerance mechanisms has been greatly increased through research on the plant genetic model, the Arabidopsis, a glycophyte. It will also enable us to effectively apply the genetic learning of one crop to another and also will yield large spillover benefits from investment in such research, like possible eventual interspecies gene transfers.

Till now, more than 2,171 salt-responsive proteins/enzymes have been discovered in different parts of 34 plant species like shoots, leaves, roots, seedlings, radicles, hypocotyls, grains, gametophytes, and unicells (Zhang et al. 2012). Most of the research has been conducted on plants like *Arabidopsis thaliana* (Jiang et al. 2007; Kim et al. 2007; Lee et al. 2004; Ndimba et al. 2005; Pang et al. 2010) and *Oryza sativa* (Abbasi and Komatsu 2004; Cheng et al. 2009; Chitteti and Peng 2007; Dooki et al. 2006; Kim et al. 2005; Li et al. 2010; Nohzadeh et al. 2007; Parker et al. 2006; Ruan et al. 2011; Salekdeh et al. 2002; Wen et al. 2010; Yan et al. 2005;

Zhang et al. 2009), *Triticum durum* (Caruso et al. 2008), *Triticum aestivum* (Huo et al. 2004; Jacoby et al. 2010; Peng et al. 2009; Wang et al. 2008a), *Hordeum vulgare* (Rasoulnia et al. 2010; Sugimoto and Takeda 2009; Witzel et al. 2009, 2010), *Zea mays* (Zörb et al. 2004, 2009, 2010), *Setaria italica* (Veeranagamallaiah et al. 2008), *Sorghum bicolor* (Kumar et al. 2011), and *Agrostis stolonifera* (Xu et al. 2010).

This chapter includes the reports of proteomics studies about plant salt response and tolerance mechanisms, especially that of cereals, and future perspectives in food security. These mechanisms include changes in photosynthesis, scavenging system of reactive oxygen species (ROS), ion homeostasis, osmotic homeostasis, membrane transport, signaling transduction, transcription, protein synthesis/turnover, cytoskeleton dynamics, and cross talks with other stresses.

## Role of Cereals in Food Security

The word cereal has been derived from *Ceres*, the name of the Roman goddess of harvest and agriculture. The common cereal crops include rice, wheat, corn, barley, sorghum, millet, oats, and rye. It also includes flours, meals, breads, and alimentary pastes or pasta. Cereals, the poor man's meat, provide staple food in almost every country and region as in the world as a whole; about 95% of starchy staple food comes from cereals and only 5% from root crops (mainly cassava, potato, and yams, depending on climate). Currently, about 50% of the world's cropland is under the cultivation of cereals. These provide about two-thirds of all human calorie intake, if we combine their direct intake (e.g., as cooked rice or bread) with their indirect consumption, in the form of foods like meat and milk (about 40% of all grain is currently fed to livestock).

Among cereals, wheat is the predominant commodity consumed for food accounting 68% of total cereal use. By 2020, total cereal consumption is projected to reach nearly 746 metric tons (Mt) (145 Mt wheat, 529 Mt rice, and the rest maize, barley, etc.) with per capita food consumption of around 66 kg per person per annum, and 2% of world wheat is utilized for biofuel production; ultimately the total consumption is expected to increase from 68 to 75% (Chand and Kumar 2006).

According to Chand and Kumar (2006), to feed a population of eight billion by 2025, average world cereal yield of about 4 metric tons/ha will be required (Evans 1998). In addition, it has been found out the reasons for food insecurity that include increasing demand from growing population, climate change, and increased linkages between energy and agricultural commodities due to the growing demand for biofuels (Chand and Kumar 2006). The emerging food insecurity condition will prevail mostly in developing countries, especially that of Asia. However, more than 90% of rice and 43% of wheat in the world is produced and consumed in Asia.

## Detrimental Effects of Salinity on Cereals

Salinity is one of the major environmental factors that adversely affect the crop growth and development processes like seed germination (Dash and Panda 2001); seedling growth (Mahajan and Tuteja 2005; Tuteja 2007; Wang et al. 2004; Wong et al. 2006); enzyme activity (Seckin et al. 2009); deoxyribonucleic acid (DNA), ribonucleic acid (RNA), and protein synthesis (Anuradha and Rao 2001); mitosis (Tabur and Demir 2010); vegetative growth (Hamed et al. 2007; Panda and Khan 2009); and flowering and fruit set (Zhu 2001a, b) and ultimately result in diminished economic yield and also quality of produce.

Since sensitivity or tolerance to salt stress is different in all plant species (Ashraf and Harris 2004), so plants are classified into two groups, namely, glycophytes or halophytes, on the basis of their ability to grow on high-salt medium. Most of the grain crops and vegetables are glycophytes and cannot tolerate salt stress as high salt concentrations decrease the osmotic potential of soil solution creating water stress in plants and ion toxicity (e.g., $Na^+$, $Cl^-$) since $Na^+$ is not readily sequestered into vacuoles as in halophytes. Ultimately, it results in nutrient imbalances and their deficiencies which can lead to plant death as a result of growth arrest and molecular damage (McCue and Hanson 1990). In addition, salinity causes oxidative stress due to the production of induced active oxygen species (Heidari 2009; Munns and Tester 2008) which disrupt the cellular metabolism through oxidative damage to membrane lipid, proteins, and nucleic acids (Mittler 2002).

## Salinity Tolerance, Multigenic Trait

Salinity is a quantitative trait, and a large number of salt-induced genes have been isolated which are concomitantly up- and downregulated (Bohnert et al. 1995). According to Meyer et al. (1990), in *M. crystallinum,* more than a 100 genes are induced, and probably transcripts, three times that number, are repressed in response to salt stress. Salinity tolerance is a mutagenic trait due to the fact that sublethal salt stress conditions cause an osmotic effect that is similar to that brought about by water deficit and to some extent by cold as well as heat stresses (Almoguera et al. 1988). So there is a high degree of similarity between salt and dehydration stress with respect to their physiological, biochemical, molecular, and genetical effects (Cushman et al. 1992). The ways in which different plants confer salt resistance are represented in the following figure (Fig. 5.1):

There is a diverse expression pattern of salt-responsive genes found in different plants (especially glycophytes and halophytes) under different salinity conditions (e.g., salt concentration and treatment time) (Nito et al. 2007). The expression pattern of genes in cereals indicates that most of the proteins related with photosynthesis in *O. sativa* (Abbasi and Komatsu 2004; Kim et al. 2005; Parker et al. 2006), *T. durum* (Caruso et al. 2008), and *T. aestivum* (Huo et al. 2004; Peng et al. 2009) and proteins involved in carbohydrate and energy metabolism in *O. sativa* (Abbasi and Komatsu 2004; Chitteti and Peng 2007; Dooki et al. 2006; Kim et al. 2005; Li et al. 2010; Nohzadeh et al. 2007; Parker et al. 2006; Ruan et al. 2011), *T. aestivum* (Peng et al. 2009; Wang et al. 2008a), *H. vulgare* (Rasoulnia et al. 2010; Witzel et al. 2010), *Z. mays* (Zörb et al. 2004, 2010), and *S. bicolor* (Kumar et al. 2011) are induced by salinity.

Most of the salt-responsive proteins are involved in basic metabolic processes like photosynthesis, energy metabolism, ROS scavenging, and ion homeostasis which make halophytes highly efficient in photosynthetic and energy metabolism, ion exclusion/compartmentalization, compatible product synthesis, induction of antioxidative enzymes and hormones, as well as modification of cell structure. Also evolution of different salt tolerance mechanisms have been found during the study of the specific proteins and/or their expression patterns in different halophytes.

## Molecular Mechanisms of Salinity Tolerance in Cereals

### Photosynthesis

In addition to osmotic, ionic, and nutrient imbalances in plants, salinity also disturbs the plant water uptake and biosynthesis of abscisic acid (ABA) in leaves (Fricke et al. 2004) which affects the stomatal conductance. This in turn affects the photosynthetic electron transport and the enzyme activities for carbon fixation during dark reaction (Parida and Das 2005; Tuteja 2007). Several genes involved in photosynthesis (encoding chlorophyll a-/b-binding proteins (CAB), ribulose-1,5-bisphosphate carboxylase/oxygenase (RuBisCO), and RuBisCO activase (RCA)) have been isolated and characterized through previous studies which are directly or indirectly involved in salinity tolerance (Wong et al. 2006; Zhang et al. 2001a, b, 2008).

Currently, through the advancement in the field of proteomics, our understanding of the photosynthetic processes underlying salinity response and tolerance has been greatly enhanced. About 367 photosynthesis-related IDs, representing 26 UPs,

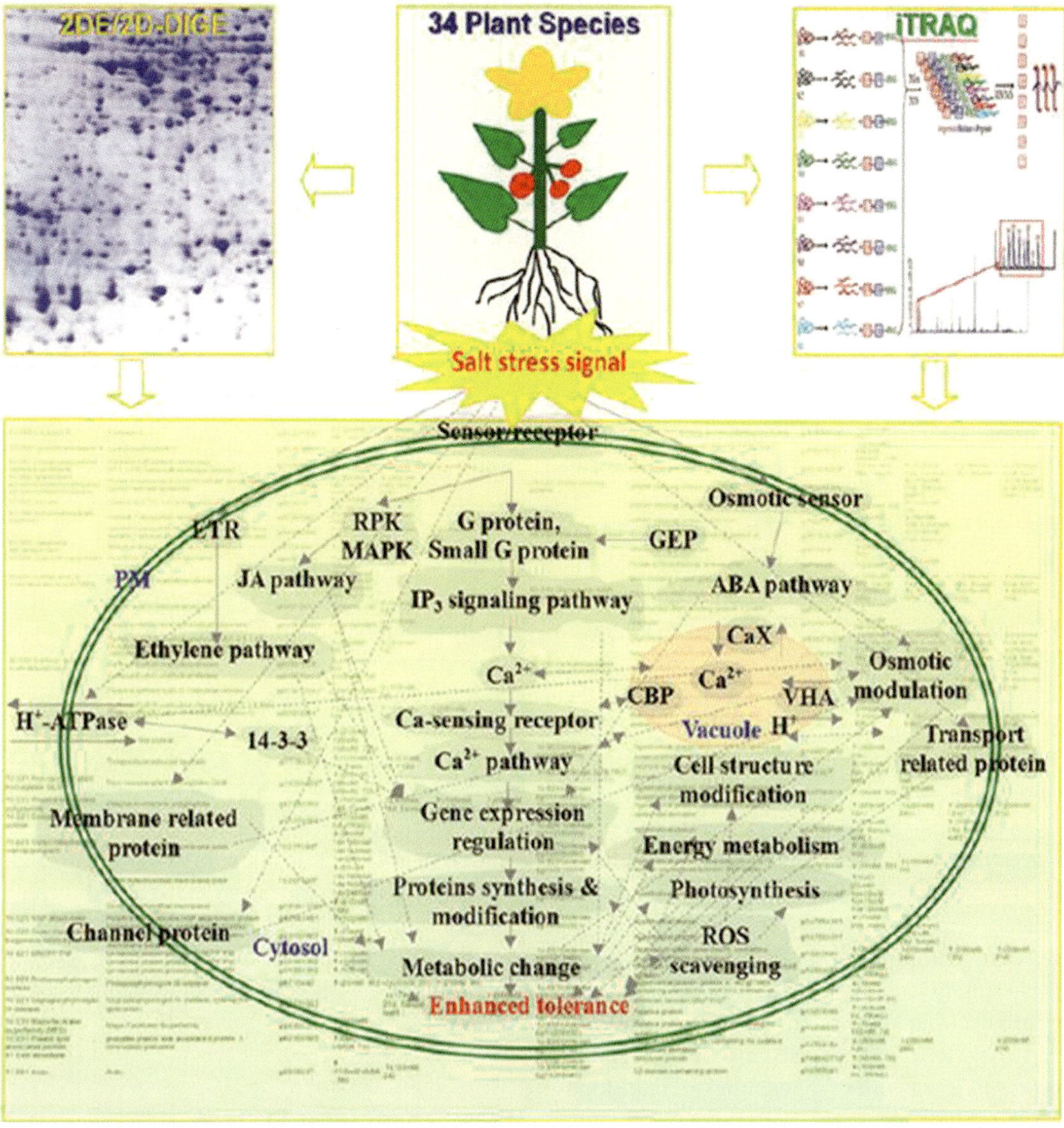


**Fig. 5.1** Different ways of salt tolerance in different plants (Zhang et al. 2012)

regulated by salinity in 27 plant species have been discovered (Nito et al. 2007). These are involved in the regulation of photosynthetic processes like that of light reaction, $CO_2$ assimilation, and other photosynthesis-related processes. Out of 26 UPs in all, 12 are related with light reaction involved in light-harvesting phase, proton gradient formation, electron transfer, and energy production.

In case of cereals, the concentration of photosystem II chlorophyll-binding protein 47 (CP47) and 33 kDa oxygen evolving complex (OEC) protein in *O. sativa* (Abbasi and Komatsu 2004; Kim et al. 2005; Sengupta and Majumder 2009) is increased under salinity stress which helps in the protection of reaction center proteins (D1 protein) from stomatal protease digestion and ensures optimal functioning of photosystem II (PSII) (Enami et al. 1997). Salinity stress affects the abundance of cytochrome b6f complex involved in the transfer of electrons from PSII to photosystem I (PSI) in *Z. mays* (Zörb et al. 2009) and PSI reaction center protein in *H. vulgare* (Rasoulnia et al. 2010). The variation in the abundance of cytochrome b6f complex disturbs the electron transfer efficiency and transmembrane electrochemical proton gradients and ultimately affects ATP synthesis and NADPH formation. Adjustment of ATP synthesis and thermal dissipation take place in halophytes due to the fact that multiple isoforms of chloroplast ATP synthases (Bandehagh et al. 2011; Caruso et al. 2008; Chen et al. 2011; Huo et al. 2004; Katz et al. 2007; Kim et al. 2005; Li et al. 2011; Liska et al. 2004; Pang et al. 2010; Parker et al. 2006; Sobhanian et al. 2010a,b; Wang et al. 2008b; Yu et al. 2011; Zörb et al. 2009) and ferredoxin NADP(H) oxidoreductases (FNR) (Bandehagh et al. 2011; Caruso et al. 2008; Li et al. 2011; Liska et al. 2004; Pang et al. 2010; Peng et al. 2009; Tanou et al. 2009; Wakeel et al. 2011; Xu et al. 2010; Yu et al. 2011; Zörb et al. 2009) are regulated by salinity.

In addition to the effects of salinity on light reaction, there are also reports about the changes in the expression pattern of 14 Calvin cycle-related enzymes. During this process, $CO_2$ is fixed in chloroplasts by RuBisCO to produce 3-phosphoglycerates (PGA). The enzymes (RuBisCO-binding protein (RBP) and RuBisCO activase (RCA), FBP, SBPase, PRK, and the other five enzymes (phosphoglycerate kinase (PGK), glyceraldehyde-3-phosphate dehydrogenase (GAPDH), triose-phosphate isomerase (TPI), fructose-bisphosphate aldolase (FBA), and transketolase (TK))) involved in this process are all affected by salinity stress and show diverse changes in different plant species (Zhang et al. 2012). Photosynthesis-related proteins affected by salt stress revealed through proteomics studies include low-$CO_2$-inducible protein (LCIC) red chlorophyll catabolite reductase (RCCR) in *D. salina* and thylakoid lumen (TL) proteins in *T. halophila* and *A. thaliana*.

Low-$CO_2$-inducible protein (LCIC) is found to be induced in *D. salina* when subjected to solution of 3 M NaCl (Yu et al. 2011). LCIC, a component of the inorganic carbon transport system in the plasma membrane, is involved in carbon-concentrating mechanism, hence important for salt tolerance in the algae *D. salina*. But the concentration of two thylakoid lumen (TL) proteins, TL18.3 and TL19, found in *T. halophila* (Pang et al. 2010) and *A. thaliana* (Kim et al. 2007), respectively, is decreased under salinity stress. TL18.3 is involved in the regulation of D1 protein turnover and the assembly of PS II monomers into dimers (Sirpio et al. 2007), while TL19 is a member of PS I subunit III and is involved in the oxidation of plastocyanin in the electron transport chain (Hippler et al. 1989).

Another protein, red chlorophyll catabolite reductase (RCCR) in *C. aurantium,* is also found to be induced when treated with NaCl (Tanou et al. 2009). During chloroplast breakdown, it helps in the conversion of an intermediary red chloroplast catabolite (RCC) into primary fluorescent catabolites (pFCCs) (Rodoni et al. 1997; Takamiya et al. 2000). Its absence causes leaf cell death due to the accumulation of RCC which leads to the production of singlet oxygen (Pruzinska et al. 2007). So the increased concentration of RCCR helps in combating the problem of reactive oxygen species (ROS) and thus plays an important role in salinity tolerance. The graphical view of salinity-responsive proteins involved in the photosynthesis in plants is given in Fig. 5.2.

## ROS Scavenging System

Due to the salinity stress, the metabolites in various processes like electron transport chain in mitochondria and chloroplasts, photorespiration, fatty acid oxidation, and various detoxification reactions, cell wall peroxidases, germin-like oxalate oxidases, and amine oxidases in the apoplast are over-reduced (Miller et al. 2010; Mittler et al. 2004). This over-reduction causes the production of reactive oxygen species (ROS) like superoxide radicals ($O_2^-$), hydrogen peroxide ($H_2O_2$), and hydroxyl radicals ($OH^-$). These ROS disturb cellular redox homeostasis and cause oxidative

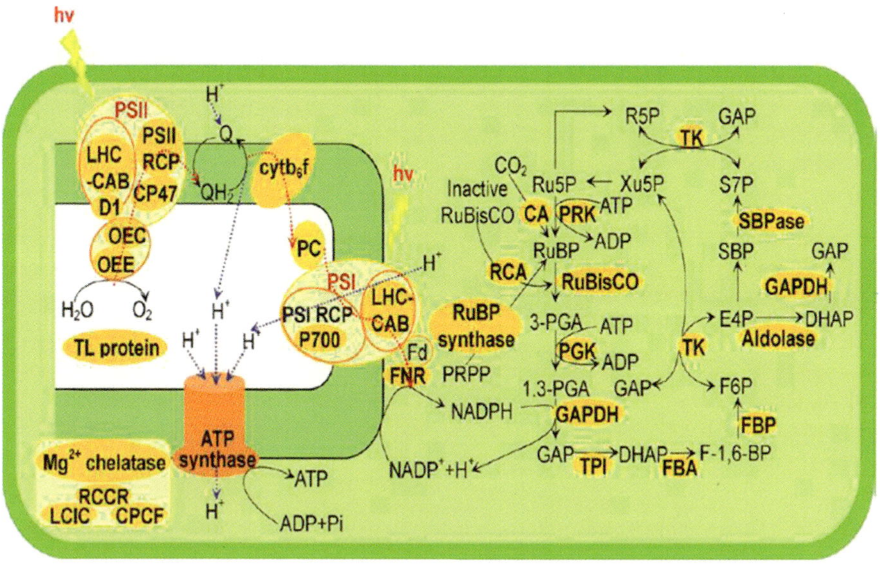


**Fig. 5.2** Schematic representation of salinity-responsive proteins involved in the photosynthesis in plants (Zhang et al. 2012)

damage to many cellular components and structures (Jithesh et al. 2006; Parida and Das 2005; Zhang et al. 2001a, b).

To cope with this problem, plants need to activate the ROS scavenging system for enhanced salt tolerance (Zhang et al. 2001a, b). Through the advancement in the field of proteomic, scientists have discovered 184 protein IDs (representing 12 UPs) as ROS scavenging-related proteins, and most of them (143 IDs) are induced by salinity in 24 plant species (Nito et al. 2007). These proteins take part in various chemical reactions like superoxide dismutation, glutathione-ascorbate cycle, catalase (CAT) pathway, peroxiredoxin/thioredoxin (PrxR/Trx) pathway, and glutathione peroxidase (GPX) pathway (Fig. 5.3). Superoxide dismutases (SOD) is the main enzyme in ROS scavenging system, usually induced by salinity, and enhances the timely dismutation of superoxide into oxygen and $H_2O_2$, which is subsequently removed through different pathways.

## Glutathione-Ascorbate Cycle

One of the most important antioxidant protection systems for removing $H_2O_2$ generated in cytosol, mitochondria, chloroplast, and peroxisomes is the glutathione-ascorbate cycle (Hasegawa et al. 2000a; Sanders et al. 1999) in which $H_2O_2$ is reduced to water by an enzyme ascorbate peroxidase (APX). In this process ascorbate (AsA) acts as the electron donor and is oxidized. As oxidized AsA (monodehydroascorbate, MDA) is still a radical, so it is converted into dehydroascorbate (DHA) spontaneously or by monodehydroascorbate reductase (MDAR). Various scientists have found that the concentration of APX, DHAR, MDAR, and GR is increased in *O. sativa* under salinity (Dooki et al. 2006; Li et al. 2010; Ruan et al. 2011; Salekdeh et al. 2002).

## Catalase (CAT) Pathway

CAT pathway mainly occurs in peroxisomes and helps in the reduction of $H_2O_2$ to $H_2O$. Through proteomics studies, it has been found that salinity stress increases the CAT levels *O. sativa* (Kim et al. 2005; Li et al. 2010) but decreases *H. vulgare* (Witzel et al. 2009).

## Peroxiredoxin/Thioredoxin (PrxR/Trx) Pathway

It is a vital antioxidant defense system in plants in which peroxiredoxins (PrxRs), a multigenic family, and thioredoxins (Trxs) are involved in ROS metabolism (Horling et al. 2003). PrxRs utilize a thiol-based catalytic mechanism to

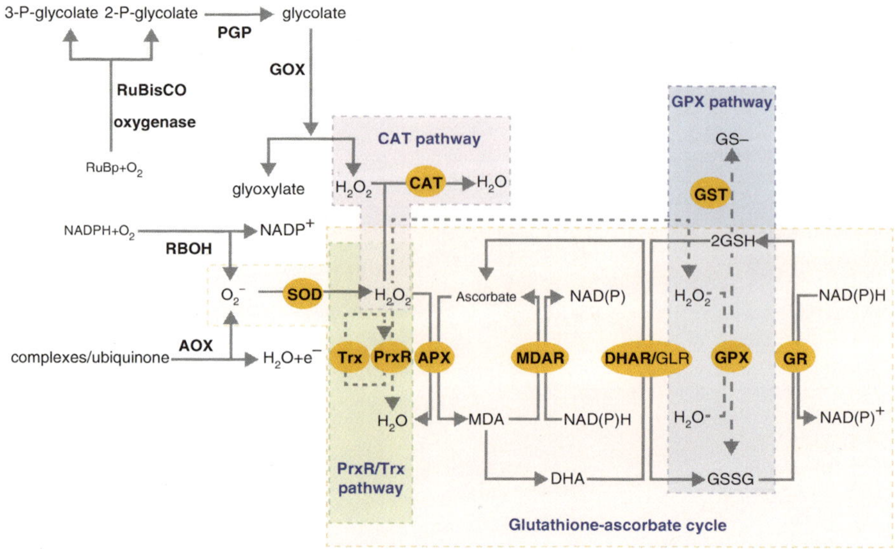


**Fig. 5.3** Schematic presentation of the salinity-responsive proteins/enzymes in ROS scavenging system in plants (Zhang et al. 2012)

reduce $H_2O_2$ and are regenerated using Trxs as electron donors (Dietz 2011). Advancement in the field of proteomics has revealed that salinity affects the two main proteins, that is, PrxRs (Askari et al. 2006; Caruso et al. 2008; Chattopadhyay et al. 2011; Du et al. 2010; Kim et al. 2007; Ndimba et al. 2005; Pang et al. 2010; Peng et al. 2009; Rasoulnia et al. 2010; Sobhanian et al. 2010b; Wang et al. 2008b, 2009; Yu et al. 2011; Zörb et al. 2010) and Trxs (Du et al. 2010; Lee et al. 2004; Nohzadeh et al. 2007; Sobhanian et al. 2010a; Wang et al. 2007), of this pathway. In *Z. mays*, the concentration of PrxRs is increased (Zörb et al. 2010).

### Glutathione Peroxidase (GPX) Pathway

This pathway is one of the major mechanisms of ROS scavenging system (Yoshimura et al. 2004) in which GPX can reduce $H_2O_2$ to the corresponding hydroxyl compounds using GSH and/or other reducing equivalents. The concentration of GPXs is increased in *S. europaea* (Li et al. 2011), while reduced in *S. aegyptiaca* (Askari et al. 2006) under salt stress conditions.

Another group of enzymes, glutathione S-transferases (GSTs), have GPX activity and can use glutathione (GSH) to reduce organic hydroperoxides of fatty acids and nucleic acids to the corresponding monohydroxy alcohols (Cummins et al. 1999; Roxas et al. 1997). These GSTs prevent the degradation of organic hydroperoxides to cytotoxic aldehyde derivatives and hence protect plants from oxidative damage under salt stress (Dixon et al. 2002). From the proteomic studies, it has been revealed that most of the GSTs are increased in salt-stressed plants (Chattopadhyay et al. 2011; Chen et al. 2011; Chitteti and Peng 2007; Jain et al. 2006; Jiang et al. 2007; Kim et al. 2007; Kumar et al. 2011; Liska et al. 2004; Pang et al. 2010; Peng et al. 2009; Rasoulnia et al. 2010; Ruan et al. 2011; Sobhanian et al. 2010b; Sugimoto and Takeda 2009; Tanou et al. 2009; Wang et al. 2008a, 2009; Witzel et al. 2009; Yu et al. 2011) but decreased in *H. vulgare* (Rasoulnia et al. 2010; Witzel et al. 2009). In addition, it has been found that the level of peroxidases (PODs) increases in all salt-stressed plants but reduces in *O. sativa* (Cheng et al. 2009; Zhang et al. 2009). PODs also play an important role in ROS scavenging system.

## Osmotic Homeostasis

As mentioned earlier, physiological water deficit and osmotic stress are the main effects of salinity in plants. In order to maintain the osmotic homeostasis, plants tend to accumulate osmolytes such as proline, soluble sugars, and glycine betaine (GB). GB is a major osmolyte which not only stabilizes the protein quaternary structure and highly ordered membrane state but also reduces lipid peroxidation during salinity stress (Chen and Murata 2008; Chinnusamy et al. 2006; Wang et al. 2004). There are a variety of proteins which are involved in osmotic homeostasis. For example, late embryogenesis abundant (LEA) proteins function to protect the steady structure of proteins, membranes, and cells (Chinnusamy et al. 2006), and their expression is increased in roots and hypocotyls of salt-treated *O. sativa* (Li et al. 2010). In *T. aestivum*, cold-regulated proteins and cold-responsive group-3 LEA-/RAB-related COR proteins are induced under stress conditions (Caruso et al. 2008). Moreover, in *O. sativa* panicles, the level of an ABA-/salt-responsive 40 kDa protein, Osr40c1s, is increased in response to salt stress (Dooki et al. 2006). This protein consists of 151 amino acids in a duplicated domain which have the ability to form amphiphilic $\alpha$-helical structures that associate with membrane proteins for salt tolerance (Moons et al. 1997).

## Salt Stress Signal Transduction

Ionic signaling, osmotic signaling, detoxification signaling, and signaling to coordinate cell division and expansion are all included in salt stress signaling (Zhu 2002). For salt tolerance in plants, the signal transduction is a burning issue, and several salt-responsive signaling pathways have been predicted which include salt overly sensitive (SOS) signaling pathway, ABA signaling pathway, $Ca^{2+}$ signal transduction pathway, protein kinase pathway, phospholipid pathway, ethylene signaling pathway, and jasmonate acid (JA)-induced signaling pathway (Cao et al. 2008; Darwish et al. 2009; Mahajan et al. 2008; Zhu 2001a, b, 2002).

According to Tuteja (2007) and Zouari et al. (2007), advancement in proteomic research has identified about 85 IDs (24 UPs) as signal transduction-related proteins in response to salt stress (Fig. 5.4). The details of the signaling pathways mentioned above are given as follows.

### G-Protein-Coupled Receptors

Under salinity conditions, kinase-mediated protein phosphorylation and/or G-proteins are involved in the transduction of stress signals (e.g., ions, ROS, and ethylene), perceived by their receptors/sensors, and helped in the regulation of corresponding signaling and metabolic pathways. From the current proteomic research, it has been clear that there are two types of receptors (the ethylene receptor (ETR) and a transforming growth factor (TGF)-beta receptor-interacting protein) which are induced in *T. aestivum* under salinity stress (Peng et al. 2009). In addition, there are also reports about the stimulation of some of G-proteins/small G-proteins and three isoforms of receptor protein kinase (RPK) identified from *T. aestivum* (Peng et al. 2009) and *O. sativa* (Dooki et al. 2006; Zhang et al. 2009) under saline conditions. The above-stated facts suggest that ethylene and ABA signaling pathways may be involved in salt response (Cao et al. 2008). In addition to the above-stated receptors, G-protein-coupled receptors are dynamically regulated to cope with salinity. It has been found through the proteomic studies that the reduced levels of a signal receiver and G-proteins/small G-proteins in *T. aestivum* (Peng et al. 2009) and *O. sativa* (Chitteti and Peng 2007), as well as two abundance-changed guanine nucleotide exchange proteins (GEP) are involved in small GTPase activation in *O. sativa* (Chitteti and Peng 2007).

### Abscisic Acid (ABA) Signaling Pathway

ABA signaling pathway is an endogenous messenger which helps in controlling plant water status and osmotic stress tolerance through its

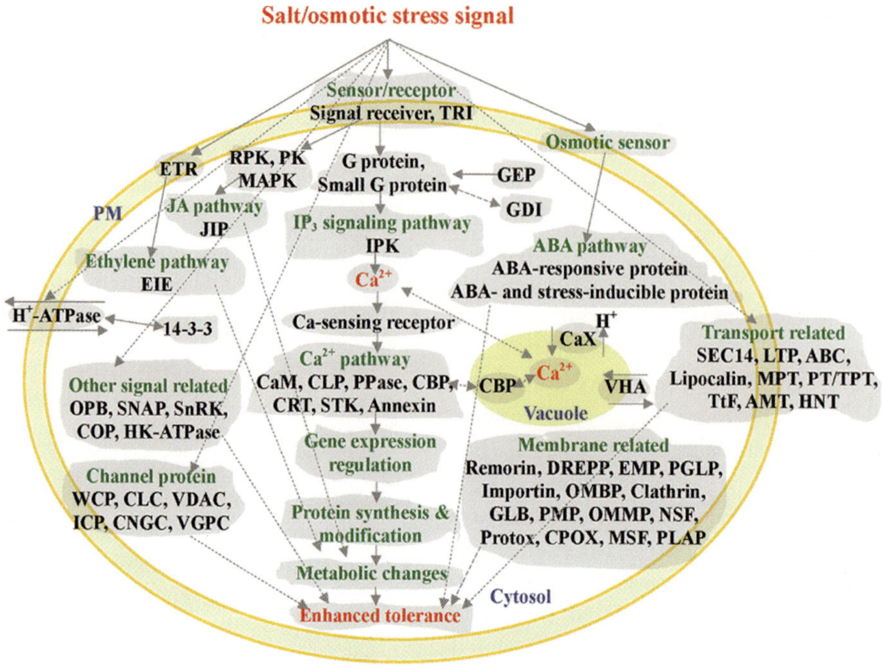


**Fig. 5.4** Graphical representation of different identified signaling pathways and transport-related proteins under salinity (Zhang et al. 2012)

role in guard cell regulation (Christmann et al. 2006; Zhu 2002). Cis-regulatory elements (i.e., ABA-responsive elements, ABREs) are shared by ABA-inducible genes which are involved in the regulation of plant stress responses. But it has been found that osmotic stress-responsive genes may be ABA independent, ABA dependent, or partially ABA dependent (Zhu 2002). Due to this, salt-/osmotic-responsive genes have been considered as either early response genes or delayed-response genes as these and their expression products are unknown. From the proteomic studies, it has been found that the level of several ABA-related proteins such as ABA-responsive proteins (ABR17 and ABR18) and ABA-/stress-inducible proteins (ASR1) is increased in *O. sativa* (Salekdeh et al. 2002) under salinity stress. But from the genetic analysis of ABA-deficient mutants, it has been found that ABA signaling is necessary in stomatal control of water loss (Schroeder et al. 2001). In conclusion, ABA signaling pathways are activated in the plants to protect themselves from water deficit associated with salinity.

## Jasmonic Acid (JA), Ethylene (ET), and Salicylic Acid (SA) Signaling Pathways

Through the advancement in proteomics studies, it has been found that a salt-inducible ethylene receptor in *T. aestivum* (Peng et al. 2009; Wang et al. 2008a), salt-responsive jasmonate-inducible proteins in the roots of *H. vulgare* (Witzel et al. 2009), as well as many members of pathogenesis-related (PR) protein in *O. sativa* (Li et al. 2010; Zhang et al. 2009) and *H. vulgare* (Sugimoto and Takeda 2009) serve as negative regulators of SA and positive regulators of JA-activated gene expression (Zhu 2002). There are also reports about the involvement of mitogen-activated protein kinase (MAPKs) in the integration of SA- and JA-dependent signals to evoke appropriate responses against pathogens and other stresses.

Moreover, some PR proteins (AtPR12) induced by JA/ET signaling, AtPR1 upregulated by SA signaling (Koornneef et al. 2008), as well as OsPR10 induced by JA/ET but suppressed by SA signaling in response to high salinity have also been found to be affected by JA/ET or SA signaling. So it has been clear that powerful applications of proteomics have helped us in unraveling the molecular mechanisms underlying hormone signaling in salt tolerance.

### $Ca^{2+}$/Calmodulin (CaM) Signaling Pathway

It is calcium ($Ca^{2+}$)-dependent signaling network that has been reported to mediate $Na^+$ homeostasis and salt resistance in many crop plants (Mahajan et al. 2008). For example, in *Z. mays* chloroplast, a $Na^+$ sensing element and $Ca^{2+}$ sensing receptor are reported to be induced at 25 mM NaCl for 1 h but decreased after 4 h (Zörb et al. 2009). Calcium-binding proteins (CBPs) are regulated by salinity in *O. sativa* (Li et al. 2010). Moreover, salinity regulates the dynamics of calmodulin in *O. sativa* (Yan et al. 2005) and in *Zea mays* (Zörb et al. 2010) and that of calcineurin-like phosphoesterase in *T. aestivum* (Peng et al. 2009) and in *O. sativa* (Li et al. 2010). These changes modulate the levels of intracellular $Ca^{2+}$ and induce specific protein kinase/phosphatase systems (Jiang et al. 2007). So $Ca^{2+}$ signaling network is closely related to the activation of the SOS signal transduction pathway, which is responsible for cellular $Na^+/K^+$ homeostasis and the osmolytes accumulation (Zhu 2002).

### 14-3-3 Proteins

*14-3-3 proteins* are a group of proteins which are commonly found and are multifunctional regulators of many cellular signaling pathways. This group of proteins interacts with a number of signaling molecules like calcium-dependent protein kinase (CDPK) and mitogen-activated protein kinase (MAPK). By interacting with the C terminus which is essential for the control of ion transport and cytoplasmic pH, these proteins act as positive regulators of plasma membrane (PM) $H^+$-ATPase (Palmgren 1998). These proteins are also known to be involved in response to salinity at multiple levels including regulating target proteins with functions including signaling, transcription activation, and defense and also working as components of transcription factor complexes associated with ABA-induced gene expression. Through the advancement in proteomics, it has been found that in *T. aestivum* (Wang et al. 2008a), *O. sativa* (Cheng et al. 2009; Nohzadeh et al. 2007), and *Z. mays* (Zhu 2003), many members of this group such as 14-3-3 protein (gi13928452, and gi12229593) (Jain et al. 2006; Zörb et al. 2010), 14-3-3-like protein (gi1168189, and gi7267542) (Wang et al. 2008a), GF14a(XP-48289), GF14b (gi50924768) (Nohzadeh et al. 2007), and GF14 kappa isoform (gi30698122) (Ndimba et al. 2005) are regulated by salinity conditions. So we can conclude that 14-3-3 proteins regulate multiple pathways involved in salt stress response.

### Ion Homeostasis and Cross-Membrane Transport

Under salinity conditions, aqueous and ionic thermodynamic equilibrium is altered through high apoplastic levels of $Na^+$ and $Cl^-$ which results in hyperosmotic stress, ionic imbalance, and toxicity. So to cope with this problem, plants reestablish cellular ion homeostasis by regulating ion uptake/exclusion and in vivo compartmentalization. Ion (e.g., $K^+$ and $Na^+$) homeostasis, a fine-tuned process, mainly relies on the proton-motive forces created by the action of $H^+$-ATPases, various ion channels, and transporters (Gong et al. 2001). From the recent studies, it has been found that for the regulation of the activity of plasma membrane $H^+$-ATPases, phototropin and 14-3-3 proteins have to work cooperatively as well as independently for opening and closing of the ion channels (e.g., $K^+$ channel) (Inoue et al. 2005; van den Wijngaard et al. 2005). Currently, various researchers through proteomic studies have found that salinity stress significantly affects $H^+$-ATPase (Barkla et al. 2009; Cheng et al. 2009; Du et al. 2010; Jiang et al. 2007; Katz et al. 2007; Manaa et al. 2011; Nat et al. 2004; Ndimba et al. 2005; Pang et al. 2010; Wang et al. 2008a, 2009; Yu et al. 2011; Zörb et al. 2004), ATP-binding cassette (ABC) transporter (Wang et al. 2008a, b, 2009), and other ion channels and transporters (Li et al. 2011; Peng et al. 2009; Sobhanian et al. 2010b; Wakeel et al.

2011; Wang et al. 2008a, b, 2009; Zörb et al. 2010).

### $H^+$-ATPase

For the maintenance of ion homeostasis in plant cells, $H^+$-ATPases are one of the most important enzymes because most of the vacuolar $H^+$-ATPases in glycophytes like *O. sativa* (Cheng et al. 2009), *T. aestivum* (Wang et al. 2008a), and *Z. mays* (Zörb et al. 2004) are induced under salinity stress. For $Na^+$ transport by salt overly sensitive (SOS1) which is essential for salt tolerance, it has been that the level of $H^+$-ATPases is increased which provides a more strong driving force for this transport (Kim et al. 2007; Zhu 2001a, b, 2003). The proton electrochemical gradient for vacuolar $Na^+/H^+$ antiporter to compartmentalize $Na^+$ in the vacuoles is generated through vacuolar $H^+$-ATPases, the major $H^+$-pumps on the tonoplast (Chinnusamy et al. 2005). For $Na^+$ sequestration and osmotic adjustment under salinity stress, increased levels and/or activities of the vacuolar $H^+$-ATPases are found to be a cost-effective strategy (Pang et al. 2010). Moreover, it has also been found that mitochondrial-/chloroplast-located $H^+$-ATPases are also involved in ion homeostasis (Huo et al. 2004; Kim et al. 2005; Ndimba et al. 2005).

### ABC Transporters and Other Transporters

In *T. aestivum*, it has been found that ABC transporters (the in charge of transporting of stress-related secondary metabolites, such as alkaloids, terpenoids, polyphenols, and quinines (Yazaki 2006)) are induced under salinity stress (Wang et al. 2008a; Peng et al. 2009). There are three proteins which cause abundant changes in plants under salinity and include ferritin (Chen et al. 2009; Parker et al. 2006; Wang et al. 2009), iron deficiency-induced protein (IDI), and iron deficiency-specific protein (IDS) (Witzel et al. 2009). The first protein, ferritin, through the Fenton reaction, helps in the sequestration of excess free irons and prevents formation of hydroxyl radicals (Laohavisit et al. 2010; Parker et al. 2006) and is reported to be induced in *O. sativa* (Parker et al. 2006) under certain salinity conditions. But in contrast, the level of other two transporters, that is, IDIs and IDSs, is decreased in *H. vulgare* under salinity stress (Witzel et al. 2009) which is beneficial in avoiding excessive ion uptake.

### Ion Channel Proteins

To maintain ion homeostasis under salinity stress, the level of different ion channels is changed, that is, increases or decreases; for example, the level of voltage-gated potassium channels in *T. aestivum* (Peng et al. 2009) is induced which is crucial for a balance of $K^+/Na^+$ in the cells, but in case of a cyclic nucleotide-gated ion channel (CNGC), the level is reduced as a nonselective cation channel (Wang et al. 2008a). The direct binding of cyclic nucleotides (cAMP and cGMP) helps in the opening of CNGC of which the activity is of little voltage dependence, but $Ca^{2+}$/calmodulin and phosphorylation help in its modulation. Another ion channel, annexin, revealed through proteomic studies, is a $Ca^{2+}$-permeable channel at endomembrane and plasma membrane for the formation of a ROS-stimulated passive $Ca^{2+}$ transport pathway (Laohavisit et al. 2010), and its level is reported to be induced in various cereal crops like *O. sativa* (Li et al. 2010) and *T. aestivum* (Peng et al. 2009). Its increased level plays a vital role in osmotic adjustment and subsequently cell expansion and exocytosis (Faurobert et al. 2007; Lee et al. 2004). Moreover, it has been found that the level of voltage-dependent anion channel protein (VDAC), a barrel protein located at the outer mitochondrial membrane and responsible for passage of small molecules (<1,000 Da) into the intermembrane space, is induced under salinity stress in *Z. mays* (Sugimoto and Takeda 2009). Its dynamic changes are found to influence the mitochondrial respiration (Rostovtseva and Bezrukov 2008; Rostovtseva et al. 2008).

### Plasma Membrane and Other Membrane-Associated Proteins

Twenty four plasma membrane proteins are reported to be found in rice which are induced under salinity stress and regulate the development of plasma membrane polypeptides containing a Glu-rich site at the C terminus, responsible for calcium binding in $Ca^{2+}$ signal transduction pathway (Yuasa and Maeshima 2000). In addition, a plant-specific PM/lipid-raft protein, remorin, helps in maintaining the membrane skeletons (Bariola et al. 2004) and thus contributes in the stabilization of damaged PM under salinity stress (Cheng et al. 2009; Nohzadeh et al. 2007). Moreover, the level of importin (a nuclear membrane transporter) and an outer mitochondrial membrane porin in *T. aestivum* (Peng et al. 2009; Wang et al. 2008a; Witzel et al. 2009) is reduced under salinity stress.

## Transcription and Protein Fates

Changes in transcriptional regulatory networks of cis-/trans-elements and transcription factors can be triggered through aforementioned signaling systems. Through the advancement in proteomic research, it has been found that the levels of transcription factors and transcription-related proteins are regulated under salinity stress and play a vital role in salt tolerance (Zhang et al. 2012). To cope with salt stress in *T. aestivum* (Peng et al. 2009; Wang et al. 2008a), the salt-induced transcription factor, basic transcription factor 3 (BTF3), is an important regulatory component and controls diverse processes. In addition, the increased levels of DNA polymerases in *Z. mays* (Zörb et al. 2004) and DNA helicases in *T. aestivum* (Wang et al. 2008a) enhance DNA replication, unwinding, and transcription under salinity. Furthermore, there are also reports about the effects of salinity on some RNA processing- and spicing-related proteins such as maturase K (Chattopadhyay et al. 2011; Wang et al. 2009; Zörb et al. 2010), nucleic acid- binding proteins (Aghaei et al. 2008; Bandehagh et al. 2011; Caruso et al. 2008; Chen et al. 2009; Jain et al. 2006; Kim et al. 2005; Pang et al. 2010; Tanou et al. 2009; Wang et al. 2008b; Witzel et al. 2009), glycine-rich RNA-binding proteins (Askari et al. 2006; Dooki et al. 2006; Jiang et al. 2007; Manaa et al. 2011), and RNA splicing factors (Wang et al. 2008a; Yan et al. 2005). In abiotic stress adaptation, protein synthesis plays a very important role, and proteomics studies have found that many of its components including different ribosomal proteins (Aghaei et al. 2008; Bandehagh et al. 2011; Chattopadhyay et al. 2011; Chen et al. 2009; Chitteti and Peng 2007; Dani et al. 2005; Du et al. 2010; Jiang et al. 2007; Kim et al. 2005; Ndimba et al. 2005; Pang et al. 2010; Peng et al. 2009; Sobhanian et al. 2010a, b; Veeranagamallaiah et al. 2008; Wang et al. 2008b; Zörb et al. 2004, 2010), translation initiation factors (Jiang et al. 2007; Ndimba et al. 2005; Pang et al. 2010; Parker et al. 2006; Peng et al. 2009; Wang et al. 2008a, b, 2009; Yu et al. 2011), poly(A)-binding proteins (Jiang et al. 2007; Witzel et al. 2009), translation elongation factors (Chen et al. 2009; Liska et al. 2004; Ndimba et al. 2005; Pang et al. 2010; Peng et al. 2009; Sobhanian et al. 2010b; Witzel et al. 2010), translationally controlled tumor proteins (Nat et al. 2004; Pang et al. 2010; Sobhanian et al. 2010b; Vincent et al. 2007; Witzel et al. 2010; Yu et al. 2011; Zörb et al. 2010), RNA recognition motif (RRM)-containing proteins (Pang et al. 2010; Yu et al. 2011), and tRNA synthases (Ndimba et al. 2005; Pang et al. 2010; Peng et al. 2009; Wen et al. 2010) are altered in expression under salinity stress conditions.

Generally, salinity stress represses protein synthesis (Tuteja 2007). However, the level of some of the above proteins is increased which shows that normal cellular processes are required for the maintenance of protein synthesis activities under salinity stress (Zhang et al. 2012). For maintaining normal cellular functions under salinity stress, correct protein folding and transport is crucial. For this, heat shock proteins (HSPs) and other molecular chaperons play a very important role in protein structure stabilization and subcellular localization (Vierling 1991). Salinity affects various HSPs/chaperonins (Aghaei et al. 2008; Chattopadhyay et al. 2011; Chen et al. 2009, 2011; Chitteti and Peng 2007; Dani et al. 2005; Du et al. 2010; Geissler et al. 2010; Jain et al. 2006; Jiang et al. 2007; Katz et al. 2007; Kim et al. 2005, 2007; Li et al. 2010; Liska et al. 2004; Manaa et al. 2011; Nat et al. 2004; Ndimba et al. 2005; Pang et al. 2010; Peng et al. 2009; Razavizadeh et al. 2009; Sengupta and Majumder 2009;

Sobhanian et al. 2010a, b; Tanou et al. 2009; Wang et al. 2008a, b, 2009; Wen et al. 2010; Witzel et al. 2010; Xu et al. 2010; Yu et al. 2011; Zörb et al. 2010), luminal-binding proteins (LBP) (Kim et al. 2007; Liska et al. 2004; Pang et al. 2010; Wang et al. 2008b, 2009), peptidyl-prolyl cis-trans isomerases (Aghaei et al. 2008; Askari et al. 2006; Chen et al. 2011; Tanou et al. 2009; Zörb et al. 2009), protein disulfide isomerases (PDI) (Chen et al. 2011; Jiang et al. 2007; Nohzadeh et al. 2007; Pang et al. 2010; Yu et al 2011), T-complex proteins (Pang et al. 2010; Wang et al. 2008a), AAA ATPase superfamily proteins (Chen et al. 2011; Wen et al. 2010; Yu et al 2011), and cold shock domain-containing proteins (Huo et al. 2004; Peng et al. 2009; Ruan et al. 2011) which are involved in maintaining normal protein folding, repair, and renaturation of the stress-damaged proteins (Fig. 5.5). In addition, for the selective degradation of proteins, plants use proteasome pathways. Some members of these pathways such as ubiquitin/poly-ubiquitin/tetraubiquitin (Du et al. 2010; Katz et al. 2007; Xu et al. 2010), SKP1 protein (Liska et al. 2004), proteasome components (Askari et al. 2006; Chattopadhyay et al. 2011; Jiang et al. 2007; Kim et al. 2005; Liska et al. 2004; Peng et al. 2009; Sobhanian et al. 2010b; Tanou et al. 2009; Vincent et al. 2007; Wang et al. 2008a, b, 2009), various proteases (Katz et al. 2007; Liska et al. 2004; Manaa et al. 2011; Pang et al. 2010; Peng et al. 2009; Veeranagamallaiah et al. 2008; Wang et al. 2008b, 2009; Xu et al. 2010; Yu et al. 2011; Zörb et al. 2009;) and peptidases (Chitteti and Peng 2007; Jiang et al. 2007; Li et al. 2010; Liska et al. 2004; Manaa et al. 2011; Ndimba et al. 2005; Pang et al. 2010; Wang et al. 2008b; Yu et al. 2011), protease inhibitors (Aghaei et al. 2009; Chen et al. 2011; Peng et al. 2009; Sobhanian et al. 2010a), and reversed protein-methionine-S-oxide reductases (Pang et al. 2010) under salinity stress exhibit tremendous changes. Degradation of proteins not only is important in protein turnover during ubiquitin-mediated degradation of proteins but also helps in the regulation of other cellular processes such as signal transduction and transcription. So, these salt-responsive proteins have a vital role to play in salinity tolerance.

## Cytoskeleton and Cell Structure

For cell turgor maintenance, a rapid remodeling of cytoskeleton for the adjustment of cell size takes place during salinity stress (Pang et al. 2010; Li et al. 2011). Through the advancement in the field of proteomics, it has been found that basic cytoskeleton components such as actin (Cheng et al. 2009; Jiang et al. 2007; Li et al. 2010; Tanou et al. 2009; Xu et al. 2010) and tubulin (Jiang et al. 2007; Katz et al. 2007; Kim et al. 2007; Liska et al. 2004; Pang et al. 2010; Peng et al. 2009) and other cytoskeleton-related proteins (some actin-binding proteins (ABPs)) (Yan et al. 2005), kinesin motor (Chitteti and Peng 2007; Sobhanian et al. 2010a; Wang et al. 2009), myosin (Cheng et al. 2009; Peng et al. 2009; Wang et al. 2009), and xyloglucan endo-transglycosylase (XET) hydrolases (Zörb et al. 2010) are changed under salinity stress. ABPs such as actin-depolymerizing factors (ADFs) (Sobhanian et al. 2010b), profilins (Askari et al. 2006; Du et al. 2010; Wang et al. 2009), and cyclase-associated proteins (CAPs) (Ndimba et al. 2005) play key roles in the remodeling as these have the ability to bind with actin cytoskeletons; for example, ADFs promote filamentous actin disassembly and thus modulate the dynamic organization of actin cytoskeletons. Profilin has the ability to join with actin monomers and cause polymerization/depolymerization of actin filaments to maintain cell structure integrity, cell mobility, tumor cell metastasis, and growth factor signaling (Staiger et al. 1997). Similarly CAPs which are multifunctional ABPs are involved in various signal transduction pathways involved in cell growth, development, vesicle trafficking, and endocytosis. Under salinity stress, it has been found that cytoskeleton dynamics is associated with other physiological

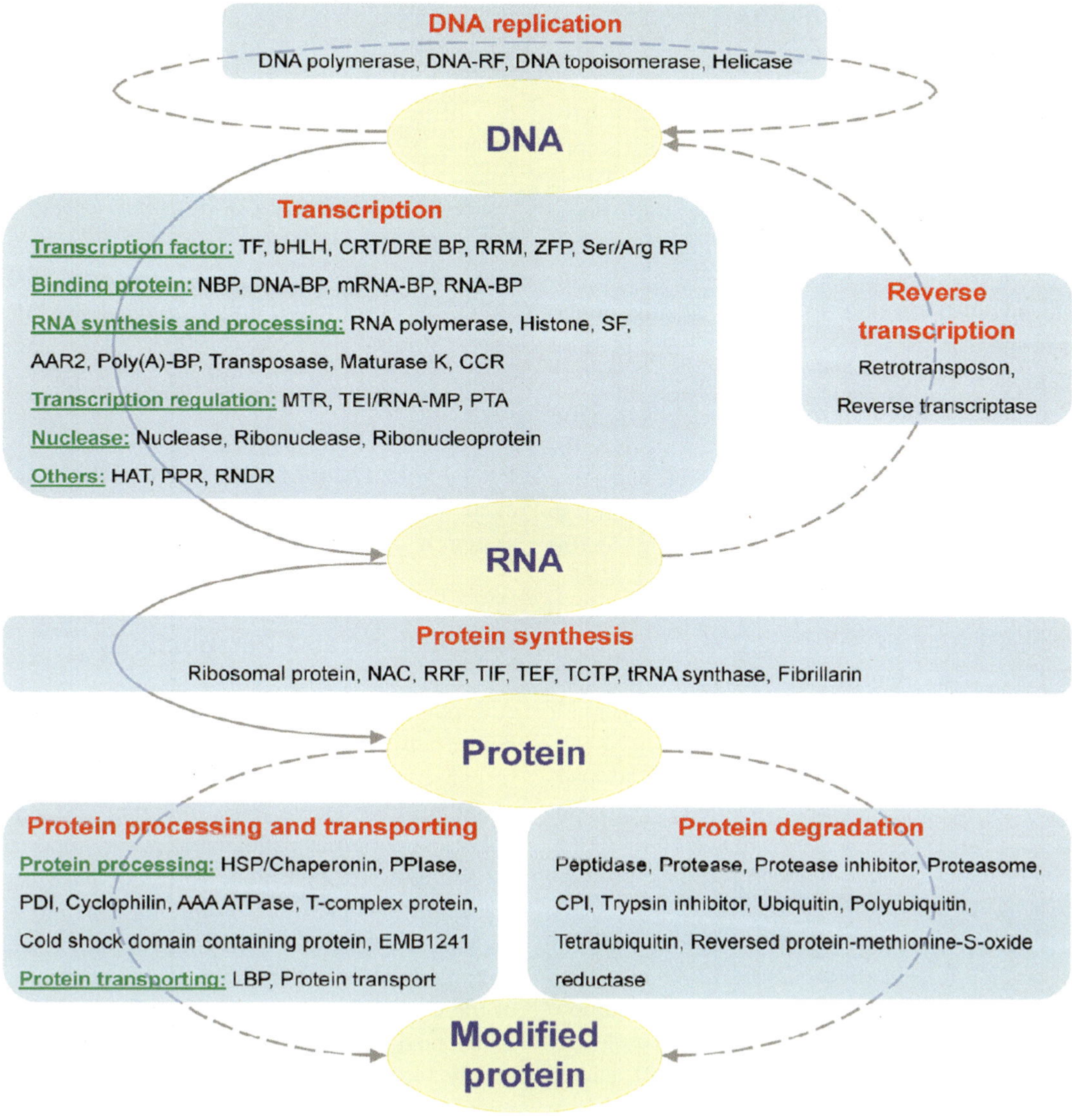


**Fig. 5.5** Schematic representation of the salinity-responsive proteins involved in transcription and protein metabolism (Zhang et al. 2012)

changes like the osmotic stress regulation of actin organization that correlates well with $K^+$ channel activity in guard cells (Luan 2002). For controlling cell expansion and morphology, tubulins may co-migrate with P-type ATPases (Campetelli et al. 2005) or connect with the plasma membrane (Drykova et al. 2003). Moreover, XETs are involved in cutting and rejoining of intermicrofibrillar xyloglucan chains and help in wall loosening and ultimately for cell expansion. So under salinity stress, growth inhibition occurs due to reduced expression of aforementioned enzymes (Fry et al. 1992).

## Cross-Tolerance to Multiple Stresses

Various signaling and metabolic pathways are connected into networks because plants have developed cross-tolerance mechanisms to cope with different stresses at the same time (Tuteja 2007); for example, in salt tolerance, some biotic stress-responsive proteins/genes play important roles. Carbohydrate-binding proteins like lectins are involved not only in defense against predators and pathogens (De Hoff et al. 2009) but also in plant salt tolerance (Van Damme et al. 2004). Researchers through proteomics studies have found that salinity stress enhances not only the levels of lectins in *O. sativa* (Chitteti and Peng 2007) and *S. bicolor* (Kumar et al. 2011) but also that of cytoplasmic mannose-binding lectins (Claes et al. 1990; Hirano et al. 2000; Zhang et al. 2000) and salt stress-induced proteins (salt proteins) (Chitteti and Peng 2007; Wen et al. 2010). The biotic stress-related proteins, revealed through proteomics, include elicitor peptides (Aghaei et al. 2009; Wang et al. 2009), disease-related/resistance proteins (Chattopadhyay et al. 2011; Chitteti and Peng 2007; Nat et al. 2004; Peng et al. 2009; Wang et al. 2008a, 2009), hypersensitive-induced response proteins (Cheng et al. 2009; Nohzadeh et al. 2007), pathogenesis-related proteins (Chattopadhyay et al. 2011; Jain et al. 2006; Li et al. 2010; Manaa et al. 2011; Pang et al. 2010; Sugimoto and Takeda 2009; Vincent et al. 2007; Wang et al. 2008b, 2009; Zhang et al. 2009), stress-inducible proteins (Dooki et al. 2006; Wang et al. 2008a; Witzel et al. 2009), and universal stress protein family (Kumar et al. 2011; Li et al. 2010; Ndimba et al. 2005; Wang et al. 2008a), and while that of abiotic stress-related proteins include cold-regulated proteins (Wang et al. 2009), cold-responsive LEA-/RAB-related COR proteins (Caruso et al. 2008), and copper homeostasis factors (Jiang et al. 2007). Salt tolerance is also contributed by the multifunctional glyoxalase system. Under salinity stress, plants tend to accumulate high amounts of methylglyoxal (MG) which is a by-product of glycolysis mainly from triose phosphate. MG has detrimental effects on plants as it involved in the inhibition of cell proliferation (Ray et al. 1994), degradation of proteins, and inactivation of antioxidant defense system (Martins et al. 2001). MG is detoxified through glyoxalase system consisting of glyoxalase I (GlyI) and glyoxalase II (GlyII). From the previous molecular studies involving transgenic plants, it has found that overexpression of GlyI and GlyII enhances plant salt tolerance (Singla-Pareek et al. 2003, 2008; Yadav et al. 2005). From the proteomics, it has been found that the level of GlyI increases in *O. sativa* (Chitteti and Peng 2007; Li et al. 2010) when treated with 150 mM NaCl for 6 or 48 h (Jiang et al. 2007) which implies that the glyoxalase system is under dynamic regulation.

## Conclusions and Perspectives

A sophisticated fine-tuned signaling and metabolic network is there under plant salinity response and tolerance. Significant discoveries of salinity-responsive genes, proteins, and metabolites in different cellular pathways important for salt stress response and tolerance have been made through previous morphological, physiological, genetic, and genomic analyses. But a large gap exists in the systematic understanding of the molecular processes and networks, as this understanding is still in the beginning. Modern highly sophisticated and state-of-the-art proteomics techniques have helped us in the acquisition of more detailed quantitative information on the temporal and spatial expression of proteins. Currently, overall 2,171 proteins have been identified in 34 plants, but the need of the hour is to fill the gaps present between the research and actual field conditions.

The dynamic nature of the proteins involved under salinity stress further provides invaluable information toward understanding of the underlying sophisticated cellular and molecular processes like photosynthesis, energy metabolism, ROS scavenging, ion/osmotic homeostasis, signaling transduction, transcription and translational regulation, and cytoskeleton dynamics. However, a large gap still exists in our knowledge of transmembrane ion transport and cellular compartmentalization, sensors/receptors in signaling transduction, molecules in long distance signaling, and metabolites in energy supply, especially in case of cereals. Moreover, molecular interactions and pathway cross talks should have to be arranged, and future targets should be set on the basis of the gaps found during these talks. New state-of-the-art technologies like nanotechnology should be used in the understanding of molecular mechanisms. Through nanotechnology, enzyme biosensors (controlled-pore glass beads with optical transducer element, polyurethane foam with photothermal transducer element, ion-selective membrane with either potentiometric or amperometric transducer element, and screen-printed electrode with amperometric transducer element) should be used for the detection of very minute changes occurring during the quantitative analysis of the salt tolerance mechanisms. Also the advanced and sophisticated proteomics approaches and technologies like multidimensional protein fractionation, isobaric tags for relative and absolute quantitation (iTRAQ), label-free quantification mass spectrometry, and phosphoprotein and glycoprotein enrichment and tagging will definitely help us in discovering low-abundance proteins (e.g., transcriptional factors, kinases, channels, and transporters) and novel regulatory mechanisms (e.g., phosphorylation) in salt stress signaling and metabolism pathways. Integration of proteomics results with findings from other large scale sources will definitely facilitate in the establishment of molecular networks underlying salt stress response and tolerance in cereals. After integration of the knowledge collected, we will be able to predict the underlying molecular mechanisms of salinity tolerance. It will prove helpful toward the ultimate goal of improving plant salt tolerance for enhanced yield and bioenergy.

## References


Abbasi FM, Komatsu S (2004) A proteomic approach to analyze salt-responsive proteins in rice leaf sheath. Proteomics 4:2072–2081

Aghaei K, Ehsanpour AA, Komatsu S (2008) Proteome analysis of potato under salt stress. J Proteome Res 7:4858–4868

Aghaei K, Ehsanpour AA, Shah AH, Komatsu S (2009) Proteome analysis of soybean hypocotyl and root under salt stress. Amino Acids 36:91–98

Almoguera C, Shibata D, Forrester K, Matin J, Arnheim N, Perucho M (1988) Most human carcinomas of the exocrine pancreas contain mutant c-K-ras genes. Cell 53:549–554

Anuradha S, Rao SSR (2001) Effect of brassinosteroids on salinity stress induced inhibition of seed germination and seedling growth of rice (*Oryza sativa* L.). Plant Growth Regul 33:151–153

Ashraf M, Harris PJC (2004) Potential biochemical indicators of salinity tolerance in plant. Plant Sci 166:3–16

Askari H, Edqvist J, Hajheidari M, Kafi M, Salekdeh GH (2006) Effects of salinity levels on proteome of *Suaeda aegyptiaca* leaves. Proteomics 6:2542–2554

Bandehagh A, Salekdeh GH, Toorchi M, Mohammadi A, Komatsu S (2011) Comparative proteomic analysis of canola leaves under salinity stress. Proteomics 11:1965–1975

Bariola PA, Retelska D, Stasiak A, Kammerer RA, Fleming A, Hijri M, Frank S, Farmer EE (2004) Remorins form a novel family of coiled coil-forming oligomeric and filamentous proteins associated with apical, vascular and embryonic tissues in plants. Plant Mol Biol 55:579–594

Barkla BJ, Vera-Estrella R, Hernandez-Coronado M, Pantoja O (2009) Quantitative proteomics of the tonoplast reveals a role for glycolytic enzymes in salt tolerance. Plant Cell 21:4044–4058

Blumwald E, Aharon GS, Apse MP (2000) Sodium transport in plant cells. Biochemica et Biophysica Acta 1465:140–151

Bohnert HJ, Nelson DE, Jensen RG (1995) Adaptations to environmental stresses. Plant Cell 7:1099–1111

Bressan RA, Hasegawa PM, Pardo JM (1998) Plants use calcium to resolve salt stress. Trends Plant Sci 3:411–412

Campetelli AN, Previtali G, Arce CA, Barra HS, Casale CH (2005) Activation of the plasma membrane H+-ATPase of *Saccharomyces cerevisiae* by glucose is mediated by dissociation of the H+-ATPase-acetylated tubulin complex. FEBS J 272:5742–5752

Cao YR, Chen SY, Zhang JS (2008) Ethylene signaling regulates salt stress response: an overview. Plant Signal Behav 3:761–763

Caruso G, Cavaliere C, Guarino C, Gubbiotti R, Foglia P, Lagana A (2008) Identification of changes in *Triticum durum* L. leaf proteome in response to salt stress by two-dimensional electrophoresis and MALDI-TOF mass spectrometry. Anal Bioanal Chem 391:381–390

Chand R, Kumar P (2006) Country case study: India. In: Thomas H (ed) Trade reforms and food security. Food and Agriculture Organization of the United Nations, Rome

Chattopadhyay A, Subba P, Pandey A, Bhushan D, Kumar R, Datta A, Chakraborty S, Chakraborty N (2011) Analysis of the grass pea proteome and identification of stress-responsive proteins upon exposure to high salinity, low temperature, and abscisic acid treatment. Phytochemistry 72:1293–1307

Chen TH, Murata N (2008) Glycinebetaine: an effective protectant against abiotic stress in plants. Trends Plant Sci 13:499–505

Chen SB, Gollop N, Heuer B (2009) Proteomic analysis of salt stressed tomato (*Solanum lycopersicum*) seedlings: effect of genotype and exogenous application of glycinebetaine. J Exp Bot 60:2005–2019

Chen FG, Zhang S, Jiang H, Ma WJ, Korpelainen H, Li C (2011) Comparative proteomics analysis of salt response reveals sex-related photosynthetic inhibition by salinity in *Populus cathayana* cuttings. J Proteome Res 10:3944–3958

Cheng YW, Qi YC, Zhu Q, Chen X, Wang N, Zhao X, Chen HY, Cui XJ, Xu LL, Zhang W (2009) New changes in the plasma membrane-associated proteome of rice roots under salt stress. Proteomics 9:3100–3114

Chinnusamy V, Jagendorf A, Zhu JK (2005) Understanding and improving salt tolerance in plants: genetic and metabolic engineering for value-added traits. Crop Sci Soc Am 45:437–448

Chinnusamy V, Zhu J, Zhu JK (2006) Salt stress signaling and mechanisms of plant salt tolerance. Genet Eng 27:141–177

Chitteti BR, Peng Z (2007) Proteome and phosphoproteome differential expression under salinity stress in rice (*Oryza sativa*) roots. J Proteome Res 6:1718–1727

Christmann A, Moes D, Himmelbach A, Yang Y, Tang Y, Grill E (2006) Integration of abscisic acid signalling into plant responses. Plant Biol 8:314–325

Claes B, Dekeyser R, Villarroel R, Van den Bulcke M, Bauw G, Van Montagu M, Caplan A (1990) Characterization of a rice gene showing organ-specific expression in response to salt stress and drought. Plant Cell 2:19–27

Cummins I, Cole DJ, Edwards R (1999) A role for glutathione transferases functioning as glutathione peroxidases in resistance to multiple herbicides in blackgrass. Plant J 18:285–292

Cushman JC, Vermon DM, Bohnert HJ (1992) ABA and the transcriptional control of CAM induction during salt stress in the common ice plant. In: Verma DPS (ed) Control of plant gene expression. CRC Press, Boca Raton, pp 287–300

Dani V, Simon WJ, Duranti M, Croy RR (2005) Changes in the tobacco leaf apoplast proteome in response to salt stress. Proteomics 5:737–745

Darwish E, Testerink C, Khalil M, El-Shihy O, Munnik T (2009) Phospholipid signaling responses in salt-stressed rice leaves. Plant Cell Physiol 50:986–997

Dash M, Panda SK (2001) Salt stress induced changes in growth and enzyme activities in germination *Phaseolus mungo* seeds. Biol Plant 44:587–589

De Hoff PL, Brill LM, Hirsch AM (2009) Plant lectins: the ties that bind in root symbiosis and plant defense. Mol Genet Genomics 282:1–15

Diedhiou CJ, Popova OV, Golldack D (2009) Transcript profiling of the salt-tolerant *Festuca rubra* ssp. litoralis reveals a regulatory network controlling salt acclimatization. J Plant Physiol 166:697–711

Dietz KJ (2011) Peroxiredoxins in plants and cyanobacteria. Antioxid Redox Signal 15:1129–1159

Dixon DP, Lapthorn A, Edwards R (2002) Plant glutathione transferases. Genome Biol 3:No. REVIEWS3004

Dooki AD, Mayer-Posner FJ, Askari H, Zaiee AA, Salekdeh GH (2006) Proteomic responses of rice young panicles to salinity. Proteomics 6:6498–6507

Drykova D, Cenklova V, Sulimenko V, Volc J, Draber P, Binarova P (2003) Plant gamma-tubulin interacts with alpha beta-tubulin dimers and forms membrane-associated complexes. Plant Cell 15:465–480

Du CX, Fan HF, Guo SR, Tezuka T, Li J (2010) Proteomic analysis of cucumber seedling roots subjected to salt stress. Phytochemistry 71:1450–1459

Enami I, Tohri A, Kamo M, Ohta H, Shen JR (1997) Identification of domains on the 43 kDa chlorophyll-carrying protein (CP43) that are shielded from tryptic attack by binding of the extrinsic 33 kDa protein with Photosystem II complex. Biochim Biophys Acta Bioenerg 1320(1):17–26

Evans LT (1998) Feeding the ten billion: plants and population growth, Cambridge, UK Cambridge University Press

Faurobert M, Mihr C, Bertin N, Pawlowski T, Negroni L, Sommerer N, Causse M (2007) Major proteome variations associated with cherry tomato pericarp development and ripening. Plant Physiol 143:1327–1346

Flowers TJ, Colmer TD (2008) Salinity tolerance in halophytes. New Phytol 179:945–963

Fricke W, Akhiyarova G, Veselov D, Kudoyarova G (2004) Rapid and tissue-specific changes in ABA and in growth rate in response to salinity in barley leaves. J Exp Bot 55:1115–1123

Fry SC, Smith RC, Renwick KF, Martin DJ, Hodge SK, Matthews KJ (1992) Xyloglucan endotransglycosylase, a new wall-loosening enzyme activity from plants. Biochem J 282:821–828

Geissler N, Hussin S, Koyro HW (2010) Elevated atmospheric $CO_2$ concentration enhances salinity tolerance in *Aster tripolium* L. Planta 231:583–594

Gong ZZ, Koiwa H, Cushman MA, Ray A, Bufford D, Kore-eda S, Matsumoto TK, Zhu JH, Cushman JC, Bressan RA, Hasegawa PM (2001) Genes that are uniquely stress regulated in salt overly sensitive (SOS) mutants. Plant Physiol 126:363–375

Hamed K, Castagna A, Salem E, Ranieri A, Abdelly C (2007) Sea Fennel (*Crithmum martimum* L.) under salinity conditions: a comparison of leaf and root antioxidant responses. Plant Growth Regul 53:185–194

Hasegawa PM, Bressan RA, Pardo JM (2000a) The dawn of plant salt to tolerance genetics. Trends Plant Sci 5:317–319

Hasegawa PM, Bressan RA, Zhu J-K, Bohnert HJ (2000b) Plant cellular and molecular responses to high salinity. Annu Rev Plant Physiol Plant Mol Biol 51:463–499

Heidari M (2009) Variation in seed germination, seedling growth, nucleic acids and biochemical component in canola (*Brassica nupus* L.) under salinity stress. Asian J Plants Sci 8:557–561

Hippler M, Ratajczak R, Haehnel W (1989) Identification of the plastocyanin binding subunit of photosystem I. FEBS Lett 250:280–284

Hirano K, Teraoka T, Yamanaka H, Harashima A, Kunisaki A, Takahashi H, Hosokawa D (2000) Novel mannose-binding rice lectin composed of some isolectins and its relation to a stress-inducible salT gene. Plant Cell Physiol 41:258–267

Horling F, Lamkemeyer P, Konig J, Finkemeier I, Kandlbinder A, Baier M, Dietz KJ (2003) Divergent light-, ascorbate-, and oxidative stress-dependent regulation of expression of the peroxiredoxin gene family in Arabidopsis. Plant Physiol 131:317–325

Huo CM, Zhao BC, Ge RC, Shen YZ, Huang ZJ (2004) Proteomic analysis of the salt tolerance mutant of wheat under salt stress. Yi Chuan Xue Bao 31:1408–1414

Inoue S, Kinoshita T, Shimazaki K (2005) Possible involvement of phototropins in leaf movement of kidney bean in response to blue light. Plant Physiol 138:1994–2004

Jacoby RP, Millar AH, Taylor NL (2010) Wheat mitochondrial proteomes provide new links between antioxidant defense and plant salinity tolerance. J Proteome Res 9:6595–6604

Jain S, Srivastava S, Sarin NB, Kav NN (2006) Proteomics reveals elevated levels of PR 10 proteins in saline-tolerant peanut (*Arachis hypogaea*) calli. Plant Physiol Biochem 44:253–259

Jha B, Agarwal PK, Reddy PS, Lal S, Sopory SK, Reddy MK (2009) Identification of salt-induced genes from *Salicornia brachiata*, an extreme halophyte through expressed sequence tags analysis. Genes Genet Syst 84:111–120

Jiang Y, Yang B, Harris NS, Deyholos MK (2007) Comparative proteomic analysis of NaCl stress-responsive proteins in Arabidopsis roots. J Exp Bot 58:3591–3607

Jithesh MN, Prashanth SR, Sivaprakash KR, Parida AK (2006) Antioxidative response mechanisms in halophytes: their role in stress defense. J Genet 85:237–254

Katz A, Waridel P, Shevchenko A, Pick U (2007) Salt-induced changes in the plasma membrane proteome of the halotolerant alga *Dunaliella salina* as revealed by blue native gel electrophoresis and nano- LC-MS/MS analysis. Mol Cell Proteomics 6:1459–1472

Kim DW, Rakwal R, Agrawal GK, Jung YH, Shibato J, Jwa NS, Iwahashi Y, Iwahashi H, Kim DH, IeS S, Usui K (2005) A hydroponic rice seedling culture model system for investigating proteome of salt stress in rice leaf. Electrophoresis 26:4521–4539

Kim YO, Pan S, Jung CH, Kang H (2007) A zinc finger containing glycine-rich RNA-binding protein, atRZ-1a, has a negative impact on seed germination and seedling growth of Arabidopsis thaliana under salt or drought stress conditions. Plant Cell Physiol 48:1170–1181

Koornneef A, Leon-Reyes A, Ritsema T, Verhage A, Den Otter FC, Van Loon LC, Pieterse CM (2008) Kinetics of salicylate mediated suppression of jasmonate signaling reveal a role for redox modulation. Plant Physiol 147:1358–1368

Kumar Swami A, Alam SI, Sengupta N, Sarin R (2011) Differential proteomic analysis of salt stress response in Sorghum bicolor leaves. Environ Exp Bot 71:321–328

Laohavisit A, Brown AT, Cicuta P, Davies JM (2010) Annexins: components of the calcium and reactive oxygen signaling network. Plant Physiol 152:1824–1829

Lee S, Lee EJ, Yang EJ, Lee JE, Park AR, Song WH, Park OK (2004) Proteomic identification of annexins, calcium-dependent membrane binding proteins that mediate osmotic stress and abscisic acid signal transduction in Arabidopsis. Plant Cell 16:1378–1391

Li XJ, Yang MF, Chen H, Qu LQ, Chen F, Shen SH (2010) Abscisic acid pretreatment enhances salt tolerance of rice seedlings: proteomic evidence. Biochim Biophys Acta 1804:929–940

Li W, Zhang CY, Lu QT, Wen XG, Lu CM (2011) The combined effect of salt stress and heat shock on proteome profiling in *Suaeda salsa*. J Plant Physiol 168:1743–1752

Liska AJ, Shevchenko A, Pick U, Katz A (2004) Enhanced photosynthesis and redox energy production contribute to salinity tolerance in Dunaliella as revealed by homology-based proteomics. Plant Physiol 136:2806–2817

Luan S (2002) Signalling drought in guard cells. Plant Cell Environ 25:229–237

Mahajan S, Tuteja N (2005) Cold, salinity and drought stresses: an overview. Arch Biochem Biophys 444:139–158

Mahajan S, Pandey GK, Tuteja N (2008) Calcium- and salt-stress signaling in plants: shedding light on SOS pathway. Arch Biochem Biophys 471:146–158

Manaa A, Ben Ahmed H, Valot B, Bouchet JP, Aschi-Smiti S, Causse M, Faurobert M (2011) Salt and genotype impact on plant physiology and root proteome variations in tomato. J Exp Bot 62:2797–2813

Martins AM, Cordeiro CA, Ponces Freire AM (2001) In situ analysis of methylglyoxal metabolism in Saccharomyces cerevisiae. FEBS Lett 499:41–44

McCue KF, Hanson AD (1990) Drought and salt tolerance: towards understanding and application. Trends Biotechnol 8:358–362

Meyer G, Schmitt JM, Bohnert HJ (1990) Direct screening of a small genome: estimation of the magnitude of plant gene expression changes during adaptation to high salt. Mol Gen Genet 224:347–356

Miller G, Suzuki N, Ciftci-Yilmaz S, Mittler R (2010) Reactive oxygen species homeostasis and signaling during drought and salinity stresses. Plant Cell Environ 33:453–467

Mittler R (2002) Oxidative stress, antioxidants and stress tolerance. Trends Plant Sci 7:405–410

Mittler R, Vanderauwera S, Gollery M, Van Breusegem F (2004) Reactive oxygen gene network of plants. Trends Plant Sci 9:490–498

Moons A, Gielen J, Vandekerckhove J, Van der Straeten D, Gheysen G, Van Montagu M (1997) An abscisic-acid- and salt-stress responsive rice cDNA froma novel plant gene family. Planta 202:443–454

Munns R, Tester M (2008) Mechanisms of salinity tolerance. Annu Rev Plant Biol 59:651–681

Nat NVK, Sanjeeva S, Laksiri G, Stanford FB (2004) Proteome level changes in the roots of *Pisum sativum* in response to salinity. Ann Appl Biol 145:217–230

Ndimba BK, Chivasa S, Simon WJ, Slabas AR (2005) Identification of Arabidopsis salt and osmotic stress responsive proteins using two-dimensional difference gel electrophoresis and mass spectrometry. Proteomics 5:4185–4196

Nito K, Kamigaki A, Kondo M, Hayashi M, Nishimura M (2007) Functional classification of Arabidopsis peroxisome biogenesis factors proposed from analyses of knockdown mutants. Plant Cell Physiol 48:763–774

Niu X, Bressan RA, Hasegawa PM, Pardo JM (1995) Ion homeostasis in NaCl stress environments. Plant Physiol 109:735–742

Nohzadeh MS, Habibi RM, Heidari M, Salekdeh GH (2007) Proteomics reveals new salt responsive proteins associated with rice plasma membrane. Biosci Biotechnol Biochem 71:2144–2154

Palmgren MG (1998) Proton gradients and plant growth: role of the plasma membrane H+-ATPase. Adv Bot Res 28:1–70

Panda SK, Khan MH (2009) Growth, oxidative damage and antioxidant responses in greengram (*Vigna radiate* L.) under short- term salinity stress and its recovery. J Agron Crop Sci 195:442–454

Pang QY, Chen SX, Dai SJ, Chen YZ, Wang Y, Yan XF (2010) Comparative proteomics of salt tolerance in *Arabidopsis thaliana* and *Thellungiella halophila*. J Proteome Res 9:2584–2599

Parida AK, Das AB (2005) Salt tolerance and salinity effects on plants: a review. Ecotoxicol Environ Saf 60:324–349

Parker R, Flowers TJ, Moore AL, Harpham NV (2006) An accurate and reproducible method for proteome profiling of the effects of salt stress in the rice leaf lamina. J Exp Bot 57:1109–1118

Peng ZY, Wang MC, Li F, Lv HJ, Li CL, Xia GM (2009) A proteomic study of the response to salinity and drought stress in an introgression strain of bread wheat. Mol Cell Proteomics 8:2676–2686

Pruzinska A, Anders I, Aubry S, Schenk N, Tapernoux-Luthi E, Muller T, Krautler B, Hortensteiner S (2007) In vivo participation of red chlorophyll catabolite reductase in chlorophyll breakdown. Plant Cell 19:369–387

Rasoulnia A, Bihamta M, Peyghambari S, Alizadeh H, Rahnama A (2011) Proteomic response of barley leaves to salinity. Mol Biol Rep 38: 5055–5063

Ray S, Dutta S, Halder J, Ray M (1994) Inhibition of electron flow through complex I of the mitochondrial respiratory chain of Ehrlich ascites carcinoma cells by methylglyoxal. Biochem J 303:69–72

Razavizadeh R, Ehsanpour AA, Ahsan N, Komatsu S (2009) Proteome analysis of tobacco leaves under salt stress. Peptides 30:1651–1659

Rodoni S, Vicentini F, Schellenberg M, Matile P, Hortensteiner S (1997) Partial purification and characterization of red chlorophyll catabolite reductase, a stroma protein involved in chlorophyll breakdown. Plant Physiol 115:677–682

Rostovtseva TK, Bezrukov SM (2008) VDAC regulation: role of cytosolic proteins and mitochondrial lipids. J Bioenerg Biomembr 40:163–170

Rostovtseva TK, Sheldon KL, Hassanzadeh E, Monge C, Saks V, Bezrukov SM, Sackett DL (2008) Tubulin binding blocks mitochondrial voltage-dependent anion channel and regulates respiration. Proc Natl Acad Sci USA 105:18746–18751

Roxas VP, Smith RK Jr, Allen ER, Allen RD (1997) Overexpression of glutathione S-transferase/glutathione peroxidase enhances the growth of transgenic tobacco seedlings during stress. Nat Biotechnol 15:988–991

Ruan SL, Ma HS, Wang SH, Fu YP, Xin Y, Liu WZ, Wang F, Tong JX, Wang SZ, Chen HZ (2011) Proteomic identification of OsCYP2, a rice cyclophilin that confers salt tolerance in rice (*Oryza sativa* L.) seedlings when overexpressed. BMC Plant Biol 11:34

Salekdeh GH, Siopomgcp J, Wade LJ, Ghareyazie B, Bennett J (2002) A proteomic approach to analyzing drought- and salt responsiveness in rice. Field Crops Res 76:199–219

Sanders D, Brownlee C, Harper JF (1999) Communicating with calcium. Plant Cell 11:691–706

Schroeder JI, Allen GJ, Hugouvieux V, Kwak JM, Waner D (2001) Guard cell signal transduction. Annu Rev Plant Physiol Plant Mol Biol 52:627–658

Seckin B, Sekmen AH, Turkan I (2009) An enhancing effect of exogenous mannitol on the antioxidant enzyme activities in roots of wheat under salt stress. J Plant Growth Regul 28:12–20

Sengupta S, Majumder AL (2009) Insight into the salt tolerance factors of a wild halophytic rice, *Porteresia coarctata*: a physiological and proteomic approach. Planta 229:911–929

Serrano R, Culiañz-Maciá A, Moreno V (1999) Genetic engineering of salt and drought tolerance with yeast regulatory genes. Sci Hortic 78:261–269

Singla-Pareek SL, Reddy MK, Sopory SK (2003) Genetic engineering of the glyoxalase pathway in tobacco leads to enhanced salinity tolerance. Proc Natl Acad Sci USA 100:14672–14677

Singla-Pareek SL, Yadav SK, Pareek A, Reddy MK, Sopory SK (2008) Enhancing salt tolerance in a crop plant by overexpression of glyoxalase II. Transgenic Res 17:171–180

Sirpio S, Allahverdiyeva Y, Suorsa M, Paakkarinen V, Vainonen J, Battchikova N, Aro EM (2007) TLP18.3, a novel thylakoid lumen protein regulating photosystem II repair cycle. Biochem J 406:415–425

Sobhanian H, Motamed N, Jazii FR, Nakamura T, Komatsu S (2010a) Salt stress induced differential proteome and metabolome response in the shoots of *Aeluropus lagopoides* (Poaceae), a halophyte C4 plant. J Proteome Res 9:2882–2897

Sobhanian H, Razavizadeh R, Nanjo Y, Ehsanpour AA, Jazii FR, Motamed N, Komatsu S (2010b) Proteome analysis of soybean leaves, hypocotyls and roots under salt stress. Proteome Sci 8:19

Staiger CJ, Gibbon BC, Kovar DR, Zonia LE (1997) Profilin and actin-depolymerizing factor: modulators of actin organization in plants. Trends Plant Sci 2:275–281

Sugimoto M, Takeda K (2009) Proteomic analysis of specific proteins in the root of salt-tolerant barley. Biosci Biotechnol Biochem 73:2762–2765

Tabur S, Demir K (2010) Role of some growth regulators on cytogenetic activity of barley under salt stress. Plant Growth Regul 60:99–104

Takamiya KI, Tsuchiya T, Ohta H (2000) Degradation pathway-(s) of chlorophyll: what has gene cloning revealed? Trends Plant Sci 5:426–431

Tang L, Kim MD, Yang KS, Kwon SY, Kim SH, Kim JS, Yun DJ, Kwak SS, Lee HS (2008) Enhanced tolerance of transgenic potato plants overexpressing nucleoside diphosphate kinase 2 against multiple environmental stresses. Transgenic Res. doi:10.1007/s11248-007-9155-2

Tanou G, Job C, Rajjou L, Arc E, Belghazi M, Diamantidis G, Molassiotis A, Job D (2009) Proteomics reveals the overlapping roles of hydrogen peroxide and nitric oxide in the acclimation of citrus plants to salinity. Plant J 60:795–804

Tuteja N (2007) Mechanisms of high salinity tolerance in plants. Methods Enzymol 428:419–438

Van Damme EJM, Barre A, Rouge P, Peumans WJ (2004) Cytoplasmic/nuclear plant lectins: a new story. Trends Plant Sci 9:484–489

van den Wijngaard PW, Sinnige MP, Roobeek I, Reumer A, Schoonheim PJ, Mol JN, Wang M, De Boer AH (2005) Abscisic acid and 14-3-3 proteins control K channel activity in barley embryonic root. Plant J 41:43–55

Veeranagamallaiah G, Jyothsnakumari G, Thippeswamy M, Chandra Obul Reddy P, Surabhi GK, Sriranganayakulu G, Mahesh Y, Rajasekhar B, Madhurarekha C, Sudhakar C (2008) Proteomic analysis of salt stress responses in foxtail millet (*Setaria italica* L. cv. Prasad) seedling. Plant Sci 175:631–641

Vierling E (1991) The roles of heat shock proteins in plants. Annu Rev Plant Physiol Plant Mol Biol 42:579–620

Vincent D, Ergul A, Bohlman MC, Tattersall EA, Tillett RL, Wheatley MD, Woolsey R, Quilici DR, Joets J, Schlauch K, Schooley DA, Cushman JC, Cramer GR (2007) Proteomic analysis reveals differences between *Vitis vinifera* L. cv. Chardonnay and cv. *Cabernet Sauvignon* and their responses to water deficit and salinity. J Exp Bot 58:1873–1892

Wakeel A, Asif AR, Pitann B, Schubert S (2011) Proteome analysis of sugar beet (*Beta vulgaris* L.) elucidates constitutive adaptation during the first phase of salt stress. J Plant Physiol 168:519–526

Wang ZL, Li PH, Fredricksen M, Gong ZM, Kim CS, Zhang C, Bohnert HJ, Zhu JK, Bressan RA, Hasegawa PM, Zhao YX, Zhang H (2004) Expressed sequence tags from *Thellungiella halophila*, a new model to study plant salt-tolerance. Plant Sci 166:609–616

Wang XC, Li XF, Deng X, Han HP, Shi WL, Li YX (2007) A protein extraction method compatible with proteomic analysis for the euhalophyte *Salicornia europaea*. Electrophoresis 28:3976–3987

Wang MC, Peng ZY, Li CL, Li F, Liu C, Xia GM (2008a) Proteomic analysis on a high salt tolerance introgression strain of *Triticum aestivum/Thinopyrum ponticum*. Proteomics 8:1470–1489

Wang XQ, Yang PF, Gao Q, Liu XL, Kuang TY, Shen SH, He YK (2008b) Proteomic analysis of the response to high-salinity stress in Physcomitrella patens. Planta 228:167–177

Wang XC, Fan PX, Song HM, Chen XY, Li XF, Li YX (2009) Comparative proteomic analysis of differentially expressed proteins in shoots of *Salicornia europaea* under different salinity. J Proteome Res 8:3331–3345

Wen FP, Zhang ZH, Bai T, Xu Q, Pan YH (2010) Proteomics reveals the effects of gibberellic acid (GA3) on salt-stressed rice (*Oryza sativa* L.) shoots. Plant Sci 178:170–175

Witzel K, Weidner A, Surabhi GK, Borner A, Mock HP (2009) Salt stress-induced alterations in the root proteome of barley genotypes with contrasting response towards salinity. J Exp Bot 60:3545–3557

Witzel K, Weidner A, Surabhi GK, Varshney RK, Kunze G, Buck-Sorlin GH, Borner A, Mock HP (2010) Comparative analysis of the grain proteome fraction in barley genotypes with contrasting salinity tolerance during germination. Plant Cell Environ 33:211–222

Wong CE, Li Y, Whitty BR, Díaz-Camino C, Akhter SR, Brandle JE, Golding GB, Weretilnyk EA, Moffatt BA, Griffith M (2005) Expressed sequence tags from the Yukon ecotype of Thellungiella reveal that gene expression in response to cold, drought and salinity shows little overlap. Plant Mol Biol 58:561–574

Wong CE, Li Y, Labbe A, Guevara D, Nuin P, Whitty B, Diaz C, Golding GB, Gray GR, Weretilnyk EA, Griffith M, Moffatt BA (2006) Transcriptional profiling implicates novel interactions between abiotic stress and hormonal responses in Thellungiella, a close relative of Arabidopsis. Plant Physiol 140:1437–1450

Xu CP, Sibicky T, Huang BR (2010) Protein profile analysis of salt responsive proteins in leaves and roots in two cultivars of creeping bentgrass differing in salinity tolerance. Plant Cell Rep 29:595–615

Yadav SK, Singla-Pareek SL, Ray M, Reddy MK, Sopory SK (2005) Methylglyoxal levels in plants under salinity stress are dependent on glyoxalase I and glutathione. Biochem Biophys Res Commun 337:61–67

Yan SP, Tang ZC, Su WA, Sun WN (2005) Proteomic analysis of salt stress-responsive proteins in rice root. Proteomics 5:235–244

Yazaki K (2006) ABC transporters involved in the transport of plant secondary metabolites. FEBS Lett 580:1183–1191

Yoshimura K, Miyao K, Gaber A, Takeda T, Kanaboshi H, Miyasaka H, Shigeoka S (2004) Enhancement of stress tolerance in transgenic tobacco plants overexpressing Chlamydomonas glutathione peroxidase in chloroplasts or cytosol. Plant J 37:21–33

Yu JJ, Chen SX, Zhao Q, Wang T, Yang CP, Diaz C, Sun GR, Dai SJ (2011) Physiological and proteomic analysis of salinity tolerance in *Puccinellia tenuiflora*. J Proteome Res 10:3852–3870

Yuasa K, Maeshima M (2000) Purification, properties, and molecular cloning of a novel $Ca^{2+}$-binding protein in radish vacuoles. Plant Physiol 124:1069–1078

Zahn DK, Moreira JA, Millar JG (2008) Identification, synthesis and bioassays of a male specific aggregation pheromone from the Harlequin bug *Murgantia histrionica*. J Chem Ecol 34:238–251

Zhang W, Peumans WJ, Barre A, Astoul CH, Rovira P, Rouge P, Proost P, Truffa-Bachi P, Jalali AA, Van Damme EJ (2000) Isolation and characterization of a jacalin-related mannose-binding lectin from salt-stressed rice (*Oryza sativa*) plants. Planta 210:970–978

Zhang HX, Hodson JN, Williams JP, Blumwald E (2001a) Engineering salt-tolerant Brassica plants: characterization of yield and seed oil quality in transgenic plants with increased vacuolar sodium accumulation. Proc Natl Acad Sci USA 98:12832–12836

Zhang L, Ma XL, Zhang Q, Ma CL, Wang PP, Sun YF, Zhao YX, Zhang H (2001b) Expressed sequence tags from a NaCl treated *Suaeda salsa* cDNA library. Gene 267:193–200

Zhang YY, Lai JB, Sun SH, Li Y, Liu YY, Liang LM, Chen MS, Xie Q (2008) Comparison analysis of transcripts from the halophyte *Thellungiella halophila*. J Integr Plant Biol 50:1327–1335

Zhang L, Tian LH, Zhao JF, Song Y, Zhang CJ, Guo Y (2009) Identification of an apoplastic protein involved in the initial phase of salt stress response in rice root by two-dimensional electrophoresis. Plant Physiol 149:916–928

Zhang H, Han B, Wang T, Chen S, Li H, Zhang Y, Dai S (2012) Mechanisms of plant salt response: insights from proteomics. J Proteome Res 11:49–67

Zhu JK (2000) Genetic analysis of plant salt tolerance using Arabidopsis. Plant Physiol 124:941–948

Zhu JK (2001a) Cell signaling under salt, water and cold stresses. Curr Opin Plant Biol 4:401–406

Zhu JK (2001b) Plant salt tolerance. Trends Plant Sci 6:66–71

Zhu JK (2002) Salt and drought stress signal transduction in plants. Annu Rev Plant Biol 53:247–273

Zhu JK (2003) Regulation of ion homeostasis under salt stress. Curr Opin Plant Biol 6:441–445

Zhu CL, Zheng Y, Jia Y (2007) A theoretical study on activation of transcription factor modulated by intracellular $Ca^{2+}$ oscillations. Biophys Chem 129:49–55

Zörb C, Schmitt S, Neeb A, Karl S, Linder M, Schubert S (2004) The biochemical reaction of maize (*Zea mays* L.) to salt stress is characterized by a mitigation of symptoms and not by a specific adaptation. Plant Sci 167:91–100

Zörb C, Herbst R, Forreiter C, Schubert S (2009) Short-term effects of salt exposure on the maize chloroplast protein pattern. Proteomics 9:4209–4220

Zörb C, Schmitt S, Muhling KH (2010) Proteomic changes in maize roots after short-term adjustment to saline growth conditions. Proteomics 10:4441–4449

Zouari N, Ben Saad R, Legavre T, Azaza J, Sabau X, Jaoua M, Masmoudi K, Hassairi A (2007) Identification and sequencing of ESTs from the halophyte grass *Aeluropus littoralis*. Gene 404:61–69

# Salt Stress: A Biochemical and Physiological Adaptation of Some Indian Halophytes of Sundarbans

6

Nirjhar Dasgupta, Paramita Nandy, and Sauren Das

**Abstract**

Experiment was conducted with five typical mangroves (*Bruguiera gymnorrhiza*, *Excoecaria agallocha*, *Heritiera fomes*, *Phoenix paludosa*, and *Xylocarpus granatum*) both from Sundarbans (*in-situ*) and grown in mesophytic condition (*ex-situ*, in the Indian Statistical Institute's premises) since 15–17 years. A comparative account on PAR utilization for maximum photosynthesis, stomatal conductance, total leaf proteins, and polymorphic expression of two antioxidative enzymes (peroxidase and superoxide dismutase) and two hydrolyzing enzymes (esterase and acid phosphatase) were estimated both qualitatively and quantitatively. The present work revealed that the net photosynthesis was higher in mangroves from mesophytic habitats than those of the native plants, but the PAR acquisitions for maximum photosynthesis were greater in most of the Sundarbans species, except *H. fomes* and *X. granatum*. At the same time, the stomatal conductance was remarkably depleted under salinity stressed habitats than those of the nonsaline counterparts and ranged between nearly 25 and 52%. Total leaf protein content form the above said taxa revealed that the increment of total protein occurred in mesophytic habitat and it was ranged between 156% (in *P. paludosa*) and 5.7% (in *X. granatum*). PAGE analysis revealed that in most of the cases there were extra numbers of protein bands expressed with relatively low molecular weight in saline habitat plants. In all salinity imposed plants, there were sharp increase in band intensity and number of isoforms of each enzyme. Peroxidase increment in

N. Dasgupta • S. Das (✉)
Agricultural and Ecological Research Unit, Indian Statistical Institute, 203, Barrackpore Trunk Road, Kolkata 700 108, India
e-mail: ndg.1009@gmail.com; sauren@isical.ac.in

P. Nandy
Department of Botany, Barasat Government College, Barasat, Kolkata 700 124, India
e-mail: nandydattap@gmail.com

G.R. Rout and A.B. Das (eds.), *Molecular Stress Physiology of Plants*,
DOI 10.1007/978-81-322-0807-5_6, © Springer India 2013

saline plants was ranged between 257% (in *Bruguiera*) and 139% (in *Excoecaria*). Similarly, superoxide dismutase (SOD) was estimated as 247% (in *Heritiera*) to 147% (in *Excoecaria*) in saline habitats. Increments of esterase and acid phosphatase were varied from 287% (in *Phoenix*) to 154% (in *Excoecaria*) and 293% (in *Bruguiera*) to 139% (in *Excoecaria*), respectively. Salinity imposed increment of antioxidant enzymes proved their efficient scavenging ability to evolved reactive oxygen species (ROS), but these increments were relatively lower in *Heritiera* and *Xylocarpus* even though the net photosynthesis was higher. This might be related to their less adaptability in elevated salinity stress than those of the other three species investigated from the same regime. Among the plants grown in *in situ* condition, some taxa have the better ability of enzyme production, which might be correlated with the efficient stress management practice. A statistical relationship was observed between the total protein content and the investigated enzyme concentration, dependent on the habitat and discussed accordingly.

## Introduction

### Mangrove Vegetation: Global Context

The environment and ecosystem of tropical and subtropical coastal zones are marked with unique geophysical characteristics like frequent sea surges with tidal influences, upland discharges, rapid sedimentation, substrate erosion, and incidence of episodic cyclones. It is estimated that about 55% of the world's population lives in coastal areas. In East Asia, more than 70% of the population depends on coastal resources for food, employment, and generation of income (Kathiresan and Bingham 2001). In South Asia, demographic compulsion is considered as greatest threat on diminishing coastal resources. The Western Pacific region, due to unplanned developmental activities and climatic change, significant degradation occurred on coastal ecology. Moreover, the increasing human habitation in coastal areas globally, predicted to be 6 billion by 2030 provides the inevitable need for biodiversity conservation (Adeel and Caroline 2002). In most of the developing countries, due to demographic need for more food production and socioeconomic developmental activities which are mostly at the expense of coastal biodiversity loss.

Mangroves, a cosmopolitan assemblage of plant families representing a converging genetic adaptation to a typical saline environment and are best developed on shorelines of tropical world particularly in vast areas of tidal influence. Mangroves are largely restricted to latitudes between 30°N and 30°S. Northern extensions of this limit occur in Japan (31°22′ N) and Bermuda (32°20′ N); southern extensions are in New Zealand (38°03′ S), Australia (38°45′ S), and on the east coast of South Africa (32°59′ S) (Spalding et al. 1997). The cradle of mangrove ecosystem is the Indo-Malaysian area as the mangroves first developed here and spread afterward to other tropical and subtropical world (Alongi 2009). The luxuriant occurrence of mangrove composition is found in the Indo-Pacific region. In Southeast Asia, the mangrove forests are the richest in species diversity. The mangrove forests are distributed over the tropical and subtropical estuaries from mean sea level to highest spring tide (Alongi 2009). The current estimate of mangrove forests of the world is less than half of what it once was (Spalding et al. 1997; Spiers 1999) and maximum of which exists in a gaunt condition. Coastal habitats across the world are presently under constant demographic and developmental pressure and periodic natural calamities. The existence crisis of this important habitat starts after industrialization and development; conversion of agricultural land, aquaculture, human habitation, industrial runoff, and over exploitation are all considered as much threats on these estuarine vegetation (Alongi 2002; Giri et al.

2008). Simultaneously, global climatic changes also have a great impact on mangrove vegetation. The destruction of mangrove forests is happening at a time when there are clear indications of potential changes in the climate (precipitation and temperature), sea level rise and incidence of UV-β radiation. Mangrove ecosystems are the front-line defense against these adverse consequences of change in the sea level. These plants are known to absorb the UV radiation to a considerable extent (Hogarth 2007). The earth's crust has warmed by 0.5–1.2°C since industrial revolution. Further warming of 0.8–2.0°C is to be expected owing to the inertia of the geosystem. Doubling of atmospheric $CO_2$ levels to about 700 ppm in the next 40–50 years expected to lead to a further increase in mean temperature of 1.5–4.5°C. Rise in relative sea level globally has been reported to be about 10 cm over the last 100 years, with an expected rise of 30–50 cm by 2005 and 100 cm by 2100 (Clough 1994). These climatic changes will result in changing patterns of rainfall, cloud cover and aridity. Duke et al. (2007) opined that the decline of the mangrove forest occurs at a faster rate than any other inland tropical forest. The relative sea level rise could be the greatest threat to mangrove formation (Gilman et al. 2008) and it is postulated that 30–40% of coastal wet lands and 100% of the mangrove forest could be lost in the next 100 years if the present rate of denudation continues (Duke et al. 2007). These losses represent about 2.0% per year since 1980–1990, and 0.7% per year within 1990–2000 (FAO 2007). These figures show the magnitude of mangrove loss and demand serious attention to the potentiality of mangrove restoration program. Indian coastline covers about 7,500 $km^2$ and it accounts for 8% of the world's mangrove area, and the eastern coast of India accounts for about 82% of the mangrove forest cover throughout India (Parida et al. 2002). Due to lack of any national plan for conservation and sustainable utilization, mangroves along the Indian coast have reached an alarming stage of depletion. It is reported that 25% reduction in mangrove forest cover has been estimated along the Indian region during the last 25 years (FAO 2007). Predictions of these features of global changes indicate an increase in frequency and intensity of episodic cyclones, hurricanes, tsunami, storms, and floods. Though mangroves do provide a protective buffer and minimize these impacts, they will have increasing impacts on mangrove ecosystem too.

Mangrove ecosystems currently cover 146,530 $km^2$ of the tropical shorelines of the world (FAO 2007). This represents a decline from 198,000 $km^2$ of mangroves in 1980, and 157,630 $km^2$ in 1990 (FAO 2007). From the recent estimate done by Giri et al. (2011), it is revealed that the total mangrove forests in the world include 137,760 $km^2$, and it is 0.7% of the total tropical forest in the world. In India, the estimated mangrove forest area includes 4,844 $km^2$ (FAO 2007). Using Satellite data, FSI (2009) reported that the Indian mangrove area covers 4,639 $km^2$ (Fig. 6.1c) which is 58 $km^2$ increments (2.4%) in comparison with the earlier estimate in 2005 (4,581 $km^2$). The largest mangrove forest area is covered in Asia (about 42%) followed by Africa (20%), North and Central America (15%), Oceania (12%), and South America (11%). Giri et al. (2011) confirmed that the mangrove area decreases with the increase of latitude, except 20°N and 25°N latitude, where the Sundarbans area is located and considered this area as the largest block of the mangrove forest in the world. Although the species diversity is the highest in this region, the conservation and sustainable management of this diversity presently need serious attention.

## Sundarbans Mangrove: Ecosystem Scenario

Sundarbans delta, a mangrove swamp with world's richest species diversity in the Indian subcontinent (extends between 21°31′–22°30′ N and 88°10′–89°51′ E), is formed by two major rivers the Ganga and the Brahmaputra and their innumerous tributaries (Fig. 6.1a). The flora comprises 36 true mangroves, 28 associates, and seven obligatory mangrove species

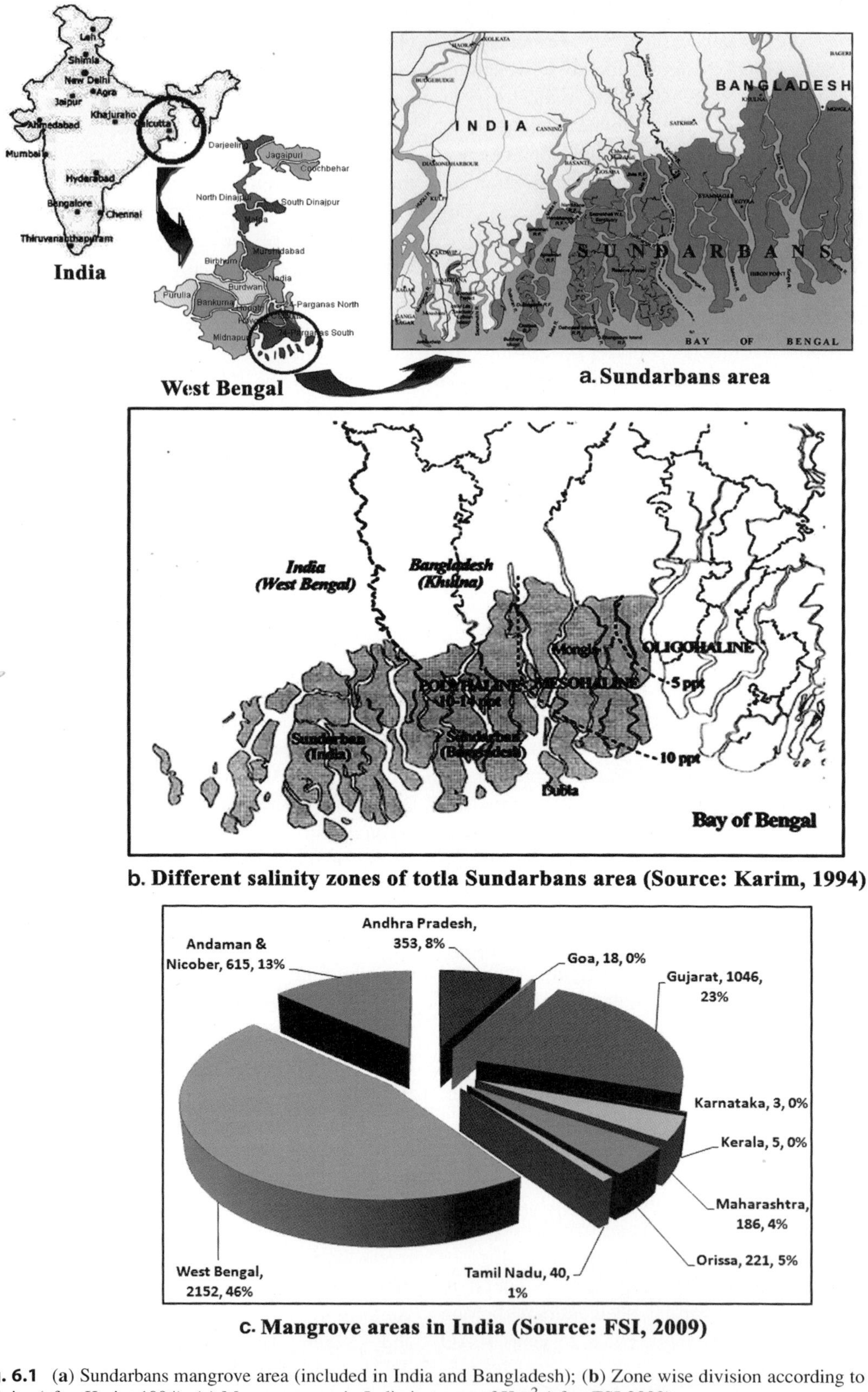


**Fig. 6.1** (**a**) Sundarbans mangrove area (included in India and Bangladesh); (**b**) Zone wise division according to soil salinity (after Karim 1994); (**c**) Mangrove area in India in terms of $Km^2$ (after FSI 2009)

representing 29 families and 49 genera (Naskar and Guha Bakshi 1983). The forest area (Indian Territory) covers approximately 2,152 $km^2$ (FSI 2009) excluding the anastomosing network of creeks and backwaters. But this scenario is changing rapidly. Due to tectonic upliftment, there has been a very slow tilting of the coast in the northwestern part (India) and subsidence in the east (Bangladesh); an increased salinity persists in the western part (India) of Sundarbans that has a direct impact on mangrove vegetation pattern. Since the twentieth century, damming on the lower Gangetic plane cut off an ample freshwater influx through Hooghly and its tributaries to the islands of Sundarbans. Moreover, urban sewage and industrial wastes expedite silt formation in the riverbed causing decreased runoff of sweet water influx through the river system leading to elevate the salinity level of water in the delta region. The soil salinity reaches between 15 and 27 PPT (Nandy (Datta) et al. 2007). Haq et al. (1999) estimated that salinity has gone up by 20% in the Sundarbans since 1990. These manual and environmental adversities have posed disastrous for some important plant species of Sundarbans forest including *Aegialitis rotundifolia*, *Heritiera fomes*, *Nypa fruticans*, *Xylocarpus granatum*, and *X. mekongensis* (Banarjee 1999; Upadhyay et al. 2002). These species are predominant in between the Raimangal and Matla rivers, where freshwater influx from the Ichamati River through Raimangal is much better (in Bangladesh part). Especially *H. fomes* prefers slightly and/or moderately saline zone and the ridges of higher elevation that are inundated only during spring tide (Alim 1979). Previously in West Bengal these trees used to be 2 m in girth, but presently beyond 1 m girth is no longer common, and top dying of *H. fomes* is very frequent in the Sundarbans forest (Curtis 1993). However, rise in water level and increasing salinity in the rivers and creeks, human interference, and changed ecology pose the biggest threat to the Sundarbans and have caused rapid destruction to the mangrove forests and even caused extinction to a few of the locally existing species (Parani et al. 2000). The soil of the Sundarbans is saline due to tidal interactions, although the salinity is low compared to soil salinity in other mangrove forests of the world (Karim 1994). Soil salinity, however, is regulated by a number of other factors including surface runoff and groundwater seepage from adjacent areas, amount and seasonality of rainfall, evaporation, groundwater recharge and depth of impervious subsoil, soil type, and topography. It is found that conductivity of subsurface soil is much higher than that of surface soil (Chaffey et al. 1985). Considering the salinity scale established by Walter (1971), the forest areas have been divided into three zones based on soil salinity (Karim 1994), such as (1) *oligohaline (ormiohaline) zone* – the zone is characterized by the soil containing less than 5 PPT of NaCl salt. The oligohaline zone occupies a small area of the northeastern part of the forest; (2) *mesohaline zone* – the zone is characterized by NaCl content within the concentration range of 5–10 PPT in soil. This zone covers the north-central to south-central part of the forest; and (3) *polyhaline zone* –NaCl content of the soil in this zone is higher than 10 PPT. This zone covers the western portion (Indian part) of the forest (Fig. 6.1b).

## The Problems Addressed

The mangroves are specially suited for the inhospitable environmental condition and thus pose a lot of challenging problems to the biologists. Salinity is the major environmental factor that is limiting plant growth and productivity worldwide. It has been well established that they are considered as an ecologically essential component in protecting adjacent land by forming a front-line barrier against tidal waves and sea storm (Hogarth 2007). Salinity stress has long been considered as potential and important factor that are regulating different physiological and biochemical processes (Lin and Sternberg 1993). Mangroves growing in the areas of high salinity, excess irradiance, lower substrate osmotic potential, and anaerobic soil and thus imposed stress on them which were confirmed by decreased $CO_2$ assimilation, higher stomatal conductance, and increased leaf water potential (Nandy Datta et al. 2005; Dasgupta et al. 2011). Differential photosynthetic gas exchange ability, extent of water potential, and stomatal conductance along with leaf anatomical characteristics (e.g., mesophyll ratio, nature of palisade cells) are reported as potential indicator of physiological status of a plant and well related to the relative adaptability of different mangrove species in the same regime (Ball and Farquhar 1984; Das 1999, Nandy Datta et al. 2007, 2009).

Unlike morphological markers, molecular markers are not prone to environmental influences and provide some vital information toward the priority areas for conservation strategies. Therefore, the use of molecular markers (enzymes, DNA) might enhance the understanding of such situation. Enzyme analysis is an added tool for detecting this diversity (Zeidler 2000). The International Union for Protection of New Varieties of Plants (UPOV) have harmonized and adopted test guidelines and procedures for the use of isozyme electrophoresis as a characteristic for establishing uniqueness of plants (UPOV 1997).

Mangroves have to cope with considerably high soil salinity and, consequently, a physiologically dry substrate. As such, they are confronted with the problem of maintaining adequate turgor pressure within the cell sap because of high salt concentrations in the growth medium and thus protecting their metabolic activity (Flowers et al. 1977). This leads to accumulation and /or synthesis of organic substances in the form of compatible solutes within the vacuole (Hasegawa et al. 2000). Cheeseman et al. (1997) experimentally showed that ascorbate peroxidase and SOD synthesis are much higher in field grown mangroves. Superoxide dismutase (SOD) and several antioxidant enzymes are potentially involved in $H_2O_2$ metabolism leading to photoprotection. Parida et al. (2002) reported that sugar, proline, and some polyphenol compounds accumulate in the cell sap of *Bruguiera parviflora* to restore the required water potential more negative. Experimental works reported that in mangroves, the synthesis of these osmolytes, specific proteins, and translatable mRNA induced and increased by salt stress (Hurkman et al. 1989; Bray 1993; Xu et al. 1996, 2001; Swire-Clark and Marcotte 1999). A Positive linear relationship between peroxidase activity and leaf tissue metal concentrations were reported in *Avicennia marina* (Macfarlane and Burchett 2001). *In-vitro* experiment on *B. parviflora* resulted the differential changes in the levels of the isoforms of antioxidative enzymes due to NaCl treatment which may be useful as markers for recognizing salt tolerance in mangroves and suggested that the elevated levels of the antioxidant enzymes protect the plants against the reactive oxygen species (ROS) thus avoiding lipid peroxidation during salt stress (Parida et al. 2004a, b). An increased level of peroxidase and SOD accumulation was reported in water logging stress in *Kandelia candel* and *Bruguiera gymnorrhiza* (Ye et al. 2003). Antioxidative enzymes in relation to salt tolerance in different plant have been evaluated much, but it is still a paradox because enhanced accumulation of these not only associated with salt tolerance but also with slat sensitivity too (Abogadalla 2010). This paradox is due to the following: (1) it is difficult to estimate the evolved amount of ROS in a plant and (2) detoxification of ROS leading to efficient upregulation of genes responsible for all antioxidative enzymes (SOD, PRX, CAT, GR, etc.), though the efficient antioxidative activity never solely means the strong upregulation of all antioxidant enzymes.

In obligate halophytes, reverse adaptation often provoke significant metabolic shifts that can be partially characterized by isozyme study. Peroxidase (in different isoforms) is widely distributed throughout the growing phase and has great biological importance. In plants, peroxidase is either bound to cell wall or located in the protoplast (Mader 1976). Cell wall bound peroxidases are probably involved in lignification while other isoenzymes have the regulatory role in plant senescence or in the destruction of auxins (Frenkel 1972; Stonier and Yang 1973). Generation of reactive oxygen species (ROS) such as superoxide, hydroxyl, and peroxyl radicals is inevitable under oxidative stress as does the level of ROS-induced oxidative damage to lipids, proteins, and nucleic acids (Meloni et al. 2003). To mitigate the extent of destruction of cellular components by ROS, a front-line defense mechanism is developed in plants with complex antioxidant enzyme mechanisms like peroxidase (PRX), superoxide dismutase (SOD), catalase (CAT), and glutathione reductase (GR). Salinity resistance is improved by elevated regulation of antioxidant enzymes leading to ROS scavenging (Alscher et al. 2002). Salinity imposed upregulation of cellular ROS accumulation leading to destruction of membrane lipids, proteins, and nucleic acids has been reported by earlier works (Hernandez et al. 2000; Mansour et al. 2005; Ben-Amor et al. 2007; Eyidenan and Oz 2007).

Due to changed ecology, isoforms of these stress related enzymes were differentially expressed. There are hardly any report dealt with a comparative account of quantitative and qualitative analysis of antioxidant and hydrolyzing enzymes in Indian context. In view of above, this work aims to understand the extent of changes of isoforms of two antioxidant enzymes (peroxidase and superoxide dismutase) and two important hydrolyzing enzymes (esterase and acid phosphates) in five true mangrove species grown in the natural field condition (in Sundarbans) and their counterparts grown in the freshwater condition in the garden of ISI Kolkata. The comparative assessment, both gel electrophoresis study and quantitative estimation of total leaf protein and enzyme, would provide some important clues toward their reverse adaptability to mesophytic condition for postulating proper conservation technique in *ex-situ* condition.

## Methodology

Five true mangroves species grown in Sundarbans mangrove forest (*Bruguiera gymnorrhiza*, *Excoecaria agallocha*, *Heritiera fomes*, *Phoenix paludosa*, and *Xylocarpus granatum*) were selected among which *Heritiera* and *Xylocarpus* are precarious in occurrence in western part of Sundarbans (Indian territory) and the remaining grown profusely in the same regime (considered as natural control). Leaf buds were collected from *in-situ* (from Sundarbans forest, where salinity ranges from 15 to 27 PPT) plants (about 15–17 years old) and their replica from *ex-situ* (grown in mesophytic condition, in the premises of the Indian Statistical Institute, Kolkata) of all most same age and the soil salinity ranging from 2 to 2.5 PPT.

## Estimation of Carbon Assimilation and Stomatal Conductance

The rate of net photosynthesis and stomatal conductance in different PAR were measured with an infrared $CO_2$ gas analyzer (PS 301 CID, USA) that uses an electronic mass flow meter to monitor airflow rate. Measurements were taken from the exposed surface of leaves from top, middle, and bottom of each plant. The rate of net photosynthesis ($P_n$) was determined measuring the rate, at which a known leaf area assimilated $CO_2$ concentration at a given time. The data were taken from randomly 20 plants of almost same age in full sunshine condition. The average data and their standard error bars were presented in the graphs.

$$\begin{aligned} P_n &= -W \times (C_{\mathrm{o}} - C_{\mathrm{I}}) \\ &= -2005.39 \times \{(V \times P)/(T_a \times A)\} \\ &\quad \times (C_{\mathrm{o}} - C_{\mathrm{I}}) \end{aligned}$$

where $W$ = mass flow rate per leaf area (mmol $m^{-2}$ $s^{-1}$); $C_{\mathrm{o}}$ ($C_{\mathrm{I}}$) = outlet (inlet) $CO_2$ conc. (μmol $m^{-2}$ $s^{-1}$); $P$ = atm. pressure (bar); and $T_{\mathrm{a}}$ = air temp. (K).

Stomatal conductance ($C_{\mathrm{leaf}}$) was calculated from the rate of water efflux and leaf surface temperature (°C).

$$\begin{aligned} C_{\mathrm{leaf}} = W/[\{(e_{\mathrm{leaf}} - e_{\mathrm{o}})/(e_{\mathrm{o}} - e_{\mathrm{I}})\} \\ \times \{(P - e_{\mathrm{o}})/P\} - R_b W] \times 1{,}000 \end{aligned}$$

where $e_{\mathrm{leaf}}$ = saturated water vapor at leaf temperature (bar); $R_{\mathrm{b}}$ = leaf boundary layer resistance ($m^2$/mol); $P$ = atm. pressure (bar); and $W$ = mass flow rate per leaf area (mmol $m^{-2}$ $s^{-1}$).

The data were downloaded and computed through RS 232 Port.

## Protein Estimation and SDS–PAGE Analysis

Total protein estimation was carried out for five mangrove taxa from both habitat following Lawrey et al. (1951). Extraction of protein for gel electrophoresis was done from 2 g of fresh leaf. Leaf samples were macerated in a mortar–pestle, add 5 ml of extraction buffer (containing 10% (w/v) SDS, 10 mM β-mercaptoethanol, 20% (v/v) glycerol, 0.2 M Tris-HCl (pH 6.8) and 0.05% Bromophenol blue). Centrifuge at 10,000 rpm for 20 min. Supernatants were used as samples. Protein samples were resolved in 12.5% SDS–PAGE gels following the procedure of Laemmli (1970) and stained with Coomassie Brilliant Blue R-250 (Sigma). Molecular weights of different protein bands were determined with respect to standard protein marker (Bioline Hyper Page prestained protein marker, 10–200 kDa) with the Kodak MI software after documentation the gel slab with Gel-Doc system (Biostep GmbH – Germany).

## Extraction of Enzymes Native Gel Electrophoresis

Two grams of young leaf buds were macerated to powder in liquid nitrogen with a mortar–pestle; then 0.1 g PVP and 5 ml of extraction buffer (consists of 1 M sucrose, 0.2 M Tris-HCl, and 0.056 M β-mercaptoethanol; pH is adjusted at 8.5) were added to it and homogenized. The extractants were centrifuged at 10,000 rpm for 20 min at 4° C; supernatants were used as samples for gel electrophoresis. Isozyme analysis of four enzymes, *viz.*, peroxidase, superoxide dismutase, esterase, and acid phosphatase, were done for the investigated five taxa. Equimolar amount of enzymes were loaded in each well. Samples from saline and nonsaline environment were loaded side by side for precision of polymorphic band expression. Slab gels were stained for definite enzymes following Das and Mukherjee (1997). Gels were documented with a Gel-Doc system (Biostep GmbH – Germany) and analysis for band intensity and relative mobility factor ($R_{mf}$) were estimated with Kodak MI software.

## Enzyme Assay

*Peroxidase (PRX, E.C.1.11.1.7)*: 200 mg fresh leaf sample was extracted in 1–1.5 ml of 0.9% KCl and centrifuged at 12,000 rpm for 15 min at 4°C; supernatant used as enzyme sample. Absorbance; was taken by Helios γ spectrophotometer (Thermo electron Corporation, USA) at 460 nm in respect to the standard curve prepared following Shannon et al. (1966) with minute modification.

*Superoxide dismutase (SOD, E.C.1.15.1.1)*: Cell sap was extracted from 200 mg of leaf in 1–1.5 ml of 50 mM phosphate buffer, ph adjusted to 7.0, centrifuged at 12,000 rpm for 15 min at 4°C. Supernatants were used for enzyme samples. Different aliquots (50, 100, 150, 200, 250 μg/ml) of the standard enzyme samples were also used for preparing the standard curve, and absorbance was measured at 550 nm (Keith et al. 1983).

*Esterase (EST, E.C.3.1.1.1):* Enzyme sample was prepared from 200 mg fresh leaf sample extracted with 1–1.5 of ml ice cold 0.1 M Tris-HCl buffer adjusted pH 8.0. Extractants were centrifuged at 12,000 rpm for 15 min at 4°C. The supernatant used as sample. Absorbance was noted at 322 nm with respect to the prepared standard curve (Balen et al. 2004).

*Acid Phosphatase (ACP, E.C.3.1.3.2):* 200 mg fresh leaf sample was extracted in 1–1.5 ml of 40 mM succinic acid/NaOH buffer, pH adjusted to 4.0, centrifuged at 12,000 rpm for 15 min at 4°C. Supernatant was taken for enzyme assay. Prepared a standard curve with the known enzyme samples and absorbance was taken at 322 nm (Huttová et al. 2002).

## Results

The salinity ranged between 15 and 27 PPT throughout the year in Sundarbans, but in garden soil it never exceeds beyond 2 PPT. The Irradiance (PAR) that measured in two ecosystems is ranged between 428 and 2,110 μmolm$^{-2}$s$^{-1}$ in the saline-habitat (Sundarbans) and 600 and 1,880 μmol m$^{-2}$ s$^{-1}$ in the nonsaline (ISI garden, Kolkata) environment.

## Carbon Assimilation and Stomatal Conductance

The net photosynthesis was higher in mangroves of nonsaline soil habitat than that of the native ones (Fig. 6.2a), but the PAR acquisition for maximum photosynthesis was greater in most of the Sundarbans species except *H. fomes* and *X. granatum* (Fig. 6.2a). In *B. gymnorrhiza*, the maximum photosynthesis (10.47 μmol m$^{-2}$ s$^{-1}$) was achieved only at 873 μmol m$^{-2}$ s$^{-1}$ PAR when grown in nonsaline soil, but as high as 1078.5 μmol m$^{-2}$ s$^{-1}$ PAR was utilized to obtain the highest assimilation rate (9.19 μmol m$^{-2}$ s$^{-1}$) under saline condition (Fig. 6.2a). In *E. agallocha* the optimum PAR required for maximum photosynthesis were 1445.8 μmol m$^{-2}$ s$^{-1}$ in Sundarbans and 1402.6 μmol m$^{-2}$ s$^{-1}$ in garden, whereas the highest assimilation rates were 12.27 and 14.69 μmol m$^{-2}$ s$^{-1}$, respectively (Fig. 6.2a). Similarly, in *P. paludosa*, the optimum PAR value was 1662.3 μmol m$^{-2}$ s$^{-1}$ in Sundarbans forest beyond which photosynthesis started declining, whereas in garden, the highest rate of net photosynthesis (6.92 μmol m$^{-2}$ s$^{-1}$) was recorded at a much lower PAR value (1012.6 μmol m$^{-2}$ s$^{-1}$) (Fig. 6.2a). On the contrary, under salt stress, the rate of assimilation in *X. granatum* dropped just beyond 827.7 μmol m$^{-2}$ s$^{-1}$ PAR, whereas in nonsaline condition, the optimum PAR was as high as 1557.6 μmol m$^{-2}$ s$^{-1}$ (Fig. 6.2a). Among the studied species, photosynthesis rate was maximal in *H. fomes* under both the environmental conditions (10.63 μmol m$^{-2}$ s$^{-1}$ in Sundarbans and 12.63 μmol m$^{-2}$ s$^{-1}$ in garden) (Fig. 6.2a). Stomatal conductance was remarkably decreased under salinity stressed habitats than those of the sweet water counterparts (Fig. 6.2b). In *B. gymnorrhiza* and *E. agallocha*, the salinity imposed restriction of stomatal conductance was noticed about 44%; in *P. paludosa* and *X. granatum*, it was nearly 52% and in *H. fomes* 25%.

## Protein Analysis

### SDS–PAGE Analysis

This analysis revealed that the numbers of protein bands were expressed differentially in the same species from two different habitats. The molecular weights of these bands were calculated with respect to standard marker run in the same gel. The result revealed that in *Bruguiera*, the saline habitat individual showed one extra band than its nonsaline replica and molecular weight ranged between 169.1 and 66.67 kDa (nonsaline) and 210.7 and 66.11 kDa (saline). *Excoecaria* showed the same number of bands in both habitats having molecular weight ranged between 205.8 and 65.55 kDa (nonsaline) and 213.2 and 77.72 kDa (saline). The highest number of protein bands appeared in *Heritiera* from both the environments, nine bands in each having molecular weight 211.2–26.71 kDa in nonsaline and 212.2–37.0 kDa in saline taxa. One extra band appeared in nonsaline *Phoenix* than its saline pair, and the molecular weight ranged between 201.3 and 46.43 kDa and 213.2 and 46.0 kDa, respectively. In *Xylocarpus*, one more band was expressed in saline plant, having 202.8–50.57 kDa (nonsaline) and 197.3–58.27 kDa (saline) (Fig. 6.2c).

### Total Protein

Total leaf protein was estimated from the five enough mature taxa, grown in both saline and freshwater environment. In all five species, the total protein content showed higher amount in freshwater grown plants than those of their Sundarbans counterparts (salt stress environment).

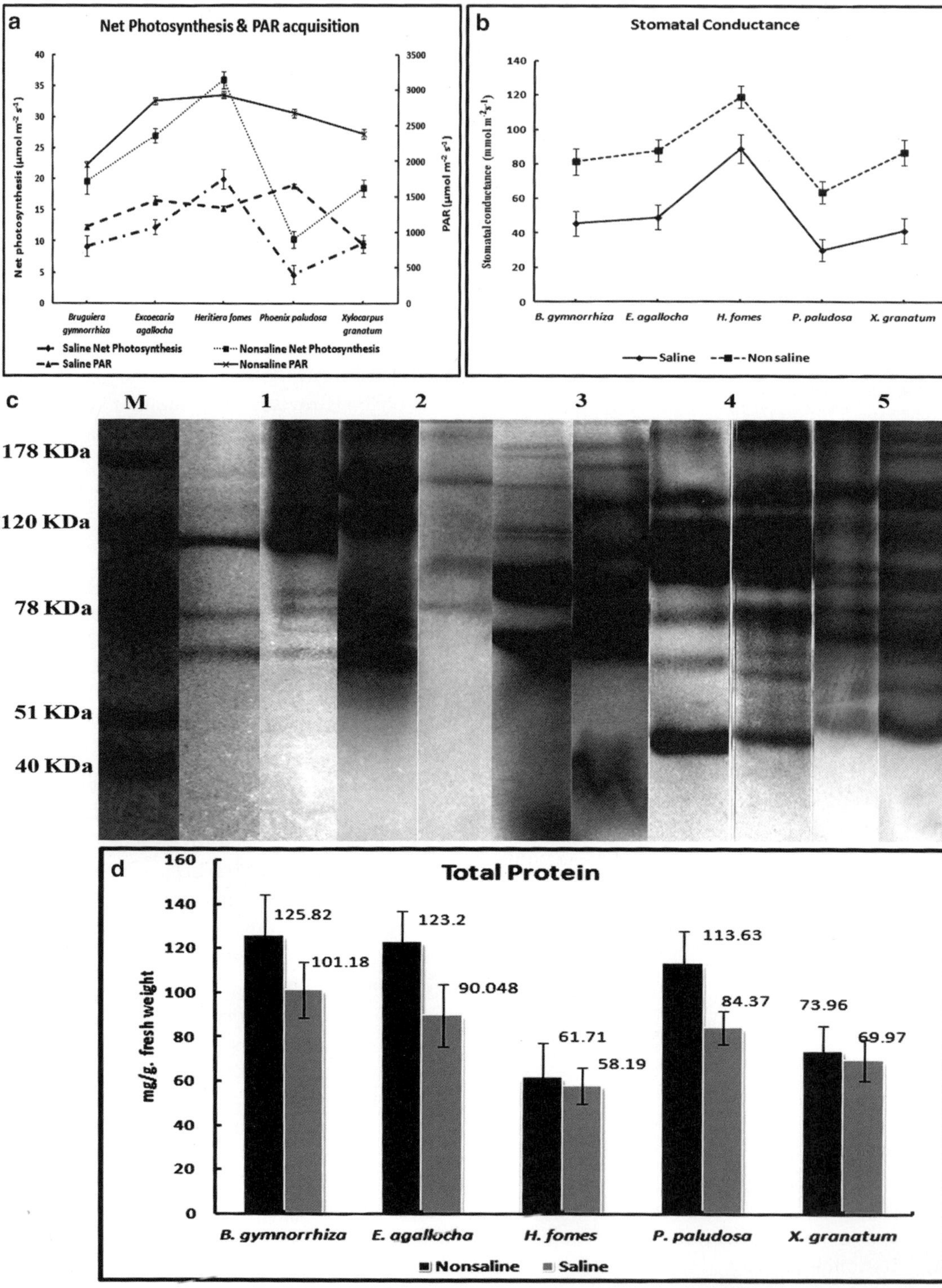


**Fig. 6.2** (**a**) Net photosynthesis among the five investigated mangrove taxa in saline and nonsaline habitat; (**b**) Stomatal conductance in two different habitats; (**c**) *SDS*-PAGE photograph of total leaf proteins; (**d**) Amount of leaf proteins in two different habitats along with standard error bars

The highest amount was estimated in *B. gymnorrhiza* (125.82 mg/g fr. wt.) and *E. agallocha* (123.2 mg/g fr. wt.), and minimum was in *X. granatum* (73.96 mg/g fr. wt.) grown in *ex-situ* condition. The increment of total protein was estimated at highest in *P. paludosa* (156%) and lowest in *X. granatum* (5.7%). In *H. fomes*, freshwater habitat showed 57% more protein content than that of the *in-situ* habitat (Fig. 6.2d).

## Native Gel Electrophoresis

### Peroxidase (PRX)

Band expression obtained from gel electrophoresis revealed that *H. fomes* and *X. granatum* showed the same number of isoforms in two different habitats, whereas in *B. gymnorrhiza*, *P. paludosa*, and *E. agallocha*, the numbers of isoforms were higher in Sundarbans species than those of their replicas from freshwater condition. But the $R_{mf}$ and band intensity were different to a large extent in all the five species. In *Bruguiera*, the saline plant showed eight isoforms with highest OD 163.5 (0.07 $R_{mf}$), whereas the freshwater individual showed five isoforms with highest OD 51.37 (0.68 $R_{mf}$). In *Heritiera* and *Xylocarpus*, the numbers of isoforms were same but highest OD obtained 206.0 (0.18 $R_{mf}$) and 180.0 (0.68 $R_{mf}$) from saline individual and from freshwater habitats highest OD values were 166.0 (0.07 $R_{mf}$) and 89.9 (0.07 $R_{mf}$), respectively. Nonsaline *Phoenix* and *Excoecaria* showed three and two isoforms of PRX, and saline partners expressed four and three isoforms, respectively (Fig. 6.3).

### Superoxide Dismutase (SOD)

The experimental data showed that in all five species, isoforms of SOD expressed in less number from the freshwater grown individuals than those of their saline replicas. All four species expressed three isoforms in nonsaline environment, except *Phoenix*, where it was two. The plants from saline habitat, *Heritiera*, *Phoenix* and *Xylocarpus* showed five isoforms, *Bruguiera* and *Excoecaria* have two. The densitometric scanning resulted that the band intensity of each isoforms were much higher in saline habitats. In *Heritiera*, highest intensity (138.7 OD) occurred with $R_{mf}$ value 0.78 in saline individual, where in reverse habitat it was much less (6.48 OD and 0.63 $R_{mf}$). Similarly, in *Bruguiera*, it was 142.0 OD with 0.73 $R_{mf}$ in saline and 41. 53 OD at 0.18 $R_{mf}$ in nonsaline habitat. In *Xylocarpus* and *Phoenix*, saline and freshwater condition showed the highest peak as 147.0 OD (0.87 $R_{mf}$) and 12.0 OD (0.49 $R_{mf}$) and 184.0 OD (0.78 $R_{mf}$) and 42.0 OD (0.26 $R_{mf}$), respectively. In *Excoecaria* highest peak of intensity were observed in saline and nonsaline habitats as 170.07 OD (0.31 $R_{mf}$) and 163.0 OD (0.33 $R_{mf}$) respectively (Fig. 6.4).

### Esterase (EST)

From the stained gel, it revealed that EST expression in all species from freshwater habitats were two isoforms, except *Xylocarpus* (single band) and *Excoecaria* (three bands). The comparative band intensities were also remarkably high from all saline habitat taxa except in *Phoenix*, where it was slightly higher (222.0 OD at 0.48 $R_{mf}$ in saline plants and 177.0 OD at 0.3 $R_{mf}$ in nonsaline habitat). In *Heritiera*, among the four expressed bands in saline habitat, the highest band intensity occurred at 226.0 OD (0.48 $R_{mf}$) and it was 53.6 OD (0.36 $R_{mf}$) in reverse habitat. *Bruguiera* showed as high as 221.0 OD (0.53 $R_{mf}$) in saline (expressed number of isoforms was three) and 20.0 OD (0.48 $R_{mf}$). In *Xylocarpus*, out of five isoforms in saline condition, the highest OD was 223.0 (0.37 $R_{mf}$) in saline, and on the other side it was 59.5 OD (0.3 $R_{mf}$). Out of three isoforms, in saline species of *Excoecaria*, highest OD obtained 214.0 (0.17 $R_{mf}$) and in nonsaline it was 102.27 OD (0.24 $R_{mf}$) (Fig. 6.5).

### Acid Phosphatase (ACP)

Among the five investigated taxa, all four species showed excess number of isoforms of ACP in saline individual except in *Excoecaria*, where it was single band in both the environment, though the band intensity was higher in saline plants (196.0 OD, 0.4 $R_{mf}$) than nonsaline partner (26.15 OD, 0.53 $R_{mf}$). In *Bruguiera*, the saline habitat expressed two isoforms of ACP with higher intensity of 157.0 (0.49 $R_{mf}$) and 148.6

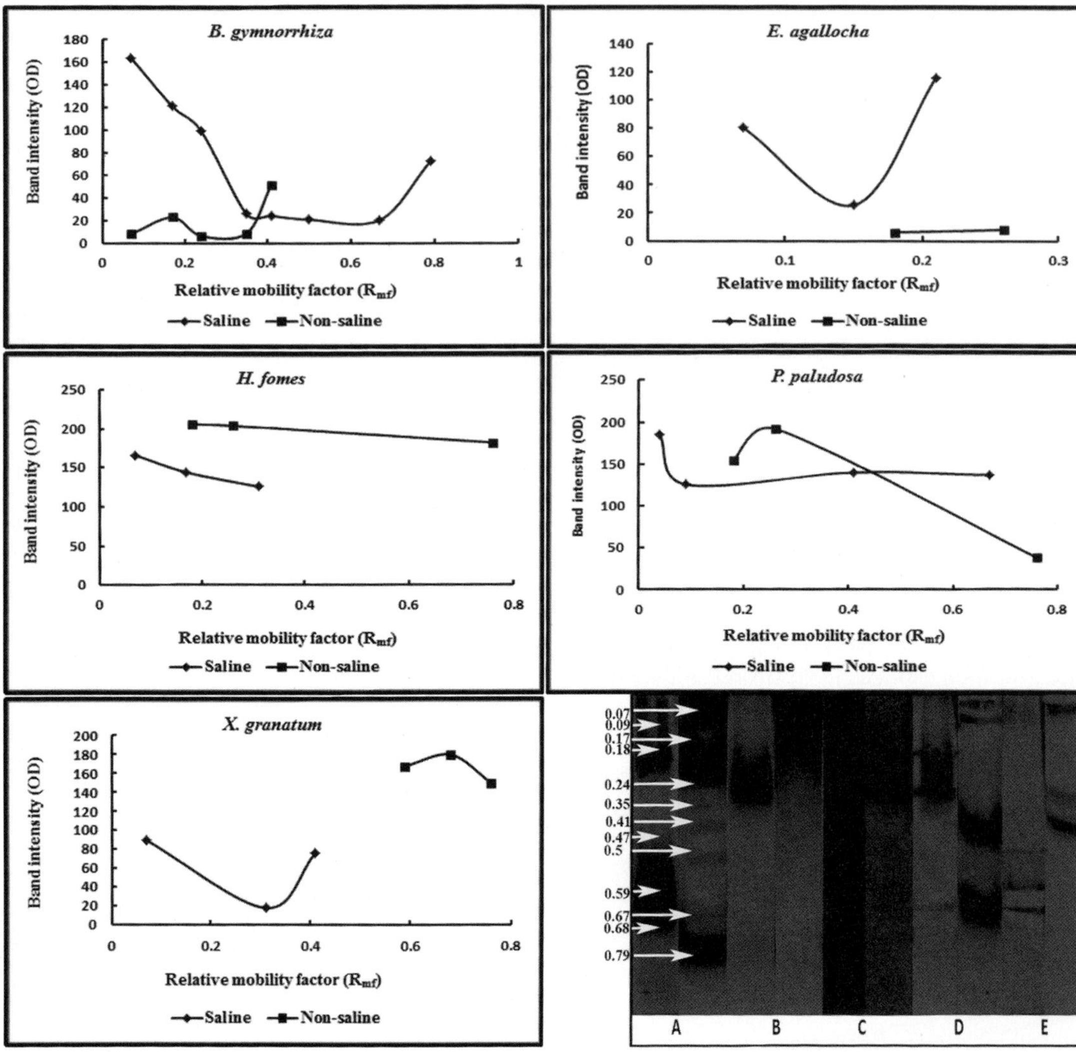


**Fig. 6.3** Band intensities and relative mobility factors ($R_{mf}$) of peroxidase enzyme among the investigated taxa: a comparative graphical representation of two habitat plants

OD (0.32 $R_{mf}$), but the freshwater plant has only one band with 124.0 OD (0.53 $R_{mf}$). In both *Xylocarpus* and *Phoenix*, saline environment expressed one more isoforms than those of their reverse habitat (three isoforms were expressed in freshwater habitat in each). The highest band intensity in *in-situ Xylocarpus* occurred with 247.0 OD (0.49 $R_{mf}$) and in reverse condition the highest band intensity and $R_{mf}$ value were almost same (248.0 OD and 0.46). In *ex-situ* plant of *Phoenix,* the highest intensity is observed at 96.03 OD (0.17 $R_{mf}$), and in counterpart it was 227.0 at 0.12 $R_{mf}$. Among the three expressed bands, highest OD value occurred as 168.4 (0.49 $R_{mf}$) in *Heritiera* (saline) and 145.0 OD (0.37 $R_{mf}$) in nonsaline plant (Fig. 6.6).

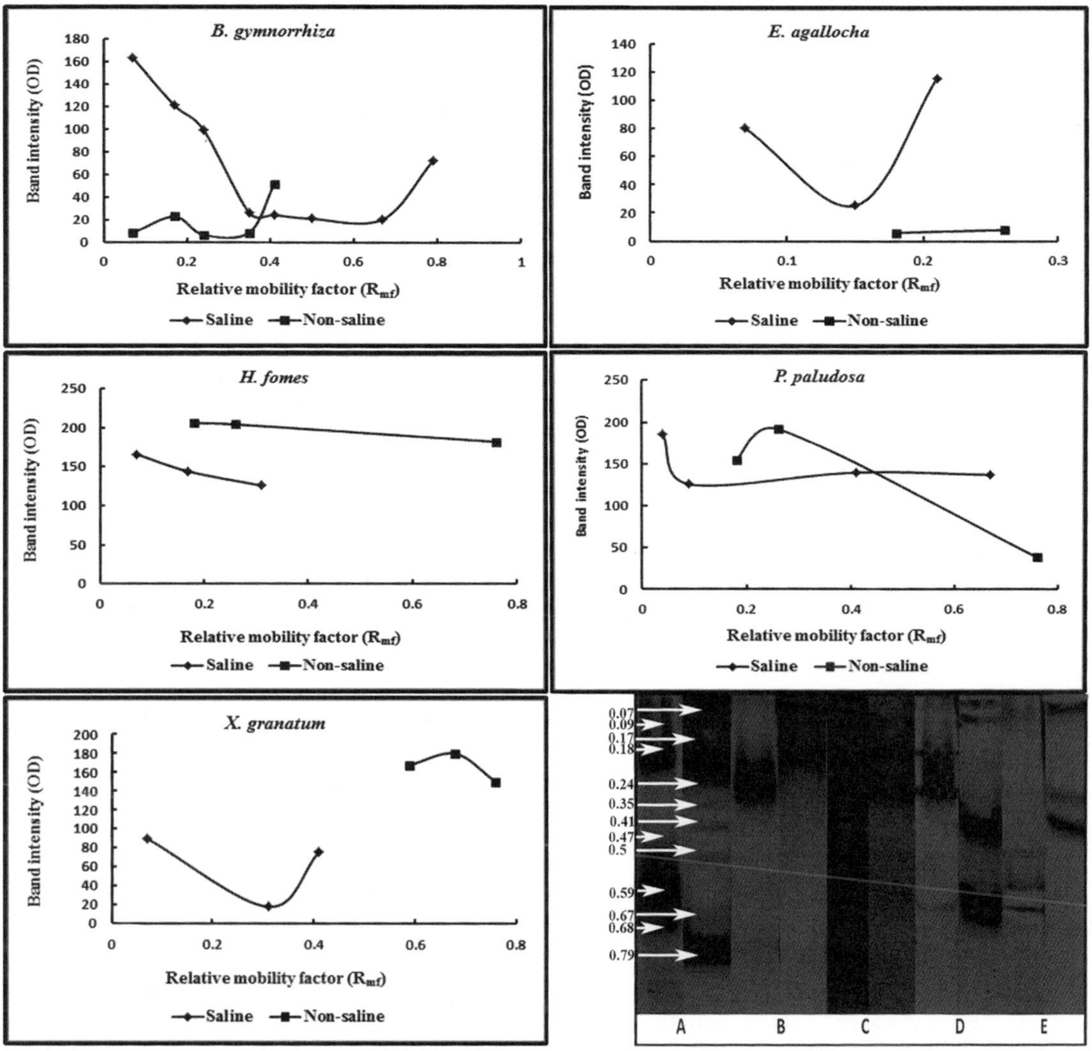


**Fig. 6.4** Band intensities and relative mobility factors ($R_{mf}$) of superoxide dismutase (*SOD*) enzyme among the investigated taxa: a comparative graphical representation of two habitat plants

## Quantitative Assay of Enzymes

The plant species from saline environment showed all four (PRX, SOD, EST, and ACP) investigated enzymes were in higher quantities than those of their freshwater grown individual. Increase in PRX quantity (μg/g) was highest in *Bruguiera* (257%), then *Xylocarpus* (209%), *Phoenix* (181%), and *Heritiera* (176%) while the increment was 139% in *Excoecaria* (Fig. 6.7a).

In case of SOD, the highest increment occurred in *Heritiera* (241%), then *Bruguiera* and *Phoenix* (229 and 224%, respectively), and lowest in *Excoecaria* (147%) (Fig. 6.7b). Similarly, EST was highest increased in *Phoenix* (287%), *Bruguiera* (257%), and *Heritiera* (241%), and lowest in *Excoecaria* (154%) (Fig. 6.7c). ACP reached its maximum increment in *Bruguiera* (293%) and *Xylocarpus* (267%) and lower in *Excoecaria* (139%) (Fig. 6.7d).

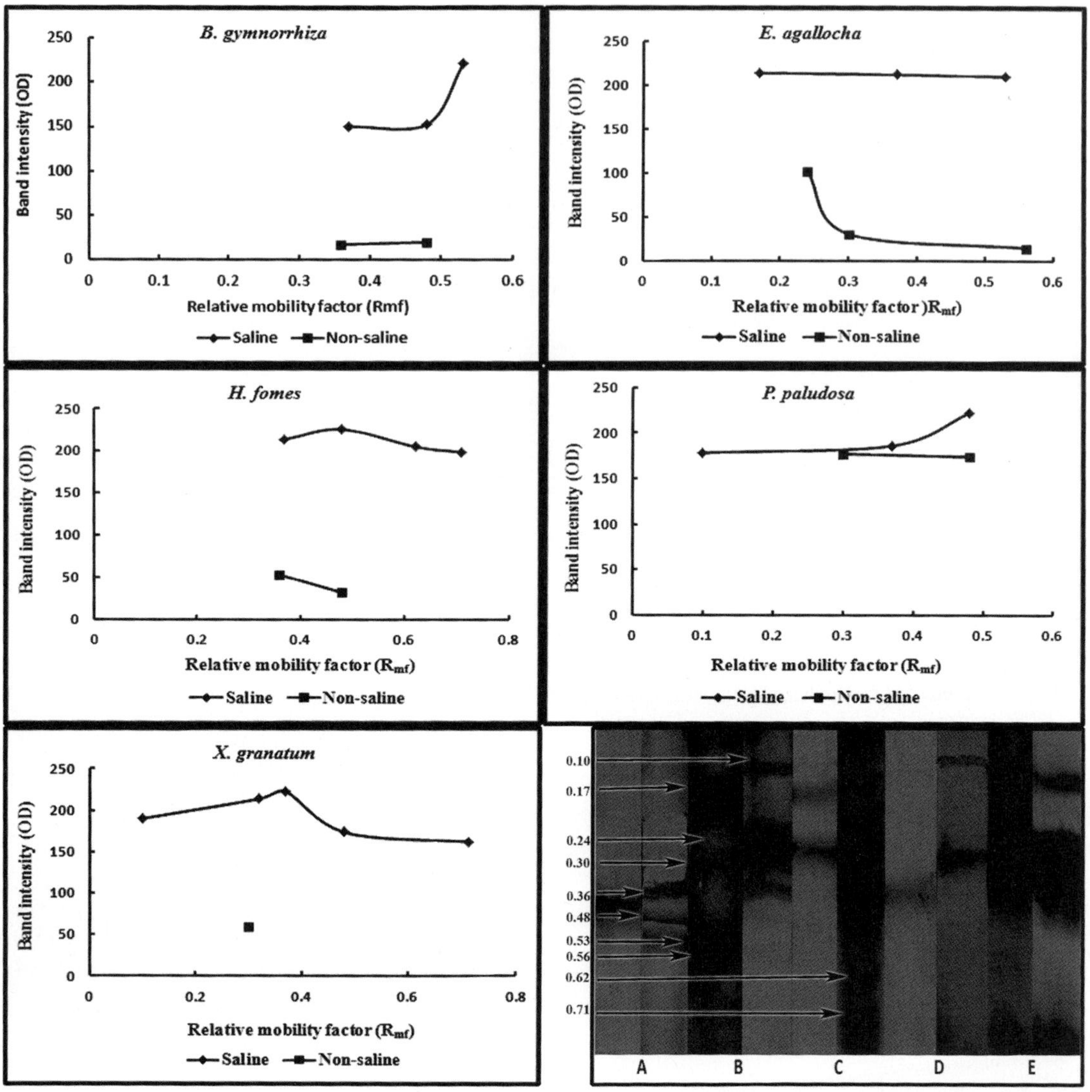


**Fig. 6.5** Band intensities and relative mobility factors ($R_{mf}$) of esterase enzyme among the investigated taxa: a comparative graphical representation of two habitat plants

## Statistical Analysis

Estimated total protein and four enzymes from two habitats were taken into account. A two-tailed bivariate correlation coefficient (Pearson coefficient) was calculated among each parameter (Table 6.1). The analysis showed that in case of the relationship between protein and SOD, all species in saline environment have inverse relationship (at 0.01% level) except of *Bruguiera*, wherein it was significant at 0.05% level. In PRO vs. PRX, significant inverse relationship was observed only in *Bruguiera* (0.05%) and *Phoenix* (0.01%), whereas the other three plants (*Excoecaria*, *Heritiera*, and *Xylocarpus*) showed no statistically significant relationship. Correlation between PRO and EST obtained a significant positive relationship at 0.01% level only in

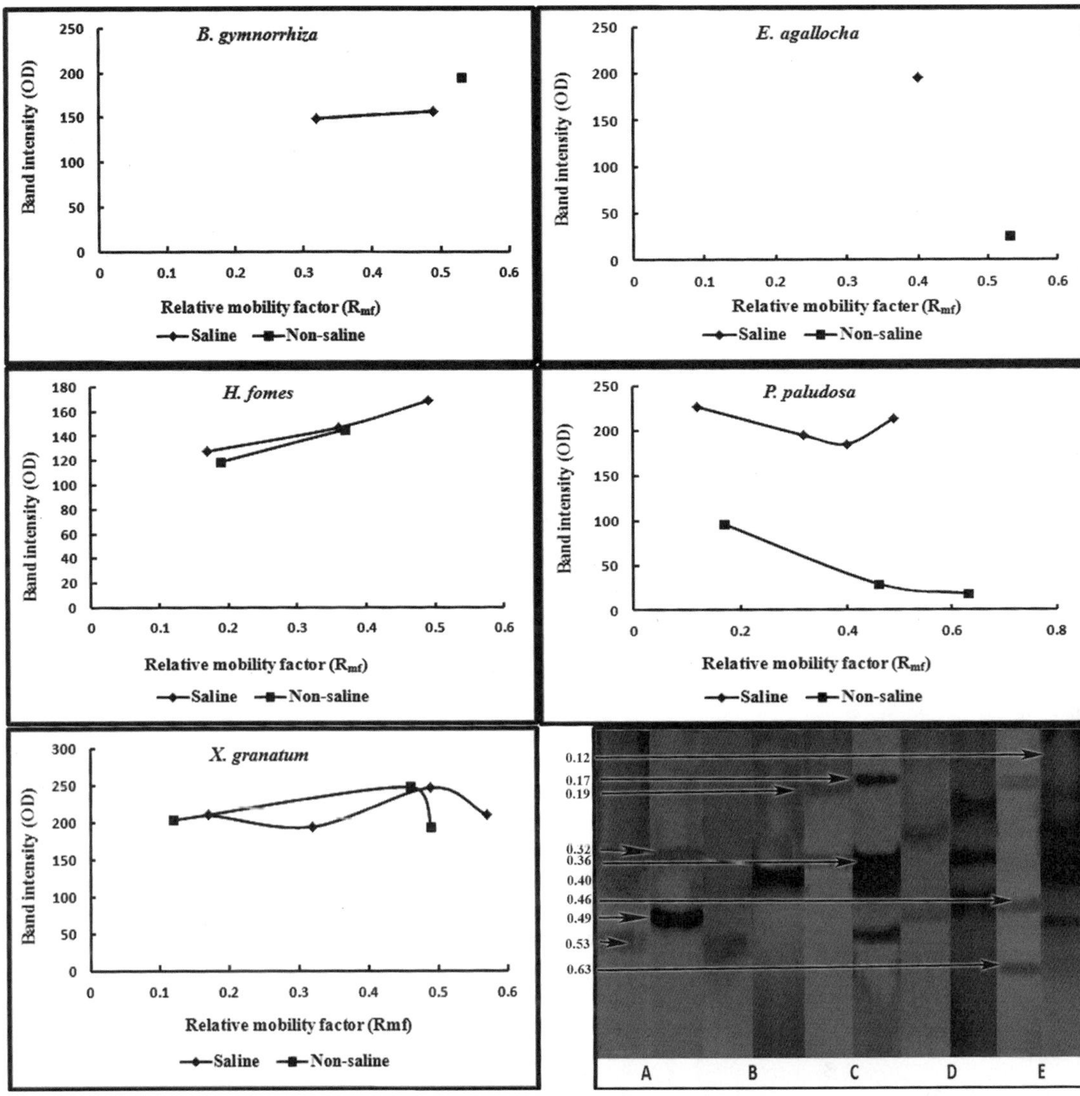


**Fig. 6.6** Band intensities and relative mobility factors ($R_{mf}$) of acid phosphatase enzyme among the investigated taxa: a comparative graphical representation of two habitat plants

*Bruguiera* and *Excoecaria* in saline inhabitants, and others did not show any relationship. The only inversely correlation was obtained in *Excoecaria* (saline plant) at 0.01% level, whereas in case of other plants, it showed no relationship.

## Discussion

Five typical mangroves (*Bruguiera gymnorrhiza*, *Excoecaria agallocha*, *Heritiera fomes*, *Phoenix paludosa*, and *Xylocarpus granatum*) from *in-situ* grown where salinity level of the substrate was quiet high (15–27 PPT) and *ex-situ* (mesophytic) habitat (salinity level was 1.8–2 PPT) were investigated with respect to their comparative approach of rate of net photosynthesis, stomatal conductance, and expression of two antioxidative enzymes, both qualitative and quantitative estimation.

Among the five investigated taxa *B. gymnorrhiza*, *E. agallocha*, and *P. paludosa*, the optimum

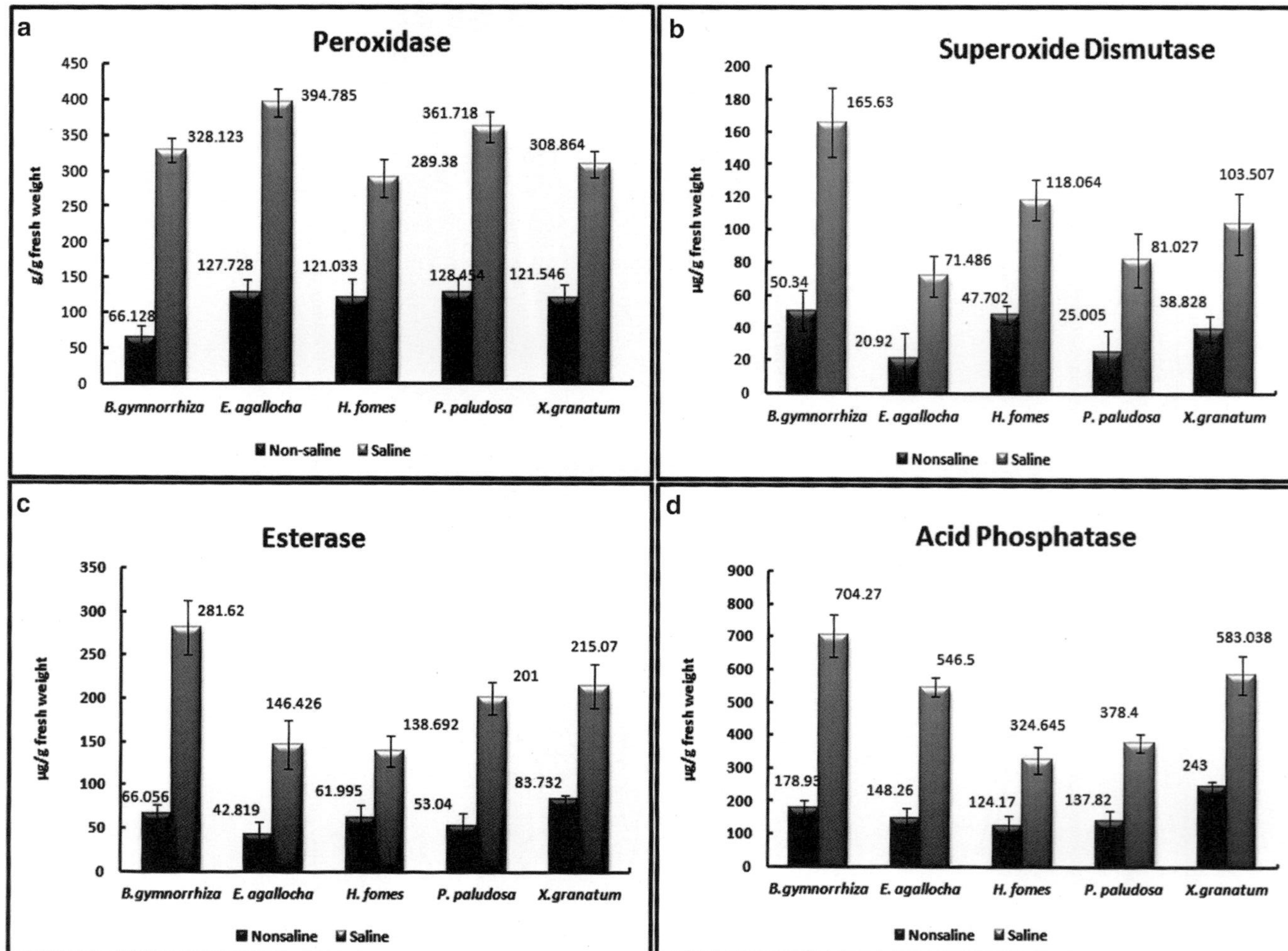


**Fig. 6.7** Quantitative estimation of four (**a**, **b**, **c**, and **d**) enzymes among the taxa, in two habitats. *Bars* indicates as standard error

**Table 6.1** Correlations among the different enzymes and total proteins in the plants of two habitats

| Species | *B. gymnorrhiza* | | *E. agallocha* | | *H. fomes* | | *P. paludosa* | | *X. granatum* | | |
|---|---|---|---|---|---|---|---|---|---|---|---|
| Treatment | Ns | S | Ns | S | Ns | S | Ns | S | Ns | S | Treatment |
| PRO | 0.186 | –0.571** | 0.380 | –0.754* | 0.045 | –0.529* | –0.383 | –0.731* | 0.145 | –0.705* | SOD |
| | 0.110 | 0.301 | 0.308 | –0.442 | 0.795** | –0.348 | 0.698* | 0.187 | 0.555 | –0.013 | POX |
| | 0.213 | 0.667* | 0.603 | –0.66* | 0.468 | 0.099 | 0.554 | –0.517 | 0.518 | –0.363 | EST |
| | –0.206 | 0.430 | –0.510 | –0.65* | 0.192 | –0.9** | –0.254 | 0.495 | –0.407 | –0.403 | ACP |

*Ns* nonsaline, *S* saline
*Significant at 0.01%; **Significant at 0.05%

PAR requirements were higher for maximum photosynthesis in Sundarbans than those in the mesophytic taxa, whereas the peak photosynthesis rates were higher in the nonsaline soil. But *H. fomes* and *X. granatum* showed the reverse phenomenon, where at comparatively low PAR the highest net photosynthetic rate occurred. Krauss and Allen (2003) pointed that *B. sexangula* prefers low salinity combined with low light intensity. Cheeseman and Lovelock (2004) experimentally proved that in *Rhizophora mangle* under low saline condition, net $CO_2$ exchange and photosynthetic electron transport become light saturated at less than 500 μmol $m^{-2}$ $s^{-1}$. In Sundarbans, however, despite tidal influence, high salinity makes the substrate physiologically dry. In order to check desiccation and xylem embolism, mangrove leaves reduce the rate of water efflux (Nandy (Datta) and Ghose 2001) that may enhance the tendency to elevate the leaf temperature with subsequent decline in photosynthesis. The present observation revealed that in all five species, stomatal conductance was reduced ranged by 25–52% under salinity stress that effectively limited $CO_2$ influx. Although reduced stomatal conductance imposed by high salinity restricts $CO_2$ diffusion, but may elevate the $CO_2$ partial pressure across the stomata, that utilized by mangrove leaves to maintain a consistently moderate rate of photosynthesis throughout the day, leading to avoid $CO_2$ starvation and photoinhibition. This result is well accord with Cowan (1982), Nandy (Datta) and Ghose (2001). Naidoo et al. (2002) also measured the optimum PAR for highest photosynthesis in *B. gymnorrhiza* at Durban Bay site that is similar (around 1,000 μmol $m^{-2}$ $s^{-1}$) to the present data. The opposite phenomenon occurred in *H. fomes* and *X. granatum* can be explained as less affinity of these species toward high salinity, irradiance, and temperature of the Sundarbans forest. Theoretically high photosynthetic efficiency can increase water use efficiency as more carbon is assimilated per unit water transpired. In mangroves, a positive correlation was reported between photosynthesis and stomatal conductance – an important determinant of water use efficiency (Nandy (Datta) et al. 2005). The effect of salinity stress on the photosynthetic enzyme activities postulated to be a secondary effect mediated by the reduced $CO_2$ partial pressure in the leaves caused by the stomatal closure (Lawlor and Cornic 2002; Meloni et al. 2003; DeRidder and Salvucci 2007). The present study also reveals that in all the mangroves grown in nonsaline soil, an increased rate of assimilation is coupled with increased stomatal conductance.

All the five investigated mangrove taxa from freshwater habitat showed an increase amount of total leaf protein than those of their saline replicas. It was noted that the percent of increment varied in a wide range from 5 to 36%, in which the highest increment occurred in *Excoecaria* and *Phoenix* while lowest in *Heritiera* and *Xylocarpus* (6.05 and 5.7%, respectively). This occurred probably as salinity imposed plants are adversely affected in their growth and metabolism due to osmotic effect of salt, nutritional imbalance, and accumulation of incompatible toxic ions. The decreased protein content in saline environment might be due to enhance activity of protease (Parida et al. 2002). The present result was well accord to Rajesh et al. (1999), where they experimentally reported that in *Ceriops*, the total leaf protein decreased under higher concentration of saline treatment. Raymond et al. (1994) opined that stress-induced protein degradation may be essential which provides amino acids for synthesis of new proteins suited for growth or survival under the modified condition. Mansour (2000) reported that protein biosynthesis declines under salt stress condition, while cells preferentially synthesize some specific stress proteins. Stress-induced proteins accumulated in the cell which might be synthesized de novo in response to salt or might be present constitutively at low level (Pareek et al. 1997). In the present investigation, the degradation of proteins in salt habitat *Heritiera* and *Xylocarpus* was lesser amount than the other three taxa investigated probably leading to synthesis of lesser amount of compatible amino acids in salt habitat. Parida et al. (2002) reported that the total soluble leaf proteins decreased in *Bruguiera parviflora* under NaCl treatment. This decreased might have the outcome of adverse effect of NaCl treatment resulted synthesis of certain low molecular weight proteins which are yet to be elucidated.

Among the various antioxidant enzymes, in this chapter we estimated two – peroxidase (PRX) and superoxide dismutase (SOD). Qualitative and quantitative study of two antioxidant enzymes (PRX and SOD) and two other important (hydrolyzing) enzymes (EST and ACP) from saline and freshwater grown plants revealed that in most of the cases number of isoforms, band intensity, and enzyme expression were higher in salt-stressed plant. It has been proved that during electron transport in the mitochondria and chloroplasts, some leakage of electrons occurs and these leaked electrons react with $O_2$ during aerobic metabolism to produce reactive oxygen species (ROS) such as superoxide ($O_2^-$), hydrogen peroxide ($H_2O_2$), and the hydroxyl radical ($^-OH$) (Halliwell and Gutteridge 1985). These cytotoxic ROS may seriously affect the normal metabolism through oxidative damage of lipids, proteins, and nucleic acids (Fridovich 1986). During photosynthesis, the internal $O_2$ level becomes high and chloroplast is prone to generate ROS at that time (Foyer and Mullineaux 1994). Plants synthesize a number of antioxidative enzymes to counteract these ROS, especially SOD converts $O_2^-$ into $H_2O_2$ and PRX catalyze $H_2O_2$ (Asada 1994). In salinity imposed plants the balance between the production of ROS and the scavenging activity of the antioxidants becomes disrupted which ultimately results in oxidative damage. Plants with high levels of antioxidants, either constitutive or induced, have been reported to provide sufficient resistance against oxidative damage (Parida et al. 2004a, b). The present work resulted that both PRX and SOD expressions were high in saline plants and the increment were ranged between 139 to 257% in case of PRX and 147 to 241 in SOD. The present result was substantiated with the earlier works (Cheeseman et al. 1997; Takemura et al. 2000). In both the cases, the increments were lower in *Heritiera* (139% in PRX, 147% in SOD) and *Xylocarpus* (142% in PRX, 166% in SOD) than those of the other three species of saline habitat. Parida et al. (2004b) opined that high salt concentration enhanced the accumulation of free amino acids and polyphenols. Thus, NaCl stress not only imposes alterations in antioxidative metabolism but also accumulation of osmolytes as adaptive measures. The numbers of isoforms were also increased in case of PRX and SOD in saline habitat plants. In *Bruguiera* (saline), the highest numbers of isoforms were expressed in case of PRX, but it was unchanged in case of *Heritiera* and *Xylocarpus* (three isoforms in each habitat). This might be due to the relatively less suitability of those plants in the saline environment. SOD showed the excess isoforms in all saline plants than their freshwater counterparts. Therefore, it is evident that the salt imposed production of toxic ROS is mostly regulated by upregulation of antioxidative enzymes like PRX and SOD. Sahu and Mishra (1987) reported changes in enzymatic activity of peroxidase during senescence of rice leaves when submitted to salt stress. They observed that NaCl increased peroxidase activity which could be related to regulation of membrane permeability, cell wall formation, and oxidation of accumulated substances due to salt stress. It was also proved that peroxidases are enzymes related to polymer synthesis in cell wall (Bowles 1990), as well as with prevention of oxidation of membrane lipids (Kalir et al. 1984).

Biosynthesis of esterase (EST) revealed that in all five species it is in higher amount in the *in-situ* taxa investigated. The freshwater grown plants synthesized esterase enzyme with less number of isoforms except *Excoecaria*, where the numbers of isoforms were same (3), but band intensity was more in saline plants. Highest number of isoforms occurred in *Heritiera* (saline – 4; in nonsaline – 2) and *Xylocarpus* (saline – 5; in nonsaline – 1). Still the percentage of increment was lower in the above two taxa than the other three from saline habitat (123 and 156%, respectively); the other three species ranged between 241 and 287% of esterase increment. This result supplemented by Hassanein (1999), where he experimentally proved that nine different esterase isoenzymes were detected in embryos of seeds germinated in 105 mM NaCl, whereas only five of them were detected in the embryos of untreated seeds. Pectins are major components of the primary plant cell wall. They can be both methylesterified and acetylesterified and de-esterification occurs by specific esterases (Cécile et al. 2006). Al-Hakimi and Hamada (2001) reported that the contents of cellulose, lignin of either shoots or roots, pectin of root, and soluble sugars of shoots were lowered with the rise of NaCl concentration. Hence, esterases play a major role to counteract the salt-induced imbalance in cell wall formation.

Acid phosphatases (ACP) are a group of enzymes that catalyze the hydrolysis of a variety of phosphate esters. These enzymes are widely distributed in plants and are related to phosphate supply and metabolism from a vast array of phosphate esters which are essential for normal growth and development of plant organs (Olczak et al. 2000). The present work revealed that the magnitude of increment in saline grown plants occurred ranging from 139 to 293%. It may be due to the fact that under stressful conditions, growth is restricted and delivery of phosphate is impaired, thus resulting in the activation of the cellular phosphatases that release soluble phosphate from its insoluble compounds inside or outside of the cells, thereby modulating osmotic adjustment by free phosphate uptake mechanism (Fincher 1989). Jain et al. (2004) also demonstrated that in the endosperm, acid and alkaline phosphatase activities were significantly higher after salt treatment than that of the control in pearl millet. Olmos and Hellin (1997) observed that acid phosphatases are known to act under salt and water stress by maintaining a certain level of inorganic phosphate which can be cotransported with $H^+$ along a gradient of proton motive force. Hence, the plants in which the ACP increments were observed lower might be less suited in higher salt environment.

The present investigation revealed that a significant inverse correlation obtained between the concentration of the antioxidative enzymes, peroxidase, and SOD with total protein in the case of *Bruguiera gymnorrhiza*, *Excoecaria agallocha*, and *Phoenix paludosa* in saline habitat. This elevation in the antioxidant enzyme concentration level may have taken place to scavenge more number of free radicals that are produced during stress (Davies 2000), and the decrease in protein concentration might be the result of formation of more compatible osmolytes to restore more negative water potential in cell sap. Both these phenomenon might provide some combat forces to the plants against salinity stress. On the other hand, no such statistical significant relationship between antioxidant enzymes and total protein concentration was found in case of *Heritiera fomes* and *Xylocarpus granatum*. This relationship, as discussed above, may provide some important clue toward the proper salt management mechanism for sustainable existence in the hostile environment. Therefore, the absence of it, might be one of the reasons toward less adaptability for the plants in present situation. Though there are scopes yet to elucidate in detail regarding the significance of increment of these enzymes in salt imposed plants, the present work might provide the baseline information and a system necessary to conduct future research in relation to the genetic basis of salt tolerance.

## References

Abogadalla GM (2010) Antioxidant defense under salt stress. Plant Signal Behav 5(4):369–374

Adeel Z, Caroline K (eds) (2002) Conserving our coastal environment: a summary of UNU's research on sustainable management of the coastal hydrosphere in the Asia Pacific region. United Nations University, Tokyo, 39

Al-Hakimi AMA, Hamada AM (2001) Counteraction of salinity stress on wheat plants by grain soaking in ascorbic acid, thiamin or sodium salicylate. Biol Plant 44(2):253–261

Alim A (1979) Instruction manual for plantations in coastal areas. In: White KL (ed) Research considerations in coastal afforestation. Food and Agricultural Organization, UNDP/FAO Project BDG/72/005. Forest Research Institute, Chittagong, pp 65–75

Alongi DM (2002) Present state and future of the world's mangrove forests. Environ Conserv 29:331–349

Alongi DM (2009) Introduction in the energetic of mangrove forests. Springer Science and Business Media BV, New York

Alscher RG, Donahue JL, Cramer CL (2002) Reactive oxygen species and antioxidants: relationship in green cells. Physiol Plant 100:224–233

Asada K (1994) Production and action of active oxygen species in photosynthetic tissues. In: Foyer CH, Mullineaux PM (eds) Causes of photooxidative stress and

amelioration of defense systems in plants. CRC Press, Boca Raton/Ann Arbor/London/Tokyo, pp 77–104
Balen B, Krsnik-Rasol M, Zadro I (2004) Eserase activity and isozymes in relation to morphogenesis in *Mammillaria gracillis* Pfeiff. tissue culture. Acta Bot Croat 63(2):83–91
Ball MC, Farquhar GD (1984) Photosynthetic and stomatal responses of the grey mangrove, *Avicennia marina*, to transient salinity conditions. Plant Physiol 74:7–11
Banarjee LK (1999) Mangroves of Orissa coast and their ecology. Bishen Singh Mohendra Pal Singh, Dehra Dun, p 41
Ben-Amor N, Jimenez A, Megdiche W, Lundqvist M, Sevilla F, Abdelly C (2007) Kinetics of the antioxidant response to salinity in the helophyte *Cackile maritime*. J Integr Plant Biol 49:982–992
Bowles DJ (1990) Defense-related proteins in higher plants. Annu Rev Biochem 59:873–907
Bray EA (1993) Molecular responses to water deficit. Plant Physiol 103:1035–1040
Cécile T, Francoise L, Pierre VC (2006) Polymorphism and modulation of cell wall esterase enzyme activities in the chicory root during the growing season. J Exp Bot 57(1):81–89
Chaffey DR, Miller FR, Sandon JH (1985) A forest inventory of Sundarbans, Bangladesh main report. Overseas Development Administration, London, 196 p
Cheeseman JM, Lovelock CE (2004) Photosynthetic characteristics of dwarf and fringe *Rhizophora mangle* L. in Belizean mangrove. Plant Cell Environ 27 (6):769–780
Chesseman JM, Herendeen LB, Cheeseman AT, Clough BF (1997) Photosynthesis and photoprotection in mangroves under field conditions. Plant Cell Environ 20:579–588
Clough BF (1994) Climatic impacts on mangrove ecosystems. In: Deshmukh SV, Balaji V (eds) Conservation of mangrove genetic resources: a training manual. ITTO- CRSARD Project. MSSRF, Madras, pp 39–43
Cowan IR (1982) Regulation of water use in relation to carbon gain in higher plants. In: Water relations and carbon assimilation, vol II, Physiological plant ecology. Springer, Berlin, pp 589–614
Curtis SJ (1993) Working plan for the forests of the Sundarbans Division for the period from 1st April 1931 to 31st March 1961. Bengal Government Press, Calcutta, 175 p
Das S (1999) An adaptive feature of some mangroves of Sundarbans, West Bengal. J Plant Biol 42: 109–116
Das S, Mukherjee KK (1997) Morphological and biochemical investigations on Ipomea seedlings and their species interrelationships. Ann Bot 79:565–571
Dasgupta N, Nandy (Datta) P, Das S (2011) Photosynthesis and antioxidative enzyme activities in five Indian mangroves with respect to their adaptability. Acta Physiol Plant 33:803–810
Davies KJA (2000) Oxidative stress, antioxidant defenses and damage removal, repair and replacement systems. IUBMB Life 50:279–289
DeRidder BP, Salvucci M (2007) Modulation pf Rubisco activase gene expression during heat stress in cotton (*Gossyoium hirsutum* L.) involves post-transcriptional mechanisms. Plant Sci 172:246–252
Duke NC, Meynecke JO, Dittmann S, Ellison AM, Anger K, Berger U, Cannicci S, Diele K, Ewel KC, Field CD, Koedam N, Lee SY, Marchand C, Nordhaus I, Dahdouh-Guebas F (2007) A world without mangroves? Science 317:41–42
Eyidogen F, Oz MT (2007) Effect of salinity on antioxidant responses on chickpea seedlings. Acta Physiol Plant 29:485–493
FAO (Food and Agricultural Organization, United Nations) (2007) The world's mangroves 1980–2005. FAO forestry paper 153, FAO, Rome
Fincher GB (1989) Molecular and cellular biology association with endosperm mobilization in germinating cereal grains. Annu Rev Plant Physiol Plant Mol Biol 40:305–346
Flowers TJ, Troke PF, Yeo AR (1977) The mechanism of salt tolerance in halophytes. Annu Rev Plant Physiol 28:89–121
Forest Survey of India (FSI) (2009) State forest report 2009. Ministry of Environment and Forests, Government of India, 226 p
Foyer CH, Mullineaux PM (1994) Causes of photooxidative stress and amelioration of defense systems in plants. CRC Press, Boca Raton/Ann Arbor/London/Tokyo, pp 343–364
Frenkel C (1972) Involvement of peroxidase and indole-3-acetic acid oxidase isozymes from pear, tomato, and blueberry fruit in ripening. Plant Physiol 49:757–763
Fridovich I (1986) Biological effects of the superoxide radical. Arch Biochem Biophys 247:1–11
Gilman E, Ellison J, Duke N, Field C (2008) Threats to mangroves from climate change and adaptation options: a review. Aquat Bot 89(2):237–250
Giri C, Zhu Z, Tieszen LL, Singh A, Gillette S, Kelmelis JA (2008) Mangrove forest distribution and dynamics (1975–2005) of the Tsunami affected region of Asia. J Biogeogr 35:519–528
Giri C, Ochieng E, Tieszen LL, Zhu Z, Singh A, Loveland T, Masek J, Duke N (2011) Status and distribution of mangrove forests of the world usin earth observation satellite data. Global Ecol Biogeogr 20:154–159
Halliwell B, Gutteridge JMC (1985) Free radicals in biology and medicine. Clarendon, Oxford
Haq S, karim Z, Asaduzzaman M, Mahtab F (1999) Vulnerability and adaptation to climate change for Bangladesh. Kluwer Academic, Dordrecht
Hasegawa PM, Bressan RA, Zhu JK, Bohnert HJ (2000) Plant cellular and molecular responses to high salinity. Annu Rev Plant Physiol Plant Mol Biol 51:463–499
Hassanein AM (1999) Alterations in protein and esterase patterns of peanut in response to salinity stress. Biol Plant 42(2):241–248
Hernandez JA, Jimenez A, Mullineaus P, Sevilla P (2000) Tolerance of pea (*Pisum sativam* L.) to long term salt

stress is associated with induction of antioxidant defenses. Plant Cell Environ 23:853–862
Hogarth PJ (2007) The biology of mangroves and seagrasses. Oxford University Press, New York, p 273
Hurkman WJ, Fornia CS, Tanaka CK (1989) A comparison of the effects of salt on polypeptides and translatable mRNAs in roots of a salt-tolerant and salt-sensitive cultivar of barley. Plant Physiol 90:1444–1456
Huttová J, Tamás L, Mistrik I (2002) Aluminium induced acid phosphatase activity in roots of Al-sensitive and Al-tolerant barley varieties. Roslinná Výroba 48 (12):556–559
Jain A, Sharmai AD, Singh K (2004) Plant growth hormones and salt stress-mediated changes in acid and alkaline phosphatase activities in the pearl millet seeds. Int J Agric Biol 6(6):960–963
Kalir A, Omri G, Poljakoff-Matber A (1984) Peroxidase and catalase activity in leaves of *Halimione portulacoides* exposed to salinity. Physiol Plant 62:238–244
Karim A (1994) Vegetation. In: Hussain Z, Acharya G (eds) Mangroves of Sundarbans, vol 2 – Bangladesh. IUCN, The World Conservation Union, Bangkok
Kathiresan K, Bingham BL (2001) Biology of mangrove and mangrove ecosystems. Adv Mar Biol 40:81–251
Keith H, Emmanuelle V, Hélène B, Charistian A (1983) Superoxide dismutase assay using alkaline dimethylsulfoxide as superoxide anion-generating system. Anal Biochem 135:280–287
Krauss KW, Allen JA (2003) Influence of salinity and shade on seedling photosynthesis and growth of two mangrove species *Rhizophora mangle* and *Bruguiera sexangula*, introduced to Hawaii. Aquat Bot 77 (4):311–324
Laemmli UK (1970) Cleavage of structural proteins during the assembly of the head of bacteriophage T4. Nature 227:680–685
Lawlor DW, Cornic G (2002) Photosynthetic carbon assimilation and associated metabolism in relation to water deficits in higher plants. Plant Cell Environ 25:275–294
Lin G, Sternberg L (1993) Effects of salinity fluctuation on photosynthetic gas exchange and plant growth of the red mangrove (*Rhizophora mangle*). J Exp Bot 44:9–16
Lowrey OH, Rosebrough N, Farr AL, Randall RJ (1951) Protein measurements with folin phenol reagent. J Biol Chem 193:265–275
Macfarlane GR, Burchett MD (2001) Photosynthetic pigments and peroxidase activity as indicators of heavy metal stress in the grey mangrove, *Avicennia marina* (Forsk.) Vierh. Mar Pollut Bullet 42(3):233–240
Mader M (1976) Die Localization der Peroxidase Iso-. enzym gruppe G, in der Zellwand von Tabak-Geweben. Planta 131:11–15
Mansour MMF (2000) Nitrogen containing compound and adaptation of plants to salinity stress. Biol Plant 43(3):491–500
Mansour MMF, Salama KHA, Ali FZM, Abou Hadid AF (2005) Cell and plant responses to NaCl in *Zea mays* L. cultivars differing in salt tolerance. Gen Appl Plant Physiol 31:29–41
Meloni DA, Oliva MA, Martinez CA, Cambraia J (2003) Photosynthesis and activity of superoxide dismutase, peroxidase and glutathione reductase in cotton under salt stress. Environ Exp Bot 49:69–76
Naidoo G, Tuffers AV, von Willert DJ (2002) Changes in gas exchange and chlorophyll fluorescence characteristics of two mangroves and a mangrove associate in response to salinity in the natural environment. Trees 16(2–3):140–146
Nandy (Datta) P, Ghose M (2001) Photosynthesis and water-use efficiency of some mangroves of Sundarbans, India. J Plant Biol 44:213–219
Nandy (Datta) P, Das S, Ghose M (2005) Relation of leaf micromorphology with photosynthesis and water efflux in some Indian mangroves. Acta Bot Croat 64 (2):331–340
Nandy (Datta) P, Das S, Ghose M, Spooner Hart R (2007) Effects of salinity on photosynthesis, leaf anatomy, ion accumulation and photosynthetic nitrogen use efficiency in five Indian mangroves. Wetl Ecol Manage 15:347–357
Nandy (Datta) P, Dasgupta N, Das S (2009) Differential expression of physiological and biochemical characters of some Indian mangroves towards salt tolerance. Physiol Mol Biol Plants 15(2):151–160
Naskar KR, Guha Bakshi DN (1983) A brief review on some less familiar plants of the Sundarbans India. J Econ Taxon Bot 4(3):699–712
Olczak M, Kobialka M, Watorek WX (2000) Characterization of diphosphonucleotide phosphatase/phosphodiesterase from yellow lupin (*Lupinus luteus*) seeds. Biochim Biophys Acta 1478(2):239–247
Olmos E, Hellin E (1997) Cytochemical localization of ATPase plasma membrane and acid phosphatase by cerium based in a salt–adapted cell line of *Pisum sativum*. J Exp Bot 48:1529–1535
Parani M, Lakshmi M, Zeigenhagen B, Fladung M, Senthilkumar P, Parida A (2000) Molecular phylogeny of mangroves VII. PCR-RFLP of trnS – pbsC and rbcL gene regions in 24 mangrove and mangrove associate species. Theor Appl Genet 100:454–460
Pareek A, Singla SL, Grover A (1997) Salt responsive proteins/genes in crop plants. In: Jaiwal PK, Singh RP, Gulati A (eds) Strategies for improving salt tolerance in higher plants. Oxford and IBH Publishing Co., New Delhi, pp 365–391
Parida A, Das AB, Das P (2002) NaCl stress causes changes in photosynthetic pigments, proteins and other metabolic components in the leaves of a true mangrove, *Bruguiera parviflora*, in hydroponic cultures. J Plant Biol 45:28–36
Parida AK, Das AB, Mohanty P (2004a) Defense potentials to NaCl in a mangrove. *Bruguiera parviflora*: differential changes of isoforms of some antioxidative enzymes. J Plant Physiol 161(5):531–542
Parida AK, Das AB, Mohanty P (2004b) Investigations on the antioxidative defence responses to NaCl stress in a mangrove, *Bruguiera parviflora*: differential

regulations of isoforms of some antioxidative enzymes. Plant Growth Regul 42(3):213–226

Rajesh A, Arumugam R, Venkatesalu V (1999) Response to *Ceriops roxburghiana* to NaCl stress. Biol Pant 42 (1):143–148

Raymond P, Broquisse R, Chevalior C, Couee I, Dieuaide M, James F, Just D, Pradet A (1994) Proteolysis and proteolytic activities in the acclimation to stress: the case of sugar starvation in maize root tips. In: Cherry JH (ed) Biochemical and cellular mechanisms of stress tolerance in plants. Springer, Berlin, pp 325–334

Sahu AC, Mishra D (1987) Changes in some enzyme activities during excised rice leaf senescence under NaCl – stress. Biochemie and Physiologie der Pflanzen 182:501–505

Shannon LM, Key E, Lew JY (1966) Peroxidase isozyme from Horseraddish root. I. Isolation and physical properties. J Biol Chem 249(9):2166–2172

Spalding M, Blasco F, Field C (1997) World mangrove atlas. The International Society for Mangrove Ecosystems, Okinawa, p 178

Spiers AG (1999) Review of international continental wetland resources. In: Finlayson CM, Spiers AG (eds) Global review of wetland resources and priorities for wetland inventory. Supervising scientist report 144. Canberra, Australia, pp 63–104

Stonier T, Yang H-M (1973) Studies on auxin protectors: XI. Inhibition of peroxidase-catalyzed oxidation of glutathione by auxin protectors and o dihydroxyphenols. Plant Physiol 51:391–395

Swire-Clark GA, Marcotte WR Jr (1999) The wheat LEA protein Em functions as an osmoprotective molecule in *Saccharomyces cerevisiae*. Plant Mol Biol 39:117–128

Takemura T, Hanagata N, Sugihara K, Baba S, Karube I, Dubinsky Z (2000) Physiological and biochemical responses to salt stress in the mangrove, *Bruguiera gymnorrhiza*. Aquat Bot 68:15–28

Upadhyay VP, Ranjan R, Singh JS (2002) The human mangrove conflicts-the way out. Curr Sci 83 (11):1328–1336

UPOV (1997) Adopted report of the Technical committee. Thirty third Session, Geneva, 16–18 Oct 1996. Union for Protection of New Varieties of Plants TC/ 33/11

Walter H (1971) Ecology of tropical and subtropical vegetation. Van Nostrand, New York, 539 pp

Xu D, Duan X, Wang B, Hong B, Ho TDH, Wu R (1996) Expression of a late embryogenesis abundant protein gene, HVA1, from barley confers tolerance to water deficit and salt stress in transgenic rice. Plant Physiol 110:249–257

Xu XY, Abo M, Okubo A, Yamazaki S (2001) Salt-stress responsive membrane proteins in *Rhodobacter sphaeroides* f. sp denitrificans IL106. J Biosci Bioeng 91:228–230

Ye Y, Tam FYN, Wong YS, Lu CY (2003) Growth and physiological responses of two mangrove species (*Bruguiera gymnorrhiza* and *Kandelia candel*) to water logging. Environ Exp Bot 49(3):209–221

Zeidler M (2000) Electrophoretic analysis of plant. Acta Univ Palacki Olomuc Fac Rerum Nat Biol 38:7–16

# Molecular Physiology of Osmotic Stress in Plants

7

Hrishikesh Upadhyaya, Lingaraj Sahoo, and Sanjib Kumar Panda

**Abstract**

Osmotic stress apparently reduces growth and productivity of crop plants. When plants are subjected to abiotic stress conditions like drought and salinity stress, most of the affected cellular processes are common and some overlap with cold stress. One important response to osmotic stress is the accumulation of the phytohormone abscisic acid (ABA), which induces several responses to osmotic stress. Osmotic stress signaling consists of an ABA-dependent and an ABA-independent pathway. Osmotic stress also caused increased ROS generation, which in turn elicits various cellular signaling networks resulting into physiological damage to plant cell. Osmotic stress signaling activates specific kinases, including one that belongs to the SNF1-related protein kinase (SnRK) 2 family, which is a plant-specific protein kinase family with 10 members (SnRK2.1-2-10) in *Arabidopsis*. ABA strongly activates SnRK2.2, 2.3, and 2.6, whereas osmotic stress activate almost all members of SnRK2s. Previous report revealed that SnRK2.2, 2.3, and 2.6 are core components of the ABA pathways whereas SnRK2s are essential kinases for osmotic stress signaling. Though the activation mechanism of SnRK2s in the ABA pathway is elucidated, regulatory mechanisms of SnRK2s in the ABA-independent pathway remain obscure. Thus, further analysis of SnRK2s must be a key study to draw the whole signaling pathways. This chapter is an insight into the molecular physiology of osmotic stress signaling and response in plants.

H. Upadhyaya
Department of Botany and Biotechnology, Karimganj College, Karimganj 788710, Assam, India
e-mail: hkupbl_au@rediffmail.com

L. Sahoo
Department of Biotechnology, Indian Institute of Technology Guwahati, Gauhati 781039, India
e-mail: ls@iitg.ernet.in

S.K. Panda (✉)
Plant Molecular Biotechnology Laboratory, Department of Life Science and Bioinformatics, Assam University, Silchar 788011, India
e-mail: drskpanda@gmail.com

G.R. Rout and A.B. Das (eds.), *Molecular Stress Physiology of Plants*,
DOI 10.1007/978-81-322-0807-5_7, © Springer India 2013

## Introduction

Plants are often subjected to adverse changes in the environment which they quickly recognize and respond with suitable reactions. Drought, heat, cold, and salinity are among the major abiotic stresses that adversely affect plant growth and productivity around the globe. Abiotic stress is the principal cause of crop yield loss worldwide, reducing normal yields of crop plant by more than 50% and thereby causing enormous economic loss as well. Osmotic stress, caused by either drought, salinity, or cold stress, is one of the important abiotic factors which had a great impact on plant evolution. Osmotic stress in its broadest sense encompasses both drought and salinity stress-induced lowering of water potential in plant cell (Hoffmann 2002). Osmotic stress is becoming particularly widespread in many regions as a consequence of serious salinization of more than 50% of all arable lands by the year 2050. In general, osmotic stress often causes a series of morphological, physiological, biochemical, and molecular changes that alters growth, development, and productivity of plant. Drought, salinity, and cold stress-induced oxidative stress are often interrelated, and these conditions singularly or in combination induce various cellular damages. Such stress stimuli inducing osmotic stress in plants are complex in nature (Huang et al. 2011). Almost all the abiotic stresses including osmotic stress lead to oxidative damage and involve the formation of reactive oxygen species (ROS) in plant cells. However, plants have evolved the mechanisms to reduce their oxidative damage by the activation of antioxidant enzymes and the accumulation of compatible solutes that effectively scavenge ROS (Upadhyaya and Panda 2004; Upadhyaya et al. 2008, 2011; Gill and Tuteza 2010). If the production of ROS exceeds the plant's ability to detoxify it, adverse reactions occurs, the typical symptoms being loss of osmotic responsiveness, wilting, and necrosis. Therefore, the balance between the production and the scavenging of ROS is needed, which is critical for the maintenance of normal growth and metabolism of the plant and overall osmotic stress tolerance in plants.

Osmotic stress signaling activates specific kinases, including one that belongs to the SNF1-related protein kinase (SnRK) 2 family, which is a plant-specific protein kinase family with 10 members (SnRK2.1-2.10) in *Arabidopsis*. ABA strongly activates SnRK2.2, 2.3, and 2.6, whereas osmotic stress activate almost all members of SnRK2s. Recent study reveals that SnRK2.2, 2.3, and 2.6 are core components of the ABA pathways whereas SnRK2s are essential kinases for osmotic stress signaling (Fujii et al. 2009, 2011; Fuji and Jhu 2009; Umezawa et al. 2010). Though the activation mechanism of SnRK2s in the ABA pathway is elucidated, regulatory mechanisms of SnRK2s in the ABA-independent pathway remain obscure. Thus, further analysis of SnRK2s must be a key study to draw the whole signaling pathways. Plant genetic engineering strategies for osmotic stress tolerance have been focused largely on the expression of genes that are involved in osmolyte biosynthesis (glycine betaine, mannitol, proline, etc.); genes encoding enzymes for scavenging ROS (superoxide dismutase (SOD), glutathione S-transferase, glutathione reductase, glyoxylases, etc); genes encoding late embryogenesis protein (LEA) (LEA, HVA1, LE25, dehydrin, etc); genes encoding transcription factors (DREB1A, CBF1, Alfin1); engineering of cell membranes; and proteins involved in ion homeostasis. These aspects have opened up the avenue to produce transgenics with improved osmotic stress tolerance. To cope with osmotic stress, it is important to understand plant responses to drought, salinity, cold, etc., stresses that disturb the cellular homeostasis in order to identify a common mechanism for multiple stress tolerance. This book chapter is an insight into molecular physiological mechanism of osmotic stress responses in plants.

## Osmotic Stress Sensing and Signaling in Plants

As other living organisms, plant cells perceive and process information with the use of various receptors exposed on the cell surface (Rodríguez et al. 2005). Membrane protein kinases of two classes—receptor-like Ser/Thr kinases (RLKs)

(Shiu and Bleecker 2001) and receptor Hiskinases (HKs) (Urao et al. 2001)—act as receptors in osmotic stress. In the late 1990s, *Arabidopsis thaliana* was found to possess AtHK1, which is homologous to *Saccharomyces cerevisiae* SLN1. In yeast cells, SLN1 senses the changes in osmolarity and triggers a two-component system, which activates MAPK (mitogen-activated protein kinase) pathways. Although direct evidence for the osmosensor role of AtHK1 in *A. thaliana* is still lacking, *AtHK1* expression is known to increase in salt stress (250 mM NaCl) or a decrease in temperature to 4°C (Urao et al. 1999). In the *sln1* Δ *sho1* Δ double mutant (the two affected proteins act as osmosensors in yeasts), *AtHK1* expression inhibited the salt-sensitive *sln1*Δ *sho1*Δ phenotype (Urao et al. 1999). It is important to note that AtHK1 did not interact with any of the five *A. thaliana* response regulators examined (Urao et al. 2001). Yeast SHO1, which acts as an osmosensor at a high osmolarity, contains four transmembrane domains and the SH3 domain exposed into the cytoplasm and lacks enzymatic activity. SHO1 initiates a signaling pathway typical of higher eukaryotic cells (Raitt et al. 2000; Reiser et al. 2000). At a high osmolarity, HK CRE1, which acts as a membrane cytokinin receptor in *A. thaliana*, substitutes HK SLN1 in *sln1*Δ yeast cells when activated with its ligand zeatin (Inoue et al. 2001). Such CRE1 activity was also observed when osmotic stress was simulated with sorbitol and, consequently, the turgor pressure and cell volume decreased (Reiser et al. 2003). CRE1 and SLN1 are similar in domain organization but are highly homologous only in the cytoplasmic kinase and sensor domains. The questions arise as to how CRE1 perceives the changes in turgor pressure and which domains are indispensable for the osmosensor function of activated CRE1. Reiser et al. (2003) assumed that CRE1 mediates a physical contact of the cell wall with the plasma membrane. The osmosensor function requires a unique combination of the periplasmatic and transmembrane domains of CRE1, as well as the integrity of the total periplasmatic domain. The above property of CRE1 is of immense importance, suggesting that plants have an osmosensing system that is structurally and functionally similar to its yeast analog. This assumption needs verification with plant cells. Recent analyses of cultured plant cells with altered levels of the putative $Ca^{2+}$-permeable mechanosensitive channel indicate that OsMCA1 is involved in regulation of plasma membrane $Ca^{2+}$ influx and ROS generation induced by hypo-osmotic stress in cultured rice cells (Kurusu et al. 2012). Such findings enlight our understanding of mechanical osmotic stress sensing pathways.

## Physiochemical Effect of Osmotic Stress in Plants

Osmotic stress results into various physiochemical responses in plants growing at different developmental stages. Accumulation of compatible solutes, such as proline (Pro) and glycine betaine (GB), in response to drought and salinity to facilitate water uptake (Ashraf and Foolad 2007) is common in plants. In addition to osmotic adjustments, these osmolytes were suggested to be important for protecting cells against increased levels of ROS accumulation under stress conditions. Pro accumulates in the cytosol and the vacuole during stress (Aubert et al. 1999; McNeil et al. 1999) and was shown to protect plant cells against damages caused by $^1O_2$ or HO· (Matysik et al. 2002). By quenching $^1O_2$ and directly scavenging HO, Pro might be able to protect proteins, DNA, and membranes (Matysik et al. 2002). In addition to directly scavenging of HO, Pro might bind to redox-active metal ions and protect biological tissues against damages caused by HO· formation (Matysik et al. 2002). Transgenic wheat plants that accumulated higher Pro than wild type exhibited less lipid peroxidation of membranes during osmotic stress, indicating a role for Pro in reducing ROS damages during osmotic stress (Vendruscolo et al. 2007). *Arabidopsis* mutants deficient in salt-induced siRNA that suppresses the expression of pyrroline-5-carboxylate dehydrogenase (functions in Pro degradation) were impaired in Pro accumulation, enhanced the accumulation ROS, and consequently enhanced plant sensitivity to salinity stress (Borsani et al. 2005). Pro synthesis from glutamate, previously documented in the cytosol, was now shown to occur, at least in part, in the chloroplast (Székely et al. 2008). During Pro synthesis, NADPH is used for the reduction of glutamate, which increases $NADP^+$ availability, required for relieving of PSI over-reduction during stress, as well as increasing $NADP^+$/NADPH ratio that prevents perturbation of redox-sensitive pathways during drought- and salt-induced osmotic stresses. Accordingly, Pro was shown to protect photochemical efficiency of PSII and prevent lipid peroxidation during drought-induced osmotic stress (Molinari et al. 2007). These results indicate that Pro functions to indirectly protect PSII as well as directly scavenge ROS during drought. In contrast to Pro, GB is not thought to directly scavenge ROS during abiotic stresses (Smirnoff and Cumbes 1989; Chen and Murata 2008). However, GB has been shown to protect cells against oxidative damage during abiotic stresses (Park et al. 2007; Chen and Murata 2008). Accumulation of GB is mainly in the chloroplast and is involved in the maintenance of PSII efficiency under stress conditions (Génard et al. 1991; Ashraf and Foolad 2007; Ben Hassine et al. 2008). A previous study showed that accumulation of GB in the chloroplast is more effective than that in other cellular compartments in protecting plants against oxidative stress and salinity (Park et al. 2007). In addition, exogenous GB treatment prevents osmotic stress-induced structural damages to ROS-producing organelles, such as chloroplasts and mitochondria (Ashraf and Foolad 2007). These results suggest that GB acts to prevent excess ROS production by protecting chloroplasts during salinity-induced osmotic stress. Activation of antioxidant mechanisms by Pro and GB during salinity has been studied using tobacco Bright Yellow-2 suspension-cultured cells (Hoque et al. 2007; Banu et al. 2009). Salinity significantly inhibited the amount of reduced AsA, reduced GSH, and the activity of AsA–GSH cycle enzymes, and exogenous application of Pro or GB increased the activity of these enzymes (Hoque et al. 2007). These results suggest a role of Pro and GB in the regulation of antioxidant enzymes during osmotic stress induced by salinity. Soluble sugars also contribute to the regulation of ROS signaling as well as osmotic adjustments during abiotic stresses (Vinocur and Altman 2005; Seki et al. 2007). Soluble sugars are involved in the metabolism and protection of both ROS-producing and ROS-scavenging pathways, such as mitochondrial respiration, photosynthesis, and oxidative-pentose-phosphate pathway (Couée et al. 2006). A number of studies have suggested that mannitol can protect against photooxidative damages to the chloroplastic apparatus caused especially by HO· during stress. Transgenic tobacco plants with increased mannitol production targeted to the chloroplast showed increased scavenging capacity of HO· enhancing their resistance to oxidative stress. Because the increase in mannitol content was insufficient to account for osmotic adjustments in transgenic wheat plants expressing the *Escherichia coli* mannitol biosynthetic gene *mtlD* during drought and salinity, mannitol function in enhancing stress tolerance in these plants was mainly attributed to its potential for scavenging of HO· and $O_2^-$ (Abebe et al. 2003). Trehalose, a disaccharide, is known as a signaling molecule that regulates carbon and ABA metabolism under osmotic stress conditions (Avonce et al. 2004). In previous studies, overexpression of trehalose biosynthetic genes enhanced tolerance of transgenic plants to drought and salinity (Garg et al. 2002; Penna 2003; Miranda et al. 2007), suggesting the involvement of trehalose in the response of plants to these stresses. Transgenic rice plants that express *E. coli* trehalose biosynthetic genes showed less photooxidative damage to PSII during drought and salinity compared with wild-type plants (Garg et al. 2002). Drought stress-induced osmotic stress is known to reduce protein synthesis, photosynthetic efficiency, nutrient accumulation, antioxidant metabolism and increased ROS level, lipid peroxidation, and oxidative metabolism in plant cell (Upadhyaya and Panda 2004; Bartels and Sunkar 2005; Shinozaki and

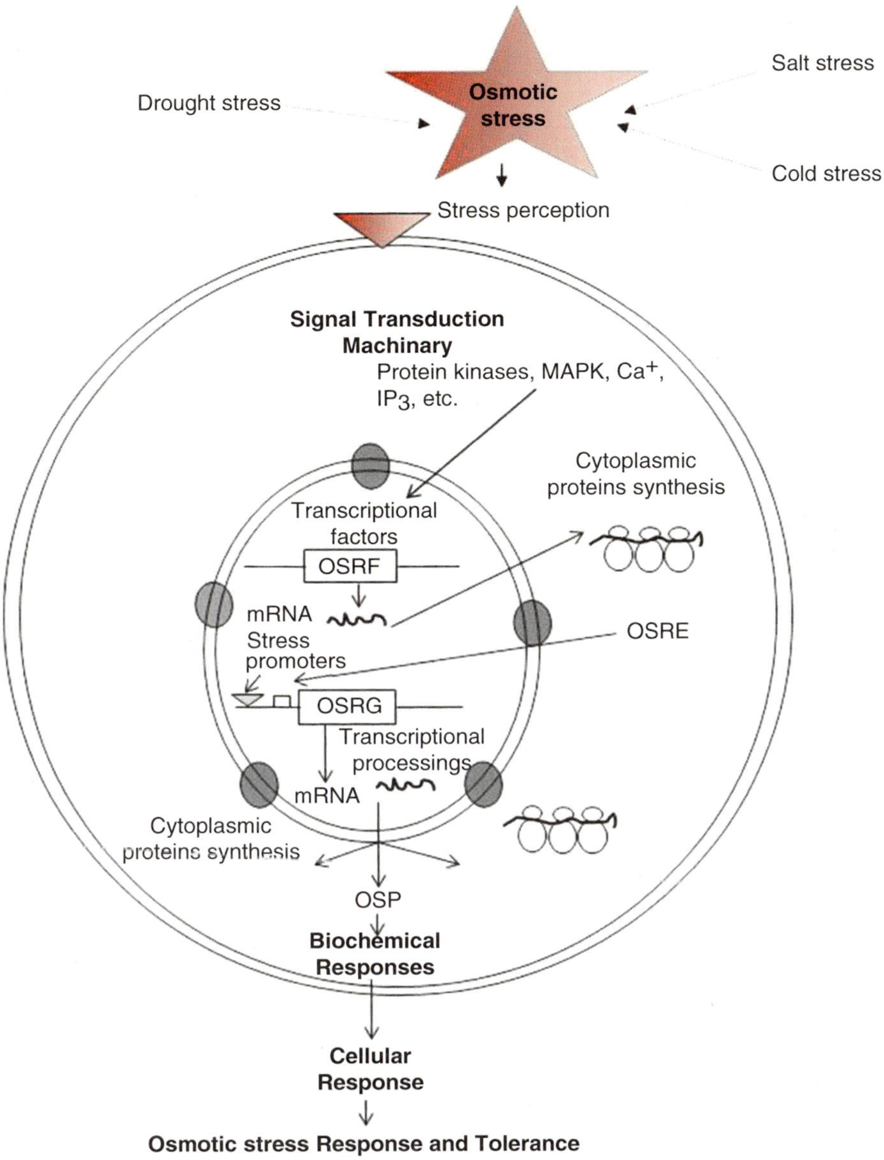


**Fig. 7.1** Schematic model depicting the generalized molecular mechanism of osmotic stress response in plants. *OSRF* osmotic stress responsive factors, *OSRE* osmotic stress responsive elements, *OSRG* osmotic stress responsive genes, *OSP* osmotic stress proteins

Yamaguchi-Shinozaki 2007; Upadhyaya et al. 2008, 2011) (Fig. 7.1).

## SOS Signaling Pathway and Osmotic Stress

Cellular ion homeostasis during salinity-induced osmotic stress is achieved by the following strategies: (1) exclusion of $Na^+$ from the cell by plasma membrane-bound Na+/H+ antiporters or by limiting the $Na^+$ entry, (2) utilization of $Na^+$ for osmotic adjustment by compartmentation of $Na^+$ into the vacuole through tonoplast $Na^+/H^+$ antiporters, and (3) $Na^+$ secretion. Thus, regulation of ion transport systems is fundamental to plant salt tolerance. Genetic analysis of *s*alt *o*verly *s*ensitive (*sos*) mutants of *Arabidopsis* led to the identification of the SOS pathway, which regulates cellular ion homeostasis and salt tolerance (Zhu 2002). The *Arabidopsis sos3* mutant is hypersensitive to salt stress. Molecular cloning revealed that the *SOS3* encodes a $Ca^{2+}$-binding protein homologous to the regulatory subunit of yeast calcineurin and animal neuronal calcium sensors. It has an N-myristoylation motif and three calcium-binding EF hands. SOS3 senses salt stress-induced increases in cytosolic $Ca^{2+}$ concentration in plants (Liu and Zhu 1998; Ishitani et al. 2000). Myristoylated SOS3 is recruited to the plasma membrane (Quintero et al. 2002). Mutations that disrupt either myristoylation (G2A) or calcium binding (*sos3-1*) cause salt stress hypersensitivity to *Arabidopsis* plants. Since myristoylation of SOS3 is essential for salt tolerance, it is likely that membrane recruitment of SOS3 is essential for its function. Membrane localization of SOS3 may help in the regulation of its target ion transporters (Ishitani et al. 2000). Identification of additional *SOS* loci (*SOS2* and *SOS1*) revealed that the SOS pathway regulates cellular ion homeostasis under salt stress. *Arabidopsis sos1* and *sos2* mutants are also hypersensitive to salt stress and *sos1*, *sos2*, and *sos3* mutations do not show an additive effect, implying that they are in the same pathway of salt stress response. SOS2 is a ser/thr protein kinase with an N-terminal kinase catalytic domain and a C-terminal regulatory domain. The SOS2 C-terminal regulatory domain consists of the SOS3-binding, autoinhibitory FISL motif (Liu et al. 2000). Binding of SOS3 activates the SOS2 protein kinase (Halfter et al. 2000). Deletion of the FISL motif from SOS2 leads to constitutive activation of the kinase. Molecular genetic analysis of the *sos1* mutant led to the identification of a target for the SOS3–SOS2 kinase complex. *SOS1* encodes a plasma membrane Na+/H+ antiporter. The *sos1* mutant accumulates high levels of Na+ in tissues under salt stress, and isolated plasma membrane vesicles from *sos1* mutants showed significantly less Na+/H+ exchange activity than the wild type, suggesting that the SOS1 Na+/H+ antiporter is located on the plasma membrane (Qiu et al. 2002). The *sos3* and *sos2* mutants accumulate higher levels of $Na^+$ than wild-type plants. Isolated plasma membrane vesicles from these mutants also showed significantly less $Na^+/H^+$ exchange activity, and this could be restored to the wild-type levels by the addition of activated SOS2. The SOS3–SOS2 kinase complex activates SOS1 by phosphorylation (Quintero et al. 2002). *SOS1* complemented yeast mutants defective in $Na^+$ transporters. Coexpression of *SOS2* and *SOS3* significantly increased *SOS1*-dependent Na+ tolerance of the yeast mutant (Quintero et al. 2002). These results show that SOS1 is a $Na^+/H^+$ antiporter involved in $Na^+$ efflux, which is activated by the SOS3-–SOS2 kinase complex ( Qiu et al. 2002; Quintero et al. 2002). Constitutive expression of a CaMV 35S promoter-driven active form of SOS2 could rescue *sos2* and *sos3* mutants under salt stress (Xiong et al. 2002). The expression of *SOS1* is stronger in cells bordering the xylem. Under salt stress (100 mM NaCl), a higher concentration of Na+ accumulates in shoots of *sos1* mutants than in those of the wild type. These results suggest that SOS1 might retrieve Na+ from the xylem, thereby preventing excess Na+ accumulation in the shoot (Shi et al. 2002). Transgenic *Arabidopsis* plants overexpressing *SOS1* showed improved salt tolerance and accumulated less Na+ in the xylem transpirational stream as well as in the shoot compared to the wild-type plants. This demonstrated that $Na^+$ efflux from the root cells and long-distance $Na^+$ transport within the plant under salt stress are regulated by SOS1 (Shi et al. 2003), which in turn is regulated by the SOS3–SOS2 kinase complex. In addition to the activation of $Na^+/H^+$ antiporter activity of SOS1, SOS3–SOS2 kinase complex also is involved in salt stress-induced upregulation of *SOS1* expression (Shi et al. 2000). In the *sos3* mutant, salt stress could not induce *SOS1* expression, while the *sos2* mutant is impaired in *SOS1* expression only in roots, but not in shoots. Interestingly, *SOS1* overexpressing transgenic *Arabidopsis* showed a significantly higher steady-state level of *SOS1* mRNA under salt stress than that grown under normal conditions. Since *SOS1* was overexpressed under the control of the CaMV 35S promoter, its higher mRNA abundance under salt stress might be due to an increase in *SOS1* transcript stability (Shi et al. 2003). In addition to positive control of $Na^+$ exclusion from the cytosol, the SOS pathway may also negatively regulate $Na^+$ influx systems. Expression of plant high-affinity $K^+$ transporters, *AtHKT1*, *EcHKT1*, and *EcHKT2*, in *Xenopus laevis* oocytes showed that they could mediate Na+ uptake. Transgenic wheat plants expressing the wheat *HKT1* in antisense orientation under control of a ubiquitin promoter showed significant downregulation of the native *HKT1* transcript. These lines showed significantly less 22Na uptake and enhanced growth under salinity when compared with the control (Laurie et al. 2002). These results suggest that HKT1 mediates sodium uptake under salinity and salt tolerance can be improved by downregulation of *HKT1* expression. Consistent with this observation, a suppressor genetic screen for the *sos3* mutation revealed that functional disruption of *AtHKT1* could suppress the salt-sensitive phenotype of *sos3*. In addition, the *athkt1* mutation alleviates the $K^+$-deficient phenotype of the *sos3* mutant (Rus et al. 2001), which suggests that the $K^+$-deficient phenotype of the *sos3* mutant might be due to an excess of cytoplasmic $Na^+$, as *sos3* impairs the $Na^+$ efflux mediated by SOS1. These results suggest that ATHKT1 might function as low-affinity $Na^+$ transporter that is involved in $Na^+$ influx under salinity. Significant amounts of $Na^+$ enter plant roots through voltage-independent channels, which are probably regulated by $Ca^{2+}$ concentrations. It is not known whether activity of these channels and their gene expression are also regulated by the calcium-dependent SOS3–SOS2 kinase complex. Thus, the SOS3–SOS2 kinase complex positively regulates $Na^+$ efflux by activating SOS1 and upregulating the *SOS1* transcript level and may negatively regulate Na+ influx by downregulating low-affinity $Na^+$ transporter (*HKT1*) genes to restore cellular ion homeostasis under salt stress in plants (Fig. 7.2; Zhu 2002).

## Hormonal Regulation of Osmotic Stress in Plant

The plant hormones play a pivotal role in a variety of developmental processes and adaptive stress responses to environmental stimuli in plants. Cellular dehydration during the seed maturation and vegetative growth stages induces an increase in endogenous ABA levels, which control many dehydration-responsive genes. In Arabidopsis plants, ABA regulates nearly 10% of the protein-coding genes, a much higher percentage than other plant hormones. Expression of the genes is mainly regulated by two different families of bZIP transcription factors (TFs), ABI5 in the seeds and AREB/ABFs in the vegetative stage, in an ABA-responsive-element (ABRE)-dependent manner (Fig. 7.2). The SnRK2-AREB/ABF pathway governs the majority of ABA-mediated ABRE-dependent gene expression in response to osmotic stress during the vegetative stage. In addition to osmotic stress, the circadian clock and light conditions also appear to participate in the regulation of ABA-mediated gene expression, likely conferring versatile tolerance and repressing growth under stress conditions. Moreover, various other TFs belonging to several classes, including AP2/ERF, MYB, NAC, and HD-ZF, have been reported to engage in ABA-mediated gene expression. It was proposed that in roots auxin and abscisic acid decrease resistance to water flow. Therefore, an adaptation of plants to osmotic stress may be an increase in root auxin concentration and a decrease in leaf auxin concentration, or auxin signaling, to minimize water loss. Indeed, Albacete et al. (2008) found an increase in root auxin content, whereas leaf auxin content of tomato stressed with 100 mM NaCl decreased. A decrease in auxin content has been shown in plants exposed to osmotic stress. This has been reported for herbaceous plants, for example, *Anastatica hierochuntica* and *Helianthus annuus*, but also in the developing xylem of poplar (Junghans et al. 2006).

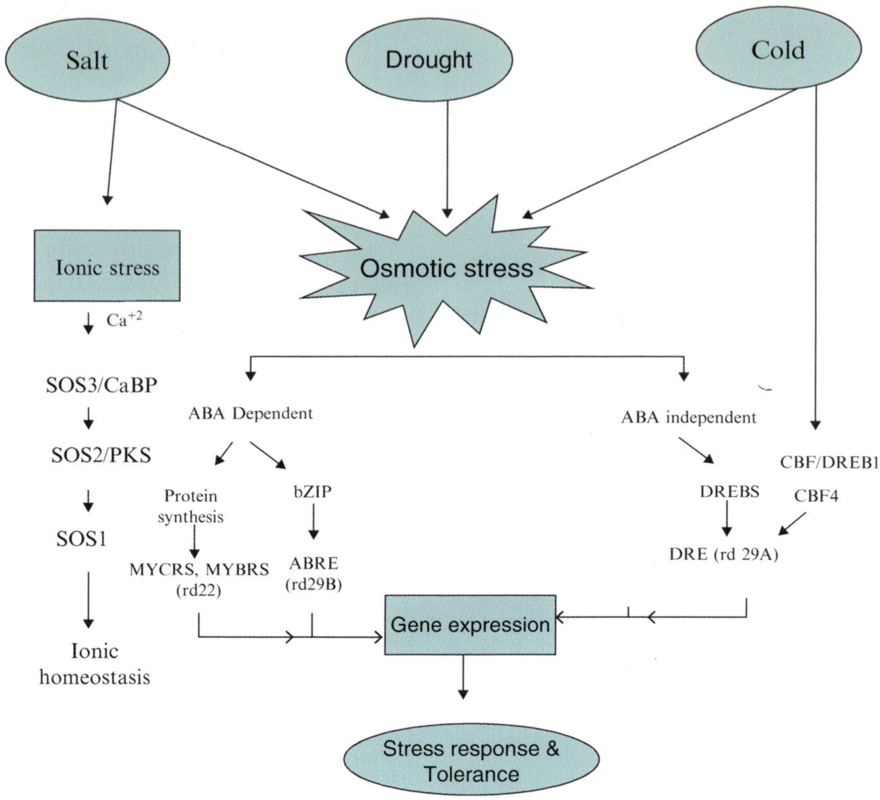


**Fig. 7.2** Osmotic stress signaling and cross talk involved in cold, drought, and salinity stress gene expression. Osmotic stress signaling generated via cold, salinity, and drought stress seems to be mediated by transcription factors such as DREB2A, DREB2B, bZIP, and MYC and MYB transcription activators, which interacts with CRT/DRE, ABRE, or MYCRE/MYBRE elements in the promoter of stress genes. Two different DRE/CRT-binding proteins, DREB1/CBF and DREB2, distinguish two different signal transduction pathways in response to cold and drought stresses, respectively. Salinity mainly works through SOS pathway reinstating cellular ionic equilibrium

Changes in auxin response can be achieved by regulation of auxin biosynthesis and auxin conjugation or by modulation of auxin signaling. In addition, local auxin concentration and auxin distribution may be regulated under stress by changes in auxin transport. It was reported that osmotic stress caused by increased salinity or drought has an impact on polar auxin transport, but response of the auxin transport machinery seems to be plant and tissue specific. In Avena mesocotyls, osmotic stress induced a drastic increase in auxin transport (Sheldrake 1979), while in *Pisum* hypocotyls a significant reduction was observed (Kaldewey et al. 1974). In the petioles of water-stressed cotton, the basipetal auxin transport capacity was reduced to 30% of that in control plants, but these authors did not investigate whether the reduced capacity was caused by decreased auxin transport or diminished auxin synthesis in the leaf blade.

IAA amidoconjugate synthases are encoded by members of the GH3 gene family (Staswick et al. 2002, 2005). Park et al. (2007) reported that environmental stresses, including high salinity and osmotic stress, induced expression of the AtGH3.5 gene. Plants overexpressing AtGH3.5 exhibited growth reduction and enhanced stress resistance, and thus, it suggests that GH3 auxin conjugate synthases are involved in adjustment of the hormonal balance in response to stress. Recent literature showed that transcriptional responses of rice to high NaCl concentrations provide evidence that auxin signaling is affected by salt stress. Jain and Khurana (2009) investigated auxin-induced gene expression in rice and found that 154 auxin-induced and 50 auxin-repressed genes also responded to abiotic stress such as drought, increased salinity, and cold. A total of 51 of the auxin-responsive genes were differentially expressed under all three of the above stress conditions. Song et al. (2009) observed that, out of the 31 predicted AuxIAA genes in rice, 15 were induced by drought while three showed decreased expression. Under salt stress, eight Aux-IAA genes were induced and eight repressed, and seven of these genes showed the same tendency under salt and drought stress, while three showed the opposite regulation pattern.

Auxin and ABA are both involved in regulation of plant water status, with opposite functions in the shoot and complementary roles in the root. To adapt root conductivity for water and allow fine-tuning of stomatal aperture, signals for plant water status have to be integrated. As a prerequisite, auxin and ABA signaling chains cannot just exist in a linear manner but must form a network that communicates by cross talk (Fig. 7.2). Knowledge on cross talk has come from studies of mutants impaired in ABA or auxin perception and transgenic plants overexpressing signaling components. Some of these mutants or transgenic lines showed pleiotropic effects that indicated that ABA and auxin signal transduction are both affected by the mutated gene or overexpressed protein. One pathway of ABA signal transduction is through a mitogen-activated protein kinase (MAPK) cascade leading to activation of transcription factors (Fig. 7.1). Possible targets are ABI3 (abscisic acid-insensitive 3) and ABI5 (Fedoroff 2002). ABA perception activates the MAP kinases AtMPK3 and AtMPK6 (Kovtun et al. 2000). These MAPKs are in a cascade with the MAP3K ANP1 and possibly the M2Ks AtMKK3 and AtMKK4. A cross-point of ABA signaling branching off to auxin signal transduction is ANP1. Activation of ANP1 interferes with activation of auxin signal transduction (Kovtun et al. 1998). More support for the hypothesis that an ABA-induced MAPK cascade negatively regulates auxin responses comes from studies of the hyl1 mutant, in which ANP1 and AtMKK3 are overexpressed. Another junction from ABA signaling to auxin signal transduction is the transcription factor ABI3, involved in ABA signal transduction. Suzuki et al. (2001) investigated VIVIPAROUS1 (VP1), the maize orthologue of ABI3. They reported that in plants overexpressing VP1 application of ABA inhibits auxin-induced lateral root formation. In contrast, loss-of-function mutants of ABI3 need higher concentrations of exogenous auxin to induce lateral roots (Brady et al. 2003). The expression of ABI3 in the root is responsive to both auxin and ABA. Brady et al. (2003) concluded that ABI3 is necessary for correct auxin signaling during lateral root formation and speculated that ABI3 may interact with ARFs or AUX IAA proteins.

During osmotic stress signaling, it is interesting to note that the stress-induced hormone level changes and, therefore, extends the network of hormonal cross talk during the stress response of plants. A major part of this cross talk may be that these hormones influence the biosynthesis of each other, for example, it has been known for a long time that auxin increases ethylene biosynthesis by upregulation of 1-aminocyclopropane-1-carboxylate (ACC) synthase (Abel et al. 1995). However, inhibition of root growth by ethylene is mediated by an increase of auxin synthesis in aerial tissues (Ruzicka et al. 2007). Newly synthesized auxin is transported to the root tip and redirected via the auxin efflux carrier PIN2 to the root elongation zone. The resulting increase of auxin in the root elongation zone causes a reduction in root growth by inhibition of root cell elongation (Ruzicka et al. 2007). In contrast, there are data indicating that ethylene and ABA negatively influence the biosynthesis of each other (Cheng et al. 2009). The ABA biosynthesis mutant aba2 exhibits upregulation of the ethylene biosynthesis gene ACC oxidase and produces significantly more ethylene than wild-type plants. Similarly, the ethylene signal transduction mutant ein2 hyperaccumulates ABA (Ghassemian et al. 2000). However, contrary to data from analysis of mutants, Cheng et al. (2002) observed that ABA enhances ethylene synthesis in water-stressed poplar.

Therefore, data from analysis of mutants have to be examined critically, since due to pleiotropic effects, mutants may not reflect the situation in wild-type plants. It will be interesting to determine the kinetics of changes in ABA, auxin, and ethylene after onset of stress. These measurements, combined with analysis of hormone biosynthesis and perception mutants, will help to unravel the high complexity of hormonal cross talk during stress. ABA and IAA both play important roles in the stress adaptation of annual, herbaceous, and perennial woody plants. The basic mechanisms on ABA and IAA biosynthesis and signaling have been elucidated in Arabidopsis. However, data on ABA biosynthesis in poplar indicate that trees may differ from annual plants with respect to localization of biosynthesis and indicate occurrence of significant root-to-shoot transport of ABA in trees. It will be worthwhile to investigate if differences in hormone biosynthesis and transport present adaptive mechanisms that are specific to trees.

## ROS Signaling During Osmotic Stress in Plants

Reactive oxygen species (ROS) are known to accumulate during abiotic stresses in different cellular compartments. It is now becoming increasingly evident that even during stress, ROS production is not necessarily a symptom of cellular dysfunction but might represent a necessary signal in adapting the cellular machinery to the stressed conditions. ROS can modulate many signal transduction pathways, such as mitogen-activated protein kinase cascades, and ultimately influence the activity of transcription factors. However, the picture of ROS-mediated signaling is still fragmentary and the issues of ROS perception as well as the signaling specificity remain open. ROS signals originating at different organelles have been shown to induce large transcriptional changes and cellular reprogramming that can either protect the plant cell or induce programmed cell death ( Davletova et al. 2005; Foyer and Noctor 2005; Gadjev et al. 2006; Rhoads et al. 2006). These types of reprogramming suggest the involvement, at least in part, of organellar retrograde signaling in mediating ROS signals in the coordination of the stress response between ROS generating organelles to the nucleus and perhaps directly between the organelles themselves. Retrograde signaling is largely divided into two phases: (1) developmental control of organelle biogenesis and (2) operational control, that is, rapid adjustments in response to environmental and developmental constraints (Pogson et al. 2008). ROS generated in these organelles are considered to be important signaling molecules that are involved in retrograde signaling under abiotic stress conditions (Rhoads and Subbaiah 2007; Pogson et al. 2008; Woodson and Chory 2008). It has been shown that all the chloroplast-to-nuclei retrograde signaling, including ROS signaling, converges into a pathway regulated by GUN1 (genome uncoupled 1) and ABI4 (ABA-insensitive 4) in *Arabidopsis* seedlings (Koussevitzky et al. 2007). Mutants of *gun1* and *abi4* were shown to be sensitive to heat stress, suggesting the importance of retrograde signaling to abiotic stress responses (Miller et al. 2007). To understand the role of organelle-to-nuclei retrograde signaling, as well as intra-organelle coordination during abiotic stress, the tolerance and transcriptome reprogramming of mutants deficient in these key regulators to the abiotic stress should be tested. ROS signaling is an integral part of the acclimation response of plants to abiotic stresses. It is used to sense stress due to enhanced ROS production caused by metabolic imbalances, as well as to actively send different signals via enhanced production of ROS at the apoplast by different RBOH (respiratory burst oxidase homolog) proteins. ROS signaling during abiotic stress, namely, drought and salinity, is highly integrated into many of the other signaling networks that regulate plant acclimation, including calcium, hormone, and protein phosphorylation. In light of this view of the integrated signaling network of plants, which is responsible for the timely activation of different acclimation pathways, it is easy to see why some changes in ROS metabolism were found to cause enhanced tolerance to stress, whereas other changes were found to cause enhanced sensitivity. It is also difficult to predict how different changes in ROS metabolism, engineered by different genetic manipulations, will affect crop tolerance to various abiotic stresses. It is likely that identifying key regulators which control stress response activation (Baena-González et al. 2007; Baena-González and Sheen 2008) will provide crucial avenues to enhance crop tolerance to abiotic stress. The way abiotic stress-induced alterations would affect ROS homeostasis and signaling will be an exciting aspect of these studies. In considering the relationships between ROS metabolism, signaling, and plant responses to various abiotic stresses including drought or salinity, we should also remember that these stresses almost always occur in nature or in the field together with other abiotic stress conditions. Such combinations could include drought and heat, or drought and cold, drought and nutrient, drought and metal, etc., and may even include triple combinations such as drought, salinity, and heat and many others. Such different stress combinations affecting ROS metabolism and signaling are subject of active research in recent years (Keles and Oncel 2002; Rizhsky et al. 2002; Hewezi et al. 2008; Koussevitzky et al. 2008), and it should be taken into consideration while attempting to engineer crops for abiotic stress tolerant to field growth conditions (Mittler 2006).

## Conclusion and Perspective

An understanding of the plant signaling pathways responding to osmotic stress is important for both basic science and agriculture, because drought and soil salinity are increasingly important problems drastically affecting crop productivity. Since the mechanisms underlying plants responses to osmotic stress are complicated, many questions remain unanswered. Osmotic stress caused reduction in growth and productivity of crop plants. Some of the events involved in the regulation of osmotic stress tolerance by ABA-dependent and independent pathways are becoming clearer. However, forward and reverse genetics approaches will continue to be imperative to dissect complex osmotic stress signaling pathways. Thorough characterization of mutant phenotypes will provide an indication whether a signaling component functions in a specific pathway or is involved in multiple pathways. Further proteomic and functional genomic analysis of expression patterns, combined with biochemical and molecular characterization of the components of signaling complexes and other regulatory proteins, will be essential to firmly establish specificity or cross talk of the signaling pathways during osmotic stress. DNA sequence of the AtMPK3 promoter for responses to osmotic stress has also been identified, which advances our understanding of the molecular mechanisms controlling AtMPK3 expression during osmotic stress in plants. Recent studies also revealed that SnRK2.2, 2.3, and 2.6 are core components of the ABA pathways whereas SnRK2s are essential kinases for osmotic stress signaling. Though the activation mechanism of SnRK2s in the ABA pathway is elucidated, regulatory mechanisms of SnRK2s in the ABA-independent pathway remain obscure. Thus, further analysis of SnRK2s must be a key study to draw the whole osmotic stress signaling pathways. Further proteomics, transcriptomics, and metabolite profiling of osmotic stress signaling components can bridge the gap in our current knowledge of osmotic stress signaling networks in plants.

## References


Abebe T, Guenzi AC, Martin B, Cushman JC (2003) Tolerance of mannitol-accumulating transgenic wheat to water stress and salinity. Plant Physiol 131:1–8

Abel T, Alberini C, Ghirardi M, Huang YY, Nguyen P, Kandel ER (1995) Steps toward a molecular definition of memory consolidation. In: Schacter DL (ed) Memory distortion. Harvard University Press, Cambridge, MA, pp 298–325

Albacete A, Ghanem ME, Martínez-Andújar C, Acosta M, Sánchez-Bravo J, Martínez V, Lutts S, Dodd IC, Pérez-Alfocea F (2008) Hormonal changes in relation to biomass partitioning and shoot growth impairment in salinized tomato (*Solanum lycopersicum* L.) plants. J Exp Bot 59(15):4119–4131

Ashraf M, Foolad MR (2007) Roles of glycine betaine and proline in improving plant abiotic stress resistance. Environ Exp Bot 59(2):206–216

Aubert S, Assard N, Boutin JP, Frenot Y, Dorne AJ (1999) Carbon metabolism in the subantarctic Kerguelen cabbage *Pringlea antiscorbutica* R. Br.: environmental controls over carbohydrates and proline contents and relation to phenology. Plant Cell Environ 22:243–254

Avonce N, Leyman B, Mascorro-Gallardo JO, Van Dijck P, Thevelein JM, Iturriaga G (2004) The Arabidopsis trehalose-6-P synthase AtTPS1 gene is a regulator of glucose, abscisic acid, and stress signaling. Plant Physiol 136:3649–3659

Baena-González E, Sheen J (2008) Convergent energy and stress signaling. Trends Plant Sci 13:474–481

Baena-González E, Rolland F, Thevelein J, Sheen J (2007) A central integrator of transcription networks in plant stress and energy signaling. Nature 448:938–942

Banu MNA, Hoque MA, Watanabe-Sugimoto M, Matsuoka K, Nakamura Y, Shimoishi Y, Murata Y (2009) Proline and glycinebetaine induce antioxidant defense gene expression and suppress cell death in cultured tobacco cells under salt stress. J Plant Physiol 166(2):146–156

Bartels D, Sunkar R (2005) Drought and salt tolerance in plants. Crit Rev Plant Sci 24:23–58

Borsani O, Zhu J, Verslues PE, Sunkar R, Zhu JK (2005) Endogenous siRNAs derived from a pair of natural cis-antisense transcripts regulate salt tolerance in Arabidopsis. Cell 123:1279–1291

Brady SM, Sarkar SF, Bonetta D, McCourt P (2003) The ABSCISIC ACID INSENSITIVE 3 (ABI3) gene is modulated by farnesylation and is involved in auxin signaling and lateral root development in Arabidopsis. Plant J 34:67–75

Chen THH, Murata N (2008) Glycinebetaine: an effective protectant against abiotic stress in plants. Trends Plant Sci 13:499–505

Cheng WH, Endo A, Zhou L, Penney J, Chen HC, Arroyo A, Leon P, Nambara E, Asami T, Seo M, Koshiba T, Sheen J (2002) A unique short-chain dehydrogenase/reductase in Arabidopsis glucose signaling and abscisic acid biosynthesis and functions. Plant Cell 14:2723–2743

Cheng WH, Chiang MH, Hwang SG, Lin PC (2009) Antagonism between abscisic acid and ethylene in *Arabidopsis* acts in parallel with the reciprocal regulation of their metabolism and signalling pathways. Plant Mol Biol 71:61–80

Couée I, Sulmon C, Gouesbet G, Amrani AE (2006) Involvement of soluble sugars in reactive oxygen species balance and responses to oxidative stress in plants. J Exp Bot 57:449–459

Davletova S, Rizhsky L, Liang HJ, Zhong SQ, Oliver DJ, Coutu J, Shulaev V, Schlauch K, Mittler R (2005) Cytosolic ascorbate peroxidase1 is a central component of the reactive oxygen gene network of Arabidopsis. Plant Cell 17:268–281

Fedoroff NV (2002) Cross-talk in abscisic acid signaling. Science's STKE, 2002, RE10

Foyer CH, Noctor G (2005) Oxidant and antioxidant signaling in plants: a re-evaluation of the concept of oxidative stress in a physiological context. Plant Cell Environ 28:1056–1071

Fujii H, Chinnusamy V, Rodrigues A, Rubio S, Antoni R, Park SY, Cutler SR, Sheen J, Rodriguez PL, Zhu JK (2009) In vitro reconstitution of an abscisic acid signaling pathway. Nature 462:660–664

Fujii H, Zhu JK (2009) An Arabidopsis mutant deficient in all three abscisic acid-activated protein kinases reveals critical roles in plant growth, reproduction and stress. Proc Natl Acad Sci USA 106:8380–8385

Fujii H, Verslues PE, Zhu JK (2011) *Arabidopsis* decuple mutant reveals the importance of SnRK2 kinases in osmotic stress responses in vivo. Proc Natl Acad Sci USA 108:1717–1722

Gadjev I, Vanderauwera S, Gechev TS, Laloi C, Minkov IN, Shulaev V, Apel K, Inzé D, Mittler R, Van Breusegem F (2006) Transcriptomic footprints disclose specificity of reactive oxygen species signaling in Arabidopsis. Plant Physiol 141:436–445

Garg AK, Kim JK, Owens TG, Ranwala AP, Choi YD, Kochian LV, Wu RJ (2002) Trehalose accumulation in rice plants confers high tolerance levels to different abiotic stresses. Proc Natl Acad Sci USA 99:15898–15903

Génard H, Lesaos J, Billard JP, Tremolieres A, Boucaud J (1991) Effect of salinity on lipid composition, glycine betaine content and photosynthetic activity in chloroplasts of *Suaeda maritima*. Plant Physiol Biochem 29:421–427

Ghassemian M, Nambara E, Cutler S, Kawaide H, Kamiya Y, McCourt P (2000) Regulation of abscisic acid signaling by the ethylene response pathway in Arabidopsis. Plant Cell 12:1117–1126

Gill SS, Tuteza N (2010) Reactive oxygen species and antioxidant machinery in abiotic stress tolerance in crop plants. Plant Physiol Biochem 48:909–930

Halfter U, Ishitani M, Zhu JK (2000) The *Arabidopsis* SOS2 protein kinase physically interacts with and is activated by the calcium-binding protein SOS3. Proc Natl Acad Sci USA 97:3735–3740

Hassine AB, Ghanem ME, Bouzid S, Lutts S (2008) An inland and a coastal population of the Mediterranean xero-halophyte species *Atriplex halimus* L. differ in their ability to accumulate proline and glycinebetaine in response to salinity and water stress. J Exp Bot 59 (6):1315–1326

Hewezi T, Leger M, Gentzbittel L (2008) A comprehensive analysis of the combined effects of high light and high temperature stresses on gene expression in sunflower. Ann Bot 102:127–140

Hoffmann S (2002) Osmotic stress signaling and osmoadaptation in yeasts. Microbiol Mol Biol Rev 66 (2):300–372

Hoque MA, Banu MN, Okuma E, Amako K, Nakamura Y, Shimoishi Y, Murata Y (2007) Exogenous proline and glycinebetaine increase NaCl-induced ascorbate-glutathione cycle enzyme activities, and proline improves salt tolerance more than glycinebetaine in tobacco bright yellow-2 suspension-cultured cells. J Plant Physiol 164:1457–1468

Huang GT, Ma SL, Bai LP, Zhang L, Ma H, Jia P, Liu J, Zhong M, Guo ZF (2011) Signal transduction during cold, salt, and drought stresses in plants. Mol Biol Rep. doi:10.1007/s11033-011-0823-1

Inoue T, Higuchi M, Hashimoto Y, Seki M, Kobayashi T, Takara S, Shinozaki K, Kakimoto T (2001) Identification of CRE1 as a cytokinin receptor from *Arabidopsis*. Nature 409:1060–1063

Ishitani M, Liu J, Halfter U, Kim CS, Wei M, Zhu JK (2000) SOS3 function in plant salt tolerance requires myristoylation and calcium-binding. Plant Cell 12:1667–1677

Jain M, Khurana JP (2009) Transcript profiling reveals diverse roles of auxin-responsive genes during reproductive development and abiotic stress in rice. FEBS J 276:3148–3162

Junghans U, Polle A, Du¨chting P, Weiler E, Kuhlman B, Gruber F, Teichmann T (2006) Effects of salt stress on the anatomy and auxin physiology of poplar xylem. Plant Cell Environ 29:1519–1531

Kaldewey H, Ginkel U, Wawczyniak G (1974) Auxin transport and water stress in pea (*Pisum sativum* L.). Berichte der Deutschen Botanischen Gesellschaft 87:563–576

Keles Y, Oncel I (2002) Response of antioxidative defence system to temperature and water stress combinations in wheat seedlings. Plant Sci 163:783–790

Koussevitzky S, Ajit Nott A, Mockler TC, Hong F, Sachetto-Martins G, Surpin M, Lim J, Mittler R, Chory J (2007) Signals from chloroplasts converge to regulate nuclear gene expression. Science 316 (5825):715–719

Koussevitzky S, Suzuki N, Huntington S, Armijo L, Sha W, Cortes D, Shulaev V, Mittler R (2008) Ascorbate peroxidase 1 plays a key role in the response of *Arabidopsis thaliana* to stress combination. J Biol Chem 283:34197–34203

Kovtun Y, Chiu WL, Zeng W, Sheen J (1998) Suppression of auxin signal transduction by a MAPK cascade in higher plants. Nature 395:716–720

Kovtun Y, Chiu WL, Tena G, Sheen J (2000) Functional analysis of oxidative stress-activated mitogen-activated protein kinase cascade in plants. Proc Natl Acad Sci USA 97:2940–2945

Kurusu T, Nishikawa D, Yamazaki Y, Gotoh M, Nakano M, Hamada H, Yamanaka T, Iida K, Nakagawa Y, Saji H, Shinozaki K, Iida H, Kuchitsu K (2012) Plasma membrane protein OsMCA1 is involved in regulation of hypo-osmotic shock-induced $Ca^{2+}$ influx and modulates generation of reactive oxygen species in cultured rice cells. BMC Plant Biol 12:11

Laurie S, Feeney KA, Maathuis FJ, Heard PJ, Brown SJ, Leigh RA (2002) A role for HKT1 in sodium uptake by wheat roots. Plant J 32:139–149

Liu J, Zhu JK (1998) A calcium sensor homolog required for plant salt tolerance. Science 280:1943–1945

Liu J, Ishitani M, Halfter U, Kim CS, Zhu JK (2000) The Arabidopsis thaliana SOS2 gene encodes a protein kinase that is required for salt tolerance. Proc Natl Acad Sci USA 97:3730–3733

Matysik J, Ali BB, Mohanty P (2002) Molecular mechanism of quenching of reactive oxygen species by proline under water stress in plants. Curr Sci 82:525–532

McNeil SD, Nuccio ML, Hanson AD (1999) Betaines and related osmoprotectants. Targets for metabolic engineering of stress resistance. Plant Physiol 120:945–949

Miller G, Suzuki N, Rizhsky L, Hegie A, Koussevitzky S, Mittler R (2007) Double mutants deficient in cytosolic and thylakoid ascorbate peroxidase reveal a complex mode of interaction between reactive oxygen species, plant development, and response to abiotic stresses. Plant Physiol 144:1777–1785

Miranda JA, Avonce N, Suarez R, Thevelein JM, Van Dijck P, Iturriaga G (2007) A bifunctional TPS-TPP enzyme from yeast confers tolerance to multiple and extreme abiotic-stress conditions in transgenic *Arabidopsis*. Planta 226:1411–1421

Mittler R (2006) Abiotic stress, the field environment and stress combination. Trends Plant Sci 11(1):16–19

Molinari HBC, Marur CJ, Daros E, Campos MKF, Carvalho JFRP, Bespalhok Filho JC, Pereira LFP, Vieira LGE (2007) Evaluation of the stress-inducible production of proline in transgenic sugarcane (*Saccharum* spp.): osmotic adjustment, chlorophyll fluorescence and oxidative stress. Physiol Plant 130:218–229

Park EJ, Jeknic Z, Pino MT, Murata N, Chen TH (2007) Glycine betaine accumulation is more effective in chloroplasts than in the cytosol for protecting transgenic tomato plants against abiotic stress. Plant Cell Environ 30:994–1005

Penna S (2003) Building stress tolerance through over-producing trehalose in transgenic plants. Trends Plant Sci 8:355–357

Pogson BJ, Woo NS, Forster B, Small ID (2008) Plastid signaling to the nucleus and beyond. Trends Plant Sci 13:602–609

Qiu QS, Guo Y, Dietrich MA, Schumaker KS, Zhu JK (2002) Regulation of SOS1, a plasma membrane $Na^+$/$H^+$ exchanger in Arabidopsis thaliana, by SOS2 and SOS3. Proc Natl Acad Sci USA 99:8436–8441

Quintero FJ, Ohta M, Shi H, Zhu J-K, Pardo JM (2002) Reconstitution in yeast of the Arabidopsis SOS signaling pathway for $Na^+$ homeostasis. Proc Natl Acad Sci USA 99:9061–9066

Raitt DC, Posas F, Saito H (2000) Yeast Cdc42 GTPase and Ste20 PAK-like kinase regulate Sho1-dependent activation of Hog1 MAPK pathway. EMBO J 19:4623–4631

Reiser V, Salah SM, Ammerer G (2000) Polarized localization of yeast Pbs2 depends on osmostress, the membrane protein Sho1 and Cdc42. Nat Cell Biol 2:620–627

Reiser V, Raitt DC, Saito H (2003) Yeast osmosensor Sln1 and plant cytokinin receptor Cre1 respond to changes in turgor pressure. J Cell Biol 161: 1035–1040

Rhoads DM, Subbaiah CC (2007) Mitochondrial retrograde regulation in plants. Mitochondrion 7:177–194

Rhoads DM, Umbach AL, Subbaiah CC, Siedow JN (2006) Mitochondrial reactive oxygen species. Contribution to oxidative stress and interorganellar signaling. Plant Physiol 141:357–366

Rizhsky L, Hallak-Herr E, Van Breusegem F, Rachmilevitch S, Barr JE, Rodermel S, Inze D, Mittler R (2002) Double antisense plants lacking ascorbate peroxidase and catalase are less sensitive to oxidative stress than single antisense plants lacking ascorbate peroxidase or catalase. Plant J 32:329–342

Rodríguez M, Canales E, Borrás-Hidalgo O (2005) Molecular aspects of abiotic stress in plants. Biotecnol Apl 22:1–10

Rus A, Yokoi S, Sharkhuu A, Reddy M, Lee B-H, Matsumoto TK, Koiwa H, Zhu J-K, Bressan RA, Hasegawa PM (2001) AtHKT1 is a salt tolerance determinant that controls $Na^+$ entry into plant roots. Proc Natl Acad Sci USA 98:14150–14155

Ruzicka K, Ljung K, Vanneste S, Podhorska R, Beeckman T, Friml J, Benkova E (2007) Ethylene regulates root growth through effects on auxin biosynthesis and transport-dependent auxin distribution. Plant Cell 19:2197–2212

Seki M, Umezawa T, Urano K, Shinozaki K (2007) Regulatory metabolic networks in drought stress responses. Curr Opin Plant Biol 10:296–302

Sheldrake AR (1979) Effect of osmotic stress on polar auxin transport in Avena mesocotyl sections. Planta 145:113–117

Shi H, Ishitani M, Kim C, Zhu JK (2000) The *Arabidopsis thaliana* salt tolerance gene SOS1 encodes a putative $Na^+/H^+$ antiporter. Proc Natl Acad Sci USA 97:6896–6901

Shi H, Xiong L, Stevenson B, Lu T, Zhu JK (2002) The Arabidopsis *salt overly sensitive 4* mutants uncover a critical role for vitamin B6 in plant salt tolerance. Plant Cell 14:575–588

Shi H, Lee BH, Wu SJ, Zhu JK (2003) Overexpression of a plasma membrane $Na^+/H^+$ antiporter gene improves salt tolerance in Arabidopsis thaliana. Nat Biotechnol 21:81–85

Shinozaki K, Yamaguchi-Shinozaki K (2007) Gene networks involved in drought stress response and tolerance. J Exp Bot 58:221–222

Shiu S-H, Bleecker AB (2001) Plant receptor-like kinase gene family: diversity, function, and signaling. Sci Signal Transduct Knowl Environ 113:RE22

Smirnoff N, Cumbes QJ (1989) Hydroxyl radical scavenging activity of compatible solutes. Phytochemistry 28:1057–1060

Song Y, Wang L, Xiong L (2009) Comprehensive expression profiling analysis of OsIAA gene family in developmental processes and in response to phytohormone and stress treatments. Planta 229:577–591

Staswick PE, Tiryaki I, Rowe ML (2002) Jasmonate response locus JAR1 and several related Arabidopsis genes encode enzymes of the firefly luciferase superfamily that show activity on jasmonic, salicylic, and indole-3- acetic acids in an assay for adenylation. Plant Cell 14:1405–1415

Staswick PE, Serban B, Rowe M, Tiryaki I, Maldonado MT, Maldonado MC, Suza W (2005) Characterisation of an Arabidopsis enzyme family that conjugates amino acids to indole-3-acetic acid. Plant Cell 17:616–627

Suzuki I, Kanesaki Y, Mikami K, Kanehisa M, Murata N (2001) Cold-regulated genes under control of the cold sensor Hik33 in *Synechocystis*. Mol Microbiol 40:235–244

Székely G, Abrahám E, Cséplo A, Rigó G, Zsigmond L, Csiszár J, Ayaydin F, Strizhov N, Jásik J, Schmelzer E (2008) Duplicated P5CS genes of Arabidopsis play distinct roles in stress regulation and developmental control of proline biosynthesis. Plant J 53:11–28

Umezawa T, Nakashima K, Miyakawa T, Takashi K, Tanokura M, Kazuo S, Yamaguchi-Shinozaki K (2010) Molecular basis of the core regulatory network in ABA responses: sensing, signaling and transport. Plant Cell Physiol 51(11):1821–1839

Upadhyaya H, Panda SK (2004) Responses of *Camellia sinensis* to drought and rehydration. Biol Plant 48 (4):597–600

Upadhyaya H, Panda SK, Dutta BK (2008) Variation of physiological and antioxidative responses in tea cultivars subjected to elevated water stress followed by rehydration recovery. Acta Physiol Plant 30:457–468

Upadhyaya H, Panda SK, Dutta BK (2011) $CaCl_2$ improves post drought recovery potential in *Camellia sinensis* (L) O. Kuntze. Plant Cell Rep 30(4):495–503

Urao T, Yakubov B, Satoh R, Yamaguchi-Shinozaki K, Seki B, Hirayama T, Shinozaki K (1999) A transmembrane hybrid-type histidine kinase in Arabidopsis functions as an osmosensor. Plant Cell 11:1743–1754

Urao T, Yamaguchi-Shinozaki K, Shinozaki K (2001) Plant histidine kinases: an emerging picture of two-component signal transduction in hormone and environmental responses. Sci Signal Transduct Knowl Environ 113:RE18

Vendruscolo ECG, Schuster I, Pileggi M, Scapim CA, Molinari HBC, Marur CJ, Vieira LGE (2007) Stress-induced synthesis of proline confers tolerance to water deficit in transgenic wheat. J Plant Physiol 164:1367–1376

Vinocur B, Altman A (2005) Recent advances in engineering plant tolerance to abiotic stress: achievements and limitations. Curr Opin Biotechnol 16:123–132

Woodson JD, Chory J (2008) Coordination of gene expression between organellar and nuclear genomes. Nat Rev Genet 9:383–395

Xiong L, Schumaker KS, Zhu JK (2002) Cell signaling during cold, drought, and salt stress. Plant Cell 14: S165–S183

Zhu JK (2002) Salt and drought stress signal transduction in plants. Annu Rev Plant Biol 53:247–273

# The Physiology of Reproductive-Stage Abiotic Stress Tolerance in Cereals

8

Rudy Dolferus, Nicola Powell, Xuemei JI, Rudabe Ravash, Jane Edlington, Sandra Oliver, Joost Van Dongen, and Behrouz Shiran

**Abstract**

In cereal crops, during the critical time of flowering, abiotic stresses cause massive yield reductions. Attempts to solve the problem using classical breeding methods have so far been unsuccessful. This is primarily due to the lack of focused, reliable screening procedures and the fact that yield components are controlled by complex gene networks. Although genetic approaches require the use of germplasm that shows a contrasting behaviour for a trait of interest, physiological and molecular studies are often based on plant lines with unknown ranking of these traits (e.g. abiotic stress tolerance). We have taken the approach to first develop a reliable and reproducible screening method to rank germplasm in terms of reproductive-stage abiotic stress tolerance. The high sensitivity of the young microspore stage of pollen development was used as a target for our screening method, enabling us to identify germplasm with higher resilience to cold (rice) and drought (wheat). This material was then used in comparative studies to identify differences in the tolerant and sensitive response to abiotic stress. In this chapter, we discuss the physiological differences between tolerant and sensitive germplasm and how this can be exploited to develop cereal lines with improved reproductive stage abiotic stress tolerance.

R. Dolferus (✉) • N. Powell • X. JI •
J. Edlington • S. Oliver
CSIRO, division of Plant Industry,
GPO Box 1600, Canberra, ACT 2601, Australia
e-mail: rudy.dolferus@csiro.au; nicola.powell@csiro.au; xuemei.ji@daff.gov.au; jane.edlington@csiro.au; sandra.oliver@csiro.au

R. Ravash • B. Shiran
Faculty of Agriculture, Shahrekord University,
P.O. Box 115, Shahrekord 8818634141, Iran
e-mail: rudabe_ravash@yahoo.com; beshiran45@gmail.com

J. Van Dongen
Max Planck Institute, division of Molecular Plant Physiology,
Am Muehlenberg 1, 14476 Golm, Germany
e-mail: dongen@mpimp-golm.mpg.de

## Introduction

### Abiotic Stresses and Cereal Grain Productivity

Abiotic stresses affect all aspects of plant growth and development. In cereals, the world's most important food crops, abiotic stress conditions strongly influence grain yield. The increasing global population, combined with the future threat of climate change, has become a pressing concern. This has created a major challenge for

G.R. Rout and A.B. Das (eds.), *Molecular Stress Physiology of Plants*,
DOI 10.1007/978-81-322-0807-5_8, © Springer India 2013

breeders to develop cereal crops that produce more grain under sustainable agricultural systems. To achieve this, future grain crops will need to be better adapted to a changing environment (Powell et al. 2012). Developing cereal varieties with higher yield potentials that are able to maintain this productivity under adverse environmental conditions will require a better understanding of how plants grow and respond to environmental cues.

Abiotic stresses such as cold, drought and heat affect cereals at particular stages of the growth cycle. In temperate climate zones where the growing season is relatively short, or at high elevation, cold affects seedling establishment at the start of the season and at the end during reproductive growth. This is problematic for an originally tropical grass such as rice, which is increasingly cultivated in temperate climate zones with shorter growing seasons. Under cold conditions, rice crops experience poor seedling establishment at the start of the season, resulting in delayed heading and huge grain losses at the end of the season when temperatures drop below a certain temperature threshold (typically around 16°; "cold spells"; Lin and Peterson 1975; Board et al. 1980; Glaszmann et al. 1990; Bertin et al. 1996). Temperate climate cereals such as wheat and barley are adapted to avoid the damaging effect of cold during the reproductive phase. Vernalisation and photoperiodism are mechanisms that sense seasonal changes in temperature and day length, preventing the plant from flowering during the coldest period of the growing season (Distelfeld et al. 2009; Greenup et al. 2009). However, cold spells can be unpredictable and flowering too early can cause severe yield losses. Adapting to drought stress is a more difficult task. Drought conditions can occur throughout the life cycle of the plant, usually taking place during the warmest periods when plants are flowering. In this case, drought stress often coincides with heat conditions, and evidence suggests that resistance to drought and heat stress has co-evolved (Jagadish et al. 2011). In temperate climate wheat-growing areas such as Australia, successful reproductive development leading to grain production requires a perfect balance between not flowering too early to avoid chilling and frost injury, and not flowering too late to avoid drought and heat damage. Although the reproductive phase is the most sensitive stage of the cereal growth cycle, stress conditions during the vegetative stage also affect crop yields. Abiotic stresses affect biomass production, causing an indirect effect on grain productivity. Several traits have been identified that improve vegetative-stage tolerance of cereals to abiotic stress conditions. For instance, in the case of drought stress, these traits include yield potential, water-use efficiency (WUE), harvest index (HI), improved transpiration efficiency and deep root penetration to access water and nutrients. Although these traits are important for successful reproductive development, they do not eliminate the effect of abiotic stresses. Even short interrupted stress periods during the reproductive stage can have catastrophic and irreversible effects on grain yield.

## Determination of Grain Number in Cereals

Grain yield in cereals is determined by grain number and grain weight. Both of these yield components are affected by abiotic stresses. It is widely recognised that grain number is the most important yield determinant of wheat, and loss in grain number is the key factor contributing to reductions in grain yield under abiotic stress conditions (Savin and Slafer 1991; Fischer 1993; Abbate et al. 1995; Sayre et al. 1997; Gonzalez et al. 2003).

Grain number is a complicated trait. Grain number per square metre can be influenced by a variety of agronomic factors (e.g. seeding rate and management practices), as well as the number of tillers per square metre. The number of grains per head varies between the main shoot and secondary tillers and also between varieties. Pre-anthesis growing conditions primarily control grain number, while grain weight is determined by post-anthesis conditions influencing grain filling (Gonzalez et al. 2003; Ji et al. 2010). Grain weight and grain number are interrelated; both depend on nitrogen availability and resource allocation from the vegetative plant parts. Water-soluble carbohydrates are important for maintaining grain size, particularly under drought conditions when photosynthesis is arrested (Gebbing et al. 1999; Yang et al. 2001; Ruuska et al. 2006, 2008; Sinclair and Jamieson 2006). Since grain number and weight are controlled at different periods of reproductive development, they may be influenced by different environmental conditions during the growing season. It has been shown that a compensation effect exists between grain number and grain weight (Acreche and Slafer 2006). Grain number is determined by the reigning conditions during floral meristem and later during spikelet differentiation. However, developing spikelets can prematurely abort and grain number can be adjusted later during floral development. Meiosis is the committed step towards gametophyte production and later fertilisation and grain development. The young microspore stage of pollen development has been shown to be particularly sensitive to a variety of abiotic stress conditions, and stress-induced sterility is a major problem in self-fertilising cereals such as wheat, barley and rice (Powell et al. 2012).

The fixation of grain number is a flexible and dynamic process that is controlled continuously by the environment throughout reproductive development. Several genes involved in inflorescence and floral development have been identified in the cereal model system rice (Kurata et al. 2005; Kellogg 2007; Yoshida and Nagato 2011). It remains unclear how these genes are affected by abiotic stresses, whether the control mechanism for grain number is the same under normal and abiotic stress conditions and whether a different mechanism is involved at different stages of reproductive development (e.g. YM stage and grain-filling stage).

## Tolerance Versus Avoidance and Escape Mechanisms

The switch from vegetative to reproductive development is controlled by seasonal factors such as photoperiod (long/short day length) and temperature (vernalisation; Jung and Müller 2009; Amasino 2010; Trevaskis 2010). Adjustment of flowering time is an adaptation mechanism that leads to avoidance of abiotic stress such as cold and frost. Breeding for short- or long-duration varieties is an effective strategy for optimising yield potential in a given environment. Although avoidance mechanisms do not prevent yield loss when unexpected and transient stress events occur (e.g. cold, heat or drought spells), a tolerance mechanism that truly protects reproductive development is required to guarantee maximum grain formation. Under field selection conditions, it is impossible to discriminate between avoidance and tolerance mechanisms. Screening for abiotic stress-tolerant germplasm under field conditions often results in selecting lines that avoid the stress through alteration of flowering time. It depends on the timing and severity of the stress in a targeted area whether selecting for stress escape/avoidance or tolerance mechanisms is the right strategy. Under unexpected or more severe stress conditions, stress-tolerant lines are more likely to safeguard grain yield. Breeding strategies targeting high yields under optimum growing conditions may select varieties that are better yielding under stress conditions. This strategy is beneficial if the genetic control of grain yield under normal and stress conditions is the same and when yield traits are constitutively expressed and are independent of environmental conditions.

Although regulation of flowering time is often associated with cold and frost conditions, adjustment of flowering time appears to be a dynamic mechanism that is also used to escape a variety of other abiotic stress conditions. Flowering before the onset of severe water deficit is a known drought escape mechanism which is particularly effective under severe terminal drought conditions (Chaves et al. 2003). Many known flowering-time regulating genes are expressed throughout the plant, not only in the flower meristems (Amasino 2010). It remains unknown how other abiotic stresses other than cold influence known flowering-time control genes.

## Mapping QTLs for Abiotic Stress Tolerance in Cereals

Breeding for reproductive-stage abiotic stress tolerance has so far not been successful for a variety of reasons (for reviews, see Collins et al. 2008; Fleury et al. 2010; Richards et al. 2010; Blum 2011; Dolferus et al. 2011). In the wild, grain production is a survival mechanism that allows the plant to survive extreme environmental conditions. Domesticated plants such as cereals have been under continuous selection for high grain productivity under "optimal" growing conditions. It is possible that this selection may have affected the capacity to produce grains under nonoptimal environmental conditions. Breeding programmes are unavoidably focusing on commercial factors such as high yield potential and grain quality and not on abiotic stress tolerance. This may have caused a bias in the selection of breeding lines that are not superior in terms of abiotic stress tolerance (Forster et al. 2000).

The development of comprehensive genetic maps was not possible until the introduction of DNA-based molecular markers; advances in marker technologies have provided virtually unlimited markers. DNA markers can be detected at any stage of the plant's development and are completely independent of environmental conditions. The first step towards locating genes or QTLs associated with abiotic stresses is the construction of a detailed genetic map with a high level of genome coverage. The continued improvement of marker systems and the identification of markers that are linked to genes or QTLs associated to the tolerant phenotype of abiotic stress traits will aid the introgression of favourable genes into new tolerant cultivars. QTL analysis can be used to identify, locate and determine the effects and interactions of contributing genes. Two fundamental components are required for QTL analysis, a genetic marker map and phenotypic data segregating for the trait of interest. Selection of lines with improved abiotic stress tolerance has proved difficult in the field, where timing, severity and even occurrence of the stress during reproductive development cannot be controlled. The lack of reliable screening methods has seriously hampered breeding progress for reproductive abiotic stress tolerance. A shift to selection under managed or controlled environment conditions, combined with a better physiological and morphological understanding of the effect of abiotic stresses on reproductive development, appears to be an essential step towards breeding cereals with improved abiotic stress tolerance. Screening under controlled environmental conditions will provide phenotypic data that is reliable and reproducible for QTL analysis. The complex issue of plant stress tolerance can be dissected through the use of QTL mapping; this approach will provide significant advancement to breeding for abiotic stress tolerance in crops.

## Pollen Development and Grain Number Control

### Pollen Sterility Is a Major Cause of Grain Loss

Our research has focused on two experimental systems: cold-induced sterility in rice and drought-induced sterility in wheat (Fig. 8.1). Both experimental systems have shown a large amount of similarity in terms of physiological and molecular underpinnings. Physiological and morphological studies in Japan have demonstrated that the problem of cold-induced sterility in rice is due to pollen sterility (Nishiyama 1984). In wheat it was known for quite some time (e.g. observation of increased frequencies of outcrossing) that drought stress affected pollen fertility (Ashton 1948; Bingham 1966; Salter and Goode 1967). In rice, cold spells (typically night temperatures below 16–17°C) can lead to yield reductions of over 40%, depending on the variety and the length of the cold event. In wheat, terminal drought stress can be equally destructive (Dolferus et al. 2011; Powell et al. 2012). We have shown that in both cases, the young microspore stage of pollen development is particularly sensitive to the stress (Oliver et al. 2005;

**Fig. 8.1** Effect on grain number of cold stress in rice (**a**) and drought stress in wheat (**b**)

Ji et al. 2010), with irreversible effects. In wheat, even moderate water stress conditions affected pollen fertility without having an irreversible effect on the vegetative plant parts (for review, see Saini 1997; Saini and Westgate 2000). Male gametophyte development is targeted by cold stress in sorghum (Brooking 1976) and so is heat stress in wheat, rice and sorghum (Saini et al. 1984; Jagadish et al. 2007; Jain et al. 2007; Barnabas et al. 2008). Even low-light conditions (e.g. heavy cloud cover) can cause high levels of sterility in wheat (Fischer and Stockman 1980). This high level of sensitivity during pollen development to abiotic stresses provides an ideal opportunity for the development of an urgently needed, more reliable and reproducible screening method. This approach may focus on only one aspect of reproductive-stage abiotic stress tolerance, but screening for improved resilience of pollen development to abiotic stresses is essential to maintain higher grain numbers under stress conditions. Elucidating what contributes to this stress tolerance mechanism in anthers may then provide an entry point for studies into the broader physiological and molecular aspects of the problem (for discussion, see Dolferus et al. 2011; Powell et al. 2012).

## Induction of Pollen Abortion Is Stage Specific

We developed a screening method to determine the sensitive stage of pollen development in rice and wheat. In wheat, the Feeke's and Zadoks' scales were created to monitor plant developmental stages (Large 1954; Zadoks et al. 1974); these scales do not however allow to precisely score pollen developmental stages. At the early stages of male gametophyte development, the panicle (rice) and ear (wheat) are still invisible and wrapped in the leaf sheaths. We therefore developed an easy nondestructive scoring system based on auricle distances (AD; the distance between the auricle of the flag leaf and the auricle of the penultimate leaf; Matsushima 1957; Satake and Hayase 1970, 1974; Heenan 1984; Fig. 8.2a). We harvested anther samples at different ADs for each rice and wheat variety being investigated and under a microscope examined the pollen developmental stages corresponding to those ADs. Plants were stress-treated for short periods (4–5 days), enabling us to scan all developmental stages ranging from before meiosis to anthesis (Fig. 8.2a, b). In the case of cold stress of the cold-sensitive Australian rice variety

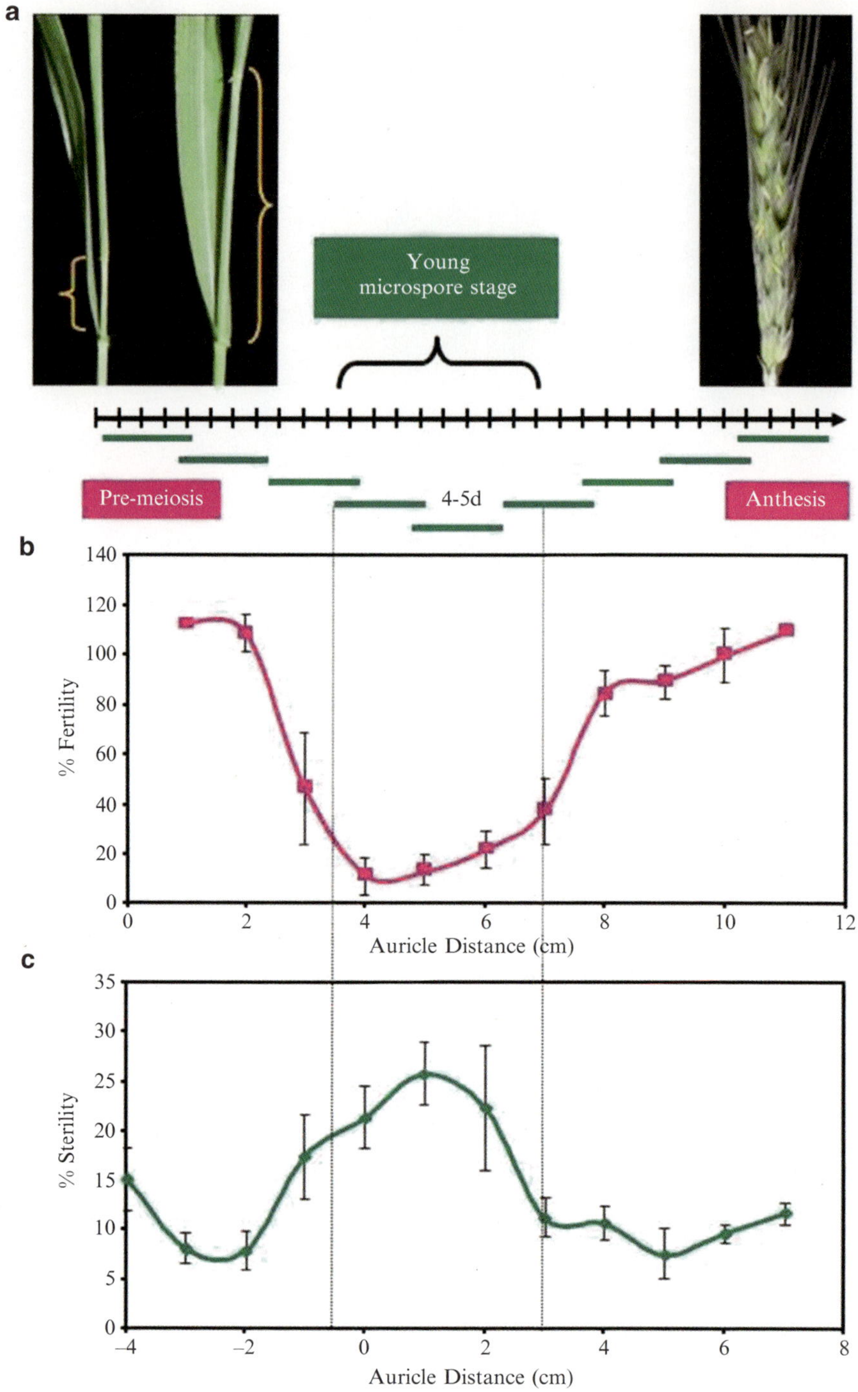


**Fig. 8.2** Procedure used for applying reproductive-stage-specific cold and drought stress in rice and wheat. (**a**) Schematic representation of the stress treatment strategy. The picture on the *left* shows how the auricle distances (AD) were determined (*brackets*). (**b**) Effect of drought stress on the % fertility in wheat (drought-sensitive variety Sundor) shows a depression around the young microspore stage. (**c**) Effect of cold stress on the % sterility in rice (cold-sensitive variety Doongara) shows a peak coinciding with YM stage

Doongara, we found that the percentage sterility was highest at ADs between 0 and +2 cm, when a large proportion of the florets at the top of the panicle were at the young microspore stage (YM) of pollen development (the stage immediately after meiosis, when tetrads separate to form the uninucleate stage young microspores; see Fig. 8.3a–f). Since a rice panicle flowers from top to bottom, we standardised our cold treatments (12°C for 5 days) to AD 0 to +2, when meiosis is starting in the top florets of the panicle. For drought stress in wheat, a similar method was developed. The difficulty with drought stress is that the treatment has to be started before the young microspore is reached. Wheat plants were drought-stressed by withholding watering for 4 days. Relative water content (RWC) measurements were carried out on the penultimate leaf to follow the plant's water status (Barr and Weatherley 1962; Morgan 1980). In contrast to rice, flowering in the wheat ear starts in the centre of the ear and proceeds to the top and bottom spikelets. We therefore focused our screening method on those florets that undergo meiosis first in the centre of the ear. To ensure drought conditions are reached around YM, the drought treatment was started at an AD of about 3 days of growth ahead of the AD where YM is reached (Ji et al. 2010). For both rice and wheat, we tagged the tillers that were at YM during the stress treatment, and only those tillers were used for scoring grain number and analytical studies. The AD at which YM is reached needs to be determined for each rice or wheat variety, as considerable differences can be observed between varieties. For cold treatment in rice, we routinely used a controlled environment chamber at a constant temperature of 12°C and a 16/8 light/dark cycle (400 μEinsteins $m^{-2}$ $sec^{-1}$). In the case of drought stress, the size of pots, the number of seeds per pot and the type of soil mix determine the rate of water loss. We use round pots (10.5 × 3.5 cm, DxH; 3 plants/pot; 0.6 kg soil/pot) or trays (36 × 25 × 13 cm, L × W × H; 10 plants/tray) filled with a relatively light soil mix (50% vermiculite, 50% composted soil with the addition of a slow release fertiliser NPK 16 + 9 + 12 + 2 Mg). Using this set-up, the RWC in leaves was around 40–50% at the end of the 5-day treatment period. Wilting of wheat leaves typically starts at a RWC of 60–70% (Morgan and King 1984). The use of trays rather than pots allows better reproducibility and more uniformity of drought stress conditions for a larger set of plants. All plants were grown in the glasshouse under natural lighting conditions and were watered twice daily before applying stress treatments. For the drought stress treatment, plants were transferred to a controlled environment chamber. This was necessary due to the variability in humidity and condensation under normal glasshouse conditions. The cold and drought stress treatments we have used in rice and wheat were relatively mild; they did not affect survival of the vegetative plant parts or the plant as a whole. Yet, using these standardised cold and drought stress treatments targeting the YM stage of pollen development, we were able to routinely induce high levels of sterility in sensitive rice and wheat lines (>40–50%). This screening method allowed us to identify rice and wheat germplasm with higher tolerance to cold and drought, respectively (Oliver et al. 2005; Ji et al. 2010).

## The Ovule Is More Resilient to the Effect of Cold and Drought Stress

Male and female gametogenesis is initiated by meiosis which occurs approximately at the same time in the anthers and the ovary of rice and wheat (Bennett et al. 1973; Xiang-Yuan and DeMason 1984; Herrero 2003). However, different florets of the ear or panicle do not flower at the same time, providing opportunities for cross-pollination. Under normal conditions, outcrossing in bread wheat is generally quite low (6%; Hucl 1996), but under cold, drought and heat stress conditions outcrossing in the field can be as high as 32% (Bingham 1966; Demotes-Mainard et al. 1996; Ferris et al. 1998). These are indications that the ovary remains receptive to pollination and that, provided enough fertile pollen is available, fertilisation and grain development can

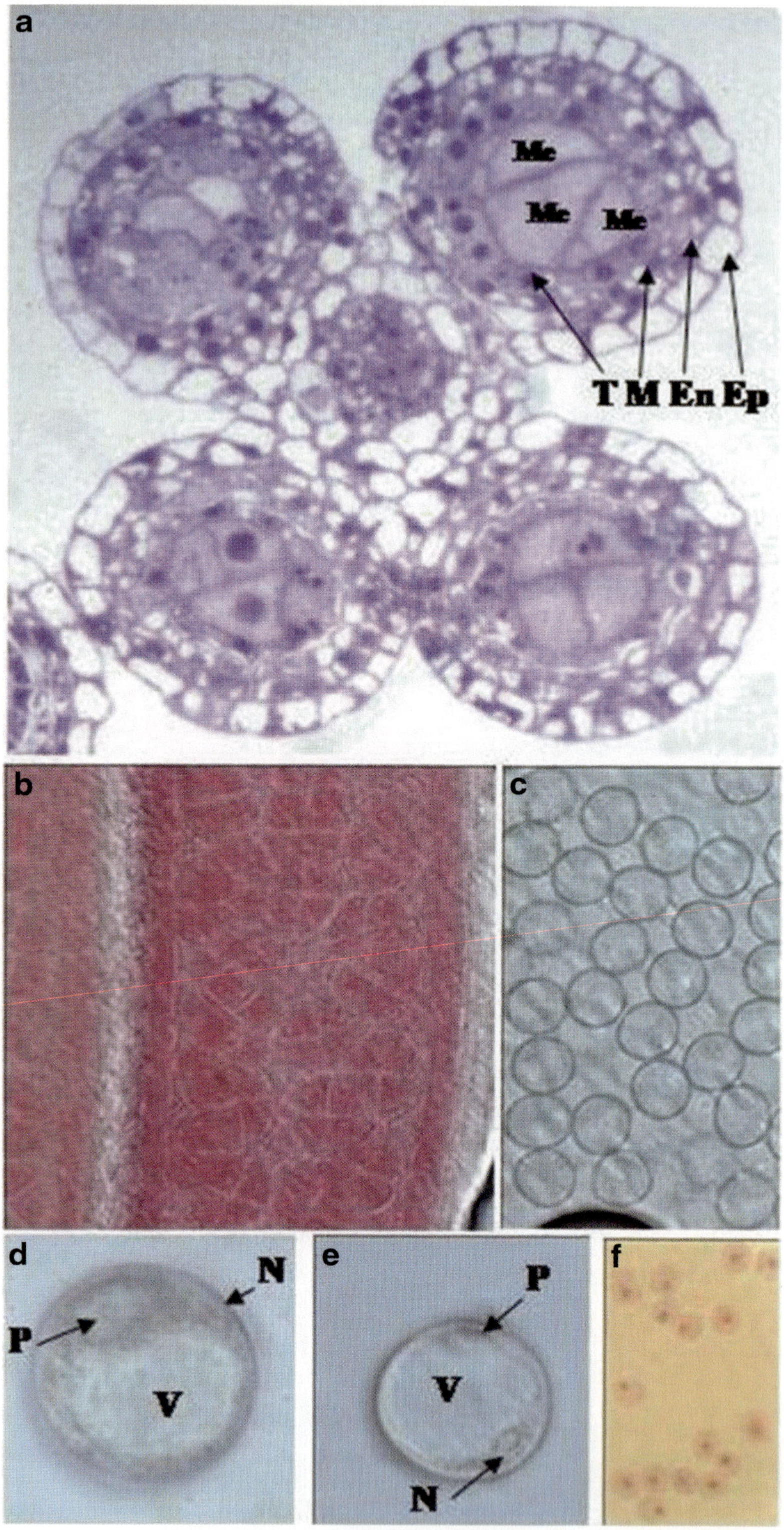


**Fig. 8.3** Determination of the young microspore stage in rice and wheat. (**a**) Cross section of a rice anther just before meiosis. The four layers of the anther wall have clearly been differentiated. From the outside to the inside, these layers are the epidermis (*Ep*), the endothecium (*En*), the middle layer (*M*) and the tapetum (*T*). In the centre are the meiocytes (*Me*) that will undergo meiosis and give rise to the microspores. (**b**) Rice anther at the tetrad stage of pollen meiosis (acetocarmine staining of squashed anther). (**c**) Rice pollen at the young microspore stage, immediately after tetrad separation. The cell wall has not yet been deposited and the pollen germination pore is not yet visible. (**d**) Rice pollen grain at the early vacuolated uninucleate stage. The vacuole (*V*) and germination pore (*P*) are clearly visible. The nucleus (*N*) is squashed against the cell wall and will move to the opposite side of the pollen germination pore. (**e**) Wheat pollen at the late vacuolated uninucleate stage. The nucleus (*N*) has moved further around the vacuole (*V*) to the opposite side of the pollen germination pore (*P*). (**f**) Vacuolated uninucleate stage wheat pollen stained with acetocarmine solution

occur. This is problematic when selection in the field is attempted, where occurrence, severity and timing of the stress are uncontrollable.

Starch staining of pollen grains prior to dehiscence revealed that pollen is sterile following YM-stage cold treatment in rice and drought treatment in wheat (Oliver et al. 2005; Ji et al. 2010). Experiments using male-sterile rice lines demonstrated that the ovule remained receptive to pollination after cold treatment (Hayase et al. 1969). Using wheat male-sterile lines, we have shown that the ovary also remained fertile following YM-stage drought treatment (Ji et al. 2010). It is possible that under extreme stress conditions (e.g. terminal drought in wheat fields), the ovary will abort and ultimately also the vegetative plant parts. For this reason, we found it important to keep the stress treatment conditions mild, so they will only affect pollen fertility and not ovary fertility. The YM-stage selection method we developed therefore targets maintenance of pollen fertility, and because rice and wheat are predominantly self-pollinating cereals, maintaining pollen fertility also means maintenance of grain number.

In maize the ovary is very sensitive to reproductive-stage drought stress. Drought-induced ovary abortion around the time of pollination is a major cause of kernel losses (Westgate and Boyer 1985; Zinselmeier et al. 1995a). Maize is different from rice and wheat not only in the fact that male and female flower structures are physically separated on the plant, making maize less dependent on self-pollination, but the ovary structure in maize is also different. Plant ovules are normally attached to the placenta and the ovary wall through a structure known as the funiculus. The maize nucellus is supported by the pedicel, the top layer of which, the placento-chalaza, is in direct contact with the nucellus. The placento-chalaza plays an important role in controlling nutrient flow to the nucellus and undergoes a cell death response after fertilisation to form the abscission zone of the seed (Kladnik et al. 2004). Drought stress induces a senescence response in the ovary (McLaughlin and Boyer 2004). In Arabidopsis, the funiculus is an umbilical cord-like structure and stress-induced ovule abortion is associated with a programmed cell death response in the chalazal cells (Sun et al. 2004). The rice and wheat ovule is of the anatropous type, which is characterised by the adhesion of the funiculus to the maternally derived ovary wall over the entire length of the ovule. The ovary wall develops into the pericarp which will ultimately enclose the seed. The ovary wall has one dorsal and two lateral vascular bundles, the latter ones continuing into the styles (Percival 1921; Fig. 8.4a, b). A cell layer equivalent to the placento-chalaza is not present in the rice and wheat ovary, and the ovule is directly in contact with the ovary wall (Fig. 8.4a–c). This higher contact surface between ovule and ovary wall in the rice and wheat ovary may provide a less vulnerable and more efficient nutrient exchange surface, making the ovary more stress-proof.

## The Young Microspore Stage Is Highly Sensitive to Abiotic Stresses

A detailed analysis of the events leading to cold-induced pollen abortion was carried out in rice (reviewed in Nishiyama 1984). Cold conditions were found to induce tapetal hypertrophy, which is observed as a swelling of the tapetal cell layer and its intrusion into the locule. The tapetal dilatation is associated with a vigorous augmentation of cytoplasmic organelles such as Golgi bodies, mitochondria, proplastids and endoplasmatic reticula (Mamun et al. 2005a, b). These observations indicate that cold stress interferes with the functioning of the tapetum and that pollen abortion may be the direct consequence of tapetal dysfunction. This was confirmed by the observation that cold stress induces a premature programmed cell death response in the tapetum (Gothandam et al. 2007). This was also shown to be the case for drought stress in rice (Nguyen et al. 2009). Premature degeneration of the tapetum is a major cause of pollen abortion in photoperiod-sensitive genic male sterility in rice (Shi et al. 2009). The tapetum layer naturally degenerates shortly after the young microspores become vacuolated (Steer 1977; Raghavan

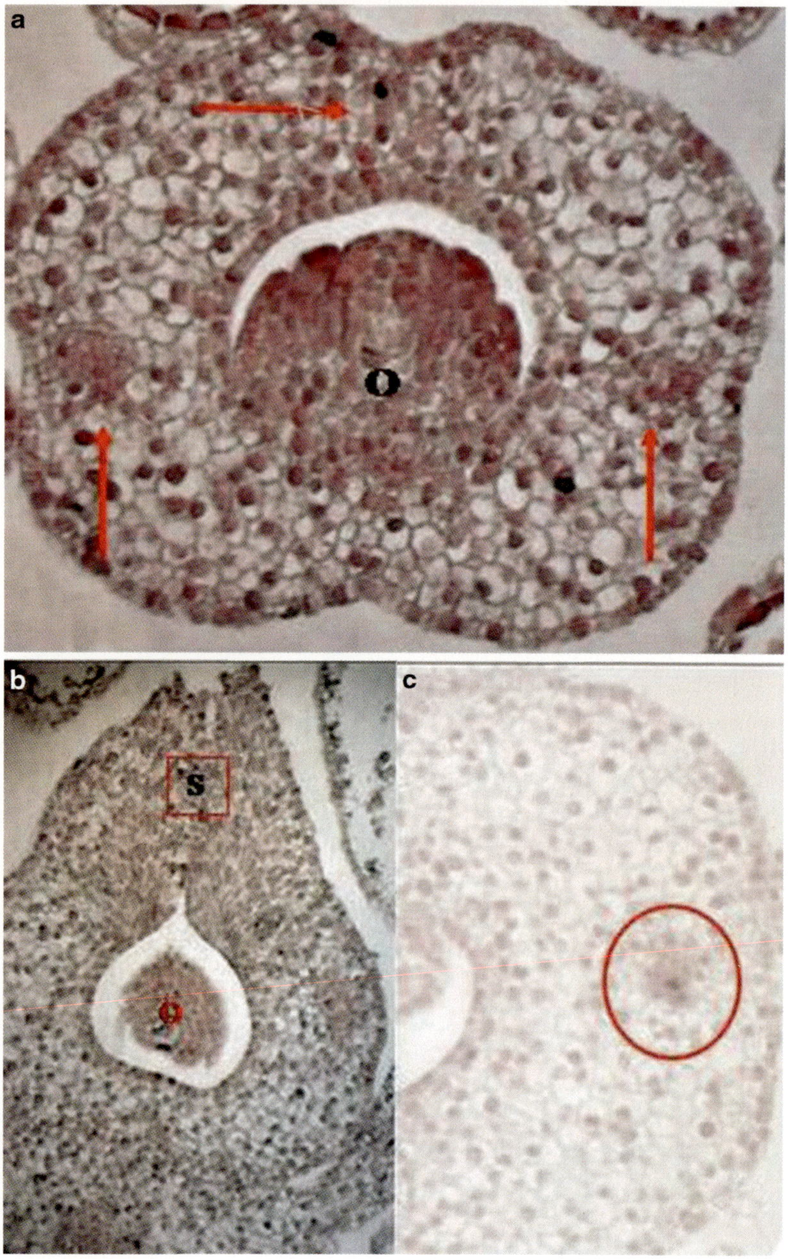


**Fig. 8.4** Cross sections of a wheat ovary at the young microspore stage of pollen development. (**a**) Cross section of the ovary showing the dorsal and lateral vascular bundles (*arrow*). The ovule (*O*) is not attached to the ovary wall via a funiculus but is over its entire length attached to the placenta and ovary wall (anatropous ovule). (**b**) Longitudinal section of a wheat ovary with the ovule (*O*) and style (*S*). (**c**) In situ hybridisation using a wheat *IVR1* cell wall invertase gene antisense probe. Weak expression is found around the lateral vascular bundle (*red circle*)

1988). The tapetum layer is a highly specialised secretory cell layer, responsible for pollen cell wall deposition and the production of the locular fluid that contains the nutrients required for pollen maturation (Steer 1977; Pacini et al. 1985; Clément and Audran 1995; Clément et al. 1998). The tapetum is very active during meiosis and at the YM stage. The young microspores are attached to the tapetum layer while their cell is deposited, and the cellular content of the tapetum is recycled for pollen nutrition. Stress-induced premature degradation of the tapetum layer is likely to interfere with these vital functions, leading to abortion of pollen development.

## Physiological Effects of Abiotic Stress on Pollen Development

### Sugars Accumulate in Stressed Anthers

One of the earliest biochemical observations of the effect of cold is interference in sugar metabolism in anthers (Fig. 8.5a). Accumulation of nonreducing sugars in panicles is observed within 12–24 h after cold treatment, well ahead of the morphological disturbances observed in the tapetum. At this time, an abnormal accumulation of starch is observed in the outer layers of the anther wall (Satake 1976). This accumulation of sugars in anthers was also observed for drought stress in rice and wheat (Saini 1997; Saini and Westgate 2000). The availability of tolerant and sensitive germplasm provided us with an incentive to reinvestigate the effect of cold and drought stress on sugar metabolism in anthers in more detail, in order to gain a better understanding of the molecular basis of pollen abortion. The accumulation of sugars in stressed anthers indicates that the abortion of pollen development is not due to repression of photosynthesis or lack of source supplies of sugars to sink tissues like anthers. Instead, sugar supply appears adequate but the sink strength of anthers is repressed by stress conditions. Sugars are therefore accumulating in the anther and they are converted to starch in the anther wall rather than in the pollen grains (Fig. 8.5b). In normal pollen grains, starch begins to accumulate only a few days before anthesis (Raghavan 1988), but sucrose, the building block for starch synthesis, needs to be supplied to the pollen before maturity. Starch accumulation is essential for pollen fertility, and starch staining can be used as a viability-staining method for pollen (Fig. 8.5c). We have shown that cold-stressed rice anthers and drought-stressed wheat anthers accumulate both nonreducing and reducing sugars (Fig. 8.5a; Oliver et al. 2005; Ji et al. 2010).

In the tolerant lines, we found that stress treatments did not affect starch accumulation in anthers. Anthers of cold-tolerant rice and drought-tolerant wheat lines did not accumulate sugars in response to stress treatment and mature pollen contained starch (Oliver et al. 2005; Ji et al. 2010). The tapetum is metabolically very active during meiosis with the anthers being the floral organ with the highest sink strength (Clément et al. 1996). The number of mitochondria per tapetum cell is increased dramatically to cope with the high-energy requirements (20- to 40-fold; Warmke and Lee 1978). The reduction in sink strength induced by cold and drought stress is therefore a likely cause for the abortion of pollen development. Elucidating the molecular mechanism of stress-induced repression of sink strength is an important step towards finding possible solutions.

### Abiotic Stresses Disturb Sugar Transport in Anthers

The finding that the nonreducing sugar sucrose accumulated in rice anthers exposed to cold at the cold-sensitive YM stage indicated that there is an obstruction in the supply mechanism of sugars to the young microspores within the anther. Shortly before meiosis, plasmodesmata between the tapetum and the outer cell layers of the anther wall disappear, and callose deposition isolates the meiocytes from the rest of the anther (Mamun 2005a, b). The symplastic isolation between the anther wall cells and the tapetum implies that sucrose supply to the tapetum and pollen grains occurs via an apoplastic transport pathway, involving cell wall invertase and a monosaccharide transporter. Cell wall invertase (β-fructofuranosidase; EC 3.2.1.26) first catalyses the hydrolysis of sucrose to glucose and fructose, and these hexoses are then transported into sink cells by membrane-bound monosaccharide transporters. Invertase plays a critical role in sugar partitioning and the regulation of the source-sink relationships (Roitsch 1999; Sturm

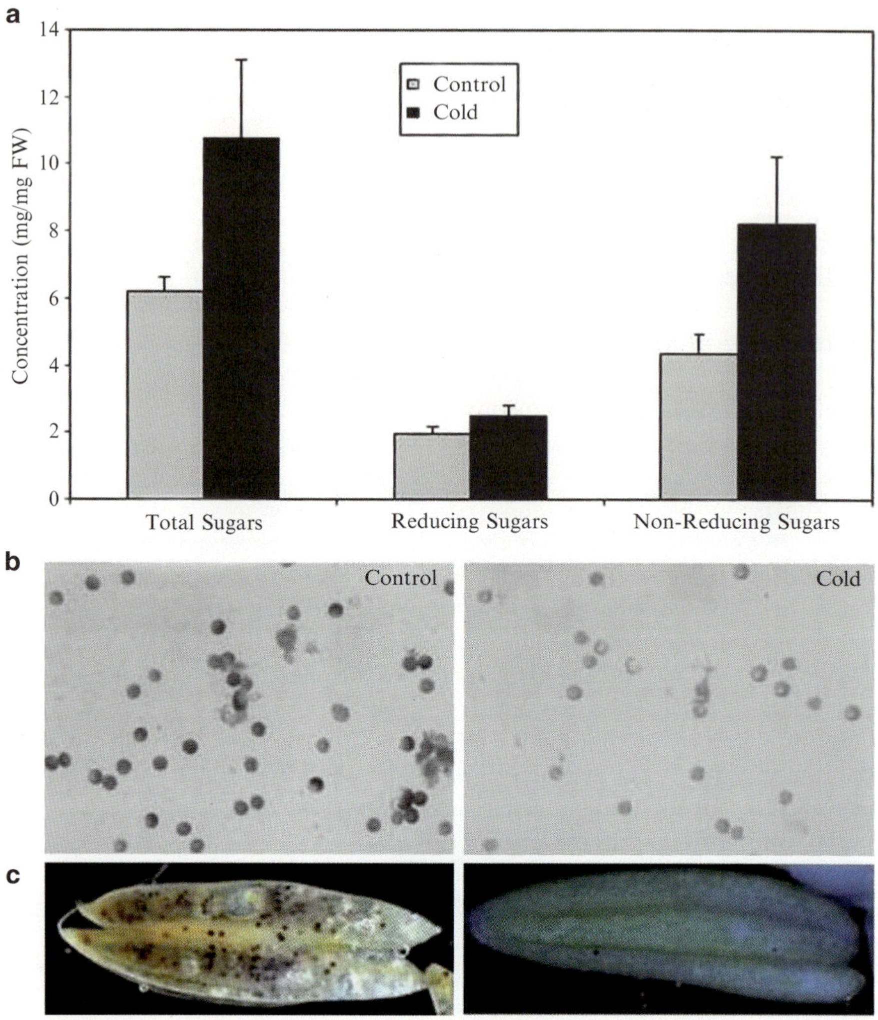


**Fig. 8.5** Effect of cold stress on sink strength in rice anthers. (**a**) Effect of YM-stage cold stress on the accumulation of total, reducing and nonreducing sugars in rice anthers (*DW* = dry weight). (**b**) Starch staining ($I_2$–KI) of control and cold-stressed pollen grains in rice anthers. Staining was carried out after grinding anthers and collecting the pollen just before reaching anthesis. Cold-stressed pollen grains do not stain dark blue and therefore lack starch. (**c**) Starch staining of control (*left*) and drought-stressed wheat anthers (*right*) show also absence of starch in the stressed anthers

1999; Sturm and Tang 1999). We have shown that cold stress reduced cell wall invertase enzyme activity in rice, indicating that sink strength is reduced by cold treatment (Oliver et al. 2005). We were able to identify the rice gene encoding cell wall invertase in anthers (*OSINV4*), and we found that this gene was also repressed by cold stress, but not in cold-tolerant rice lines (Fig. 8.6a; Oliver et al. 2005). The expression of *OSINV4* correlated well with the sugar and cell wall invertase activity measurements in cold-sensitive and tolerant rice lines. *OSINV4* is closely related to the wheat *IVR1* gene (Oliver et al. 2005). Drought stress in wheat anthers was shown to repress *IVR1* (Koonjul et al. 2005). Our results confirm this finding as we found that wheat *IVR1* expression was maintained in drought-tolerant wheat lines (Ji et al. 2010). In situ

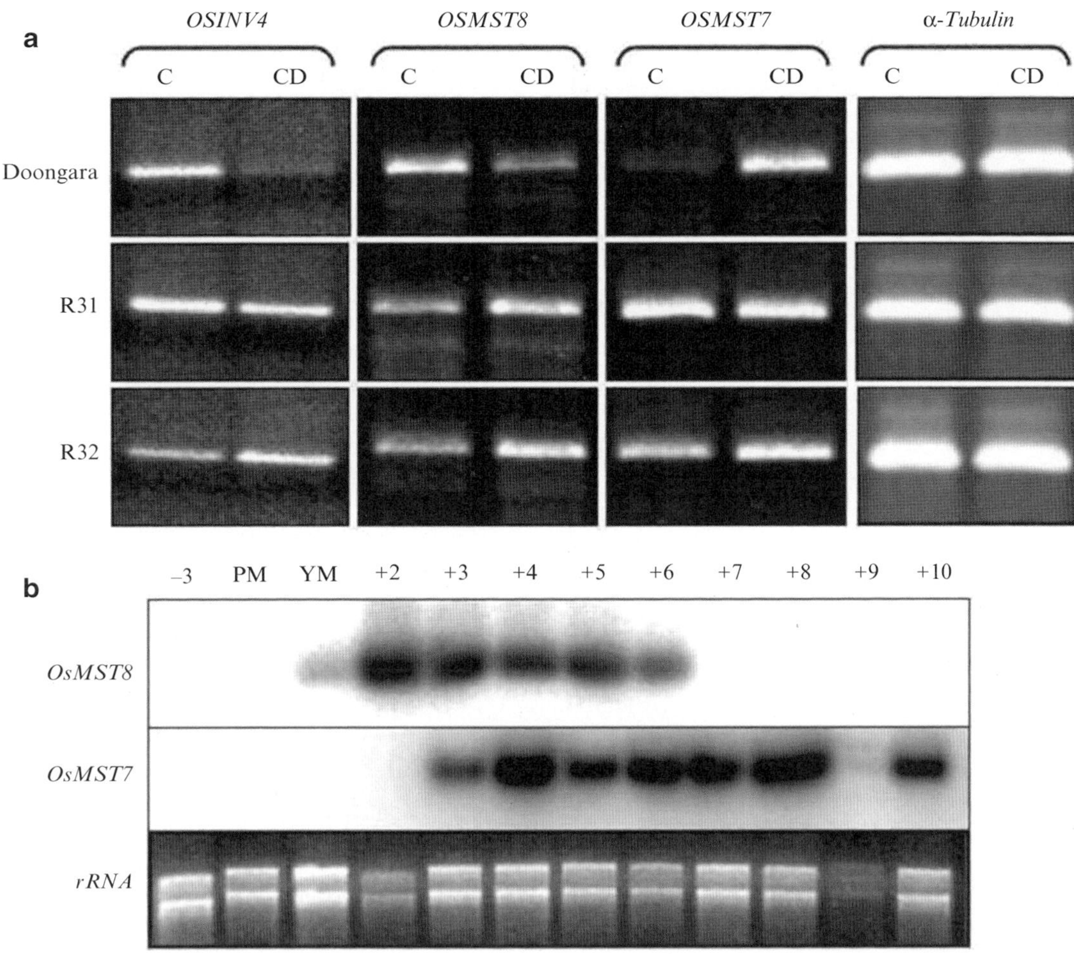


**Fig. 8.6** Effect of cold stress on sink strength in rice anthers. (**a**) Semi-quantitative RT-PCR showing the effect of cold stress at the young microspore stage on the expression of cell wall invertase OSINV4 and two monosaccharide transporter genes (OSMST7 and 8). The rice varieties tested were Doongara (cold-sensitive) and two cold-tolerant lines (R31 and R32). α-Tubulin was used as a standard for the cDNA concentration. (**b**) Temporal expression pattern of the rice OSMST7 and OSMST8 genes. OSMST8 is expressed from the YM stage onwards and is more similar to OSINV4 expression, while OSMST7 expression is mainly later in pollen development (*PM* = pre-meiosis)

hybridisation experiments and promoter-GUS constructs showed that rice *OSINV4* is expressed in the tapetum at the cold-sensitive YM stage and in the pollen grains from the binucleate stage onwards (Oliver et al. 2005). The wheat *IVR1* gene is also expressed in the tapetum at the YM stage (Koonjul et al. 2005). Tobacco cell wall invertase gene *NIN88* displayed a similar expression pattern, and antisense repression of this gene was shown to lead to male sterility (Goetz et al. 2001).

In wheat, fructans are nonreducing sugars and an important storage form of carbohydrate. In addition, fructans can play a protective role under abiotic stress conditions (Valluru and Van den Ende 2008). We estimated the effect of drought stress on the expression of two fructan biosynthesis genes in wheat anthers: sucrose 1-fructosyl-transferase (1-SST) and sucrose: fructan 6-fructosyl-transferase (6-SFT; Ruuska et al. 2008). Our findings showed that drought stress repressed 1-SST and especially 6-SFT gene expression in anthers of drought-sensitive wheat, but both genes were not repressed or even slightly up-regulated in drought-tolerant wheat lines (Ji et al. 2010). This suggests that cell wall invertase and the fructan biosynthesis genes are subject to a common regulatory mechanism that differentially affects their expression in anthers of sensitive and tolerant lines.

Our sugar measurements indicated that hexoses (glucose and fructose) accumulated in anthers in response to cold treatment (Oliver et al. 2005), suggesting that monosaccharide transporters may also be down-regulated by cold. The expression of cell wall invertase and monosaccharide transporter genes is synchronised in plants (Ehness et al. 1997; Dimou et al. 2005). Monosaccharide transporters (MSTs) in plants are encoded by large, functionally diverse multigene families. Arabidopsis has 14 *MST* genes; most are expressed in the later stages of pollen development, but only one (*AtSTP2*) is expressed at meiosis and YM (Truernit et al. 1999). *OSMST5* is also known to be expressed in rice pollen (Ngampanya et al. 2003). In an attempt to identify anther *MST* genes that were affected by cold in rice, we screened an anther cDNA library using the rice *OSMST1* and *OSMST3* cDNAs as probes (Toyofuku et al. 2000). This led to the identification of the *OSMST7* cDNA clone that corresponded to rice genomic locus Os01g38680 (AY643749). We then discovered that the neighbouring gene 2 kb upstream of *OSMST7* is another *MST* gene, *OSMST8* (AY822464; Os01g38670). Both *OSMST7* and *OSMST8* were expressed in rice anthers but during slightly different stages of pollen development (Oliver et al. 2007). *OSMST7* is expressed predominantly later in pollen development (Fig. 8.6b), and *OSMST8* expression more closely resembles *OSINV4* expression. In situ hybridisation results have shown that, like *OSINV4*, the gene is expressed in the tapetum at YM (Mamun et al. 2006). Like *OSINV4*, *OSMST8* expression is repressed by cold in YM anthers of cold-sensitive rice lines, and expression of the gene is maintained in cold-tolerant lines (Fig. 8.6a; Oliver et al. 2007). In addition, *OSINV4* and *OSMST8* appear to be regulated by the same rice R2R3 Myb family transcription factor CSA (*c*arbon *s*tarved *a*nther), a key transcriptional regulator for sugar partitioning in rice during male reproductive development (Zhang et al. 2010). Expression of *OSMST7* is induced in cold-sensitive rice, but expression is constitutively high and not affected by cold in cold-tolerant rice (Fig. 8.5a; Oliver et al. 2007). These results confirm that sugar partitioning in anthers is affected by a common regulatory mechanism that functions differently in stress-sensitive and stress-tolerant lines.

## Abiotic Stresses and Ovary Sink Strength

Sink strength of the ovary is highest after fertilisation and during grain filling (Zinselmeier et al. 1995a). In maize, drought-induced ovary abortion around the time of pollination is associated with a reduction in cell wall invertase levels (Westgate and Boyer 1985; Zinselmeier et al. 1995a, b). Due to their strategic position between the ovule and mother plant, the maize placento-chalazal cells express cell wall invertase activity and control sugar flow for grain filling. After repression of sugar transport, drought stress induces a senescence response in the maize ovary. It remains unclear where precisely this senescence response occurs, but its late induction may be the cause of the irreversibility of the drought effect (McLaughlin and Boyer 2004). At the YM stage, we found that sink strength is repressed by drought stress in wheat, similar to sink strength repression in anthers. In contrast to anthers, this response appears to be reversible in the ovary. This was shown using starch staining in both drought-stressed wheat and cold-stressed rice ovaries. Upon reversion of the stress stimulus, starch accumulation returned in rice and wheat ovaries (Ji et al. 2010). As suggested above, this may have something to do with the different structure of the rice and wheat ovaries (Fig. 8.4a, b). The *OSINV4* cell wall invertase gene is also expressed at low levels in the rice ovary. This expression is mainly located in cells around the

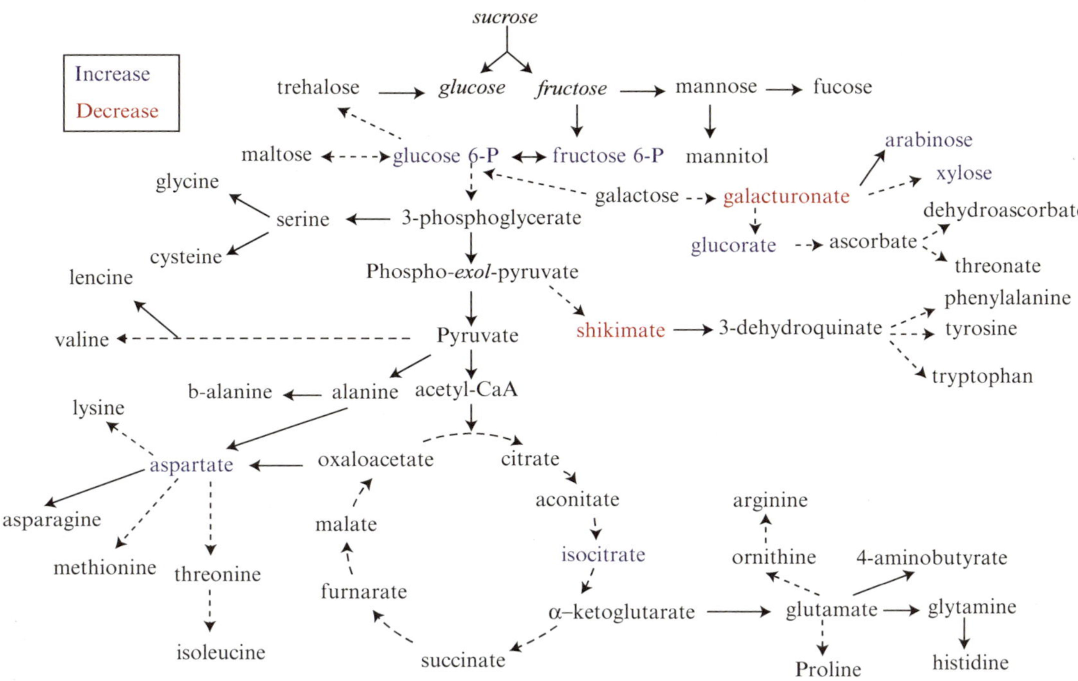


**Fig. 8.7** Effect of YM-stage cold stress on the accumulation of metabolites of the main metabolic pathways in rice anthers

lateral vascular bundles of the ovary wall (Fig. 8.4c). The rice cell wall invertase gene *OSCIN1* is expressed in the vascular parenchyma cells mainly around the dorsal vein of the ovary in the early stages of grain filling (Hirose et al. 2002). This suggests that apoplastic unloading of sugars also plays an essential role during ovule development and grain filling in rice.

## Effect of Abiotic Stress on Other Spike Metabolites

We have attempted to determine metabolite changes other than sucrose, fructose and glucose in dissected anther material of cold-stressed and control YM-stage rice anthers (Fig. 8.7). Organic acids, amino acids and sugars were extracted and measured according to Roessner et al. (2000). Relatively few metabolites were significantly affected by cold treatment in two biological repeat experiments. We found that glucose 6-P and fructose 6-P, the phosphorylated forms of glucose and fructose which are the starting points of glycolysis, were significantly increased. Whether this reflects a reduction of the flux through the glycolytic pathway remains to be established. Shikimate, a substrate for the synthesis of lignin and various aromatic compounds (Hermann and Weaver 1999), was decreased. The enzymes of this pathway have been shown to be involved in defence responses (Kasai et al. 2005). Auxin biosynthesis from tryptophan is also coupled to this pathway. In one of the repeat experiments, the aromatic amino acids tyrosine and tryptophan were found to be increased. Gluconate, which feeds into ascorbate biosynthesis, was also increased. In one of the repeat experiments, we found increased levels of dehydro-ascorbate and threonate. It is possible that this pathway is induced by cold stress as a protection against antioxidants (Foyer and Noctor 2011). Also coupled to galacturone and gluconate are the synthesis of sugars for pectin and cell wall synthesis (Reboul et al. 2011). Both arabinose and xylose synthesis are increased in

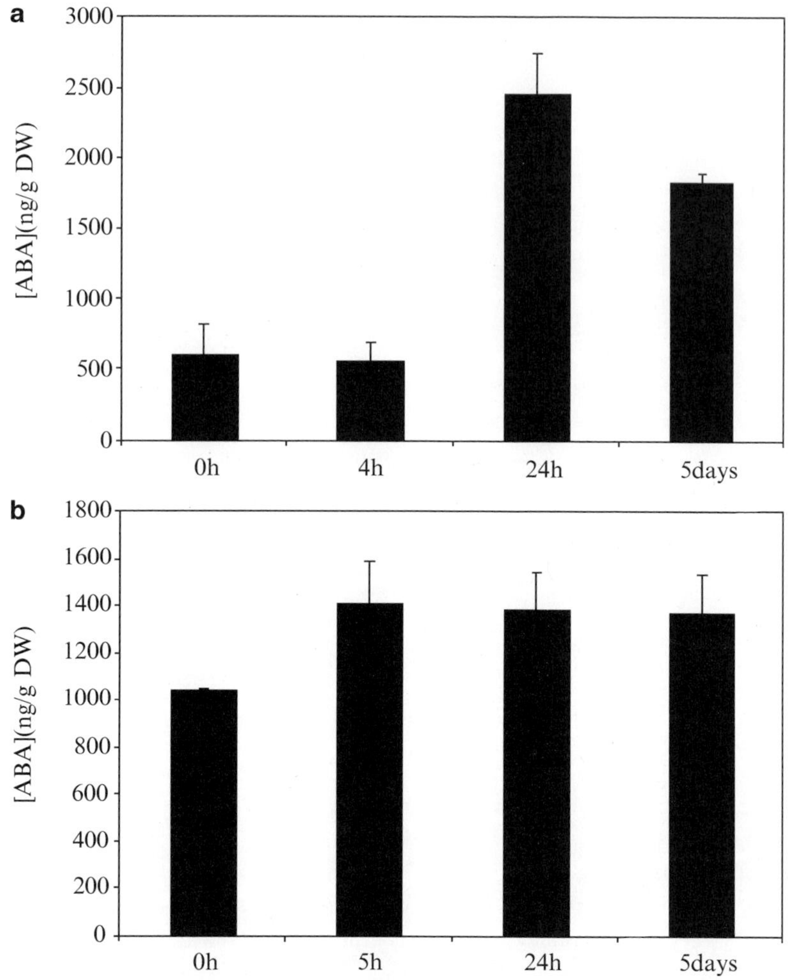


**Fig. 8.8** Effect of cold stress on ABA levels in leaves (**a**) and YM-stage rice anthers (**b**) (*DW* = dry weight)

cold-stressed anthers. Xylose is known to play a role in pollen tube growth (Suen and Huang 2007). We also found increased levels of aspartate, which is the starting point for synthesis of various other amino acids, and the TCA cycle intermediate isocitrate (Fig. 8.7).

## Control of Sink Strength in Cereal Reproductive Organs

### The Role of ABA

ABA is important for a variety of abiotic stress responses in plants, including drought and cold stress (Thomashow 1999; Shinozaki and Yamaguchi-Shinozaki 2000; Shinozaki et al. 2003; Chinnusamy et al. 2004). The similarity between sterility induced by cold and drought stress in rice and wheat, respectively, pointed to the stress hormone ABA as common denominator. We found that ABA indeed accumulates in anthers of both cold-stressed rice (Oliver et al. 2007; Fig. 8.8) and drought-stressed wheat (Ji et al. 2011). Interestingly, ABA did not accumulate in response to stress treatments in the anthers of cold and drought-tolerant rice and wheat lines (Oliver et al. 2007; Ji et al. 2011). The lower ABA levels in cold- and drought-tolerant lines were the result of both lower ABA biosynthesis levels and increased levels of ABA catabolism.

The negative correlation with stress tolerance at the reproductive stage is the opposite of what has been observed at the vegetative stage. ABA treatment at the vegetative stage can improve acclimation to abiotic stresses (Chen et al. 1983; Larosa et al. 1985; Robertson et al. 1994; Lu et al. 2009). This suggests that ABA plays different roles in stress tolerance in the vegetative and reproductive plant parts. ABA treatments mimic the effect of stress treatment; rice and wheat lines that are more tolerant to cold and drought stress, respectively, were more sensitive to exogenous ABA treatment. This suggests that tolerant germplasm reacts to different threshold levels of ABA (Oliver et al. 2007; Ji et al. 2011). Cold- and drought-tolerant rice and wheat are not noticeably disadvantaged at the vegetative stage, and it is not clear whether this inverted relationship has anything to do with the fact that reproductive tissues are sink tissues for sugars. Exogenous treatment of reproductive structures with ABA revealed that the hormone is able to repress cell wall invertase gene expression in both rice and wheat anthers, suggesting that ABA accumulation under cold and drought conditions is responsible for the observed reduction in anther sink strength (Oliver et al. 2007; Ji et al. 2011). The rice *OSINV4* promoter contains ABA response elements (ABRE; Kang et al. 2002), as well as cold/drought response elements (CBF/DREB transcription factor binding sites; Xue 2002; Dubouzet et al. 2003). It remains to be established whether ABA directly regulates the *OSINV4* promoter via the ABRE cis-acting elements, or whether the regulation by ABA is indirect through antagonism with GA. GA has also been shown to affect cell wall invertase gene expression (Wu et al. 1993; Proels et al. 2003; Ji et al. 2005).

ABA biosynthesis is localised in the anther vascular parenchyma cells in rice and wheat (Fig. 8.9a; Ji et al. 2011). Genes encoding the 9-cis-epoxycarotenoid dioxygenase (NCED) ABA biosynthetic gene are expressed in leave vascular parenchyma cells (Endo et al. 2008). ABA synthesis can be induced around the vascular bundles and can then diffuse to the surrounding tissues (Christmann et al. 2005). In anthers, this can then lead to repression of cell wall invertase expression. In the ovary, NCED expression is located in the tissue surrounding the dorsal vascular bundle (Fig. 8.9b). We have shown that transgenic rice plants over-expressing the wheat ABA-8′-hydroxylase gene under the control of a strong tapetum-specific promoter results in maintenance of anther sink strength and improved cold tolerance (Ji et al. 2011). This result illustrates the importance of ABA in regulating both anther sink strength and pollen fertility.

## Role of Other Plant Hormones

The antagonistic interaction between ABA and gibberellic acid (GA) has been known for some time (Jones and Jacobsen 1991). In a comparison of several tolerant and susceptible rice varieties, cold tolerance was found to be associated with greater anther length and pollen number, a trait that may be controlled by GA (Saito et al. 2001; Gunawardena et al. 2003). We have tried to counteract the effect of cold stress and ABA accumulation in anthers by treating the panicle with GA solutions ($10^{-5}$ and $10^{-4}$ M). This resulted in much higher sterility levels, even under control temperatures (Dolferus et al. unpublished result). This was a surprising finding, considering the fact that GA is required for stimulating flower development and a normal ABA/GA balance is required for pollen and floral development (Weiss et al. 1995; Wassom et al. 2001). GA biosynthesis genes are expressed in the tapetum (Kaneko et al. 2003; 2004; Murray et al. 2003). It is possible that the ABA-GA interactions in anthers do not involve higher GA levels but occur at the level of the GA signalling pathway.

The fact that cold and drought stress induce premature tapetal cell death (Gothandam et al. 2007; Nguyen et al. 2009) suggests that ethylene may be involved (Fig. 8.10). It was previously suggested that induction of pollen sterility by ABA injection is caused by ethylene (Morgan 1980). ABA and ethylene act antagonistically in plants (Cheng et al. 2009), and both ABA and ethylene induce plant senescence (Van Doorn 2004; Wingler and Roitsch 2008; Lin et al. 2009). Ethylene and ABA have overlapping functions in controlling sink strength, as mutant genes that confer sugar insensitivity encode components of ABA metabolism and signalling, as well as genes involved in ethylene signalling

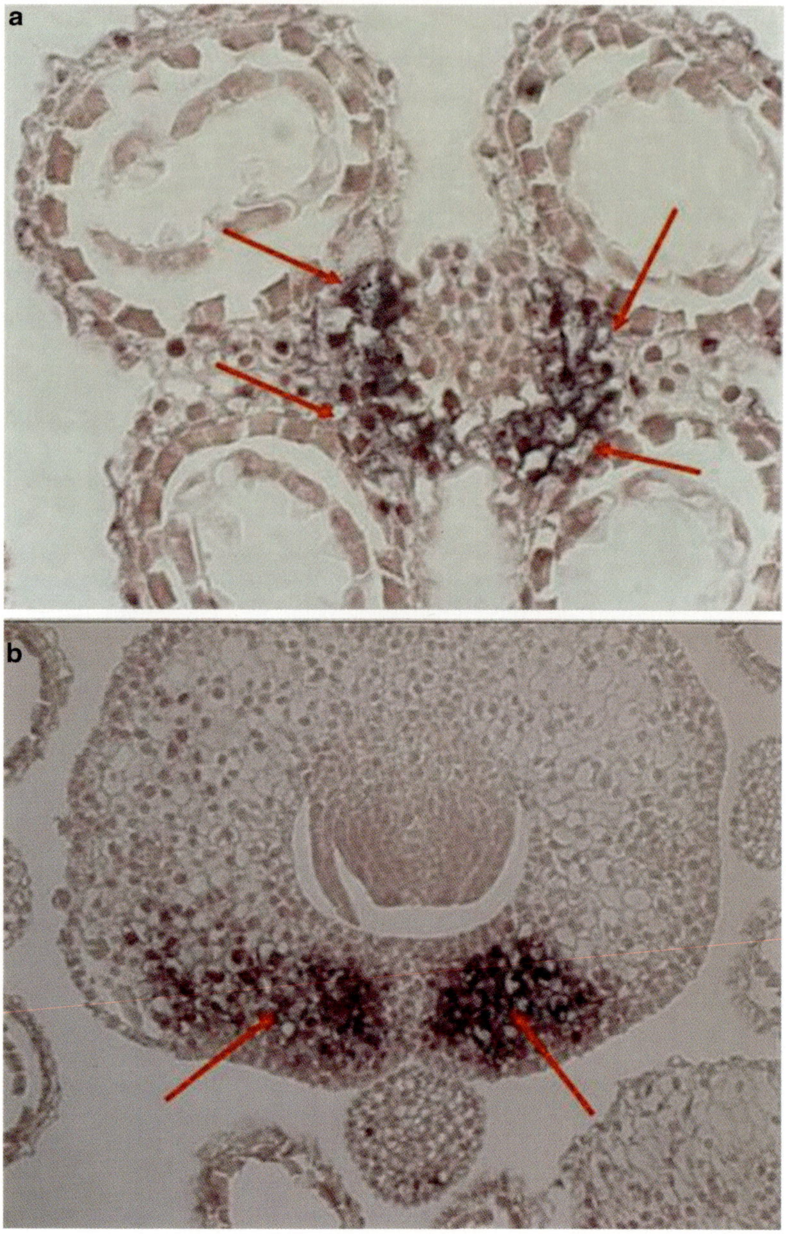


**Fig. 8.9** In situ hybridisation using the wheat *TaNCED2* ABA biosynthetic gene on cross sections of wheat anthers (**a**) and ovules (**b**). The deep purple colour (*red arrows*) indicates areas where the ABA biosynthetic gene is expressed in both tissues

(Laby et al. 2000; Gibson et al. 2001; Pourtau et al. 2004; Rook et al. 2006; Huang et al. 2008). In leaves senescence is tightly linked to sugar signalling, metabolic activity and the hormone cytokinin (Wingler and Roitsch 2008). The overlapping functions of ABA and ethylene signalling are a possible convergence point for abiotic stress response pathways (Fig. 8.10). Ethylene interacts with DELLA proteins in the GA signalling pathway (Achard et al. 2003).

Another plant hormone that may play a role in controlling abiotic stress tolerance is auxin (Fig. 8.8; Jain and Khurana 2009; Shen et al. 2010; Tognetti et al. 2010; Wang et al. 2010). Auxins play a key role in pollen development (Feng et al. 2006; Cecchetti et al. 2008). They

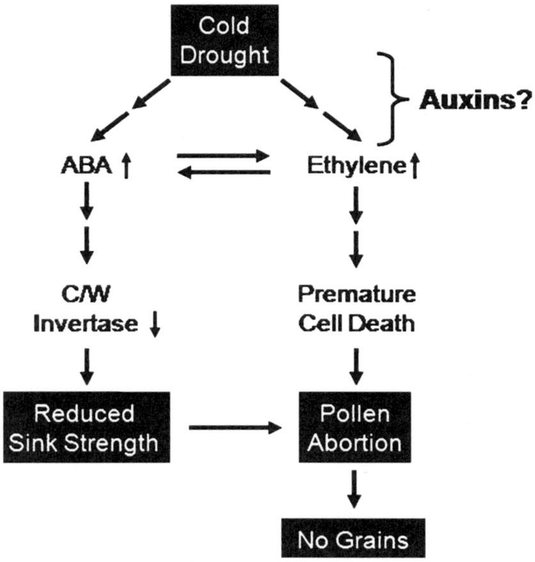


**Fig. 8.10** Model for the effect of cold and drought stress on the regulation of sink strength in YM-stage anthers

regulate anther GA biosynthesis (Wolbang et al. 2004; Yin et al. 2007), and recently it was shown that auxin treatment could reverse the effect of heat stress on pollen fertility in barley (Sakata et al. 2010). This effect of auxins still remains to be demonstrated in the case of cold stress in rice and drought stress in wheat.

## Conclusions

The establishment of a screening system that targets pollen fertility and not ovule fertility has allowed us to identify germplasm in wheat and rice that is more capable of maintaining grain number under cold and drought conditions. This germplasm has allowed us to carry out comparative physiological and molecular studies into the effect of cold and drought stress on pollen development. This approach has revealed interesting differences in the capacity to maintain sink strength in anthers, and this is providing us with an insight into the hormonal signalling events that control sugar supply to the nascent pollen grains. Tolerant germplasm maintains sink strength through a different capacity of controlling ABA homeostasis. Keeping ABA levels low in anthers is important to maintain sink strength, and we have shown that transgenic rice lines with lower ABA levels under stress conditions have higher sink strength and improved cold tolerance. Controlling ABA homeostasis and maintaining pollen fertility under stress conditions are likely to involve interactions with other plant hormones, e.g. ethylene, GA and auxins. Further molecular and physiological studies are under way to identify candidate genes in the upstream regulatory mechanism that controls pollen fertility under abiotic stress conditions. Our work has provided germplasm and screening methods that can form the basis of QTL mapping and marker design. QTL mapping approaches are currently in progress in both rice and wheat in order to further unravel the mechanism of abiotic stress tolerance in cereals. With the rice (and soon also the wheat) genome sequence available, together with the availability of state-of-the-art genomics technologies (SNPs), the identification of genes from QTLs has become more feasible. Our work on the physiological and molecular characterisation of cold and drought tolerance will not only provide potential to improve reproductive abiotic stress tolerance via targeted transgenic approaches, but the improved knowledge may provide helpful information for the identification of candidate genes within mapped QTL regions.

**Acknowledgements** The research presented in this chapter was supported by the Australian Grains Research and Development Corporation (wheat; GRDC grants CSP00089, CSP00130 and CSP00143) and the Australian Rural Industries Research and Development Corporation (rice; RIRDC grant US-143A). The authors thank Dr Alisdair Fernie (Max Planck Institute for Molecular Plant Physiology, Germany) for the use of the gas chromatography-mass spectrometry platform to measure rice anther metabolites.

## References

Abbate PE, Andrade FH, Culot JP (1995) The effects of radiation and nitrogen on number of grains in wheat. J Agric Sci 124:352–360

Achard P, Vriezen WH, Van Der Straeten D, Harberd NP (2003) Ethylene regulates Arabidopsis development via the modulation of DELLA protein growth repressor function. Plant Cell 15:2816–2825

Acreche MM, Slafer GA (2006) Grain weight response to increases in number of grains in wheat in Mediterranean area. Field Crops Res 98:52–59

Amasino R (2010) Seasonal and developmental timing of flowering. Plant J 61:1001–1013

Ashton T (1948) Technique of breeding for drought resistance in crops. Commonwealth Bureau of Plant Breeding, Technical Communication #14, pp 8–9

Barnabas B, Jäger K, Feher A (2008) The effect of drought and heat stress on reproductive processes in cereals. Plant Cell Environ 31:11–38

Barr HD, Weatherley PE (1962) A re-examination of the relative turgidity technique for estimating water deficit in leaves. Aust J Biol Sci 15:413–428

Bennett MD, Finch JB, Smith JB, Rao MK (1973) The time and duration of female meiosis in wheat, rye and barley. Proc R Soc Lond B 183:301–319

Bertin P, Kinet JM, Bouharmont J (1996) Evaluation of chilling sensitivity in different rice varieties. Relationship between screening procedures applied during germination and vegetative growth. Euphytica 89:201–210

Bingham J (1966) Varietal response in wheat to water supply in the field, and male sterility caused by a period of drought in a glasshouse experiment. Ann Appl Biol 57:365–377

Blum A (2011) Drought resistance – is it really a complex trait? Funct Plant Biol 38:753–757

Board JE, Peterson ML, Ng E (1980) Floret sterility in rice in a cool environment. Agron J 72:483–487

Brooking IR (1976) Male sterility in *Sorghum bicolor* (L.) Moench induced by low temperature. I. Timing of the stage sensitivity. Aust J Plant Physiol 3:589–596

Cecchetti V, Altamura MM, Falasca G, Costantino P, Cardarelli M (2008) Auxin regulates Arabidopsis anther dehiscence, pollen maturation, and filament elongation. Plant Cell 20:1760–1774

Chaves MM, Maroco JP, Pereira JS (2003) Understanding plant responses to drought: from genes to the whole plant. Funct Plant Biol 30:239–264

Chen HH, Li PH, Brenner ML (1983) Involvement of abscisic acid in potato cold acclimation. Plant Physiol 71:362–365

Cheng WH, Chiang MH, Hwang SG, Lin PC (2009) Antagonism between abscisic acid and ethylene in Arabidopsis acts in parallel with the reciprocal regulation of their metabolism and signaling pathways. Plant Mol Biol 71:61–80

Chinnusamy V, Schumaker K, Zhu JK (2004) Molecular genetic perspectives on cross-talk and specificity in abiotic stress signalling in plants. J Exp Bot 55:225–236

Christmann A, Hoffmann T, Teplova I, Grill E, Müller A (2005) Generation of active pools of abscisic acid revealed by *in vivo* imaging of water-stressed Arabidopsis. Plant Physiol 137:209–219

Clément C, Audran JC (1995) Anther wall layers control pollen sugar nutrition in *Lilium*. Protoplasma 187:172–181

Clément C, Burrus M, Audran JC (1996) Floral organ growth and carbohydrate content during pollen development in *Lilium*. Am J Bot 83:459–469

Clément C, Laporte P, Audran JC (1998) The loculus content and tapetum during pollen development in *Lilium*. Sex Plant Reprod 11:94–106

Collins NC, Tardieu F, Tuberosa R (2008) Quantitative trait loci and crop performance under abiotic stress: where do we stand? Plant Physiol 147:469–486

Demotes-Mainard S, Doussinault G, Meynard JM (1996) Abnormalities in the male developmental programme of winter wheat induced by climatic stress at meiosis. Agronomie 16:505–515

Dimou M, Flemetakis E, Delis C, Aivalakis G, Spyropoulos KG, Katinakis P (2005) Genes coding for a putative cell-wall invertase and two putative monosaccharide/H + transporters are expressed in roots of etiolated *Glycine max* seedlings. Plant Sci 169:798–804

Distelfeld A, Li C, Dubcovsky J (2009) Regulation of flowering in temperate cereals. Curr Opin Plant Biol 12:178–184

Dolferus R, Ji X, Richards RA (2011) Abiotic stress and control of grain number in cereals. Plant Sci 181:331–341

Dubouzet JG, Sakuma Y, Ito Y, Kasuga M, Dubouzet EG, Miura S, Seki M, Shinozaki K, Yamaguchi-Shinozaki K (2003) *OsDREB* genes in rice, *Oryza sativa* L., encode transcription activators that function in drought-, high-salt- and cold-responsive gene expression. Plant J 33:751–763

Ehness R, Ecker M, Godt DE, Roitsch T (1997) Glucose and stress independently regulate source and sink metabolism and defense mechanisms via signal transduction pathways involving protein phosphorylation. Plant Cell 9:1825–1841

Endo A, Sawada Y, Takahashi H, Okamoto M, Ikegami K, Koiwai H, Seo M, Toyomasu T, Mitsuhashi W, Shinozaki K et al (2008) Drought induction of Arabidopsis 9-cis-epoxycarotenoid dioxygenase occurs in vascular parenchyma cells. Plant Physiol 147:1984–1993

Feng XL, Ni WM, Elge S, Mueller-Roeber B, Xu ZH, Xue HW (2006) Auxin flow in anther filaments is critical for pollen grain development through regulating pollen mitosis. Plant Mol Biol 61:215–226

Ferris R, Wheeler TR, Hadley P, Ellis RH (1998) Recovery of photosynthesis after environmental stress in soybean grown under elevated $CO_2$. Crop Sci 38:948–955

Fischer RA (1993) Irrigated spring wheat and timing and amount of nitrogen fertilizer. II. Physiology of grain yield response. Field Crops Res 33:57–80

Fischer RA, Stockman YM (1980) Kernel number per spike in wheat (*Triticum aestivum* L.): responses to pre anthesis shading. Aust J Plant Physiol 7:169–180

Fleury D, Jefferies S, Kuchel H, Langridge P (2010) Genetic and genomic tools to improve drought tolerance in wheat. J Exp Bot 61:3211–3222

Forster BP, Ellis RP, Thomas WTB, Newton AC, Tuberosa R, This D, El-Enein RA, Bahri MH, Ben Salem M

(2000) The development and application of molecular markers for abiotic stress tolerance in barley. J Exp Bot 51:19–27

Foyer CH, Noctor G (2011) Ascorbate and glutathione: the heart of the redox hub. Plant Physiol 155:2–18

Gebbing T, Schnyder H, Kuhbauch W (1999) The utilization of pre-anthesis reserves in grain filling of wheat. Assessment by steady-state $^{13}C–CO_2/^{12}C–CO_2$ labelling. Plant Cell Environ 22:851–858

Gibson SI, Laby RJ, Kim D (2001) The sugar-insensitive1 (*sis1*) mutant of Arabidopsis is allelic to *ctr1*. Biochem Biophys Res Commun 280:196–203

Glaszmann JC, Kaw RN, Khush GS (1990) Genetic divergence among cold tolerant rices (*Oryza sativa* L.). Euphytica 45:95–104

Goetz M, Godt DE, Guivarch A, Kahmann U, Chriqui D, Roitsch T (2001) Induction of male sterility in plants by metabolic engineering of the carbohydrate supply. Proc Natl Acad Sci USA 98:6522–6657

Gonzalez FG, Slafer GA, Miralles DJ (2003) Grain and floret number in response to photoperiod during stem elongation in fully and slightly vernalized wheats. Field Crops Res 81:17–27

Gothandam KM, Kim ES, Chung YY (2007) Ultrastructural study of rice tapetum under low-temperature stress. J Plant Biol 50:396–402

Greenup A, Peacock WJ, Dennis ES, Trevaskis B (2009) The molecular biology of seasonal flowering-responses in Arabidopsis and the cereals. Ann Bot 103:1165–1172

Gunawardena RA, Fukai S, Blamey FPC (2003) Low temperature induced spikelet sterility in rice. II. Effects of panicle and root temperatures. Aust J Agric Res 54:947–956

Hayase H, Satake T, Nishiyama I, Ito N (1969) Male sterility caused by cooling treatment at the meiotic stage in rice plants. II. The most sensitive stage to cooling and the fertilising ability of pistils. Proc Crop Sci Soc Jpn 38:706–711

Heenan DP (1984) Low-temperature induced floret sterility in the rice cultivars *Calrose* and *Inga* as influenced by nitrogen supply. Aust J Agric Anim Husb 24:255–259

Hermann KM, Weaver LM (1999) The shikimate pathway. Annu Rev Plant Physiol Mol Biol 50:473–503

Herrero M (2003) Male and female synchrony and the regulation of mating in flowering plants. Philos Trans R Soc Lond Biol Sci 358:1019–1024

Hirose T, Takano M, Terao T (2002) Cell wall invertase in developing rice caryopsis: molecular cloning of *OsCIN1* and analysis of its expression in relation to its role in grain filling. Plant Cell Physiol 43:452–459

Huang Y, Li CY, Biddle KD, Gibson SI (2008) Identification, cloning and characterization of *sis7* and *sis10* sugar-insensitive mutants of Arabidopsis. BMC Plant Biol 8:104–128

Hucl P (1996) Out-crossing rates for 10 Canadian spring wheat cultivars. Can J Plant Sci 76:423–427

Jagadish SVK, Craufurd PQ, Wheeler TR (2007) High temperature stress and spikelet fertility in rice (*Oryza sativa* L.). J Exp Bot 58:1627–1635

Jagadish KSV, Cairns JE, Kumar A, Somayanda IM, Craufurd PQ (2011) Does susceptibility to heat stress confound screening for drought stress in rice? Funct Plant Biol 38:261–269

Jain M, Khurana JP (2009) Transcript profiling reveals diverse roles of auxin-responsive genes during reproductive development and abiotic stress in rice. FEBS J 276:3148–3162

Jain M, Vara Prasad PV, Boote KJ, Hartwell AL Jr, Chourey PS (2007) Effects of season-long high temperature growth conditions on sugar-to-starch metabolism in developing microspores of grain sorghum (*Sorghum bicolor* L. Moench). Planta 227:67–79

Ji X, Raveendran M, Oane R, Lafitte R, Bruskiewich R, Cheng SH, Bennett J (2005) Tissue-specific expression and drought responsiveness of cell-wall invertase genes of rice at flowering. Plant Mol Biol 59:947–967

Ji X, Shiran B, Wan J, Lewis DC, Jenkins CLD, Condon AG, Richards RA, Dolferus R (2010) Importance of pre-anthesis anther sink strength for maintenance of grain number during reproductive stage water stress in wheat. Plant Cell Environ 33:926–942

Ji X, Dong B, Shiran B, Talbot MJ, Edlington JE, Hughes T, White RG, Gubler F, Dolferus R (2011) Control of ABA catabolism and ABA homeostasis is important for reproductive stage stress tolerance in cereals. Plant Physiol 156:647–662

Jones RL, Jacobsen JV (1991) Regulation of synthesis and transport of secreted proteins in cereal aleurone. Int Rev Cytol 126:49–88

Jung C, Müller AE (2009) Flowering time control and applications in plant breeding. Trends Plant Sci 14:563–573

Kaneko M, Itoh H, Inukai Y, Sakamoto T, Ueguchi-Tanaka M, Ashikari M, Matsuoka M (2003) Where do gibberellin biosynthesis and gibberellin signaling occur in rice plants? Plant J 35:104–115

Kaneko M, Inukai Y, Ueguchi-Tanaka M, Itoh H, Izawa T, Kobayashi Y, Hattori T, Miyao A, Hirochika H, Ashikari M, Matsuoka M (2004) Loss-of-function mutations of the rice GAMYB gene impair alpha-amylase expression in aleurone and flower development. Plant Cell 16:33–44

Kang JY, Choi HI, Im MY, Kim SY (2002) Arabidopsis basic leucine zipper proteins that mediate stress-responsive abscisic acid signaling. Plant Cell 14:343–357

Kasai K, Kanno T, Akita M, Ikejiri-Kanno Y, Wakasa K, Tozawa Y (2005) Identification of three shikimate kinase genes in rice: characterization of their differential expression during panicle development and of the enzymatic activities of the encoded proteins. Planta 222:438–447

Kellogg EA (2007) Floral displays: genetic control of grass inflorescences. Curr Opin Plant Biol 10:26–31

Kladnik A, Chamusco K, Dermastia M, Chourey P (2004) Evidence of programmed cell death in post-phloem transport cells of the maternal pedicel tissue in developing caryopsis of maize. Plant Physiol 136:3572–3581

Koonjul PK, Minhas JS, Nunes C, Sheoran IS, Saini HS (2005) Selective transcriptional down-regulation of anther invertases precedes the failure of pollen development in water-stressed wheat. J Exp Bot 56:179–190

Kurata N, Miyoshi K, Nonomura KI, Yamazaki Y, Ito Y (2005) Rice mutants and genes related to organ development, morphogenesis and physiological traits. Plant Cell Physiol 46:48–62

Laby RJ, Kincaid MS, Kim D, Gibson SI (2000) The Arabidopsis sugar-insensitive mutants *sis4* and *sis5* are defective in abscisic acid synthesis and response. Plant J 23:587–596

Large EG (1954) Growth stages in cereals: illustration of the Feeke's scale. Plant Pathol 3:128–129

Larosa PC, Handa AK, Hasegawa PM, Bressan RA (1985) Abscisic acid accelerates adaptation of cultured tobacco cells to salt. Plant Physiol 79:138–142

Lin SS, Peterson ML (1975) Low temperature-induced floret sterility in rice. Crop Sci 15:657–660

Lin Z, Zhong S, Grierson D (2009) Recent advances in ethylene research. J Exp Bot 60:3311–3336

Lu S, Su W, Li H, Guo Z (2009) Abscisic acid improves drought tolerance of triploid Bermuda grass and involves $H_2O_2$- and NO-induced antioxidant enzyme activities. Plant Physiol Biochem 47:132–138

Mamun EA, Cantrill LC, Overall RL, Sutton BG (2005a) Cellular organization and differentiation of organelles in pre-meiotic rice anthers. Cell Biol Int 29:792–802

Mamun EA, Cantrill LC, Overall RL, Sutton BG (2005b) Cellular organization in meiotic and early post-meiotic rice anthers. Cell Biol Int 29:903–913

Mamun EA, Alfred S, Cantrill LC, Overall RL, Sutton BG (2006) Effects of chilling on male gametophyte development in rice. Cell Biol Int 30:583–591

Matsushima S (1957) Analysis of developmental factors determining yield and yield prediction in lowland rice. Bull Natl InstAgric Sci Jpn A5:1–271

McLaughlin JE, Boyer JS (2004) Sugar-responsive gene expression, invertase activity, and senescence in aborting maize ovaries at low water potentials. Ann Bot 94:675–689

Morgan JM (1980) Possible role of abscisic acid in reducing seed set in water-stressed wheat plants. Nature 285:655–657

Morgan JM, King RW (1984) Association between loss of leaf turgor, abscisic acid levels and seed set in two wheat cultivars. Aust J Plant Physiol 11:143–150

Murray F, Kalla R, Jacobsen J, Gubler F (2003) A role for *HvGAMYB* in anther development. Plant J 33:481–491

Ngampanya B, Sobolewska A, Takeda T, Toyofuku K, Narangajavana J, Ikeda A, Yamaguchi J (2003) Characterisation of rice functional monosaccharide transporter, *OsMST5*. Biosci Biotechnol Biochem 67:556–562

Nguyen GN, Hailstones DL, Wilkes M, Sutton BG (2009) Drought-induced oxidative conditions in rice anthers leading to a programmed cell death and pollen abortion. J Agron Crop Sci 195:157–164

Nishiyama I (1984) Climatic influence on pollen formation and fertilization. In: Tsunoda S, Takahashi N (eds) Biology of rice. Japan Scientific Societies Press, Tokyo, pp 153–171

Oliver SN, Van Dongen JT, Alfred SC, Mamun EA, Zhao X, Saini HS, Fernandes SF, Blanchard CL, Sutton BG, Geigenberger P, Dennis ES, Dolferus R (2005) Cold-induced repression of the rice anther-specific cell wall invertase gene *OSINV4* is correlated with sucrose accumulation and pollen sterility. Plant Cell Environ 28:1534–1551

Oliver SN, Dennis ES, Dolferus R (2007) ABA regulates apoplastic sugar transport and is a potential signal for cold-induced pollen sterility in rice. Plant Cell Physiol 48:1319–1330

Pacini E, Franchi GG, Hesse M (1985) The tapetum: its form, function, and possible phylogeny. In Embryophyta. Plant Syst Evol 149:155–185

Percival J (1921) The wheat plant, a monograph. Duckworth and Co., London

Pourtau N, Marès M, Purdy S, Quentin N, Ruël A, Wingler A (2004) Interactions of abscisic acid and sugar signalling in the regulation of leaf senescence. Planta 219:765–772

Powell N, Ji X, Ravash R, Edlington J, Dolferus R (2012) Yield stability for cereals in a changing climate. Funct Plant Biol 39:539–552

Proels RK, Hause B, Berger S, Roitsch T (2003) Novel mode of hormone induction of tandem tomato invertase genes in floral tissues. Plant Mol Biol 52:191–201

Raghavan V (1988) Anther and pollen development in rice (*Oryza sativa*). Am J Bot 75:183–186

Reboul R, Geserick C, Pabst M, Frey B, Wittmann D, Lütz-Meindl U, Léonard R, Tenhaken R (2011) Down-regulation of UDP-glucuronic acid biosynthesis leads to swollen plant cell walls and severe developmental defects associated with changes in pectic polysaccharides. J Biol Chem 286:39982–39992

Richards RA, Rebetzke GJ, Watt M, Condon AG, Spielmeyer W, Dolferus R (2010) Breeding for improved water productivity in temperate cereals: phenotyping, quantitative trait loci, markers and the selection environment. Funct Plant Biol 37:85–97

Robertson AJ, Ishikawa M, Gusta LV, MacKenzie SL (1994) Abscisic acid-induced heat tolerance in *Bromus inermis* Lcyss cell-suspension cultures. Heat-stable, abscisic acid-responsive polypeptides in combination with sucrose confer enhanced thermostability. Plant Physiol 105:181–190

Roessner U, Wagner C, Kopka J, Trethewey RN, Willmitzer L (2000) Simultaneous analysis of

metabolites in potato tuber by gas chromatography-mass spectrometry. Plant J 23:131–142
Roitsch T (1999) Source-sink regulation by sugar and stress. Curr Opin Plant Biol 2:198–206
Rook F, Hadingham SA, Li Y, Bevan MW (2006) Sugar and ABA response pathways and the control of gene expression. Plant Cell Environ 29:426–434
Ruuska SA, Rebetzke GJ, van Herwaarden AF, Richards RA, Fettell NA, Tabe L, Jenkins C (2006) Genotypic variation for water soluble carbohydrate accumulation in wheat. Funct Plant Biol 33:799–809
Ruuska SA, Lewis DC, Kennedy G, Furbank RT, Jenkins CL, Tabe LM (2008) Large scale transcriptome analysis of the effects of nitrogen nutrition on accumulation of stem carbohydrate reserves in reproductive stage wheat. Plant Mol Biol 66:15–32
Saini HS (1997) Effects of water stress on male gametophyte development in plants. Sex Plant Reprod 10:67–73
Saini HS, Westgate ME (2000) Reproductive development in grain crops during drought. In: Sparks DL (ed) Advances in agronomy, vol 68. Academic, New York, pp 59–96
Saini HS, Sedgley M, Aspinall D (1984) Developmental anatomy in wheat of male sterility induced by heat stress, water deficit or abscisic acid. Aust J Plant Physiol 11:243–253
Saito K, Miura K, Nagano K, Hayano-Saito Y, Araki H, Kato A (2001) Identification of two closely linked quantitative trait loci for cold tolerance on chromosome 4 of rice and their association with anther length. Theor Appl Genet 103:862–868
Sakata T, Oshino T, Miura S, Tomabechi M, Tsunaga Y, Higashitani N, Miyazawa Y, Takahashi H, Watanabe M, Higashitani A (2010) Auxins reverse plant male sterility caused by high temperatures. Proc Natl Acad Sci USA 107:8569–8574
Salter RJ, Goode JB (1967) Crop responses to water at different stages of growth. Commonwealth Agricultural Bureau, Farnham Royal, Bucks, England, pp 22–26
Satake T (1976) Determination of the most sensitive stage to sterile-type cool injury in rice plants. Res Bull Hokkaido Natl Agric Exp Stat 113:1–40
Satake T, Hayase H (1970) Male sterility caused by cooling treatment at the young microspore stage in rice plants. V. Estimation of pollen developmental stage and the most sensitive stage to coolness. Proc Crop Sci Soc Jpn 39:468–473
Satake T, Hayase H (1974) Male sterility caused by cooling treatment at the young microspore stage in rice plants. IX. Revision of the classification and terminology of pollen developmental stages. Proc Crop Sci Soc Jpn 43:36–39
Savin R, Slafer GA (1991) Shading effects on the yield of an Argentinean wheat cultivar. J Agric Sci Camb 116:1–7
Sayre KD, Rajaram S, Fischer RA (1997) Yield potential progress in short bread wheats in northwest Mexico. Crop Sci 37:36–42
Shen C, Bai Y, Wang S, Zhang S, Wu Y, Chen M, Jiang D, Qi Y (2010) Expression profile of PIN, AUX/LAX and PGP auxin transporter gene families in *Sorghum bicolor* under phytohormone and abiotic stress. FEBS J 277:2954–2969
Shi Y, Zhao S, Yao J (2009) Premature tapetum degeneration: a major cause of abortive pollen development in photoperiod sensitive genic male sterility in rice. J Integr Plant Biol 51:774–781
Shinozaki K, Yamaguchi-Shinozaki K (2000) Molecular responses to dehydration and low temperature: differences and cross-talk between two stress signaling pathways. Curr Opin Plant Biol 3:217–223
Shinozaki K, Yamaguchi-Shinozaki K, Seki M (2003) Regulatory network of gene expression in the drought and cold stress responses. Curr Opin Plant Biol 6:410–417
Sinclair TR, Jamieson PD (2006) Grain number, wheat yield, and bottling beer: an analysis. Field Crops Res 98:60–67
Steer MW (1977) Differentiation of the tapetum in Avena. II. The endoplasmatic reticulum and Golgi apparatus. J Cell Sci 28:71–86
Sturm A (1999) Invertases. Primary structures, functions, and roles in plant development and sucrose partitioning. Plant Physiol 121:1–8
Sturm A, Tang GQ (1999) The sucrose-cleaving enzymes of plants are crucial for development, growth and carbon partitioning. Trends Plant Sci 4:401–407
Suen DF, Huang AHC (2007) Maize pollen coat xylanase facilitates pollen tube penetration into silk during sexual reproduction. J Biol Chem 282:625–636
Sun K, Hunt K, Hauser BA (2004) Ovule abortion in Arabidopsis triggered by stress. Plant Physiol 135:2358–2367
Thomashow MF (1999) Plant cold acclimation: freezing tolerance genes and regulatory mechanisms. Annu Rev Plant Physiol Plant Mol Biol 50:571–599
Tognetti VB, Van Aken O, Morreel K, Vandenbroucke K, van de Cotte B, De Clercq I, Chiwocha S, Fenske R, Prinsen E, Boerjan W, Genty B, Stubbs KA, Inzé D, Van Breusegem F (2010) Perturbation of indole-3-butyric acid homeostasis by the UDP-glucosyltransferase UGT74E2 modulates Arabidopsis architecture and water stress tolerance. Plant Cell 22:2660–2679
Toyofuku K, Kasahara M, Yamaguchi J (2000) Characterization and expression of monosaccharide transporters (OSMSTs) in rice. Plant Cell Physiol 41:940–947
Trevaskis B (2010) The central role of the *VERNALIZATION1* gene in the vernalization response of cereals. Funct Plant Biol 37:479–487
Truernit E, Stadler R, Baier K, Sauer N (1999) A male gametophyte-specific monosaccharide transporter in Arabidopsis. Plant J 17:191–201
Valluru R, Van den Ende W (2008) Plant fructans in stress environments: emerging concepts and future prospects. J Exp Bot 59:2905–2916
Van Doorn WG (2004) Is petal senescence due to sugar starvation? Plant Physiol 134:35–42

Wang S, Bai Y, Shen C, Wu Y, Zhang S, Jiang D, Guilfoyle TJ, Chen M, Qi Y (2010) Auxin-related gene families in abiotic stress response in *Sorghum bicolor*. Funct Integr Genomics 10:533–546

Warmke HE, Lee SLJ (1978) Pollen abortion in T cytoplasmic male-sterile corn (*Zea mays*): a suggested mechanism. Science 200:561–563

Wassom JJ, Mei C, Rocheford TR, Widholm JM (2001) Interaction of environment and ABA and GA treatments on the maize anther culture response. Plant Cell Tissue Org Cult 64:69–72

Weiss D, Van Der Luit A, Knegt E, Vermeer E, Mol J, Kooter JM (1995) Identification of endogenous gibberellins in petunia flowers (induction of anthocyanin biosynthetic gene expression and the antagonistic effect of abscisic acid). Plant Physiol 107:695–702

Westgate ME, Boyer JS (1985) Carbohydrate reserves and reproductive development at low water potentials in maize. Crop Sci 25:762–769

Wingler A, Roitsch T (2008) Metabolic regulation of leaf senescence: interactions of sugar signalling with biotic and abiotic stress responses. Plant Biol 10(Suppl 1):50–62

Wolbang CM, Chandler PM, Smith JJ, Ross JJ (2004) Auxin from the developing inflorescence is required for the biosynthesis of active gibberellins in barley stems. Plant Physiol 134:769–776

Wu LL, Mitchell JP, Cohn NS, Kaufman PB (1993) Gibberellin (GA3) enhances cell wall invertase activity and mRNA levels in elongating dwarf pea (*Pisum sativum*) shoots. Int J Plant Sci 154:280–289

Xiang-Yuan X, DeMason DA (1984) Relationship between male and female gametophyte development in rye. Am J Bot 71:1067–1079

Xue GP (2002) Characterisation of the DNA-binding profile of barley HvCBF1 using an enzymatic method for rapid, quantitative and high-throughput analysis of the DNA-binding activity. Nucleic Acids Res 30:e77

Yang J, Zhang J, Wang Z, Zhu Q, Liu L (2001) Water deficit induced senescence and its relationship to the remobilization of pre-stored carbon in wheat during grain filling. Agron J 93:196–206

Yin C, Gan L, Ng D, Zhou X, Xia K (2007) Decreased panicle-derived indole-3-acetic acid reduces gibberellin A1 level in the uppermost internode, causing panicle enclosure in male sterile rice Zhenshan 97A. J Exp Bot 58:2441–2449

Yoshida H, Nagato Y (2011) Flower development in rice. J Exp Bot 62:4719–4730

Zadoks JC, Chang TT, Konzak CF (1974) A decimal code for the growth stages of cereals. Weed Res 14:415–421

Zhang H, Liang W, Yang X, Luo X, Jiang N, Ma H, Zhang D (2010) Carbon starved anther encodes a MYB domain protein that regulates sugar partitioning required for rice pollen development. Plant Cell 22:672–689

Zinselmeier C, Westgate ME, Schlusser JR, Jones RJ (1995a) Low water potential disrupts carbohydrate metabolism in maize (*Zea mays* L.) ovaries. Plant Physiol 107:385–391

Zinselmeier C, Lauer MJ, Boyer JS (1995b) Reversing drought-induced losses in grain yield: sucrose maintains embryo growth in maize. Crop Sci 35:1390–1400

# Salicylic Acid: Role in Plant Physiology and Stress Tolerance

# 9

Gopal K. Sahu

**Abstract**

The plant phenolic salicylic acid (SA) plays a regulatory role in plant physiology and metabolism. An updated review has been made based on previous and recent publications related to influence of SA on physiological processes of plants such as flowering, thermogenesis, membrane function, ion uptake, and photosynthesis. The role of SA as a mediator in disease resistance in plants is also described. This chapter deals with the role of SA in plant response to abiotic stress factors of various natures, i.e., temperature, heavy metal, ozone, salinity, and drought. The exact mechanism of SA action in plants as a signal molecule in stress mitigation and other biological processes needs to be understood.

## Introduction

### Salicylic Acid: A Brief History

Salicylic acid (SA) is one of the numerous plant phenolics. The compound was discovered in the fourth century B.C. and was prescribed by Hippocrates as a pain reliever during child birth. Native Americans are known to have widely used SA in the form of willow bark for treatment of pain, fever, and inflammation. In 1928, Johann Buchner, at the University of Munich, Germany, was the first to isolate SA from willow bark in minute quantities as salicin. Charles Gerhardt, a Frenchman, invented aspirin (acetylsalicylic acid) in 1853, and the product was marketed until the Bayer Company patented the formula in 1899. The compound was named as salicylic acid from Latin Salix, a willow tree by Raffaele Piria in 1938. During the nineteenth century, SA and other salicylates were extracted from a variety of plant sources.

### Chemistry of SA

Chemically SA is known as *o*-hydroxybenzoic acid. In free state, it is colorless and a crystalline powder with melting temperature 159°C. It is poorly soluble in water (0.2 g/100 ml of $H_2O$ at 20°C) and is well soluble in polar organic solvents. The pH and p*K*a of aqueous solution are nearly 2.4 and 2.9, respectively (Minnick and

G.K. Sahu (✉)
School of Life Sciences, MATS University,
Raipur 493 441, India
e-mail: drgksahu@matsuniversity.ac.in

G.R. Rout and A.B. Das (eds.), *Molecular Stress Physiology of Plants*,
DOI 10.1007/978-81-322-0807-5_9, © Springer India 2013

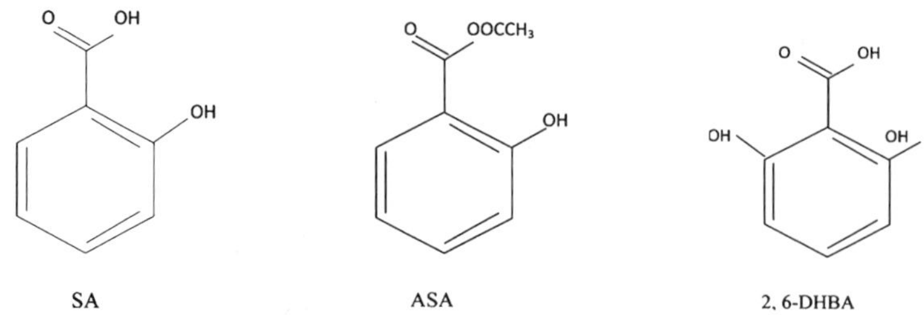


**Fig. 9.1** Chemical structure of salicylic acid and its active analogs. *SA* salicylic acid, *ASA* acetylsalicylic acid, 2,6-*DHBA* 2,6-dihydrobenzoic acid

Kilpatrick 1939). SA fluoresces at 412 nm with 301-nm excitation, a property that is used for the detection of the compound in the plants (Raskin et al. 1987). SA is transported to other parts when applied to one part of the plant and is metabolized into different other compounds. It can also be conjugated to other compounds. Acetylsalicylic acid (ASA), 2,6-dichloroisonicotinic acid (INA), and benzothiadiazole S-methyl ester (BTH) are the compounds that impart functions similar to SA in plants. The structure of SA and some of its derivatives are given in Fig. 9.1.

## Biosynthesis and Endogenous Level of SA in Plants

SA is synthesized by both plants and microorganisms. In plants, SA is synthesized from phenylalanine (Phe) in two different pathways. In one of the pathways, Phe is converted to cinnamic acid (CA) by phenylalanine ammonia-lyase (PAL), a regulatory enzyme in the phenylpropanoid pathway. Further, CA is hydroxylated to form *ortho*-coumaric acid followed by oxidation of side chain (El-Basyouni et al. 1964; Chadha and Brown 1974) into SA. In the second pathway, the side chain of CA is oxidized to give benzoic acid (BA) which is hydroxylated in the *ortho* position and gives rise to SA in the presence of the enzyme benzoic acid 2-hydroxylase (Ellis and Amrhein 1971). Independent operation of these pathways has been observed in tomato seedlings under infection with *Agrobacterium tumefaciens* and without infections, respectively (Chadha and Brown 1974). Labeling studies in tobacco (Yalpani et al. 1993a) and rice (Silverman et al. 1995) suggested that SA is synthesized from CA via the formation of BA. Further, BA 2-hydroxylase, the enzyme that causes hydroxylation of BA into SA, was detected in the leaves of healthy tobacco plants (León et al. 1993).

In microorganisms, an alternative pathway for the synthesis of SA has been proposed. Most of these organisms synthesize SA from chorismate, an intermediate metabolite of shikimic acid pathway. Chorismate is converted to isochorismate and then to salicylic acid catalyzed by isochorismate synthase (ICS) and isochorismate pyruvate lyase (IPL), respectively (Serino et al. 1995). Overproduction of these two enzymes through the incorporation of bacterial transgene expression causes enhanced accumulation of SA (Verberne et al. 2000; Mauch et al. 2001). Presence of an analogous pathway is believed to operate in plants which has been studied in *Arabidopsis* (Wildermuth et al. 2001).

After its synthesis, SA usually conjugates with glucose and/or methyl groups. The conjugation with glucose at the hydroxyl group forms SA 2-*O*-β-D-glucoside as a major conjugate and at carboxyl group produces the SA glucose ester. These reactions are catalyzed by cytosolic SA glucosyltransferase that are known to be induced by SA application or pathogen attack in tobacco and *Arabidopsis* (Lee and Raskin 1999; Song 2006). SA glucosides are transported from the cytoplasm into the vacuole of soybean and tobacco cells where it may function as inactive storage form that can release free SA (Dean et al. 2005). SA is methylated at carboxyl group, forming methyl salicylate (MeSA) that acts as an important long-distance signal in systemic

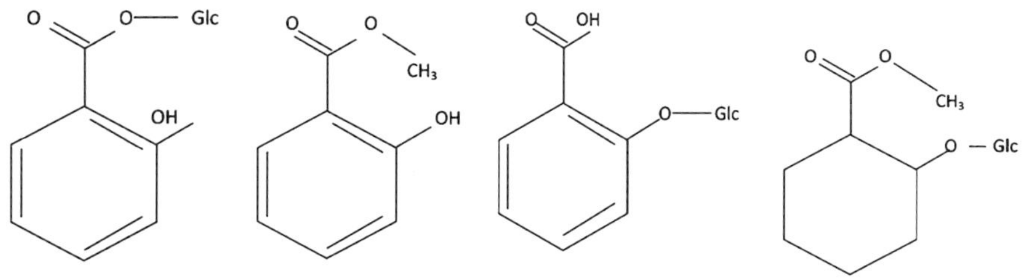


**Fig. 9.2** Glucosylated and methylated conjugates of SA. *SGE* SA glucose ester, *SAG* SA 2-*O*-β-D-glucoside, *MeSA* methyl salicylate, *MeSAG* methyl salicylate 2-*O*-β-D-glucose

acquired resistance (SAR) in tobacco and *Arabidopsis* (Chen et al. 2003; Vlot et al. 2008). Further, MeSA can be glucosylated to produce MeSA 2-*O*-β-D-glucose. The conjugated form of SA has been shown in Fig. 9.2.

Salicylates are known to exist in various plant species (Procter 1843; Griffiths 1958). The endogenous level of SA among different species varies widely. The basal level of SA in tobacco leaves has been reported to be less than 100 ng $g^{-1}$ f.w. (Yalpani et al. 1991), whereas in potato, it varied to 10 μg $g^{-1}$ f.w. (Navarre and Mayo 2004), though both the species belong to the family *Solanaceae*. The level of total SA in *Arabidopsis* varies from 0.250 to 1 μg $g^{-1}$ f.w. Endogenous level of SA has been quantified in the leaves, flowers, and inflorescences of a wide variety of plants. A survey for the detection of SA level in thermogenic and non-thermogenic plants using spectrofluorometric techniques confirmed the wide distribution of the compound in plants (Raskin et al. 1990). Among the various non-thermogenic agricultural important plant species examined, *Oryza sativa* possessed the largest quantity of SA. The level of SA in different cultivars of rice varied between 24 and 68 μg $g^{-1}$ f.w. SA level in crab grass, barley, green foxtail, and soybean has been estimated to be more than 1 μg $g^{-1}$ f.w. The lowest detectable concentration of SA was 10 ng $g^{-1}$ f.w., whereas the highest level of SA was recorded in the inflorescences of thermogenic plants. Though the function of endogenous SA in thermogenic plants is to act as a regulatory molecule in respiratory metabolism, its role in non-thermogenic plants was not fully understood. The level of SA in response to pathogen infection and various abiotic stress increases by several folds in plant tissues that will be discussed in the forthcoming sections.

## SA and Plant Physiological Processes

Plant phenolics generally play their role as growth regulators such as flowering, thermogenesis, and disease resistance (Aberg 1981). Exogenous application of SA regulates a variety of plant processes. These compounds have drawn the attention of researchers as important factors in normal plants in cell wall biosynthesis, regulation of plant growth, and plant-microbe interactions (Harborne 1980; Hahlbrock and Scheel 1989). Various studies have demonstrated that salicylates act as endogenous signal molecules participating in a number of plant processes. Some of these effects might have been caused by the general chemical properties of SA as iron chelator (Raskin 1992).

### SA and Flowering in Plants

The role of salicylates in flowering has been well documented in quite a good number of plants. Lee and Skog (1965) have experimentally demonstrated the induction of flowering by SA in tobacco tissue culture containing kinetin and indole acetic acid (IAA). SA could stimulate the process at minimum concentration of 4 μM. Later on, the process of flowering induction with response to SA was also observed in

*Lemna gibba* and other species of *Lemna* (Cleland and Ajami 1974). The factor responsible for flowering was isolated and identified as SA which induces flowering in 5.6 μM concentration in *L. gibba*. Lack of flowering induction activity in honeydew collected from synthetic diet-feeding aphids confirmed the origin of these regulators from plants. The induction of flowering by SA, aspirin, and other plant phenolics under noninductive photoperiod was also noticed in *Spirodela polyrrhiza* (Khurana and Maheswari 1980), *Spirodela punctata* (Scharfetter et al. 1978), and *Wolffia microscopica* (Khurana and Maheswari 1987) belonging to family *Lemnaceae*. Though exogenous application of SA could induce flowering in plants, the level of SA did not show appreciable variation in honeydew obtained from vegetative and flowering plants of *Xanthium strumarum* and *Lemna* as well. Various other compounds including benzoic acids (Watanabe et al. 1981), ferricyanide (Tanaka and Cleland 1980), nicotinic acid, and cytokinins (Fujioka et al. 1983, 1986) were observed to induce flowering in *Lemna* under noninductive photoperiodic condition. Thus, the possible and specific role of SA as an endogenous regulator of flower induction in these plants could not be fully established. Watanabe et al. (1981) studied the relationship between structure and flowering induction activity of BA derivatives in *Lemna paucicostata* and found that there is a correlation between the electron-withdrawing ability and decrease in size of the benzyl ring constituent with the ability to induce flowering.

The flower-inducing effect of SA was later reported in other plant species of other families. Aspirin in combination with sucrose was able to induce the opening of flower in *Oncidium* (Hew 1987). Nanda et al. (1976) have reported the formation of floral buds in *Impatiens balsamina*, a qualitatively short-day plant, by the application of 1 mg $l^{-1}$ SA applied to the apices of the plants. Moreover, it has been recently demonstrated that SA also regulates flowering time in non-stressed plants as SA deficient are late flowering. Interestingly, the regulation of flowering time by SA does not require the functioning of the flowering time genes (Martinez et al. 2003).

The exact mechanism lying behind SA-induced flowering is not yet been thoroughly resolved. According to Oota (1975), the chelating properties of the *o*-hydroxy group of SA might be the reason for induction of flowering. The fact that chelating agent EDTA can induce flowering in *Lemna* (Seth et al. 1970; Oota 1972) that is similar to the flower induction effect of SA (Pieterse and Muller 1977) supported the above hypothesis. However, the flowering induction ability of non-chelating phenolics including BA (Fujioka et al. 1985; Watanabe et al. 1981) suggests the existence of an alternate mechanism of induction process. Maximum flowering was recorded in *Lemna gibba* with the presence of SA in the medium (Ben-Tal and Cleland 1982). Based on this finding, the role of SA in flowering is suggested to be probably noninductive. The noninductive nature of SA has been further evidenced in *Lemna paucicostata* (Takimoto and Kaihara 1986). Short-term treatment of plants with SA had no effect on floral induction. Using [$^{14}$C] SA and [$^{14}$C] BA, Khurana and Cleland (1992) compared the effect of SA to that of BA in terms of interaction with cytokinin and whether the continuous presence of SA and BA is required for the occurrence of maximum flowering in *Lemna paucicostata* and revealed that the effect of these compounds can be of inductive nature, if treated at pH 4.0 followed by the treatment of 6-aminobenzylpurine (BAP) treatment. In spite of these effects on flowering, the nature of the flowering induction signal generated by salicylates remains to be elucidated.

## SA and Thermogenesis in Plants

Thermogenesis in plants is the phenomenon of heat production whereby the temperature increases beyond the normal level. The generation of heat in the inflorescences of *Arum* lilies during blooming was first of all described by Lamarck (1778). Thermogenesis is known to occur in the flowers and inflorescences of plants of the families *Anonaceae*, *Araceae*, *Aristolochiaceae*, *Cyclanthaceae*, *Nymphaeaceae*, and *Palmae* (Meeuse and Raskin 1988) and is associated with an increase in alternative respiratory (cyanide-insensitive) electron transport pathway (James and Beevers 1950; Meeuse 1985). The alternative pathway diverges from the cytochrome pathway after the ubiquinone pool (Bendal and Bonner 1971; Storey 1976), and therefore, the flow of electrons is not coupled to phosphorylation of ADP at the two sites of proton gradient formation (complexes III and IV) that are downstream of the ubiquinone pool (Moore and Bonner 1982). In this pathway, alternative oxidase (AOX) acts as the terminal oxidase catalyzing the transfer electrons to molecular oxygen. The energy of the electron flow through this pathway is not conserved as chemical energy, but is dissipated as heat (Storey and Bahr 1969). The inflorescences of these aroid plants develop the level of alternative pathway to such extent that they become thermogenic (Day et al. 1980). The heat so produced volatilizes amines and indoles and thus attracts insect pollinators (Smith and Meeuse 1966; Chen and Meeuse 1971). The hypothesis that calorigen, a water-soluble substance produced in male flowers and afterward transported to thermogenic tissues, is responsible for heat production as proposed by Van Herk (1937) came to an end after the purified calorigen from the male flowers of *Sauromatum guttatum* (voodoo lily) was identified as SA (Raskin et al. 1987). Application of 0.13 μg SA $g^{-1}$ f.w. to the immature appendix increased the endogenous temperature by 12°C. This elevation of temperature was very much similar to that produced by the crude extracts of calorigen and showed that SA is calorigen playing its role as endogenous regulator of heat production.

The role SA as an endogenous inducer of heat production has been reported in the inflorescences of voodoo lily (Raskin et al. 1989). Analytical assay for the quantification of SA demonstrated an increase in SA level in the thermogenic organs of voodoo lily. Endogenous SA increased in late afternoon proceeding the day of blooming and was almost 100-fold more as compared to the level 3 days before the blooming. This increase was associated with heat and odor production in the appendix (a thermogenic part of the inflorescences) in next morning. The temperature of the appendix reached to the normal level in late afternoon with concomitant decrease in SA content to the pre-blooming level, thus indicating the direct role of SA in triggering heat production in thermogenic tissues (Raskin et al. 1989). Further, a comparative study was carried out by the same workers to find the thermogenic activity of SA analogs. Among the 33 compounds tested for thermogenicity, only two of them, 2,6-dihydroxybenzoic acid (2,6-DHBA) and ASA, could induce heat production in thin slices of voodoo lily appendix. The thermogenicity by 2,6-DHBA was relatively more, and that of ASA was half as compared to SA. Structural modification at any other position made the compound thermogenically inactive.

Observation of three mitochondrial proteins of molecular masses 35, 36, and 37 kD correlating with the activity of AOX in voodoo lily appendix tissue (Elthon and Mclntosh 1987) gave an insight for further investigation of the molecular basis of SA-induced physiology of thermogenesis. A protein (42 kD) that acts as the precursor of all the three AOX proteins has been identified by Rhoads and McIntosh (1991). Application of SA increases alternative pathway activity and the levels of the 35-, 36-, and 37-kD AOX protein in voodoo lily appendix tissue sections (Elthon et al. 1989). Both SA and calorigen extract induced the accumulation of a 1.6-kb transcript (encodes the 42-kD precursor protein), leading to the accumulation of AOX with molecular weight of 38.9 kD (Rhoads and McIntosh 1992, 1993). The level of the transcript was relatively high in the day of blooming as compared to the transcript level of the preceding days. In the day of blooming, the high level of the transcript corresponded to that of the highest AOX protein when thermogenesis was at its peak. This further confirmed the chemical nature of calorigen as SA and also the involvement of SA in inducing alternative respiratory pathway. Increased rate in alternative pathway was also observed by incubating the cells of unicellular algae *Chlamydomonas reinhardtii* with SA (Goyal and Tolbert 1989). For the first time, SA-induced alternate pathway leading to heat production in non-thermogenic vegetative plant tissues was reported by Kapulnik et al. (1992). They have detected a significant increase in heat evolution in tobacco cell-suspension cultures incubated with 2–20 μM SA. Though, in tobacco cells, 2,6-DHBA stimulated the alternative pathway similar to that of SA, the analog 4-hydroxybenzoic acid that was not able to induce thermogenesis in thermogenic plants induced alternative pathway in tobacco cells in contrast to thermogenic plants.

## SA in Membrane Function and Ion Uptake

Literature survey indicates that only few studies have been devoted to the involvement of SA in the regulation of membrane function and ion uptake. SA-mediated inhibition of potassium ion absorption with respect to varied pH and concentration has been studied in excised oat roots (Harper and Balke 1981). The amount of inhibition was more at acidic pH indicating that SA is more active in protonated form as compared to its dissociated form. Role of SA in membrane function of animal system has also been studied. In molluscan neuron, it could increase membrane potential by the increase of potassium conductance and decrease of chloride conductance (Levitan and Barker 1972). SA has been reported to collapse the electrochemical potential across mitochondrial membrane and the ATP-dependent $H^+$ gradient of vesicles enriched with tonoplasts (Macri et al. 1986). The uncoupling of oxidative phosphorylation in the presence of diflunisal (a derivative of SA) that acts as an ionophore has been recorded by Gutknecht (1990). Katz et al. (2002) have observed the SA-mediated early $K^+$/pH response in parsley cells with subsequent coumarin (phytoalexin) secretion. The inhibition of sugar and amino acid uptake under the influence of SA in a dose-dependent manner is known in leaf discs of sugar beet during aging period (Bourbouloux et al. 1998) and suggested that SA may affect the uptake of sugars and amino acids by indirect inhibition of the plasma membrane $H^+$-ATPase. Though the inhibition of phosphate uptake (Glass 1973) and reduction in potassium ion absorption (Glass 1974) in plant system were observed, the measurement of radiolabeled $PO_4^{3-}$ uptake in the leaves of wheat plants hydroponically grown in the presence of SA revealed that the uptake is stimulated at lower concentration (50 μM) of SA and is drastically inhibited at higher concentrations (Sahu et al. 2010). In this study, the autoradiogram of SA-treated wheat plants showed an increase in intensity of radioactivity of the leaves both in light and in dark with 50 μM SA that further decreased at higher concentration of SA. Yang et al. (2003) have assumed that activation of SA-mediated components might be involved in the modification of anion channel, which promotes the Al-induced citrate release in *Cassia tora* L.

## SA and Stomata Movement

There have been a number of reports which suggest that SA could inhibit stomatal opening (Larque-Saavedra 1978; Bhatia et al. 1986; Manthe et al. 1992; Lee 1995). Application of 1 μM SA has been shown to inhibit stomatal opening by 67% at pH 5.0 in detached epidermis of *Commelina* (Manthe et al. 1992). Contrary to this, SA-mediated opening of stomatal aperture has been documented by Rai et al. (1986) at pH 7.0. Lee (1998) had explained two possible mechanisms of SA-induced stomatal closing. One mode is SA-dependent catalase (CAT) inhibition, which accumulates $H_2O_2$ in the cytoplasm of guard cells. $H_2O_2$ oxidizes the plasma membrane and increases the membrane permeability of $K^+$, leading to its mass efflux. This causes the loss of turgor pressure and thus the closing of stomata. The second mode is the combined effect of $K^+$ efflux and the elevated concentration of intercellular $CO_2$ concentration. Involvement of extracellular superoxide anion ($O_2^{\cdot -}$) in SA-induced stomatal closure has also been proposed in *Vicia faba* (Mori et al. 2001).

## SA and Photosynthesis

Photosynthesis is one of the major physiological processes in plants, the rate of which determines the yield of the plants. Data with regard to the effect of SA on growth and photosynthesis have shown that long-term treatment with SA decreases the rate of photosynthesis and activity of ribulose 1,5-bisphosphate carboxylase/oxygenase (RuBisCo) in barley plants (Pancheva et al. 1996). The effect was concentration dependent with a maximum of 30% inhibition of enzymatic activity at 1 mM SA. As compared to control plants, an increase in phosphoenolpyruvate carboxylase (PEPcase) activity was recorded in SA-treated plants. Short-term application of SA caused no significant change in rate of photosynthesis. Measure of intracellular $CO_2$ concentration in control and SA-treated plants implied the reduction in photosynthesis as non-stomatal. The decrease in total soluble protein in general and RuBisCo synthesis in particular has been explained as the possible reason for the inhibition of photosynthesis and RuBisCo activity. It has also been assumed that similar to other stress factors, SA application retards chloroplast photosynthetic activity as a result of effects on the thylakoid membranes and photochemical reactions connected with them, thus playing an indirect role in regulation of RuBisCo activity (Pancheva and Popova 1998). To substantiate these assumptions, Sahu et al. (2002) have investigated the alteration in thylakoid photofunction and photophosphorylation activity of wheat plants grown with different concentrations of SA in the nutrient solution. Treatment of plants with SA for 7 days affected its growth, leaf photochemical activities including phosphorylation efficiency, and thylakoid organization. A low concentration of the compound (50 μM) has a stimulatory effect on both PS-I and PS-II catalyzed electron transport reaction including photophosphorylation activity (Fig. 9.3). The stimulation of electron transport was also found to be accompanied by marginal elevation of cytochrome $b_{563}$ and $P_{700}$. The lowering of cytochrome $f_{554}$ (Fig. 9.4) level at high concentration of SA (0.5 and 1 mM) in this study may be viewed as rate-limiting step contributing towards the attenuation of PS-I catalyzed electron flow. These observations lend support to the suggestion that long-term treatment of wheat seedlings with a relatively high concentration of SA affects the thylakoid composition so as to bring about impairment of ATP synthesis mostly due to membrane disorganization, thus inhibiting the selective translocation of $H^+$ into the lumen. The acid also has a specific role to play in

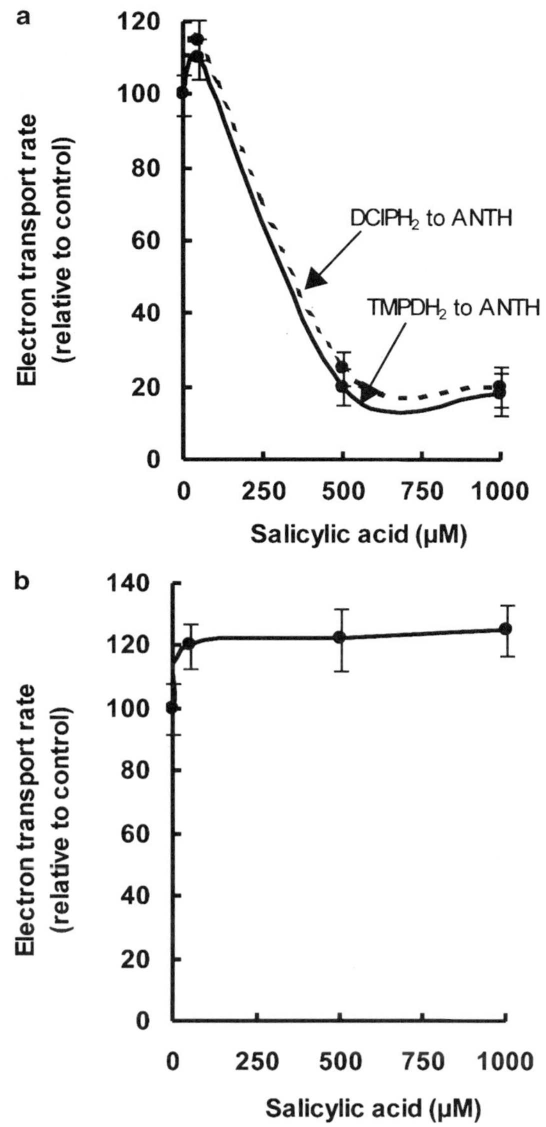


**Fig. 9.3** (**a**) PS-I electron transport activities of thylakoids isolated from control and SA-cocultivated wheat plants with reduced DCIP ($DCIPH_2$) and TMPD ($TMPDH_2$) as electron donors. (**b**) PS-II electron transport activity of thylakoids isolated from control and SA-treated wheat plants. 100% activity refers to 195 ± 20 μmol $O_2$-evolved mg $chl^{-1}$ $h^{-1}$ (Sahu et al. 2002)

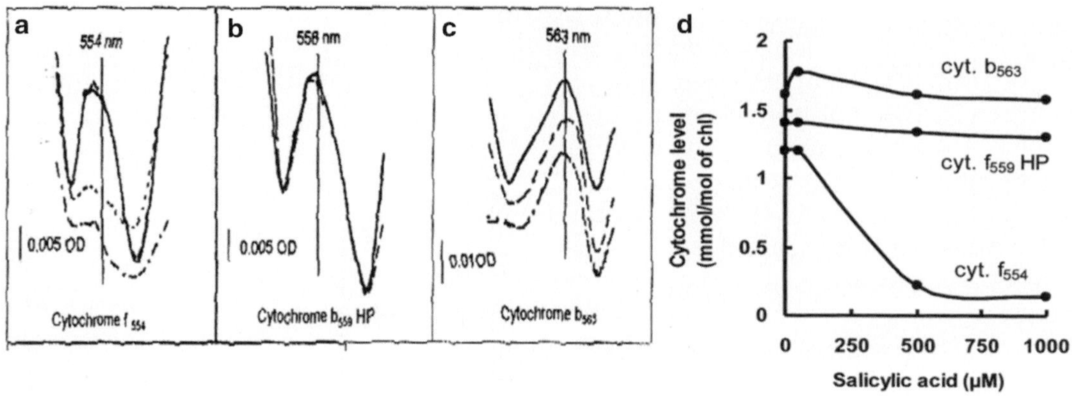


**Fig. 9.4** (**a–c**) Spectral resolution of different thylakoid cytochromes. The individual cytochrome traces from different SA treatments are shown as (-) control, (–) 50 μM, (– – –) 500 μM, and (– - - -) 1,000 μM. (**d**) Quantitative determination of cytochromes $b_{559}$ HP, $b_{563}$, and $f_{554}$ in thylakoids isolated from control SA-treated plants (Sahu et al. 2002)

reducing the level of cytochrome $f_{554}$. Such inhibitions together with SA-mediated impairment of carboxylating enzymes reported by Pancheva et al. (1996, 1998) could lead to severe down-regulation of photosynthetic activity.

Application of low concentration of SA (10 μM) to mustard seedlings stimulates photosynthetic net $CO_2$ assimilation accompanied with increased carboxylation efficiency, chlorophyll content, and carbonic anhydrase and nitrate reductase activity (Fariduddin et al. 2003). The effect of exogenous SA on the pigment content of plants is dose dependent. Foliar spray of low concentration of SA ($10^{-5}$ M) increases the pigment content in *Brassica napus* (Ghai et al. 2002) and *Brassica juncea* (Fariduddin et al. 2003). Higher concentration of SA reduced the level of the pigments. A reduction in chlorophyll quantity has been noticed in barley (Pancheva et al. 1996) and wheat seedlings grown with SA (Sahu et al. 2002). The leaf and chloroplast ultrastructure is also known to be affected by SA (Uzunova and Popova 2000).

## SA and Seed Germination

Germination of seed is an important physiological aspect in the life of a plant. The process is regulated by various external and internal factors including phytohormones. SA can either inhibit or induce the processes of seed germination, depending on the concentration of the compound employed. Inhibition of seed germination has been reported in many cases. SA concentrations greater than 1 mM either delay or inhibit seed germination in *Arabidopsis* (Rajjou et al. 2006). SA concentration >0.250 mM inhibit seed germination in barley (Xie et al. 2007), whereas in maize, complete inhibition of germination was noticed by SA at doses ranging from 3 to 5 mM (Guan and Scandiolios 1995). SA-generated oxidative stress may be attributed to the negative effect on seed germination as it has been observed that $H_2O_2$ level increases due to increase in superoxide dismutase (SOD) and inactivation of CAT and ascorbate peroxidase (APX) in *Arabidopsis* (Rao et al. 1997).

Though SA is known to negatively affect seed germination usually at high doses in unstressed plants, exogenous application of the compound at low doses improves the process under various abiotic stress conditions (Alonso-Ramírez et al. 2009). Significant increase in germination percentage in *Arabidopsis* was observed under NaCl stress in the presence of 0.05–0.5 mM SA. Exogenous SA could also remove the inhibitory effect of heat and paraquat-generated oxidative stress. This is evidenced in *Arabidopsis* mutant *sid2* (defective ICS gene), a delayed germination phenotype containing low level of SA. In *Nah*G transgenic plants, seed germination was not affected by salt stress (Borsani et al. 2001; Lee et al. 2010). This might happen due to the antioxidant effect of catechol, which is produced from SA through the expression of transgene salicylate hydroxylase. Rajjou et al. (2006) have reported the induction of two SOD isoforms during germination of *Arabidopsis* seeds that thereby reduces the oxidative stress.

Salicylic acid interacts with both abscisic acid (ABA) and gibberellins (GAs) during seed germination. Synthesis of SA is stimulated in *Arabidopsis* seeds by imbibitions with $GA_3$ (50 μM). Overexpression of *$FcGASA_4$*, a GA-stimulated gene in *Arabidopsis*, induced an increase in SA level. The relation between SA and GA with respect to seed germination is both synergistic and antagonistic. Exogenous application of 50 μM SA partially induces seed germination in *gal-3*, a GA-deficient mutant, whereas exogenous $GA_3$ slightly improved the germination of *sid2* mutants under NaCl stress (Alonso-Ramírez et al. 2009). Suppression of GA-induced *Amy32b* expression (codes for alpha amylase) caused the inhibition of barley seed germination by SA (Xie et al. 2007). The interaction between SA and ABA has also been evidenced. Increased synthesis of dehydrins, heat shock proteins, and late embryogenesis abundant (LEA) proteins that are ABA regulated was observed in *Arabidopsis* in presence of 0.5 mM SA (Rajjou et al. 2006).

## SA and Environmental Stress

Plants grow under diverse environmental conditions. Extreme and also mild changes in the environment parameters result in stress of various forms that become unfavorable for plant growth and productivity, development, and hence the survival. Depending on its nature, stress may be biotic or abiotic. Biotic stress includes pathogen and insect attacks, wounding, and plant competition, whereas abiotic stresses include high and low temperature, water in excess or deficit, low or excess light, excess salt, toxic metals, environmental pollutants like sulfur dioxide, ozone, and UV light. Plants being sessile, in order to sustain and survive, respond to these stresses by activating elaborate defense mechanisms. To cope with the stress conditions, plants may undergo whole-sum physiological changes that reduce the severity of the stress. This could often be linked with changes in gene expression marked by synthesis of new protein(s) or some other metabolites. Some of the protective mechanisms adopted by plants against insects and pathogen attacks involve the biosynthesis of various secondary metabolites, changes in composition and physical properties of cell walls, and formation of necrotic lesions at the site of infection. Synthesis of various heat shock proteins (HSPs) in response to high temperature conferring thermotolerance has been well documented (Lindquist 1986; Nagao et al. 1986).

Plant hormones like ABA, auxins, GAs, cytokinin, and ethylene are influenced in response to these environmental stresses and thus provide various forms of adaptations (Swamy and Smith 1999). Accumulation of ABA in response to NaCl-imposed osmotic stress has also been observed in various plants (Boussiba et al. 1975; Swamy and Smith 1999). The role of ABA as a messenger molecule in the signal transduction pathway in response to various stresses has been reported (Jeffrey and Giraudt 1998). In addition, some other plant hormones such as brassinosteroids, jasmonic acid, SA, or salicylates have been identified as new faces and play major roles in plant response to both biotic and abiotic stresses. Of these, the role of SA in recent years has been widely investigated in both biotic and abiotic stresses.

## SA and Biotic Stress in Plants

Even though plants lack an immune system like vertebrate animals, they are surprisingly resistant to diseases caused by the fungi, bacteria, viruses, and nematodes that are ever present in the environment. Some plants protect against pathogenic organisms by the formation of a necrotic lesion due to the death of cells at the site of infection. This rapid, localized cell death due to pathogen attack is called the hypersensitive response (HR). When a plant survives the infection of a pathogen at one site, it often develops increased resistance to subsequent attacks at sites throughout the plant and thus protects a wide range of pathogen species. This phenomenon of acquired resistance may be local acquired resistance (LAR) when developed in the vicinity of the HR lesions (Ross 1961) and SAR detected in uninoculated, pathogen-free parts of plants (Ryals et al. 1996). SAR develops after a period of days following the initial infection. Both HR and SAR are known to be associated with synthesis of several families of pathogenesis-related (PR) proteins (Bol and van Kan 1988; Carr and Klessig 1989).

The role of SA as a mediator in plant defense against pathogen attack has been well documented in a variety of plants. Application of 0.01% SA and 0.02% aspirin solution to tobacco leaves enhanced the resistance to subsequent infection of tobacco mosaic virus (TMV) with less number of necrotic lesions (White 1979). In addition to enhancing resistance to TMV in tobacco, SA-mediated induction of SAR has also been observed against many other viral, bacterial, and fungal pathogens (Weete 1992). Application of SA also provides resistance of blue mold (*Peronospora tabacina*) infection to tobacco (Ye et al. 1989) and *Colletotrichum lagenarium* infection to cucumber (Mills and Woods 1984). *Arabidopsis* mutants expressing a *NahG* transgene or treated with a PAL inhibitor showed enhanced symptoms of *Botrytis cinerea,* a necrotic fungal pathogen, and suggested that SA synthesized via PAL, and not via ICS, mediates lesion development (Ferrari et al. 2003).

The effect of SA on pathogen resistance is probably due to the expression of pathogen-related genes (PR genes) encoding PR proteins. About 13 sets of PR gene induction have been identified in response to SA in various plant species. The induction of PR proteins parallel to pathogen resistance with exogenous application of aspirin was investigated by Antoniw and White (1980) in different cultivars of tobacco. An increase in mRNA level of both basic and acidic isozymes of β-1,3-glucanase, one of the PR-2 group proteins, was reported after TMV inoculation or SA treatment of tobacco (Linthorst et al. 1990). In cucumber, extracellular endochitinase, one of the PR-3 group proteins, was induced by SA at the level of RNA accumulation (Metraux et al. 1989). Ward et al. (1991) have reported the induction of pathogenesis-related genes in tobacco with the application of SA, which are also associated with both HR and SAR of TMV-infected plants. Interestingly, the SA analog 2,6-DHBA that could stimulate the generation of heat in voodoo lily also induced PR proteins, leading to virus resistance (Van Loon 1983).

Accumulation of significant amount of SA in plants in response to pathogen attack has also been observed. Malamy et al. (1990) have noticed a several hundredfold increase in the endogenous level of SA after TMV inoculation of resistant, but not susceptible, tobacco cultivars with simultaneous induction of PR genes. Rise in SA level in the phloem of tobacco necrosis virus (or *Colletotrichum lagenarium*)-infected leaves before development of SAR in distal tissues has been noticed in cucumber (Metraux et al. 1990). Further evidence for the accumulation and involvement of SA in plant defense was contributed by Gaffney et al. (1993). These authors have shown that *NahG* gene from the bacterium *Pseudomonas putida* encoding salicylate hydroxylase that converts SA into catechol, when expressed in transgenic tobacco, prevents the accumulation of SA, alters PR gene expression, and suppressed the establishment of SAR with the development of comparatively larger viral lesions than those produced by wild plants. Several mutants of *Arabidopsis* have been identified that are both defective in SA signal transduction and susceptible to pathogen infection (Cao et al. 1994, 1997; Delaney et al. 1995; Shah et al. 1997). Though the accumulation of SA causing PR protein synthesis in response to pathogen infection and exogenous application of SA leading to synthesis of PR proteins and thereby prevention of pathogen infection have been widely studied in plants like tobacco, cucumber, and *Arabidopsis*, its involvement in plant defense activation in healthy potato plants containing comparatively high basal level of SA is not able to activate the defense mechanism against *Phytophthora infestans*, the causative organism of late blight in potato (Coquoz et al. 1995). This view was experimentally demonstrated in transgenic *NahG* potato plants that were not significantly susceptible to primary infection by *P. infestans*, suggesting that high basal level of SA in healthy potato plants does not constitutively activate defense mechanism against the pathogen *P. infestans* (Yu et al. 1997). These observations indicated that there exists a difference in SA responsiveness among different plant species and can be correlated with endogenous level of SA (Enyedi et al. 1992; Vernooij et al. 1994). Furthermore, in certain hybrids of tobacco (Yalpani et al. 1993b) and *Arabidopsis* mutants (Bowling et al. 1994; Dietrich et al. 1994; Greenberg et al. 1994), stunted growth and the development of necrotic lesions are associated with alleviated SA levels.

The critical role of SA as a signal molecule in plant defense after pathogen attack is accomplished in a complex network (Shah 2003). Experimental analyses based on advanced techniques in the field of molecular biology have identified several components of SA signaling pathway. These include (1) $H_2O_2$-degrading enzymes CAT and APX, (2) a high-affinity SA-binding protein (SABP), (3) a SA-inducible protein kinase (SIPK), (4) non-expressor of PR1 (NPR1), and (5) members of TGA/OBF family of bZIP transcription factors. These bZIP factors physically interact with NPR1 and bind the SA-responsive element in the promoters of several defense genes such as PR1 gene (Shah and Klessig 1999).

## SA and Abiotic Stress in Plants

Though SA plays an important role as a signal molecule in plant disease resistance, thermogenesis, flowering, and other physiological process, in the last decade much attention has been given to find out its role in preventing the deleterious effect of various stresses of abiotic nature such as heat, extreme temperature, ozone, UV radiation, heavy metals, salinity, and drought. SA responsiveness to these stress factors are discussed below.

### SA and Temperature Stress

Heat stress due to increased temperature is an agricultural problem in many areas in the world. Transitory or constantly high temperatures cause an array of morpho-anatomical, physiological, and biochemical changes in plants, affecting plant growth and development. High temperature may adversely affect photosynthesis, respiration, water relations, membrane stability and also modulate the level of hormones and primary and secondary metabolites (Wahid et al. 2007). Hormones are known to play a significant role in plant response to abiotic stress including temperature stress. It is well established that the stress hormones ABA and ethylene act as signal molecules regulating various physiological process, thus imparting temperature stress tolerance to plants. Among the other hormones, SA has been suggested to be involved in heat-stress responses elicited by plants. Foliar spray of mustard (*Sinapis alba* L.) seedlings with 10 500 μM SA significantly increased their tolerance to a subsequent heat shock (Dat et al. 1998b). It has been proposed that high temperature increases endogenous SA, whereas SA treatment and heat acclimation induced changes in ascorbate and glutathione pool and antioxidant enzymes, suggesting a possible role of endogenous SA and antioxidants in heat tolerance of mustard seedlings (Dat et al. 1998a). The induction of thermotolerance in potato microplants by ASA was also reported by Lopez-Delago et al. (1998). The effective concentration of ASA for induction of thermotolerance was 1–10 μM, similar to that of SA that induced thermogenesis in *Arum* lilies (Raskin et al. 1987). SA may involve in thermotolerance of plants by reducing oxidative damage (Larkindale and Knight 2002; Larkindale and Huang 2004). They have investigated that pretreatment of plants with SA did not affect the peroxidase (POX) activity, whereas the CAT activity declined significantly compared to control plants. In contrast to the above report, He et al. (2005) have observed increased activity of CAT and SOD in heat-stressed *Poa pratensis* after treatment with SA. In another study, it has been reported that the membrane damage due to heat stress in *Cicer arietinum* was significantly diminished by the application of SA (Chakraborty and Tongden 2005). SA application, in this plant, enhanced the total protein and proline content along with the induction of POX and APX and decline of CAT activity. Long-term thermotolerance in young grape plants can be induced by SA, in which both $Ca^{2+}$ homeostasis and antioxidant systems are assumed to be involved (Wang and Li 2006a, b).

Sulfosalicylic acid (SSA), a SA derivative, removed $H_2O_2$ in cucumber seedlings and decreased heat stress. Increase in CAT activity plays a key role in removing $H_2O_2$. The enzymatic antioxidants GPX, APX, and glutathione reductase (GR) though showed higher activities in SA treatments under heat stress, they were not the key enzymes in removing $H_2O_2$ (Shi et al. 2006). Foliar accumulation of MeSA gives thermotolerance to holm oak (*Quercus ilex*) which might partly be attributed to enhanced xanthophyll de-epoxidation and increases in ascorbate and α-tocopherol (Llusia et al. 2005). Pretreatment of grape leaves with SA alleviates the decrease of net photosynthesis under heat stress, maintaining a higher RuBisCo activation state and greater PS-II efficiency. These effects of SA may partly be related to expression of higher levels of HSP21 in chloroplast (Wang and Li 2007; Wang et al. 2010).

Experimental data with respect to involvement of SA in plant response to low temperature are also available. SA treatment is known to protect the plants from chilling injury in maize (Janda et al. 1999). Addition of 0.5 mM SA to the hydroponic growth solutions of young maize plants protected the plants from subsequent low-temperature stress. As compared to the control plants, an increase in chlorophyll fluorescence parameters ($F_m/F_v$) and decrease in electrolytic leakage were observed in plants exposed to low-temperature stress that were pretreated with SA. In addition to the above changed parameters, a sharp decline in CAT activity and slight increase in GR and POX activity were observed in the treated plants. Other phenolics such as benzaldehyde, ASA, and coumaric acid are also known to induce chilling tolerance in maize plants (Janda et al. 2000; Horvath et al. 2002). Inhibition of the isoform CAT-1 has been implicated as the possible mechanism for SA-induced chilling tolerance in maize (Horvath et al. 2002). Following exposure of germinating seeds of maize, rice, and cucumber to aqueous solution of 0.5 mM SA for 24 h before chilling at 2.5°C for 14 days, electrolytic leakage from maize and rice leaves and cucumber hypocotyls reduced (Kang and Saltveit 2002). The percentage germination of SA pretreated seeds of carrot (Rajasekaran et al. 2002) and *Capsicum annum* (Korkmaz 2005) has been found to increase at low temperature as compared to the untreated control seeds. The ameliorating effect of SA to low-temperature stress was also investigated in bean (Senaratna et al. 2000), wheat (Tasgin et al. 2003), and rice (Szalai et al. 2002).

## SA and Ozone-UV Radiation Stress

Tropospheric ozone ($O_3$) is a major component of photochemical air pollution responsible for causing significant change in both natural and cultivated plants (Pell et al. 1997). Many studies have demonstrated the damaging effects of $O_3$ on several crop and forest species. Exposure of plants to $O_3$ causes photosynthetic inhibition (Reich 1987), accelerates senescence of leaves (Wardlaw 1968; Black et al. 2000; Saitanis and Karandinos 2002), and reduces the growth and yield of many plant species (Cooley and Manning 1987). Plants develop several physiological, molecular, and biochemical protective mechanisms to deal with the damaging effects induced by $O_3$. The molecular basis of $O_3$-induced responses in plants, based on the studies in model plant system, *Arabidopsis thaliana*, demonstrated that $O_3$-induced responses are caused in part by the activation of a SA-dependent signaling pathway (Sharma and Davis 1997). Accumulation of SA in response to $O_3$ stress has been reported in few cases. Increase in endogenous level of SA was noticed in plants treated with $O_3$ or UV light (Yalpani et al. 1994; Rao and Davis 1999). Their findings have revealed that $O_3$-mediated induction of several defense genes is correlated with accumulation of SA. Experimentally they have proved that *NahG* plants are $O_3$ sensitive. Sharma et al. (1996) have found $O_3$-induced rapid accumulation of both free and glycosylated form of SA in *Arabidopsis thaliana*. Low level of SA is required to maintain the redox state of glutathione and activate the antioxidant defense responses required to minimize $O_3$-induced oxidative stress. In contrast, a high level of SA accumulation causes an oxidative burst leading to cell death, resulting in $O_3$ sensitivity (Rao and Davis 1999). The mediation of SA in UV radiation is least studied. Accumulation of SA in tobacco plants exposed to UV radiation has been reported which is considered to be due to the higher activity of BA hydroxylase, an enzyme of SA biosynthetic pathway (Yalpani et al. 1994).

The mechanism of $O_3$ stress tolerance in plants induced by SA has been studied in few cases. Analyses of the responses of molecular markers of specific signaling pathways indicated a relationship between SA and ethylene signaling pathways and $O_3$ sensitivity. Ethylene-overproducing $O_3$-sensitive mutants accumulated higher level of SA and ethylene prior to lesion formation in response to acute $O_3$ exposure. *NahG* and *NPR1* plants that do not produce SA failed to produce ethylene in response to $O_3$ and other stress factors, suggesting that SA is required for stress-induced ethylene production (Rao et al. 2002). SA-deficient *NahG* plants reduced the expression FAD7 gene which is responsible for generating oxidative responses that consequently activates various defense-related genes induced by $O_3$ exposure (Yaeno et al. 2006).

## SA and Metal Stress in Plants

Plants exposed to heavy metals generate oxidative stress leading to cellular damage. Exogenous application of SA is known to reduce the toxic effect of certain heavy metals in plants. Alleviation of inhibitory effects of heavy metals on germination of seeds of rice was studied by Mishra and Choudhuri (1997). Lead (Pb)- and mercury (Hg)-induced membrane damage mediated by lipoxygenase was found to be reduced in presence of SA (Mishra and Choudhuri 1999). Role of SA in mitigation of cadmium (Cd)-induced stress has been investigated in various crop plants. The DF/$F_m$, an index for PS-II quantum efficiency, has revealed that in maize, there was hardly any change in PS-II efficiency when Cd and SA were applied simultaneously, whereas in treatment with 0.5 mM Cd alone, the decrease in PS-II quantum efficiency was well pronounced (Pal et al. 2002). Metwally et al. (2003) have observed the accumulation of free SA in *Hordeum vulgare* following Cd exposure. Presoaking of the seeds with SA or hydroponic growth of plants with 0.5 mM of SA protected the seedlings from Cd toxicity through the suppression of Cd-induced upregulation of activities of antioxidant enzymes such as CAT and APX. A decline in quantity of thiobarbituric acid-reactive substances and $O^{2-}$ and $H_2O_2$ generated under Cd stress was found in rice treated with SA and suggested the protective role of SA to oxidative stress induced by Cd (Panda and Patra 2007). Simultaneous addition of Cd and SA to seedlings of soybean lessened the toxic effects of Cd such as leaf desiccation. The ionic homeostasis disturbed by Cd exposure was found maintained by SA in this plant (Drazic and Mihailovic 2005). Protection of photosynthesis and oxidative damage in Cd-stressed condition has been suggested in maize plants (Krantev et al. 2008). The authors have observed that presoaking maize seeds with SA for 6 h alleviated the inhibitory effect of RuBisCo and PEPcase. Pretreatment of SA also caused a decrease in lipid peroxidation and the electrolytic leakage induced by Cd. The activity of APX and SOD was enhanced with a significant reduction in CAT activity as compared to Cd-treated plants. Similar changes in biochemical parameters mediated by SA in protecting the plants from Cd toxicity were found in pea plants (Popova et al. 2009). Endogenous SA potentiates Cd-induced oxidative stress in *Arabidopsis thaliana*. The seedlings of *A. thaliana* exposed to Cd, coupled to SA biosynthesis that was not seen in *NahG* mutants. The protective role of SA to stress though extensively known in several plants, the endogenous SA in *Arabidopsis* may serve as a signal molecule for generation and further amplification of Cd-induced oxidative stress (Zawoznik et al. 2007).

Though the role of SA in Cd stress is well studied, its role in alleviation of toxicity of other metals has been reported in few cases. It was found that exogenous application of SA increases the Al-induced citrate efflux from roots of *Cassia tora* and is associated with increased Al tolerance. The accumulation of endogenous SA in *C. tora* root tips in response to Al suggested that Al-induced stress could be mediated by SA-involved signal transduction pathway (Yang et al. 2003). Freeman et al. (2005) have reported SA as strong predictor of Ni hyperaccumulation in six species of *Thlaspi* and signals glutathione-mediated Ni tolerance. Role of SA in mitigation of toxicity of other metals and the molecular basis of metal tolerance mediated by SA warrants further investigation.

## SA and Water Stress in Plants

Effect of drought that arises due to water deficit is one of the important limiting factors of crop production. Drought stress triggers various changes in plants, leading to altered gene expression and plant metabolism and growth. One of the early responses to drought is the reduction in leaf expansion, which is usually followed by stomatal closure and reductions in photosynthesis (Munne-Bosch et al. 2007). Plants adapt drought avoidance mechanisms through the involvement of several signal molecules. As SA is known to participate in regulating plant responses to abiotic stress, it has fascinated many workers to find the role of SA in mitigation of water stress. Drought tolerance in wheat plants was observed when the seeds were presoaked with ASA (Hamada 1998). Low concentration of exogenous SA protected the damaging effects of drought stress in tomato and bean (Senaratna et al. 2000). Singh and Usha (2003) reported that wheat seedlings under water stress when treated with SA possessed higher moisture content, dry matter accumulation, carboxylase activity of RuBisCo, SOD, and chlorophyll content compared to untreated control plants. Application of SA is also known to alleviate the damaging effects of water stress in barley with increased content of ABA in the leaves that might have contributed the water stress tolerance (Bandurska and Stroinski 2005). Munne-Bosch and Penuelas (2003) have evaluated the relationship between mechanisms of photo- and antioxidative protection and SA accumulation in stressed *Phillyrea angustifolia*. Endogenous SA increased to five-fold in water-stressed plants and was associated with loss of chlorophyll and carotenoid content. However, the water-stressed *NahG* transgenic *Arabidopsis* plants suppressed the water stress-induced loss of chlorophyll and carotenoid observed in wild plants (Munne-Bosch et al. 2007). Growth of tomato plants (*Lycopersicon esculentum*) with low concentration of SA under water stress increased photosynthetic parameters, leaf water potential, chlorophyll content, membrane stability index, and enzymatic activities of nitrate reductase, carbonic anhydrase, CAT, POX, and SOD as compared to plants grown without SA (Hayat et al. 2008). SA-induced physiological and biochemical changes in lemongrass under water stress have been studied. Foliar application of SA ($10^{-5}$ M) improved the growth parameters that were significantly reduced in stress-affected plants. Also, SA application alleviated the level of total chlorophyll, carotenoid, and the activities of nitrate reductase and carbonic anhydrase and reduced the adverse effects of electrolytic damage due to water stress (Idrees et al. 2010). Very recently, Mardani et al. (2012) have found that 1 mM SA could improve the water stress tolerance in cucumber seedlings. As an exception to the above studies, Nemeth et al. (2002) have observed that the concentration of SA (0.5 mM) inducing chilling tolerance in maize and other plants caused an increased sensitivity to drought.

## SA and Salt Stress in Plants

Detailed survey of literature indicates the role of SA in salinity stress in plants. Grains of wheat presoaked with SA solution enhanced the salinity stress in wheat and tomato plant seedlings (Hamada and Al-Hakimi 2001; Tari et al. 2002; Szepesi et al. 2005). Shakirova et al. (2003) have demonstrated that SA treatment to wheat seedlings reduced the damaging actions of salinity and water deficit on seedling growth. In this study, SA treatment caused an increase in concentration of proline, an effective osmoprotectant (Kuznetsov and Shevyakova 1999; Pesci 1987), and was preceded by a transitory accumulation of ABA that is known to induce the synthesis of a range of stress proteins (Leung and Giraudat 1998; Rock 2000) and PR proteins (Moons et al. 1997). This indicated the involvement of SA induced increase of endogenous ABA, contributing to preparedness of the plants to salinity stress. This was further substantiated that SA treatment led to prolonged ABA accumulation and to enhanced activity of aldehyde oxidase, an enzyme responsible for the conversion of ABA-aldehyde to ABA, both in root and leaf tissues of *Solanum lycopersicum* L. cv. Rio Fuego (Szepesi et al. 2005). Yusuf et al. (2008) have noticed that 30-day-old NaCl-stressed mustard seedlings when sprayed with 10 μM SA improved the growth, photosynthetic parameters, and activities of nitrate reductase and carbonic anhydrase. The antioxidant enzymes (CAT, POX, and SOD) and proline contents were enhanced in response to NaCl stress that further increased with SA treatment. Recently Karlidag et al. (2009) have reported the mediation of SA in ameliorating the adverse effects of SA in strawberry, a NaCl salinity-sensitive plant species. In contrast, SA potentiates the generation of reactive oxygen species (ROS) in photosynthetic tissues of *Arabidopsis* during salt and osmotic stress and thus plays a role in the oxidative damage generated by salt stress (Borsani et al. 2001).

## Mechanism of SA Action in Plants

The foregoing discussions claim that SA acts as a signal molecule in various plant physiological processes including biotic and abiotic stress. The mode of action of SA is process specific and varies from plant to plant. The acting of SA in response to plant stress signaling is contradictory. The discovery of its targets and the understanding of its molecular modes of action in physiological processes could help in the dissection of the complex SA signaling network, confirming its important role in both plant health and disease (Vicente and Plasencia 2011).

## Interaction Between SA and $H_2O_2$

Existence of complex interaction between SA and $H_2O_2$ in signal mechanism during pathogenesis (Chen et al. 1993; Conrath et al. 1995; Kauss and Jeblick 1995, 1996; León et al. 1995) and abiotic stress (Foyer et al. 1997) has been identified. Salicylates can increase the level of $H_2O_2$ in plant tissues (Dat et al. 1998b; Lopez-Delgado et al. 1998). In *Arabidopsis* plants treated with SA (1–5 mM), the $H_2O_2$ level increases up to threefold as a result of increased activities of Cu and Zn-SOD and inactivation of the $H_2O_2$-degrading enzymes CAT and APX (Rao et al. 1997; Durner and Klessig 1995, 1996). This SA-mediated elevation of $H_2O_2$ could be presumed as a secondary messenger in the signaling pathway of stress adaptation. On the other hand, high concentrations of $H_2O_2$ have been shown to induce SA biosynthesis and accumulation (León et al. 1995; Chamnongpol et al. 1998). Increased SA level generates more $H_2O_2$, and thus, an interaction between $H_2O_2$ and SA results in high free-radical concentrations that alter the cellular redox state, resulting in the activation of various defense-related genes (Hammond-Kosack and Jones 1996) including PR gene expression (Chen et al. 1993). Thus, the acting of $H_2O_2$ in the up- or downstream of SA-mediated stress response is controversial and seems to be stress specific.

## Mitogen-Activated Protein Kinases in SA-Mediated Stress Responses

Mitogen-activated protein kinases (MAPKs) are a specific class of serine/threonine protein kinases that play a central role in the transduction of various extra- and intracellular signals and are conserved throughout eukaryotes. MAPKs generally function via a cascade of networks, where MAPK is phosphorylated and activated by MAPK kinase (MAPKK), which itself is activated by MAPKK kinase (MAPKKK). Evidences suggest that MAPKs play important signaling roles in plants (Jonak et al. 1994; Hirt 1997). Signaling through MAP kinase cascade leads to cellular responses including cell division, differentiation, as well as response to various stresses. In plants, MAPKs are represented by multigene families and are organized into a complex network for efficient transmission of specific stimuli (Mishra et al. 2006).

SA-induced activation of a 48-kD kinase known as p48 SIP kinase that uses myelin basic protein (MBP) as a substrate has been reported in suspension cells of tobacco (Zhang and Klessig 1997). As SIP kinase is activated by SA that acts as a signal molecule in defense responses, it was hypothesized that SA kinase is also involved in the activation of these responses. A 48-kD kinase that uses MBP as a substrate was also activated by both SA treatment and TMV infection. This SA-/TMV-activated kinases and the p48 SIP kinase from suspension cells could be considered as same (Zhang and Klessig 1997). Moreover, MAPK cascades are also important mediators of the interplay between SA, other phytohormones, and ROS signaling in cell growth regulation (Vicente and Plasencia 2011).

## Future Prospectives

SA acts as a plant hormone regulating various processes in plants including growth and development. It acts in coordination with other phytohormones in plants. The physiological processes like flowering, thermogenesis, ion uptake, and photosynthesis are induced by exogenous application of SA. However, the applied concentration of SA is both plant and process specific. SA when applied at low concentrations usually induces photosynthesis and other biochemical process in plants. With response to biotic stress, SA accumulates and establishes SAR. SA-responsive genes to abiotic stress have been identified, and their expression is studied to a little extent. Effect of SA to abiotic stress tolerance has been studied only in selected plants. An extensive investigation of SA responsiveness to abiotic stress including xenobiotics in a wide number of plants needs to be undertaken along with the identification of underlying molecular mechanisms. Tissue-specific expression of desired concentration of SA in plants with very low/lack of SA biosynthesis might play a significant role in agricultural crop productivity. At the same time, the role of SA and its functional analogs in unrevealed aspects of plant physiology and biochemistry needs to be explored.

## References


Aberg B (1981) Plant growth regulators. XII mono-sub stituted benzoic acids. Swed J Agric Res 11:93–105

Alonso-Ramírez A, Rodrı'guez D, Reyes D, Jime' nez JA, Nicola's G et al (2009) Evidence for a role of gibberellins in salicylic acid-modulated early plant responses to abiotic stress in Arabidopsis seeds. Plant Physiol 150:1335–1344

Antoniw JF, White RF (1980) The effects of aspirin and polyacrylic acid on soluble leaf proteins and resistance to virus infection in five cultivars of tobacco. Phytopathol Z 98:331–341

Bandurska H, Stroinski A (2005) The effect of salicylic acid on barley response to water deficit. Acta Physiol Plant 27:379–386

Bendal DS, Bonner WD Jr (1971) Cyanide-insensitive respiration in plant mitochondria. Plant Physiol 47:236–245

Ben-Tal Y, Cleland CF (1982) Uptake and metabolism of [$^{14}$C]-salicylic acid in *Lemna Gibba* G3. Plant Physiol 70:291–296

Bhatia DS, Jindal V, Malke CP (1986) Effects of salicylic acid and tannic acid on stomatal aperture and some enzyme changes in isolated epidermal peelings of *Euphorbia hirta* L. Biochem Physiol Pflanz 181:262–264

Black VJ, Black CR, Roberts JA, Stewart CA (2000) Impact of ozone on the reproductive development of plants. New Phytol 147:421–447

Bol JF, van Kan JAL (1988) The synthesis and possible functions of virus-induced proteins in plants. Microbiol Sci 5:47–52

Borsani O, Valpuesta V, Botella MA (2001) Evidence for a role of salicylic acid in the oxidative damage generated by NaCl and osmotic stress in *Arabidopsis* seedlings. Plant Physiol 126:1024–1030

Bourbouloux A, Raymond P, Delrot S (1998) Effects of salicylic acid on sugar and amino acid uptake. J Exp Bot 49:239–247

Boussiba S, Rikin A, Richmond AE (1975) The role of ABA in cross-adaptation of tobacco plants. Plant Physiol 56:337–339

Bowling SA, Guo A, Cao H, Gordon AS, Klessig DF, Dong X (1994) A mutation in *Arabidopsis* that leads to constitutive expression of systemic acquired resistance. Plant Cell 6:1845–1857

Cao H, Bowling SA, Gordon AS, Dong X (1994) Characterization of an Arabidopsis mutant that is nonresponsive to inducers of systemic acquired resistance. Plant Cell 6:1583–1592

Cao H, Glazebrook J, Clarke JD, Volko S, Dong X (1997) The Arabidopsis NPR1 gene that controls systemic acquired resistance encodes a novel protein containing ankyrin repeats. Cell 88:57–63

Carr JP, Klessig DF (1989) The pathogenesis-related proteins of plants. In: Setlow JK (ed) Genetic engineering. Principles and methods. Plenum, New York, pp 65–109

Chadha KC, Brown SA (1974) Biosynthesis of phenolic acids in tomato plants infected with *Agrobacterium tumefaciens*. Can J Bot 52:2041–2046

Chakraborty U, Tongden C (2005) Evaluation of heat acclimation and salicylic acid treatments as potent inducers of thermotolerance in *Cicer arietinum* L. Curr Sci 89:384–389

Chamnongpol S, Wilieekens H, Moeder W, Langebartels C, Sandermann H et al (1998) Defense activation and enhanced pathogen tolerance induced by $H_2O_2$ in transgenic tobacco. Proc Natl Acad Sci USA 95:5818–5823

Chen J, Meeuse BJD (1971) Production of free indole by some Aroids. Acta Bot Neerl 20:627–635

Chen Z, Ricigliano J, Klessig DF (1993) Purification and characterization of a soluble salicylic acid-binding protein from tobacco. Proc Natl Acad Sci USA 90:9533–9537

Chen F, D'Auria JC, Tholl D, Ross JR, Gershenzon J, Noel JP, Pichersky E (2003) An *Arabidopsis thaliana* gene for methylsalicylate biosynthesis, identified by a biochemical genomics approach, has a role in defense. Plant J 36:577–588

Cleland CF, Ajami A (1974) Identification of the flower-inducing factor isolated from aphid honeydews as being salicylic acid. Plant Physiol 54:904–906

Conrath U, Chen Z, Ricigliano J, Klessig DF (1995) Two inducers of plant defense responses, 2,6-dichloroisonicotinic acid and salicylic inhibit catalase activity in tobacco. Proc Natl Acad Sci USA 92:7143–7147

Cooley DR, Manning WJ (1987) The impact of ozone on assimilate partitioning in plants: a review. Environ Pollut 47:95–113

Coquoz JL, Buchala AJ, Meuwly PH, Metrux JP (1995) Arachidonic acid induces local but not systemic synthesis of salicylic acid and confers systemic resistance in potato plants to *Phytophthora infestans* and *Alternaria solani*. Phytopathology 85:1219–1224

Dat JF, Foyer CH, Scott IM (1998a) Changes in salicylic acid and antioxidants during induction of thermo tolerance in mustard seedlings. Plant Physiol 118:1455–1461

Dat JF, Lopez-Delgado H, Foyer CH, Scott IM (1998b) Parallel changes in $H_2O_2$ and catalase during thermotolerance induced by salicylic acid or heat acclimation in mustard seedlings. Plant Physiol 116:1351–1357

Day DA, Arron GP, Laties GG (1980) Nature and control of respiratory pathway in plants: the interaction of cyanide-resistant respiration with cyanide-sensitive pathway. In: Davies DD (ed) The biochemistry of plants: a comprehensive treaties, vol 2. Academic, New York, pp 197–241

Dean JV, Mohammed LA, Fitzpatrick T (2005) The formation, vacuolar localization, and tonoplast transport of salicylic acid glucose conjugates in tobacco cell suspension cultures. Planta 221:287–296

Delaney TP, Friedrich L, Ryals JA (1995) Arabidopsis signal transduction mutant defective in chemically and biologically induced disease resistance. Proc Natl Acad Sci USA 92:6602–6606

Dietrich RA, Delaney TP, Uknes SJ, Ward ER, Ryals JA, Dang JL (1994) Arabidopsis mutants stimulating disease resistance response. Cell 77:565–577

Drazic G, Mihailovic N (2005) Modification of cadmium toxicity in soybean seedlings by salicylic acid. Plant Physiol 168:511–517

Durner J, Klessig DF (1995) Inhibition of ascorbate peroxidase by salicylic acid and 2,6 dichloroisonicotinic acid, two inducers of plant defense responses. Proc Natl Acad Sci USA 92:11312–11316

Durner J, Klessig DF (1996) Salicylic acid is a modulator of tobacco and mammalian catalases. J BiolChem 271:28492–28501

El-Basyouni S, Chen D, Ibrahim RK, Neish AC, Towers GHN (1964) The biosynthesis of hydroxy benzoic acids in higher plants. Phytochemistry 3:485–492

Ellis BE, Amrhein N (1971) The "NIH shift" during aromatic ortho hydroxylation in higher plants. Phytochemistry 10:3069–3072

Elthon TE, Mclntosh L (1987) Identification of the alternative terminal oxidase of higher plant mitochondria. Proc Natl Acad Sci USA 84:8399–8403

Elthon TE, Nickels RL, McIntosh L (1989) Mitochondrial events during the development of thermogenesis in *Sauromatum guttatum* (Schott). Planta 180:82–89

Enyedi AJ, Yalpani N, Sliverman P, Raskin I (1992) Signal molecule in systemic plant resistance to pathogens and pests. Cell 70:879–886

Fariduddin Q, Hayat S, Ahmad A (2003) Salicylic acid influences net photosynthetic rate, carboxylation efficiency, nitrate reductase activity and seed yield in *Brassica juncea*. Photosynthetica 41:281–284

Ferrari S, Plotnikova JM, De Lorenzo G, Ausubel FM (2003) *Arabidopsis* local resistance to *Botrytis cinerea* involves salicylic acid and camalexin and requires EDS4 and PAD2, but not SID2, EDS5 or PAD4. Plant J 35:193–205

Foyer CH, Lopez-Delago H, Dat JF, Scott IM (1997) Hydrogen peroxide- and glutathione-associated mechanisms of acclimatory stress tolerance and signalling. Physiol Plant 100:241–254

Freeman JL, Garcia D, Kim D, Hopf A, Salt DE (2005) Constitutively elevated salicylic acid signals glutathione-mediated nickel tolerance in *Thlaspi* nickel hyperaccumulators. Plant Physiol 137: 1082–1091

Fujioka S, Yamaguchi I, Muorfushi N, Takahashi N, Kaihara S, Takimoto A (1983) The role of plant hormones and benzoic acid in flowering of *Lemna paucicostata 151* and *381*. Plant Cell Physiol 24:241–246

Fujioka S, Yamaguchi I, Murofushi N, Takahashi N, Kaihara S et al (1985) The role of benzoic acid and plant hormones in flowering of *Lemna gibba* G3. Plant Cell Physiol 26:655–659

Fujioka S, Yamaguchi I, Muorfushi N, Takahashi N, Kaihara S et al (1986) The influence of nicotinic acid and plant hormones on flowering in *Lemna*. Plant Cell Physiol 27:109–116

Gaffney T, Friedrich L, Vernooij B, Negrotto D, Nye G et al (1993) Requirement of salicylic acid for the induction of systemic acquired resistance. Science 261:754–756

Ghai N, Setia RC, Setia N (2002) Effects of paclobutrazol and salicylic acid on chlorophyll content, hill activity and yield components in *Brassica napus* L. (cv. GSL-1). Phytomorphology 52:83–87

Glass ADM (1973) Influence of phenolic acids on ion uptake: I. Inhibition of phosphate uptake. Plant Physiol 51:1037–1041

Glass ADM (1974) Influence of phenolic acids upon ion uptake: III. Inhibition of potassium absorption. J Exp Bot 25:1104–1113

Goyal A, Tolbert NE (1989) Variations in the alternative oxidase in *Chlamydomonas* grown in air or high $CO_2$. Plant Physiol 89:958–962

Greenberg JT, Guo A, Klessig DF, Ausubel FM (1994) Programmed cell death in plants: a pathogen-triggered response activated with multiple defense functions. Cell 77:551–563

Griffiths LA (1958) Occurrence of gentisic acid in plant tissues. Nature 182:733–734

Guan L, Scandalios JG (1995) Developmentally related responses of maize catalase genes to salicylic acid. Proc Natl Acad Sci USA 92:5930–5934

Gutknecht J (1990) Salicylate and proton transport through lipid bilayer membranes: a model for salicylate-induced uncoupling and swelling in mitochondria. J Membr Biol 115:253–260

Hahlbrock K, Scheel D (1989) Physiology and molecular biology of phenylpropanoid metabolism. Annu Rev Plant Physiol Plant Mol Biol 40:347–369

Hamada AM (1998) Effects of exogenously added ascorbic acid, thiamin or aspirin on photosynthesis and some related activities of drought-stressed wheat plants. In: Garab G (ed) Photosynthesis: mechanisms and effects. Kluwer Academic, Dordrecht, pp 2581–2584

Hamada AM, Al-Hakimi AMA (2001) Salicylic acid versus salinity-drought induced stress on wheat seedlings. Rostl Vyr 47:444–450

Hammond-Kosack KE, Jones JDG (1996) Resistance gene-dependent plant defense responses. Plant Cell 8:1773–1791

Harborne JB (1980) Plant phenolics. In: Bell EA, Charlwood BV (eds) Secondary plant products. Springer, Berlin, pp 329–402

Harper JP, Balke NE (1981) Characterization of the inhibition of $K^+$ absorption ion in oat roots by salicylic acid. Plant Physiol 68:1349–1351

Hayat S, Hasan SA, Fariduddin Q, Ahmad A (2008) Growth of tomato (*Lycopersicon esculentum*) in response to salicylic acid under water stress. J Plant Interact 3:297–304

He YL, Liu YL, Cao WX, Huai MF, Xu BG, Huang BG (2005) Effects of salicylic acid on heat tolerance associated with antioxidant metabolism in Kentucky bluegrass. Crop Sci 45:988–995

Hew CS (1987) The effect of 8-hydroxyquinoline sulphate, acetylsalicylic acid and sucrose on bud opening of *Oncidium* flowers. J Hort Sci 62:75–78

Hirt H (1997) Multiple roles of MAP kinases in plant signal transduction. Trends Plant Sci 2:11–15

Horvath E, Janda T, Szalai G, Paldi E (2002) In vitro salicylic acid inhibition of catalase activity in maize: differences between the isozymes and a possible role in the induction of chilling tolerance. Plant Sci 163:1129–1135

Idrees M, Khan MMA, Aftab T, Naeem M, Hashmi N (2010) Salicylic acid induced physiological and biochemical changes in lemongrass varieties under water stress. J Plant Interact 5:293–303

James WO, Beevers H (1950) The respiration of *Arum spadix*. A rapid respiration, resistant to cyanide. New Phytol 49:453–474

Janda T, Szalai G, Tari I, Paldi E (1999) Hydroponic treatment with salicylic acid decreases the effect of chilling injury in maize (*Zea mays* L.) plants. Planta 208:175–180

Janda T, Szalai G, Antunovics ZS, Horvath E, Paldi E (2000) Effect of benzoic acid and aspirin on chilling tolerance and photosynthesis in young maize plants. Maydica 45:29–33

Jeffrey L, Giraudt J (1998) Abscisic acid and signal transduction. Annu Rev Plant Physiol Plant Mol Biol 49:199–222

Jonak C, Heberle-Bors E, Hirt H (1994) MAP kinases: universal multipurpose signaling tools. Plant Mol Biol 24:407–416

Kang HM, Saltveit ME (2002) Chilling tolerance of maize, cucumber and rice seedling leaves and roots are differentially affected by salicylic acid. Physiol Plant 115:571–576

Kapulnik Y, Yalpani N, Raskin I (1992) Salicylic induces cyanide-resistant respiration in tobacco cell-suspension cultures. Plant Physiol 100:1921–1926

Karlidag H, Yildirim E, Turan M (2009) Salicylic acid ameliorates the adverse effect of salt stress on strawberry. Sci Agric 66:180–187

Katz V, Fuchs A, Conrath U (2002) Pretreatment with salicylic acid primes parsley cells for enhanced ion transport following elicitation. FEBS Lett 520:53–57

Kauss H, Jeblick W (1995) Pretreatment of parsley suspension cultures with salicylic acid enhances spontaneous and elicited production of $H_2O_2$. Plant Physiol 108:1171–1178

Kauss H, Jeblick W (1996) Influence of salicylic acid on the induction of competence for $H_2O_2$ elicitation. Plant Physiol 111:755–763

Khurana JP, Cleland CF (1992) Role of salicylic acid and benzoic acid in flowering of a photoperiod-insensitive strain, *Lemna Paucicostata* LP6. Plant Physiol 100:1541–1546

Khurana JP, Maheswari SC (1980) Some effects of salicylic acid on growth and flowering in *Spirodela polyrrhiza* SP20. Plant Cell Physiol 21:923–927

Khurana JP, Maheswari SC (1987) Floral induction in *Wolffia microscopica* by non-inductive long days. Plant Cell Physiol 24:907–912

Korkmaz A (2005) Inclusion of acetyl salicylic acid and methyl jasmonate into the priming solution improves low temperature germination and emergence of sweet pepper. Hort Sci 40:197–200

Krantev A, Yordanova R, Janda T, Szalai G, Popova L (2008) Treatment with salicylic acid decreases the effect of cadmium on photosynthesis in maize plants. J Plant Physiol 165:920–931

Kuznetsov VV, Shevyakova NI (1999) Proline under stress conditions: biological role, metabolism, and regulation. Russ J Plant Physiol 46:274–289

Lamarck JB (1778) Flore Francaise 3. L'Emprimerie Royale, Paris, pp 537–539

Larkindale J, Huang B (2004) Thermotolerance and antioxidant systems in *Agrostis stolonifera*: involvement of salicylic acid, abscisic acid, calcium, hydrogen peroxide, and ethylene. J Plant Physiol 161:405–413

Larkindale J, Knight MR (2002) Protection against heat stress-induced oxidative damage in *Arabidopsis* involves calcium, abscisic acid, ethylene and salicylic acid. Plant Physiol 128:682–695

Larque-Saavedra A (1978) The antitranspirant effect of acetylsalicylic acid on *Phaseolus vulgaris* L. Physiol Plant 43:126–128

Lee JS (1995) Effect of salicylic acid and its analogues on stomatal closing in *Commelina communis* L. J Korean Environ Sci Soc 4:317–321

Lee JS (1998) The mechanism of stomatal closing by salicylic acid in *Commelina communis* L. J Plant Biol 41:97–102

Lee HI, Raskin I (1999) Purification, cloning and expression of a pathogen inducible UDP-glucose:salicylic acid glucosyltransferase from tobacco. J Biol Chem 274:36637–36642

Lee TT, Skog F (1965) Effect of substituted phenols on bud formation and growth of tobacco tissue culture. Physiol Plant 18:386–402

Lee S, Kim S-G, Park C-M (2010) Salicylic acid promotes seed germination under high salinity by modulating antioxidant activity in *Arabidopsis*. New Phytol 188:626–637

León J, Yalpani N, Raskin I, Lawton MA (1993) Induction of benzoic acid 2-hydroxylase in virus-inoculated tobacco. Plant Physiol 103:323–328

León J, Lawton MA, Raskin I (1995) Hydrogen peroxide stimulates salicylic acid biosynthesis in tobacco. Plant Physiol 108(1673–1):678

Leung J, Giraudat J (1998) Abscisic acid signal transduction. Annu Rev Plant Physiol Plant Mol Biol 49:166–222

Levitan H, Barker JL (1972) Salicylate: a structure-activity study of its effects on membrane permeability. Science 126:1423–1425

Lindquist S (1986) The heat shock response. Annu Rev Biochem 45:39–72

Linthorst HJM, Melchers LS, Mayer A, van Roekel JSC, Cornelissen BJC, Bol JF (1990) Analysis of gene families encoding acidic and basic β-1, 3-glucanase of tobacco. Proc Natl Acad Sci USA 87:8756–8760

Llusia J, Penuelas J, Munne-Bosch S (2005) Sustained accumulation of methyl salicylate alters antioxidant protection and reduces tolerance of holm oak to heat stress. Physiol Plant 124:353–361

Lopez-Delgado H, Dat JF, Foyer CH, Scott IM (1998) Induction of thermotolerance in potato microplants by acetylsalicylic acid and $H_2O_2$. J Exp Bot 49:713–720

Macri F, Vianello A, Pennazio S (1986) Salicylate-collapsed membrane potential in pea stem mitochondria. Physiol Plant 67:136–140

Malamy J, Carr JP, Klessig DF, Raskin I (1990) Salicylic acid: a likely endogenous signal in resistance the response of tobacco to viral infection. Science 250:1004–1006

Manthe B, Schultz M, Schnable H (1992) Effects of salicylic acid on growth and stomatal movements of *Vicia faba* L.: evidence for SA metabolism. J Chem Ecol 18:1525–1539

Mardani H, Bayat H, Saeidnejad AH, Rezaie EE (2012) Assessment of salicylic acid impacts on seedling characteristic of cucumber (*Cucumis sativus* L.) under water stress. Not Sci Biol 4:112–115

Martinez de Ilarduya O, Xie Q, Kaloshian I (2003) Aphid-induced defense responses in Mi-1-mediated compatible and incompatible tomato interactions. Mol Plant Microbe Interact 16:699–708

Mauch F, Mausch-Mani B, Gaille C, Kull B, Haas D, Reimmann C (2001) Manipulation of salicylate content in *Arabidopsis thaliana* by the expression of an engineered bacterial salicylate synthase. Plant J 25:62–77

Meeuse BJD (1985) Sauromatum. In: Halevy HA (ed) Handbook of flowering. CRC Press, Boca Raton, pp 321–327

Meeuse BJD, Raskin I (1988) Sexual reproduction in the arum lily family, with emphasis on thermogenicity. Sex Plant Reprod 1:3–15

Metraux JP, Burkhart W, Moyer M, Dincher S et al (1989) Isolation of a complementary DNA encoding a chitinase with structural homology to a bifunctional lysozyme/chitinase. Proc Natl Acad Sci USA 86:896–900

Metraux JP, Signer H, Ryals JA, Ward E, Wyss-Benz M et al (1990) Increase in salicylic acid at the onset of systemic acquired resistance in cucumber. Science 250:1004–1006

Metwally A, Finkemeier I, Georgi M, Dietz K-J (2003) Salicylic acid alleviates the cadmium toxicity in barley seedlings. Plant Physiol 132:272–281

Mills PR, Woods RKS (1984) The effects of polyacrylic acid, acetylsalicylic acid and salicylic acid on resistance of cucumber to *Colletotrichum lagenarium*. Phytopathol Z 111:209–216

Minnick L, Kilpatrick M (1939) Acid-base equilibria in aqueous and nonaqueous solutions. J Phys Chem 43:259–268

Mishra A, Choudhuri MA (1997) Ameliorating effects of salicylic acid on lead and mercury induced inhibition of germination and early seedling growth of two rice cultivars. Seed Sci Technol 25:263–270

Mishra A, Choudhuri MA (1999) Effects of salicylic acid on heavy metal-induced membrane deterioration mediated by lipoxygenase in rice. Biol Plant 42:409–415

Mishra NS, Tuteja R, Tuteja N (2006) Signaling through MAP kinase networks in plants signaling through MAP kinase networks in plants. Arch Biophys Biochem 452:55–68

Moons A, Prinsen E, Van Montagu M (1997) Antagonistic effect of abscisic acid and jasmonates on salt stress-inducible transcripts in rice roots. Plant Cell 9:2243–2259

Moore AL, Bonner WD Jr (1982) Measurements of membrane potentials in plant mitochondria with the safranine method. Plant Physiol 70:1271–1276

Mori IC, Pinontoan R, Kawano T, Muto S (2001) Involvement of superoxide generation in salicylic acid induced stomata closure in *Vicia faba*. Plant Cell Physiol 42:1383–1388

Munne-Bosch S, Penuelas J (2003) Photo- and antioxidative protection, and a role for salicylic acid during drought and recovery in field-grown *Phillyrea angustifolia* plants. Planta 217:758–766

Munne-Bosch S, Penuelas J, Llusia J (2007) A deficiency in salicylic acid alters isoprenoid accumulation in water-stressed NahG transgenic Arabidopsis plants. Plant Sci 172:756–762

Nagao RT, Kimpel JA, Vierling E, Key JL (1986) The heat shock response: a comparative analysis. In: Miflin BJ (ed) Oxford surveys of plant molecular and cell biology. Oxford University Press, Oxford, pp 384–438

Nanda KK, Kumar S, Sood V (1976) Effect of gibberellic acid and some phenols on flowering of *Impatiens balsamina*, a qualitative short-day plant. Physiol Plant 38:53–56

Navarre DA, Mayo D (2004) Differential characteristics of salicylic acid-mediated signaling in potato. Physiol Mol Plant Pathol 64:179–188

Nemeth M, Janda T, Eszter J, Horvath E, Paldi E, Szalai G (2002) Exogenous salicylic acid increases polyamine content but may decrease drought tolerance in maize. Plant Sci 162:569–574

Oota Y (1972) The response of *Lemna Gibba* G3 to a single long day in the presence of EDTA. Plant Cell Physiol 13:375–380

Oota Y (1975) Short-day flowering of *Lemna gibba* G3 induced by salicylic acid. Plant Cell Physiol 13:1131–1135

Pal M, Szalai G, Hovarth E, Janda T, Paldi E (2002) Effect of salicylic acid during heavy metal stress. In: Proceedings of the 7th Hungarian congress on plant physiology, pp 119–120

Pancheva TV, Popova LP (1998) Effect of salicylic acid on the synthesis of ribulose-1, 5-bisphosphate carboxylase/oxygenase in barley leaves. J Plant Physiol 152:381–386

Pancheva TV, Popova LP, Uzunova AN (1996) Effects of salicylic acid on growth and photosynthesis in barley plants. J Plant Physiol 149:57–63

Panda SK, Patra HK (2007) Effect of salicylic acid potentiates cadmium-induced oxidative damage in *Oryza sativa* L. leaves. Acta Physiol Plant 29:567–575

Pell EJ, Schlagnhaufer CD, Arteca RN (1997) Ozone-induced oxidative stress: mechanisms of action and reaction. Physiol Plant 100:264–273

Pesci L (1987) ABA induced proline accumulation in barley leaf segments: dependence on protein synthesis. Physiol Plant 71:287–291

Pieterse AH, Muller LJ (1977) Induction of flowering in *Lemna gibba* G3 under short-day conditions. Plant Cell Physiol 18:45–53

Popova LP, Maslenkova LT, Yordonava RY, Ivanova AP, Krantev AP, Szalai G, Janda T (2009) Exogenous treatment with salicylic acid attenuates cadmium toxicity in pea seedlings. Plant Physiol Biochem 47:224–231

Procter W (1843) Observations on the volatile oil of *Betula lenta*, and on gaultherin, a substance which, by its decomposition, yield that oil. Am J Pharm 15:241

Rai VK, Sharma SS, Sharma S (1986) Reversal of ABA-induced stomatal closure by phenolic compounds. J Exp Bot 37:129–134

Rajasekaran LR, Stiles A, Caldwell CD (2002) Stand establishment in processing carrots- Effects of various temperature regimes on germination and the role of

salicylates in promoting germination at low temperatures. Can J Plant Sci 82:443–450
Rajjou L, Belghazi M, Huguet R, Robin C, Moreau A, Job C, Job D (2006) Proteomic investigation of the effect of salicylic acid on *Arabidopsis* seed germination and establishment of early defense mechanisms. Plant Physiol 141:910–923
Rao MV, Davis KR (1999) Ozone-induced cell death occurs via two distinct mechanisms in *Arabidopsis*: the role of salicylic acid. Plant J 17:603–614
Rao MV, Paliyath G, Ormrod DP, Murr DP, Watkins CB (1997) Influence of salicylic acid on $H_2O_2$ production, oxidative stress, and $H_2O_2$-metabolizing enzymes. Plant Physiol 115:137–149
Rao MV, Lee HI, Davis KR (2002) Ozone-induced ethylene production is dependent on salicylic acid, and both salicylic acid and ethylene act in concert to regulate ozone induced cell death. Plant J 32:447–456
Raskin I (1992) Role of salicylic acid in plants. Annu Rev Plant Physiol Plant Mol Biol 43:439–463
Raskin I, Turner IM, Melander WR (1989) Regulation of heat production in the inflorescences of an *Arum* lily by endogenous salicylic acid. Proc Natl Acad Sci USA 86:2214–2218
Raskin I, Skubatz H, Tang W, Meeuse BJD (1990) Salicylic acid levels in thermogenic and non-thermogenic plants. Ann Bot 66:369–373
Raskin I, Ehmann A, Melander WR, Meeuse BJD (1987) Salicylic acid-a natural inducer of heat production in *Arum* lilies. Science 237:1545–1556
Reich PB (1987) Quantifying plant response to ozone: a unifying theory. Tree Physiol 3:63–91
Rhoads DM, McIntossh L (1992) Salicylic acid regulation of respiration in higher plants: alternative oxidase expression. Plant Cell 4:1131–1139
Rhoads DM, McIntossh L (1993) Cytochrome and alternative pathway respiration in tobacco: effect of salicylic acid. Plant Physiol 103:877–883
Rhoads DM, McIintosh L (1991) Isolation and characterization of a cDNA clone encoding an alternative oxidase protein of *Sauromatum guttatum* (Schott). Proc Natl Acad Sci USA 88:2122–2126
Rock CD (2000) Pathways to abscisic acid-regulated gene expression. New Phytol 148:357–396
Ross AF (1961) Localized acquired resistance to plant virus infection in hypersensitive hosts. Virology 14:329–339
Ryals JA, Neuenschwander UH, Willits MG, Molina A, Steiner HY, Hunt MD (1996) Systemic acquired resistance. Plant Cell 8:1809–1819
Sahu GK, Kar M, Sabat SC (2002) Electron transport activities of isolated thylakoids from wheat plants grown in salicylic acid. Plant Biol 4:321–328
Sahu GK, Kar M, Sabat SC (2010) Alteration in phosphate uptake potential of wheat plants co-cultivated with salicylic acid. J Plant Physiol 167:326–328
Saitanis CJ, Karandinos MG (2002) Effects of ozone on tobacco (*Nicotiana tabacum* L.) varieties. J Agron Crop Sci 188:51–58
Scharfetter E, Rottenburg T, Kandeler R (1978) The effect of EDDHA and salicylic acid on flowering and vegetative development in *Spirodela punctata*. Zeitschrift fur Pflanzenphysiol 87:445–454
Senaratna T, Touchell D, Bunn E, Dixon K (2000) Acetyl salicylic acid (aspirin) and salicylic acid induce multiple stress tolerance in bean and tomato plants. Plant Growth Regul 30:157–161
Serino L, Reimman C, Baur H, Beyeler M, Visca P, Hass D (1995) Structural genes for salicylates biosynthesis from chorismate in *Pseudomonas aeruginosa*. Mol Gen Genet 249:217–228
Seth PN, Venketraman R, Maheshwari SC (1970) Studies on the growth and flowering of a short day plant. Wolfia microscopic. II. Role of metal ions and chelates. Planta 90:349–359
Shah J (2003) The salicylic acid loop in plant defense. Curr Opin Plant Biol 6:365–371
Shah J, Klessig DF (1999) Salicylic acid: signal perception and transduction. In: Lebbenga K, Hall M, Hooykass PJJ (eds) Biochemistry and molecular biology of plant hormones. Elsevier, London, pp 513–541
Shah J, Tsui F, Klessig DF (1997) Characterization of a salicylic acid-insensitive mutant (Sai 1) of *Arabidopsis thaliana* identified in a selective screening utilizing the SA-inducible expression of the tms2gene. Mol Plant Microbe Interact 10:69–76
Shakirova FM, Sakhabutdinova AR, Bezrukova MV, Fatkhutdinova RA, Fatkhutdinova DR (2003) Changes in hormonal status of wheat seedlings induced by salicylic acid and salinity. Plant Sci 164:317–322
Sharma YK, Davis KR (1997) The effects of ozone on antioxidant responses in plants. Free Radical Biol Med 23:480–488
Sharma YK, Leon J, Raskin I, Davis KR (1996) Ozone-induced responses in *Arabidopsis thaliana*: the role of salicylic acid in the accumulation of defense-related transcripts and induced resistance. Proc Natl Acad Sci USA 93:5099–5104
Shi Q, Bao Z, Zhu Z, Ying Q, Qian Q (2006) Effects of different treatments of salicylic acid on heat tolerance, chlorophyll fluorescence, and antioxidant enzyme activity in seedlings of *Cucumis sativa* L. Plant Growth Regul 48:127–135
Silverman P, Seskar M, Kanter D, Schweizer P, Metraux JP, Raskin I (1995) Salicylic acid in rice (Biosynthesis, conjugation and possible role). Plant Physiol 108:633–639
Singh B, Usha K (2003) Salicylic acid induced physiological and biochemical changes in wheat seedlings under water stress. Plant Growth Regul 39:137–141
Smith BN, Meeuse BJD (1966) Production of volatile amines and skatole at anthesis in some arum lily species. Plant Physiol 41:343–374
Song JT (2006) Induction of a salicylic acid glucosyltransferase, AtSGT1, is an early disease response in *Arabidopsis thaliana*. Mol Cells 22:233–238

Storey BT (1976) Respiratory chain of plant mitochondria. XVIII. Point of interaction of the alternate oxidase with the respiratory chain. Plant Physiol 58:521–525

Storey BT, Bahr JT (1969) The respiratory chain of plant mitochondria. II. Oxidative phosphorylation in skunk cabbage mitochondria. Plant Physiol 44:126–134

Swamy PM, Smith BN (1999) Role of abscisic acid in plant stress tolerance. Curr Sci 9:1220–1227

Szalai G, Tari I, Janda T, Pestenacz A, Paldi E (2002) Effects of cold acclimation and salicylic acid on changes in ACC and MACC contents in maize during chilling. Biol Plant 43:637–640

Szepesi A, Csiszar J, Bajkan SZ, Gemes K, Horvath F, Erdei L et al (2005) Role of salicylic acid pretreatment on the acclimation of tomato plants to salt- and osmotic stress. Acta Biol Szegediensis 49: 123–125

Takimoto A, Kaihara S (1986) The mode of action of benzoic acid and some related compounds on flowering in *Lemna Paucicostata*. Plant Cell Physiol 27:1309–1316

Tanaka O, Cleland F (1980) Comparison of the ability of salicylic acid and ferricyanide to induce flowering in the long day plant *Lemna Gibba* G3. Plant Physiol 65:1058–1061

Tari I, Csiszar J, Szalai G, Horvath F, Pecsvaradi A, Kiss G et al (2002) Acclimation of tomato plants to salinity stress after a salicylic acid pre-treatment. Acta Biol Szegediensis 46:55–56

Tasgin E, Atici O, Nalbantoglu B (2003) Effects of salicylic acid and cold on freezing tolerance in winter wheat leaves. Plant Growth Regul 41:231–236

Uzunova AN, Popova LP (2000) Effect of salicylic acid on leaf anatomy and chloroplast ultrastructure of barley plants. Photosynthetica 38:243–250

Van Herk AWH (1937) Die chemischen Vorgange im *Sauromatum*-Kolben. III. Mitteilung. Proc K Ned Akad Wet 40:709–719

Van Loon LC (1983) The induction of pathogenesis-related proteins by pathogens and specific chemicals. Neth J Plant Pathol 88:265–273

Verberne MC, Verpoorte R, Bol JF, Mercado-Blanco J, Linthorst HJ (2000) Overproduction of salicylic acid in plants by bacterial transgenes enhances pathogen resistance. Nat Biotechnol 18:779–783

Vernooij B, Friedrich L, Morse A, Reist R, Kolditz-Jawahar R, Ward E, Uknes S, Kessmann H, Ryals J (1994) Salicylic acid is not the translocated signal responsible for inducing systemic acquired resistance but is required in signal transduction. Plant Cell 6:959–969

Vicente R-S, Plasencia J (2011) Salicylic acid beyond defence: its role in plant growth and development. J Exp Bot 62:3321–3338

Vlot AC, Liu PP, Cameron RK et al (2008) Identification of likely orthologs of tobacco salicylic acid-binding protein 2 and their role in systemic acquired resistance in *Arabidopsis thaliana*. Plant J 56:445–456

Wahid A, Gelani S, Ashraf M, Foolad MR (2007) Heat tolerance in plants: an overview. Environ Exp Bot 61:199–223

Wang LJ, Li SH (2006a) Thermotolerance and related antioxidant enzyme activities induced by heat acclimation and salicylic acid in grape (*Vitis vinifera* L.) leaves. Plant Growth Regul 48:137–144

Wang LJ, Li SH (2006b) Salicylic acid-induced heat or cold tolerance in relation to $Ca^{2+}$ homeostasis and antioxidant systems in young grape plants. Plant Sci 170:685–694

Wang LJ, Li SH (2007) The effects of salicylic acid on distribution of $^{14}C$-assimilation and photosynthesis in young grape plants under heat stress. Acta Hort 738:779–7851

Wang LJ, Fan L, Loescher W, Duan W et al (2010) Salicylic acid alleviates decreases in photosynthesis under heat stress and accelerates recovery in grapevine leaves. BMC Plant Biol 10:34

Ward ER, Uknes SJ, Williams SC, Dincher SS, Wiederhold DL et al (1991) Cooordinate gene activity in response to agents that induce systemic acquired resistance. Plant Cell 3:1085–1094

Wardlaw IF (1968) The control and pattern of movement of carbohydrates in plants. Bot Rev 34:79–105

Watanabe K, Fujita T, Takimoto A (1981) Relation between structure and flowering-inducing activity of benzoic acid derivatives in *Lemna paucicostata* 151. Plant Cell Physiol 20:847–850

Weete JD (1992) Induced systemic resistance to *Alternaria cassiae* in sicklepod. Physiol Mol Plant Pathol 40:437–445

White RF (1979) Acetylsalicylic acid (aspirin) induces resistance to tobacco mosaic virus in tobacco. Virology 99:410–412

Wildermuth MC, Dewdney J, Wu G, Ausubel FM (2001) Isochorismate synthase is required to synthesize salicylic acid for plant defense. Nature 414:562–565

Xie Z, Zhang ZL, Hanzlik S, Cook E, Shen QJ (2007) Salicylic acid inhibits gibberellin-induced alpha-amylase expression and seed germination via a pathway involving an abscisic-acid inducible WRKY gene. Plant Mol Biol 64:293–303

Yaeno T, Saito B, Katsuki T, Iba K (2006) Salicylic acid activates a 48-kD MAP kinase in tobacco. Plant Cell 9:809–824

Yalpani N, Silverman P, Wilson TMA, Kleier DA, Raskin I (1991) Salicylic acid is a systemic signal and an inducer of pathogenesis-related proteins in virus-infected tobacco. Plant Cell 3:809–818

Yalpani N, Leon J, Lawton MA, Raskin I (1993a) Pathway of salicylic acid biosynthesis in healthy and virus inoculated tobacco. Plant Physiol 103:315–321

Yalpani N, Shulaev V, Raskin I (1993b) Endogenous salicylic acid levels correlate with accumulation of pathogenesis-related proteins and virus resistance in tobacco. Phytopathology 83:702–708

Yalpani N, Enyedi AJ, Leon J, Raskin I (1994) Ultraviolet light and ozone stimulate accumulation of salicylic acid, pathogenesis-related proteins and virus resistance in tobacco. Planta 193:372–376

Yang ZM, Wang J, Wang SH, Xu LL (2003) Salicylic acid-induced aluminum tolerance by modulation of citrate efflux from roots of *Cassia tora* L. Planta 217:168–174

Ye XS, Pan SQ, Kuc J (1989) Pathogenesis-related proteins and systemic resistance to blue mold and tobacco mosaic virus induced by tobacco mosaic virus, *Perenospora tabacina* and aspirin. Physiol Mol Plant Pathol 35:161–175

Yu D, Liu Y, Fan B, Klessig DF, Chen Z (1997) Is the High basal level of salicylic important for disease resistance in potato. Plant Physiol 115:343–349

Yusuf M, Hasan SA, Ali B, Hayat S, Fariduddin Q, Ahmad A (2008) Effect of salicylic acid on salinity induced changes in *Brassica juncea*. J Integr Plant Biol 50:1–4

Zawoznik M, Groppa MD, Tomaro ML, Benavides MP (2007) Endogenous salicylic acid potentiates cadmium-induced oxidative stress in *Arabidopsis thaliana*. Plant Sci 173:190–197

Zhang S, Klessig DF (1997) Salicylic acid activates a 48-kD MAP kinase in tobacco. Plant Cell 9:809–824

# Role of Calcium-Mediated CBL–CIPK Network in Plant Mineral Nutrition and Abiotic Stress

10

Indu Tokas, Amita Pandey, and Girdhar K. Pandey

**Abstract**

In plants like other organisms, nutrition plays a vital role in biological processes such as growth, development and reproduction. Extensive studies done in the field of plant nutrition signalling using *Arabidopsis thaliana* as a model system have unravelled the calcium-mediated regulation of various ion transporters and channels involved in mineral nutrient acquisition and assimilation. Unlike animals, the immobile nature of plants makes them more vulnerable to the unavoidable environmental conditions in which they grow and are exposed to numerous biotic as well as abiotic stresses. Nutrition deprivation severely affects soil productivity, crop yield, and quality and stress resistance. So, a rapid and efficient signalling mechanism in response to disturbances in nutrient levels is crucial for the survival of organisms from bacteria to humans. Plants have, therefore, evolved a host of molecular pathways that can sense nutrient concentrations, both intracellular and extracellular, and quickly regulate gene expression and protein modifications to respond to any such changes. Ion channels and transporters present in the plasma membrane aid in acquisition of these nutrients. Nutrient deprivation acts as a trigger for the activation of calcium-mediated CBL–CIPK complex signalling pathways that integrate adaptive responses in plants. Several members of CBL–CIPK family such as CBL1, CBL9, CIPK6, CIPK8, CIPK16 and CIPK23 work in combination in multiple nutrient-sensing pathways to confer specific responses during uptake and transport of minerals especially nitrate and potassium. The regulatory circuit of these ion channels and transporters involves multiple post-translational modifications like phosphorylation by CBL–CIPK complex, which modulates the nutrient uptake properties in response to changes in soil nutrient

I. Tokas • A. Pandey • G.K. Pandey (✉)
Department of Plant Molecular Biology, University of Delhi South Campus, Benito Juarez Road, Dhaula Kuan, New Delhi 110021, India
e-mail: indutokas16@yahoo.co.in; amitap04@gmail.com; gkpandey@south.du.ac.in

G.R. Rout and A.B. Das (eds.), *Molecular Stress Physiology of Plants*,
DOI 10.1007/978-81-322-0807-5_10, © Springer India 2013

concentration. This chapter concisely discusses the imperative role of newly identified CBL–CIPK members as crucial components of potassium- and nitrate-sensing mechanisms during nutrition uptake, allocation and signalling.

## Introduction

Food is required by every living organism to grow and develop and is considered as one of the important aspects for defining the living versus the nonliving. The term 'nutrition' refers to the interrelated steps by which a living organism assimilates food and uses it for growth, reproduction and sustenance. It is one of the indispensable processes required for organism's survival and existence (Epstein 1972, 1973). The major role of nutrients in any living organism is to provide energy for metabolic reactions. Nutrients are classified as organic and inorganic based on their chemical compositions. Inorganic nutrients include components such as mineral elements, which are constituted by both metal and nonmetal elements.

Plants acquire most of their nutrients and water from soil by absorption through roots. As plants are sessile in nature, they are more vulnerable to environmental conditions such as nutrition, light, temperature, water as well as biotic factors like herbivores, animals and pathogens (Trewavas and Malho 1997). Sensing and responding to soil nutrient fluctuations are vital for the survival of higher plants (Tsay et al. 2011). Just like other organisms, plants also need nutrients in right combination and optimal amounts for healthy growth. They form the first trophic level in the ecological food chain and are necessary for the sustenance of all other living organisms on earth. There are two aspects of plant development: the first one is optimal growth where plants are growing in nutrient-enriched and water-sufficient soil, whereas the second one is limitation of nutrients in the soil, which leads to hampering of plant growth. The primary responses to nutrient deficiency are reduction in plant growth and alteration of nutrient acquisition, and uptake from roots as well as the improper distribution and assimilation to different organs and tissues of plants. Balance and homeostasis of these minerals for their physiological functions influence many aspects of plant growth leading to a healthy plant (Lee et al. 2009). There are 20 essential mineral elements required for plant growth (Arnon and Stout 1939). These fall into two categories depending upon the requirements of the plant. *Macronutrients* are those elements which are required in relatively large amounts and include nitrogen, potassium, phosphorous, calcium and magnesium. Others are needed in small or trace amounts and are called *micronutrients* such as iron, boron, manganese, zinc, copper, chlorine and molybdenum. With the exception of carbon, hydrogen and oxygen, all remaining nutrients are provided by the soil through plant roots. Roots are the initial sites for perceiving the nutrient status of the soil and assist in the absorption and transport of these mineral ions into the plants through active and passive transport, mass flow or diffusion. Various factors such as water content, temperature and pH alter the availability of nutrients in the soil. Deficiencies of these nutrients lead to improper growth and abnormalities in the plants. Because of less than ideal nutrition conditions in the soil, plants respond by activating several signalling components, which leads to adaptation under these conditions through involvement of plethora of genes and gene networks. Recently, calcium signalling has been implicated in sensing and responding under nutrient deprivation conditions for a few essential plant nutrients.

Calcium is one of the important signalling molecules, which regulates diverse physiological and developmental processes and is also shown to be implicated in nutrient signalling pathways in plants. In plants, there are numerous calcium decoding proteins, which are known as 'calcium sensor' involved in sensing the 'calcium signatures', which typically are the transient spatial and temporal changes in the concentration of calcium ions generated in response to a stimuli. There are three major calcium sensors identified in plants such as calmodulin (Snedden and Fromm 1998, 2001; Zielinski 1998; Reddy et al. 2002; Zhang et al. 2003), CDPK (calcium-dependent protein kinase) (Harmon et al. 2000) and calcinuerin B-like protein (CBL) (Luan et al. 2002; Batistic and Kudla 2004; Pandey 2008). CDPKs are the calcium sensors possessing kinase activity and hence act as sensor effectors. Whereas calmodulin and CBLs are typical calcium sensors, which do not possess enzymatic activity and act as sensor relay in signalling pathways to regulate their targets such as kinases and phosphatases (Luan et al. 2002,2009; Batistic and Kudla 2004; Pandey 2008 ). CBL proteins interact with CIPKs (CBL-interacting protein kinases), and most of the higher plants have multiple numbers of CBLs and CIPKs forming a complex network regulating diverse signalling pathways in phytohormones like ABA, auxin and GA and in abiotic stress management (Luan et al. 2002; Batistic and Kudla 2004; Pandey 2008; Luan 2009). Besides regulation of abiotic stress signalling, CBL–CIPK module has been implicated in mineral nutrient signalling and uptake pathways (Li et al. 2006; Xu et al. 2006; Cheong et al. 2007; Pandey et al. 2007; Pandey 2008; Luan et al. 2009; Ho et al. 2009). Regulation by CBL–CIPK module is achieved mostly by phosphorylation of the various targets such as transport proteins and other downstream components involved in various signalling pathways.

Deficiency of a particular mineral nutrient such as potassium and nitrate in the soil triggers reactive oxygen species, which in turn activate the calcium signature, and this calcium signature is decoded by CBL–CIPK pathway to modulate the transport mechanism as an adaptive response in the nutrient-deprived soil (Kim et al. 2010). Thus, this chapter mainly emphasises on the emerging roles of CBL–CIPK network during mineral nutrition in plants. Recently, the mechanism of nutrition signalling mediated by CBL–CIPK in plants has been unearthed mainly in Arabidopsis by extensive genetic and biochemical analysis; however, it is speculated to be conserved in other plant species such as crop plants and require further investigations.

## Calcium-Mediated CBL–CIPK Network During Potassium Nutrition and Signalling in Arabidopsis

Potassium apart from being a macronutrient is the most abundant inorganic cation present in the cytosol and constitutes up to 10% of a plant's dry weight (Leigh and Jones 1984). Both at the cellular and whole organism level, $K^+$ is involved in carrying out essential functions ranging from signal transduction, metabolism, growth and stress adaptations to plant sustenance (Armengaud et al. 2004; Cherel 2004; Schachtman and Shin 2007; Wang and Wu 2010). Deficiency of $K^+$ in the soil holds great agricultural importance (Laegreid et al. 1999). Plants growing on $K^+$-starved soil show major developmental abnormalities and growth arrest due to impairment in nitrogen and sugar balance, inhibition of photosynthesis and long-distance transport (Marschner 1995; Ashley et al. 2006). Short-term shortage of the $K^+$ supply in the soil is transient in nature, and plants have developed adaptive measures to cope under such conditions (Véry and Sentenac 2003; Armengaud et al. 2004). A concerted and coordinated adaptive response during $K^+$ signalling is mediated by CBL–CIPK module. This involves a synchronised interplay of sensors, transducers and effectors in the generation of effective response and eventually leading to adaptation reactions (Armengaud et al. 2004; Luan et al. 2009). $K^+$ perception as well as sensing is believed to take place in the roots, which then activates several components leading to generation of ROS and eventually calcium transient (Shin and Schachtman 2004; Kim et al. 2010); however, a clear understanding of potassium sensing and perception at the plasma membrane still requires detailed investigation. The status of potassium in the soil is sensed through reactive oxygen species (ROS) (Shin and Schachtman 2004; Shin et al. 2005) and calcium signalling (Pandey 2008; Luan et al. 2009). Under potassium deprivation conditions in plants, reactive oxygen species (ROS) are generated and shown as a critical signalling molecule for plant responses to potassium deficiency conditions (Shin and Schachtman 2004). Interestingly, ROS in root cells has been shown to elicit changes in cellular calcium and is required for root hair growth and mineral uptake (Foreman et al. 2003). Moreover, these studies suggest that ROS signal is produced under low-potassium conditions, and ROS-induced calcium changes may be a crucial messenger for downstream responses. One of the earlier works of global gene expression analysis in Arabidopsis under potassium deprivation conditions has identified several genes related to phytohormone JA, cell wall, ion transport and calcium signalling (Armengaud et al. 2004).

*Arabidopsis thaliana* genome encodes for 10 calcium sensors (CBLs) and 26 Ser/Thr protein kinases (CIPKs), which form a complex signalling network for specific and overlapping stimulus–-response coupling (Luan et al. 2002; Batistic and Kudla 2004; Pandey 2008; Luan 2009; Batistič and Kudla 2012).

## CBL1/9-CIPK23 Module Regulates Potassium Uptake and Nutrition

Two groups (Luan and co-workers and Wu and co-workers) were parallely working on identifying the molecular mechanisms of potassium nutrition signalling and have done breakthrough discovery by dissecting the signalling pathways regulating the potassium uptake and nutrition under potassium deprivation conditions in Arabidopsis. In the forward genetic screen, Xu et al. (2006) for the first time identified the *lks* (low-potassium sensitive) mutant, which shows severe sensitivity as compared to wild type based on the chlorosis of the seedling in low-potassium-containing media. Molecular cloning of this *lks* mutant leads to the identification of a loss of function of a Ser/Thr protein kinase, CIPK23. To understand the molecular mechanism for the action of CIPK23 in low-potassium condition, they have performed extensive genetic and biochemical analysis to identify the upstream and downstream interacting targets of this kinase, which turned out to be the calcium sensors CBL1 and CBL9 and high-affinity voltage-dependent potassium channel AKT1, respectively (Xu et al. 2006). By detailed phenotypic analysis of *lks* (*cipk23*), *cbl1*, *cbl9* and *akt1* mutants, they observed the similar sensitive phenotype, that is, chlorosis of leaves in low-potassium-containing media for *lks* and *akt1*; however, single mutants of *cbl1* and *cbl9* did not show any chlorosis on low-potassium-containing media. Therefore, they generated the double mutants of *cbl1cbl9* and tested the phenotype on low-potassium media, which turned out to be sensitive as well. Similarly, by gain-of-function approach, they have also confirmed the better growth phenotype of CIPK23, and CBL1 and CBL9 overexpressing transgenic plants, which also suggested them to be involved in potassium nutrition and uptake. The detailed molecular mechanism of LKS (CIPK23) in potassium uptake and nutrition has been dissected by extensive biochemical, cell biological and electrophysiological experiments. Cell biologically, the CBL1/CBL9–CIPK23 complex was targeted to be localised to plasma membrane and shown to phosphorylate the high-affinity potassium channel, AKT1. Moreover, by electrophysiological analysis, the CBL1/CBL9–CIPK23 module was shown to regulate the channel activity in *Xenopus* oocyte expression system as well as in planta (Xu et al. 2006).

Parallely, the work done by Luan group also found out CIPK23 as the major player regulating potassium nutrition by gene expression and systematic reverse genetic analysis (Li et al. 2006; Cheong et al. 2007). In the gene expression profiling of all the CBLs and CIPKs, CIPK23 transcripts were found to be highly upregulated under potassium deficiency or deprivation conditions, which provide a clue for its function in low-potassium conditions (Cheong et al. 2007). Therefore, a systematic reverse genetic approach was adopted to isolate the T-DNA-inserted mutant alleles of CIPK23, and detailed phenotypic analysis under potassium-deficient conditions was performed. One of the important phenotypes observed for *cipk23* mutant alleles was sensitive behaviour of seedling growth (mostly root growth) in potassium-deficient conditions (Cheong et al. 2007) unlike the chlorosis-based phenotype observed by Xu et al. (2006). It is important to note that different phenotypic measures were adopted by Wu and Luan group. As reported by Wu and co-workers, the chlorosis detected after post-germination-based assay under low-potassium- and high-$NH_4^+$-containing media was the major parameter; however, in the case of Luan and co-workers, they have monitored root and seedling growth sensitivity (germination-based whole seedling growth impairment) as the major parameter on diluted Murashige and Skoog (MS) media where macronutrients were diluted (1/20 times) to reduce the concentration of $NH_4^+$, since high $NH_4^+$ ion inhibit root growth drastically under low-potassium conditions (Cheong et al. 2007). When the seeds of *cipk23* mutant alleles and wild type were plated on different concentrations of potassium-containing media and a vertical germination-based growth assay was performed in optimal growth conditions, the *cipk23* mutant alleles were found to be sensitive based on reduced seedling and root growth in micromolar ranges (Cheong et al. 2007).

Similarly, the genetic analysis of CIPK23-interacting upstream calcium sensors, CBL1 and CBL9, identified by comprehensive yeast two-hybrid analysis showed them to be involved in CIPK23 mediated potassium uptake and nutrition (Cheong et al. 2007). Both CBL1 and CBL9 interact with CIPK23 and are required synergistically as identified by single and double mutant analysis. The single mutants of *CBL1* and *CBL9, that is, cbl1* and *cbl9*, did not yield any significant phenotype in potassium-deprived condition, whereas the double mutants of *cbl1cbl2* yielded a strong sensitive phenotype similar to *cipk23* mutant alleles. Thus, there is requirement of both CBL1 and CBL9 synergistically to interact and regulate CIPK23 mediated low-potassium nutrition and uptake (Cheong et al. 2007).

At molecular level, the total potassium content and uptake under low-potassium conditions were also found to be reduced in the *cipk23* mutant as well as *cbl1cbl9* double mutant but not in single mutant *cbl1* and *cbl9* (Cheong et al. 2007). This also corroborates the involvement of CBL1/9–CIPK23 complex in potassium uptake and signalling. The detailed molecular mechanism involved in regulation of potassium uptake and signalling was uncovered by heterologous expression analysis of CBL1/CBL9–CIPK23 and AKT1 channel in Xenopus oocyte system and assessing the channels opening and closing by electrophysiological TEV (two-electrode voltage clamp) analytical tools (Li et al. 2006; Xu et al. 2006). In the TEVC experiment, when AKT1 was expressed alone in the oocyte, the inherent inward-rectifying current was not detected even after applying higher negative voltage. However, when AKT1 was co-expressed along with either CBL1 or CBL9 and CIPK23, a tremendous increase in inward-rectifying current was detected (Li et al. 2006; Xu et al. 2006). By in vitro phosphorylation analysis, Li et al. (2006) have confirmed that CIPK23 phosphorylates AKT1 channels at the C-terminal region. Similarly, the channel activity was shown to be regulated by phosphorylation in vivo in oocyte where only the active CIPK23 and not inactive (ATP binding site mutant CIPK23 variant, CIPK23T60N) could lead to the activation of AKT1 channel as monitored by detection of inward-rectifying current. By TEVC-based electrophysiological analysis, the calcium-dependent activation and reconstitution of CBL–CIPK23–AKT1 pathway have been established in *Xenopus* oocytes. Since this CBL–CIPK23–AKT1 pathway has been reconstituted in heterologous system, that us, *Xenopus* oocyte, it is important to determine the functional relevance pathway in plants; therefore, the AKT1-derived inward-rectifying current was recorded in root hair cell of wild type, *cipk23* mutant alleles and *cbl1cbl9* double mutant (Li et al. 2006; Xu et al. 2006). Interestingly, the AKT1 channel activity was reduced in both *cipk23* and *cbl1cbl9* double mutants as compared to wild type, which further confirmed the functional role of CBL1/CBL9–CIPK23 module in regulating AKT1 *in planta* (Li et al. 2006; Xu et al. 2006).

## CBL1/9–CIPK23 Module Also Regulate ABA-Dependent Transpirational Streaming of Nutrients

One of the important aspects of absorption of nutrients and water from soil and transport from root to shoot of the plants is believed to be powered by transpirational streaming regulated by opening and closing of stomata (Blatt 2000; Fan et al. 2004; Schroeder et al. 2001a, b). Therefore, stomatal opening and closing regulate not only gaseous exchange ($CO_2$ and $O_2$) for photosynthesis but also orchestrate the xylem streaming and long distant transport of nutrients and water from roots to other parts of the plants. There are several factors which regulate this critical physiological process, and phytohormone abscisic acid (ABA) is one of the candidate (Schroeder et al. 2010). The role of ABA as a regulator of stomatal movement is extensively studied at molecular level (Esser et al. 1997; Wang and Song 2008). Under drought or water limitation conditions, ABA is also known to regulate several genes, and gene networks, and also induces closing of stomata to prevent the water loss through transpiration and enabling plants to cope up with the water limitation and hence adapt under such conditions (MacRobbie 1998; Wang and Song 2008). One of the major molecular mechanisms regulating ABA-mediated opening and closing of guard cell under drought conditions is through triggering the generation of hydrogen peroxide ($H_2O_2$) (Schroeder et al. 1987, 2000; Schroeder 1988, 1989; Wang and Song 2008). Interestingly, $H_2O_2$ leads to the release of calcium through activation of calcium channels in the membrane (Hamilton et al. 2000; Kwak et al. 2003; Pei et al. 2000; Peiter et al. 2005) and also from vacuoles and other endomembrane systems (McAinsh et al. 1996) through a variety of channels (Leckie et al. 1998; Lee et al. 1990). The transient increase in calcium concentration (calcium signature) is further decoded by calcium sensors and effectors, which causes activation of SLAC1 (slow activating anion channel1) leading to the closure of stomata (Vahisalu et al. 2008, 2010; Geiger et al. 2009; Roelfsema and Hedrich 2010). $H_2O_2$ is also shown to be a major signalling molecule involved in triggering the responses under nutrient-deficient conditions in roots cells (Shin and Schachtman 2004; Shin et al. 2005; Schachtman 2007), and like in leaf guard cells, $H_2O_2$ production in root cells also results in specific spatial and temporal changes in $Ca^{2+}$ concentration (Evans et al. 2001, 2005; Foreman et al. 2003).

Mechanistically, the transient level of calcium generated in both guard and root cells is further decoded by several calcium sensors such as calmodulin, CBLs and CDPKs (Snedden and Fromm 1998; Zielinski 1998; Harmon et al. 2000; Snedden and Fromm 2001; Luan 2002). Recently, CBL1/9-CIPK23 module has been identified to be involved in regulation of stomatal closure and transpirational streaming as well as mediating uptake of potassium ions from roots (Cheong et al. 2007). By systematic reverse genetic analysis, Cheong et al. (2007) have unravelled the functional role of CIPK23 in regulating drought responses since the *cipk23* mutant alleles were tolerant to dehydration stress as compared to wild type. Upon detailed molecular investigations, the stomatal movement (opening and closing of stomata) was found to be hypersensitive to ABA. Since CBL1/CBL9 targets CIPK23 to the plasma membrane in the root cell to regulate the function of voltage-gated potassium channel, AKT1 (Li et al. 2006; Xu et al. 2006), the target of CIPK23 in guard cell was also speculated to be a potassium channel because influx and efflux of $K^+$ also regulate the turgor of guard cell and ultimately lead to opening and closure of stomata. Several potassium channels such as KAT1, GORK, and AKT1/AKT2/AKT3 were also shown to be expressing in guard cell (Schroeder et al. 1987; Schroeder 1988, 1989; MacRobbie 1998; Hosy et al. 2003; Pilot et al. 2003; Lebaudy et al. 2007); the target of CIPK23 in regulating the transpiration pull through regulation of stomatal opening and closure requires further investigation.

## Other CBLs and CIPKs Regulating AKT1 in Potassium Nutrition

The functional role of calcium-mediated CBL–CIPK pathway to modulate the potassium channel activity of AKT1 at plasma membrane of root has opened up new frontiers in understanding the mechanisms of post-translationalregulation of potassium uptake under potassium-deficient conditions (Hedrich and Kudla 2006; Li et al. 2006; Xu et al. 2006). However, these studies raise number of important questions regarding the mechanisms underlying the regulation of AKT1 channel by the CBL–CIPK complexes. Based on the interaction analysis of CIPK23 with AKT1, it is quite imperative to determine the domain in AKT1 responsible for interaction with CIPK23. Therefore, deciphering the domains that mediate the AKT1–CIPK23 interaction is critical for understanding the structural basis of specificity of interaction. Moreover, the AKT1 activity is not abolished completely in *cbl1cbl9* double or *cipk23* mutant (Li et al. 2006); it is quite possible that multiple CBL–CIPK complexes might be regulating the AKT1 channel activity. To answer this question, Luan and co-workers extensively performed yeast two-hybrid analysis between 10 CBLs and 26 CIPKs and C-terminal of AKT1 and 26 CIPKs to determine multiple interacting partners of AKT1. In their comprehensive yeast two-hybrid studies, two more CIPKs, CIPK6 and CIPK16 were also found to interact with AKT1 in addition to CIPK23. Similarly, two more CBLs, CBL2 and CBL3 in addition to CBL1 and CBL9, which were already reported in earlier studies (Li et al. 2006; Xu et al. 2006), were found to interact with all three CIPKs such as CIPK6, CIPK16 and CIPK23 (Lee et al. 2007). In total, 4 CBLs and 3 CIPKs were found to be interacting with CIPK23 and AKT1, respectively (Lee et al. 2007), forming a multivalent interacting network. Based on the multiple interaction networks of CBL–CIPK and AKT1, it is logical to anticipate that a complicated network involving more CBLs and CIPKs might be cooperatively regulating AKT1 activity.

*In planta* analysis of this complex network (12 combinations of CBL–CIPK) is a difficult task; therefore, Luan and colleagues took advantage of the *Xenopus* oocytes model systems (Li et al. 2006) to investigate the mechanistic action of CBL–CIPK pair in the regulation of AKT1 potassium channel activity (Lee et al. 2007). Such a molecular network may be functionally relevant in the regulation of AKT1 activity. Based on the AKT1 channel activity measurement in *Xenopus* oocytes, differential activation of AKT1 channel was observed between these four different CBLs and three CIPKs, for example, interaction of CBL1–CIPK23 with AKT1 produces the strongest channel activity, whereas interaction combination of CBL2, 3 or 9 with CIPK6 or CIPK16 with AKT1 produces weaker channel activity (Lee et al. 2007) and hence indicating the degree of overlap and complexity in the cellular regulation by CBL–CIPK signalling network.

A key factor that governs the specificity of AKT1 activation by CBL–CIPK complexes is the physical interaction between CIPKs and AKT1 (Lee et al. 2007). The ankyrin repeat domain present in the AKT1 protein acts as a docking site, which appears to interact with the kinase domain of the CIPKs thereby determining the specificity of interaction of AKT1 channel protein with different CIPKs. Moreover, this study by Lee et al. (2007) revealed a new component in AKT1 channel regulation, a specific PP2C-type protein phosphatase known as AIP1 that interacted and inactivated AKT1. Thus, providing an important aspect of regulatory switch i.e. dephosphorylation mediated inactivation of AKT1 channel in potassium uptake (Lee et al. 2007). Overall, these finding unearthed a crucial calcium signalling pathway involved in regulation of AKT1 channel activity by phosphorylation–dephosphorylation switch in acquisition of potassium from the root.

In a recent study, the complex involvement of CBL–CIPK–PP2C-type phosphatase in regulation of AKT1 channel has been elaborated by yeast two-hybrid and electrophysiological analysis (Lan et al. 2011). In their study, Lan et al. (2011) have dissected the involvement of multiple A-type protein phosphatase 2C (PP2C) members such as AIP1, PP2CA and AHG1 in inhibition of CIPK6-mediated activation of AKT1 channels. The yeast two-hybrid and electrophysiological results showed that PP2CA specifically interacts with CIPK6 and also directly interacts with the kinase domain of CIPK6 to inactivate channel activity of AKT1. In addition, several CBLs were found to interact with and inhibit the activity of PP2CA, thereby enhancing AKT1 activation induced by CIPK6. As suggested by the authors, a working hypothetical model has been proposed for the regulation of AKT1 channel activity by CBL–CIPK–PP2C complex. According to this model, in the absence of CBLs, the kinase domain is enclosed by the NAF/FISL motif, and it cannot interact with the AKT1 ankyrin domain to activate AKT1. When CBLs are present, they interact with the NAF/FISL motif of CIPK and liberate the catalytic domain and phosphorylate AKT1. The PP2Cs interact with the PPI motif and the kinase domain of CIPK to inhibit AKT1 activation by CIPK-CBLs complex, in which CBLs do not interact with PP2Cs, through physical binding and dephosphorylation. There is reduction in the activity of phosphatase and the recovery of CIPK kinase activity upon specific interaction of CBLs with PP2Cs leading to the efficient phosphorylation of AKT1 channel (Lan et al. 2011). These findings provide new insights into the regulation of AKT1 activity by the CBL–CIPK–PP2CA network.

While understanding the detailed regulation of CBL–CIPK network function, Kudla and coworkers (Hashimoto et al. 2012) have shown the requirement of phosphorylation of CBL1 by CIPK23. CBL1 is phosphorylated at a conserved serine residue at the C-terminus in the wheat germ extract-based in vitro transcription/translation protocol. They have identified the conserved serine residue in other CBLs being phosphorylated by their respective interacting kinase. The phosphorylation status of CBLs does not appear to influence the stability, localisation or CIPK interaction of these calcium sensor proteins in general. However, proper phosphorylation of CBL1 is essential for in vivo activation of the AKT1 $K^+$ channel by CBL1–CIPK23 and CBL9–CIPK23 complexes in *Xenopus* oocytes. Moreover, the phosphorylation of CBL1 by CIPK23 is supposed to enhance the kinase activity towards its substrate, that is, AKT1 (Hashimoto et al. 2012). These findings unravelled another layer of complexity in regulation of sensor–effector component with its substrate in catalysing phosphorylation reaction, which acts as a switch in mediating a physiological process.

## CBL4–CIPK6 Regulation of AKT2 in Potassium Nutrition

In plants a large number of proteins are responsible for the transport, allocation and uptake of potassium. These proteins are classified as transporters or carriers and channels (Walker et al. 1996; Schachtman 2000; Véry and Sentenac 2003; Gierth et al. 2005; Gierth and Maser 2007; Schachtman and Shin 2007). The potassium transporters or carrier is KT/KUP/HAK (potassium transporter/potassium uptake transporter/high-affinity potassium transporters), which includes the transcriptionally highly inducible HAK5 under potassium deprivation condition (Wang and Crawford 1996; Santa-Maria et al. 1997; Kim et al. 1998; Ahn et al. 2004; Gierth et al. 2005; Rubio et al. 2008). The major potassium channels in plants are comprised of three families based on the homologous members from animal system (Lebaudy et al. 2007). The first major family of potassium channel is Shaker type as designated from the *Drosophila* potassium transporter 'Shaker', which is voltage-gated ion channel. The most well-characterised Shaker family members are AKT1 and KAT1 (both of these are inward-conducting channels) (Anderson et al. 1992; Very and Sentenac 1992; Very and Sentenac 2002). The other Shaker family of channels extensively studied is GORK and SKOR, which are outward-conducting potassium channels (Gaymard et al. 1998; Hosy et al. 2003). The second family of channels includes TPK (tandem-pore $K^+$ channels), which are voltage independent and mostly present in the membrane of vacuoles (Czempinski et al. 1999, 2002). The third family of potassium channels is Kir channel from animals also known as two-transmembrane domain inward-rectifying channels (Lebaudy et al. 2007).

After the discovery of calcium-mediated CBL–CIPK–PP2C regulation of AKT1 channel in potassium uptake and nutrition signalling, the interaction between other potassium channels beside AKT1 was investigated. However, most of the other channels were not found to interact with CIPK23 in yeast two-hybrid assays (YHC, GKP and SL unpublished data). An interesting work from J. Kudla lab (Held et al. 2011) unfolded non-phosphorylation-dependent regulation of potassium channel by CBL–CIPKs module. In this study, yeast two-hybrid analysis showed an interaction between CBL4 and CIPK6 and also confirmed between CIPK6 and AKT2. Through cell biological and electrophysiological analysis they showed, CBL4–CIPK6 complex interacts with C-terminal of AKT2 and functionally constitutes the activation of AKT2 channel activity in *Xenopus* oocyte. The interesting result of this study opens a new dimension of regulation of AKT2 channel activity by kinase-dependent interaction and translocation of AKT2 from ER to plasma membrane by CBL4–CIPK6 complex. Unlike AKT1 channel regulation by kinase activity, that is, phosphorylation mediated by CBLs–CIPKs complex (Li et al. 2006; Xu et al. 2006; Lee et al. 2007; Lan et al. 2011), AKT2 phosphorylation by CBL4–CIPK6 is not required for the functional activation of AKT2. However, the interaction of kinase with the AKT2 C-terminal and its targeting from ER to plasma membrane is critical for proper regulation of AKT2 activity (Held et al. 2011). Moreover, the dual lipid modifications of CBL4 by myristoylation as well as palmitoylation are also crucial for the translocation of AKT2 channels to plasma membrane by CBL4–CIPK6 complex (Held et al. 2011). This study suggests existence of multiple $K^+$ channels trafficking to membrane where calcium-dependent targeting of channels or transporters to the plasma membrane can be efficiently mediated by CBL–CIPK ensemble. This efficient trafficking is achieved by interaction with CIPK6 kinase, which along with CBL4 act as scaffolding protein for proper translocation of the channels to plasma membrane rather than employing phosphorylation switch to regulate the activity of channels.

Thus, new facets still need to be explored in understanding the complex array of CBL–CIPK mediated regulation of ion channels/transporters during mineral nutrition.

## Role of CIPK9 in Regulating Potassium Nutrition

Detail complexity of regulation of potassium uptake, assimilation and signalling has greatly enhanced the understanding of molecular mechanisms as one of the important aspects of plant physiology. Several members of CBLs and CIPKs have been implicated in regulation of Shaker family voltage ion-gated high affinity potassium channels, AKT1 (Li et al. 2006; Xu et al. 2006; Lee et al. 2007; Lan et al. 2011) and AKT2 (Held et al. 2011) which has significantly contributed in understanding the molecular mechanism of calcium mediated potassium nutrient uptake and signalling. However, it is inevitable that the trait of potassium uptake and nutrition is only regulated by the above discovered mechanisms. Moreover, there are multiple potassium channels and transporters, which are also involved in the uptake and assimilation of potassium from root to shoot and distribution into different parts of the plants (Walker et al. 1996; Véry and Sentenac 2003; Gierth et al. 2005; Gierth and Maser 2007). At the same time, there are multiple signalling pathways, which also regulate these transport and sensor proteins both transcriptionally and post-translationally (Shin and Schachtman 2004; Luan et al. 2009). Therefore, it is important to explore and investigate the new components involved in regulating the potassium nutrition and signalling pathways.

For this reason, Luan and colleagues have adopted gene expression and reverse genetics as major systematic approaches for understanding the potassium nutrition and signalling. In their comprehensive analysis of CBL–CIPK gene family, they have identified another member of CIPK gene family, *CIPK9*, whose transcripts were also found to be strongly inducible under potassium deprivation conditions (Pandey et al. 2007). To determine the function of CIPK9 *in planta*, they have isolated the T-DNA-inserted mutant alleles (*cipk9-1* and *cipk9-2*) and performed the germination-based phenotypic assays on potassium-deficient conditions. Both the alleles of *CIPK9* exhibited hypersensitive growth impairment phenotype as corroborated by reduced root growth and seedling development (Pandey et al. 2007). This growth-based hypersensitive phenotype is specifically observed under potassium deficiency at very low concentration of $K^+$ in the growth media. Unlike the hypersensitive growth phenotype of *cipk23* mutant alleles (Cheong et al. 2007), *cipk9* mutant exhibits the impairment of growth phenotype at lower (less than 20 μM $K^+$) $K^+$ concentration. As, upon the measurement of $K^+$ uptake and content in the *cipk9* mutant did not yield any significant difference in both potassium-sufficient (20 mM $K^+$) and potassium-deficient (20 μM $K^+$) conditions. This result suggested that CIPK9 might be regulating an alternative pathway other than CBL–CIPK23–AKT1 pathway (Li et al. 2006; Xu et al. 2006). Moreover, the authors of this study (Pandey et al. 2007) have also analysed the interaction of CIPK9 with C-terminal of AKT1 and found that these did not interact in yeast two-hybrid assays.

By genetic analysis, CIPK9 was implicated in $K^+$ deprivation condition; however, the mechanism of action of CIPK9 was not investigated (Pandey et al. 2007) and hence raises several questions like what are the targets of CIPK9? Is CIPK9 involved in regulating the other $K^+$ transporters or channels, which might be involved in loading and distribution of $K^+$ in the other parts of the plants? Is CIPK9 also involved in $K^+$ sensing and signalling by indirectly regulating ion-homeostasis pathway or generation of ROS secondary messengers in the cell? To answer all these questions, a detailed exploration of targets of CIPK9 by microarray analysis as well as proteomics-based approach is required, which will provide a holistic picture by removing the layers of complexity in potassium nutrition signalling.

## Regulation of Nitrate Signalling by CBL–CIPK Network in Arabidopsis

Nitrogen is present in the biosphere in abundant form comprising of both mineral and organic compounds. $NO_3^-$ (nitrate) and $NH_4^+$ (ammonium) are the most important nitrogen sources for plant nutrition under agricultural conditions depending upon their abundance and accessibility (Von Wirén et al. 1997a). Minerals, in particular nitrogen, serve important functions as the building blocks of organic matter, cofactors or as signalling molecules (Forde and Clarkson 1999; Forde 2000; Coruzzi and Bush 2001). Numerous physiological studies have signified the role of nitrate in higher plants as a signalling molecule. High degree of physiological and developmental plasticity is displayed by plants growing on soil with fluctuating nutritional components (Forde and Lorenzo 2001). Most of the land plants use nitrate as the major source of nitrogen from aerobic soil.

Initial uptake of minerals and organic nitrogen forms occurs across the plasma membrane of root epidermal and cortical cells (Larsson and Ingemarsson 1989). Efficient utilisation and uptake of nitrogen by roots require two kinetically distinct components – high-affinity transport system (HATS) and low-affinity transport system (LATS) (Aslam et al. 1992; Doddema and Telkamp 1979; Hole et al. 1990; Siddiqi et al. 1989, 1990). Upon sensing the nutritional status of nitrogen in the soil, there is a tight regulation of nitrogen transport system in terms of expression and activity of several components involved in the nitrogen acquisition, distribution and signalling network (Von Wiren et al. 1997a, b). Extensive work has been done in the field of nitrogen nutrition and uptake physiology (Aslam et al. 1984; Aguera et al. 1990; Crawford and Glass 1998; De Angeli et al. 2006, 2009; Miller et al. 2007; Li et al. 2007; Ho et al. 2009).

There are several transporters present in the roots for the uptake of nitrate from the soil. These transporters such as CHL1 (AtNRT1.1) (Tsay et al. 1993; Guo et al. 2001), AtNRT1.2 (Huang et al. 1999), AtNRT2.1 (Little et al. 2005) and AtNRT2.2 (Li et al. 2007) are responsible for nitrate acquisition. Three gene families of nitrate transporters have been identified in the model dicot plant *Arabidopsis thaliana*. These are *AtNRT1* consisting of 53 members, *AtNRT2* family and *AtCLC* with 7 members each (De Angeli et al. 2009; Forde 2000; Tsay et al. 1993). Based on the mode of nitrate uptake, these transporters are further categorised. Of these AtNRT2.1 and AtNRT2.2 are involved in high-affinity transport (Wang and Crawford 1996; Li et al. 2007; Little et al. 2005), AtNRT1.2 mediates low-affinity uptake (Huang et al. 1999) and AtNRT1.1 (CHL1) functions as a dual-affinity nitrate transporter during both high- and low-affinity uptake of nitrate (Liu et al. 1999; Wang et al. 1998). Assimilation of nitrate into organic nitrogen is catalysed through series of enzymatic reactions, which involve participation of nitrate reductase in the cytosol and nitrite reductase and glutamate synthase in plastids and chloroplasts (Crawford and Glass 1998). Nitrate apart from being a nutritional source also serves as a signalling molecule by inducing the expression of several genes involved in nitrate transport and assimilation (Crawford 1995; Stitt 1999; Hu et al. 2009).

Primary nitrate response is defined as rapid transcriptional induction of genes involved in nitrate transport, nitrate assimilation enzymes and carbon assimilation enzymes involved in nitrogen/carbon balance during nitrate nutrition sensing (Redinbaugh and Campbell 1991; Gowri et al. 1992; Scheible et al. 1997a, b; Lejay et al. 1999). Even in the absence of nitrate reductase activity, the genes involved in primary nitrate response are regulated by nitrate, henceforth suggesting that nitrate is the signalling molecule, which is responsible for triggering the induction of downstream targets in the response mediated by nitrate uptake (Dang et al. 1989; Pouteau et al. 1989). Thus, nitrate ion serves as the signalling molecule regulating the primary transcriptional response (Hu et al. 2009). Nitrate-sensing and signalling mechanisms involve the modulation of downstream metabolic processes in response to the external nitrate concentration. In *E. coli*, several nitrate-signalling components have been studied till date. There is presence of a two-component signalling mechanism, which comprises of membrane-spanning histidine kinases Nar X and Nar Q (Nitrate/nitrite sensor proteins) in addition to response regulators such as Nar L and Nar P, which transduce the signal to the downstream effectors modulating expression of target genes (Cavicchioli et al. 1996; Chiang et al. 1997). In *Neurospora*, localisation of transcription factors like NIT4 (positive regulator of nitrate reductase) to the nucleus is regulated by nitrate (Fu et al. 1989), and NIRA (nitrite reductase transcription regulator) discovered in *Aspergillus* mediates the transcriptional induction of nitrate assimilation genes (Berger et al. 2006; Burger et al. 1991; Bernreiter et al. 2007).

In higher plants, the mechanism of nitrate sensing and signalling involves well-characterised components such as ANR1 transcription factor belonging to MADS-box family (Zhang and Forde 1998), dual-affinity nitrate transporter CHL1 (Liu et al. 1999) and the high-affinity nitrate transporter AtNRT2.1 (Little et al. 2005). Identification of mutants *chl1* and *atnrt2.1* exhibiting altered root architecture even in the absence of external nitrate has unravelled the role of CHL1 as a nitrate sensor or signal transducer in plant mineral nutrition (Guo et al. 2001; Little et al. 2005; Walch-Liu and Forde 2008).

## Involvement of CIPK8 in the Primary Nitrate Response

Developing sophisticated adaptive mechanisms to cope with the constantly changing environmental conditions is indispensable for plant survival. Plant roots harbour a number of specialised transporters on their surface for efficient nutrient uptake and acquisition (Chen et al. 2008). Uptake of nitrate from soil is a critical process controlled by complex regulatory networks targeting nitrate transporters in the roots (Von Wirén et al. 1997a, b; De Angeli et al. 2006; Dechorgnat et al. 2011;

Tsay et al. 2011; Ho and Tsay 2011; Wang et al. 2012). As reported previously, the absorption of nitrate from soil is done by several transporters present in root cell including high-affinity NRT2.1 and dual-affinity CHL1 (Tsay et al. 2007, 2011; Ho et al. 2011; Wang et al. 2012). Nitrate directly induces the expression of these transporters as part of primary nitrate response (Tsay et al. 2007).

A differential transcriptomic study by Hu et al. (2009) has revealed the downregulation of a Ser/Thr kinase CIPK8 in the *chl1-5* mutant, thus indicating the involvement of a novel CBL–CIPK network in nitrate response. The rapid induction of *CIPK8* transcript by nitrate was shown to be concentration dependent and was speculated to be the positive regulator of primary nitrate response (Hu et al. 2009). To understand the role of CIPK8 in nitrate response, genetic analysis involving isolation of two mutant alleles of *CIPK8* was undertaken (Hu et al. 2009). In the *cipk8* mutants, the increased primary root length was observed indicating the role of CIPK8 in modulation of long-term nitrate-regulated root growth. A detailed expression analyses of the *cipk8* mutants led to the identification of two nitrate-induced nitrate uptake transporter genes, *CHL1* and *NRT2.1*, and three nitrate assimilation genes, nitrate reductase1 (*N1A1*), nitrate reductase2 (*N1A2*) and nitrite reductase (*NiR*), were found to be reduced in the *cipk8* mutant when compared with wild type (Hu et al. 2009). Interestingly, there are two other primary nitrate responsive genes namely glutamine synthetase2 (*GLN2*), glucose-6-phosphate dehydrogenase3 (*G6PDH3*) whose expression was also reduced in the *cipk8* mutant (Hu et al. 2009). This nitrate-responsive expression analysis suggested the positive regulation of nitrate transporters and nitrate assimilation genes by CIPK8 (Hu et al. 2009).

The uptake of nitrate from root is dependent on two kinetically different response phases, high affinity phase and low affinity phase of nitrate uptake and shown to be genetically distinct mechanistically (Hu et al 2009). Regulation of low-affinity phase of primary nitrate response was mediated by the Ser/Thr kinase, CIPK8. There is a profound effect caused by *CIPK8* on regulation of CHL1 and NRT2.1 in a concentration-dependent manner during low-affinity phase of nitrate-sensing system (Hu et al. 2009). Moreover, a boron transporter (BOR1) expression was downregulated in the *cipk8* mutant, which is responsible for translocation and distribution of boron ions to xylem vessels (Hu et al. 2009). This suggested that CIPK8 also participates in maintaining the homeostasis between nitrate and other anions such as borate.

The concentration of nitrate in the soil can vary from 1 μM in case of high-affinity uptake system to 10 mM for low-affinity transport. Thus, in plants two major nitrate uptake systems have evolved for efficient utilisation of the nitrate under such fluctuating conditions (Miller et al. 2007). These are high-affinity system with a $K_m$ in micromolar range and low-affinity system with a $K_m$ in millimolar range. Nitrate itself is responsible for biphasic response patterns rather than downstream metabolic products. This is a unique property of nitrate response depending upon uptake potential. As there is similarity in the $K_m$ values of nitrate-sensing and uptake system, there is possibility of sharing some common signalling components between these responses. Identification of CIPK8 as a signalling component during nitrate uptake has facilitated the elucidation of CBL–CIPK sensing network in higher plants (Hu et al. 2009). There is a direct and rapid regulation of CIPK8 by nitrate, which suggests that it is autoregulated and displays a biphasic pattern with similar $K_m$ values in both the low- and high-affinity phases. Although CIPK8 has been genetically implicated to be the regulator of primary nitrate response by low-affinity nitrate uptake process in Arabidopsis, the direct physical interaction of CIPK8 with any of the nitrate transporter such as CHL1 (NRT1.1) or NRT2.1 has not been established. Moreover, there is no evidence of phosphorylation by CIPK8 of these transporters. Therefore, in the future a detailed biochemical investigation needs to be undertaken by employing phosphorylation-based studies of nitrate transporters like CHL1 and NRT2.1 in both low- and high-affinity nitrate acquisitions from root.

## CBL1/9-CIPK23 Regulate Nitrate Sensing and Uptake

With the discovery of a kinase, CIPK8 in regulation of primary nitrate response by acting as positive regulator of low-affinity transport suggested the involvement of calcium-mediated signalling in regulation of nitrate uptake in plants (Hu et al. 2009). Despite of the alterations in transcripts of several nitrate transporters and assimilation genes in *cipk8* mutant, direct relationship of CIPK8 with any transporters has not been established by phosphorylation or direct physical interaction. In quest for detailed understanding of nitrate-sensing and response pathway, Ho et al. (2009) made a remarkable discovery by highlighting the molecular mechanisms of nitrate signalling by CHL1 in Arabidopsis. Studies on loss-of-function mutants of *chl1* revealed the regulation of NRT2.1 by CHL1 in nitrate response. In the presence of high concentration of nitrogen, there was no downregulation of NRT2.1 in the *chl1* mutants (Munos et al. 2004) and absence of lateral root proliferation in nitrate-rich zones in the soil (Remans et al. 2006).

By extensive genetic and biochemical approaches, they have revealed the regulation mediated through phosphorylation by another CIPK family member, CIPK23 (Ho et al. 2009). CIPK23 a Ser/Thr kinase phosphorylates the threonine residue present at position 101 in CHL1. Phosphorylation at this particular residue T101 acts as a 'toggle switch' in converting the low- to high-affinity modes of nitrate transporter CHL1 (Vert and Chory 2009; Ho et al. 2009). The generation of transgenic plants expressing the CHL1 wild type, mutant variants such as T101D (phosphorylated CHL1) and T101A (dephosphorylated CHL1), reflecting a shift in high affinity to low affinity nitrate response has been established (Ho et al. 2009). Dephosphorylation leads to the conversion of CHL1 to the original form mediating low-affinity transport during high concentration of nitrate in the soil (Ho et al. 2009). As suggested by Ho et al. (2009), CHL1 acts as a bona fide nitrate sensor, which can sense the low and high concentration of nitrate in the soil and accordingly leads to the activation by calcium-mediated phosphorylation switch. Biphasic nitrate uptake response has a range from less than 1 mM of nitrate concentration (high-affinity uptake) to more than 1 mM of external nitrate. Apparently, CIPK8 acts as positive regulator of low-affinity phase of nitrate uptake, and CIPK23 acts as a negative regulator of high-affinity phase (Hu et al. 2009; Ho et al. 2009; Vert and Chory 2009). Low-nitrate content in the soil activates the CIPK23–CHL1 complex (Ho et al. 2009; Vert and Chory 2009). The amino acid T-101 gets phosphorylated in CHL1 and acts as a high-affinity nitrate transporter, whereas unphosphorylated T-101 (T101N) CHLI mediates low-affinity nitrate transport.

Indeed, calcium signalling regulates both potassium and nitrate uptake and responses and requires CBL–CIPK23-dependent phosphorylation of plant channel/transporter (Li et al. 2006; Xu et al. 2006; Ho et al. 2009).

## Concept of 'Transceptor' in Plant Nutrient Signalling

Ion transporters, which also serve as sensors and receptors of both extracellular and intracellular nutrients, are known as 'transceptors'. They act as gatekeeper of nutrient exchange by employing switch mechanisms regulated by phosphorylation and thereby modulating the uptake properties quickly in response to changes in soil nutrient concentration in time-dependent manner (Hundal and Taylor 2009). CHL1 also functions as a transceptor in nitrate signalling and sensing. Topology comprises of 12 transmembrane domains in CHL1 (Forde 2000). Genetic studies were done in the *chl1-9* mutant, which exhibited point mutation caused by the replacement of leucine with proline at 492 position Pro492 in the cytosolic loop region between 10 and 11 transmembrane domains (Liu et al. 1999). *chl1-9* mutant has normal levels of CHL1 transcript and protein. It showed lesser uptake of nitrate than wild type, but when compared with *chl1-5* deletion mutant, the uptake capacity was similar in both the low- and high-affinity range

(Ho et al. 2009). There was also presence of the biphasic primary nitrate response typical for nitrate sensing in the *chl1-9* mutants (Ho et al. 2009). Moreover, introduction of the chl1-9 genomic fragment into the *chl1-5* mutant rescued the signalling defect, but the uptake activity was not reverted back. Thus, these findings corroborate the results that sensory function of CHL1 is not dependent on the transporter activity.

Phosphorylation/dephosphorylation is an essential mode of regulation of transporter activity (Lee et al. 2007). In the case of CHL1, it acts as a molecular switch in converting the two modes of action in nitrate transport (Liu and Tsay 2003). CHL1 T101D is the phosphorylated form mediating high-affinity uptake of nitrate and also maintains low levels of primary nitrate response. In contrast, the unphosphorylated form CHL1T101A enhanced both the high-affinity and low-affinity nitrate response phases. Thus, during nitrogen-deficient conditions, phosphorylation of CHL1 T101 is crucial for enhanced nitrate uptake and reduction of primary nitrate response to low levels (Ho et al. 2009).

The phosphorylation-mediated switch is accomplished by CIPK23, which phosphorylates CHL1 T101 in response to low nitrate concentrations. BiFC experiments confirmed the physical and direct interaction of CIPK23 with CHL1 in the plasma membrane. The *cipk23* mutants showed the similar primary nitrate response as that of transgenic CHL1 T101A plants. Also these mutants of *cipk23* exhibited elevation in the high-affinity nitrate primary response. Hence CHL1 is a target of CIPK23 in nitrate uptake and response pathway. Further in vitro kinase assays and *Xenopus* oocyte co-injection revealed a regulatory complex formed by CBL9–CIPK23, which phosphorylates CHL1 at residue threonine 101 and switches on the high-affinity mode of the transporter. Henceforth, low concentrations of nitrate in the soil act as a signal perceived by CHL1 and trigger the high-affinity transport (Ho et al. 2009). Mechanistically, it has been suggested that the binding of nitrate to CHL1 leads to the recruitment of CBL9–CIPK23 complex, which phosphorylates it and prevents the increased primary nitrate response. In contrast, there is inhibition of phosphorylation of CHL1 when concentration of nitrate exceed in the soil (Ho et al. 2009). Nitrate signalling network thus involves several molecular players working in a coordinated and synchronised manner. Complex interplay of different CIPKs and CBLs has been deciphered in nutrient sensing and signalling where two CIPKs work antagonistically to each other in mediating the specific response, such as CIPK8 acting as a positive regulator of low-affinity response, whereas CIPK23 generates high-affinity nitrate uptake response and work as a negative regulator of low-affinity nitrate uptake and signalling.

## Conclusion and Future Prospective

Environment and nutrition in particular determine the growth of the plants. Soil has a fluctuating nutrient profile, which varies by several orders of magnitude and imposes nutrient-related stress on plants (Jackson and Caldwell 1993). Hence, various adaptive measures including a rapid detection of nutrients in the soil by different mechanisms become imperative for the survival of plants, better growth and sustenance. There are numerous breakthrough discoveries made in the field of calcium signalling mediated by CBL–CIPK regulation of nutrient sensing and signalling in plants. Two major mineral nutrients, that is, potassium and nitrate uptake and sensing, are regulated by calcium-mediated CBL–CIPK signalling pathways. Interestingly, CIPK23 is the common kinase regulating two diverse mineral nutrients, that is, $K^+$ and $NO_3^-$ uptake from soil, by employing phosphorylation switch. A comprehensive view of potassium sensing, uptake and signalling is summarised as a hypothetical depiction in Fig. 10.1. In the case of $K^+$ nutrient uptake from root, a major CBL–CIPK23 pathway has been implicated to regulate the voltage-gated high-affinity $K^+$ channel, AKT1. A complete mechanistic interplay of multiple CBLs and CIPKs regulating the AKT1 channel by phosphorylation and dephosphorylation (involvement of PP2C in dephosphorylation of AKT1) has been shown in Arabidopsis. However, in

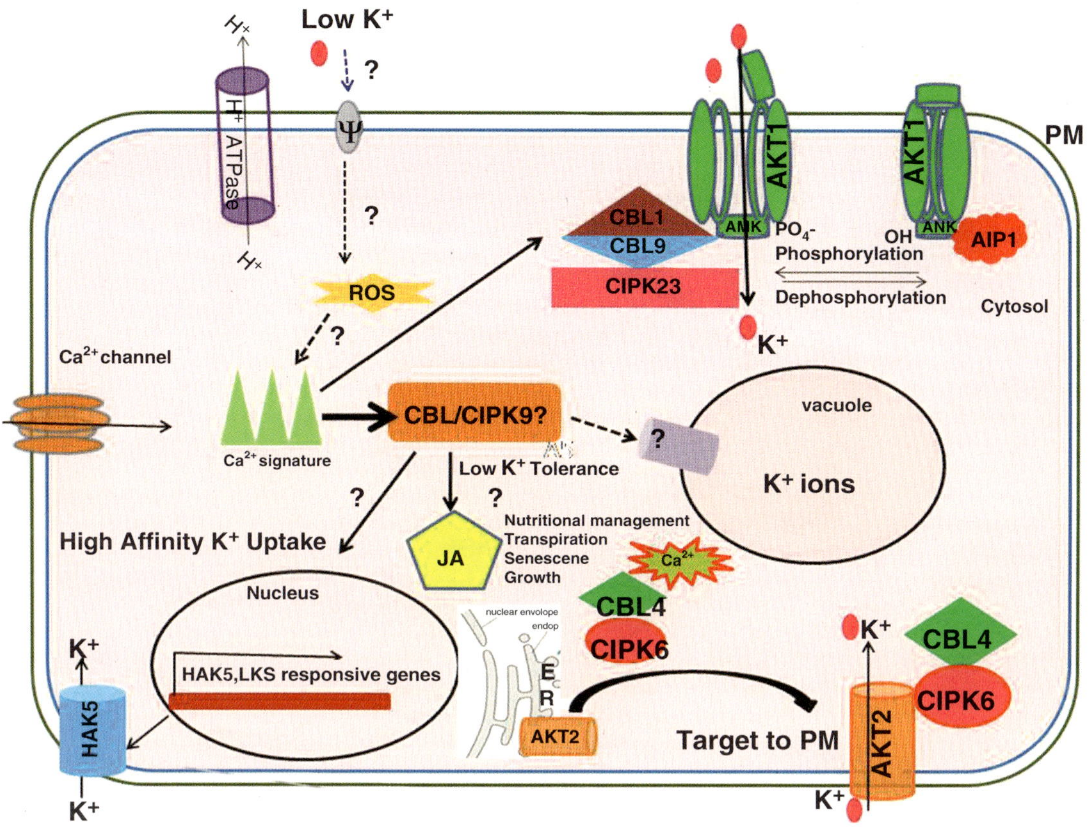


**Fig. 10.1** CBL–CIPK mediated potassium signalling in Arabidopsis. Low-$K^+$ conditions in soil is believed to cause hyperpolarisation of plasma membrane (PM, hyperpolarisation designated as Ψ) by $H^+$ conductance through $H^+$-ATPases and generation of reactive oxygen species (ROS), which lead to activation of voltage-dependent $Ca^{2+}$ channels as well as transcriptional activation of high-affinity $K^+$ uptake responsive genes (HAK5 and LKS). CBL are calcium sensor, which senses the $Ca^{2+}$ signatures and activates CIPKs. CBL1/9–CIPK23–AIP1 complex regulates the AKT1 channel activity causing influx of $K^+$ into the cell in a phosphorylation–dephosphorylation-dependent manner. CIPK9 mediates low-$K^+$ stress tolerance by targeting unknown components hypothesised as vacuolar-localised $K^+$ transporters, channels and downstream transcription factors regulating high-affinity $K^+$ uptake gene expression. Upregulation of JA (jasmonic acid)-mediated response genes under $K^+$ starvation conditions hypothesised to be activated by unknown mechanism through $Ca^{2+}$–CBL–CIPK. Another, CBL4/CIPK6 module translocates AKT2, a $K^+$ channel from ER (endoplasmic reticulum) to plasma membrane (PM) of the cell in a phosphorylation-independent manner leading to $K^+$ acquisition under potassium-deficient conditions

nitrate uptake responses primarily, two CIPKs, that is, CIPK8 and CIPK23, were shown to regulate the low-affinity versus high-affinity nitrate uptake. Phosphorylation mediated by CBL9–CIPK23 module of CHL1 has been shown to change the activity of CHL1 from low-affinity nitrate transporter to high-affinity transporter. Hypothetical presentation in Fig. 10.2 describes the overall mechanism of nitrate sensing involving nitrate transporters and calcium signalling components.

Although these studies revealed a detailed mechanism of mineral nutrient uptake and signalling, there are several questions, which need to be answered in the future. Firstly, how does the sensing of $NO_3^-$ by CHL1 triggers the calcium signalling to regulate the phosphorylation of CHL1? Secondly, which phosphatases are responsible for dephosphorylation of CHL1, and how does the interplay of kinases and phosphatase regulate the conversion of CHL1 from high-affinity to low-affinity nitrate uptake mode

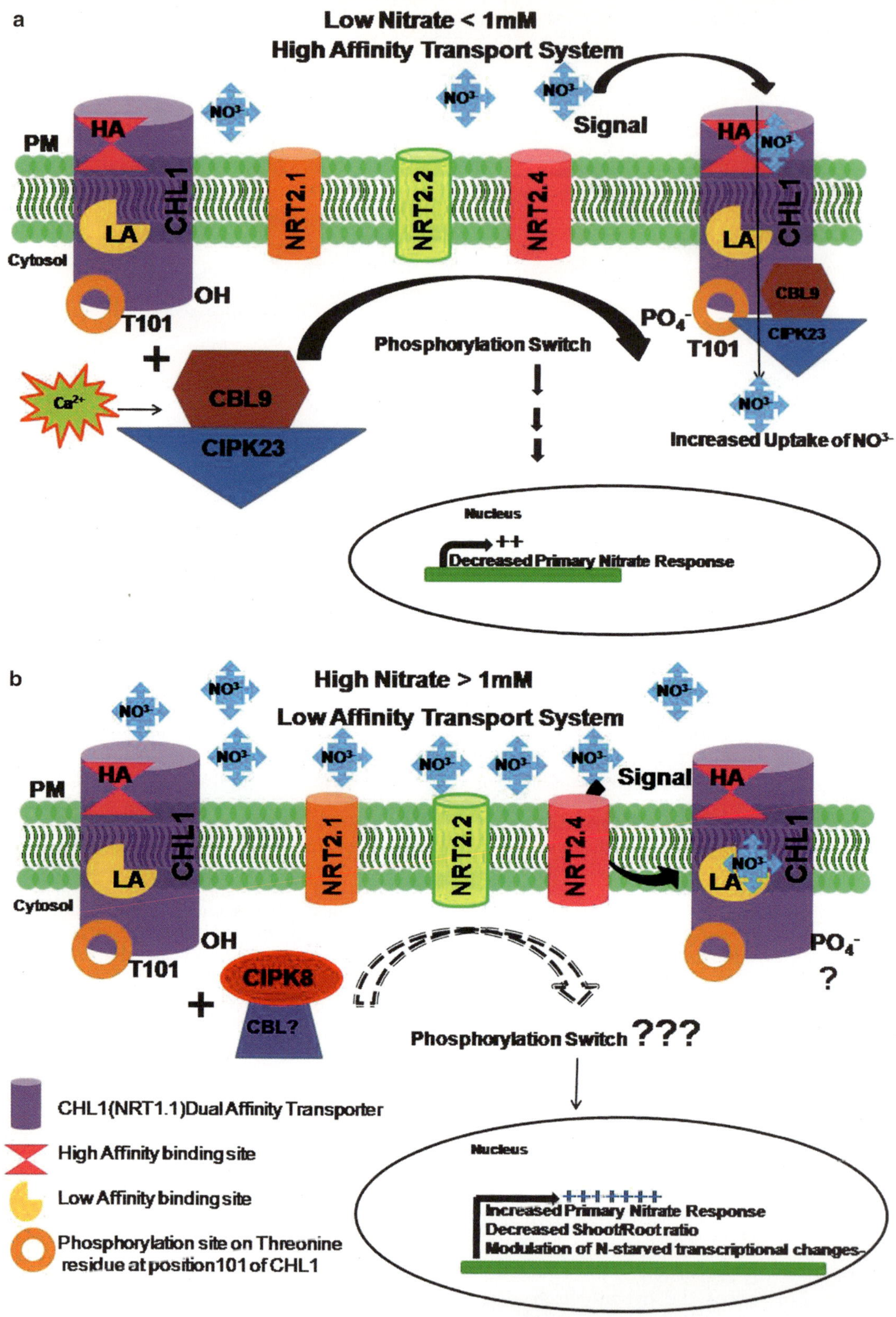


**Fig. 10.2** Schematic model representing nitrate uptake and signalling mediated by CBL–CIPK complex in Arabidopsis. (**a**) CHL1 can act as 'transceptor' where it can sense and uptake nitrate from soil. Low nitrate concentration (<1 mM) activates high-affinity transport system in roots. CBL9–CIPK23 module triggers the phosphorylation switch in CHL1 (NRT1.1), which acts as a sensor by binding of nitrate to its high-affinity site and then

at site of action, that is, root cell? In the case of $K^+$ uptake and signalling by AKT1 from root, a bona fide $K^+$ ion sensor has not been identified. There are no evidences yet how the $K^+$ sensing takes place at the root cell plasma membrane and again how the sensing of $K^+$ leads to generation of calcium signal and activation of CBL–CIPK pathway. Although, ROS is implicated to be involved in connecting ion sensing and activation of calcium transient, however, a clear link has not been established. Therefore, a holistic future analysis involving cell biology and biochemistry is required to investigate these important questions in regulation of this crucial physiological process at molecular level.

Indeed, the identification of these novel mineral nutrient uptake and signalling mechanism in plant has opened up doors for further dissecting the molecular mechanisms underlying the complex nutrient signalling networks and improving the efficiency of nutrient utilisation by plants especially in crop plants growing under nutrition-deficit conditions. The future development of tools and technology for the manipulation of this trait would also lead to the reduction in the usage of fertilisers in agricultural practices and eventually aid in saving the environment from pollution such as eutrophication, global warming and preserving the renewable as well as nonrenewable energy resources. Thus, the challenge lies in protecting and conserving the resources present on earth from premature extinction.

**Acknowledgement** Research in GKP's lab is supported by funding from Department of Biotechnology (DBT), Department of Science and Technology (DST), India. IT acknowledges Council for Scientific and Industrial Research (CSIR), India, for her research fellowship.

## References

Aguera E, Haba P, Fontes AG, Maldonado JM (1990) Nitrate and nitrite uptake and reduction in intact sunflower. Planta 182:149–154

Ahn SJ, Shin R, Schachtman DP (2004) Expression of KT/KUP genes in Arabidopsis and the role of root hairs in $K^+$ uptake. Plant Physiol 134:1135–1145

Anderson JA, Huprikar SS, Kochian LV, Lucas WJ, Gaber RF (1992) Functional expression of a probable *Arabidopsis thaliana* potassium channel in *Saccharomyces cerevisiae*. Proc Natl Acad Sci USA 89:3736–3740

Armengaud P, Breitling R, Amtmann A (2004) The potassium dependent transcriptome of *Arabidopsis* reveals a prominent role of jasmonic acid in nutrient signalling. Plant Physiol 136:2556–2576

Arnon DI, Stout PR (1939) Molybdenum as an essential element for higher plants. Plant Physiol 14:599–602

Ashley MK, Grant M, Grabov A (2006) Plant responses to potassium deficiencies: a role for potassium transport proteins. J Exp Bot 57:425–436

Aslam M, HuVaker RC, Rains DW (1984) Early effects of salinity on nitrate assimilation in barley seedlings. Plant Physiol 76:321–325

Aslam M, Travis RL, HuVaker RC (1992) Comparative kinetics and reciprocal inhibition of uninduced and induced barley (*Hordeum vulgare* L.) seedlings. Plant Physiol 99:1124–1133

Batistic O, Kudla J (2004) Integration and channeling of calcium signaling through the CBL calcium sensor/CIPK protein kinase network. Planta 219:915–924

Batistič O, Kudla J (2012) Analysis of calcium signaling pathways in plants. Biochim Biophys Acta 1820:1283–1293

Berger H, Pachlinger R, Morozov I, Goller S, Narendja F, Caddick M, Strauss J (2006) The GATA factor AreA regulates localization and in vivo binding site occupancy of the nitrate activator NirA. Mol Microbiol 59:433–446

Burger G, Tilburn J, Scazzocchio C (1991) Molecular cloning and functional characterization of the pathway-specific regulatory gene nirA, which controls nitrate assimilation in *Aspergillus nidulans*. Mol Cell Biol 11(2):795–802

**Fig. 10.2** (continued) subsequent transport into the cytosol by activation of calcium-mediated phosphorylation of CHL1 at T101 residue. This leads to the low level of primary nitrate response gene expression and makes CHL1 a high-affinity nitrate transporter (**b**) Under high external nitrate conditions, there is binding of the nitrate to the low-affinity site of CHL1 and activation of CIPK8, which also physically interacts with it. This interaction stimulates the expression of primary nitrate response modulated genes. The interaction and phosphorylation mediated by CBL–CIPK8 with any of the nitrate transporters have not been established, which is represented as question mark (?) in this hypothetical model. Thus, CHLI acts as a dual-affinity transporter and generates different levels of the responses depending upon the nitrate availability in the soil

Bernreiter A, Ramon A, Fernandez-Martinez J, Berger H, Araujo-Bazan L, Espeso EA, Pachlinger R, Gallmetzer A, Anderl I, Scazzocchio C, Strauss J (2007) Nuclear export of the transcription factor NirA is a regulatory checkpoint for nitrate induction in *Aspergillus nidulans*. Mol Cell Biol 27:791–802

Blatt MR (2000) Cellular signaling and volume control in stomatal movements of plants. Annu Rev Cell Dev Biol 16:221–241

Cavicchioli R, Chiang RC, Kalman LV, Gunsalus RP (1996) Role of the periplasmic domain of the *Escherichia coli* NarX sensor-transmitter protein in nitrate-dependent signal transduction and gene regulation. Mol Microbiol 21:901–911

Chen YF, Wang Y, Wu WH (2008) Membrane transporters for nitrogen, phosphate and potassium uptake in plants. J Integr Plant Biol 50:835–848

Cheong YH, Pandey GK, Grant JJ, Batistic O, Li L, Kim BG, Lee SC, Kudla J, Luan S (2007) Two calcineurin B-like calcium sensors, interacting with protein kinase CIPK23, regulate. Plant J 52:223–239

Cherel I (2004) Regulation of $K^+$ channel activities in plants: from physiological to molecular aspects. J Exp Bot 55:337–351

Chiang RC, Cavicchioli R, Gunsalus RP (1997) 'Locked-on' and 'locked-off' signal transduction mutations in the periplasmic domain of the *Escherichia coli* NarQ and NarX sensors affect nitrate- and nitrite-dependent regulation by NarL and NarP. Mol Microbiol 24:1049–1060

Coruzzi GM, Bush DR (2001) Nitrogen and carbon nutrient and metabolite signaling in plants. Plant Physiol 125:61–64

Crawford NM (1995) Nitrate: nutrient and signal for plant growth. Plant Cell 7:859–868

Crawford NM, Glass ADM (1998) Molecular and physiological aspects of nitrate uptake in plants. Trends Plant Sci 3:389–395

Czempinski K, Gaedeke N, Zimmermann S, MuellerRoeber B (1999) Molecular mechanisms and regulation of plant ion channels. J Exp Bot 50:955–966

Czempinski K, Frachisse JM, Maurel C, Barbier-Brygoo H, Mu¨ller-Ro¨ber B (2002) Vacuolar membrane localization of the Arabidopsis 'two-pore' $K^+$channel KCO1. Plant J 29:809–820

Dang J, Harvey D, Jobbagy A, Grady C Jr (1989) Evaluation of biodegradation kinetic with respirometric data. Res J Water Pollut Control Federation 61:1711–1721

De Angeli A, Monachello D, Ephritikhine G, Frachisse JM, Thomine S, Gambale F, Barbier-Brygoo H (2006) The nitrate/proton antiporter AtCLCa mediates nitrate accumulation in plant vacuoles. Nature 442:939–942

De Angeli A, Moran O, Wege S, Filleur S, Ephritikhine G, Thomine S, BarbierBrygoo H, Gambale F (2009) ATP binding to the C terminus of the *Arabidopsis thaliana* nitrate/proton antiporter, AtCLCa, regulates nitrate transport into plant vacuoles. J Biol Chem 284:26526–26532

Dechorgnat J, Nguyen CT, Armengaud P, Jossier M, Diatloff E, Filleur S, Daniel-Vedele F (2011) From the soil to the seeds: the long journey of nitrate in plants. J Exp Bot 62:1349–1359

Doddema H, Telkamp GP (1979) Uptake of nitrate by mutants of *Arabidopsis thaliana* disturbed in uptake or reduction of nitrate. II. Kinetics. Physiol Plant 45:332–338

Epstein E (1972) In: Epstein E (ed) Mineral nutrition of plants: principles and perspectives. Wiley, New York, pp 325–344

Epstein E (1973) Mechanisms of ion transport through plant cell membranes. Int Rev Cytol 34:123–167

Esser JE, Liao YJ, Schroeder JI (1997) Characterization of ion channel modulator effects on ABA- and malate-induced stomatal movements: strong regulation by kinase and phosphatase inhibitors, and relative insensitivity to mastoparans. J Exp Bot 48:539–550

Evans NH, McAinsh MR, Hetherington AM (2001) Calcium oscillations in higher plants. Curr Opin Plant Biol 4:415–420

Evans AM, Mustard KJ, Wyatt CN, Peers C, Dipp M, Kumar P, Kinnear NP, Hardie DG (2005) Does AMP-activated protein kinase couple inhibition of mitochondrial oxidative phosphorylation by hypoxia to calcium signaling in O2-sensing cells? J Biol Chem 280:41504–41511

Fan L-M, Zhao Z, Assmann SM (2004) Guard cells: a dynamic signalling model. Curr Opin Plant Biol 7:537–546

Forde BG (2000) Nitrate transporters in plants: structure, function and regulation. Biochem Biophys Acta 1465:219–235

Forde BG, Clarkson DT (1999) Nitrate and ammonium nutrition of plants: physiological and molecular perspectives. Adv Bot Res 30:1–90

Forde BG, Lorenzo H (2001) The nutritional control of root development. Plant Soil 232:57–68

Foreman J, Demidchik V, Bothwell JHF, Mylona P, Miedema H, Torres MA, Linstead P, Costa S, Brownlee C, Jones JDG et al (2003) Reactive oxygen species produced by NADPH oxidase regulate plant cell growth. Nature 422:442–446

Fu Y-H, Kneesi JY, Marzluf GA (1989) Isolation of ni t - 4, the minor nitrogen regulatory gene which mediates nitrate induction in *Neurospora crassa*. J Bacteriol 171:4067–4070

Gaymard F, Pilot G, Lacombe B, Bouchez D, Bruneau D, Boucherez J, Michaux-Ferriere N, Thibaud JB, Sentenac H (1998) Identification and disruption of a plant shaker-like outward channel involved in K + release into the xylem sap. Cell 94:647–655

Geiger D, Scherzer S, Mumm P, Stange A, Marten I, Bauer H, Ache P, Matschi S, Liese A, Al-Rasheid KA, Romeis T, Hedrich R (2009) Activity of guard cell anion channel SLAC1 is controlled by drought-stress signaling kinase-phosphatase pair. Proc Natl Acad Sci USA 106:21425–21430

Gierth M, Maser P (2007) Potassium transporters in plants–Involvement in K+ acquisition, redistribution and homeostasis. FEBS Lett 581:2348–2356

Gierth M, Maser P, Schroeder JI (2005) The potassium transporter AtHAK5 functions in $K^+$ deprivation induced high-affinity K+ uptake and AKT1 $K^+$ channel contribution to K+ uptake kinetics in *Arabidopsis* roots. Plant Physiol 137:1105–1114

Gowri G, Kenis JD, Ingemarsson B, Redinbaugh MG, Campbell WH (1992) Nitrate reductase transcript is expressed in the primary response of maize to environmental nitrate. Plant Mol Biol 18:55–64

Guo FQ et al (2001) The *Arabidopsis* dual-affinity nitrate transporter gene AtNRT1.1 (CHL1) is activated and functions in nascent organ development during vegetative and reproductive growth. Plant Cell 13:1761–1777

Hamilton DW, Hills A, Kohler B, Blatt MR (2000) $Ca^{2+}$ channels at the plasma membrane of stomatal guard cells are activated by hyperpolarization and abscisic acid. Proc Natl Acad Sci USA 97:4967–4972

Harmon AC, Gribskov M, Harper JF (2000) CDPKs. A kinase for every $Ca^{2+}$ signal? Trends Plant Sci 5:154–159

Hashimoto K, Eckert C, Anschuetz U, Scholz M, Held K, Waadt R, Reyer A et al (2012) Phosphorylation of Calcineurin B-like (CBL) calcium sensor proteins by their CBL-interacting protein kinases (CIPKs) is required for full activity of CBL-CIPK complexes towards their target proteins. J Biol Chem 287:7956–7968

Hedrich R, Kudla J (2006) Calcium signaling networks channel plant $K^+$ uptake. Cell 125:1221–1223

Held K, Pascaud F, Eckert C, Gajdanowicz P, Hashimoto K, Corratgé-Faillie C, Offenborn JN et al (2011) Calcium-dependent modulation and plasma membrane targeting of the AKT2 potassium channel by the CBL4/CIPK6 calcium sensor/protein kinase complex. Cell Res 21:1116–1130

Ho C-H, Tsay YF (2011) Nitrate, ammonium, and potassium sensing and signaling. Curr Opin Plant Biol 13:604–610

Ho CH, Lin SH, Hu HC, Tsay YF (2009) CHL1 functions as a nitrate sensor in plants. Cell 138:1184–1194

Hole DJ, Emran AM, Fares Y, Drew MC (1990) Induction of nitrate transport in maize roots, and kinetics of influx, measured with nitrogen-13. Plant Physiol 93:642–647

Hosy E, Vavasseur A, Mouline K, Dreyer I, Gaymard F, Poree F, Boucherez J, Lebaudy A, Bouchez D, Very AA et al (2003) The *Arabidopsis* outward $K^+$ channel GORK is involved in regulation of stomatal movements and plant transpiration. Proc Natl Acad Sci USA 100:5549–5554

Hu HC, Wang YY, Tsay YF (2009) AtCIPK8, a CBL-interacting protein kinase, regulates the low-affinity phase of the primary nitrate response. Plant J 57:264–278

Huang NC et al (1999) Cloning and functional characterization of an *Arabidopsis* nitrate transporter gene that encodes a constitutive component of low-affinity uptake. Plant Cell 11:1381–1392

Hundal HS, Taylor PM (2009) Amino acid transceptors: gate keepers of nutrient exchange and regulators of nutrient signalling. Am J Physiol Endocrinol Metab 296:E603–E613

Jackson RB, Caldwell MM (1993) The scale of nutrient heterogeneity around individual plants and its qualification with geostatistics. Ecology 74:612–614

Kim EJ, Kwak JM, Uozumi N, Schroeder JI (1998) AtKUP1: an Arabidopsis gene encoding high-affinity potassium transport activity. Plant Cell 10:51–62

Kim MJ, Ciani S, Schachtman DP (2010) A peroxidase contributes to ROS production during *Arabidopsis* root response to potassium deficiency. Mol Plant 3:420–427

Kim TH, Böhmer M, Hu H, Nishimura N, Schroeder JI (2010) Guard cell signal transduction network: advances in understanding abscisic acid, $CO_2$, and $Ca^{2+}$ signaling. Annu Rev Plant Biol 61:561–591

Kwak JM, Mori I, Pei Z-M, Leonhardt N, Torres MA, Dangl JL, Bloom R, Bodde S, Jones JDG, Schroeder JI (2003) NADPH oxidase AtrbohD and AtrbohF genes function in ROS-dependent ABA signaling in *Arabidopsis*. EMBO J 22:2623–2633

Laegreid M, Bockman OC, Kaarstad O (1999) Agriculture, fertilizers and the environment. CABI (CABI Publishing in association with Norsk Hydro ASA), Oxon, UK, 294 pp

Lan WZ, Lee SC, Che YF, Jiang YQ, Luan S (2011) Mechanistic analysis of AKT1regulation by the CBL-CIPK-PP2CA interactions. Mol Plant 4:527–536

Larsson CM, Ingemarsson B (1989) Molecular aspects of nitrate uptake in higher plants. In: Wray J, Kinghorn J (eds) Molecular and genetic aspects of nitrate assimilation. Oxford University Press, Oxford, pp 3–24

Lebaudy A, Ve'ry AA, Sentenac H (2007) $K^+$ channel activity in plants: genes, regulations and functions. FEBS Lett 581:2357–2366

Leckie CP, McAinsh MR, Montgomery L, Priestley AJ, Staxen I, Webb AAR, Hetherington AM (1998) Second messengers in guard cells. J Exp Bot 49:339–349

Lee SC, Lan WZ, Kim BG, Li L, Cheong YH, Pandey GK, Lu G, Buchanan BB, Luan S (2007) A protein phosphorylation/dephosphorylation network regulates a plant potassium channel. Proc Natl Acad Sci USA 104:15959–15964

Lee SC, Lan W, Buchanan BB, Luan S (2009) A protein kinase-phosphatase pair interacts with an ion channel to regulate ABA signaling in plant guard cells. Proc Natl Acad Sci USA 106(50):21419–21424

Leigh RA, Jones RGW (1984) A hypothesis relating critical potassium concentrations for growth to the distribution and functions of this ion in the plant-cell. New Phytol 97:1–13

Lejay L, Tillard P, Lepetit M, Olive F, Filleur S et al (1999) Molecular and functional regulation of two

NO3− uptake systems by N- and C-status of *Arabidopsis* plants. Plant J 18:509–519

Li L, Kim BG, Cheong YH, Pandey GK, Luan S (2006) A $Ca^{(2)+}$ signaling pathway regulates a $K^{(+)}$ channel for low-K response in *Arabidopsis*. Proc Natl Acad Sci USA 103:12625–12630

Li W, Wang Y, Okamoto M, Nigel M, Crawford NM, Siddiqui MY, Glass ADM (2007) Dissection of the AtNRT2.1:AtNRT2.2 inducible high-affinity nitrate transporter gene cluster. Plant Physiol 143:425–433

Little DY et al (2005) The putative high-affinity nitrate transporter NRT2.1 represses lateral root initiation in response to nutritional cues. Proc Natl Acad Sci USA 102:13693–13698

Liu KH, Tsay YF (2003) Switching between the two action modes of the dual-affinity nitrate transporter CHL1 by phosphorylation. EMBO J 22:1005–1013

Liu KH, Huang CY, Tsay YF (1999) CHL1 is a dual-affinity nitrate transporter of *Arabidopsis* involved in multiple phases of nitrate uptake. Plant Cell 11:865–874

Luan S (2009) The CBL–CIPK network in plant calcium signaling. Trends Plant Sci 14:37–42

Luan S, Kudla J, Rodriguez-Concepcion M, Yalovsky S, Gruissem W (2002) Calmodulins and calcineurin B-like proteins: calcium sensors for specific signal response coupling in plants. Plant Cell 14(Suppl): S389–S400

Luan S, Lan W, Chul Lee S (2009) Potassium nutrition, sodium toxicity, and calcium signaling: connections through the CBL–CIPK network. Curr Opin Plant Biol 12:339–346

MacRobbie EAC (1998) Signal transduction and ion channels in guard cells. Philos Trans R Soc Lond B Biol Sci 353:1475–1488

Marschner H (1995) Functions of mineral nutrients: macronutirents. In: Mineral nutrition of higher plants, 2nd edn. Academic, New York, pp 299–312

McAinsh MR, Clayton H, Mans®eld TA, Hetherington AM (1996) Changes in stomatal behavior and guard cell cytosolic free calcium in response to oxidative stress. Plant Physiol 111:1031–1042

Miller AJ, Fan X, Orsel M, Smith SJ, Wells DM (2007) Nitrate transport and signalling. J Exp Bot 58:2297–2306

Munos S, Cazettes C, Fizames C, Gaymard F, Tillard P, Lepetit M, Lejay L, Gojon A (2004) Transcript profiling in the chl1-5 mutant of *Arabidopsis* reveals a role of the nitrate transporter NRT1.1 in the regulation of another nitrate transporter, NRT2.1. Plant Cell 16:2433–2447

Pandey GK (2008) Emergence of a novel calcium signalling pathway in plants: CBL-CIPK signaling network. Physiol Mol Biol Plants 14:51–68

Pandey GK, Cheong YH, Kim BG, Grant JJ, Li L, Luan S (2007) CIPK9: a calcium sensor-interacting protein kinase required for low-potassium tolerance in *Arabidopsis*. Cell Res 17:411–421

Pei ZM, Murata Y, Benning G, Thomine S, Klusener B, Allen GJ, Grill E, Schroeder JI (2000) Calcium channels activated by hydrogen peroxide mediate abscisic acid signaling in guard cells. Nature 406:731–734

Peiter E, Maathuis FJM, Mills LN, Knight H, Pelloux M, Hetherington AM et al (2005) The vacuolar $Ca^{2+}$ -activated channel TPC1 regulates germination and stomatal movement. Nature 434:404–408

Pilot G, Gaymard F, Mouline K, Cherel I, Sentenac H (2003) Regulated expression of Arabidopsis shaker K1 channel genes involved in K1 uptake and distribution in the plant. Plant Mol Biol 51:773–787

Pouteau S, Cherel I, Vaucheret H, Caboche M (1989) Nitrate reductase mRNA regulation in *Nicotiana plumbaginifolia* nitrate reductase–deficient mutants. Plant Cell 1:1111–1120

Reddy VS, Ali GS, Reddy AS (2002) Genes encoding calmodulin-binding proteins in the *Arabidopsis* genome. J Biol Chem 277:9840–9852

Redinbaugh MG, Campbell WH (1991) Higher plant responses to environmental nitrate. Physiol Plant 82:640–650

Remans T, Nacry P, Pervent M, Filleur S, Diatloff E, Mounier E, Tillard P, Forde BG, Gojon A (2006) The *Arabidopsis* NRT1.1 transporter participates in the signaling pathway triggering root colonization of nitrate-rich patches. Proc Natl Acad Sci USA 103:19206–19211

Roelfsema MR and Hedrich R (2010) Making sense out of $Ca^{2+}$ signals: their role in regulating stomatal movements. Plant Cell Environ 33(3):305–21

Rubio F, Nieves-Cordones M, Aleman F, Martinez V (2008) Relative contribution of AtHAK5 and AtAKT1 to K+ uptake in the high-affinity range of concentrations. Physiol Plant 134:598–608

Santa-Maria GE, Rubio F, Dubcovsky J, Rodriguez-Navarro A (1997) The HAK1 gene of barley is a member of a large gene family and encodes a high-affinity potassium transporter. Plant Cell 9:2281–2289

Schachtman DP (2000) Molecular insights into the structure and function of plant $K^+$ transport mechanisms. Biochim Biophys Acta 1465(1–2):127–139

Schachtman DP, Shin R (2007) Nutrient sensing and signalling: NPKS. Annu Rev Plant Biol 58:47–69

Scheible WR, Gonzalez Fontes A, Lauerer M, Muller Rober B, Caboche M, Stitt M (1997a) Nitrate acts as a signal to induce organic acid metabolism and repress starch metabolism in tobacco. Plant Cell 9:783–798

Scheible WR, Lauerer M, Schulze ED, Caboche M, Stitt M (1997b) Accumulation of nitrate in the shoot acts as a signal to regulate shoot root allocation in tobacco. Plant J 11:671–691

Schroeder JI (1988) K+ transport properties of K+ channels in plasma membrane of *Vicia faba* guard cells. J Gen Physiol 92:667–683

Schroeder JI (1989) Quantitative analysis of outward rectifying $K^+$ channel currents in guard cell protoplasts from *Vicia faba*. J Membr Biol 7:229–235

Schroeder JI, Raschke K, Neher E (1987) Voltage dependence of $K^+$ channels in guard cell protoplasts. Proc Natl Acad Sci USA 84:4108–4112

Schroeder BC, Hechenberger M, Weinreich F, Kubisch C, Jentsch TJ (2000) KCNQ5, a novel potassium channel broadly expressed in brain, mediates M-type currents. J Biol Chem 275:24089–24095

Schroeder JI, Allen GJ, Hugouvieux V, Kwak JM, Waner D (2001a) Guard cell signal transduction. Annu Rev Plant Physiol Plant Mol Biol 52:627–658

Schroeder JI, Kwak JM, Allen GJ (2001b) Guard cell abscisic acid signalling and engineering drought hardiness in plants. Nature 410:327–330

Shin R, Schachtman DP (2004) Hydrogen peroxide mediates plant root cell response to nutrient deprivation. Proc Natl Acad Sci USA 101(23):8827–8832

Shin R, Berg RH, Schachtman DP (2005) Reactive oxygen species and root hairs in *Arabidopsis* root response to nitrogen, phosphorus and potassium deficiency. Oxford journals. Plant Cell Physiol 46 (8):1350–1357

Siddiqi MY, Glass ADM, Ruth TJ, Fernando M (1989) Studies of the regulation of nitrate influx by barley seedlings using $^{13}NO_3^-$. Plant Physiol 90:806–813

Siddiqi MY, Glass ADM, Ruth TJ, Rufty TW (1990) Studies of the uptake of nitrate in barley. I. Kinetics of $^{13}$NO, – influx. Plant Physiol 93:1426–1432

Snedden WA, Fromm H (1998) Calmodulin, calmodulin-related proteins and plant responses to the environment. Trends Plant Sci 3:299–304

Snedden WA, Fromm H (2001) Calmodulin as a versatile calcium signal transducer in plants. New Phytol 151:35–66

Stitt M (1999) Nitrate regulation of metabolism and growth. Curr Opin Plant Biol 2:178–186

Trewavas AJ, Malho R (1997) Signal perception and transduction: the origin of the phenotype. Plant Cell 9(7):1181–1195

Tsay YF, Schroeder JI, Feldmann KA, Crawford NM (1993) The herbicide sensitivity gene CHL1 of *Arabidopsis* encodes a nitrate-inducible nitrate transporter. Cell 72:705–713

Tsay YF, Chiu CC, Tsai CB, Ho CH, Hsu PK (2007) Nitrate transporters and peptide transporters. FEBS Lett 581:2290–2300

Tsay YF et al (2011) Integration of nitrogen and potassium signaling. Annu Rev Plant Biol 62:207–226

Vahisalu T, Puzõrjova I, Brosché M, Valk E, Lepiku M, Moldau H, Pechter P, Wang YS, Lindgren O, Salojärvi J, Loog M, Kangasjärvi J, Kollist H (2008) Ozone-triggered rapid stomatal response involves the production of reactive oxygen species, and is controlled by SLAC1 and OST1. Plant J 62(3):442–453

Vahisalu T, Kollist H, Wang YF, Nishimura N, Chan WY, Valerio G, Lamminmäki A, Brosché M, Moldau H, Desikan R, Schroeder JI, Kangasjärvi J (2010) SLAC1 is required for plant guard cell S-type anion channel function in stomatal signalling. Nature 452 (7186):487–491

Vert G, Chory J (2009) A toggle switch in plant nitrate uptake. Cell 138(6):1064–1066

Véry AA, Sentenac H (2002) Cation channels in the Arabidopsis plasma membrane. Trends Plant Sci 7 (4):168–175

Véry AA, Sentenac H (2003) Molecular mechanisms and regulation of $K^+$ transport in higher plants. Annu Rev Plant Biol 54:575–603

Von Wirén N, Bergfeld A, Ninnemann O, Frommer WB (1997a) OsAMT1-1: A high-affinity ammonium transporter fromrice (*Oryza sativa* cv. Nipponbare). Plant Mol Biol 3:681

Von Wirén N, Gazzarrini S, Frommer WB (1997b) Regulation of mineral nitrogen uptake in plants. Plant Soil 196:191–199

Walch-Liu P, Forde BG (2008) Nitrate signalling mediated by the NRT1.1 nitrate transporter antagonises l-glutamate-induced changes in root architecture. Plant J 54:820–828

Walker DJ, Leigh RA, Miller AJ (1996) Potassium homeostasis in vacuolated plant cells. Proc Natl Acad Sci USA 93:10510–10514

Wang R, Crawford NM (1996) Genetic identification of a gene involved in constitutive, high-affinity nitrate transport in higher plants. Proc Natl Acad Sci USA 93:9297–9301

Wang P, Song CP (2008) Guard-cell signalling for hydrogen peroxide and abscisic acid. New Phytol 178:703–718

Wang Y, Wu WH (2010) Plant sensing and signaling in response to $K^+$- deficiency. Mol Plant 3:280–287

Wang R et al (1998) The Arabidopsis CHL1 protein plays a major role in high-affinity nitrate uptake. Proc Natl Acad Sci USA 95:15134–15139

Wang YY, Hsu PK, Tsay YF (2012) Uptake, allocation and signaling of nitrate. Trends Plant Sci 971:1–10

Xu J, Li HD, Chen LQ, Wang Y, Liu LL, He L, Wu WH (2006) A protein kinase, interacting with two calcineurin B-like proteins, regulates K+ transporter AKT1 in Arabidopsis. Cell 125:1347–1360

Zhang H, Forde BG (1998) An *Arabidopsis* MADS box gene that controls nutrient-induced changes in root architecture. Science 279:407–409

Zhang YJ, Lynch JP, Brown KM (2003) Ethylene and phosphorus availability have interacting yet distinct effects on root hair development. J Exp Bot 54:2351–2361

Zielinski RE (1998) Calmodulin and calmodulin binding proteins in plants. Annu Rev Plant Physiol Plant Mol Biol 49:697–725

# Isothermal Calorimetry and Raman Spectroscopy to Study Response of Plants to Abiotic and Biotic Stresses

# 11

Andrzej Skoczowski and Magdalena Troć

**Abstract**

This review presents applying a typical chemical methods such as isothermal calorimetry and Raman sectroscopy to study response of plants to aboiotic and biotic stress, so how can be used analytical techniques to better understunding of plant phisiology.

Plants produce heat during metabolism, and measurements of these metabolic heat could lead understanding of plant physiology. Calorimetric measurements have proven to be useful as monitors for many types of biological processes such as seed germination, seedlings growth or plant tissue vitality. Differences in amount of heat production give the information about how big is the impact of stress factors on plant.

Raman spectroscopy can be used for in situ analysis of valuable substances in living plant tissue. Chemical compounds (primary and secondary metabolites) content could be done as point-measurements and distribution as a Raman mapping. This techniques allow to obtain the data as a Raman-spectra, which present some characteristic bands of plants components. These bands give informations about the specific chemical composition of plant tissue. Two-dimensional Raman maps provide insight not only into the distribution, but also into relative content of the compounds in the specified area of the plant tissue.

Combination of spectroscopy and hierarchical cluster analysis enables a simple, quickly and reliable method for chemotaxonomy charakterization.

A. Skoczowski
Department of Plant Physiology, Institute of Biology, Pedagogical University of Cracow, Podchorążych 2 Street, 30-084 Cracow, Poland
e-mail: skoczows@up.krakow.pl

M. Troć (✉)
The Franciszek Górski Institute of Plant Physiology, Polish Academy of Sciences, Niezapominajek 21 Street, 30-239 Cracow, Poland
e-mail: mtroc@ifr-pan.krakow.pl

G.R. Rout and A.B. Das (eds.), *Molecular Stress Physiology of Plants*,
DOI 10.1007/978-81-322-0807-5_11, 

Isothermal calorimetry and Raman spectroscopy, especially combination of these two analitical techniques is a very good tool to describe the susceptibility and response of plant to abiotic and biotic stress factors.

## Isothermal Calorimetry

### Calorimetric Methods in Plant Stress Physiology

#### Introduction

According to Criddle and Hansen (1999), calorimetry plays a unique role in the measurement of metabolic plant properties. These authors recognised that "calorimetry provides more than just another means for measuring metabolic rate because it measures a fundamentally different property (energy) while other methods measure mass". Although at the beginning of the twentieth century it was recognised that plants produce heat during metabolism and that measurements of the metabolic heat of plants could guide understanding of plant physiology, only recently has metabolic heat in plants become a rapidly expanding area of investigation (for review see Criddle and Hansen 1999).

#### Theoretical Basis

Criddle et al. (1991a) recognised that "heat production is an inevitable by-product of all metabolic processes". Equations describing metabolic heat production were published, with minor differences, in several papers (Criddle and Hansen 1999; Smith et al. 1999, 2001; Hansen et al. 2002, 2004; Thygerson et al. 2002). We present below equations for catabolic and anabolic reactions according to Thygerson et al. (2002).

Aerobic respiration covers catabolic as well as anabolic processes. In catabolic reactions, organic substrates are oxidised to produce $CO_2$. Part of the energy produced by oxidation is used to produce ATP from ADP and inorganic phosphate ($P_i$). The rest of the energy is lost as heat.

$$\text{Organic substrate} + O_2 + \text{xADP} + \text{xP}_i \rightarrow CO_2 + H_2O + \text{xATP} + \text{heat}$$

ATP produced in catabolic reactions is used for cellular work, including anabolism:

$$\text{Substrate} + \text{yATP} \rightarrow \text{growth} + \text{yADP} + \text{yP}_i + \text{heat}$$

In anabolism, heat and new plant tissue are produced, and ATP is hydrolyzed back to ADP and $P_i$. A calorimeter measures the rate of heat loss ($R_q$) from both catabolism and anabolism. For that reason, in most cases, calorimetric measurements of metabolic heat production alone are not sufficient to analyse the metabolic efficiency of plant tissues.

As early as 1991, Criddle and co-workers postulated that only the combined measurements of heat rates, $CO_2$, and $O_2$ fluxes provide important information on the bioenergetic efficiency of cell metabolism (Criddle et al. 1991a). This problem was widely presented in several papers (Criddle et al. 1997; Criddle and Hansen 1999; Hansen et al. 2002, 2004; Smith et al. 2001). Roughly, respiration processes in living tissues are described by three parameters: the rate of heat production ($R_q$), the rate of oxygen consumption ($R_{o_2}$), and the rate of carbon dioxide production ($R_{CO_2}$). From these three measures of respiratory rate, three ratios can be calculated, i.e. $R_q/R_{CO_2}, R_q/R_{O_2}$, and $R_{CO_2}/R_{O_2}$. There are differences in the information content of $R_q/R_{CO_2}$ and $R_q/R_{O_2}$, so that each ratio provides unique information on metabolic activities and efficiency. Measurement of both ratios provides information not available from either alone (for details see Hansen et al. 2004).

#### Some Practical Remarks

Unfortunately, terminology concerning calorimetry is not homogeneous with respect to both instruments and measured values. For example, the poorly defined term "microcalorimeter" is used to indicate that the instrument's sensitivity is in the range of 1 μW or better. Consequently, in this chapter, we use the term isothermal calorimeter. Not long ago, Hansen (2001) proposed a systematic nomenclature for describing calorimeters. According to Hansen, the instrument's name and description must include four parts: the method of heat measurement, a description of the temperature control of the surroundings, a description of the means of initiating the heat effect, and a description of the operation of the calorimeter, including data analysis.

The problem is also with the uniform names for measured values. In the literature, many synonyms for metabolic heat production exist, such as heat production rate, heat flow, heat emission, heat evolution, heat output, and, finally, metabolic heat rate. Recently the term specific thermal power was introduced.

How can the best calorimeter for the investigation of living plants materials be chosen? The answer is not simple because, unfortunately, no single source adequately describes the many applications or current capabilities of calorimeters. Applications of calorimetry to metabolism are not widely known or fully understood. Furthermore, most books on calorimetry are too basic or too specific for the plant physiologist, and physical chemistry textbooks include only the basics on traditional applications. The literature is also a poor guide for selecting the optimum calorimeter for a given task. Hansen and Russell (2006) recently described a procedure for choosing the best calorimeter for a given task.

Practically all calorimetric experiments conducted on living plant systems have been carried out under dark condition. Isothermal photocalorimeters are not available on the market, so they are constructed individually by particular researchers (Petrov et al. 1995; Janssen et al. 2005; Mukhanov and Kemp 2006, Alyabyev et al. 2007, Oroszi et al. 2011).

New trend in calorimetric measurements was described by Lamprecht and Schmolz (2000). In speculative paper, they presented idea that modern biological calorimetry may go afield in cases where biological system is fixed in their place (e.g. plants).

Isothermal calorimetric techniques used in measurements on living systems are very well described by Wadsö (1995) and Criddle and Hansen (1999).

### Plant Material

Recent calorimetric investigations in the plant field have been conducted on cell suspensions (Rank et al. 1991; Janeczko et al. 2007b), callus tissues (Płażek et al. 2000a), algae (Loseva et al. 1995, 1998, 2002, 2003, 2004; Alyabyev et al. 2007), mosses (Możdżeń and Skoczowski – in this chapter), seeds (Schabes and Sigstad 2005), seedlings (Płażek and Rapacz 2000; Alyabyev et al. 2002, Skoczowski et al. 2011b), cotyledons (Janeczko et al. 2007a; Skoczowski et al. 2011a), leaves (Ádám et al. 1993; Fodor et al. 2007; Baltruschat et al. 2008), roots and tuber tissue (Minibayeva et al. 1998; Gordon et al. 2004; Wadsö et al. 2004; Alyabyev et al. 2007), and fruits (Criddle et al. 1991b).

## Applications in Stress Physiology

Response of plants to temperature was extensively discussed by Criddle and Hansen (1999) and Smith et al. (1999, 2001). Recently, a lot of papers described influence of the climate at the place of origin (e.g. temperature) on the intensity of plant metabolism and respiration (Smith et al 2002; Thygerson et al. 2002; Anekonda et al. 2004; Keller et al. 2004).

Below we presented our selected studies about influence of low temperatures on plant metabolism.

### Low and High Temperature

#### Response to Low Temperature

Regenerable cell cultures offer an in vitro system for mutant selection, propagation of varieties, gene transfer experiments, and biochemical studies (Janeczko et al. 2007b). The main aim of the investigation conducted by Janeczko and coworkers (2007b) was to determine the impact of zearalenone and 24-epibrasinolide on differentiation of winter wheat callus developed from cell suspension cultured for 0–9 weeks at 5°C.

Nevertheless, because the survival of winter wheat suspension for a longer period of cold had not been studied earlier, the changes in metabolic activity (heat production rate) of wheat suspension over 9 weeks at low temperature will be also characterised. Using isothermal calorimetry, it was shown that the cell suspension of winter wheat may be cultured for 9 weeks without the necessity of changing the medium. Its metabolic activity is slowed down, and there is observed a stage of acclimation to cold and subsequently of fast growth and stabilisation. Suspension may be used to study the response to cold of isolated cells (Janeczko et al. 2007b).

The precise time and temperature dependence of the decrease of metabolism of cultured cells of tomato (*Lycopersicon esculentum* L.) resulting from exposures to high and low temperatures were determined by Rank et al. (1991). Analysis of metabolic heat rates, $O_2$-consumption rates, and $CO_2$-evolution rates demonstrated a simultaneous shift in metabolic pathways and metabolic activities towards more anaerobic metabolism below about 12°C and at high temperatures that stress the growth of tomato cells.

Skoczowski and Troć (data not yet published) are probably the first researchers to comprehensively investigate the energy dependence associated with the early stages of the growth of seedlings of different varieties of wheat at chilling temperatures. Papers relating to the calorimetric measurement of heat energy accompanying the germination of seeds are relatively numerous (Sigstad and Prado 1999; Sigstad and Garcia 2001; Schabes and Sigstad 2005, 2006). However, the authors did not encounter any work done at temperatures near 0°C for seeds or seedlings of plants. This is probably due to lack of access to isothermal calorimeters allowing this type of measurement. The study's aim was to determine the differences in the response of different cultivars of winter wheat seedlings to changes of growth temperature (from 20 to 0.5°C). The specific thermal power accompanying the growth of seedlings at a temperature of 0.5°C is shown below. Tests were performed on seedlings of winter wheat (*Triticum aestivum* L.) cultivars: Trend, Kobiera, Alcazar, and Ludwig. The seeds were germinated at 20°C for 24 h. Then 20 seedlings, which were morphologically similar to each other, were placed in a ventilated calorimetric ampoule. The cell with distilled water was placed on the bottom of ampoule and covered with plastic mesh in order to isolate seedlings from direct contact with the water, which were also in an atmosphere of saturated water vapour. The reference ampoule contained water only.

Measurements were carried out in an isothermal calorimeter TAM I (Thermometric AB, Jarfalla, Sweden) at 0.5°C for 6 days, with 5 replicates for each variety. The monitoring of the heat flow was by the computer programme DIGITAM (Thermometric AB, Sweden). After measurement the seedlings were lyophilised and weighed, and the specific thermal power was determined based on the dry weight.

In Fig. 11.1, the specific thermal power–time curves obtained for the tested winter wheat cultivars are shown. Due to the large differences in thermal power emitted by the seedling cultivars (from about 70 $mW \cdot g_{DW}^{-1}$ in the case of cv. Trend up to about 180 $mW \cdot g_{DW}^{-1}$ for cv. Ludwig), a detailed analysis of the changes in the various pattern curves required individual adjustment of the Y scale.

Based on the data, tested cultivars can be divided into those with high heat production (cv. Ludwig and cv. Kobiera) and those with a much lower heat production (cv. Trend and cv. Alcazar). The variety Trend is characterised by low values of specific thermal power (max. about 75 $mW \cdot g_{DW}^{-1}$), and the characteristic mark of this variety is its fast acclimatisation to the new thermal conditions. The increase of specific thermal power, after an initial decrease caused by thermal shock, appears on the second day at 0.5°C (Fig. 11.1a). The Kobiera variety seems to be very resistant to rapid temperature changes, because the decrease of thermal power caused by the change of seedlings' growth temperature is little (about a few $mW \cdot g_{DW}^{-1}$), and then a rapid increase of the thermal power is observed (Fig. 11.1b). In the case of the Alcazar variety, a decrease of specific thermal power, which occurs on day 6 of growth at 0.5°C, is

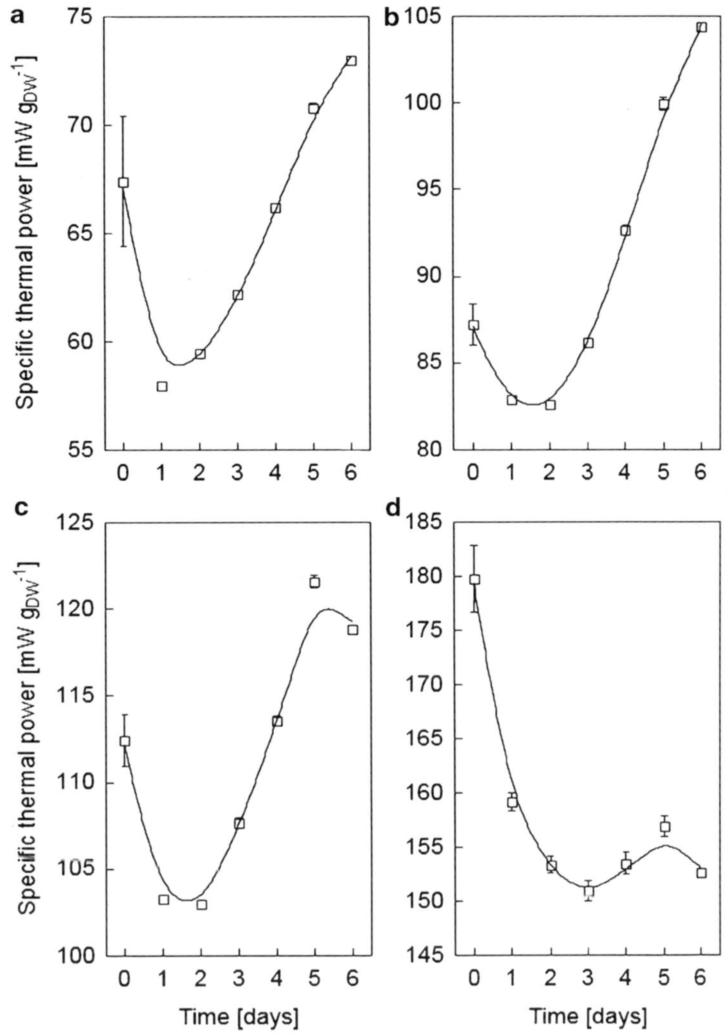


**Fig. 11.1** Specific thermal power–time curves for winter wheat seedlings cv. Trend (**a**), cv. Kobiera (**b**), cv. Alcazar (**c**), and cv. Ludwig (**d**) transferred, after 24 h of growth at 20–0.5°C on 6 days. Mean values from five repetitions ± SD. Interpolation by the method of least squares

characteristic, as this does not occur in other varieties (Fig. 11.1c).

The variety Ludwig has a high emission of heat in the cold. At the same time, processes of acclimatisation to the new thermal conditions proceed slowly in this variety, as evidenced by the absence of a significant increase of thermal power during the investigated period of time (Fig. 11.1d).

## Heat Shock

Loseva et al. (1998) investigated both the effect of extreme temperature (45°C) and high salt concentration (450 mM) on the rate of heat evolution by *Chlorella* cells. In fact, this work is connected with adaptive changes in energy of plant cells under stress. On the other hand, Alyabyev et al. (2002) examined the influence of blue and red light on the resistance of energetic processes of wheat seedlings to treatment by stress temperature (45°C). The authors concluded that blue light has a positive effect on the resistance of wheat seedlings to high temperature. In light of what has been said above, it seems that high-temperature stress (heat shock) is still an open field for research using isothermal calorimetry.

## Salinity

Most crops in saline environments are negatively affected in their rate of growth. This effect is attributed either to osmotic causes or to ion toxicity, depending on the plant species, salt composition, and salt concentration. A serious problem affecting arable land is its increasing salinity. Moreover, there is still a lack of well-defined plant indicators that could be used by plant breeders to improve agricultural crops for their tolerance to saline environments.

The effect of salinity stress on metabolic heat production by barley (*Hordeum vulgare* L.) roots has been measured by isothermal calorimetry (Criddle et al. 1989). Experiments were performed on three barley varieties differing in tolerance to salinity. Root tips were put on moistened filter paper in a calorimeter ampoule, and heat efflux was measured. The roots were then flushed with solutions in increasing concentrations of NaCl, and metabolic heat rates were remeasured at each concentration. Two levels of inhibition by increasing salt concentrations were found. The first, occurring at a concentration of NaCl up to 150 mM, inhibiting metabolic rate at about 50%, was cultivar dependent. At a higher concentration of NaCl (>150 mM), metabolism was further decreased, and this decrease was not cultivar dependent. Inhibition of root tip metabolism was not reversed by washing with NaCl-free solution.

In order to elucidate the physiological responses of *Piriformospora indica*-colonised barley plants to salinisation, Baltruschat et al. (2008) measured important indicators of salt stress, such as metabolic heat production, lipid peroxidation, fatty acid composition, and analysed antioxidant activities. The fungus *Piriformospora indica* colonises roots and increases the biomass of both monocot and dicot plants. The above-mentioned researchers provided clear evidence that salt-induced responses indicated by heat emission in the *P. indica*-infected salt-sensitive barley cv. Ingrid resemble those found in salinity-tolerant plants (cv. California Mariout). Calorimetric studies indicated that the rate of metabolic activity increased in leaves of *P. indica*-infected plants after salt treatment. Therefore, the endophyte seemed to overcompensate the salt-induced inhibition of metabolic leaf activity. This suggests that enhanced tolerance to salt stress can be associated with higher metabolic activity in *P. indica*-colonised barley.

Schabes and Sigstad (2005) evaluated two cultivars of quinoa (*Chenopodium quinoa* Willd.) for tolerance to saline stress by isothermal calorimetric measurements of seed germination in various salt solutions (NaCl, KCl, $Na_2SO_4$, $K_2SO_4$, $Na_2CO_3$). $HgCl_2$ was used in combination with the salts to evaluate the possible existence of channels blocked by the mercurial reagent involved in the transport of ions. They conclude that seeds of cv. Robura are less tolerant to saline stress than are seeds of cv. Sajama, with a tolerance limit for seeds of the former cultivar of 100 mM NaCl.

The salt inhibition of *Chlorella* has been studied by calorimetric methods by Loseva et al. (1995). The addition of low salt concentrations (450 mM of NaCl) causes an increase in metabolic rate. The increased metabolic rate is not sufficient to offset energetic losses, and the growth rate of *Chlorella* is reduced.

The unicellular green halotolerant microalga *Dunaliella maritime* and the freshwater microalga *Chlorella vulgaris* were used by Alyabyev et al. (2007) as suitable model systems in studies on the adaptation of energy mechanisms to changes in salinity. The study of the alteration of energy-yielding processes of cells with different sensitivity to salt stress may be a key to the understanding of the complex mechanisms that play an important role in the adaptation of algae to unfavourable conditions. *Chlorella* was unable to adapt to concentrations of 500 mM NaCl and higher. This microalga has a metabolic rate significantly lower than that of *Dunaliella*. This halotolerant and cell wall-less microalga adapts to very high salinities. This tolerance is connected with the ability to maintain high rates of energy-yielding processes. Alyabyev et al. (2007) showed that the heat production, $O_2$ uptake, and $O_2$ evolution rates increased in *Dunaliella* cells in conditions up to 2M salt in the medium. These data may support the opinion that an important feature of salt-sensitive and salt-tolerant algae is the increased rate of energy dissipation, which ensures quick and effective microalgal adaptation.

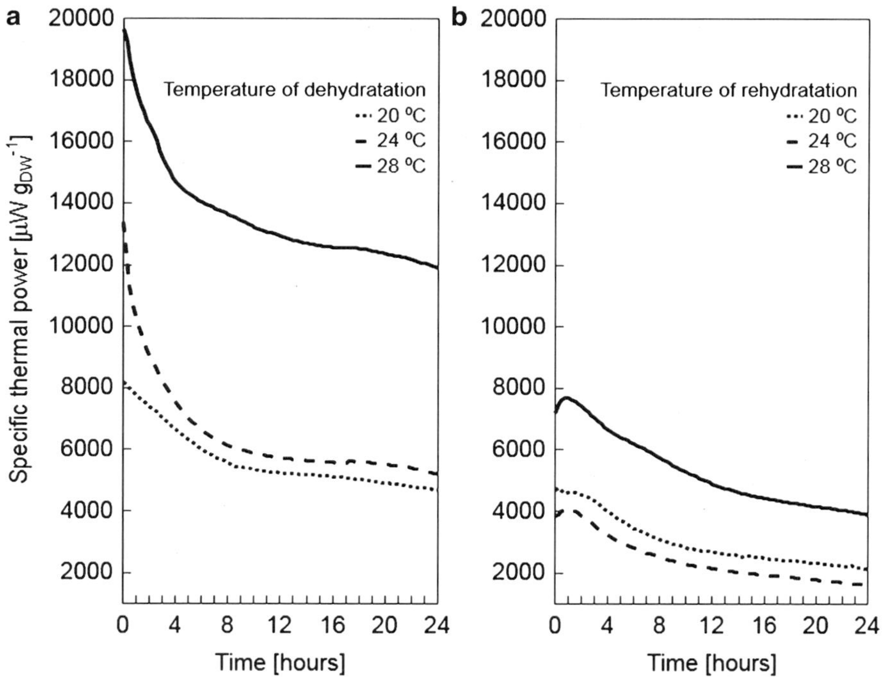


**Fig. 11.2** Typical specific thermal power–time curves ($\mu W \cdot g_{DW}^{-1}$) for plants of mosses *Mnium undulatum* obtained during dehydration (**a**) and rehydration (**b**) processes at 20, 24, and 28°C (According Możdżeń and Skoczowski)

## Drought

Calorimetry may be used also for estimation of plant tolerance to drought. For example, relative degree of drought tolerance was tested for some populations of small burnet (*Sanguisorba minor* Scop.) and few cultivar of alfalfa (*Medicago sativa* L.). Small burnet was more drought tolerant than alfalfa (Jones et al. 1999, cited in Smith et al. 2001). Mosses are known as plants with very high water demand. However, many mosses can survive desiccation, sometimes for months, returning to life even within a few hours of rehydration. Skoczowski and Możdżeń tested, by isothermal calorimetry, plants of the moss *Mnium undulatum* under their response to dehydration and rehydration process (not published data). Measurements were carried out at three temperatures: 20, 24, and 28°C. For dehydration experiments, five plants were put into 20 ml ampoules equipped with lids that enabled air exchange, and specific thermal power curves were recorded during 24 h. After this time, 1 ml of water was added and measurements of metabolic response to rehydration were continued for additional 24 h. Specific thermal power values accompanying dehydration process were much higher than for rehydration process (Fig. 11.2a, b). At both cases, the highest amount of metabolic heat at 28°C was observed. At 20 and 24°C, the metabolic heat production both in de- or rehydration was similar. The most interesting metabolic changes were observed in first hours of de- and rehydration. So, in the case of dehydration, rapid decrease of metabolic activity was observed. On the other hand, the thermal power–time curves showed that the beginning of rehydration is connected with significant increase of metabolic heat production. However, after short time, metabolic heat production significantly decreased. Processes related to drought stress in mousses are still under intensive investigation in our laboratory.

## Herbicides

The continuous employment of herbicides with the same mode of action may lead to resistance in weeds. Herbicide resistance denotes the inherited ability of a plant to survive and reproduce following exposure to a dose of herbicide normally lethal to a wild type (http://www.weedscience.com/).

The most popular methods for detecting herbicide resistance are glasshouse pot experiments and laboratory tests (http://www.plantprotection.org/HRAC/). But these methods are time and space consuming. Nowadays, when resistance to herbicides in many countries has become a serious problem, rapid tests for herbicide resistance detection are needed.

In 1995, Suwanagul (quoted in Criddle and Hansen 1999), studying three species of weeds and three herbicides with different mechanisms of action (atrazine, metsulphuron-methyl, and diclofop-methyl), showed that the metabolic activity (metabolic rate) of sensitive biotypes was inhibited by lower concentrations of herbicide than resistant biotypes.

In the described research, herbicides were applied to young plants of weeds, from which, after a certain time, meristematic tissue was collected and metabolic heat rate measured. A few years later, Stokłosa et al. (2006) proposed a much faster method for testing weeds' resistance to herbicides using isothermal calorimetry. Three-day-old seedlings of wild oat (*Avena fatua* L.) were put into calorimetric ampoules on filter paper moistened with herbicide solution (152 and 40% of the field dose for fenoxaprop and diclofop, respectively). Rate of heat flow was measured for 72 h. However, differences were already visible in the first hours of growth on each herbicide. Rate of heat flow for seedlings' resistant to both herbicides was higher than for susceptible ones. The most evident differences between susceptible and resistant biotypes were noticed after 10–20 h and 25–40 h (of the seedlings' growth) on fenoxaprop and diclofop, respectively, when a sharp increase in the rate of heat flow was observed. In conclusion, isothermal calorimetry may be used as a rapid test for the detection of the resistance of weed biotypes to herbicides, with a large practical application in agriculture.

## Allelopathy

Allelopathy is the direct or indirect influence of chemical compounds (allelochemicals) released from one living plant on the growth and development of another plant (Rice 1984). Many allelochemical compounds can be toxic to higher plants, and therefore, they can be used as a natural herbicide. Nowadays, a new method of weed control is searched for, and allelochemicals are considered to be environmentally friendly herbicides (Vyvyan 2002). This interest has been stimulated by the negative effects of using herbicides. On the one hand, applying herbicides caused the pollution of water and soil. The next problem is that there is a rapid increase in the number of biotypes of weed which are resistant to herbicides (Heap 2006). On the other hand, at the moment, there is, essentially, stagnation in the elaboration of methods of the synthesis of new active herbicidal substances (Steinrücken and Hermann 2000). A source of natural phytotoxins may be products of the secondary metabolism of plants. These may be compounds which have allelopathic potential (Weston and Duke 2003) and which are secreted as substances to aid competition between species (Inderjit and Weiner 2001).

Kupidłowska et al. (2006) probably first used microcalorimetric methods for the study of allelophatic interactions. We investigated the influence of water extracts of sunflower leaves on catabolic processes taking place during mustard (*Sinapis alba* L.) seed germination. Besides metabolic heat rate production, $O_2$ uptake by intact seeds, and energy metabolism by determining adenylate (ATP, ADP, AMP), concentrations in tissues of germinating seeds were also measured. The heat efflux from seeds germinating on water was expressed as a continuous increase during the 40th hour of the culture. Sunflower phytotoxins altered the typical heat emission pattern observed in control seeds. During first 12 h, sharp increases in metabolic heat emission were detected, being twofold higher than in the control. In control seeds, heat emission was correlated with an increased respiration rate, whereas in seeds imbibing sunflower extracts, this correlation was not detected. Maximum heat emission after 12 h was accompanied neither by high oxygen uptake nor by any increase in ATP concentration. Therefore, we cannot exclude the possibility that the transient increase in heat production may be a result of oxidative reactions taking place in seeds exposed to sunflower phytotoxins.

Since undisturbed reserve mobilisation is an important requirement for germination, as well as for subsequent plant growth and development, the influence of various extracts from sunflower and mustard leaves (as autoallelopathy) on catabolic activity during mustard seed germination was investigated further (Troć et al. 2009). Total catabolic activity was measured as the heat production rate (in mW) of germinated seeds in an isothermal calorimeter. It was shown that the heat production rate–time curves of mustard seeds can be correlated with different extracts used for seed germination and, consequently, with different allelopathic compounds which influence the metabolism of seeds. The seeds germinated on the mustard extract have higher values of metabolic activity than those germinated on the sunflower extract. Later investigations confirmed that sunflower and mustard extracts have varied impact on the growth and heat production rate of mustard seedlings (Skoczowski et al. 2011b).

In order to better understand the allelophatic interactions, and to elucidate the impact of various herbal extracts on seedling growth, investigations were initiated using isothermal calorimetry as a monitor (Troć et al. 2011). Seeds of wheat, mustard, rape, and clover were germinated on aqueous herbal extracts from arnica, hypericum, milfoil, ribwort, sage, and sunflower (until the root was visible). Then, seedlings were put into a calorimeter ampoule on moistened filter paper discs containing herbal extracts. The specific thermal power (heat production rate) of the seedlings during their growth was measured over 48 h. As a control seedlings were grown on water. The patterns of the thermal power–time curve during seedling growth on the herbal extracts and on water were completely different. In comparison with the water control, seedling growth on the herbal extracts was accompanied by a strong exothermic peak (first phase), whereas in the second phase distinct endothermic peaks were observed. The time after which the maxima of exo- and endothermic peaks occurred strongly depended on the seedling species and the origin of the herbal extract. Similarly, the total thermal effect connected with seedling growth was correlated with the seedling species and herbal extract type.

Schabes and Sigstad (2007) used isothermal calorimetry to measure the effect of "cnicin" on the germination of the soybean (*Glycine max* L.) Merr. and the radish (*Raphanus sativus* L.). The sesquiterpene lactone cnicin was isolated from a highly invasive plant, the diffuse knapweed *Centaurea diffusa* Lam. Calorimetric experiments were performed with seeds on moistened filter paper discs or in agar, both containing varying concentrations of cnicin. Results indicate that this substance blocks the water uptake by roots, inhibiting subsequent seedling growth, but has no effect during germination.

Numerous studies suggest that plants warn neighbours about stresses and danger through airborne communication via the release of volatiles. Some plant's volatiles are known to be phytotoxic. Recently (Skoczowski – unpublished data) isothermal calorimetry has been used to study the effect of volatile substances released from plant tissues on the metabolism of white mustard seedlings (*Sinapis alba* L.) and winter wheat (*Triticum aestivum* L.). Day-old seedlings were placed in a calorimetric ampoule (volume 20 $cm^3$) on filter paper moistened with 300 μl of distilled water. Into an Eppendorf tube, placed next to the seedlings in the measuring ampoule was introduced 200 μl of juice extruded from fresh garlic. Ampoules were placed in the calorimeter for 24 h, and the emission of thermal power was measured. Control seedlings were grown in the presence of distilled water. Garlic bulb (*Allium sativum* L.) contains numerous sulphur-containing compounds, most of which are derived from allicin. Allicin is generated from alliin by the enzyme alliinase, which is released when garlic is crushed. Allicin is an unstable volatile and is transformed into other more stable sulphur-containing products that are believed to be the bioactive agents. The pattern of curves for particular types of seedlings is different, which may indicate that garlic has a different influence on the metabolism of mustard and wheat. Depending on the storage materials, garlic caused either a reduction or an increase of specific thermal power. The specific thermal

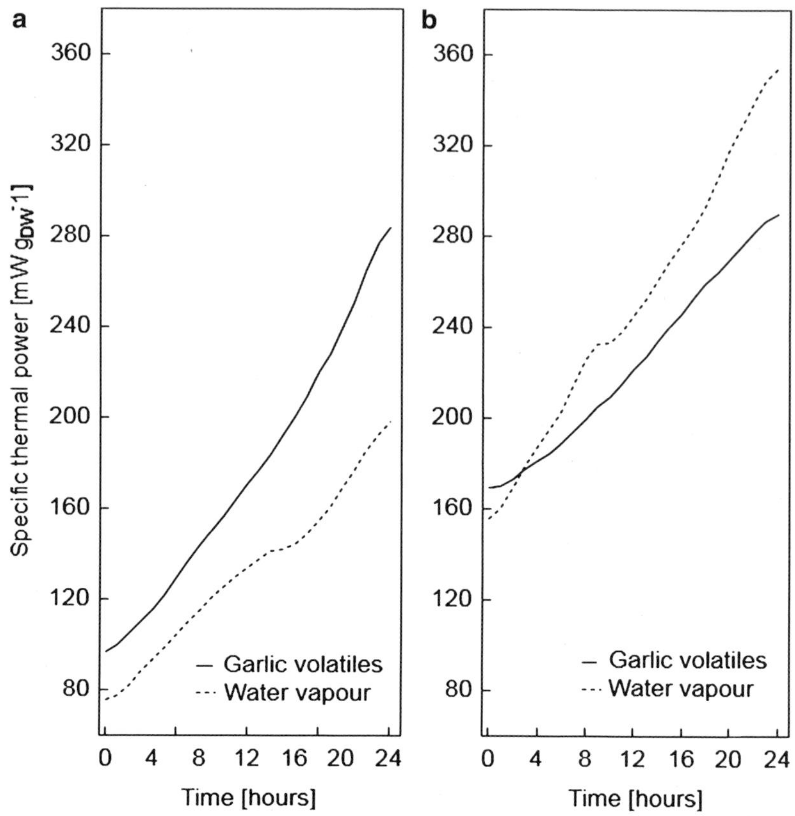


**Fig. 11.3** Specific thermal power–time curves ($mW \cdot g_{DW}^{-1}$) for mustard (**a**) and winter wheat seedlings (**b**), which were grown in air contained volatiles from fresh garlic and in water vapour (control). Measurements were conducted by 24 h at 20°C

power of wheat seedlings is reduced in relation to the control, while the growth of mustard seedlings caused an increase in the specific heat capacity compared to controls (Fig. 11.3).

The presented results demonstrate the power of isothermal calorimetry as a tool to investigate the phenomena of allelopathy. The other methods used for this purpose mainly show inhibition of growth (length, weight), but they do not show changes of metabolic activity. Only with the use of calorimetry can we see the endothermic reaction during seedling growth. It is not possible to use any other method. Moreover, this method precisely showed the moment when additional detailed analytical methods should be used. In other words, it might be desirable to combine isothermal calorimetry with specific analytical techniques (Sigstad and Prado 1999). This trend was proposed earlier by Wadsö (1995, 1997) and is also recommended by Skoczowski et al. (2011a b).

## Wounding Response

In most cases, information about the metabolic response of tissues to wounding stress derived from investigations connected directly with food sciences and technology (Criddle et al. 1991b; Smith et al. 2000; Wadsö et al. 2004, 2005; Rocculi et al. 2005). Studying the wound response in cut vegetables (to quantify metabolic response after cutting) may be used for the optimisation of process parameters, which is of importance for the large-scale market in minimally processed, ready-to-eat vegetables (for a review see Gómez Galindo et al. 2005). Smith et al. (2000) studied metabolic changes in 1 mm slices of potato and found that the heat production rate increased for up to 30 h. They also found a change from lipid to starch as the substrate.

Wadsö et al. (2004) conducted a study by isothermal calorimetry of the heat production response of root and tuber tissue to wounding. Samples with different surface-to-volume ratios were prepared from carrots, potatoes, and rutabaga, and total metabolic heat was measured in an isothermal calorimeter. Authors showed that the total measured heat of cut samples is the sum of normal metabolism and wounding stress heat emission. The results showed that the wound response part was very high. In some cases, almost half of the overall heat derived from the wound response and not from the normal metabolic activity of tissue. Results obtained by Wadsö et al. (2004) are, generally, very important from the methodological point of view because they give a recommendation on how samples should be cut (prepared) for calorimetric measurements.

## Pathogenesis

Pathogenic processes could be monitored by using calorimetry. Calorimetric measurement of heat emission by cotyledons or leaves can be used to complement other tests in studying plant–pathogen interactions. This technique has an advantage in that measurements are continuous and do not interfere with the investigated processes. Moreover, calorimetric changes in plant metabolism caused by external stressors are visible very rapidly. As a result, there have been a lot of investigations designed to reveal the mechanisms employed in the response of plant cells to infection by viruses (Ádám et al. 1993; Fodor et al. 2007), fungus (Płażek et al. 2000b; Płażek and Rapacz 2000), mycoplasma (Loseva et al. 2002, 2003, 2004), and bacteria (Janeczko et al. 2007b; Skoczowski et al. 2011b).

### Virus Infection

Tobacco (*Nicotiana tabacum* L. cv. Xanthi-nc) plants carry the N resistance gene against the *Tobacco mosaic virus* (TMV) and localise the infection to cells adjacent to the site of viral entry, developing a hypersensitive response in the form of local necrotic lesions. Studies performed by Ádám and co-workers (1993) have shown an enhanced heat production in Xanthi-nc plants during the formation of TMV-induced necrotic lesions. A much more detailed analysis of these changes was carried out in wild-type (WT) Xanthi-nc and salicylic acid (SA)-deficient phenotype NahG tobacco plants during TMV-induced cell death (Fodor et al. 2007). TMV-inoculated WT plants develop systemic acquired resistance (SAR), while NahG tobacco plants are not able to establish SAR against a second inoculation by TMV. Heat efflux from uninoculated leaves of NahG plants was found to be significantly lower than from WT tobacco. However, the rates of heat efflux increased up to similar levels in TMV-inoculated WT plants either expressing or not expressing SAR and in NahG plants. The elevated level of heat efflux detected in the TMV-inoculated NahG leaves indicates that heat production can be independent of salicylic acid.

### Fungus

Płażek and Rapacz (2000) checked whether the difference in susceptibility to *Bipolaris sorokiniana* (Sacc.) Shoem between barley and fescue leaves is also reflected in heat emission and respiration courses. Seedlings of spring barley and meadow fescue were inoculated with conidia and mycelium. Leaf respiration was measured using a Clark-type electrode, while heat emission was evaluated by means of an isothermal calorimeter. Leaves of meadow fescue were characterised by a higher respiration rate at the 6th hour, while barley leaves showed an elevated rate at 24th hours after inoculation. In the case of meadow fescue, the greatest heat emission was noted between 24 and 168 h after inoculation, whereas barley leaves emitted the greatest amount of heat only during the first 3 h of the pathogenesis. The observed opposing reaction of respiration rate and heat emission in the infected seedlings of both species may illustrate a disorder in metabolic processes in plants during pathogenesis (Płażek and Rapacz 2000).

Sensitivity of spring barley and meadow fescue to the leaf spot pathogen *Bipolaris sorokiniana* was also investigated on the callus tissues (Płażek et al. 2000a, b). Callus reaction to fungus phytotoxins was examined on the basis of the amount of total phenolics and heat emission. The results obtained show that spring barley and meadow fescue differ in the rate of response to *Bipolaris sorokiniana* phytotoxins with regard to changes in metabolic activity and phenolic content. The authors suggested a different degree of the plants' sensitivity to *B. sorokiniana* at tissue level.

### Mycoplasma

A calorimetric study has demonstrated that oxygen uptake and superoxide burst are also in strong correlation with heat flow rate in *Chlorella vulgaris* alga inoculated by a mycoplasma, *Acholeplasma laidlawii* (Loseva et al. 2002). Chlorella is extensively used as a suitable model for investigating pathogenesis (Loseva et al. 2003, 2004) because of the structural and physiological similarity of these protists to higher plants.

### Bacteria

Studies by Janeczko et al. (2007a) and Skoczowski et al. (2011a) illustrated the usefulness of isothermal calorimetry in investigations on plant response to bacterial infection. These researchers showed that heat emission (metabolic activity) was markedly increased in cotyledons of spring oilseed rape (*Brassica napus* L.) infected with the incompatible bacterium *Pseudomonas syringae* pv. *syringae*. Heat emission continued to increase as the hypersensitive response progressed, reflecting an increase in the metabolic rate in the infected tissue (Janeczko et al. 2007a). The authors also investigated the possibly protective role of 24-epibrasinolide ($BR_{27}$) in defence of the reactions of plants against bacterial infection. Pretreatment with $BR_{27}$ markedly increased the metabolic activity (heat production rate) in infected cotyledons and increased host plant resistance to necrosis by caused bacterial infection.

The study on the protective effect of $BR_{27}$ in oilseed rape cotyledons against infection by an incompatible wild type of, a hypersensitive response mutant of, and saprophytic *Pseudomonas* bacteria was continued by Skoczowski et al. (2011a). In this study, changes in metabolic activity (heat flow) during pathogenesis were also analysed. The following strains of *Pseudomonas* were used: *P. syringae pv. syringae* (*Ps*), *P. syringae pv. syringae hrcC* mutant (*Pm*), and *P. fluorescence* (*Pf*). The study was carried out using two cultivars of spring oilseed rape (*Brassica napus* L.): 'Licosmos' and 'Huzar'. The values of heat flow in all treatments, except for cotyledons infected with *Ps*, decreased during 20 h after inoculation. However, the curves of heat flow for *Ps*-infected cotyledons showed a completely different pattern, with at least two peaks. $BR_{27}$ pretreated cotyledons infected with *Ps* had higher heat flow in comparison to *Ps*-infected ones. As shown, the first effects of pathogenesis occurred 4 h after inoculation. The large exothermic peaks observed in *Ps* and $BR_{27}$ *Ps* calorimetric curves were an indicator of tissue response to an incompatible pathogen (*Ps*). $BR_{27}$ treatment did not change the specific enthalpy of cotyledon growth ($\Delta gh$) for both cultivars if compared with absolute control. However, infection with *Ps* markedly increased $\Delta gh$ values by about 200% for both cultivars. In conclusion the authors suggested the protective action of $BR_{27}$ in oilseed rape cotyledons after bacterial infection with *Pseudomonas* (Skoczowski et al. 2011a).

### Ozone Stress

Tropospheric ozone ($O_3$) has become one of the strongest factors affecting agricultural and forest plants and other types of vegetation (Kraft et al. 1996). The concentration of tropospheric $O_3$ has risen at least twice in the last 100 years, mainly because of the industrial revolution; it will continue to rise if anthropogenic activity remains at the same level (Wieser et al. 2001). The negative influence of ozone on plants depends on the concentration of $O_3$ in the genetic background and on the efficiency of plant defence mechanisms (Calatayud and Barreno 2004). This particle is responsible for the destruction of the structure and function of biological membranes and for changes in the properties of thylakoids (Płażek et al. 2000b). At the same time, there exist some reports which indicate that higher ozone concentration can lead to enhanced efficacy of protective systems through an increase in the levels of apoplastic and symplastic ascorbates and glutathione (Wieser et al. 2001; Kronfuß et al. 1998).

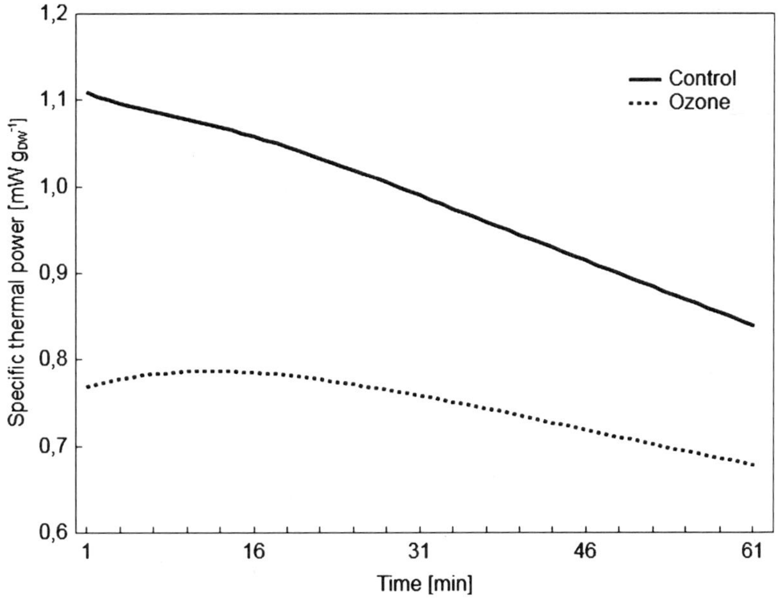


**Fig. 11.4** Specific thermal power–time curves ($mW \cdot g_{DW}^{-1}$) for cv. Optico F1 Chinese cabbage leaves treated with ozone and for control ones. Mean values from six determinations

Recently, Wdowin and Skoczowski (unpublished data) investigated the impact of ozone on metabolic activity (heat production rate) by Chinese cabbage leaves (*Brassica pekinensis* lour.). Five-week-old cabbage plants were fumigated by ozone (100 ppb) for 18 days (8 h $day^{-1}$). The control constituted plants which were grown in similar conditions but not exposed to ozone (Fig. 11.4). After 18 days, the seventh leaves from the treated and control plants were harvested for analysis of metabolic activity. Measurements were carried out over 60 min on a TAM III isothermal calorimeter at 20°C. The above-mentioned researchers have shown, among other findings, that ozone causes a significant decrease of leaves' metabolic activity.

## Raman Spectroscopy

### Basis of the Method

Raman scattering is dedicated to inelastic scattering of photons, where the Raman shift of the scattered light is measured in two ways. If the frequency of the scattered photon is less than the incident photon, then Stokes scattering occurs. If the frequency of the scattered photon is greater than the incident photon, then anti-Stokes scattering occurs. The invention of the laser makes Raman spectroscopy a very useful and versatile technique in probing low-frequency elementary excitation (Szczepaniak 2004).

Raman spectroscopy can be used to obtain information about the vibrational spectrum of a sample, and this can lead to an understanding of its chemical composition. This is possible for many organic compounds in which different chemical bonds have very characteristic vibrational frequencies. The intensity or power of a Raman peak depends in a complex way upon the polarisability of the molecule, the intensity of the source, and the concentration of the active group. Raman intensity is usually directly proportional to the concentration of the active substance in the sample (Twardowski and Anzenbacher 1994; Cygański 2002).

### Applying Raman Spectroscopy in Plant Physiology

Recently there has been considerable development in Raman spectroscopy with Fourier transformation (FT-Raman). It makes this technique useful for characterising and identifying chemical compositions in living tissue. This allows the study of analytes in situ in their natural environment, which is not possible when using other analytical techniques. Raman measurements with the excitation in the near-infrared range can usually be performed without any mechanical, chemical, photochemical, or thermal decomposition of the sample (Schrader et al. 1999, 2000). In these conditions, plant tissues do not absorb light, so fluorescence can also be avoided. The major advantage of the Raman technique in comparison to other standard analytical methods is its non-destructive effect on plant tissue. Therefore, when using Raman spectroscopy, information on chemical composition can be obtained without any need to disrupt samples. The first applications of NIR-FT-Raman spectroscopy for non-destructive measurements of various plant tissues were found to be very promising, and therefore, this method has been used extensively for this purpose in the last few years (Baranska et al. 2005, 2006a, b; Baranski et al. 2005). Samples can be analysed at ambient temperature and pressure without any special preparation required.

Raman spectroscopy can be successfully applied for identifying various plant components if characteristic key bands of individual analyte molecules are found in the spectrum. This technique can also be considered as a potential analytical tool in taxonomy. The usefulness of Raman spectroscopy for general plant classification was reported previously (Reitzenstein et al. 2007; Baranski and Baranska 2008). Raman measurements of single seeds enabled the correct separation of accessions in accordance with their taxonomical classification (Baranski et al. 2006).

The Raman spectrum, the dependence of the intensity of scattered radiation from the quantum energy gap of the incident and scattered radiation, contains information about the characteristic vibration of the studied molecules, which is then used to identify and determine the total amount of the chemical compound in the sample. In the Raman spectrum, the intensity of radiation is given in arbitrary units, referring to the selected standard (Kęcki 1998).

Using a modern FT-Raman spectrometer, it is possible not only to non-destructively detect chemical compounds which are contained in plants but also to study their distribution in the plant. In this way, it can create two-dimensional maps showing the chemical composition of the tested samples, and what is more, the coexistence of various plant components in the same tissue can be studied in a single analysis. Analysed materials can be both primary metabolites (mono- and oligosaccharides, fatty acids, amino acids, proteins) and secondary metabolites: alkaloids (Schulz et al. 2005), flavonoids (Baranska et al. 2006a), carotenoids (Baranski et al. 2005, Baranska et al. 2006b), terpenoids (Baranska et al. 2005, Rösch et al. 1999), and polyacetylenes (Baranska et al. 2005).

## Applications in Stress Physiology

### Allelopathy

Allelopathy is a direct or indirect influence of chemical compounds (allelochemicals) released from one living plant on the growth and development of another plant (see section "Allelopathy"). Investigations into allelopathic potential usually begin with testing the activity of phytotoxicity crude extracts from plant donors. The next step is determining the participation of individual phytotoxic compounds in investigated extracts. So, we prepared crude extracts from sunflower (*Helianthus annuus* L.) and mustard (*Sinapis alba* L.) leaves, and we wanted to check the allelopathic effect and the autoallelopathic effect of these extracts on the growth of mustard seedlings (Skoczowski et al. 2011b).

In Fig. 11.5 are presented the FT-Raman spectra obtained from mustard cotyledons germinated on water – the control – and sunflower and mustard leaf extracts. The band at 1,655 and 1,630 $cm^{-1}$, together with signals at 1,300 and 1,266 $cm^{-1}$, may be associated with fatty acids. The first two Raman bands are related to the content of individual *cis* and *trans* isomers present in various edible oils (Sadeghi-Jorabchi et al. 1991). The total degree of unsaturation can also be determined by calculating the ratio of the intensity of a band at 1,266 $cm^{-1}$ to that of a band at 1,300 $cm^{-1}$ (Sadeghi-Jorabchi et al.

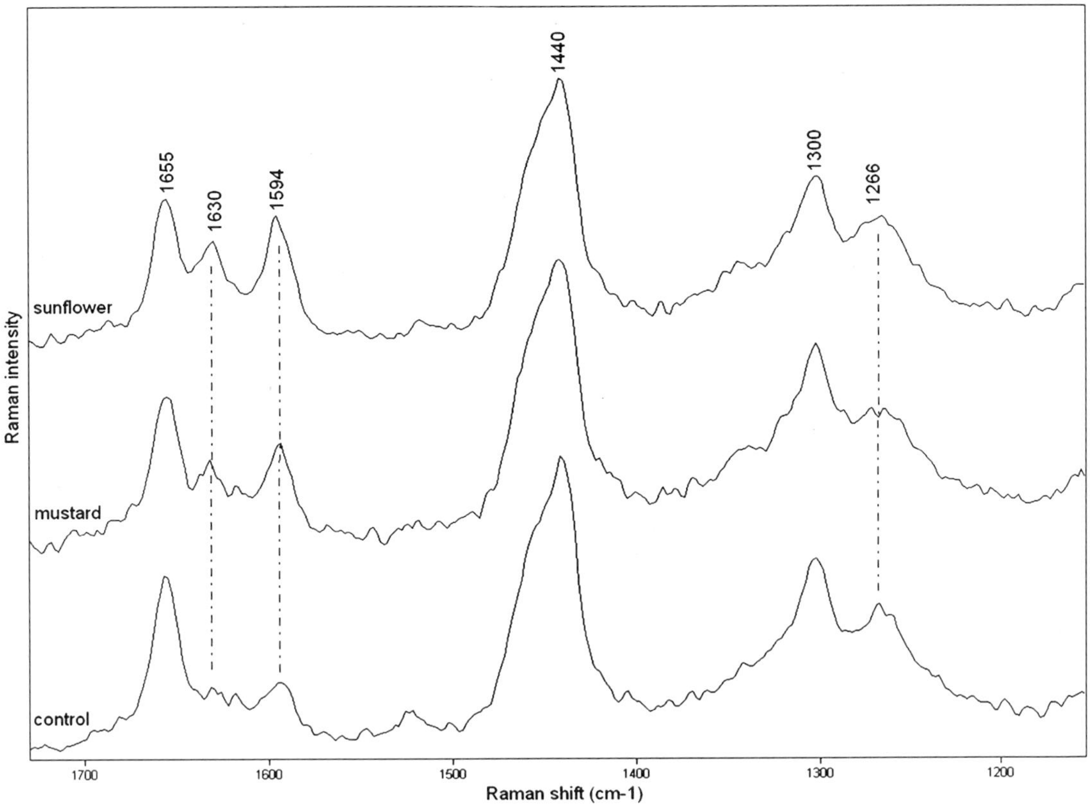


**Fig. 11.5** FT-Raman spectra of mustard cotyledons germinated on extracts from sunflower (*top*), mustard leaves (*middle*), and control (*bottom*) (From Skoczowski et al. 2011b)

1990; Sadeghi-Jorabchi et al. 1991). On the spectra, a characteristic flavonoid band near 1,590 $cm^{-1}$ can be detected (Fig. 11.5).

Some changes can be seen in the intensities of Raman bands originating from various plant components. Because there is a correlation between signal intensity and the concentration of analyte, the observed differences indicate various contents of the components, i.e. fatty acids, carotenoids, flavonoids. To obtain this information, some chemometric methods should be used. In order to find meaningful and systematic differences among the measured spectra of mustard cotyledons, cluster analysis was applied. When the analysis was limited to 1,320–1,680 $cm^{-1}$ (Fig. 11.6), a distinct discrimination between cotyledons germinated on extracts from sunflower and mustard leaves and the control was achieved. These results indicate that changes of plant constituents caused by chemical compounds from extract are related to characteristic Raman signals in this range. The fact is that samples treated by sunflower leaf extract are in a completely different group. This indicates that the allelocompounds from this extract much more strongly influence the metabolism of seedlings as compared with extract from mustard leaves.

FT-Raman spectroscopy with hierarchical cluster analysis is a very good tool to compare changes in the chemical composition of cotyledons (or other plant tissues) caused by allelochemicals. For example, the spectra of mustard seed cotyledons showed that allelochemicals mainly influenced the degree of saturation of fatty acids but also in the same way the metabolism of flavonoids.

## Ozone Stress

Tropospheric ozone ($O_3$) has become one of the strongest factors affecting agricultural and forest

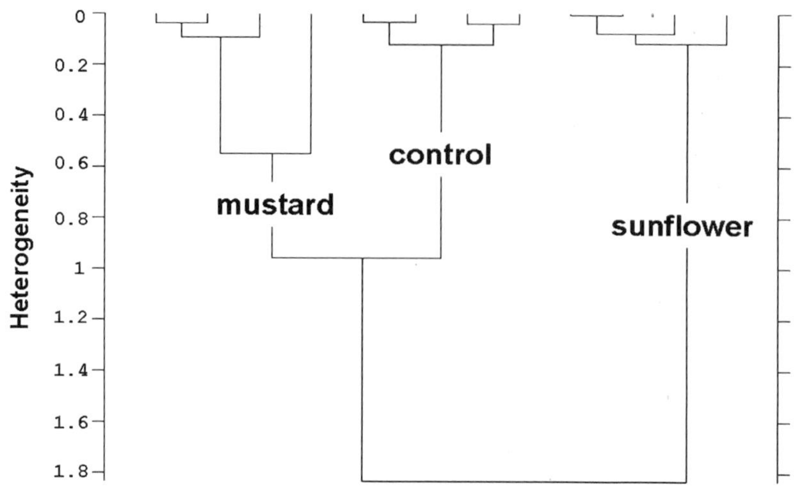


**Fig. 11.6** Dendrogram showing classification of mustard cotyledons treated with mustard and sunflower extracts and control cotyledons after cluster analysis of the FT-Raman spectra at the wave number range of 1,320–1,680 $cm^{-1}$ using the vector normalisation and factorisation (three factors) algorithm for calculating spectral distances and Ward's algorithm. "Mustard", cotyledons germinating on mustard extract; "Control", cotyledons germinating on water; "Sunflower", cotyledons germinating on sunflower extract (From Skoczowski et al. 2011b)

plants and other types of vegetation (see section "Ozone Stress").

A considerable decrease in the content of assimilation pigments in cabbage plants, especially carotenoids, can have great significance for plant quality. Carotenoids are widespread natural pigments fulfilling two essential biological functions: light energy collection and photoprotection (Astorg 1997). The photoprotection role of carotenoids originates from their ability to quench and to deactivate reactive oxygen species, such as singlet oxygen formed from exposure to light and air. This role is also associated with its antioxidant activity for human health, i.e. protecting biological membrane systems against free radicals and inhibiting a relatively early stage of carcinogenesis (Kim et al. 2001). The reactivity of carotenoids depends on the length of the chain of conjugated double bonds and the characteristics of terminal groups. Astaxanthin, zeaxanthin, and lutein are excellent lipid-soluble antioxidants that scavenge free radicals, especially in a lipid-soluble environment. Carotenoids in sufficient concentrations can prevent lipid oxidation and related oxidative stress (Liu 2004). Although carotenoids occur in plants as minor components at the ppm level (Ozaki et al. 1992), a very sensitive detection can be achieved by resonance Raman in the visible region when the wave number of laser excitation coincides with the electronic transition of an individual carotenoid (Withnall et al. 2003; Schoefs 2002). NIR-FT-Raman spectroscopy also gives a strong enhancement of carotenoids due to the known pre-resonance effect; furthermore, disturbance of the fluorescence effect of biological material usually observed when laser excitation is performed in the visible range is avoided (Ozaki et al. 1992).

In section "Ozone Stress", the calorimetric method for investigating the impact of ozone on the metabolism of plant tissue is described. Here we demonstrate the application of FT-Raman spectroscopy for the analysis of plants under ozone stress. For this purpose, broccoli (*Brassica oleracea* var. *botrytis cymosa*) and white cabbage (*B. oleracea* var. *capitata* f. *alba*) plants were exposed (at the stage of 5–6 leaves) to elevated ozone concentration (70 ppb, 6 h daily), whereas non-treated plants were taken as the control. After 6 days, leaves from 5 different plants of each kind of treatment (broccoli and

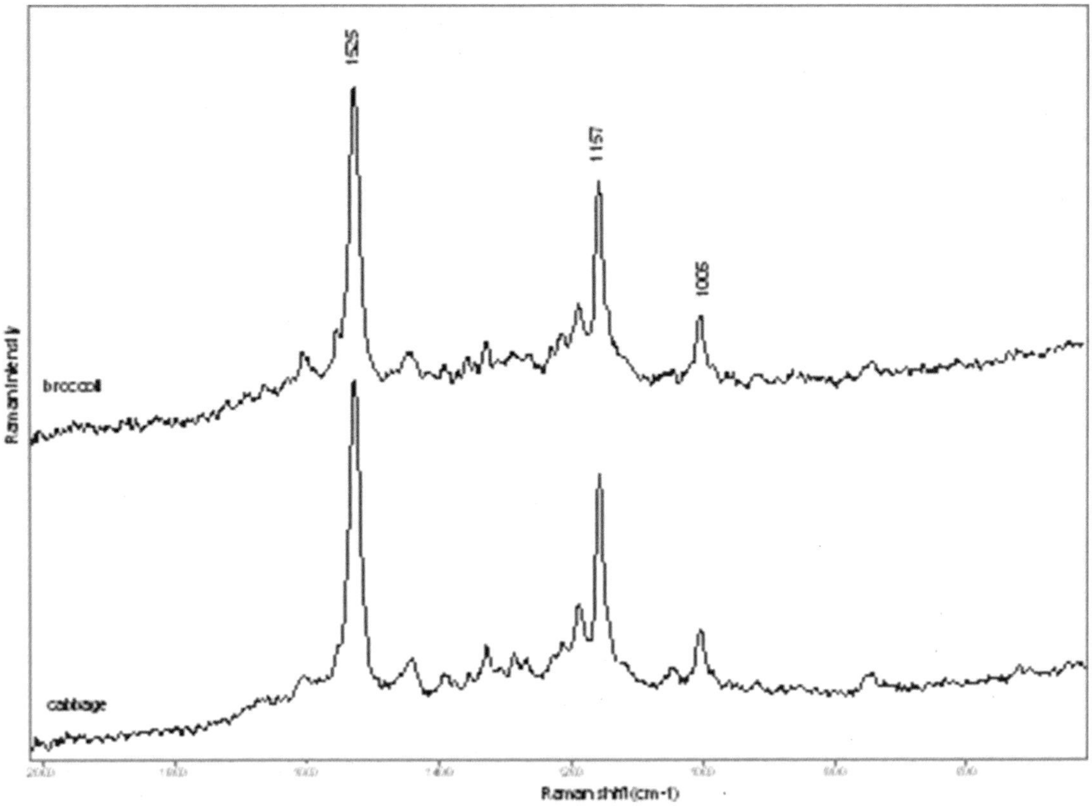


**Fig. 11.7** Typical FT-Raman spectra of broccoli and white cabbage leaves fumigated by ozone over 6 days (70 ppb, 6 h daily)

white cabbage ozone fumigated and control) were taken for FT-Raman measurements. We show that Raman spectroscopy combined with chemometrics can be considered as a potential analytical technique allowing for discriminating the plants stressed by ozone from the non-processed control group.

The cluster analysis was performed separately for each sample type, for the whole, as well as for specific wave number ranges using Ward's algorithm. Spectral distances for broccoli were calculated with the standard algorithm after applying vector normalisation. For white cabbage, the cluster analysis was carried out with the factorisation algorithm using the first three factors.

Single Raman spectra obtained from leaves did not show any distinctive differences which could be used for analytical purposes (Fig. 11.7). Yet it does not mean that ozone did not result in any change in the chemical composition of plants. This information can be spread in the broad range of the spectrum, and it overlaps with other signals assigned to various plant components. To extract this information, some chemometric methods should be used.

The most intensive carotenoid bands can be found in the Raman spectrum within the 1,500–1,550 $cm^{-1}$ range due to in-phase CC ($\nu_1$) stretching vibrations of the polyene chain. The medium signal of carotenoids at 1,150–1,170 $cm^{-1}$ can be assigned to C–C ($\nu_2$) stretching vibration. Additionally, the in-plane rocking mode of $CH_3$ groups attached to the polyene chain and coupled with C–C bonds is seen as a peak of weak intensity in the 1,000–1,020 $cm^{-1}$ region (Baranski et al. 2005, Baranska et al. 2006). A special advantage of using FT-Raman spectroscopy with laser excitation at 1,064 nm is the possibility to investigate the content of chlorophylls in green parts of plants at room

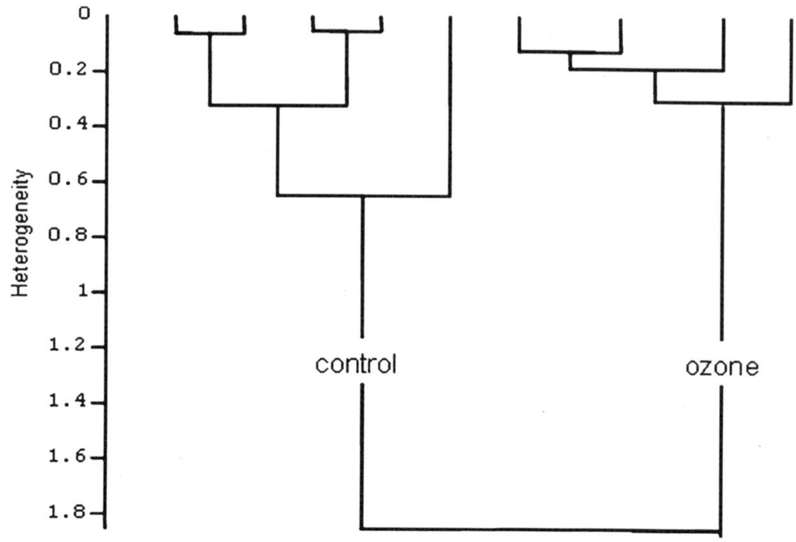


**Fig. 11.8** Dendrogram showing the classification of white cabbage leaves after cluster analysis of spectra at the wave number range of 1,570–1,470 $cm^{-1}$ with the use of the first three factors for calculating spectral distances and Ward's algorithm. "Control", non-treated plants; "Ozone", plants under exposure to elevated concentration of ozone. For details see Fig. 11.7

temperature. Characteristic Raman bands due to chlorophyll (Chl) can be seen at 1,606 $cm^{-1}$ (relatively strong), 1,326, 1,286 $cm^{-1}$ (medium to weak), and 740 $cm^{-1}$ (weak) (Schrader et al. 1998; Schulz et al. 2005). The distinction between Chl *a* and Chl *b* is also possible if this technique is used when pigments occur in the isolated form or are dissolved in sufficient solvent (Schulz and Baranska 2007). Spectra of Chl *a* and *b* contain a number of bands with similar frequencies; however, a few of them may be considered as fingerprints. The characteristic band for Chl *b* in the FT-Raman spectrum at 1,660 $cm^{-1}$ arises from the stretching mode of its formyl group (this group is absent in Chl *a*). The other key band of lower intensity located at about 1,565 $cm^{-1}$ for Chl *b* and about 1,550 $cm^{-1}$ for Chl *a* is sensitive to the coordination state of the central Mg atom of chlorophyll molecules (Pascal et al. 2000). However, when FT-Raman measurement is performed from the green tissue of the plant, the characteristic bands for Chl *a* and Chl *b* cannot be seen, and only the total content of both pigments can be analysed. The best markers are the bands at 1,326 or 740 $cm^{-1}$ since they do not overlap with other Raman signals associated with the remaining plant component, but their intensity is relatively weak in comparison to carotenoid signals.

In order to find meaningful and systematic differences among the measured spectra of white cabbage under elevated ozone concentration and in normal growing conditions, cluster analysis was applied. Despite the application of the standard procedures of spectra pretreatment, as well as various mathematical algorithms, no discrimination could be achieved if the whole spectral range was considered. Analyses based on various narrow wave number ranges gave a variety of outputs, with randomly distributed plants from both experimental groups, except for one specific spectral range. When the analysis was limited to 1,570–1,470 $cm^{-1}$ (Fig. 11.8), a distinct discrimination between ozone-fumigated and non-treated plants was achieved. This is the range in which the most intense carotenoid band can be found. These results indicated that changes of plant constituents caused by ozone exposure were related to the characteristic Raman signals in this range, i.e. correlated with carotenoid content. It can be concluded that the content of this pigment in white cabbage leaves is strongly influenced by ozone.

Raman spectra obtained for broccoli leaves were also subjected to cluster analysis. The best discrimination between ozonated and non-treated plants was achieved when almost the whole registered wave number range was considered

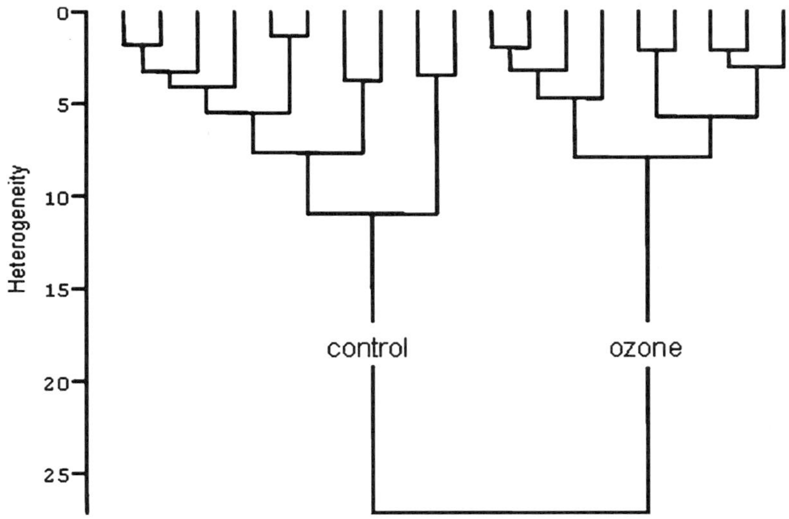


**Fig. 11.9** Dendrogram showing the classification of broccoli leaves in the cluster analysis of the spectra at the wave number range of 4,000–220 $cm^{-1}$ using the standard algorithm for calculating spectral distances and Ward's algorithm. "Control", non-treated plants; "Ozone", plants under exposure to elevated concentration of ozone. For details see Fig. 11.7

(Fig. 11.9). Carotenoid bands are naturally the most intense signals in the Raman spectra of broccoli leaves, so the main weight of the analysis was put on these components, but other plant constituents were also included. Generally, a clear discrimination can be observed between both analysed groups, named as "control" and "ozone" (Fig. 11.9).

Results show that some chemical changes in plant composition upon the stress caused by exposure to elevated ozone concentration can be detected by using Raman spectroscopy. It is very sensitive to carotenoids, and it can be successfully analysed even when this pigment is present at the level of ppm. This is due to the high Raman activity of this compound and the resonance effect resulting in a strong enhancement of the carotenoid band. Sometimes individual Raman spectra do not show any distinctive visual differences, but the application of cluster analysis allowed us to divide the investigated plants into two groups – ozone-fumigated samples and non-processed control samples. Finally, FT-Raman spectrometry seems to be a good tool to describe the susceptibility of particular plant genotypes to ozone stress.

## Response to Light Quality

Light quality plays an important role in morphogenesis and photosynthesis (Hoenecke et al. 1992; Saebo et al. 1995; Tripathy and Brown 1995), but rapid changes in the spectral quality of light are stressful for plants.

Blue light is important in the formation of chlorophyll (Senger 1982; Pushnik et al. 1987), chloroplast development (Akoyunoglou and Anni 1984), stomata opening (Zeiger 1984), enzyme synthesis (Senger 1982), activation of the circadian rhythm of photosynthesis (Senger 1982), and photomorphogenesis (Cosgrove 1981, 1982; Senger 1982; Wheeler et al. 1991).

One light source is light-emitting diodes (LEDs). The advantages of using light-emitting diodes (LED) as an artificial light source for controlled-environment plant growth applications include high energy-conversion efficiency, the use of DC power, small mass and volume, longer life, wavelength specificity, adjustable light intensity/quality, and low thermal energy output (Barta et al. 1992; Tennessen et al. 1994; Okamoto et al. 1997; Schuerger et al. 1997). Physiological responses to spectral changes can vary among different plant species (Deutch and Rasmussen 1974; Boardman 1977; Senger 1982). Heo et al. (2002) reported that 1st internode elongation in salvia and marigold was strictly dependent on red light. In the case of salvia, the authors observed the inhibiting (and in the case of marigold, the promoting) effect of red light on internode elongations.

Some crops and flowers have been cultured by LED light radiation, such as lettuce (Hoenecke et al. 1992; Okamoto et al. 1996), pepper (Brown et al. 1995), wheat (Goins et al. 1997; Tripathy and Brown 1995), spinach (Yanagi and Okamoto

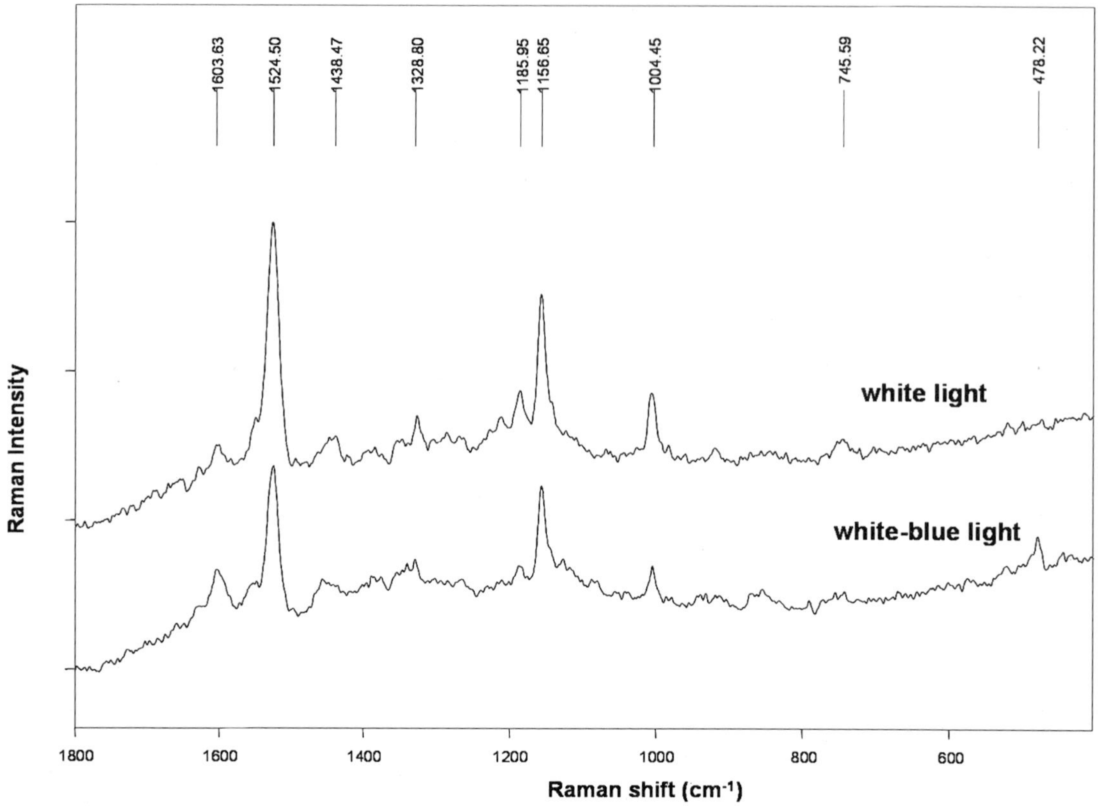


**Fig. 11.10** FT-Raman spectra of white cabbage leaves growing for 6 days in *white light* or in *white-blue light* (white light supplemented by blue light). Photoperiod, 12/12 (light/darkness); temperature, 18/14°C (day/night); relative humidity, 80%

1997; Yanagi et al. 1996), mustard (Cosgrove 1981), and banana (Duong et al. 2002).

In this experiment, we examined the influence of rapid changes of spectral light quality (derived from LED) on selected metabolic properties of white cabbage (*B. oleracea* var. *capitata* f. *alba*) leaves. Cabbage plants grown in white light (photoperiod 12/12 – light/darkness) at 18/14°C (day/night) were rapidly transferred to blue light-rich spectral conditions for 6 days.

On the FT-Raman spectra obtained for the leaves, three bands are visible, which are connected with the presence of carotenoid compounds in the sample. There are, respectively, 1,004 (in-plane rocking mode of $CH_3$ groups attached to the polyene chain and coupled with C–C bonds), 1,155 (C–C ($\nu_2$) stretching vibration), and 1,525 cm$^{-1}$ (C=C ($\nu_1$) stretching vibrations of the polyene chain) (Baranski et al. 2005; Baranska et al. 2006a, b). In addition, some bands of low intensity, at 1,604, 1,329, and 745 cm$^{-1,}$ can be assigned to chlorophyll (Schrader et al. 1998; Schulz et al. 2005) (Fig. 11.10).

In order to find meaningful and systematic differences among the measured spectra of white cabbage leaf growth in white and white + blue light, cluster analysis was applied. The range of wave number was limited to 1,556–1,115 cm$^{-1}$ (Fig. 11.11), and a distinct discrimination between leaves from white and white + blue light conditions was achieved. These results indicate that changes of plant constituents caused by different light conditions are related to characteristic Raman signals in this range – carotenoids compounds. To the first group of investigated objects belong the leaves of plants growing under white light. The chemical composition (especially carotenoids) of these

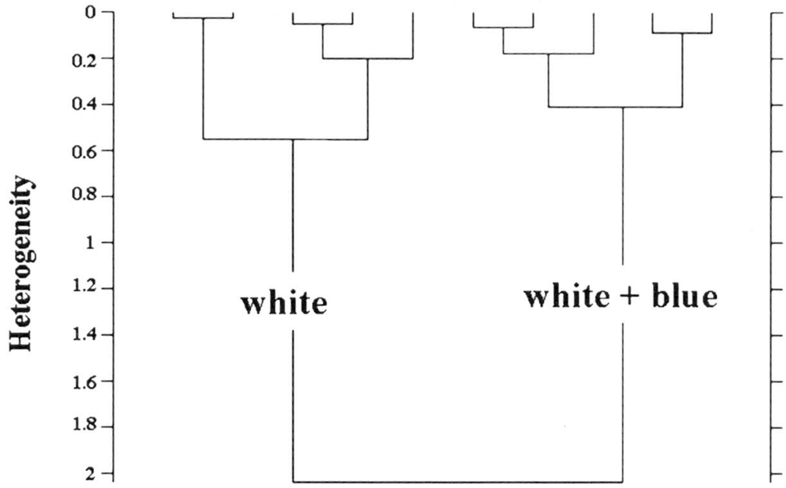


**Fig. 11.11** Dendrogram showing the classification of white cabbage leaves after cluster analysis of spectra at the wave number range of 1,556–1,115 $cm^{-1}$ with the use of the first three factors for calculating spectral distances and Ward's algorithm. "white", plants growing under white light; "white + blue", plants growing under blue light-rich spectrum. For details see Fig. 11.10

plants is very similar. In the second group are the leaves of plants which were grown under white + blue light.

The performed cluster analysis showed that the range between 1,115 and 1,556 $cm^{-1}$ could be used for discriminating leaves which were grown under different light condition – white and white-supplemented blue light.

## Pathogenesis

Raman mapping is performed by taking measurements point by point from the specified area of the leaf. For each measured point, full spectral information is obtained that could be analysed at any desired wave number. Later, a map can be constructed and compared to a visual picture of the leaf. Raman maps provide insight into the distribution and relative content of the analyte in the specified area (Baranski et al. 2005). The Raman spectra (data not shown) obtained for the leaves of pepper inoculated by Pepper mild mottle virus (PMMoV) and Obuda pepper virus (ObPV) show three characteristic bands of carotenoids at 1,004, 1,155, and 1,525 $cm^{-1}$ (Withnall et al. 2003; Schulz et al. 2005). The first, most intense C=C stretching vibration ($\nu_1$) of β-carotene is observed at 1,525 $cm^{-1}$. The second, medium in intensity at 1,155 $cm^{-1}$ is assigned with C–C stretching vibration. The third, low intensity, at 1,004 $cm^{-1}$, came from $CH_3$ groups attached to the polyene chain and coupled with C–C bonds. Some bands of low intensity at 1,575, 1,328, 1,290, and 746 $cm^{-1}$ can be assigned to chlorophyll (Schrader et al. 1998; Schulz et al. 2005).

The obtained data are presented as two-dimensional Raman maps coloured according to the band intensity of carotenoids (Fig. 11.12).

On the Raman maps, it can be seen that the distribution of carotenoids on the leaf surface is very uneven as a result of infection. As the result, areas where the number of carotenoids is high or much lower are visible. In the case of pepper leaf infected by an incompatible virus, PMMoV, the decline in the number of carotenoids occurred (42 h after inoculation) first in the upper part of the leaf blade. On the other hand, inoculation by compatible virus, ObPV, caused a "mosaic" distribution of carotenoids, i.e. areas with a high concentration bordering areas where the content of these compounds is very low. The results of our last investigations showed that necrotic lesions are observed on the leaf blade where the concentrations of carotenoids were very high, according to the Raman maps (data not shown).

These results indicate that Raman mapping is a credible tool with which to investigate plant response to stress factors.

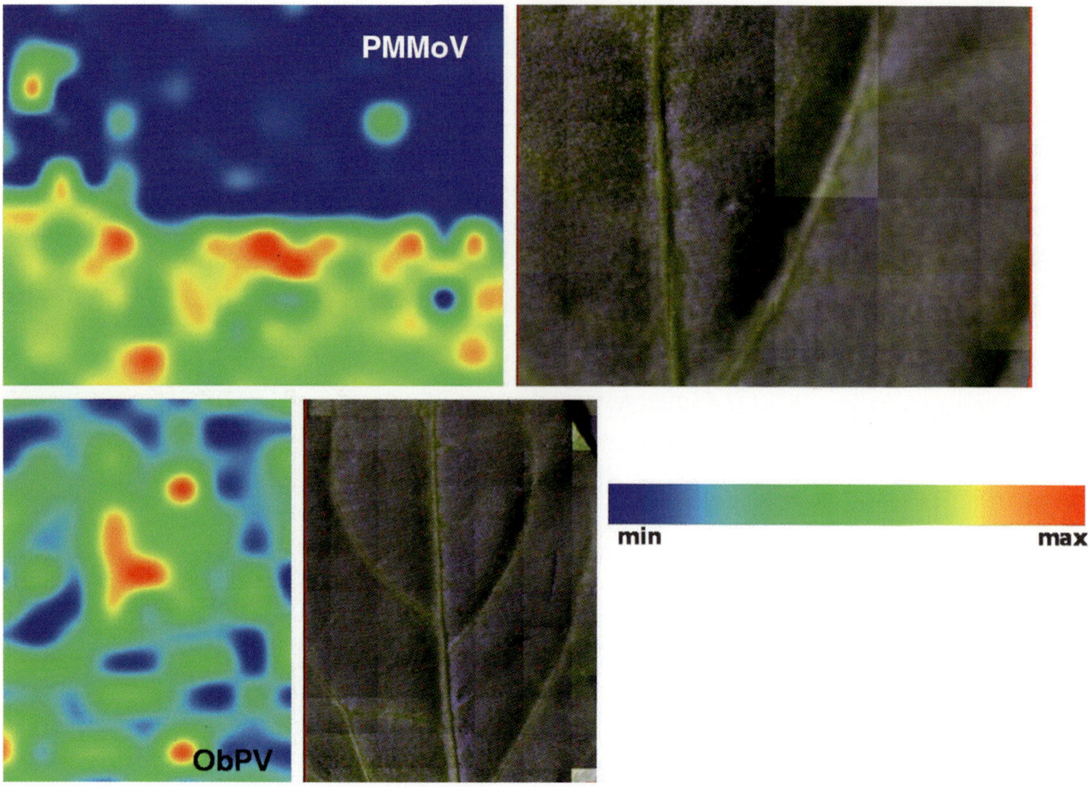


**Fig. 11.12** The two-dimensional Raman maps for pepper leaves 42 h after inoculation with PMMoV or ObPV viruses. Colour intensity corresponds with carotenoid content (carotenoid band intensity). On the *right-hand side*, analysed leaf areas are shown. The experiment was conducted within the framework of cooperation with the Plant Protection Institute of the Hungarian Academy of Sciences in Budapest

## References

Ádám AL, Dubert F, Gilly A, Barna B, Skoczowski A (1993) Measurement of exothermic heat flow and leaf temperature of TMV- and CMV-infected plants by microcalorimetry and infrared detection technique. Acta Phytopath et Entomol Hung 28:195–200

Akoyunoglou G, Anni H (1984) Blue light effect on chloroplast development in higher plants. In: Senger H (ed) Blue light effects in biological systems. Springer, Berlin

Alyabyev AJ, Loseva NL, Jakushenkova TP, Rachimova GG, Tribunskih VI, Estrina RI, Petrov VY (2002) Comparative effects of blue light and red light on the rates of oxygen metabolism and heat production in wheat seedlings stressed by heat shock. Thermochim Acta 394:227–231

Alyabyev AJ, Loseva NL, Gordon LK, Andreyeva IN, Rachimova GG, Tribunskih VI, Ponomareva AA, Kempb RB (2007) The effect of changes in salinity on the energy yielding processes of *Chlorella vulgaris* and *Dunaliella maritima* cells. Thermochim Acta 458:65–70

Anekonda T, Jones C, Smith BN, Hansen LD (2004) Differences in physiology and growth between coastal and inland varieties of Douglas-fir seedlings in a common garden. Thermochim Acta 422:75–79

Astorg P (1997) Food carotenoids and cancer prevention: an overview of current research. Trends Food Sci Technol 8:406–413

Baltruschat H, Fodor J, Harrach BD, Niemczyk E, Barna B, Gullner G, Janeczko A, Kogel K-H, Schäfer P, Schwarczinger I, Zuccaro A, Skoczowski A (2008) Salt tolerance of barley induced by the root endophyte *Piriformospora indica* is associated with a strong increase in antioxidants. New Phytol 180:501–510

Baranska M, Schulz H, Baranski R, Nothnagel T, Chrstensen L (2005) In situ simultaneous analysis of polyacetylenes, carotenoids and polysaccharides in carrot roots. J Agric Food Chem 53:6565–6571

Baranska M, Schulz H, Joubert E, Manley M (2006a) In situ flavonoid analysis by FT-Raman spectroscopy: identification, distribution, and quantification of aspalathin in green rooibos (*Aspalathus linearis*). Anal Chem 78:7716–7721

Baranska M, Baranski R, Schulz H, Nothnagel T (2006b) Tissue-specific accumulation of carotenoids in carrot roots. Planta 224:1028–1037

Baranski R, Baranska M (2008) Discrimination between nongenetically modified (non-GM) and GM plant tissue expressing cysteine-rich polypeptide using FT-Raman spectroscopy. J Agric Food Chem 56:4491–4496

Baranski R, Baranska M, Schulz H (2005) Changes in carotenoid content and distribution in living plant tissue can be observed and mapped in situ using NIR-FT-Raman spectroscopy. Planta 222:448–457

Baranski R, Baranska M, Schulz H, Simon PW, Nothnagel T (2006) Single seed Raman measurements allow taxonomical discrimination of Apiaceae accessions collected in gene banks. Biopolymers 81:497–505

Barta DJ, Tibbitis TW, Bula RJ, Morrow RC (1992) Evaluation of light emitting diode characteristics for a space-based plant irradiation source. Adv Space Res 12(5):141–149

Boardman NK (1977) Comparative photosynthesis of sun and shade plants. Annu Rev Plant Physiol 28:355–377

Brown CS, Schuerger AC, Sager JC (1995) Growth and photomorphogenesis of pepper plants under red light-emitting diodes with supplemental blue or far-red lighting. J Am Soc Hortic Sci 120:808–813

Calatayud A, Barreno E (2004) Response to ozone in two lettuce varieties on chlorophyll *a* fluorescence, photosynthetic pigments and lipid peroxidation. Plant Physiol Biochem 42:549–555

Cosgrove DJ (1981) Rapid suppression of growth by blue light: occurrence, time course, and general characteristics. Plant Physiol 67:584–590

Cosgrove DJ (1982) Rapid inhibition of hypocotyl growth by blue light in *Sinapis alba*. Plant Sci Lett 25:305–312

Criddle RS, Hansen LD (1999) Calorimetric methods for analysis of plant metabolism. In: Kemp RB (ed) Handbook of thermal analysis and calorimetry. Elsevier Science B.V, Amsterdam, pp 711–763

Criddle RS, Hansen LD, Breidenbach RW, Ward MR, Huffaker RC (1989) Effects of NaCl on metabolic heat evolution rates by barley roots. Plant Physiol 90:53–58. doi:10.1104/pp. 90.1.53

Criddle RS, Fontana AJ, Rank DR, Paige D, Hansen LD, Breidenbach RW (1991a) Simultaneous measurements of metabolic heat rate, $CO_2$ production, and $O_2$ consumption by microcalorimetry. Anal Biochem 194:413–417

Criddle RS, Breidenbach RW, Hansen LD (1991b) Plant calorimetry: how to quantitatively compare apples and oranges. Thermochim Acta 193:67–90

Criddle RS, Smith BN, Hansen LD (1997) A respiration based description of plant growth rate response to temperature. Planta 201:441–445

Cygański A (2002) Metody spektroskopowe w chemii analitycznej. Wydawnictwa Naukowo-Techniczne, Warsaw (in polish)

Deutch B, Rasmussen O (1974) Growth chamber illumination and photomorphogenetic efficacy I. Physiological action of infrared radiation beyond 750 mm. Physiol Plant 30:64–71

Duong TN, Hong LTA, Watanabe H, Goi M, Tanaka M (2002) Growth of banana plant lets cultured in vitro under red and blue light-emitting diode (LED) irradiation source. Acta Hortic 575:117–124

Fodor J, Harrach BD, Janeczko A, Barna B, Skoczowski A (2007) Metabolic responses of tobacco to induction of systemic acquired resistance. Thermochim Acta 466:29–34

Goins GD, Yorio NC, Sanwo MM, Brown CS (1997) Photomorphogenesis, photosynthesis, and seed yield of wheat plants grown under red light-emitting diodes (LEDs) with and without supplemental blue lighting. J Exp Bot 48:1407–1413

Gómez Galindo F, Rocculi P, Wadsö L, Sjöholm I (2005) The potential of isothermal calorimetry in monitoring and predicting quality changes during processing and storage of minimally processed fruits and vegetables. Trends Food Sci Technol 16:325–331

Gordon LK, Minibayeva FV, Rakhmatullina DF, Alyabyev AJ, Ogorodnikova TI, Loseva NL, Valitova YN (2004) Heat production of wheat roots induced by the disruption of proton gradient by salicylic acid. Thermochim Acta 422:101–104

Hansen LD (2001) Toward a standard nomenclature for calorimetry. Thermochim Acta 371:19–22

Hansen LD, Russell DJ (2006) Which calorimeter is best? A guide for choosing the best calorimeter for a given task. Thermochim Acta 450:71–72

Hansen LD, Macfarlane C, McKinnon N, Smith BN, Criddle RS (2004) Use of calorespirometric ratios, heat per $CO_2$ and heat per $O_2$, to quantify metabolic paths and energetics of growing cells. Thermochim Acta 422:55–61

Hansen LD, Church JN, Matheson S, McCarlie VW, Thygerson T, Criddle RS, Smith BN (2002) Kinetics of plant growth and metabolism. Thermochim Acta 388:415–425

Heap I (2006) International survey of herbicide-resistant weeds. http://www.weedscience.org

Heo J, Lee C, Chakrabarty D, Paek K (2002) Growth responses of marigold and salvia bedding plants as affected by monochromic or mixture radiation provided by a Light-Emitting Diode (LED). Plant Growth Regul 38:225–230

Hoenecke M, Bula RJ, Tibbitts TW (1992) Importance of 'blue' photon levels for lettuce seedlings grown under red light-emitting diodes. Hortic Sci 27: 427–430

Inderjit WJ (2001) Plant allelochemical interference or soil chemical ecology? Perspect Plant Ecol 4(1): 3–12

Janeczko A, Tóbias I, Skoczowski A, Dubert F, Gullner G, Barna B (2007a) Bacterial infection and pretreatment with 24-epibrassinolide markedly affect the heat emission and membrane permeability of rape cotyledons. Thermochim Acta 458:88–91

Janeczko A, Biesaga-Kościelniak J, Skoczowski A, Marcińska I (2007b) Differentiation of wheat cell suspension cultured in cold – impact of zearalenone and 24-epibrassinolide. Zeszyty Problemowe Postępów Nauk Rolniczych Z 523:127–136

Janssen M, Patiño R, Von Stockar U (2005) Application of bench-scale biocalorimetry to photoautotrophic cultures. Thermochim Acta 435:18–27

Kęcki Z (1998) Podstawy spektroskopii molekularnej. Wydawnictwo Naukowe PWN, Warsaw (in polish)

Keller EA, Anekonda TS, Smith BN, Hansen LD, Clair JBS, Criddle RS (2004) Stress and respiration traits differ among four geographically distinct *Pinus ponderosa* seed sources. Thermochim Acta 422:69–74

Kim MK, Ahn SH, Lee-Kim YC (2001) Relationship of serum α-tocopherol, carotenoids and retinol with risk of breast cancer. Nutr Res 21:797–809

Kraft M, Weigel H-J, Mejer G-J, Brandes F (1996) Reflectance measurements of leaves for detecting visible and non-visible ozone damage to crops. J Plant Physiol 148:148–154

Kronfuß G, Polle A, Tausz M, Havranek W, Wieser G (1998) Effects of ozone and mild drought stress on gas exchange, antioxidants and chloroplast pigments in current-year needles of young Norway spruce [*Picea abies* (L.) Karst]. Trees 12:482–489

Kupidłowska E, Gniazdowska A, Stępień J, Corbineau F, Vinci D, Skoczowski A, Janeczko A, Bogatek R (2006) Impact of sunflower (*Helianthus annuus* L.) extracts upon reserve mobilization and energy metabolism in germinating mustard (*Sinapis alba* L.) seeds. J Chem Ecol 32:2569–2583

Lamprecht I, Schmolz E (2000) Calorimetry goes afield. Thermochim Acta 355:95–106

Liu RH (2004) Potential synergy of phytochemicals in cancer prevention: mechanism of action. In: International research conference on food, nutrition and cancer, Washington

Loseva NL, Alyabyev AJ, Rachimova GG, Estrina RI (1995) Aspects of the energetic balance of plant cells under different salt conditions. Thermochim Acta 251:357–362

Loseva NL, Alyabyev AJ, Rachimova GG, Estrina RI (1998) The effect of high temperature and salt on the metabolic heat rate of *Chlorella* cells. Thermochim Acta 309:129–131

Loseva NL, Gordon LK, Minibayeva FV, Alyabyev AJ, Chernov VM, Chernowa OA, Andreyeva FV, Rachimova GG, Tribunskih VI, Estrina RI, Gogolev JB, Kemp RB (2002) Direct and indirect calorimetric studies of stress responses of chlorella cells to infection with the mycoplasma, *Acholeplasma laidlawii*. Thermochim Acta 390:39–46

Loseva NL, Alyabyev AJ, Gordon LK, Andreyeva IN, Kolesnikov OP, Ponomareva AA, Chernov VM, Kemp RB (2003) The energetic stress response of the microalgal *Chlorella vulgaris* to the mycoplasma, *Acholeplasma laidlawii* as a model system for plant pathogen interaction. Thermochim Acta 397:37–47

Loseva N, Gordon L, Alyabyev A, Andreyeva I, Kolesnikova O, Chernov V, Ponomareva A, Kemp RB (2004) Effect of induced changes in membrane permeability on the defence response of *Chlorella vulgaris* to infection by *Acholeplasma laidlawii*. Thermochima Acta 422:95–100

Minibayeva FV, Gordon LK, Alyabyev AJ, Rakhmatullina DF, Loseva NL (1998) Heat production of root cells upon the dissipation of ion gradients on plasma membrane. Thermochim Acta 309:139–143

Mukhanov VS, Kemp RB (2006) Simultaneous photocalorimetric and oxygen polarographic measurements on *Dunaliella maritima* cells reveal a thermal discrepancy that could be due to nonphotochemical quenching. Thermochim Acta 446:11–19

Okamoto K, Yanagi T, Takita S (1996) Development of plant growth apparatus using blue and red led as artificial light source. Acta Hortic 440:111–116

Okamoto K, Yanagi T, Kondo S (1997) Growth and morphogenesis of lettuce seedlings eaisde under different combinations of red and blue light. Acta Hortic 435:149–157

Oroszi S, Jakob T, Wilhelm C, Harms H, Maskow T (2011) Photosynthetic energy conversion in the diatom *Phaeodactylum tricornutum* Measuring by calorimetry, oxygen evolution and pulse-amplitude modulated fluorescence. J Therm Anal Calorim 104:223–231. doi:10.1007/s10973-011-1358-3

Ozaki Y, Cho R, Ikegawa K, Muraishi S, Kawauchi K (1992) Potential of near-infrared Fourier transform Raman spectroscopy in food analysis. Appl Spectrosc 46:1503–1507

Pascal A, Peterman E, Gradinaru C, Van Amerongen H, Van Grondelle R, Robert B (2000) Structure and interactions of the chlorophyll *a* molecules in the higher plant Lhcb4 antenna protein. J Phys Chem 104:9317–9321

Petrov VY, Alyabyev AJ, Loseva NL, Klementyeva GS, Tribunskich VI (1995) A differential photocalorimetric method for investigating the rate of energy storage in plants. Thermochim Acta 251:351–356

Płażek A, Rapacz M (2000) The intensity of respiration and heat emission from seedlings of *Festuca pratensis* (Hud.) and *Hordeum vulgare* L. during pathogenesis caused by *Bipolaris sorokiniana* (Sacc.) Shoem. Acta Physiol Plant 22:25–30

Płażek A, Skrzypek I, Żur I (2000a) The changes of heat emission and phenolic compound level in *Hordeum vulgare* (L.) and *Festuca pratensis* (Huds.) calli treated with *Bipolaris sorokiniana* (Sacc.) Shoem. phytotoxins. J Agron Crop Sci 184:17–21

Płażek A, Rapacz M, Skoczowski A (2000b) Effects of ozone fumigation on photosynthesis and membrane permeability in leaves of spring barley, meadow fescue and winter rape. Photosynthetica 38:409–413

Pushnik JC, Miller GW, Jolley VD, Brown GC, Davis TD, Burnes AM (1987) Influences of ultra-violet (UV)-blue light radiation on the growth of cotton. II. Photosynthesis, leaf anatomy and iron reduction. J Plant Nutr 10:2283–2297

Rank DR, Breidenbach RW, Fontana AJ, Hansen LD, Cirddle RS (1991) Time-temperature response of tomato cells during high-temperature and low-temperature inactivation. Planta 185:576–582

Reitzenstein S, Rösch P, Strehle MA, Berg D, Baranska M, Schulz H, Rudloff E, Popp J (2007) Nondestructive analysis of single rapeseeds by means of Raman spectroscopy. J Raman Spectrosc 38:301–308

Rice EL (1984) Allelopathy. Academic, New York

Rocculi P, Romani S, Rosa MD, Gómez F, Sjöholm M, Wadsö L (2005) Influence of pre-treatment on metabolism and wounding response of fresh cut potatoes evaluated with isothermal calorimetry. In: Mencarelli F, Tonutti P (eds) Proceeding of the 5th international postharvest symposium, Acta horticulturae 682., pp 1833–1838

Rösch P, Popp J, Kiefer W (1999) Raman and surface enhanced Raman spectroscopic investigation on *Lamiaceae* plants. J Mol Struct 480–481:121–124

Sadeghi-Jorabchi H, Hendra PJ, Wilson RH, Belton PS (1990) Determination of the total unsaturation in oils and margarines by Fourier transform Raman spectroscopy. J Am Oil Chem Soc 67:483–486

Sadeghi-Jorabchi H, Wilson RH, Belton PS, Edward-Webb JD, Coxon DT (1991) Quantitative analysis of oils and fats by Fourier-transform Raman spectroscopy. Spectrochim Acta A 47:1449–1458

Saebo A, Krekling T, Appelgren M (1995) Light quality affects photosynthesis and leaf anatomy of birch plantlets in vitro. Plant Cell Tissue Organ Cult 41:177–185

Schabes FI, Sigstad EE (2005) Calorimetric studies of quinoa (*Chenopodium quinoa* Willd.) seed germination under saline stress conditions. Thermochim Acta 428:71–75

Schabes FI, Sigstad EE (2006) Optimizing conditions to study seed germination by calorimetry using soybean (*Glycine max* [L.] Merr.) seeds. Thermochim Acta 450:96–101

Schabes FI, Sigstad EE (2007) A calorimetric study of the allelopathic effect of cnicin isolated from *Centaurea diffusa* Lam. on the germination of soybean (*Glicine max*) and radish (*Raphanus sativus*). Thermochim Acta 458:84–87

Schoefs B (2002) Chlorophyll and carotenoid analysis in food products. Properties of the pigments and methods of analysis. Trends Food Sci Technol 13:361–371

Schrader B, Erb I, Löchte T (1998) Differentiation of conifers by NIR-FT-Raman spectroscopy. Asian J Phys 7:259–264

Schrader B, Klump HH, Schenzel K, Schulz H (1999) Non-destructive NIR FT Raman analysis of plants. J Mol Struct 509:201–212

Schrader B, Schulz H, Andreev GN, Klump HH, Sawatzki J (2000) Non-destructive NIR-FT-Raman spectroscopy of plant and animal tissues, of food and works of art. Talanta 53:35–45

Schuerger AC, Brown C, Stryjewski EC (1997) Anatomical Features of pepper plants (AnalsofB ntt ff B L.) grown under red light-emitting diodes supplemented with blue or far-red light. Ann Bot 79:273–282

Schulz H, Baranska M (2007) Identification and quantification of valuable plant substances by IR and Raman spectroscopy. Vib Spectrosc 43:13–25

Schulz H, Baranska M, Baranski R (2005) Potential of NIR-FT-Raman spectroscopy in natural carotenoid analysis. Biopolymers 77:212–221

Senger H (1982) The effect of blue light on plants and microorganisms. Photochem Photobiol 35:911–920

Sigstad EE, Garcia CI (2001) A microcalorimetric analysis of quinoa seeds with different initial water content during germination at 25°C. Thermochim Acta 366:149–155

Sigstad EE, Prado FE (1999) A microcalorimetric study of *Chenopodium quinoa* Willd. seed germination. Thermochim Acta 326:159–164

Skoczowski A, Janeczko A, Gullner G, Tóbias I, Kornas A, Barna B (2011a) Response of brassinosteroid-treated oilseed rape cotyledons to infection with the wild type and HR-mutant of *Pseudomonas syringae* or with *P. fluorescence*. J Therm Anal Calorim 104:131–139

Skoczowski A, Troć M, Baran A, Baranska M (2011b) Impact of sunflower and mustard leave extracts on the growth and dark respiration of mustard seedlings. J Therm Anal Calorim 104:187–192

Smith BN, Jones AR, Hansen LD, Criddle RS (1999) Growth, respiration rate, and efficiency response to temperature. In: Pessarakli M (ed) Handbook of plant and crop stress, 2nd edn. Revised and expanded. Marcel Dekker Inc., New York/Basel

Smith BN, Hansen LD, Breidenbach RW, Criddle RS, Rank DR, Fontana AJ, Paige D (2000) Metabolic heat rate and respiratory substrate changes in aging potato slices. Thermochim Acta 349:121 124

Smith BN, Harris LC, McCarlie VW, Stradling DL, Thygerson T, Walker J, Criddle RS, Hansen LD (2001) Time, plant growth, respiration, and temperature. In: Pessarakli M (ed) Handbook of plant and crop physiology, 2nd edn. Marcel Dekker, New York/Basel

Smith BN, Monaco TA, Jones C, Holmes RA, Hansen LD, McArthur ED, Freeman DC (2002) Stress-induced metabolic differences between populations and subspecies of *Artemisia tridentata* (sagebrush) from a single hillside. Thermochim Acta 394:205–210

Steinrücken HCM, Hermann D (2000) Spreading the search for crop chemicals. Chem Ind 7(4):246–249

Stokłosa A, Janeczko A, Skoczowski A, Kieć J (2006) Isothermal calorimetry as a tool for estimating resistance of wild oat (*Avena fatua* L.) to aryloxyphenoxypropionate herbicides. Thermochim Acta 441:203–206

Suwanagul D (1995) Early detection of weed resistance: pattern-thinking and rapid microcalorimetric analysis. Dissertation, Oregon State University, Corvallis

Szczepaniak W (2004) Metody instrumentalne w analizie chemicznej. Wydawnictwo Naukowe PWN, Warsaw (in Polish)

Tennessen DJ, Singsaas EL, Sharkey TD (1994) Light-emitting diodes as a light source for photosynthesis research. Photosynth Res 39:85–92

Thygerson T, Harris JM, Smith BN, Hansen LD, Pendleton RL, Booth DT (2002) Metabolic response to temperature for six populations of winterfat (*Eurotia lanata*). Thermochim Acta 394:211–217

Tripathy BC, Brown CS (1995) Root–shoot interaction in the greening of wheat seedlings grown under red light. Plant Physiol 107:407–411

Troć M, Skoczowski A, Barańska M (2009) The influence of sunflower and mustard leaf extracts on the germination of mustard seeds. J Therm Anal Calorim 95 (3):727–730

Troć M, Saja D, Kornas A, Żuraw A, Skoczowski A (2011) Strong endothermic effects caused by allelopathic interactions during growth of mustard, rape, wheat and clover seedlings. J Therm Anal Calorim 104:141–148

Twardowski J, Anzenbacher P (1994) Raman and IR spectroscopy in biology biochemistry. Wydawnictwo Naukowe PWN, Warsaw

Vyvyan JR (2002) Allelochemicals as leads for new herbicides and agrochemicals. Tetrahedron 58:1631–1646

Wadsö I (1995) Microcalorimetric techniques for the investigation of living plants materials. Thermochim Acta 250:285–304

Wadsö I (1997) Trends in isothermal microcalorimetry. Chem Soc Rev 26:79–86

Wadsö L, Gómez F, Sjöholm I, Rocculi P (2004) Effect of tissue wounding on the results from calorimetric measurements of vegetable respiration. Thermochim Acta 422:89–93

Wadsö L, Rocculi P, Romani S, Gómez Galindo F, Sjöholm I (2005) Isothermal calorimetry approach to evaluate the effect of tissue wounding on vegetable metabolism. In: Mencarelli F, Tonutti P (eds) Proceeding of the 5th international postharvest symposium, Acta horticulturae 682., pp 1833–1838

Weston LA, Duke SO (2003) Weed and crop allelopathy. Crit Rev Plant Sci 22(3–4):367–389

Wheeler RM, Mackowiak CL, Sager JC (1991) Soybean stem growth under high-pressure sodium with supplemental blue lighting. Agron J 83:903–906

Wieser G, Tausz M, Wonish A, Havranek WM (2001) Free radical scavengers and photosynthetic pigments in *Pinus cembra* L. needles as affected by ozone exposure. Biol Plant 44:225–232

Withnall R, Chowdhry BZ, Silver J, Edwards HGM, De Oliveira LFC (2003) Raman spectra of carotenoids in natural products. Spectrochim Acta A 59: 2207–2212

Yanagi T, Okamoto K (1997) Utilization of super-bright light emitting diodes as an artificial light source for plant growth. Acta Hortic 418:223–228

Yanagi T, Okamoto K, Takita S (1996) Effects of blue, red and blue/red lights of two different PPF levels on growth and morphogenesis of lettuce plants. Acta Hortic 440:117–122

Zeiger E (1984) Blue light and stomata function. In: Senger H (ed) Blue light effects in biological systems. Springer, Berlin/Heidelberg/New York

# Mechanism of Plant Tolerance in Response to Heavy Metals

# 12

Jot Sharma and Nivedita Chakraverty

**Abstract**

Heavy metal (HM) toxicity is one of the major abiotic stresses leading to hazardous effects in plants. A common consequence of heavy metal toxicity is the excessive accumulation of reactive oxygen species (ROS) and methylglyoxal (MG). These events are a cascade which finally leads to peroxidation of lipids, oxidation of protein, inactivation of enzymes, DNA damage, and/or interaction with other vital constituents of plant cells. Plants respond to heavy metal toxicity in a variety of different ways. Such responses include immobilization, exclusion, chelation and compartmentalization of the metal ions, and the expression of more general stress response mechanisms such as ethylene and stress proteins. This potential of plants to sustain in toxic conditions by detoxifying the heavy metals has been explored to mitigate the damage done to nature by emerging technology known as phytoremediation. Phytorem means those plants which are recommended to detoxify metals and metalloid from environment and soil. Growing plants at the contaminated sites over a number of years aim either to remove pollutants from the contaminated sites or to alter the chemical and physical nature of the contaminant so that they no longer present a risk to human health and environment.

## Introduction

Elements that are characterized with properties like malleability, ductility, shiny luster, conductance of heat and electricity and chemically their atom have the tendency to lose the outer electron to achieve stability are classified as metals (Muller 2007; Housecroft and Sharpe 2008). Whereas the term heavy metals refers to elements that resemble to metals in their physical and chemical properties, their specific gravity is at least five times more than water, approximately around eight grams per cubic centimeter ($g/cm^3$) at 4°C. Fifty-three of the ninety naturally occurring elements are heavy metals which have density very high as compared to other metals (http://www.humet.com/acatalog/heavymetals).

J. Sharma (✉) • N. Chakraverty
Bim College of Professional Studies, Surya Mandir Road, Residency, Gwalior, Madhya Pradesh, India 474005,
e-mail: jotsharma_68@yahoo.com; nivedita.chakraverty@gmail.com

G.R. Rout and A.B. Das (eds.), *Molecular Stress Physiology of Plants*,
DOI 10.1007/978-81-322-0807-5_12, © Springer India 2013

Based on their solubility under physiological conditions, 17 heavy metals are of biological importance as they are available for living cells (Weast 1984). Among these metals, Fe, Mo, and Mn are micronutrients to plants. Zn, Ni, Cu, V, Co, W, and Cr are toxic elements with high or low importance as trace elements. As, Hg, Ag, Sb, Cd, Pb, and U have no known function as nutrients and seem to be more or less toxic to plants and microorganisms (Breckle 1991; Nies 1999; Godbold and Hüttermann 1985).

In general, the term heavy metal is linked to toxicity; apart it is a feeling rather than based on scientific evidence. There are two major facts: (1) No substance is always toxic, its effect depends upon concentration that is available to the cell (bioavailability). (2) Heavy metals show a bell-shaped dose-response relationship (Marschner 1995), that is, they are crucial to metabolism at low concentration and toxic at higher concentration. Based on the average of all tested parameters, the following phytotoxicity series was obtained: $Ag^{+} > Cd^{2+} > Hg^{2+} > Tl^{+} > Cu^{2+} > Ni^{2+} > Zn^{2+} > Co^{6+} > Cr^{6+} > As^{3+} > As^{5+}$.

## Toxicity in Plants

As stated above heavy metals that are essential for plant growth at lower quantity become toxic at higher concentrations and show adverse effects on plant growth and their productivity (Gomes-Junior et al. 2006). Once the heavy metal enters the plant system from soil, they accumulate and later may enter the food chain and cause damage to various herbivores (Vitoria et al. 2001).

Soil in the most terrestrial ecosystems is mainly at high risk of heavy metal contamination. The main sources of heavy metals are the underlying parent material, and their concentrations in soils depend on the weathering of the bedrock and on atmospheric inputs of metals. Natural sources are volcanoes and continental dusts (Lantsy and Mackensie 1979; Angelone and Bini 1992; Galloway et al. 1982), and anthropogenic activities are like mining, combustion of fossil fuels, metal-working industries, use of leaded gasoline and paints, land application of fertilizers, animal manures, sewage sludge, pesticides, wastewater irrigation, coal combustion residues, spillage of petrochemicals, and atmospheric deposition (Khan et al. 2008; Zhang et al. 2010). Heavy metals constitute an ill-defined group of inorganic chemical hazards, and those most commonly found at contaminated sites are lead (Pb), chromium (Cr), arsenic (As), zinc (Zn), cadmium (Cd), copper (Cu), mercury (Hg), and nickel (Ni) which lead to the emission of heavy metals and the accumulation of these compounds in ecosystems (Fig. 12.1).

The type of the soil plays a significant role in the bioavailability of any heavy metal. Mobile and immobilized fractions are marked since heavy metals bind to inorganic and organic soil compounds and to the humus. Soil type determines the extent to which a particular heavy metal gets adsorbed and complexes with any of the above-mentioned compounds and thus influences its solubility and mobility of metals. The availability of heavy metals to plants and their toxicity depends on interaction between soil and plants roots. Access of heavy metals to bare roots is confined to the first few millimeters of the root tip. Within the cortex the metals are transported in the apoplastic space according to their concentration gradient and also accumulate in the cell walls. Toxic effects are exerted at the plasma membrane and within the cell. Two different uptake routes have been reported: (a) passive uptake, only driven by the concentration gradient across the membrane and (b) inducible substrate-specific and energy-dependent uptake (Nies 1999; Williams et al. 2000). Transport systems have also been reported for Cd and Ni in roots of spruce and soybean (Godbold 1991; Cataldo et al. 1978, 1981). Steady-state flux between import and export rates was acquired relatively quickly (Schützendübel et al. 2001). The mode of toxicity of heavy metals in plants is based on their redox activity. So it is important to understand the biochemistry of heavy metals. On general account all the elements listed under the term heavy metal have incompletely filled δ-orbital present. So they can either

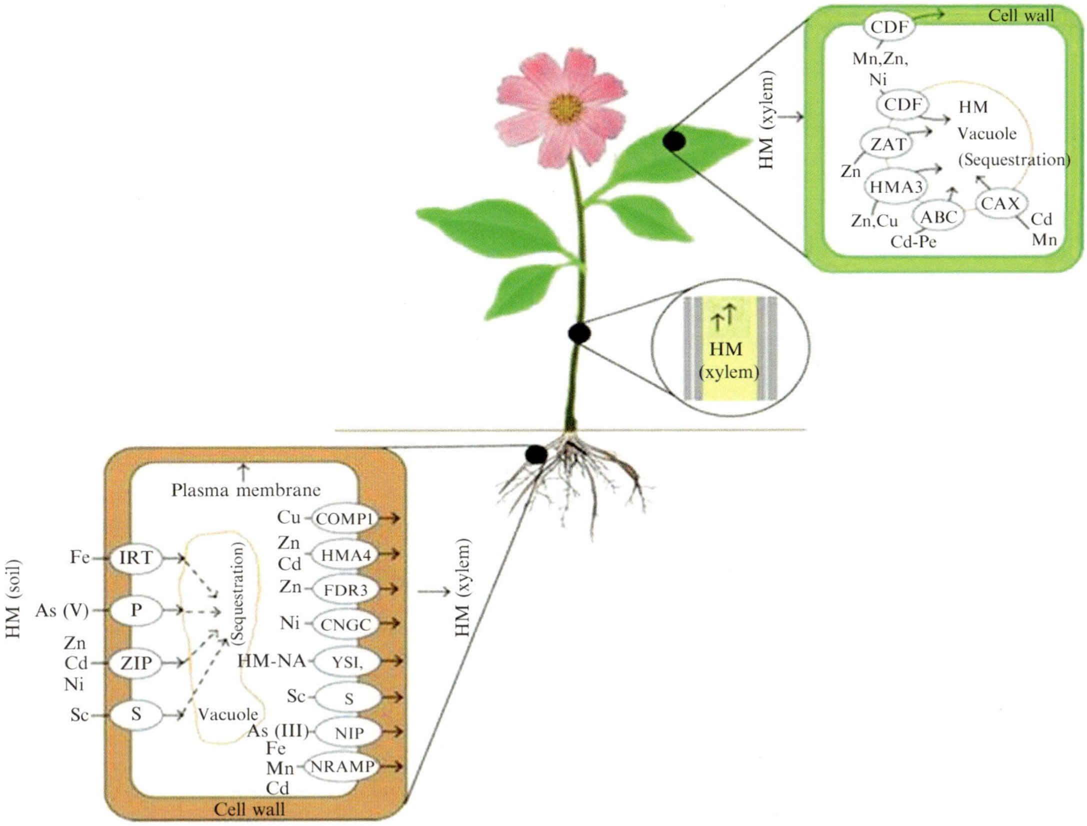


**Fig. 12.1** Diagrammatic representation of uptake and transport of heavy metals in plants through metal transporters (Hossain et al. 2012b)

receive or donate electrons and participate in redox reactions. The physiological redox range of aerobic living cells stretches from −420 to +800 mV. Therefore, heavy metals with higher redox potentials than those of biological molecules are biologically active. Those who have lower redox potentials than those of biological molecules cannot participate in biological redox reactions. Autoxidation of redox-active metals, such as $Fe^{2+}$ or $Cu^{+}$, results in $O_2^{\bullet -}$ formation and subsequently in $H_2O_2$ and $OH^{-}$ production via Fenton-type reactions. Cellular injury by this type of mechanism is well documented for iron (Imlay et al. 1988; Halliwell and Gutteridge 1986), copper (Li and Trush 1993a, b), as well as other metals (Shi and Dalal 1993; Jones et al. 1991; Lund et al. 1991; Shi et al. 1993).

Another important mechanism of heavy metal toxicity is their potential to bind strongly to electronegative oxygen, nitrogen, and sulfur atoms. This binding affinity is related to free enthalpy of the formation of the product of metal and ligand. Attributed to the above-mentioned characteristic, heavy metals can inhibit enzymes. Enzymes with amino acid residue containing sulfur (thiol) like cysteine are poisoned with heavy metals like mercury and silver. Direct effects of cadmium on the sulfhydryl homeostasis of cells and inhibition of enzymes have been reported for mammalian and animal cells (Canesi et al. 1998; Chrestensen et al. 2000).

Many enzymes contain metals in positions important for their activity. The displacement of one metal by another will normally also lead to inhibition or loss of enzyme activities. Divalent cations such as $Co^{2+}$, $Ni^{2+}$, and $Zn^{2+}$ were found to displace $Mg^{2+}$ in ribulose-1,5-bisphosphate carboxylase/oxygenase and resulted in loss of activity. Displacement of $Ca^{2+}$ by $Cd^{2+}$ in the protein calmodulin, important in cellular signaling, led to an inhibition in the calmodulin-dependent phosphodiesterase activity in radish (Rivetta et al. 1997).

The phytotoxicity of Zn is exhibited by decrease in growth and development and metabolism and an induction of oxidative damage in various plant species such as *Phaseolus vulgaris* (Cakmak and Marshner 1993). Zinc causes chlorosis in the younger leaves, which can extend to older leaves after prolonged exposure to high soil Zn levels (Ebbs and Kochian 1997). The appearance of a purplish-red color in leaves, which is ascribed to phosphorus (P) deficiency, is also caused by high Zn concentrations (Lee et al. 1996). Cadmium inhibits the nitrate reductase activity in the shoots thus negatively effecting absorption of nitrate and its transport from roots to shoots (Hernandez et al. 1996). Appreciable inhibition of the nitrate reductase activity was also found in plants of Silene cucubalus (Mathys 1975). Excess of copper in soil impedes plant growth, causes leaf chlorosis (Lewis et al. 2001), and induces oxidative stress and ROS (Stadtman and Oliver 1991). Toxic level of mercury can induce a cascade of events like it first binds to water channel proteins; this induces the closure of leaf stomata causing physical obstruction of water flow in plants (Zhang and Tyerman 1999). Also it interferes with the mitochondrial activity and induces oxidative stress by triggering the generation of ROS. This leads to the disruption of biomembrane lipids and cellular metabolism in plants (Messer et al. 2005; Cargnelutti et al. 2006). Chromium stress is one of the important factors that affect photosynthesis in terms of $CO_2$ fixation, electron transport, photophosphorylation, and enzyme activities (Clijsters and Van Assche 1985). The reduced germination of seeds under Cr stress could be a depressive effect of Cr on the activity of amylases and on the subsequent transport of sugars to the embryo axes (Zeid 2001). Inhibitory effects of lead on metabolic plant processes causes cell growth inhibition arises by simulation of indole-3-acetic acid (IAA) oxidation and inhibiting photosynthesis carboxylating enzymes activity (Stiborova et al. 1987). The appearance of iron toxicity in plants is related to high iron uptake by roots and its transportation to leaves and via transpiration stream. Excess iron causes free radical production that damages cellular structure irreversibly and impairs membranes, DNA, and proteins (Arora et al. 2002; de Dorlodot et al. 2005).

These examples show that, according to their chemical and physical properties, three different molecular mechanisms of metal toxicity can be distinguished: (a) production of reactive oxygen species by autoxidation and Fenton reaction, (b) blocking of essential functional groups in biomolecules, and (c) displacement of essential metal ions from biomolecules (Fig. 12.2).

## Mechanism of Detoxification

Plant tolerance to particular heavy metals is governed by an interrelated network of physiological and molecular mechanisms, and understanding of these mechanisms and their genetic basis is an important aspect to developing plants as agents of phytoremediation (Clemens et al. 2002; Dalcorso et al. 2008, 2010; Hossain et al. 2010). Physiological, biochemical, and molecular approaches continue to be employed to identify the underlying mechanisms of heavy metal accumulation, tolerance, and adaptive mechanisms to cope with heavy metal stress. Different plant species may have evolved different mechanisms to tolerate excess heavy metals, and even within the one plant species, more than one mechanism could be in operation. Plants have both constitutive and adaptive mechanisms to withstand excess heavy metals (Meharg 1994).

### Phytoremediation

Some adaptive mechanisms evolved by plants include immobilization, plasma membrane exclusion, restriction of uptake, and transport. Once the metal pollutant is taken up by plant cells by specific HM transporters, heavy metals can have varied fates depending on the type of metal and plant. It can be stored (phytoextraction), volatalized

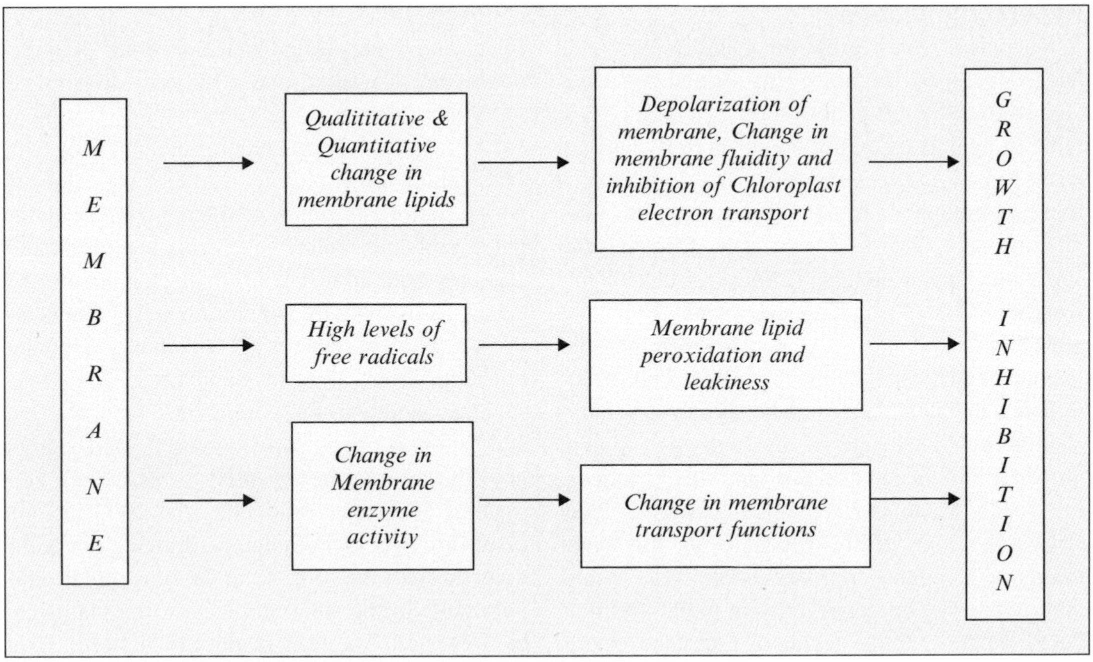


**Fig. 12.2** Flow chart depicting heavy metal induces inhibition of membrane function and growth (Nagajyoti et al. 2010)

(phytovolatalized), metabolized (phytodegradation), or chelated (phytochelatin synthesis) by the plant. The manipulation of phytochelatin (PC) expression is one of the potential mechanisms for increasing the capacity of plants for phytoremediation. The heavy metal also induces mechanisms such as upregulation of antioxidant and glyoxalase systems, production of stress proteins, and the biosynthesis of Pro, polyamines, and signaling molecule such as salicylic acid and nitric oxide (Cobbett 2000; Dalcorso et al. 2010; Toppi and Gabbrielli 1999; Sharma and Dietz 2009; Hossain and Fujita 2009; Yang et al. 2005; Pareek et al. 2006; Dalcorso et al. 2008; Hossain et al. 2009, 2010, 2011a, b, 2012a, b; Zhu et al. 1999).

Phytoextraction is the uptake and storage of pollutants in the stem or leaves. Living plants can be compared to solar-driven pumps that can extract and concentrate several elements from their environment. Generally, all the plants have the ability to accumulate heavy metals that are essential for their growth and development and at the same time certain plants have the ability to accumulate heavy metals with no known biological function. These include Cd, Cr, Pb, Co, Ag, Se, and Hg. Effective phytoextraction eventually requires that plants efficiently take up and translocate the metal to be extracted, sequester the metals in their tissues so that they can be tolerated, and have a high annual production of aboveground biomass. Plants have developed three basic strategies for growing on contaminated and metalliferous soils:

1. *Metal excluders*: These plants effectively sequester a large amount of metals in their roots and prevent metal from entering their aerial parts over a broad range of metal concentrations in the soil.
2. *Metal indicators*: These plants accumulate metals in the aboveground tissues and the metal levels in the tissues of these plants generally reflect metal levels in the soil.
3. *Metal accumulators*: These plant species can concentrate metals in their aboveground tissues to levels far exceeding those present in the soil or in the nonaccumulating species growing nearby.

## Hyperaccumulation

Some plants which grow on metalliferous soils have developed the ability to accumulate massive amounts of indigenous metals in their tissues without exhibiting symptoms of toxicity and are termed as *hyperaccumulators*. By definition, a hyperaccumulator must accumulate at least 100 mg $g^{-1}$ (0.01% dry weight) Cd, As, and some other trace metals; 1,000 mg $g^{-1}$ (0.1 dry weight) Co, Cu, Cr, Ni, and Pb; and 10,000 mg $g^{-1}$ (15 dry weight) Mn and Ni (Reeves and Baker 2000). To effectively accumulate a metal, a plant must be able to efficiently absorb it, translocate it through the xylem, unload it into the shoot tissue, and finally sequester it into vacuoles. Evolutionary studies have shown that one of the mechanisms for metal tolerance is uptake, not exclusion. This means that metal-tolerant genotypes are metal accumulators. In the extreme forms of tolerance, plants exhibit hypertolerance and often accumulate exceptionally high concentrations of heavy metals in their tissues. Most of the hyperaccumulators seem to have evolved from widespread species that surround or once surrounded metal-rich patches without actually colonizing them. Hyperaccumulating plant species perhaps present the ultimate tolerance to extremely hostile edaphic environments that would kill many other species. Hyperaccumulators can be found in a wide range of plant taxa but to date about 400 plants that hyperaccumulate metals are reported. The families dominating these members are Asteraceae, Brassicaceae, Caryophyllaceae, Cyperaceae, Fabaceae, Lamiaceae, Poaceae, and Euphorbiaceae. But many of the most commonly studied species are found in Brassicaceae which have the largest number of taxa, that is, 11 genera and 87 species. Some plants can accumulate up to 30,000 ppm of their shoot dry weight at Zn while growing in their native habitat. This accumulation is 300-fold higher than what most *normal* plants require for healthy growth. Remarkably, plants can also hyperaccumulate As, Mn, Ni, and Se (Salt 2006) (Table 12.1).

Plants which have the potential to successfully amass very high concentrations of heavy metals are classified under the term hyperaccumulator. Following are some salient feature of such plants: (a) Plants have roots and root hairs that create an enormous surface area through which pollutants can be extracted from contaminated soil and water. (b) Plants are autotrophs and they take up nearly all their elemental nutrients directly from the environment. (c) Plants use their leaves to extract $CO_2$, other gases, and nutrients from air and rain water. (d) Plants secrete fixed carbon compounds into the soil and support necessary bacterial and fungal growth that may be essential to recovering ecosystem (Subhadra and Sharma 2007). Some plants are hyperaccumulators, drawing the pollutants through the root. After the accumulation of pollutants in the stem and leaves, the plants can be harvested. Then the plants can be burnt. Even if the plants cannot be used, incineration and disposal of the plants is still cheaper than the traditional remediation methods. As a comparison, it is estimated that a site containing 5,000 tons of contaminated soils will produce only 20–30 tons of ash. This method is particularly used in remediating metals.

## Mechanisms for Hyperaccumulation

Mercury is a worldwide problem as a result of its many diverse uses in industries. Mercury has been used in bleaching operation, as a catalyst, as a pigment for paints, for gold mining, as well as a fungicide and antibacterial agent in seeds and bulbs. Elemental mercury, Hg(0), is oxidized to $Hg_2$ by biological systems and subsequently is leached into lands, waterways, and estuaries. Mercury can accumulate in animals as methylmercury ($CH_3Hg$), dimethylmercury $(CH_3)_2Hg$, or other organomercury salts. Organic mercury is produced by some anaerobic bacteria more toxic in some eukaryotes, and it efficiently permeates biological membrane. Monomethylmercury is responsible for severe neurological degeneration in birds and humans (Prasad and de Oliveira Freitas 2005). Plant tolerance to mercury is

**Table 12.1** Metals accumulated by natural hyperaccumulators

| Name of the plant | Metal(s) accumulated | Reference |
|---|---|---|
| *Pistia stratiotes* | Ag, Cd, Cr, Cu, Zn, Hg, Ni Pb, and Zn | Subhadra and Sharma (2007) |
| *Azolla caroliniana* | Hg and Cr | Subhadra and Sharma (2007) |
| *Helianthus annus* | Pb | Subhadra and Sharma (2007) |
| *Hemidesmus indicus* | Pb | Subhadra and Sharma (2007) |
| *Sesbania drummondii* | Pb | Subhadra and Sharma (2007) |
| *Pteris vittata* | As | Subhadra and Sharma (2007) |
| *Solanum nigrum* | Cd, Pb, Cu, and Zn | Subhadra and Sharma (2007) |
| *Phytolacca acinosa* | Mn | Subhadra and Sharma (2007) |

quite low, and therefore, phytoremediation can be limited by plant tolerance. An approach has been set out to introduce bacterial genes that convert methylmercury to volatile elemental mercury in plants (Bizely 2000). This involves sequential action of two enzymes: organomercurial lyase encoded by merB gene which converts methylmercury to $Hg_2$ and another mercuric reductase enzyme, encoded by merA gene (reduces $Hg_2$ to elemental mercury using NADPH as electron donor). Plants expressing merB and merA genes are resistant to extremely high levels of environmental toxin, methylmercury. Mineralization of mercuric compounds in soil can be done by expressing bacterial genes for mercuric reductase in yellow poplar seedlings. Mustard family serves as potent Hg hyperaccumulator (Prasad and de Oliveira Freitas 2005). Plants and humans require adequate amounts of micronutrients like iron and zinc, but accumulation of an excess uptake of nonessential metals like cadmium (Cd) and plumbum (Pb) can be extremely harmful. Proteins of the CDF (cation diffusion facilitator) family or efflux family are involved in the homeostasis of $Cd^{+2}$, $Co^{+2}$, $Fe^{+2}$, and $Zn^{+2}$ in microbes, animals, and plants. Therefore, the elucidation of the role of CDF proteins in *Arabidopsis thaliana* would be advantageous to the success of phytoremediation.

Nickel (Ni) hyperaccumulation in *Brassica* species and *Thlaspi goesingense* has been reported (Persans et al. 2001). This accumulation is due to constitutive expression of putative vacuolar transporters known as TgMTP1. *Thlaspi goesingense* displays an impressive ability to accumulate Ni up to 1–2% of its shoot dry mass. *Niemeyera acaminata* (New Calodenian tree) can accumulate Ni (25% of its dry weight). This is locally named *Sens bleue* because its sap is blue green due to excess Ni (Salt 2006). Recent research suggests that ZAT, a member of cation efflux (CE) family recently identified in *A. thaliana*, is involved in Zn efflux into the vacuole. Persons et al. used PCR and primers based on the sequence of the ZAT gene to clone an ortholog from *T. goesingense*, which they designated TgMTP1 for *T. goesingense* metal tolerance protein. Further research has proved the fact that Ni hyperaccumulation is known to be a constitute trait in *T. goesingense*, but the extent to which the high level of TgMTP1 is responsible for hyperaccumulation phenotype, however, remains to be determined. Scientists have developed cultivars of naturally occurring Ni hyperaccumulating plants for commercial Ni phytomining. Mining engineers in the USA planted one million Streptanthus plants in 1 ha of wasteland rich in nickel; this yielded 15–20% Ni by weight which can be reused (Salt 2006). Indian mustard and pickleweed volatilizes uptake of selenium (Se) completely and converts it to selenate and selenite into volatile nontoxic Se such as dimethyl selenide. The role of Se volatilization can be genetically enhanced (Sharma et al. 2004). Other crops like onion, garlic, mustard, cabbage, and pea family are also found to accumulate selenium.

Therefore, the hyperaccumulating ability to *T. crassulaceans* was linked to Zn transport in a number of plant species (Pence 2000).

Phytovolatilization is the uptake and vaporization of pollutants by a plant. This mechanism uptakes a solid or liquid contaminant and transforms it to an airborne vapor. The vapor can either be the pure pollutant or the metabolized pollutant by the plant before it is vaporized, as in the case of mercury, lead, and selenium (Subhadra and Sharma 2007). Phytodegradation is a process where plants metabolize pollutants. After the contaminant has been drawn into the plants, it assimilates into plant tissue, where the plant then degrades the pollutant. This metabolic activity was carried out by plant-derived enzymes such as nitroreductase, laccase, and nitrilase (Olson et al. 2003). Development of phytoremediation technologies requires thorough understanding of the underlying processes at the genetic, molecular, biochemical, physiological, and agronomic levels (Kramer 2005). Therefore, the most effective current phytoremediation sites in practice combine phytoextraction, phytovolatilization, phytodegradation, and phytochelatin synthesis mechanisms to clean up a contaminated site. Certain genetically modified plants used in phytoremediation program are indicated in Table 12.2.

## Phytochelatins

Chelating of metals by high-affinity ligands in the cytosol is potentially a very important mechanism of heavy metal detoxification and tolerance. Potential ligands include amino acids, organic acids, and two classes of peptides, the phytochelatins (PCs) and metallothionein (MT) (Rauser 1999; Clemens 2001). MTs and PCs were primarily identified as cadmium (Cd)-binding proteins in mammalian tissues and plants, respectively. Early reports of metal-binding proteins in plants were generally assumed to be MTs.

The general structure of phytochelatins is $(\gamma\text{-Glu-Cys})_n$-Gly. Phytochelatins (PCs) consisted of solitary three amino acids such as glutamine (Glu), cystine (Cys), and glycine (Gly) with the Glu and Cys residues being linked through a-carboxyl amide bond. PCs form a family of structures with increasing repetitions of the -Glu-Cys dipeptide followed by a terminal Gly, $(\text{-Glu-Cys})_n$-Gly, where *n* has been reported as being as high as 11 but is generally in the range of 2–5. PCs have been identified in a wide variety of plant species and in some microorganisms. They are structurally related to glutathione (GSH; -Glu-Cys-Gly) and were accepted to be the products of a biosynthetic pathway. In addition, a number of structural variants, for example, $(\text{-Glu-Cys})_n$-β-Ala, $(\text{-Glu-Cys})_n$-Ser, and $(\text{-Glu-Cys})_n$-Glu, have been identified in some plant species (Rauser 1995, 1999; Zenk 1996).

### Synthesis of PCs

Physiological, biochemical, and genetic studies on intact plants and in in vitro cell culture have established that glutathione (GSH) serve as substrate for the PCs biosynthesis (Rauser 1995, 1999; Zenk 1996). Cell culture studies have demonstrated that the induction of PCs in presence of Cd concur with transient decrease in GSH level. This was further deep rooted on exposure of intact plants and cell cultures to an inhibitor of GSH biosynthesis, thiamine sulfoximine (BSO), which increased sensitivity of Cd with corresponding inhibition of PC biosynthesis, and this inhibition could be reversed by addition of GSH to the growth medium. In the same way genetic studies have confirmed that GSH deficient mutants of the fission yeast and *Arabidopsis* which are PCs deficient are hypersensitive to Cd.

It was identified for the first time from cultured cells of *Silene cucubalus* that an enzyme activity synthesized PCs from GSH by transferring a γ-Glu-Cys moiety from a donor to an acceptor molecule. The enzyme mediates a bisubstrate reaction in which the thiol groups of at least one substrate are usually blocked (but not necessarily throughout the formation of heavy metal thiolates). This γ-Glu-Cys dipeptidyl transpeptidase (EC 2.3.2.15) was named PCs syntheses. The enzyme is a 95,000-Mr tetramer with a Km of 6.7 mM for GSH. Heavy metal activates the enzyme, which acts upon glutathione substrate to produce PCs.

**Table 12.2** Genetically modified plants used for phytoremediation

| Host plant | Transgenic protein expressed | Plant phenotype |
|---|---|---|
| *Brassica juncea* (Indian mustard) | β -Glutamylcysteine synthetase | Cadmium accumulation |
| *B. juncea* | ATP sulfurylase | Accumulation of selenium and other metals |
| *B. juncea* | Gamma-glutamylcysteine synthetase, glutathione synthetase | Accumulation of cadmium and other metals |
| *B. juncea* | Cystathionine-it *c*-synthase | Selenium volatilization |
| *Arabidopsis thaliana* (thale cress) | Selenocysteine lyase | Selenium accumulation |
| *A. thaliana* | Nitroreductase | TNT degradation |
| *A. thaliana* | Yeast cadmium factor (YFC1) | Cd(II) and Sb(II) |
| *A. thaliana* | Mercuric reductase | Hg(II) volatilization |
| *A. thaliana* | Organomercurial lyase | Methylmercury detoxification |
| *A. thaliana* | Arsenate reductase and γ -glutamylcysteine synthetase | Arsenic accumulation |
| *Liriodendron tulipifera* (yellow poplar) | Organomercurial lyase | Methylmercury detoxification |
| *Nicotiana tabacum* (tobacco) | Human cytochrome P450 2IE1 | Trichloroethylene degradation |
| *N. tabacum* | Citrate synthetase | Al(III) tolerance |
| *N. tabacum* | Metallothionein | Cadmium partitioning |
| *Oryza sativa* (rice) | Mercuric reductase | Mercury volatilization |

In vitro reactions have shown that PCs biosynthesis continued till metal ions are chelated by either PC or metal chelators like EDTA (Loeffler et al. 1989). This provides a mechanism to autoregulate the biosynthesis of PCs where the product chelates the activating metal ion and terminates the reaction (Fig. 12.1). PC syntheses activity has been detected in tomato (Chen et al. 1997) and *Arabidopsis* (Howden et al. 1995). After the structures of PCs had been elucidated and it was found that these peptides are distributed widely in the plant kingdom, it was proposed that PCs were the functional equivalent of MTs (Grill et al. 1989). Subsequently, numerous examples of MT-like genes, and in some cases MT proteins, have been isolated from a variety of plant species, and it is now apparent that plants express both of these Cys-containing metal-binding ligands. Furthermore, it is likely that the two play relatively independent functions in metal detoxification and/or metabolism. However, the extent to which this is true is not yet clear and will not become apparent until a complete set of MT-deficient mutants have been identified in *Arabidopsis*. PCs have not been reported in an animal species, supporting the notion that in animals, MTs may well perform some of the functions normally contributed by PCs in plants. However, the isolation of the PC syntheses gene from plants and the consequent identification of similar genes in animal species, described below, suggest that, at least in some animal species, both of these mechanisms contribute to metal detoxification and/or metabolism. Despite considerable and rapid progress in recent years, a lack of basic understanding of metal handling in plants is still limiting the design of phytoremediation approaches. Research should be further focused on how metal-sensitive crop plants can be explored for phytoremediation.

The most significant recent advances in our understanding of PCs biosynthesis and function have come from molecular genetic studies using a variety of model systems. These will continue to provide a wealth of mutants for biochemical, molecular, and physiological analysis. The isolation of PC syntheses genes from a number of species will allow a considerably greater understanding of the mechanism of metal activation of PC biosynthesis and the catalytic mechanism itself. There is considerable potential for the

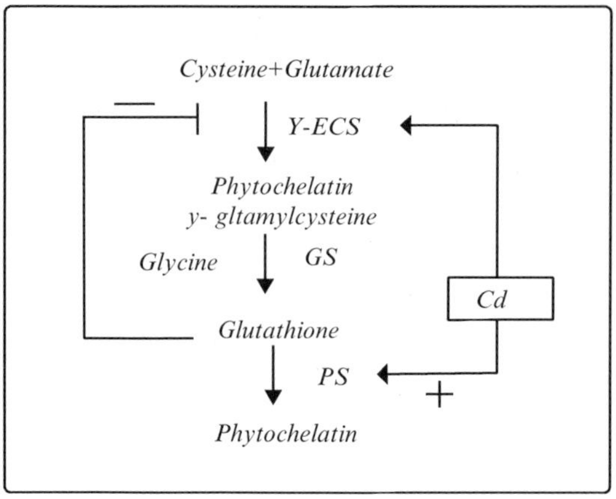


**Fig. 12.3** Synthesis of phytochelatins (PCs)

application of that understanding to optimize the process of phytoremediation (Subhadra and Sharma 2007) (Fig. 12.3).

## Cellular Mechanisms for Heavy Metal Detoxification

Elevated concentrations of both essential and nonessential metals can result in growth inhibition and toxicity symptoms. Plants possess a range of potential cellular mechanisms that may be involved in the detoxification of heavy metals and thus tolerance to metal stress. These include reduced uptake or efflux pumping of metals at the plasma membrane, the repair of stress-damaged proteins, and the compartmentation of metals in the vacuole by tonoplast located transporters.

### Mycorrhiza Symbiosis as a Mode of Tolerance

Although not always considered in general reviews of plant metal tolerance mechanisms, mycorrhizas, and particularly ectomycorrhizas that are characteristic of trees and shrubs, can be effective in ameliorating the effects of metal toxicity on the host plant (Marschner 1995; Huttermann et al. 1999; Jentschke and Godbold 2000). The mechanisms employed by the fungi at the cellular level to tolcrate heavy metals are probably similar to some of the strategies employed by higher plants, namely, binding to extracellular materials or sequestration in the vacuolar compartment. Thus, in the fungus *Pisolithus tinctorius*, tolerance to Cu and Zn was achieved by binding to extrahyphal slime (Tam 1995), whereas detoxification of Cd in *Paxillus involutus* involved binding of Cd to the cell walls and accumulation of Cd in the vacuole (Blaudez et al. 2000). In relation to the role of ectomycorrhizas in metal tolerance by the host plant, most mechanisms that have been proposed involve various exclusion processes that restrict metal movement to the host roots. These have been extensively reviewed and assessed (Jentschke and Godbold 2000) and include absorption of metals by the hyphal sheath, reduced access to the apoplast due to the hydrophobicity of the fungal sheath, chelating by fungal exudates, and adsorption onto the external mycelium. Clearly, from the variation between species described above, these different exclusion mechanisms are likely to vary in significance between different plant/fungal interactions.

### The Cell Wall and Root Exudates

The binding properties of the cell wall and its role as a mechanism of metal tolerance have been a controversial one. Although the root cell wall is directly in contact with metals in the soil solution, adsorption onto the cell wall is limited and thus have a limited effect on metal activity at the surface of the plasma membrane. Root exudates have an array of roles (Marschner 1995) including that of metal chelators which increases the uptake of certain metals. Salt et al. in 1999 observed the role of Ni-chelating exudates in the roots and accumulated amount of Ni-chelating histidine and citrate in Ni nonhyperaccumulating plants which could help to reduce Ni uptake and so play a role in a Ni detoxification strategy. Since the range of heavy metal compounds and exudes are very big, different exudates have detoxifying effect against different metals in a variety of plants. Buckwheat secretes oxalic acid from the roots in response to Al stress and accumulates nontoxic Al oxalate in the leaves; thus, detoxification occurs both externally and internally (Ma et al. 1997).

### Plasma Membrane as a Selective Barrier

The heavy metal once through from cell wall encounters the first *living* structure, the plasma membrane. In the presence of high concentrations of metals, its function is negatively affected as seen by an increased leakage from cells. Thus, tolerance may involve the protection of plasma membrane integrity against heavy metal damage that would produce increased leakage of solutes from cells (Meharg 1993; Strange and Macnair 1991; De Vos et al. 1991). The cell membrane may play an important role in metal homeostasis, by impeding the entry of metal into the cell as plasma membrane has a very interesting property of selective permeability. It also effluxes the toxic metals selectively. In most of the prokaryotic systems like bacteria, the selective efflux is coupled with ATP hydrolysis. It appears that the metabolic penalty for having more specific uptake mechanisms, and thus restricting the entry of toxic ions, is greater than that of having inducible efflux systems (Silver 1996).

Heat shock proteins (HSPs) are the proteins that show increased expression in response to temperatures above the optimal growth temperature of organism. HSPs are found in all living beings and act as molecular chaperones in normal protein modeling and also function in the protection and repair of proteins under stress conditions (Lewis et al. 1999). Increased expression of HSPs has been reported in response to a variety of heavy metal stress. Tseng et al. in 1993 showed that, in rice, both heat stress and heavy metal stress increased the levels of mRNAs for low molecular mass HSPs (16–20 kDa). Interestingly, it was reported that a short heat stress given prior to heavy metal stress induces a tolerance effect by preventing membrane damage, as judged by ultrastructural studies by Neumann et al. (1994). But the exact molecular mechanism is not yet clear. Heavy metals have a high affinity for biomolecule like carboxylic acids and amino acids such as citric acid, malic acid, and histidine. These could play a role in tolerance and detoxification as reviewed (Rauser 1999; Clemens 2001). A strong evidence for correlation between amounts of amino acid produced and exposure to a metal have been reported by Kramer et al. in (1996) to mark their widespread role in tolerance. A 36-fold increase was reported in the histidine content of the xylem sap on exposure to Ni in the Ni-hyperaccumulating plant *Alyssum lesbiacum*. In addition, supplying histidine to a nonaccumulating species greatly increased both its Ni tolerance and the capacity for Ni transport to the shoot. A possible Ni detoxification of histidine found in root exudates has been discussed earlier, as has the role of organic acids (oxalic acid) in Al tolerance.

### Vacuolar Compartmentalization

The vacuole is the site for the accumulation of a number of heavy metals including Zn and Cd (Ernst et al. 1992; De 2000). The role of vacuolar accumulation in relation to metal tolerance is for cadmium to have been substantiated best in the *Arabidopsis* plant. Two genes, *CAX1* and *CAX2*, have been studied which are localized in vacuolar membrane high- and low-efficiency $H^+/Ca^{2+}$ exchangers. *CAX1* is thought to be involved in vacuolar calcium ion ($Ca^{2+}$) accumulation; it was suggested that *CAX2* could be a high-capacity $H^+$/heavy metal cation transporter (Hirschi et al. 1996). Studies on meristematic cells of *Festuca rubra* roots showed that when these cells were treated with zinc, the cells responded by increasing vacuolation (Davies et al. 1991). Although there is evidence for $H^+$ antiport systems for Ca and Cd in oat root tonoplasts (Gries and Wagner 1998), a detailed information on other heavy metal transport systems at the tonoplast is limited

### Glyoxalase and Its Role in Heavy Metal Detoxification

The most common outcome of heavy metal toxicity is the excessive accumulation of methylglyoxal (MG) which can cause inactivation of enzymes, deoxyribonucleic acid (DNA) damage, and/or interaction with other vital constituents of plant cells. In higher plants the *glyoxalase* system eliminates methylglyoxal (MG). Being a central molecule, GSH is involved both directly and indirectly in control of MG and their reaction products in plant cells, thus protecting the plant from HM (Yadav et al. 2005a, b; Pareek et al. 2006; Hossain et al. 2009, 2012a, b). Cellular exclusion of heavy metals is an important adaptive strategy for HM tolerance in plants. A large fraction of heavy metals in plant roots are found in the free diffusion space outside the plasma membrane (apoplastic space). Tice et al. (1992) found that when wheat cultivar is treated with equal external Al concentrations, a tolerant had less symplastic Al than the sensitive cultivar substantiating the exclusion mechanism. HM transporter proteins are potentially involved in the exclusion of toxic HM ions from the symplastic to the apoplastic space.

## Heavy Metal Complex at the Cell Wall Plasma Membrane Interface

The apoplastic space, as recently discussed above, sequesters a large amount of heavy metals suggesting that the cell wall interface could be the potential site of metal tolerance. Cation exchange capacity (CEC) characterizes the number of fixed negative charges of plant cell walls and is an important parameter in studies dealing with the uptake of ions into plant tissues, especially in the roots (Fritz 2007). Sensitive wheat cultivars have much lower cell wall CECs than tolerant cultivars indicating that tolerant cultivars use a high CEC to complex HMs at the cell wall and prevent entry to the cell (Hossain et al. 2012a, b).

### Distribution of Heavy Metal

*Plants* cope up higher levels of metals by regulating the translocation of heavy metal. The evidence to this approach comes from the fact that roots in the tolerant varieties, have shown to contain much larger amount of heavy metal than the shoot system thus protecting the sensitive shoot from getting exposed to the much higher amounts of heavy metal. Heavy metals are found in plant roots than in shoots, except for hyperaccumulators; thus, plants are able to lessen the adverse effects of excess heavy metal by restricting the translocation of heavy metals within their organs or cells (Wheeler and Power 1995; Zheng et al. 1998). Cellular distribution of heavy metal in plant is indicated in Fig. 12.4.

## Enzyme Activity (ROS) Under Heavy Metal Stress

Aerobic organisms are exposed to ROS (*reactive oxygen species*) formation. These incomplete reduced oxygen species are toxic by-products, generated at low levels in nonstressed plant cells in chloroplasts and mitochondria and also by cytoplasmic membrane-bound or exocellular enzymes involved in redox reactions (especially photosynthetic electron transport processes and respiration). Extra amounts of ROS occur under stressful conditions such as pathogen attacks, wounding, herbivore feeding, UV light, heavy metals, and others (Wojtaszek 1997; Diaz et al. 2001).

ROS has some very important positive roles in several metabolic processes like it is involved in lignin formation in cell walls, participates in an oxidative burst, and protects against invading pathogens; it also acts as signals for activating further reactions (HR-hypersensitive response or phytoalexin biosynthesis) (Inez and Van 1995; Wojtaszek 1997). The high reactivity of ROS is based on the specificity of their electronic configuration. ROSs are known to damage cellular membranes by inducing lipid peroxidation (Ramadevi and Prasad 1998). They also can damage DNA, proteins, lipids, and chlorophyll (Mittova et al. 2000). The most popular ROS are $O_2^-$, *superoxide radical*; $H_2O_2$, *hydrogen peroxide*; and $OH^-$, *hydroxyl radical* originating from one, two, or three electron transfers to dioxygen ($O_2$). $H_2O_2$ is relatively reactive but is very hazardous because it can permeate through plasma membranes and reaches cell compartments far from the site of its formation (Wojtaszek 1997).

Therefore, the general mechanism of heavy metal toxicity in plants can be summed up in four broad ways. Firstly, heavy metals have unpaired electrons in their orbitals to accept and donate single electrons, thus promoting

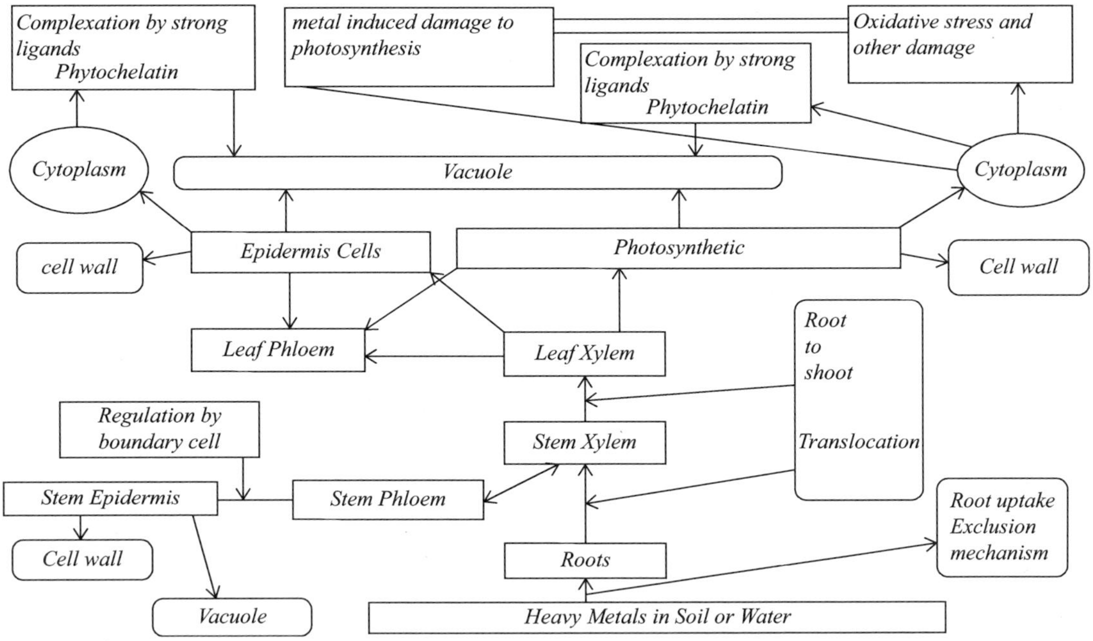


**Fig. 12.4** Cellular distribution of heavy metal in plant

monoelectron transfers to $O_2$ and generally ROS interconversion and oxireduction phenomena. Secondly, these metals reach metabolically active organelles, especially in the thylakoid membrane, and disturb the electron balance creating free radicals and reactive oxygen species. Thirdly, as already discussed, heavy metals mainly inactivate the antioxidant enzymes (peroxidases, catalases, superoxide dismutases) responsible for free radical detoxification and switch off the defense systems. And finally, heavy metal accumulation results in the depletion of low molecular weight antioxidants, such as glutathione, which is consumed under phytochelate formation (Dietz et al. 1999; Sahw et al. 2004). In the presence of redox-active transition metals such as $Cu^+$ and $Fe^{2+}$, $H_2O_2$ can be converted to OH molecule in a metal-catalyzed reaction via the Fenton reaction (Wojtaszek 1997; Mithofer et al. 2004).

*Scheme 1.* Haber-Weiss reaction

$$H_2O_2 + \cdot O_2^- \rightarrow \cdot OH + OH^- + O_2$$

Fenton reaction

$$H_2O_2 + Fe^{2+}(Cu^+) \rightarrow Fe_3{}^+(Cu^{2+}) + \cdot OH + OH^-$$

$$O_2 + Fe^{3+}(Cu^{2+}) \rightarrow Fe^{2+}(Cu^+) + O_2$$

## Plant's Defense Systems

Oxidative stress can lead to inhibition of the photosynthesis and respiration processes and, thus, plant growth. Plants have evolved enzymatic and nonenzymatic systems to scavenge active oxygen species. The antioxidant enzymes in the plant cell include superoxide dismutase (SOD), catalase (CAT), ascorbate peroxidase (APX), monodehydroascorbate reductase, dehydroascorbate reductase, and glutathione reductase. At least four of them participate in a highly developed detoxification system named the ascorbate-glutathione cycle (Halliwell-Asada cycle) (Morabito and Guerrier 2000; Rucinska et al. 1999; Noctor et al. 2002). Nonenzymatic antioxidants are ascorbate, glutathione, and tocopherol. As a major scavenger SOD catalyzes the dismutation of superoxide ($O^{-2}$) to hydrogen peroxide ($H_2O_2$) and oxygen ($O_2$). However, $H_2O_2$ is also toxic to cells and has to be further detoxified by CAT and/or peroxidase (POD) to water and oxygen (Zhu et al. 2004). Ascorbate peroxidase (APX) plays a key role in the ascorbate-glutathione cycle by reducing $H_2O_2$ to water at the expense of oxidizing ascorbate to monodehydroascorbate (MDHA). The balance between the production of ROS and their detoxification by the antioxidative system maintains homeostasis, but when this balance is disturbed to a large extent by heavy metal stress, then plants bear the oxidative insult (Hernandez et al. 1993, 1995; Gomez et al. 1999). Increasing the capacity to scavenge reactive oxygen species corresponds to tolerance (Smirnoff 1993; Bowler et al. 1992; Foyer et al. 1994). APX uses ascorbic acid as a reductant in the first step of the ascorbate-glutathione cycle. This is the most important peroxidase in $H_2O_2$ detoxification operating both in cytosol and chloroplasts (Smirnoff 2000; Mittova et al. 2000). *Apx* gene expression is rapidly induced by various stress conditions (Inez and Van 1995). The conception of antioxidant action of phenolic compounds also evolved from the idea that when plants are treated with high concentrations of phenolic compounds and peroxidase, activity rises in proportion (Bors et al. 1990). The aromatic rings present in the phenols act as nucleophile that attract the heavy metal and chelate them (Morgan et al. 1997).

Peroxidase (APX) is mainly localized in chloroplasts, cytosol, and peroxisomes and its function is to scavenge the $H_2O_2$ which is formed in these organelles. Two main types of this enzyme are (a) peroxidases (APX) which use ASC as the preferential electron donor and (b) peroxidases (APX) which use phenolics as the donor. The latter can be either soluble or cell wall-bound apoplastic POXs and vacuolar ones (Takahama and Oniki 2000; Diaz et al. 2001; De et al. 2005; Kilpelainen et al. 2002; Passardi et al. 2004). According to Rai et al. (2004), peroxidase can play significant role in sequestering heavy metals, especially as a physical barrier in cadmium toxicity, wounding, and pathogen response. It has been proposed that phytophenolics, especially flavonols and phenylopropanoids of vacuoles and the apoplast, can detoxify $H_2O_2$ as electron donors for phenol peroxidases (guaiacol peroxidases) localized in these compartments, which results in the formation of respective phenoxyl radicals (Takahama and Oniki 1997, 2000; Yamasaki et al. 1997). The similarity between APX action and POX Cycle (Sakihama et al. 2000) is indicated in Fig. 12.5.

## Glutathione (GSH), Phytochelatins (PCs), Metallothioneins (MTs), Reactive Oxygen Species (ROS), and Glyoxalase System-Related Gene Expression Under Heavy Metal Stress

Plant adaptation to stress is controlled by a well-balanced, genetically determined signaling system. The identification and characterization of plant genes which control responses to heavy metal stresses is an essential step to elucidate the complex regulatory network, which determines the tolerance of plants to heavy metal. Although molecules related to heavy metal transport, chelation, and sequestration such as transporters and heavy metal chelators are good candidates for the function of tolerance and accumulation, knowledge about global gene expression remains limited. The molecular genetics of model organisms can lend much to our understanding of the basic components of essential heavy metal metabolism in plants. With the exception of *PCS*, *GPX*, and *GS* genes, which were expressed at similar levels at all periods analyzed, the genes were induced in a transient manner. Mercury produces a clear induction of *GR1* (cytosolic) and *GR2* (plastidic) genes, with maximum expression being reached at 3 h after metal supply and decreasing thereafter. Similarly, *ECS* and *hGS* genes were expressed transiently, also reaching maximum transcript accumulation after 3 h. Notably, this pattern can be linked to GSH/hGSH depletion. The response was specific for that heavy metal whose toxicity is thought to be mitigated through PCs and other toxic and nontoxic heavy metal, which did not alter mRNA levels. JA also activated the same suite of genes, which suggests that it might be involved in the signal transduction pathway for

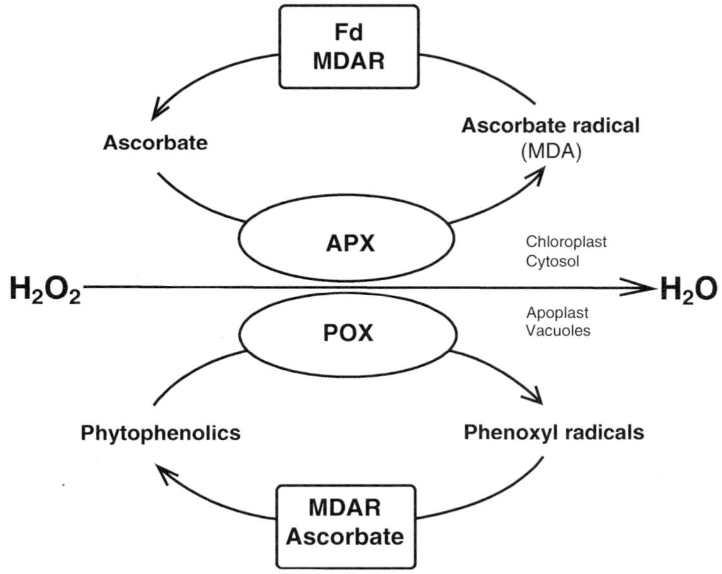


**Fig. 12.5** Similarity between APX action and POX Cycle (Sakihama et al. 2000)

Cu and Cd (Xiang and Oliver 1998). PCs are synthesized from GSH by the action of the enzyme $\gamma$-glutamylcysteine dipeptidyl transpeptidase, trivially named phytochelatin synthase (PCS). The enzyme is constitutively expressed but may be regulated at transcriptional and translational levels by heavy metal and metalloids (Heiss et al. 2003). The relative contribution of PCs and MTs in heavy metal (Cd or Cu) tolerance was examined in black mangrove, *Avicennia germinans*.

## Transgenic Plants and Heavy Metal Tolerance with Reference to Glutathione

Molecular biology and genetic engineering are being increasingly considered as effective tools for better understanding and improving the phytoremediation capability of plants because several key steps have been identified at the molecular level, through molecular-genetic technologies (Yang et al. 2005). Classic genetic studies demonstrated that many genes are involved in heavy metal uptake, translocation, sequestration, chemical modification, and tolerance. Unless regulatory genes are identified that simultaneously induce many heavy metal-related genes, it is feasible that more than one gene will need to be upregulated in order to substantially enhance heavy metal phytoremediation capacity. Few research findings showed that transgenic plants overexpressing the GSH synthesis genes, $\gamma$-ECS or GS, showed enhanced heavy metal tolerance. Two *Escherichia coli* genes involved in GSH biosynthesis, GS or $\gamma$-ECS, were constitutively expressed in *Brassica juncea* (Zhu et al. 1999a, b). The expression of either GS or $\gamma$-ECS in plant species resulted in higher accumulation of GSH and PCs as well as enhanced tolerance to Cd-induced oxidative stress. Additionally, transgenic poplar hybrid overexpressing a cytosol-targeted $\gamma$-*ECS* gene showed a three- to fourfold increase in GSH levels. However, transgenic tobacco overexpressing chloroplast-targeted $\gamma$- ECS contained elevated GSH levels but exhibited light-dependent chlorosis and necrosis. This was attributed to the increased levels of GSSG and a failure of the redox-sensing mechanism in chloroplasts of the transgenic plants. However, Li et al. 2006 showed that over expression of either $\gamma$-*ECS* or *GS* in *A. thaliana* was not sufficient to increase accumulation of As or Hg in above-ground tissues, even though the plants were more resistant to these contaminants. Coexpression of two genes, $\gamma$-*ECS* and *GS*, showed an increase in the accumulation of Hg governed by enhanced concentrations of PCs as compared to single-gene transformants. Based on their finding, it can be concluded that for the accumulation of some heavy metal, both GSH and PCs are a prerequisite for higher tolerance. The aforementioned results clearly indicate that use of transgenic plants overexpressing cysteine, GSH, PC, MT, or glyoxalase pathway-related genes showed substantial tolerance against heavy metal stress. Therefore, transgenic approaches can be effectively used for the cleanup of heavy metal pollution and as a promising approach for phytoremediation.

## Conclusion

The numerous deleterious health effects upon exposure to toxic heavy metals (HMs) in the environment are a matter of serious concern and a global issue. Therefore, molecular and cellular adaptations of plant cells in response to heavy metal stress appears to be necessary to improve plant heavy metal tolerance that ultimately reduces the chance of entering heavy metal into the food chain. Plant heavy metal tolerance is a multigenic trait and controlled at multiple levels. A wealth of recent studies demonstrated that plants protect themselves from heavy metal toxicity, besides other mechanisms, through an elevated level of nonenzymatic and enzymatic components of antioxidant and glyoxalase defense systems. Significant progress has been achieved in regard to heavy metal toxicity and tolerance, and different key components ensuring heavy metal tolerance in plants have been identified; however, many key questions remain unanswered. Additionally, different heavy metals appear to have different mechanisms to elicit toxicity symptoms, and plants employ different mechanisms as resistant reactions to particular heavy metal. Likewise, the reaction response and tolerance mechanisms are also different when a plant is grown under excess heavy metal in hydroponic or actual field conditions. Therefore, it is very difficult to hypothesize a common resistance mechanism against all heavy metal and metalloids. It would be interesting to see in the next few years how research on heavy metals and oxidative stress evolved and which approaches were adopted and found to be most useful.

## References


Angelone M, Bini C (1992) Trace elements concentrations in soils and plants of Western Europe. In: Adriano DC (ed) Biogeochemistry of trace metals. Lewis Publishers, Boca Raton, pp 19–60

Arora A, Sairam RK, Srivastava GC (2002) Oxidative stress and antioxidative system in plants. Cur Sci 82:1227–1338

Bizely SP (2000) Phytodetoxification of hazardous organomercurials by genetically engineered plants. Nat Biotechnol 18:213–217

Blaudez D, Botton B, Chalot M (2000) Cadmium uptake and subcellular compartmentation in the ectomycorrhizal fungus Paxillus involutus. Microbiology 146:1109–1117

Bors W, Heller W, Michel C, Saran M (1990) Flavonoids as antioxidants: determination of radical-scavenging efficiency. Methods Enzymol 186:343

Bowler C, Van MM, Inze D (1992) Superoxide dismutase and stress tolerance. Annu Rev Plant Physiol Plant Mol Biol 43:83–116

Breckle CW (1991) Growth under heavy metals. In: Waisel Y, Eshel A, Kafkafi U (eds) Plant roots: the hidden half. Marcel Dekker, New York, pp 351–373

Cakmak I, Marshner H (1993) Effect of zinc nutritional status on superoxide radical and hydrogen peroxide scavenging enzymes in bean leaves. In: Barrow NJ (ed) Plant nutrition-from genetic engineering field practice. Kluwer, Dordrecht, pp 133–137

Canesi L, Ciacci C, Piccoli G, Stocchi V, Viarengo A, Gallo G (1998) In vitro and in vivo effects of heavy metals on mussel digestive gland hexokinase activity: the role of glutathione. Comp Biochem Physiol C Pharmacol Toxicol Endocrinol 120:261–268

Cargnelutti D, Tabaldi LA, Spanevello RM, Jucoski GO, Battisti V, Redin M, Linares CEB, Dressler VL, Flores MM, Nicoloso FT, Morsch VM, Schetinger MRC (2006) Mercury toxicity induces oxidative stress in growing cucumber seedlings. Chemosphere 65:999–1106

Cataldo DA, Garland TR, Wildung RE (1978) Nickel in plants. I. Uptake kinetics using intact soybean seedlings. Plant Physiol Biochem 62:563–565

Cataldo DA, Garland TR, Wildung RE (1981) Cadmium uptake kinetics in intact soybean plants. Plant Physiol Biochem 73:844–848

Chen J, Zhou J, Goldsbrough PB (1997) Characterization of phytochelatin synthase from tomato. Physiol Plant 101:165–172

Chrestensen CA, Starke DW, Mieyal JJ (2000) Acute cadmium exposure inactivates thioltransferase (glutaredoxin), inhibits intracellular reduction of protein-glutathionyl-mixed disulfides, and initiates apoptosis. J Biol Chem 275:26556–26565
Clemens S (2001) Molecular mechanisms of plant metal tolerance and homeostasis. Planta 212:475–486
Clemens S, Palmgren MG, Kramer U (2002) A long way ahead: understanding and engineering plant metal accumulation. Trends Plant Sci 7(7):309–315
Clijsters H, Van Assche F (1985) Inhibition of photosynthesis by heavy metals. Photosynth Res 7:31–40
Cobbett CS (2000) Phytochelatins and their roles in heavy metal detoxification. Plant Physiol 123(3):825–832
Dalcorso G, Farinati S, Maistri S, Furini A (2008) How plants cope with cadmium: staking all on metabolism and gene expression. J Integr Plant Biol 50 (10):1268–1280
Dalcorso G, Farinati S, Furini A (2010) Regulatory networks of cadmium stress in plants. Plant Signal Behav 5(6):1–5
Davies KL, Davies MS, Francis D (1991) The influence of an inhibitor of phytochelatin synthesis on root growth and root meristematic activity in Festuca rubra L. in response to zinc. New Phytol 118:565–570
De DN (2000) Plant cell vacuoles. CSIRO Publishing, Collingwood
de Dorlodot S, Lutts S, Bertin P (2005) Effects of ferrous iron toxicity on the growth and mineral composition of an inter specific rice. J Plant Nutr 28:1–20
De Vos CHR, Schat H, De Waal MAM, Vooijs R, Ernst WHO (1991) Increased resistance to copper induced damage of the root cell plasma lemma in copper tolerant Silene cucubalus. Physiol Plant 82:523–528
Diaz J, Bernal A, Pomar F, Merino F (2001) Induction of shikimate dehydrogenase and peroxidase in pepper (Capsicum annuum L.) seedlings in response to copper stress and its relation to lignification. Plant Sci 161:179
Dietz KJ, Baier M, Kramer U (1999) Free radicals and active oxygen species as mediators of heavy metal toxicity in plants. In: Prasad MNV, Hagemeyer J (eds) Heavy metal stress in plants: from molecule to ecosystems. Springer, Berlin, pp 73–97
Ebbs SD, Kochian LV (1997) Toxicity of zinc and copper to Brassica species: implications for phytoremediation. J Environ Qual 26:776–781
Ernst WHO, Verkleij JAC, Schat H (1992) Metal tolerance in plants. Acta Botanica Neerlandica 41:229–248
Foyer CH, Lelandais M, Kunert KJ (1994) Photooxidative stress in plants. Physiol Plant 92:696–717
Fritz E (2007) Measurement of cation exchange capacity (CEC) of plant cell walls by X-ray microanalysis (EDX) in the transmission electron microscope. Microsc Microanal 13(4):233–244
Galloway JN, Thornton JD, Norton SA, Volcho HL, McLean RA (1982) Trace metals in atmospheric deposition: a review and assessment. Atmos Environ 16:1677
Godbold DL (1991) Cadmium uptake in Norway spruce. Tree Physiol 9:349–357
Godbold DL, Hüttermann A (1985) Effect of zinc, cadmium and mercury on root elongation of *Picea abies* (Karst.) seedlings and the significance of these metals to forest dieback. Environ Pollut 38:375–381
Gomes-Junior RA, Moldes CA, Delite FS, Pompeu GB, Gratão PL, Mazzafera P, Lea PJ, Azevedo RA (2006) Antioxidant metabolism of coffee cell suspension cultures in response to cadmium. Chemosphere 65:1330–1337
Gomez JM, Hernandez JA, Jimenez A, del Rıo LA, Sevilla F (1999) Differential response of antioxidative enzymes of chloroplasts and mitochondria to long-term NaCl stress of pea plants. Free Radic Res 31:11–18
Gries GE, Wagner GJ (1998) Association of nickel versus transport of cadmium and calcium in tonoplast vesicles of oat roots. Planta 204:390–396
Grill E, Loffler S, Winnacker EL, Zenk MH (1989) Phytochelatins, the heavy-metal-binding peptides of plants, are synthesized from glutathione by a specific g-glutamylcysteine dipeptidyl transpeptidase (phytochelatin synthase). Proc Natl Acad Sci U S A 86:6838–6842
Halliwell B, Gutteridge JMC (1986) Iron and free radical reactions: two aspects of antioxidant protection. Trends Biochem Sci 11:375
Heiss S, Wachter A, Bogs J, Cobbett C, Rausch T (2003) Phytochelatin synthase (PCS) protein is induced in *Brassica juncea* leaves after prolonged Cd exposure. J Exp Bot 54(389):1833–1839
Hernandez JA, Corpas FJ, Gomez M, del Rıo LA, Sevilla F (1993) Salt-induced oxidative stress mediated by activated oxygen species in pea leaf mitochondria. Physiol Plant 89(1):103–110
Hernandez JA, Olmos E, Corpas FJ, Sevilla F, del Rıo LA (1995) Salt-induced oxidative stress in chloroplast of pea plants. Plant Sci 105:151–167
Hernandez LE, Carpena-Ruiz R, Garate A (1996) Alterations in the mineral nutrition of pea seedlings exposed to cadmium. J Plant Nutr 19:1581–1598
Hirschi KD, Zhen RG, Cunningham KW, Rea PA, Fink GR (1996) CAXI, an $H^+/Ca^{2+}$ antiporter from Arabidopsis. Proc Natl Acad Sci U S A 93:8782–8786
Hossain MA, Fujita M (2009) Purification of glyoxalase I from onion bulbs and molecular cloning of its cDNA. Biosci Biotechnol Biochem 73(9):2007–2013
Hossain MA, Hossain MZ, Fujita M (2009) Stress-induced changes of methylglyoxal level and glyoxalase I activity in pumpkin seedlings and cDNA cloning of glyoxalase I gene. Aust J Crop Sci 3(2):53–64
Hossain MA, Hasanuzzaman M, Fujita M (2010) Up-regulation of antioxidant and glyoxalase systems by exogenous glycinebetaine and proline in mung bean confer tolerance to cadmium stress. Physiol Mol Biol Plant 16(3):259–272
Hossain MA, da Silva JAT, Fujita M (2011a) Glyoxalase system and reactive oxygen species detoxification system in plant abiotic stress response and tolerance: an intimate relationship. In: Shanker AK, Venkateswarlu B (eds) Abiotic stress in plants-mechanisms and

adaptations. INTECH-Open Access Publisher, Rijeka, Croatia, pp 235–266
Hossain MA, Hasanuzzaman M, Fujita M (2011b) Coordinate induction of antioxidant defense and glyoxalase system by exogenous proline and glycinebetaine is correlated with salt tolerance in mung bean. Front Agric China 5(1):1–14
Hossain MA, Hossain MD, Rohman MM, da Silva JAT, Fujita M (2012a) Onion major compounds (flavonoids, organosulfurs) and highly expressed glutathione-related enzymes: possible physiological interaction, gene cloning and abiotic stress response. In: Aguirre CB, Jaramillo LM (eds) Onion consumption and health. Nova, New York
Hossain MA, Piyatida P, da Silva JAT, Fujita Z (2012b) Molecular mechanism of heavy metal toxicity and tolerance in plants: central role of glutathione in detoxification of reactive oxygen species and methylglyoxal and in heavy metal chelation. J Bot 2012:1–37
Housecroft CE, Sharpe AG (2008) Inorganic chemistry. Prentice hall, Harlow
Howden R, Goldsbrough PB, Andersen CR, Cobbett CS (1995) Cadmium-sensitive, cad1, mutants of Arabidopsis thaliana are phytochelatin deficient. Plant Physiol 107:1059–1066
Huttermann A, Arduini I, Godbold DL (1999) Metal pollution and forest decline. In: Prasad NMV, Hagemeyer J (eds) Heavy metal stress in plants: from molecules to ecosystems. Springer, Berlin, pp 253–272
Imlay JA, Chin SM, Linn S (1988) Toxic DNA damage by hydrogen peroxide through the Fenton reaction in vivo and in vitro. Science 240:640–642
Inez D, Van MM (1995) Oxidative stress in plants. Curr Opin Biotechnol 6:153
Jentschke G, Godbold DL (2000) Metal toxicity and ectomycorrhizas. Physiol Plant 109:107–116
Jones P, Kortenkamp A, O'Brien P, Wang G, Yang G (1991) Evidence for the generation of hydroxyl radicals from a chromium (V) intermediate isolated from the reaction of chromate with glutathione. Biochim Biophys Acta 286:652–655
Khan S, Cao Q, Zheng YM, Huang YZ, Zhu YG (2008) Health risks of heavy metals in contaminated soils and food crops irrigated with wastewater in Beijing, China. Environ Pollut 152(3):686–692
Kilpelainen I, Teeri TH, Simola LK (2002) Lignification related enzymes in Picea abies suspension cultures. Physiol Plant 114:343
Kramer U (2005) Phytoremediation: novel approaches to cleaning up polluted soil. Curr Opin Biotechnol 16 (2):133–141
Kramer U, Cotter Howells JD, Charnock JM, Baker AJM, Smith JAC (1996) Free histidine as a metal chelator in plants that accumulate nickel. Nature 379:635–638
Lantzy RJ, Mackenzie FT (1979) Atmospheric trace metals: global cycles and assessment of man's impact. Geochim Cosmochim Acta 43:511
Lee CW, Choi JM, Pak CH (1996) Micronutrient toxicity in seed geranium (Pelargonium 9 hortorum Bailey). J Am Soc Hortic Sci 121:77–82
Lewis S, Handy RD, Cordi B, Billinghurst Z, Depledge MH (1999) Stress proteins (HSPs): methods of detection and their use as an environmental biomarker. Ecotoxicology 8:351–368
Lewis S, Donkin ME, Depledge MH (2001) Hsp 70 expression in Enteromorpha intestinalis (Chlorophyta) exposed to environmental stressors. Aquat Toxicol 51:277–291
Li Y, Trush MA (1993a) DNA damage resulting from the oxidation of hydroquinone by copper: role for a Cu(II)/Cu/(I) redox cycle and reactive oxygen generation. Carcinogenesis 7:1303–1311
Li Y, Trush MA (1993b) Oxidation of hydroquinone by copper: chemical mechanism and biological effects. Biochim Biophys Acta 300:346–355
Li Y, Heaton ACP, Carreira L, Meagher RB (2006) Enhanced tolerance to and accumulation of mercury, but not arsenic, in plants over expressing two enzymes required for thiol peptide synthesis. Physiol Plant 128 (1):48–57
Loeffler S, Hochberger A, Grill E, Winnacker EL, Zenk MH (1989) Termination of the phytochelatin synthase reaction through sequestration of heavy metals by the reaction product. FEBS Lett 258:42–46
Lund BO, Miller DM, Woods JS (1991) Mercury-induced $H_2O_2$ production and lipid peroxidation in vitro in rat kidney mitochondria. Biochem Pharmacol 42:181–187
Ma JF, Zheng SJ, Matsumoto H (1997) Detoxifying aluminum with buckwheat. Nature 390:569–570
Marschner H (1995) Mineral nutrition of higher plants, 2nd edn. Academic, London
Mathys W (1975) Enzymes of heavy metal-resistant and non-resistant populations of Silene cucubalus and their interactions with some heavy metals in vitro and in vivo. Physiol Plant 33:161–165
Meharg AA (1993) The role of the plasmalemma in metal tolerance in angiosperms. Physiol Plant 88:191–198
Meharg AA (1994) Integrated tolerance mechanisms: constitutive and adaptive plant responses to elevated metal concentrations in the environment. Plant Cell Environ 17(9):989–993
Messer RL, Lockwood PE, Tseng WY, Edwards K, Shaw M, Caughman GB, Lewis JB, Wataha JC (2005) Mercury (II) alters mitochondrial activity of monocytes at sublethal doses via oxidative stress mechanisms. J Biomed Mat Res B 75:257–263
Mithofer A, Schulze B, Boland W (2004) Biotic and heavy metal stress response in plants: evidence for common signals. FEBS Lett 566:1
Mittova V, Volokita M, Guy M, Tal M (2000) Activities of SOD and the ascorbate-glutathione cycle enzymes in subcellular compartments in leaves and roots of the cultivated tomato and its wild salt-tolerant relative *Lycopersicon pennellii*. Physiol Plant 110:45

Morabito D, Guerrier G (2000) The free oxygen radical scavenging enzymes and redox status in roots and leaves of Populus × euramericana in response to osmotic stress, desiccation and rehydration. J Plant Physiol 157:74

Morgan JF, Klucas RV, Grayer RJ, Abian J, Becana M (1997) Complexes of iron with phenolic compounds from soybean nodules and other legume tissues: prooxidant and antioxidant properties. Free Radic Biol Med 22:861

Muller U (2007) Inorganic structural chemistry. Wiley, Chichefeter

Nagajyoti PC, Lee KD, Sreekanth TVM (2010) Heavy metals, occurrence and toxicity for plants: a review. Environ Chem Lett 8:199–216

Neumann D, Lichtenberger O, Gunther D, Tschiersch K, Nover L (1994) Heat shock proteins induce heavy metal tolerance in higher plants. Planta 194: 360–367

Nies DH (1999) Microbial heavy metal resistance. Appl Microbiol Biotechnol 51:730–750

Noctor G, Gomez L, Vanacker H, Foyer CH (2002) Interactions between biosynthesis, compartmentation and transport in the control of glutathione homeostasis and signaling. J Exp Bot 53:1283

Olson PE, Reardon KF, Pilon-Smits EAM (2003) Ecology of rhizosphere bioremediation. In: McCutcheon SC, Schnoore JL (eds) Phytoremediation: transformation and control of contaminants. Wiley-Interscience, Inc: NJ, pp 317–353

Pareek SLS, Yadav SK, Pareek A, Reddy MK, Sopory SK (2006) Transgenic tobacco over expressing glyoxalase pathway enzymes grow and set viable seeds in zinc-spiked soils. Plant Physiol 140(2):613–623

Passardi F, Penel C, Dunand C (2004) Performing the paradoxical: how plant peroxidases modify the cell wall. Trends Plant Sci 9(11):534–540

Pence NS (2000) The molecular physiology of heavy metal transport in Zn/Cd hyperaccumulator *Thlaspi caerulescens*. Proc Natl Acad Sci U S A 97: 4956–4960

Persans MW, Nieman K, Salt DE (2001) Functional activity and role of cation efflux family members in Ni hyperaccumulation in *Thlaspi goesingense*. Proc Natl Acad Sci U S A 98:9995–10000

Prasad MNV, de Oliveira Freitas HM (2005) Metal hyperaccumulation in plants: biodiversity prospecting for phytoremediation technology. Nat Biotechnol 6:1–27

Rai V, Vajpayee P, Singh SN, Mehrotra S (2004) Effect of chromium accumulation on photosynthetic pigments, oxidative stress defense system, nitrate reduction, proline level and eugenol content of *Ocimum tenuiflorum* L. Plant Sci 167:1159

Ramadevi S, Prasad MNV (1998) Copper toxicity in *Ceratophyllum demersum* L. (Coontail), a free floating macrophyte: Response of antioxidant enzymes and antioxidants. Plant Sci 138:157

Rauser WE (1995) Phytochelatins and related peptides: structure, biosynthesis, and function. Plant Physiol 109:1141–1149

Rauser WE (1999) Structure and function of metal chelators produced by plants: the case for organic acids, amino acids, phytin and metallothioneins. Cell Biochem Biophys 31:19–48

Reeves R, Baker A (2000) Metal-accumulating plants. In: Raskin I, Ensley BD (eds) Phytoremediation of toxic metals: using plants to clean up the environment. Wiley, New York

Rivetta A, Negrini N, Cocucci M (1997) Involvement of $Ca^{2+}$-calmodulin in $Cd^{2+}$ toxicity during the early phases of radish (*Raphanus sativus* L.) seed germination. Plant Cell Environ 20:600–608

Rucinska R, Waplak S, Gwozdz E (1999) Free radical formation and activity of antioxidant enzymes in lupin roots exposed to lead. Plant Physiol Biochem 37:187

Sahw BP, Sahu SK, Mishra RK (2004) Heavy metal induced oxidative damage in terrestrial plants. In: Presad MNV (ed) Heavy metal stress in plants: from biomolecules to ecosystems, 2nd edn. Springer, Berlin, pp 84–126

Sakihama Y, Mano J, Sano S, Asada K, Yamasaki H (2000) Reduction of phenoxyl radicals mediated by monodehydroascorbate reductase. Biochem Biophys Res Commun 279:949

Salt DE (2006) From meals to metals and back. In: Taiz L, Zeiger E (eds) Plant physiology. A companion to plant physiology, 5th edn. Sinauer Associates Inc, Sunderland

Salt DE, Kato N, Kramer U, Smith RD, Raskin I (1999) The role of root exudates in nickel hyper accumulation and tolerance in accumulator and nonaccumulator species of Thlaspi. In: Terry N, Banuelos G (eds) Phytoremediation of contaminated soil and water. Terry N, and Banuelos GS (eds), chapter 10, pp 191–202, CRC Press LLC, Boca Raton, FL, USA.

Schützendübel A, Schwanz P, Teichmann T, Gross K, Langenfeld-Heyser R, Godbold DL, Polle A (2001) Cadmium-induced changes in antioxidative systems, $H_2O_2$ content and differentiation in pine (*Pinus sylvestris*) roots. Plant Physiol 127:887–892

Sharma SS, Dietz KJ (2009) The relationship between metal toxicity and cellular redox imbalance. Trends Plant Sci 14(1):43–50

Sharma NC, Gardea Torresdey JL, Parson J, Sahi SV (2004) Chemical speciation and cellular deposition of lead in Sesbania drummondii. Environ Toxicol Chem 23:2068–2073

Shi X, Dalal NS (1993) Vanadate-mediated hydroxyl radical generation from superoxide radical in the presence of NADH: Haber-Weiss versus Fenton mechanism. Biochim Biophys Acta 307:336–341

Shi X, Dalal NS, Kasprzak KS (1993) Generation of free radicals from hydrogen peroxide and lipid hydroperoxides in the presence of Cr(III). Biochim Biophys Acta 302:294–299

Silver S (1996) Bacterial resistance to toxic metal ions- a review. Gene 179:9–19

Smirnoff N (1993) The role of active oxygen in the response of plants to water deficit and desiccation. New Phytol 125:27–58

Smirnoff N (2000) Ascorbic acid: metabolism and functions of multi-facetted molecule. Curr Opin Plant Biol 3:229

Stadtman ER, Oliver CN (1991) Metal-catalyzed oxidation of proteins. Physiological consequences. J Biol Chem 266:2005–2008

Stiborova M, Pitrichova M, Brezinova A (1987) Effect of heavy metal ions in growth and biochemical characteristic of photosynthesis of barley and maize seedlings. Biol Plant 29:453–467

Strange J, Macnair MR (1991) Evidence for a role for the cell membrane in copper tolerance of *Mimulus guttatus* Fischer ex DC. New Phytol 119:383–388

Subhadra AV, Sharma J (2007) Phytoremediation technology – a nature's bliss. Res J Biotechnol 2 (2):52–57

Takahama U, Oniki T (1997) A peroxidase / phenolics / ascorbate system can scavenge hydrogen peroxide in plant cells. Physiol Plant 101:845

Takahama U, Oniki T (2000) Flavonoid and some other phenolics as substrates of peroxidase: physiological significance of the redox reactions. J Plant Res 113:301

Tam PCF (1995) Heavy metal tolerance by ectomycorrhizal fungi and metal amelioration by *Pisolithus tinctorius*. Mycorrhiza 5:181–187

Tice KR, Parker DR, DeMason DA (1992) Operationally defined apoplastic and symplastic aluminum fractions in root tips of aluminum-intoxicated wheat. Plant Physiol 100(1):309–318

Toppi LSD, Gabbrielli R (1999) Response to cadmium in higher plants. Environ Exp Bot 41(2):105–130

Tseng TS, Tzeng SS, Yeh CH, Chang FC, Chen YM, Lin CY (1993) The heat-shock response in rice seedlings—isolation and expression of cDNAs that encode class-I low-molecular-weight heat-shock proteins. Plant Cell Physiol 34:165–168

Vitoria AP, Lea PJ, Azevedo RA (2001) Antioxidant enzymes responses to cadmium in radish tissues. Phytochemistry 5:701–710

Weast RC (1984) CRC Handbook of chemistry and physics, 64th edn. CRC Press, Boca Raton

Wheeler DM, Power IL (1995) Comparison of plant uptake and plant toxicity of various ions in wheat. Plant Soil 172(2):167–173

Williams LE, Pittman JK, Hall JL (2000) Emerging mechanisms for heavy metal transport in plants. Biochim Biophys Acta 1465:104–126

Wojtaszek P (1997) Oxidative burst: an early plant response to pathogen infection. Biochem J 322:681

Xiang C, Oliver DJ (1998) Glutathione metabolic genes coordinately respond to heavy metals and jasmonic acid in *Arabidopsis*. Plant Cell 10(9):1539–1550

Yadav SK, Pareek SLS, Ray M, Reddy MK, Sopory SK (2005a) Transgenic tobacco plants overexpressing glyoxalase enzymes resist an increase in methylglyoxal and maintain higher reduced glutathione levels under salinity stress. FEBS Lett 579(27):6265–6271

Yadav SK, Pareek SLS, Ray M, Reddy MK, Sopory SK (2005b) Methylglyoxal levels in plants under salinity stress are dependent on glyoxalase I and glutathione. Biochem Biophys Res Commun 337(1):61–67

Yamasaki H, Sakihama Y, Ikehara N (1997) Flavonoid peroxidase reaction as a detoxification mechanism of plant cell against $H_2O_2$. Plant Physiol 115:1405

Yang XE, Jin XF, Feng Y, Islam E (2005) Molecular mechanisms and genetic basis of heavy metal tolerance/ hyper accumulation in plants. J Integr Plant Biol 47(9):1025–1035

Zeid IM (2001) Responses of Phaseolus vulgaris to chromium and cobalt treatments. Biol Plant 44:111–115

Zenk MH (1996) Heavy metal detoxification in higher plants: a review. Gene 179:21–30

Zhang WH, Tyerman SD (1999) Inhibition of water channels by HgCl2 in intact wheat root cells. Plant Physiol 120:849–857

Zhang MK, Liu ZY, Wang H (2010) Use of single extraction methods to predict bioavailability of heavy metals in polluted soils to rice. Commun Soil Sci Plant Anal 41(7):820–831

Zheng SJ, Ma JF, Matsumoto H (1998) High aluminum resistance in buckwheat: I. Al- induced specific secretion of oxalic acid from root tips. Plant Physiol 117 (3):745–751

Zhu YL, Pilon-Smits EAH, Jouanin L, Terry N (1999a) Overexpression of glutathione synthetase in Indian mustard enhances cadmium accumulation and tolerance. Plant Physiol 119(1):73–79

Zhu YL, Pilon-Smits EAH, Tarun AS, Weber SU, Jouanin L, Terry N (1999b) Cadmium tolerance and accumulation in Indian mustard is enhanced by overexpressing $\gamma$- glutamylcysteine synthetase. Plant Physiol 121 (4):1169–1177

Zhu Z, Wei G, Li J, Qian Q, Yu J (2004) Silicon alleviates salt stress and increases antioxidant enzymes activity in leaves of salt-stressed cucumber (Cucumis sativus). Plant Sci 167:527–533

Zhu XF, Zheng C, Hu YT et al (2009) Cadmium-induced oxalate secretion from root apex is associated with cadmium exclusion and resistance in Lycopersicon esulentum. Plant Cell Environ 34(7):1055–1064

# Brassinosteroids: Biosynthesis and Role in Growth, Development, and Thermotolerance Responses

13

Geetika Sirhindi

**Abstract**

Brassinosteroids are new and unique class of plant growth regulators that constitutes the sixth class of phytohormones. Till date more than 70 analogues of these natural occurring and nonnatural analogues of brassinolides (BL) have been identified from different plant species. Brassinoslides due to their ubiquitous presence in plant species in almost all parts of plant orchestrate myriad growth and developmental processes. Owing to their peculiar structural features, their extremely low abundance in natural sources, and potent biological activity, BLs are being studied intensively to understand their role in plant metabolism under normal and inadequate growth conditions. Their main physiological effects in plants include cell division control, germination and growth promotion, regulation of hormonal balance, activation of protein and nucleic acid synthesis, enzyme activity, and most interestingly increased resistance to abiotic and biotic stresses. Exogenous application of BLs to plants at seed level or as foliar spray enhances antioxidant defense activities, and accumulation of osmoprotectants such as proline and glycine betaine under stress conditions illustrated antistress properties of brassinosteroids. BLs reported to play a regulatory role in the control of cell-cycle progression and differentiation in the *Arabidopsis*, and other plants may offer a novel therapeutic strategy for various diseases.

## Introduction

Brassinosteroids are ubiquitously occurring plant growth regulators which are assumed to function as a "master switch" in abiotic and biotic responses (Bajguz 2010). Plants have to adapt to abiotic factors such as high and low temperatures, ultraviolet (UV) light, light conditions, photoperiodicity, oxidative stress, hypoxia, anoxia, desiccation, salt stress, and metal stress or to biotic factors such as pathogens or herbivores by complex alterations and modulations in biochemical behavior at cellular level and/or at molecular level by regulating gene expressions and protein expression

G. Sirhindi (✉)
Department of Botany, Punjabi University, Patiala 147 002, Punjab, India
e-mail: geetikasir123@gmail.com; geetika@pbi.ac.in

G.R. Rout and A.B. Das (eds.), *Molecular Stress Physiology of Plants*,
DOI 10.1007/978-81-322-0807-5_13, © Springer India 2013

particularly those related to antioxidant defense system and heat shock proteins. Many facets of these alterations occur under inadequate conditions of environment, which is known to be promoted by brassinosteroids. Because of the unique structure of brassinosteroids (BRs) and their resemblance with animal steroid cholesterol in their basic structure, BLs orchestrate myriad growth and developmental processes in plants (Vleesschauwer et al. 2012) and are the latest growth regulators to be implicated in plant immunity against various abiotic and biotic stresses. BLs are polyhydroxylated steroid hormones that play an imperative role in regulating array of physiological and developmental processes, including seed germination, growth promotion, reproductive potential, and senescence (Clouse and Sasse 1998). BL induced molecular changes that are related to stress tolerance including enhanced expression of stress responsive genes (Kagale et al. 2007), protection of translational machinery (Dhaubhadel et al. 2002), potentiated accumulation of osmoprotectants (Divi and Krishna 2009; Sirhindi et al. 2009, 2011), NADPH-oxidase-mediated accumulation of hydrogen peroxide (Xia et al. 2009) and enhanced photosynthetic efficiency (Xia et al. 2009), and under abiotic stress (Divi et al. 2010). Recent biochemical studies provided riveting insights into the various aspects of plant steroid signaling. Divi et al. (2010) using mutant and transgenic *Arabidopsis* demonstrated that 24-epibrassinolide (epiBL)-induced tolerance to salt and temperature stress is reliant on the SA, master regulatory protein nonexpressor of pathogenesis-related genes 1 (NPR1), implicating a crucial role of the SA signaling pathway in BR-mediated stress responses. Kim and Wang (2010) illustrated BR signaling pathway ranging from BR receptor at the cell surface to activation of transcription factors in the nucleus. In this chapter, various roles of brassinosteroids, their biosynthesis, and their ubiquitous role in the modulation of protein and gene expression upon environmental and developmental stimuli will be discussed.

## Brassinosteroids and Related Compounds

Brassinosteroids are polyhydroxy steroidal lactone with the structure of brassinolide (BS) and the structure of steroids having the same carbon skeleton of animal steroids as cholestane, ergostane, and stigmastane which are also used as the basis for BS nomenclature (Fig. 13.1).

The term "brassinosteroids" originates from the Latin name for rape – *Brassica napus* L. – and most of the compounds of this class have received their names in a similar manner, derived from the name of the plant from which they were isolated or identified for the first time. Mostly, the nomenclature of a new class of compounds is a mix of established nomenclature rules combined with names derived from the sources from which the compound has been isolated (Khripach et al. 1999).

However, the chemists and plant physiologists used an approach in which the most active and first identified representative of this class of compounds, i.e., brassinolide (BL), is taken as the basis structure of the system. A great diversity in the basic structure at cyclic and side chain is found which is responsible for important metabolic transformations (Figs. 13.2 and 13.3).

Till date, about 70 BS have been identified from various plants and more can be expected that many more members of this class will be discovered in near future (Ikekawa et al. 1984; Yokota et al. 1986). The rules for nomenclature of various analogues of brassinolides are summarized in Table 13.1.

While the chemical classification for arranging BS is not only bases, the classification also needs further analysis based on the biosynthesis which may provide insight into their mutual relationship. On the basis of recent studies and analysis done by various workers for the structure and biosynthesis of BS and level of their biological activity, the following groups can be distinguished:

1. Compounds in which one or more function characteristic for BL are absent. These compounds are not as active as others but are considered to be precursors of other groups.

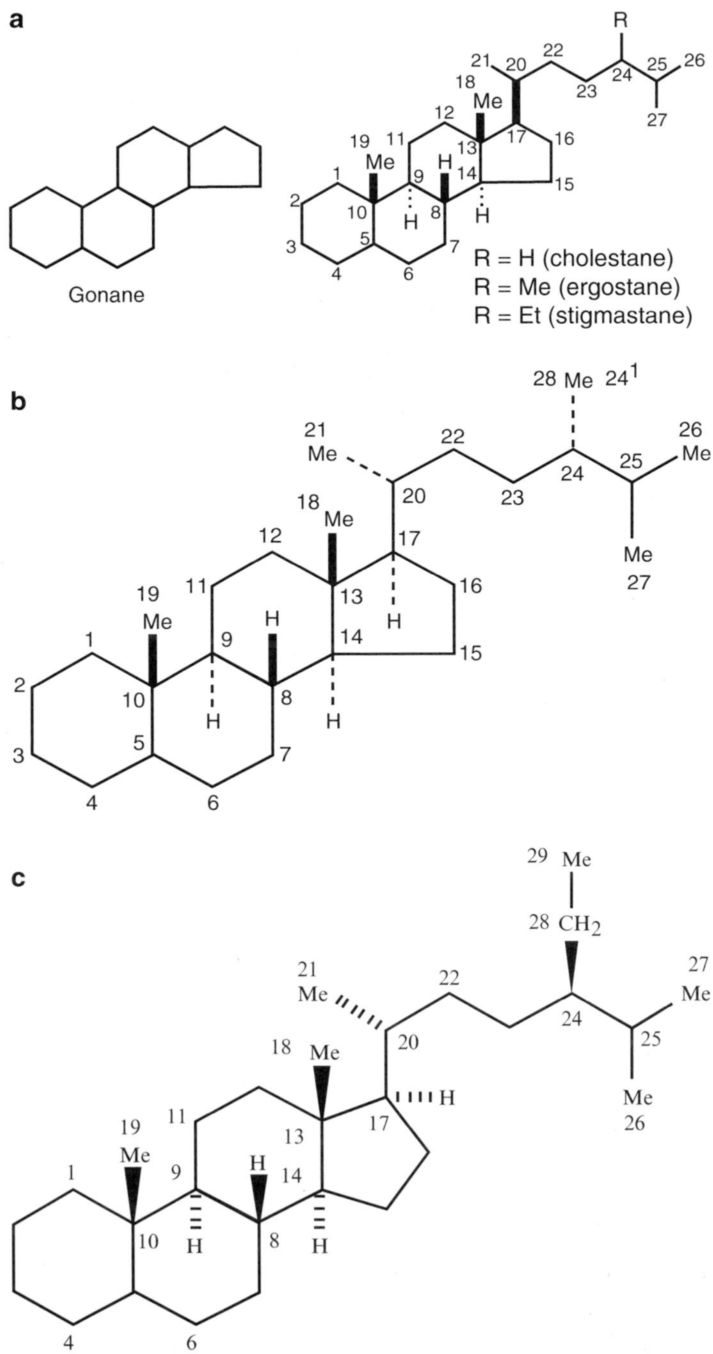


**Fig. 13.1** The basic chemical structures used for naming brassinosteroids (**a**) Cholestane (**b**) Ergostane (**c**) Stigmastane

2. BS with a full set of functional groups in the molecules (2α-, 3α- and 22*R*-, 23*R*-diol functions, B-homo-7-oxa-6-keto or 6-ketone). These compounds divulge the highest level of biological activity. Epibrassinolide and homobrassinolides belong to this group of BS.
3. Metabolites of BS including compounds with trans-diol group in ring A and BS conjugates belong to this group.

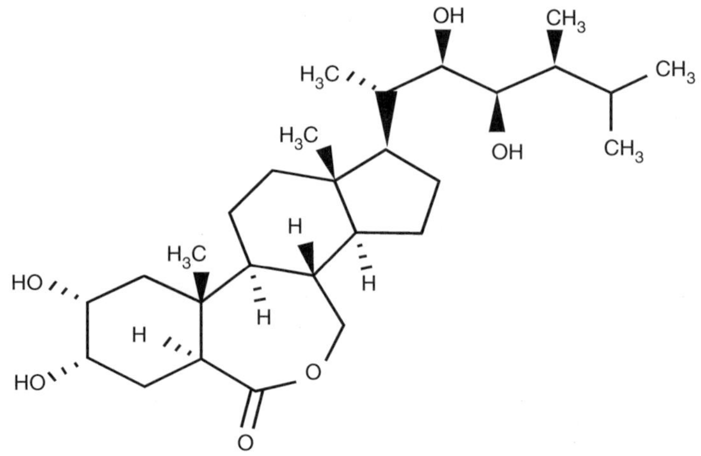


**Fig. 13.2** Brassinolide (BL)

**Table 13.1** Codes for shorthand description of Brassinosteroids

| Code | Meaning |
|---|---|
| *In capitals* | *Side chain* |
| **NB** | 27C norbrassinolide-like side chain |
| **B** | 28C brassinolide-like side chain |
| **HB** | 29C 28-homobrassinolide-like side chain |
| **E** | Epi at C-24 |
| **D** | Dehydro, $\Delta^{24(28)}$ |
| **25HB** | 29C 25-homobrassinolide |
| *In normal letters* | *Cyclic part* |
| **l** | Lactone, brassinolide-like cyclic part |
| **k** | Ketone, castasterone-like cyclic part |
| **3k** | Ketone at C-3 |
| **d** | 6-Deoxocastasterone-like cyclic part |
| **2d, 3d** | 2-Deoxy, 3-deoxy, absence of the indicated hydroxyl group |
| **2β, 3β** | β-Position of the indicated hydroxyl group |

## Biosynthesis of Brassinosteroids

The first report of brassinolide biosynthesis has been reported in 1981 by Wada and Marumo, who mentioned possible pathways to 2α-, 3α-diols. Second biosynthetic route was suggested by Mandava (1988), who proposed a route to brassinolide starting from plant sterols. Kim et al. (1988) in their experiments on *Phaseolus vulgaris* showed that the major sterols of this plant were those carrying a 24-ethyl or a 24-ethylidene group (sitosterol, stigmasterol, isofucosterol). The existence of similar BS with full set of functional groups but with a different carbon skeleton in the side chain (epibrassinolide, homobrassinolide, epicastasterone, (24*S*)-ethylbrassinone, norbrassinolide, brassinone) may be explained by the co-occurrence of campesterol as starting material for the biosynthesis of brassinolide together with other sterols such as brassicasterol, β-sitosterol, and 22-dehydrocholesterol (Khripach et al. 1999). Different experiments done by various workers for analysis of biosynthetic pathways involved in brassinolide synthesis established two pathways for BS biosynthesis. Feeding experiments on cultured cells of *Catharanthus roseus* established biosynthetic route from campesterol to brassinolide and referred it as "early C6-oxidation pathway" (Sakurai and Fujioka 1997a, b; Yokota 1997) (Fig. 13.4).

An alternative biosynthetic pathway of brassinolide which may operate independently was proposed and called the "late C6-oxidation pathway" (Choi et al. 1997; Fujioka et al. 1998).

However, with present knowledge, we cannot exclude the existence of other routes for biosynthesis of brassinolide as the identification of "early" and "late" C6-oxidation pathway is mainly based on studies carried out in cultured cells of *Catharanthus roseus* (Schneider et al.

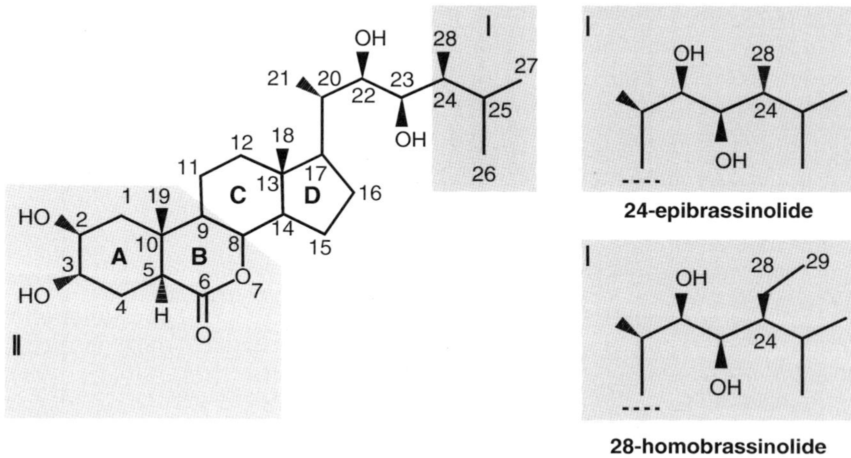


**Fig. 13.3** 24-Epibrassinolide and 28-homobrassinolide

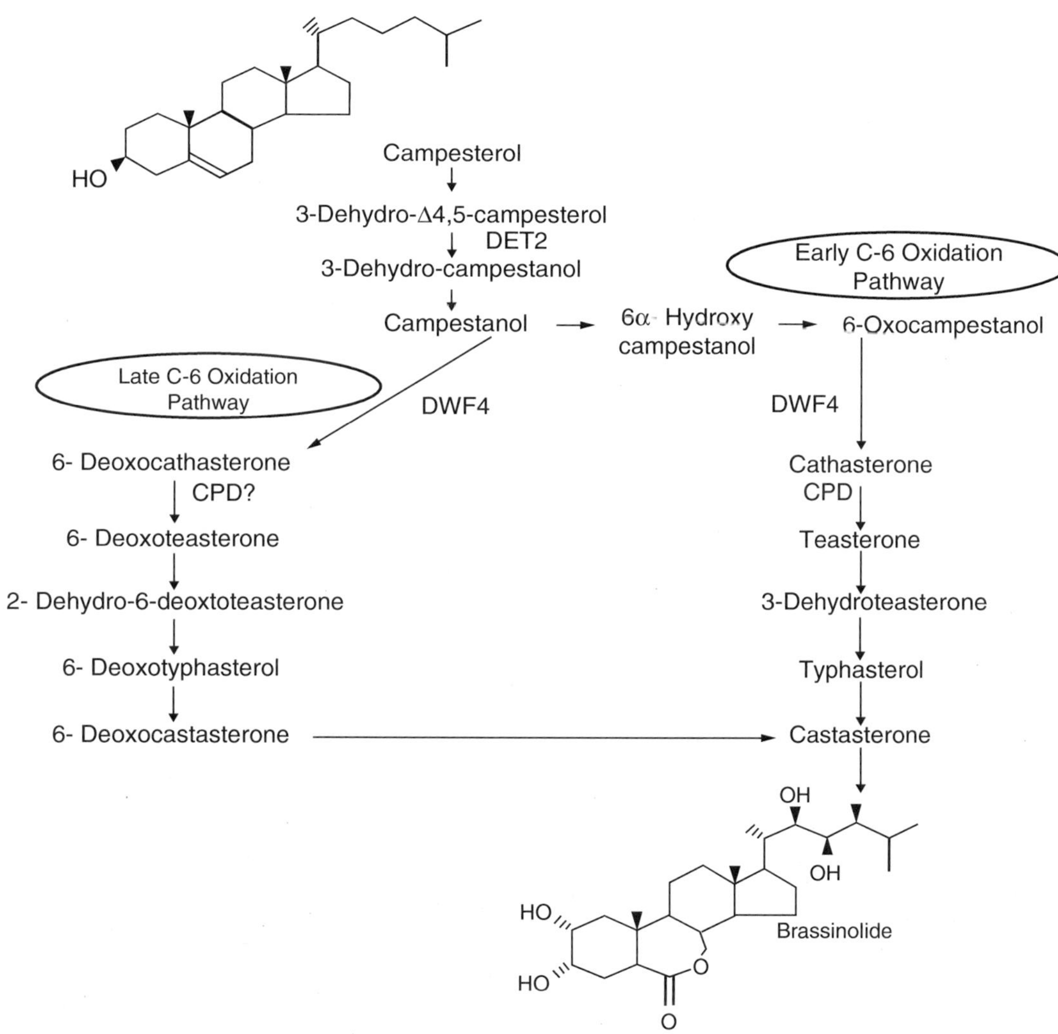


**Fig. 13.4** Scheme for early and late C6-oxidation pathway

1997). For instance, the pathway involved in transformation of castasterone into brassinolide is not found in tissues of *Phaseolus vulgaris* (Yokota et al. 1991). Co-occurrence of 24*R*- and 24*S*-BS in several plants creates one interesting question about the biosynthetic pathway involved in BS biosynthesis (Schmidt et al. 1993a, b; Schneider et al. 1997; Kauschmann et al. 1997). The best possible explanation available till date in the literature regarding this is that there may be inversion of the configuration at C-24 and/or the formation of both isomers from a common $\Delta^{24(28)}$ precursor.

## Brassinosteroids Physiological Roles

In all plants scrutinized so far, to ascertain role of various active forms of BLs underpins the importance of this phytohormone in regulating physiological, biochemical, and molecular properties of plant system under normal and inadequate conditions. Seed germination is enhanced to decisive high level by exogenous application of 24-EBL and 28-HBL in *Brassica juncea* L. (Sirhindi et al. 2009, 2011; Kumar et al. 2012), *Nicotiana tabacum* L. (Leubner-Metzger 2001), *Triticum aestivum* L. cv. HD2204 and *Lens culinaris* cv. Pusa-6 (Hayat and Ahmad 2003a, b), and *Orobanche* spp. (Song et al. 2005, 2006). Exogenous application of BL remarkably promoted growth of hypocotyls and cotyledons, leaf blades (Tanaka et al. 2003), root elongation (Müssig et al. 2003), lateral root development (Bao et al. 2004), and apical hook differentiation (De Grauwe et al. 2005). However, this amelioration is very much dependent on the concentration of BL used, on the time of treatment, and also on the phase or level of growth at which treatment was introduced. Seed priming with different concentrations of BL (24-EBL and 28-HBL) in *Vigna radiate* L. and *Brassica juncea* L. enhanced net photosynthetic rate, total chlorophyll, Chl a and b, carotenoids, shoot length, fresh weight, and dry weight (Fariduddin et al. 2011; Sirhindi et al. 2011). However, attribute regarding the root growth modulation influenced by BL illustrated contradictory results. Root growth promoted or inhibited by BL is dependent upon the concentration, the time and stage of treatment, and also the way of treatment, i.e., spray treatment or supplemented in the growing medium. The root growth in *Brassica juncea* L. was dwindled to significant levels as compared to control by both HBL and EBL (Kumar et al. 2010, 2012; Sirhindi et al. 2009, 2011). Root growth promotes effects in requisites of root weight, rooting ability, and number of adventitious roots in rice, potato, and Norway spruce (Rönsch et al. 1993; Wang and Deng 1992; Chen et al. 1990; Bobrick 1995). BL application is also responsible in modulating photosynthetic activity by changing the phosphorylation of various enzymes involved in Calvin cycle in detached pea (*Pisum sativum* L.) shoot incubated in a medium supplemented with 0.1 μM EBL (Fedina et al. 2008). The effect of exogenous application of 24-EBL on gas exchange, chlorophyll fluorescence, RuBisCO activity and carbohydrate metabolism in cucumber (*Cucumis sativus* L.) plants demonstrated that the maximum effect of BR application was occurred between day 1 to 3 after EBL treatment both at light saturated and at growing light intensity (Yu et al., 2004). EBR is also decisive in increasing the quantum yield of PSII and capacity of $CO_2$ assimilation in Calvin cycle attributed to initial increase in Rubisco activity of *Brassica oleraceae* L. (Cag et al. 2007). Presowing soaking treatment of various concentrations of EBL and HBL to seeds of *Lupinus angustifolius*, *L. luteus* L., and *L. albus* L. caused an increase in protein content and also made changes in proportion of some amino acids (Kandenlinskaya et al. 2007). Remarkable increase in quantity of nitrate reductase and carbonic anhydrase and content in leaves of tomato (*Lycopersicon esculentum* Mill.) was observed whose roots were dipped in 28-HBL and fruit number also found to be increased in such treated plants than in control (Ali et al. 2006). Significant increase in activities of nitrate reductase and carbonic anhydrase was also observed in *Cucumis sativus* L. under normal and chilling temperature conditions (Fariduddin et al. 2011). Brassinosteroid spraying at treatment at cotyledons and/or shoot apices in vitro inhibited flowering in *Pharbitis nil* (Kesy et al. 2003) and promoted panicle ripening in *Oryza sativa* L. cv. Nipponbare (Saka et al. 2003). Exogenous application of BL in *Vitis vinifera* L. cv. Cabernet Sauvignon significantly promoted fruit ripening (Symons et al. 2006). EBL application in cucumber, melon, and zucchini caused increase in ethylene production (Papadopoulou and Grumet 2005). BL application also reported to induce and/or increase femaleness in Cucurbitaceae family (Papadopoulou and Grumet 2005) and parthenocarpy in *Cucumis sativus* L. (Fu et al. 2008). Brassinosteroids also control male fertility in *Arabidopsis* (Ye et al. 2010). BL treatment also accelerated senescence in *Triticum aestivum* L. (Saglam-Cag 2007).

## Brassinosteroid Signaling Properties

Substantial empirical observations in plants established brassinosteroids as the sixth class of plant growth regulators which complement cellular and biochemical homeostasis in plants leading to increase plant growth and productivity. The obligatory precondition of growth, cell elongation, and cell division is an active and ameliorative biosynthesis of proteins. This functioning of plants is well deliberated in plants treated with various active forms of BL. Such amelioration in rate of protein synthesis has to be preceded by activation of transcribing nuclear DNA into RNA. Li et al. (1996) reported the first BR-insensitive *Arabidopsis* mutant *bri1*, which clear the path that leads to identification of the BR receptors. Shiu and Bleecker (2001) revealed another plant receptor having kinase-like activity from BR-insensitive mutants of *Arabidopsis* named as *bin2*. Identification of various components of the BR signal transduction pathway identified and established *BRI1* and *BAK1* as surface receptor proteins (Tang et al. 2008b). Proteomic and biochemical studies assembled these components into a phosphorylation cascade that connects BR perception at the cell surface for activation of transcription factors in the nucleus (Kim et al. 2009; Kim and Wang 2010). Brassinosteroid regulation of various gene expressions is mediated by transcription factor *BZR* identified from *Arabidopsis thaliana* (Gampala et al. 2007; Ryu et al. 2007). 24-EBL involved in upregulation of transcription of the *CycD3*, a D-type plant cycling gene through which cytokinin activates cell division, was identified in *Arabidopsis* using cDNA array (Hu et al. 2000). Overexpression of BZR1 (brassinazole-resistant 1) reported to increase hypocotyl length in *Arabidopsis* (He et al. 2005; Wang et al. 2002). BZR1 played a dual role in feedback inhibition of BR biosynthesis, and BR promotion of plant growth revealed that BZR1 is likely to activate certain promoters which might be possible by interacting with different partners (He et al. 2005). Although subtle, reports are available divulging the signaling pathways and TFs involved in BLs' role in plant growth and development, and more studies on these lines are needed to unveil the BLs' role in plant life. Conversely, a recent study identified two *Jumonji* domain-containing proteins, ELF6 and REF6, in addition to BZR2/BES1-interacting proteins which are found earlier to regulate flowering (Yu et al. 2008). Furthermore, BIM1 was shown to interact with two AP2/ERF transcription factors, DORNROSCHEN and DORNROSCHEN-LIKE, which are reported to be involved in patterning during embryo development, and the *bim1* mutant also showed embryo-patterning defects at low penetrance (Chandler et al. 2009). These results suggested that BZR2/BES1 recruited other transcriptional regulators to modulate the expression of subsets of target genes and specific developmental processes in plants (Kim and Wang 2010).

## Brassinosteroids and Plant Stress Responses

Plants due to their sessile nature are always challenged by swift internal and external environmental changes that many abiotic and biotic stresses can impinge upon a plant. These observations have created much interest in investigating the possibility of generating plants with catchall alterations involving at cellular and molecular level that ultimately through the signaling pathways transpire accountable early responses that are common to several abiotic and biotic stresses (Seki et al. 2002). This could be done by altering levels or patterns of expression of higher level transcription factors involved in the early responses to stresses, as has been described to some extent in *Arabidopsis* (Dubouzet et al. 2003). Manipulation in preemptive defense against abiotic and biotic stresses has the benefit of potentially facilitating the coordinated response to a stress since many stresses require more than one response for tolerance to occur. The possibilities for increasing tolerance to stresses are enormous, although it is notable that the actual production of transgenic plants with demonstrably improved abiotic stress tolerance has been slow. Traditional approaches to breeding crop plants with improved abiotic stress tolerances have so far met limited success (Richards 1996) because of the contribution made by several factors, comprising the focus on yield rather than on specific traits; the difficulties in breeding for tolerance traits, due to the complexities introduced by genotype by environment, or genome to environmental interactions; and the relatively infrequent use of simple physiological traits as measures of tolerance, and moreover the desired traits can only be introduced from closely related species as large number of inherent difficulties are involved in such processes.

An alternate to this long-term endeavor to fabricate transgenic crops, having potential to cope with changing environment, is the use of plant growth regulators (PGRs) to induce stress tolerance in plant through exogenous application of these eco-friendly chemicals. This approach has achieved a good response in agriculture crops as it is easier and simpler to make crops more tolerant to changing environment and also help in increasing crop productivity without encumbering the genomic traits. The use of PGRs in improving crop productivity, along with potential to fluctuating sudden changes in environment, is becoming increasingly more common (Brosa 1999; Steber and McCourt 2001; Wang et al. 2005; Arora et al. 2008; Bajguz 2009; Bajguz and Hayat 2009; Sirhindi et al. 2009, 2011; Ashraf et al. 2010; Kumar et al. 2010, 2011, 2012). However, the success of this approach is both beneficial and adverse effects on growth and development which has been addressed extensively in several reviews and investigations. Such results may be due to the reason that endogenous concentrations and ratios of different PGRs are impacted by numerous internal and external stimuli along with exogenous applications. Although a number of evidences have been found till date regarding the ameliorative potential of various PGRs under inadequate and normal growth and developmental conditions, till date, it has not been clear whether exogenous application of PGRs overcomes the imbalance of regulatory substances caused by the stress (Khan et al. 2000; Debez et al. 2001), generates specific defense mechanism against the stress, or just improves plant vitality (Ashraf et al. 2010; Sirhindi et al. 2011). Hitherto, practically exogenous application of PGRs tenders an impending loom to mitigate the detrimental effects of various stresses plants imposed to have from external and internal environment on plant growth and crop productivity.

Since the discovery and isolation of brassinosteroids and their substantiate presence in all orders of plants raging from lower fungi, pteridophytes, gymnosperms to higher plants, continuous efforts to explore the functioning of this compound in the plants at cellular, molecular, and signaling level under ample and inadequate conditions have been going on. Exogenous application of BRs can effectively ameliorate the adverse effects of abiotic and biotic stresses by homeostasis of various growth and developmental activities and modulating oxidative damage caused as a result of normal and traumatic growth conditions. BRs reported to have positive role in modulating plants from oxidative damage of reactive oxygen species (ROS), detrimental effects on pigments and photosynthesis, amelioration of various components of antioxidant defense system, osmoprotectant regulation and the aptitude to assist plants to synthesis protective substances, and expression of genes involved in defense responses as well as biosynthesis of other PGRs.

## Biotic Stress

All these protective roles of BRs have been well documented. For example, exogenous application of BR (24-EBL) in 0.1 μM to 10 ml per plant of *Cucumis sativus* at root and leaves level helps to reduce *Fusarium* pathogen-induced accumulation of ROS along with protective substances such as flavonoids and total phenolic compounds. In the same study, it was further added that BL treatment ameliorates the activities of various defense-related enzymes and ROS-scavenging enzymes (SOD, APOX, GPOX, CAT, PAL, and PPO) and thus responsible for reduction in disease severity (Ding et al. 2009a, b). BL treatment (2, 20, 100 μg/pot) at root level in *Oryza sativa* reduced disease symptoms caused by *Xanthomonas oryzae* cv. oryzae and *Magnaporthe grisea* (Nakashita et al. 2003). They further verified that wild tobacco (*Nicotiana tabacum*) when treated with different concentrations of BL (20, 40, 200 μM) at selected leaves or whole plant level enhanced resistance level against pathogen attacks of TMV, *Pseudomonas syringae* cv. tabaci, and *Odium* sp. BR involved in plant defense by regulating thionin protein which is low molecular weight, basic cysteine-rich antimicrobial protein is expressed in a range of plant species. BL application did not induce either acidic or basic pathogenesis-related (PR) gene expression, but it is suggested that BL resistance induction is distinct from systemic acquired resistance (SAR) and wound-inducible disease resistance. From various investigations, it has been suggested that BL functions as one of the common signaling molecules which cross talk with other PGRs in the innate system of higher plants. Kitanaga et al. (2006) suggested that transcripts of thionin genes encoding antimicrobial peptides were present at a high level in rice coleoptiles just after germination and decreased to undetectable levels after 3 days which was suppressed by brassinosteroids supplemented with gibberellic acid. The results illustrated that this collaborative cross talk between signaling molecules, viz., BR, GA, and JA, is very sequential and light dependent leading to control of the thionin transcript levels. 24-EBL at 0.1-μM concentration when sprayed at whole seedling of *Cucumis sativus* promoted the expression of genes involved in defense responses against cucumber mosaic virus (CMV). Untreated controlled distilled water seedlings developed typical CMV symptoms along with increased MDA content which lowered to significant levels in seedlings primed with 24-EBL (Xia et al. 2009).

## Water Stress

Water stress is another inadequate condition plants subjected to have as a consequence of high or low water availability in the growing medium. A decisive adaptation mechanism of plants aligned with water stress is the osmotic adjustments at cellular level prop up by other metabolic and physiological changes like amelioration in antioxidant's activities and photosynthetic apparatus enzyme's activities which consequently increased ABA content when there is paucity of water to the plant system. Under stress conditions, the first function plants started at cellular and molecular level is synthesis and increase in enzyme activities which are associated with maintenance of cell membrane. Treatment of 24EBL and/or 28-HBL strengthened all these properties of plants exposed to water stress by ameliorating various growth-related traits such as improvement in resistance level to dehydration in *Cucumis sativus* when plants were sprayed with 24-EBR (Pustovoitova et al. 2001). BR also initiated increase in tolerance to water stress in two varieties of wheat that resulted in increase in relative water content (RWC), nitrate reductase activity, chlorophyll content, and photosynthesis (Sairam 1994). EBR (1 μM) application at seed level in *Arabidopsis* and *Brassica napus* ascertained higher survival rate when seedlings are subjected to drought (Kagale et al. 2007). Seed treated with different concentrations of BR in *Sorghum vulgare* showed increased germination and seedling growth under osmotic stress (Vardhini and Rao 2003). BR (28-HBL and 24-EBL) applied in concentrations of 1 and 5 μM at flowering stage by foliar spray prepared the plants to resist against water stress by closing stomatal aperture to decrease water loss in plants exposed to water stress in *Phaseolus vulgaris* along with increased in integer of root nodulation in this legume, accumulation of ABA and cytokinin, and amelioration of nitrogenase activity (Upreti and Murti 2004). Stomatal closure in presence of BR treatment was also observed in *Vicia faba* rendered to water stress (Haubrick et al. 2006). Jager et al. (2008) observed in wild-type *WT* and BR mutants for BR-deficient *lkb* and BR-perception mutant *lka* of *Pisum sativum* L. that water stress caused increase in ABA level but did not show any relation with endogenous level of BR. Exogenous application of BL (0.1 mg/L) to leaves of *Glycine max* through foliar spray at the beginning of bloom resulted in accumulation of biomass, sugar and proline content along with amelioration of enzyme activities corollary to grain yield and quantum yield of PSII (Zhang et al., 2008). Similarly exogenous application of BL (0.4 mg/L) to roots and leaves of 1 year old seedlings of water stressed *Robinia pseudoacacia* showed accumulation of soluble sugars and proline content along with increased activities of various antioxidant enzymes (POD, SOD and CAT) as compared to untreated control plants (Li et al. 2008). Antioxidant enzyme activities are also reported to ameliorate in *Sorghum vulgare* seedlings raised from seeds primed with 2 and 3 μM 28-HBL and 24-EBL along with accumulation of soluble proteins and proline content (Vardhini and Rao 2003).

## Salt Stress

Saline stress is another abiotic stress plants frequently come in contact with, which is one of the consequences of water stress. Accumulation of proteins particularly those which play vital role in osmolarity homeostasis as proline and glycine betaine is a customary technique plant adopted to protect from inadequate environment. The physiological, biochemical, and molecular role of BRs in plant processes toward salinity tolerance was reviewed by Ashraf et al. (2010). BR reported to enhance the accumulation of these osmoprotectants to further higher levels. Seed soaking treatment with various concentrations of 24-EBR and 28-HBR is widespread practice for building high tolerance in large number of plants. Two varieties of wheat (*Triticum aestivum*) cultivars S-24 (salt tolerant) and MH-97 (moderately salt sensitive) when exposed to 120 mM NaCl in continuously aerated Hoagland's nutrient solution after treated with 24-EBL (0.052, 0.104, 0.156 μM) at seed and/or seedling levels improved growth and yield when compared with growth and yield of plants exposed to salt stress only which was associated with improved photosynthetic capacity and translocation of more photoassimilates toward grain (Ali et al. 2008a). Shahbaz et al. (2008) confirmed these improvements in growth yield in same wheat cultivars (S-24 and MH-97) under nonsaline and salinity stress (150 mM NaCl) by foliar spray of 24-EBL (0.0125, 0.025, and 0.0375 mg/L) and by increasing plant biomass and leaf area per plant in both cultivars. Saygideger and Deniz (2008) inveterate in *Spirulina platensis* cultures that 24-EBL (0.5, 1.0, and 3.0 μM) supplemented with NaCl (50, 100, 150, and 200 mM) increased biomass accumulation, growth, and free proline concentration over 5 days. 24-EBL treatment reported to alter cytogenetic responses in root meristem cells of barley (*Hordeum vulgare* L. cv. Bülbül 89) under different salt (NaCl) conditions (0.3, 0.35, and 0.4 M NaCl) and at 0.4 M NaCl supplemented with EBR caused total inhibition of mitotic activity in root tip cells (Tabur and Demir 2009). Conversely, when all concentrations of NaCl and control were evaluated with or without EBR, it was revealed that EBR treatment mitigate the detrimental effects of salinity stress on chromosomal abnormalities. Currently, ample empirical evidences are available in literature and exemplify ameliorative potential of BR for growth, productivity, and metabolic mechanisms operating in plants under normal and stress conditions. Arora et al. (2008) avowed that 28-HBL ($10^{-7}$, $10^{-9}$, and $10^{-11}$ M) at seed presowing soaking treatment in *Zea mays* coapplied with NaCl mitigates the stress effect and ameliorated growth, lipid peroxidation, and antioxidant enzymes as POD, SOD, CAT, and APOX. The MDA content of seedlings treated with different concentrations of salt supplemented with different concentrations of HBR and maximum decrease in MDA content was observed in $10^{-9}$ M HBR treatment. Similar results were observed in three lucrene (*Medicago sativa* L.) varieties (cv. Victor, Victoria and Golden Empress) (Zhang et al. 2007) and in *Oryza sativa* (Nüñez et al. 2003; Ozdemir et al. 2004) treated with 5 mM/L BL on MDA content in which it was found that MDA content was decreased in BL treated seedlings as compared to untreated salinity stressed and unstressed control seedlings which may be due amelioration in antioxidant enzyme (POD, CAT and SOD) activities which also improved seed germination and seedling growth. 28-HBL in different concentrations at seed or seedling levels in *Brassica juncea* and *Cicer arietnum* ameliorates nitrate reductase and carbonic anhydrase activities in seedlings raised under high saline condition (Ali et al. 2007; Hayat et al. 2007a). Shahid et al. (2011) evaluated role of 24 EBL and NaCl alone and in combination with pea (*Pisum sativum*) when applied prior to sowing. In the same study, it was observed that EBL treatment in pea reduced deleterious effects induced by salinity by ameliorating fresh and dry biomass; seedling growth; photosynthetic rate (Pn); stomatal conductance ($g_s$); total chlorophyll contents; proline contents; activity of SOD, POD, and CAT; nitrate reductase activity; and nitrite reductase activity as compared to control water-grown seedlings. Application of 24-EBL significantly ameliorated the adverse effects of salinity on growth, chlorophyll, and proline content along with reduction in electrolyte leakage in pepper plants (Houimli et al. 2010). From the current knowledge about BL effects under saline stress, it is recommended that BL modulates plant growth and development and develops saline tolerance in plants at specific concentrations and time development at which it is applied which may induce cellular and metabolic sensitivity to tolerate salinity.

## Heavy Metal Stress

In present developing world of industries, soil pollution due to heavy metal accumulation in soil is a very widespread problem to which plants particularly agricultural crops are bound to expose without any preference. BRs have shown an ability to modulate the uptake of ions into the plant cells, and they can be used to reduce the accumulation of heavy metals and radioactive elements in plants. Ameliorative properties of 28-HBL were deliberated in *Raphanus sativus* L. under chromium toxicity mediated by regulating the antioxidant enzymes and help in mitigating the toxic effect of chromium (Sharma and Bhardwaj 2007; Sharma et al. 2011). In the same study, they affirmed that 28-HBL treatments in radish seedlings lowered Cr (VI) uptake. Similar results were reported in *Brassica juncea* L. under nickel toxicity in which 28-HBL mitigates the Ni ion uptake (Sharma et al. 2008). Hayat et al. (2007b) avowed that foliar spray of 0.1 μM HBL to 30-day-old seedlings of *B. juncea* L. fed with cadmium (50, 100, and 150 μM) was surmounted. This mitigation in cadmium toxicity was enhanced by HBL through improvement of antioxidative enzymes (viz., CAT, POD, SOD), proline content, carbonic anhydrase activity, and nitrate reductase. Protective role of 28-HBL in cultivars of *Triticum aestivum* cultivars PBW-373, UP-2338, DL-LOK-01, DL-373, and HD-2338 observed under different levels of nickel was studied by Yusuf et al. (2011). The treatment of these Ni-stressed plants with HBL enhanced the activities of carbonic anhydrase and nitrate reductase, catalase, peroxidase, and superoxide dismutase along with proline content which acts as osmolyte in such stressed plants. The improvement of photosynthetic parameters and growth characteristics in such Ni-stressed wheat plants strengthened the concept of ameliorative potential of BL. Ameliorative potential of BRs against various heavy metal stress was studied in aluminum stress in *Vigna radiata* L. (Ali et al. 2008b), Ni stress (Alam et al. 2007), and copper stress in *Brassica juncea*

(Fariduddin et al. 2009). Sharma and Bhardwaj (2007) investigated the ameliorative effects of 24-EBL on plant growth and metal uptake in *Brassica juncea* L. plants under copper metal stress. The protective effect of 24-EBL in winter rape plants under Cd stress was investigated by Janeckzo et al. (2005). Bajguz (2002) revealed the mechanism involved for reducing the toxicity may be the chelation of the metal ion by a ligand that includes organic acids, amino acids, peptides, or polypeptides. Further, the reduction of metal toxicity by BRs was associated with lesser uptake of ions and enhanced levels of soluble proteins and nucleic acids with the increasing activity of ATPase, an enzyme responsible for acid secretion and changes in membrane level (Bajguz 2000). The effect of 24-EBL and 28-HBL in *Raphanus sativus* L. seedlings growing under cadmium stress was evaluated by Anuradha and Rao (2007a, b). They affirmed that 28-HBL was more effective in mitigating the toxic effect of Cd metal than 24-EBL. The amelioration of seedling growth by BRs under metal toxicity was associated with enhanced level of free proline. The activities of antioxidant enzymes CAT, SOD, APOX, and GPX were increased in the seedlings raised from treatments with Cd along with BRs.

## Brassinosteroids in Plant Development and Thermotolerance

Brassinosteroids are steroidal hormones which have been reported from more than 60 plant species including 51 angiosperms (12 monocots and 39 dicots), 6 gymnosperms, 1 pteridophyte (*Equisetum ravens* L.), 1 bryophyte (*Marchantia polymorpha* L.), and 1 alga (*Hydrodictyon reticulatum* (Linn.) Lagerh) and are now established to be ubiquitous in nature (Rao et al. 2002; Bajguz and Tretyn 2003; Sasse 1999). BRs participated over an array of physiological and metabolic processes and modulated them according to the environment and thus help plants in acclimatization and adjustments of their growth and development according to the circumstances. BRs' participation in regulation of many cellular and physiological processes occurring in plants is well documented, such as cell division and elongation of shoot and root (Sirhindi et al. 2011; Kumar et al. 2010, 2012; Sharma et al. 2010); biosynthesis of cell wall components; synthesis of DNA, RNA, and various proteins; microtubule organization; nitrogen fixation; distribution of assimilates to plant organs; pollen tube growth; differentiation of plant vascular system; formation of adventitious roots; flowering and reproduction; germination of seeds; sugar accumulation; and total carbohydrate increase (Khripach et al. 1999, 2000; Castle et al. 2003; Krishna 2003; Haubrick and Assmann 2006; Bajguz and Hayat 2009; Divi and Krishna 2009; Sirhindi et al. 2011). Indian mustard (*Brassica juncea* L.) 10-day-old seedlings treated with 24-EBL and 28-HBL ($10^{-6}$, $10^{-8}$, $10^{-10}$ M) showed increase in percentage germination, growth in form of shoot length, and total protein content to significant high level as compared to untreated control seedlings (Sirhindi et al. 2009). In the same study, it was observed that presowing soaking treatment of 24-EBL and 28-HBL improved various enzyme activities such as auxinase (IAAO), polyphenol oxidase (PPO), SOD, CAT, and APOX, which help in increasing the growth potential of plants and intensify the homeostasis. Exogenous application of 24-EBL and 28-HBL diverge in their magnitude to activate the various enzyme activities, $10^{-8}$ M 28-HBL augmented IAAO and PPO more as compared to all other concentrations of 28-HBL and all 24-EBL. 24-EBL is unable to enhance the APOX activity as compared to control and 28-HBL-treated seedlings. It was suggested that 28-HBL is more effective at germination stage, whereas 24-EBL showed this potential at initial and further growth and development of seedlings. However, the explanation for this is still unknown.

In addition to their vital function in plant growth and development, BRs protected plants from various abiotic and biotic stresses by inducing tolerance as a consequence of changes in cellular level ROS production and expression of genes encoding both structural and regulatory proteins (Kagale et al. 2007). BRs are reported to trigger the generation of $H_2O_2$ in plants exposed to various stresses, which act as a signaling molecule for initializing cascade of cellular and molecular changes to induce tolerance in plants. Cui et al. (2011) suggested that BR induces tolerance in cucumber leaves against abiotic stress by triggering the generation of $H_2O_2$. In *Brassica napus* and tomato seedlings, BR application improved the basic thermotolerance through increasing synthesis and accumulation of heat shock proteins as well as components of translational machinery (Dhaubhadel et al. 1999, 2002). The BRs induced and elevated antioxidant levels associated with various stress tolerance mechanisms.

Temperature is the foremost abiotic factor limiting the growth and productivity. Use of plant growth regulators is common practice nowadays all over the world to increase plant productivity under normal environmental conditions and induce tolerance in plants by triggering various cellular and molecular cascades. Revelation of plants to extreme temperature of chilling and heat, consequenced to inhibition of growth by effecting working of essential metabolic enzymes of photosynthesis and respiration. *Brassica juncea* L. plants pre-treated with 24-EBL are more tolerant to chilling injuries and dreadful affects of $H_2O_2$, which are produced as end product of photochemical reactions taking place in the environment under foggy cold environment, than untreated control plants (Sirhindi et al. 2011). Exposure of *B. juncea* L. plants to 15 mM $H_2O_2$ under field conditions showed reduction in germination rate to detrimental level. In this study, they found that 24-EBL proved to have affirmative role in protecting the plants from decisive effect of $H_2O_2$ which at higher concentration when accumulated in the plant acts as reactive oxygen species and 24-EBL facilitates its mitigation through antioxidant enzyme activation amelioration to significant high level. In another study, the effect of exogenous $H_2O_2$ on *B. juncea* L. under low temperature on antioxidant enzymes activity with or without 24-EBL treatment confirmed the previous results that BR treatment at seed level makes the plants more hardy to tolerate low-temperature injuries by ameliorating various antioxidant enzyme activities such as catalase (CAT, E.C. 1.11.1.6), ascorbate peroxidase (APOX, E.C. 1.11.1.11), and superoxide dismutase (SOD, E.C. 1.15.1.1) (Kumar et al. 2010). CAT and APOX are the two enzymes which directly dismutase the $H_2O_2$ but increase in activity of SOD in $H_2O_2$-treated seeds with or without 24-EBL is not understandable. One possible elucidation for this might be that exogenous applications of $H_2O_2$ at higher concentrations degrade various membrane proteins by increasing lipid peroxidation which integrated proteins of ETC and released various ions and amino acids including $Fe^{3+}/Fe^{2+}$ ions. These ferrous/ferric ions responsible for starting of Fenton's reaction channelized the production of more free super radicals ($O_2^{\bullet}$, $OH^{\bullet}$, etc.), and to dismutase these free radicals' SOD activity increased. Under chilling stress, significant increase in reactive oxygen species (ROS) and lipid peroxidation was reported to increase in suspension cultured cells of *Chorispora bungeana* after exposure to 4 and 0°C. But when 24-EBL application was done, the rate of ROS production and lipid peroxidation decreased to significant low level. The activities of antioxidant enzymes such as APOX, CAT, POD, and SOD were increased under chilling exposure which increased to further levels when cultures were supplemented with 24-EBL (Liu et al. 2009). The EBR treatment also significantly enhanced contents of ascorbic acid and GR under chilling stress. They concluded from these results that EBR could play the positive roles in the alleviation of oxidative damage caused by ROS overproduction through enhancing antioxidant defense system, resulting in improvement of chilling tolerance in suspension cultures of *C. bungeana*. BRs reduced lipid peroxidation and ROS production to decisive levels thus protecting the plant cell membranes and help in maintaining the structural integrity of the membranes under chilling stress and thus enhancing the chilling tolerance.

Indian mustard seedlings (*Brassica juncea* L.) whose growth was suppressed by chilling and $H_2O_2$ treatments recovered by treatment of 24-EBL by ameliorating shoot and root growth along with total protein contents (Kumar et al. 2010). Huang et al. (2006) reported that in mung bean epicotyls (*Vigna radiate* L.), growth was initially suppressed by chilling which partly recovered their ability to elongate after treatment with 24-EBL (10 μM). They affirmed upregulation of 17 proteins which were downregulated under chilling stress, and these upregulated proteins are involved in methionine assimilation, ATP synthesis, cell wall construction, and the stress response. Another study on *Brassica napus* L. cv. Lycosmos seedlings using 24-EBL under cold stress verified that at 2°C, BR (0.05 and 1.00 μM) treatment to cotyledons or primary leaves eradicated the effect of cold injuries on permeability (Janeckzo et al. 2007). Seedlings exposed to 2°C showed significant increase in pigment content and alleviation of elevated ion leakage in BR-treated seedlings as compared to untreated stressed plants. Xia et al. (2009) demonstrated that in cucumber (*Cucumis sativus* L.) cv. Jinyan No. 4 seedlings, significant decrease in electron transport rate under chilling stress was alleviated by 24-EBL (0.1 μM) and helps in protection of plants from chilling injuries. In *Brassica napus* and *Arabidopsis thaliana* grown on a nutrient solution containing 1 μM 24-EBL when exposed to cold stress, they illustrated accumulation of cold-related genes to higher levels as compared to untreated stressed plants (Kagale et al. 2007).

*Brassica juncea* L. (RCM 619) treated with 24-EBL ($10^{-6}$, $10^{-8}$, $10^{-10}$ M) before exposed to heat shock prepared the seedlings tolerant to heat shocks (40°C) by elevating the antioxidant enzyme activity such as SOD, CAT, and APOX to higher levels. Total proteins also increased in such seedlings as compared to untreated heat-exposed seedlings (Kumar et al. 2012). Singh and Shono (2005) avowed that tomato plants (*Lycopersicon esculentum* Mill.) treated with 24-EBL are more tolerant to high-temperature stress than untreated plants. Accumulation of mitochondrial small heat shock proteins (Mt-sHSP) in higher amount as response to heat stress at 38°C in 24-EBL treated plants as compared to 24-EBL treated plants at 25°C is possibly the mechanism by which EBL induced thermotolerance in tomato plants. Mazorra et al. (2002) also studied the effect of 24-EBL (10.6 and 2.12 nM) on tomato (*Lycopersicon esculentum* Mill.) plants and verified that treatment of 24-EBL stimulated the activity of SOD at 25 and 40°C. SOD is a key enzyme in the detoxification of superoxide radicals which might mitigate the toxic free-radical concentration to decisive levels to protect plant from its detrimental effects and thus build up the thermotolerance of plant. Peroxidase activity was unaffected at 25°C, while at 40°C the activity enhanced; the dynamics of CAT activity observed was markedly dependent on the structure, dose, and temperature characteristics. There are evidences available that BL refined its effect independently and also through cross talk with other hormones in signal transduction pathway. When exogenous application of 24-EBL (0.01, 0.1, and 1.0 mg/L) were applied in tomato (*Lycopersicon esculentum* Mill. cv. 9021) plants exposed to high temperature (40/30°C), the net photosynthetic rate, stomatal conductance, and maximum carboxylation rate of Rubisco were ameliorated significantly as compared to plants without EBR treatment. In EBR-treated plants, activities of antioxidant enzymes such as SOD, APOX, guaiacol peroxidase, and CAT improved to greater levels when exposed to heat stress as compared to without EBR heat stress-exposed plants. EBR application also strengthens the membrane integrity by decreasing the MDA and $H_2O_2$ content to detrimental low level under heat stress and also at normal temperature. It could be concluded that EBR alleviates the detrimental effects of high temperatures on plant enzyme systems in leaves which is EBR concentration dependent by affecting the net $CO_2$ assimilation at high temperature by modulating and/or protecting the rate-limiting enzyme Rubisco in the Calvin cycle and other enzymes involved in RuBP regeneration well under high temperature (Ogweno et al. 2008). Protection of plants from high-temperature stress is also done by plants by modulating the cell osmolarity and bleaching level to certain extent. Kagale et al. (2007) reported that *Arabidopsis thaliana* seedlings when exposed to 43°C in the

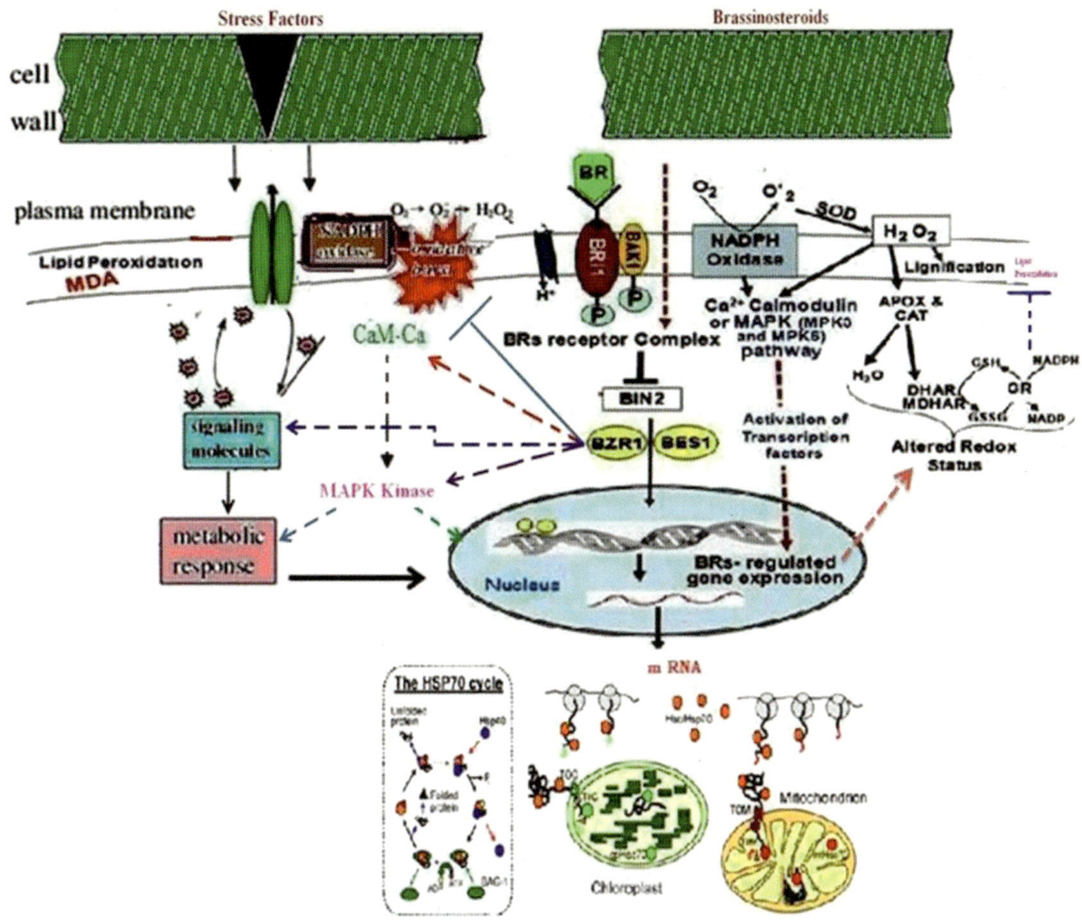


**Fig. 13.5** Possible mechanism of brassinosteroid-regulated stress tolerance in plants. *Dotted lines* represent the proposed possible mechanisms, which remains to be explored. *BR* brassinosteroid, *BRI1* brassinosteroid-insensitive 1; *BAK1* BRI1-associated kinase 1, *P* phosphate, *BIN2* brassinosteroid-insensitive 2, *BZR1* brassinozole-resistant 1, *BES1* BRI1-ems-suppressor 1, *APOX* ascorbate peroxidase, *CAT* catalase, *POD* guaiacol peroxidase, *DHAR* dehydroascorbate reductase, *MDHAR* monodehydroascorbate reductase, *GR* glutathione reductase

presence of 1 μM 24-EBL for varied hours (1, 2, 3, or 4 h) and then allowed them to recover at 22°C for 7 days showed decrease level of bleaching as compared to untreated heat stress-exposed seedlings. Kurepin et al. (2008) confirmed the involvement of abscisic acid (ABA) in thermotolerance construction of *Brassica napus* L. cv. Westar plants by BL. They suggested that when BL (0.1% aqueous ethanol plus BL at $10^{-6}$ M) was applied to canola seedlings at 45°C, heat stress endogenous concentration of ABA increased to significant high level as compared to seedlings at normal temperature.

## Stress Mechanism

Over the past decade, in biochemical and molecular genetic studies using *Arabidopsis* and rice as model plants, numerous genes have been identified which are involved in BR-regulated growth, development, and stress management potential, which provided fascinating insights into the various aspects of plant steroid signaling under normal and inadequate growth conditions. According to current concepts, BRs directly bind to the extracellular domain of the receptor-like kinase brassinosteroid-insensitive 1 (BRI1; She et al. 2011), thereby inducing a series of biochemical and metabolic responses, including heterodimerization of BRI1 with and activation of another receptor kinase, BRI1-associated kinase 1 (BAK1; Li et al. 2002; Yun et al. 2009), phosphorylation of BRI1-interacting signal kinase (Tang et al. 2008a), and activation of the protein phosphatase BRI1 suppressor protein 1 (Kim et al. 2009). These events eventually culminate in inhibition of the shaggy-like kinase BRI1-insensitive 2 (Vert and Chory 2006) and resultant activation of the transcription factors BRASSINOZOLE-RESISTANT 1 (BZR1) and BRI1-ems-suppressor 1 (BES 1)/BZR2 that orchestrate downstream gene expression (Sun et al. 2010; Yu et al. 2011) (Fig. 13.5).

## Conclusion

The end of twentieth century gave its departed gift to plant physiologists, the gift of discovery of the sixth class of steroidal hormones which start gaining its youthfulness at the start of twenty-first century. At present more than 70 brassinosteroid types have been isolated naturally or synthesized artificially which shows bountiful of properties for growth and development enhancement under normal growth conditions and ameliorative potential under inadequate environment. Exogenous application of BRs proved to be a useful approach in improving plant growth and productivity under a variety of growth conditions. However, the mechanism underlying this improvement is still undiscovered. Well-established evidences are available regarding enhancement of growth and productivity in plants grown with treatment of BRs as compared to untreated plants, but it is not clear whether BR could compensate for the imbalance of other plant growth substances which are required for growth improvement under normal environment or the imbalance caused due to any stress conditions or it is responsible for upregulation of defense-related genes or mechanism against stress or growth pressure or it merely enhances growth vigor. Such determination still requires extensive molecular and physiological examination of BR-treated and BR-untreated plants grown under different environmental conditions. Although BRs have been shown good potential under any normal and inadequate environment for plants to adjust and acclimatized at its level best, there are still gaps in knowledge between the perception of this hormone, the resulting signal transduction pathways, and the physiological responses in plants growing under changing environment. Likewise, a superior knowledge of the fundamental mechanisms of action of exogenously applied BRs and also the stage of growth they should apply to get maximum assistance of these hormones to ameliorate plant growth and development will certainly promote their efficient use in crop production and improvement.

**Acknowledgments** The author is grateful to University Grants Commission, New Delhi, India, which supported this research on the role of brassinosteroids under stress and growth and development by grant No. 33-194 (2007)/ SR under major research project. Immense gratitude also to Prof. Renu Bhardwaj from Department of Botanical and Environmental Sciences, Guru Nanak Dev University, Amritsar, who introduced me to the fascinating world of brassinosteroids.

## References

Alam MM, Hayat S, Ali B, Ahmad A (2007) Effect of 28-homobrassinolide treatment on nickel toxicity in *Brassica juncea*. Photosynthetica 45:139–142

Ali B, Hayat S, Hasan SA, Ahmad A (2006) Effect of root applied 28-homobrassinolide on the performance of *Lycopersicon esculentum*. Sci Hortic 110:267–273

Ali B, Hayat S, Ahmad A (2007) 28-Homobrassinolide ameliorates the saline stress in chickpea (*Cicer arietinum* L). Environ Exp Bot 59:217–223

Ali B, Hasan SA, Hayat S, Hayat Q, Yadav S, Fariduddin Q, Ahmad A (2008a) A role for brassinosteroids in the amelioration of aluminium stress through antioxidant system in mung bean (*Vigna radiate* L. Wilczek). Environ Exp Bot 62:153–159

Ali Q, Athar HUR, Ashraf M (2008b) Modulation of growth, photosynthetic capacity and water relations in salt stressed wheat plants by exogenously applied 240epibrassinolide. Plant Growth Regul 56:107–116

Anuradha S, Rao SSR (2007a) The effect of brassinosteroids on radish (*Raphanus sativus* L) seedlings growing under cadmium stress. Plant Soil Environ 53:465–472

Anuradha S, Rao SSR (2007b) Effect of 24-epibrassinolide on the growth and antioxidant enzyme activities in radish seedlings under lead toxicity. Indian J Plant Physiol 12:396–400

Arora N, Bhrdwaj R, Sharma P, Arora HK (2008) 28-Homobrassinolide alleviates oxidative stress in salt treated maize (*Zea mays* L.) plants. Braz J Plant Physiol 20:153–157

Ashraf M, Akram NA, Arteca RN, Foolad MR (2010) The physiological, biochemical and molecular roles of brassinosteroids and salicylic acid in plant processes and salt tolerance. Crit Rev Plant Sci 29:162–190

Bajguz A (2000) Effects of brassinosteroids on nucleic acids and protein in cultured cells in *Chlorella vulgaris*. Plant Physiol Biochem 38:209–215

Bajguz A (2002) Brassinosteroids and lead as stimulators of phytochelatins synthesis in *Chlorella vulgaris*. J Plant Physiol 159:321–324

Bajguz A (2009) Blockage of heavy metal accumulation in *Chlorella vulgaris* cells by 24-epibrassinolide. Plant Physiol Biochem 38:797–801

Bajguz A (2010) An enhancing effect of exogenous brassinolide on the growth and antioxidant activity in *Chlorella vulgaris* cultures under heavy metals stress. Environ Exp Bot 68:175–179

Bajguz A, Hayat S (2009) Effect of brassinosteroids on the plant responses to environmental stresses. Plant Physiol Biochem 47:1–8

Bajguz A, Tretyn A (2003) The chemical characteristic and distribution of brassinosteroids in plants. Phytochemistry 62:1027–1046

Bao F, Shen J, Brady SR, Muday GK, Asami T, Yang Z (2004) Brassinosteroids interact with auxin to promote lateral root development in Arabidopsis. Plant Physiol 134(4):1624–1631

Bobrick AO (1995) Application of brassinosteroids in the potato breeding. In: Brassinosteroids – biorational, ecologically safe regulators of growth and productivity of plants, 4th, Minsk, p 23

Brosa C (1999) Structure activity relationship. In: Sakurai A, Yokota T, Clouse SD (eds) Brassinosteroids: steroidal plant hormones. Springer, Tokyo, pp 191–222

Cag S, Goren-Saglam N, Cingil-Baris C, Kaplan E (2007) The effect of different concentration of epibrassinolide on chlorophyll, protein and anthocyanin content and peroxidase activity in excised red cabbage (*Brassica oleraceae* L.) cotyledons. Biotech Biotechnol Equip 21:422–425

Castle J, Montoya T, Bishop GJ (2003) Selected physiological responses of brassinosteroids: a historical approach. In: Hayat S, Ahmad A (eds) Brassinosteroids: bioactivity and crop productivity. Kluwer Academic Publishers, Dordrecht/Boston/London, pp 45–68

Chandler JW, Cole M, Filer A, Werr W (2009) BIM1, a bHLH protein involved in brassinosteroid signaling, controls *Arabidopsis* embryonic patterning via interaction with DORNROSCHEN and DORNROSCHEN–LIKE. Plant Mol Biol 69:57–68

Chen JC, Wang LF, Zhao YJ (1990) Effects of 24-epibrassinolide on growth of tobacco root explants. Acta Agric Shanghai 6:89–92

Choi YH, Fujioka S, Nomura T, Harada A, Yokota T, Takatsuto S, Sakurai A (1997) An alternative brassinolide biosynthetic pathway via late C-6 oxidation. Phytochemistry 44:609–613

Clouse SD, Sasse JM (1998) Brassinosteroids: essential regulators of plant growth and development. Annu Rev Plant Physiol Plant Mol Biol 49:427–451

Cui JX, Zhou YH, Ding JG, Xia XJ, Shi K, Chen SC, Asami T, Chen Z, Yu JQ (2011) Role of nitric oxide in hydrogen peroxide-dependent induction of abiotic stress tolerance by brassinosteroids in cucumber. Plant Cell Environ 34:347–358

De Grauwe L, Vandenbussche F, Tietz O, Palme K, Van Der Straeten D (2005) Auxin, ethylene and brassinosteroids: tripartite control of growth in the *Arabidopsis* hypocotyls. Plant Cell Physiol 46(6):827–836

Debez A, Chaibi W, Bouzid S (2001) Effect of NaCl and growth regulators on germination of *Atriplex halimus* L. Cah Agric 10:135–138

Dhaubhadel S, Chaudhary S, Dobinson KF, Krishna P (1999) Treatment with 24- epibrassinolide (a brassinosteroid) increases the basic thermotolerance of *Brassica napus* and tomato seedlings. Plant Mol Biol 40:332–342

Dhaubhadel S, Browning KS, Gallie DR, Krishna P (2002) Brassinosteroid functions to protect the translational machinery and heat shock protein synthesis following thermal stress. Plant J 29(6):681–691

Ding J, Shi K, Zhou YH, Yu JQ (2009a) Effects of root and foliar applications of 24-epibrassinolide on fusarium wilt and antioxidant metabolism in cucumber roots. Hortic Sci 44(5):1340–1349

Ding J, Shi K, Zhou YH, Yu JQ (2009b) Microbial community responses associated with the development of *Fusarium oxysporum* f. sp. *cucumerinum* after 24-epibrassinolidee applications to shoots and roots in cucumber. Eur J Plant Pathol 124:141–150

Divi UK, Krishna P (2009) Brassinosteroids: a biotechnological target for enhancing crop yield and stress tolerance. New Biotechnol 26:131–136

Divi UK, Rahman T, Krishna P (2010) Brassinosteroid – mediated stress tolerance in Arabidopsis shows interactions with abscisic acid, ethylene and salicylic acid pathways. BMC Plant Biol 10:151

Dubouzet JG, Sakuma Y, Ito Y, Kasuga M, Dubouzet EG, Miura S, Seki M, Shinozaki K, Yamaguchi-Shinozaki K (2003) OsDREB genes in rice, Oryza sativa L, encode transcription activators that function in drought-, high-salt- and cold-responsive gene expression. Plant J 33:751–763

Fariduddin Q, Yusuf M, Hayat S, Ahmad A (2009) Effect of 28-homobrassinolide on antioxidant capacity and photosynthesis in *Brassica juncea* plants exposed to different levels of copper. Environ Exp Bot 66:418–424

Fariduddin Q, Yusuf M, Chalkoo S, Hayat S, Ahmad A (2011) 28-Homobrassinolide improves growth and photosynthesis in *Cucumis sativus* L. through an enhanced antioxidant system in the presence of chilling stress. Photosynthetica 49(1):55–64

Fedina EO, Karimova FG, Tarchevsky IA, Khripach VA (2008) Effect of epibrassinolide on tyrosine phosphorylation of the Calvin cycle enzymes. Russ J Plant Physiol 55:193–200

Fu FQ, Mao WH, Shi K, Zhou YH, Asami T, Yu JQ (2008) A role of brassinosteroids in early fruit development in cucumber. J Exp Bot 59:2299–2308

Fujioka S, Noguchi T, Yokota T, Takatsuto S, Yoshida S (1998) Brassinosteroids in *Arabidopsis thaliana*. Phytochemistry 48:595–599

Gampala SS, Kim TW, He JX, Tang W, Deng Z (2007) An essential role of 14-3-3 proteins in brassinosteroid signal transduction in *Arabidopsis*. Dev Cell 13:177–189

Haubrick LL, Assmann SM (2006) Brassinosteroids and plant function: some clues, more puzzles. Plant Cell Environ 29:446–457

Haubrick LL, Torsethaugen G, Assmann SM (2006) Effect of brassinolide, alone and in concert with abscisic acid, on control of stomatal aperture and potassium currents of *Vicia faba* guard cell protoplasts. Physiol Plant 128:134–143

Hayat S, Ahmad A (2003a) 28-Homobrassinolide induced changes favoured germinability of wheat seeds. Bulg J Plant Physiol 29(1–2):55–62

Hayat S, Ahmad A (2003b) Soaking seeds of *Lens culinaris* with 28-homobrassinolide increased nitrate reductase activity and grain yield in the field of India. Ann Appl Biol 143:121–124

Hayat S, Ali B, Hasan SA, Ahmad A (2007a) Effect of 28-homobrassinolide on salinity induced changes in *Brassica juncea*. Turk J Biol 31:141–146

Hayat S, Ali B, Hasan SA, Ahmad A (2007b) Brassinosteroid enhanced the level of antioxidants under cadmium stress in *Brassica juncea*. Environ Exp Bot 60:33–41

He JX, Gendron JM, Sun Y, Gampala SSL, Gendron N, Sun CQ, Wang ZY (2005) BZR1 is a transcriptional repressor with dual roles in brassinosteroid homeostasis and growth responses. Science 307:1634–1638

Houimli SIM, Denden M, Mouhandes BD (2010) Effects of 24-epibrassinolide on growth, chlorophyll, electrolyte leakage and proline by pepper plants under NaCl stress. Eur Asia J Bio Sci 4:96–104

Hu Y, Bao F, Li J (2000) Promotive effect of brassinosteroids on cell division involves a distinct CycD3-induction pathway in *Arabidopsis*. Plant J 24 (5):693–701

Huang B, Chu CH, Chen SL, Juan HF, Chen YM (2006) A proteomics study of the mung bean epicotyls regulated by brassinosteroids under conditions of chilling stress. Cell Mol Biol Lett 11:264–278

Ikekawa N, Takatsuto S, Kitsuwa T, Saito H, Morishita T, Abe H (1984) Analysis of natural brassinosteroids by gas chromatography and gas chromatography-mass spectrometry. J Chromatogr 290:289–302

Jager CE, Gregory MS, John JR, James BR (2008) Do brassinosteroids mediate the water stress response? Physiol Plant 133:417–425

Janeckzo A, Koscielniak J, Pilipowiez M, Szarek-Lukaszewska G, Skoczowski A (2005) Protection of winter rape phosystem II by 24-epibrassinolide under cadmium stress. Photosynthetica 43:293–298

Janeckzo A, Gullner G, Skoczowski A, Dubert F, Barna B (2007) Effects of brassinosteroid infiltration prior to cold treatment on ion leakage and pigment contents in rape leaves. Biol Plant 51:335–358

Kagale S, Divi UK, Krochko JE, Keller WA, Krishna P (2007) Brassinosteroid confers tolerance in *Arabidopsis thaliana* and *Brassica napus* to a range of abiotic stresses. Planta 225:353–364

Kandenlinskaya OL, Topunov AF, Grishehenko ER (2007) Biochemical aspects of growth-stimulating effects of steroid phytohormones on lupine plants. Appl Biochem Microbiol 43:324–331

Kauschmann A, Adam G, Lichtblau D, Mussig C, Schmidt J, Voigt B, Willmitzer L, Altmann T (1997) Molecular/genetic analysis of brassinosteroid synthesis and action. Proc Plant Growth Regul Soc Am 24:95–96

Kesy J, Trzaskalska A, Galoch E, Kopeewiez J (2003) Inhibitory effect of brassinostreoids on the flowering of the short-day plant *Pharbitis nil*. Biol Plant 47 (4):597–600

Khan MA, Gul B, Weber DJ (2000) Germination responses of *Salicornia rubra* to temperature and salinity. J Arid Environ 45:207–214

Khripach VA, Zhabinski VN, Groot AE (1999) Physiological mode of action of BS. In: Khripach VA, Zhabinskii VN, Groot AE (eds) Brassinosteroids: a new class of plant hormones. Academic, San Diego/London/Boston/New York/Sydney/Tokyo/Toronto, pp 219–299

Khripach V, Ahabinskii V, de Groot A (2000) Twenty years of brassinsoteroids: steroidal plant hormones warrant better crops for the XXI century. Ann Bot 86:441–447

Kim TW, Wang ZY (2010) Brassinosteroid signal transduction from receptor kinases to transcription factors. Annu Rev Plant Biol 61:681–704

Kim SK, Akihisa T, Tamura T, Matsumoto T, Yokota T, Takahashi N (1988) 24-Methylene-25-methylcholesterol in *Phaseolus vulgaris* seed: structural relation to brassinosteroids. Phytochemistry 27:629–631

Kim TW, Guan S, Sun Y, Deng Z, Tang W (2009) Brassinosteroid signal transduction from cell surface receptor kinases to nuclear transcription factors. Nat Cell Biol 11:1254–1260

Kitanaga Y, Jian C, Hasegawa M, Yazaki J, Kishimoto N, Kikuchi S, Nakamura H, Ichikawa H, Asami T, Yoshida S, Yamaguchi I, Suzuki Y (2006) Sequential regulation of gibberellins, brassinosteroid and jasmonic acid biosynthesis occurs in rice coleoptiles to control the

transcript levels of anti-microbial thionin genes. Biosci Biotechnol Biochem 70(10):2410–2419
Krishna P (2003) Brassinosteroid-mediated stress responses. J Plant Growth Regul 22:289–297
Kumar M, Sirhindi G, Bhardwaj R, Kumar S, Jain G (2010) Effect of exogenous $H_2O_2$ on antioxidant enzymes of *Brassica juncea* L. seedlings in relation top 24-epibrassinolide under chilling stress. Indian J Biochem Biophys 47:378–382
Kumar S, Sirhindi G, Bhardwaj R, Kumar M (2011) Evaluation of brassinolide effect on growth, proteins and antioxidative enzyme activities in *Brassica juncea* L. J Indian Bot Soc 90(1&2):154–158
Kumar S, Sirhindi G, Bhardwaj R, Kimar M, Arora P (2012) Role of 24-epibrassinolide in amelioration of high temperature stress through antioxidant defense system in *Brassica juncea* L. Plant Stress, 6(1):55–58
Kurepin LV, Qaderi MM, Back TG, Reid DM, Pharis RP (2008) A rapid effect of applied brassinolide on abscisic acid concentrations in *Brassica napus* leaf tissue subjected to short-term heat stress. Plant Growth Regul 55:165–167
Leubner-Metzger G (2001) Brassinosteroids and gibberellins promote tobacco seed germination by distinct pathway. Planta 213:758–763
Li JM, Nagpal P, Viart V, McMorris TC, Chory J (1996) A role of brassinosteroids in light-dependent development of Arabidopsis. Science 272:398–401
Li J, Wen JQ, Lease KA, Doke JT, Tax FE, Walker JC (2002) BAK1, an Arabidopsis LRR receptor like protein kinase, interacts with BRI1 and modulates brassinosteroid signaling. Cell 110:213–222
Li KR, Wang HH, Han G, Wang QJ, Fan J (2008) Effects of brassinolide on the survival, growth and drought resistance of *Robinia pseudoacacia* seedlings under water-stress. New For 35:255–266
Liu Y, Zhao Z, Si J, Di C, Han J, An L (2009) Brassinosteroids alleviate chilling oxidative damage by enhancing antioxidant defense system in suspension cultured cells of *Chorispora bungeana*. Plant Growth Regul 59 (3):207–214
Mandava NB (1988) Plant growth promoting brassinosteroids. Annu Rev Plant Physiol Plant Mol Biol 39:23–52
Mazorra LM, Núñez M, Hechavarria M, Coll F, Sánchez-Blanco MJ (2002) Influence of brassinosteroid on antioxidant enzymes activity in tomato under different temperatures. Biol Plant 45(4):593–596
Müssig C, Shin GH, Altman T (2003) Brassinosteroids promote root growth in *Arabidopsis*. Plant Physiol 133 (3):1261–1271
Nakashita H, Yasuda M, Nitta T, Asami T, Fujioka S, Arai Y, Sekimata K, Takatsuto S, Yamaguchi I, Yoshida S (2003) Brassinosteroid functions in a broad range of disease resistance in tobacco and rice. Plant J 33(5):887–898
Nüñez M, Mazzafera P, Mazorra LM, Siqueira WJ, Zullo MAT (2003) Influence of a brassinosteroid analogue on antioxidant enzymes in rice grown in culture medium with NaCl. Biol Plant 47(1):67–70
Ogweno JO, Song XS, Shi K, Hu WH, Mao WH, Zhou YH, Yu JQ, Nogues S (2008) Brassinosteroids alleviate heat-induced inhibition of photosynthesis by increasing carboxylation efficiency and enhancing antioxidant systems in *Lycopersicon esculentum*. J Plant Growth Regul 27:49–57
Ozdemir F, Bor M, Demiral T, Turkan I (2004) Effects of 24-epibrassinolide on seed germination, seedling growth, lipid peroxidation, proline content and antioxidative system of rice (*Oryza sativa* L) under salinity stress. Plant Growth Regul 42 (3):203–211
Papadopoulou E, Grumet R (2005) Brassinosteroid – induced femaleness in cucumber and relationship to ethylene production. Hortic Sci 49:1763–1767
Pustovoitova TN, Zhdanova NE, Zholkevich VN (2001) Epibrassinolide increases plant water resistance. Dokl Biochem Biophys 376:36–38
Rao SSR, Vardhini BVV, Sujatha E, Anuradha S (2002) Brassinosteroids – a new class of phytohormones. Curr Sci 82:1239–1245
Richards RA (1996) Defining selection criteria to improve yield under drought. Plant Growth Regul 20:57–166
Rönsch H, Adam G, Matschke J, Schachler G (1993) Influence of (22S, 23S)-homobrassinolide on rooting capacity and survival of adult Norway spruce cuttings. Tree Physiol 12:71–80
Ryu H, Kim K, Cho H, Park J, Choe S, Hwang I (2007) Nucleocytoplasmic shuttling of BZR1 mediated by phosphorylation is essential in *Arabidopsis* brassinosteroid signaling. Plant Cell 19:2749–2762
Saglam-Cag S (2007) The effects of epibrassinolide on senescence in wheat leaves. Biotech Biotechnol Equip 21:63–65
Sairam RK (1994) Effects of homobrassinolide application on plant metabolism and grain yield under irrigated and moisture-stress conditions of two wheat varieties. Plant Growth Regul 14:173–181
Saka H, Fujii S, Imakawa AM, Kato N, Watanabe S, Nishizawa T, Yonekawa S (2003) Effect of brassinolide applied at the meiosis and flowering stages on the levels of endogenous plant hormones during grain-filling in rice plant (*Oryza sativa* L). Plant Prod Sci 6 (1):36–42
Sakurai A, Fujioka S (1997a) Biosynthesis of steroidal plant hormones, brassinosteroids. Baiosaiensu to Indasutori 55:38–40
Sakurai A, Fujioka S (1997b) Studies on biosynthesis of brassinosteroids. Biosci Biotech Biochem 61:757–762
Sasse JM (1999) Physiological actions of brassinosteroids. In: Sakurai A, Yokota T, Clouse SD (eds) Brassinosteroids: steroidal plant hormones. Springer, Tokyo, pp 137–161
Saygideger S, Deniz F (2008) Effect of 24-epibrassinolide on biomass, growth and free proline concentration in

*Spirulina platensis* (Cyanophyta) under NaCl stress. Plant Growth Regul 56:219–223

Schmidt J, Spengler B, Yokota T, Adam G (1993a) The co-occurrence of 24-epi-castasterone and castasterone in seeds of *Ornithopus sativus*. Photochemistry 32:1614–1615

Schmidt J, Yokota T, Spengler B, Adam G (1993b) 28-Homoteasterone a naturally occurring brassinosteroid from seeds of *Raphanus sativus*. Phytochemistry 34:391–392

Schneider B, Kolbe A, Winter J, Porzel A, Schmidt J, Strack D, Adam G (1997) Pathways and enzymology of brassinosteroid metabolism. Proc Plant Growth Regul Soc Am 24:91–93

Seki M, Narusaka M, Ishida J, Nanjo T, Fujita M, Oono Y, Kamiya A, Nakajima M, Enju A, Sakurai T, Satou M, Akiyama K, Taji T, Yamaguchi-Shinozaki K, Carninci P, Kawai J, Hayashizaki Y, Shinozaki K (2002) Monitoring the expression profiles of 7000 *Arabidopsis* genes under drought, cold and high salinity stresses using a full length cDNA microaaray. Plant J 31:279–292

Shahbaz M, Ashraf M, Athar HUR (2008) Does exogenous application of 24-epibrassinolide ameliorate salt induced growth inhibition in wheat (*Triticum aestivum* L)? Plant Growth Regul 55:51–64

Shahid MA, Pervez MA, Balal RM, Mattson RA, Ahmad R, Ayyub CM, Abbas T (2011) Brassinosteroid (24-epibrassinolide) enhances growth and alleviates the deleterious effects induced by salt stress in pea (*Pisum sativum* L). Aust J Crop Sci 5 (5):500–510

Sharma P, Bhardwaj R (2007) Effects of 24-epibrassinolide on growth and metal uptake in *Brassica juncea* L under copper metal stress. Acta Physiol Plant 29:259–263

Sharma P, Bhardwaj R, Arora N, Arora HK, Kumar A (2008) Effects of 28-homobrassinolide on nickel uptake, protein content and antioxidative defence system in *Brassica juncea*. Biol Plant 52:767–770

Sharma I, Pati PK, Bhardwaj R (2010) Regulation of growth and antioxidant enzyme activities by 28-homobrassinolide in seedlings of *Raphanus sativus* L under cadmium stress. Indian J Biochem Biophys 47 (3):172–177

Sharma I, Pati PK, Bhardwaj R (2011) Effect of 28-homobrassinolide on antioxidant defence system in *Raphanus sativus* L under chromium toxicity. Ecotoxicology 20:862–874

She J, Han Z, Kim TW, Wang J, Cheng W, Chang J, Shi S, Wang J, Yang M, Wang ZY et al (2011) Structural insight into brassinosteroid perception by BRI1. Nature 474:472–476

Shiu SH, Bleecker AB (2001) Receptor-like kinases from *Arabidopsis* form a monophyletic gene family related to animal receptor kinases. Proc Natl Acad Sci U S A 98:10763–10768

Singh I, Shono M (2005) Physiological and molecular effects of 24-epibrassinolide, a brassinosteroid on thermotolerance of tomato. Plant Growth Regul 47:111–119

Sirhindi G, Kumar S, Bhardwaj R, Kumar M (2009) Effects of 24-epibrassinolide and 28-homobrassinolide on the growth and antioxidant enzyme activities in the seedlings of *Brassica juncea* L. Physiol Mol Biol Plant 15(4):335–341

Sirhindi G, Kumar M, Bhardwaj R, Kumar S, Pradhan SK (2011) Effect of 24-epibrassinolide on activity of antioxidant enzymes in *Brassica juncea* L. under $H_2O_2$ stress. Indian J Plant Physiol 16(1):68–71

Song WJ, Zhou WJ, Jin ZL, Cao DD, Joel DM, Takeuchi Y, Yoneyama K (2005) Germination response of *Orobanche* seeds subjected to conditioning temperature, water potential and growth regulator treatments. Weed Res 45:467–476

Song WJ, Zhou WJ, Jin ZL, Zhang D, Yoneyama K, Takeuchi Y, Joel DM (2006) Growth regulators restore germination of *Orobanche* seeds that are conditioned under water stress and suboptimal temperature. Aust J Agric Res 57:1195–1201

Steber CM, McCourt P (2001) A role for brassinosteroids in germination in *Arabidopsis*. Plant Physiol 125:763–769

Sun Y, Fan X-Y, Cao D-M, Tang W, He K, Zhu J-Y, He J-X, Bai M-Y, Zhu S, Oh E et al (2010) Integration of brassinosteroid signal transduction with the transcription network for plant growth regulation in Arabidopsis. Dev Cell 19:765–777

Symons GM, Davies C, Shavrukov Y, Dry IB, Reid JB, Thomas MR (2006) Grapes on steroids. Brassinosteroids are involved in grape berry ripening. Plant Physiol 140:150–158

Tabur S, Demir K (2009) Cytogenetic response of 24-epibrassinolide on the root meristem cells of barley seeds under salinity. Plant Growth Regul 58:119–123

Tanaka K, Nakamura Y, Asami T, Yoshida S, Matsuo T, Okamoto S (2003) Physiological roles of brassinosteroids in early growth of *Arabidopsis*: brassinosteroids have a synergistic relationship with gibberellins as well as auxin in light-grown hypocotyl elongation. J Plant Growth Regul 22(3):259–271

Tang W, Deng Z, Oses-Prieto JA, Suzuki N, Zhu SW (2008a) Proteomics studies of brassinosteroid signal transduction using perfactionation and two-dimensional DIGE. Mol Cell Proteomics 7:728–738

Tang W, Kim TW, Oses-Prieto JA, Sun Y, Deng Z (2008b) BSKs mediate signal transduction from the receptor kinase BRI1 in *Arabidopsis*. Science 321:557–560

Upreti KK, Murti GSR (2004) Effects of brassinosteroids on growth, nodulation, phytohormone content and nitrogenase activity in French bean under water stress. Biol Plant 48:407–411

Vardhini BV, Rao SSR (2003) Amelioration of osmotic stress by brassinosteroids on seed germination and seedling growth of three varieties of sorghum. Plant Growth Regul 41(1):25–31

Vert G, Chory J (2006) Downstream nuclear events in brassinosteroid signalling. Nature 441:96–100

Vleesschauwer DD, Buyten EV, Satoh K, Balidion J, Mauleon R, Choi IR, Vrea-Cruz C, Kikuchi S, Höfte M (2012) Brassinosteroids antagonize gibberellin – and salicylate – mediated root immunity in rice. Plant Physiol 158:1833–1846

Wada K, Marumo S (1981) Synthesis and plant growth-promoting activity of brassinolide analogues. Agric Biol Chem 45:2579–2585

Wang SG, Deng RF (1992) Effects of brassinosteroid (BR) on root metabolism in rice. J Southwest Agric Univ 14:177–181

Wang ZY, Nakano T, Gendron J (2002) Nuclear-localized BZR1 mediates brassinosteroid-induced growth and feedback suppression of brassinosteroid biosynthesis. Dev Cell 2:505–513

Wang X, Li X, Meisenhelder J, Hunter T, Yoshida S, Asami T, Chory J (2005) Autoregulation and homodimerization are involved in the activation of the plant steroid receptor BRI1. Dev Cell 8:855–865

Xia XJ, Wang YJ, Zhou YH, Tao Y, Mao WH, Shi K, Asami T, Chen Z, Yu JQ (2009) Reactive oxygen species are involved in brassinosteroid-induced stress tolerance in cucumber. Plant Physiol 150:801–814

Ye Q, Zhu W, Li L, Zhang S, Yin Y, Ma H, Wang X (2010) Brassinosteroids control male fertility by regulating the expression of key genes involved in *Arabidopsis* anther and pollen development. Proc Natl Acad Sci U S A 13:6100–6105

Yokota T (1997) The structure, biosynthesis and function of brassinosteroids. Trends Plant Sci 2:137–143

Yokota T, Kim SK, Kosaka Y, Ogino Y, Takahashi N (1986) Conjugation of brassinostreoids. In: Schreiber K, Schütte HR, Sembdner G (eds) Conjugated plant hormones. Structure, metabolism and function. Proceedings of the international symposium. VEB Deutscher Verlag der Wissenschaften, Berlin, pp 288–296

Yokota T, Ogino Y, Suzuki H, Takahashi N, Saimoto H, Fujiko S, Sakurai A et al (1991) Metabolism and biosynthesis of brassinosteroids. In: Cutler HG, Yokota T, Adam G (eds) Brassinosteroids chemistry bioactivity and applications, vol 474, ACS symposium series. American Chemical Society, Washington, pp 86–96

Yu JQ, Huang LF, Hu WH, Zhou YH, Mao WH, Ye SF, Nogues S (2004) A role for brassinosteroids in the regulation of photosynthesis in *Cucumis sativus*. J Exp Bot 55(399):1135–1143

Yu X, Li L, Guo M, Chory J, Yin Y (2008) Modulation of brassinosteroid-regulated gene expression by Jumonji domain-containing proteins ELF6 and REF6 in *Arabidopsis*. Proc Natl Acad Sci U S A 105:7618–7623

Yu X, Li L, Zola J, Aluru M, Ye H, Foudree A, Guo H, Anderson S, Aluru S, Liu P et al (2011) A brassinosteroid transcriptional network revealed by genome-wide identification of BESI target genes in *Arabidopsis thaliana*. Plant J 65:634–646

Yun HS, Bae YH, Lee YJ, Chang SC, Kim S-K, Li J, Nam KH (2009) Analysis of phosphorylation of the BRI1/BAK1 complex in Arabidopsis reveals amino acid residues critical for receptor formation and activation of BR signaling. Mol Cells 27:183–190

Yusuf M, Fariduddin Q, Hayat S, Hasan SA, Ahmad A (2011) Protective response of 28-homobrassinolide in cultivars of *Triticum aestivum* with different levels of nickel. Arch Environ Contam Toxicol 60:68–76

Zhang S, Hu J, Zhang Y, Xie XJ, Knapp A (2007) Seed priming with brassinolide improves Lucerne (*Medicago sativa* L) seed germination and seedling growth in relation to physiological changes under salinity stress. Aust J Agric Res 58:811–815

Zhang MC, Zhai ZX, Tian XL, Duan LS, Li ZH (2008) Brassinolide alleviated the adverse effect of water deficits on photosynthesis and the antioxidant of soybean (*Glycine max* L). Plant Growth Regul 56:257–264

# Submergence Stress: Responses and adaptations in crop plants

# 14

Chinmay Pradhan and Monalisa Mohanty

**Abstract**

Submergence stress frequently encountered in crop plants is a widespread limiting factor for crop production throughout the world especially in irrigated and high-rainfall environments which results in huge economic losses. This chapter covers various features of submergence stress with special reference to crop plants, viz. causes of submergence and biophysical and biochemical alterations in crops, and various defence mechanisms adopted by crop plants. A brief discussion on different types of naturally or artificially developed tolerance mechanisms are presented here.

## Introduction

Environmental stresses are numerous and often crop or location specific which causes significant crop losses. They include increased UV-B radiation, water stress, high salinity, temperature extremes, hypoxia (restricted oxygen supply in waterlogged and compacted soil), mineral nutrient deficiency, metal toxicity, herbicides, fungicides, air pollutants, light, temperature and topography. There are two types of water stresses that plants experience in general. One is when water is not available in sufficient quantity, hence referred to as water deficit, while the second one is that when water is available but in excess called waterlogging/submergence. Water deficit affects plants through decrease of leaf water potential, which in turn brings about loss of cell turgor and stomatal closure resulting in decreased transpiration and photosynthesis and subsequently leads to reduced growth as well as wilting. On the other hand, waterlogging occurs when a large proportion of the pore spaces in the soil are occupied by water which limits the diffusion of oxygen and gas exchange between the soil, plants and atmosphere and resulted in decreased growth of roots and their functioning, thus negatively affecting the plant growth and survival.

Waterlogging/submergence/flooding is considered to be one of the major constraints for

C. Pradhan (✉)
Laboratory of Plant Biotechnology, Post Graduate Department of Botany, Utkal University, Bhubaneswar 751004, Odisha, India
e-mail: chinmay_pr@yahoo.com

M. Mohanty
Laboratory of Environmental Physiology and Biochemistry, Post Graduate Department of Botany, Utkal University, Bhubaneswar 751004, Odisha, India
e-mail: 18.monalisa@gmail.com

G.R. Rout and A.B. Das (eds.), *Molecular Stress Physiology of Plants*,
DOI 10.1007/978-81-322-0807-5_14, © Springer India 2013

crop production in many areas of the world (Kozlowski 1984; Pang et al. 2004; Conaty et al. 2008) which adversely affect approximately 10% of the global land area (FAO 2002). Soil waterlogging and submergence (collectively termed as flooding) are abiotic stresses that influence species composition and productivity in numerous plant communities, worldwide. Flooding is a complex stress that imposes several often-concurrent challenges to normal plant functioning. Starvation of oxygen and carbon dioxide is imposed by extremely slow rates of diffusion through the floodwater compared to that in air. Deficiency (hypoxia) or complete absence (anoxia) of oxygen in soil environment restricting the growth, development and crop yield is an important biological consequence of waterlogging or submergence stress. Submergence occurs when rainfall or irrigation water deposits on the soil surface or subsoil for prolonged period of time and can also occur when the amount of water added through rainfall or irrigation is more than what can percolate into the soil within 1 or 2 days. Waterlogging of field-grown crop plants can occur either as 'surface waterlogging' when the surface of poorly drained soils is flooded or 'root-zone waterlogging' when the water table rises to saturate a part or entire root zone with water. Thus, partial to complete flooding has detrimental effects for most terrestrial crop plants, except for some tolerant species, because it hampers growth and can result in premature death (Blom and Voesenek 1996; Bailey-Serres and Voesenek 2008) due to the rapid development of anoxic or hypoxic conditions in waterlogged soils. For most crops, excess water is a major constraint to productivity in many regions and situations (Jackson 2004), adversely affecting grain yields (Settler and Waters 2003) and growth of pasture species (Gibberd and Cocks 1997; Gibberd et al. 2001).

Recent progress has been made in the isolation and functional analyses of genes controlling yield and tolerance to submergence stress. In addition, promising new methods are being developed for identifying additional genes and variants of interest and putting these to practical use in crop improvement. Rice plants are much damaged by several days of total submergence. The effect can be a serious problem for rice farmers in the rain-fed lowlands of Asia and runs contrary to a widespread belief amongst plant biologists that rice is highly tolerant of submergence. This chapter assesses damaging effects of submergence to various crop plants, examines various physiological mechanisms of injury and reviews recent progress achieved using advanced genetic engineering process.

## Types of Submergence Stress

### Hypoxia and Anoxia

In plant physiological studies, the term 'hypoxia' is referred as situations in which the oxygen concentration is a limiting factor (Morard and Silvestre 1996) that occurs in soil environments as oxygen in soil decreased to a point below optimum level. It is the most common form of stress in wet soils and occurs during short-term flooding when the roots are submerged under water but the shoot remains in the atmosphere. It may also occur in roots near the surface of long-term floodwater (Sairam et al. 2008). This type of flood injury may be described as moisture injury.

Anoxia, the extreme form of hypoxia, also known as flooding injury is used to qualify the complete lack of oxygen in physiological experiments (Morard and Silvestre 1996) and occurs during long-term flooding or waterlogging condition when plants are completely submerged by water and in deep roots below floodwaters (Sairam et al. 2008).

## Plant Response to Water Stress

Plants respond to water stress in two ways, viz. by avoidance of the stress or by tolerating it. Stress avoidance is accomplished when plants alter their growth schedule to escape the exposure to damaging stress. Well-known examples in this category include completing the life cycle while conditions are optimal or using strategies to maximise water uptake from the environment and/or conservation.

On the other hand, the tolerance response to water stress occurs when plants develop certain biochemical and morphological characteristics to minimise the potential damage from stress. An example of the latter could be additions to photosynthetic pathways such as crassulacean acid metabolism (Scott 2000) for drying stress.

One of the best characterised plant responses to soil waterlogging is the metabolic switch from aerobic respiration to anaerobic fermentation. In fact, most proteins induced during hypoxic conditions are enzymes involved in the establishment of this fermentative pathway. Because the plant cells need to keep a continuous ATP supply, the use of alternative electron acceptors and/or alternative pathways may be key elements of survival under soil waterlogging. Plants react to an absence of oxygen by switching from an oxidative to a solely substrate-level phosphorylation of ADP to ATP; the latter reactions predominantly involve glycolysis and fermentation. An important adaptive response is formation of aerenchyma, specialised tissues in roots, which allow diffusion of gases like $O_2$ from aerobic shoot to hypoxic/anoxic roots. The plant response may also include a reduction in stomatal conductance and photosynthesis, as well as root hydraulic conductivity. These physiological modifications may in turn affect carbohydrate reserves and translocation. In fact, efficient use of carbohydrates may discriminate between tolerant and intolerant species.

## Impact of Submergence Stress on Crop Plants

Growth is greatly inhibited in the deficiency (hypoxia) or complete absence (anoxia) of oxygen (Visser et al. 2003). In water-saturated soils, roots grow only in a small region near the surface and do not exploit a large soil volume. Plants invariably wilt within few hours to days (2–4) of imposing a flooding stress (Jackson and Drew 1984) as a consequence of higher resistance to mass flow of water through the root. Wilting is caused by the inhibition of respiration and loss of ATP synthesis in the roots. This blocks the ion transport systems that normally create the gradient in water potential across the root endodermis. When flooding extends to submergence of the shoot, photosynthesis becomes severely restricted by a deficiency of external $CO_2$ and light. Furthermore, total submergence can interfere with flowering and pollination essential for completion of reproductive cycle. Under submergence stress, various physico-chemical events in rhizospheric environment occur along with several physiological and metabolic modifications followed by initiation of adaptive response (Fig. 14.1).

## Changes in Soil Physico-Chemical Properties

Flooding drastically influences the soil physico-chemical properties, most notably soil redox potential, pH and $O_2$ level. Hypoxia or anoxia limits aerobic respiration and energy productivity, leading to the decrease in soil redox potential and the accumulation in toxicants and threatening plant survival and crop productivity (Waters et al. 1991a).

As water saturates the soil, air spaces are filled, leading to the modification of several soils' physico-chemical characteristics (Kirk et al. 2003; Dat et al. 2004). The first event that takes place is the increased presence of $H_2O$ in soil; this water saturation characterises flooding. Nevertheless, the mechanisms which trigger a plant response are often presumed by products of root-zone flooding. Alterations in soil physical properties, viz. soil compaction, have demonstrated that the soil bulk density and porosity greatly affect soil hydraulic and chemical properties (i.e. water and nutrients); gaseous diffusion, root exploration and function and stomatal conductance are commonly observed during flooding.

Waterlogging causes less than 10% air-filled pore space in the soil (0% is achieved at soil saturation), which hampers root activity and hence induces aeration stress (Wesseling and van Wijk 1957). A major constraint resulting from excess water is inadequate supply of oxygen to submerged tissues; diffusion of oxygen

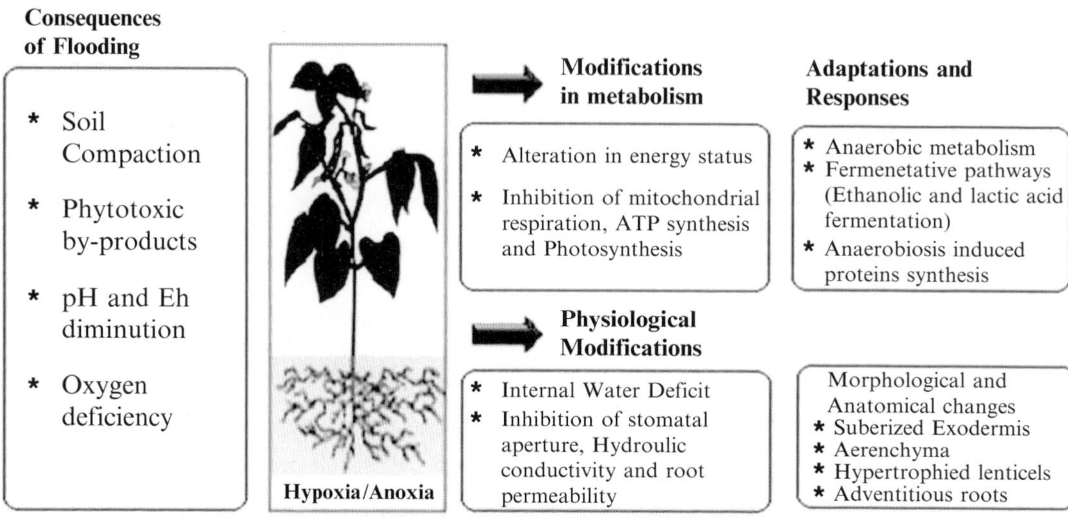


**Fig. 14.1** Physico-chemical changes in rhizospheric environment and physiological and metabolic modifications followed by initiation of adaptive response under submergence stress

through water is 104-fold slower than in air (Armstrong and Drew 2002). Excess water also leads to other changes in the soil that influence levels of the plant hormone ethylene (Smith and Russell 1969; Jackson 1982) and products of anaerobic metabolism by soil microorganisms (e.g. $Mn^{2+}$, $Fe^{2+}$, $S^{2-}$, $H_2S$ and carboxylic acids) which can accumulate (Ponnamperuma 1984; McKee and McKevlin 1993). Moreover, when flooding results in complete submergence, and in normally submersed aquatic plants, availability to the shoots of carbon dioxide, light and oxygen typically diminish (Jackson and Ram 2003).

Other changes affecting soil chemical characteristics during flooding include variations in soil pH and redox potential (Eh). As soil becomes reduced, iron and iron oxides can also become reduced leading to a modification of proton (i.e. pH) and cation balances. This process is heightened by the fact that the partial pressure of $CO_2$ will buffer carbonate, thus lowering pH. The pH medications may not directly affect plant growth; however, undesirable effects through aluminium or manganese phytotoxicity, calcium deficiency, reduced mineralisation or reduced turnover of soil organic matter will drastically alter plant metabolism (Probert and Keating 2000).

Soil redox potential (Eh) is often considered the most appropriate indicator of the chemical changes taking place during soil flooding (Pezeshki and DeLaune 1998). Eh generally declines during soil waterlogging (Pezeshki and DeLaune 1998; Pezeshki 2001; Boivin et al. 2002; Lu et al. 2004). It is not only an indicator of $O_2$ level (Eh around +350 mV under anaerobic conditions) (Pezeshki and DeLaune 1998) as reducing conditions lead to a high competitive demand for $O_2$, but it also critically affects the availability and concentration of different plant nutrients (Pezeshki 2001). However, changes in Eh are influenced by the presence of organic matter as well as Fe and Mn (Lu et al. 2004). Soil reduction induces the release of cations and phosphorous through adsorption of ferrous ion and dissolution of oxides (Boivin et al. 2002). Soil reducing conditions also favour the production of ethanol, lactic acid, acetaldehyde, acetic acid and formic acid.

The reduction in soil redox resulting in consequent reduction in uptake capacity will tend to decrease plant nitrogen and phosphorous contents. In contrast, soil iron and manganese availability will increase, as ferric and manganic forms are reduced to soluble ferrous and manganous forms. Sudden availability or restriction of various element forms will drastically affect root metabolism. There are numerous studies on the adverse effects of increased levels of phosphorous (P), potassium (K), copper (Cu) and iron ($Fe^{3+}$) combined with decreased bioavailability of nitrogen (N), sulphur (S) and zinc (Zn) on plant growth (Drew 1997). Finally, slowed diffusion of gases in water will hasten the accumulation of potentially phytotoxic by-products of anaerobic metabolism such as ethanol, lactic acid, $CO_2$, $N_2$, $H^+$ and methane. These may accumulate intracellularly and/or be released in the soil solution and adversely alter soil chemical properties.

## Plant Sensitivity

Soil flooding and submergence stress is one of the major abiotic constraints imposed on plant growth, species' distribution and agricultural productivity. Flooding stress that is highly damaging to the majority of plant species has resulted in a wide range of biochemical, molecular and morphological adaptations (Fig. 14.1) (Jackson and Colmer 2005).

Schematic diagram of the main metabolic pathways proposed during plant flooding stress is presented in Fig. 14.2. Hypoxia causes a decrease in mitochondrial respiration, which is partly compensated by increases in both the glycolytic flux and fermentation pathways. Nitrate has been proposed as an intermediate electron acceptor under low $O_2$ tensions and may participate in NAD(P)H oxidation during hypoxia (Igamberdiev et al. 2005). NO can be oxygenated to nitrate with the tightly bound $O_2$ of class-1 haemoglobin [Hb($Fe^{2+}$)$O_2$ ], which is oxidised to metHb [Hb($Fe^{3+}$)]. The alanine aminotransferase enzyme which converts pyruvate to alanine is strongly induced in hypoxic conditions (Fig. 14.3). However, unlike ethanol formation, there is no consumption of NAD(P)H in the process (Gibbs and Greenway 2003).

### Submergence Leads to Low-Oxygen Stress

Oxygen serves as an electron acceptor in the oxidative phosphorylation pathway, which generates ATP by regenerating essential $NAD^+$ cofactor from NADH. Many microorganisms react to low oxygen tensions by inducing genes encoding alternative enzymes with a higher affinity for oxygen and are therefore able to utilise limiting oxygen concentrations more efficiently (Zitomer and Lowry 1992).

Exposure of crop plants to hypoxia or anoxia causes oxidative stress, which affects plant growth due to the production of reactive oxygen species (ROS) such as superoxide radicals, hydroxyl radicals and hydrogen peroxide (Mittler et al. 2004). These ROS are very reactive and cause severe damage to membranes, DNA and proteins (Bowler et al. 1992; Foyer et al. 1997). Hypoxia stress triggers the formation of ROS and induces oxidative stress in plants (Yan et al. 1996; Geigenberger 2003; Narayanan et al. 2005). Moreover, reexposure to air after a period of oxygen deprivation can induce such stress causing serious injury (Monk et al. 1987; Crawford 1992).

The induction by low-oxygen stress of pyrophosphate-dependent phosphofructokinase (Botha and Botha 1991; Mertens 1991) and a vacuolar $H^+$- translocating pyrophosphatase replacing $H^+$-ATPase (Carystinos et al. 1995) suggests that plants might have adaptations to cope with limited availability of ATP. Hydrogen peroxide accumulations under hypoxic conditions have been shown in the roots and leaves of barley (Kalashnikov et al. 1994) and in wheat roots (Biemelt et al. 2000). Indirect evidence of ROS formation such as thiobarbituric acid reactive substances (TBARS; lipid peroxidation products) under low oxygen has also been detected (Chirkova et al. 1998; Blokhina et al. 1999).

## Effect of Waterlogging on Growth, Physiology and Metabolism of Crop Plant

Waterlogging (hypoxia) has a range of effects on plants. Firstly, it rapidly decreases growth, initially of roots and subsequently of shoots (Barrett-Lennard 1986; Drew 1983, Trought and Drew 1980a), and it increases the senescence of roots, beginning at the tips (Barrett-Lennard et al. 1988; Huck 1970; Webb and Armstrong 1983). Secondly, it affects processes associated with solute movement across membranes, such as the

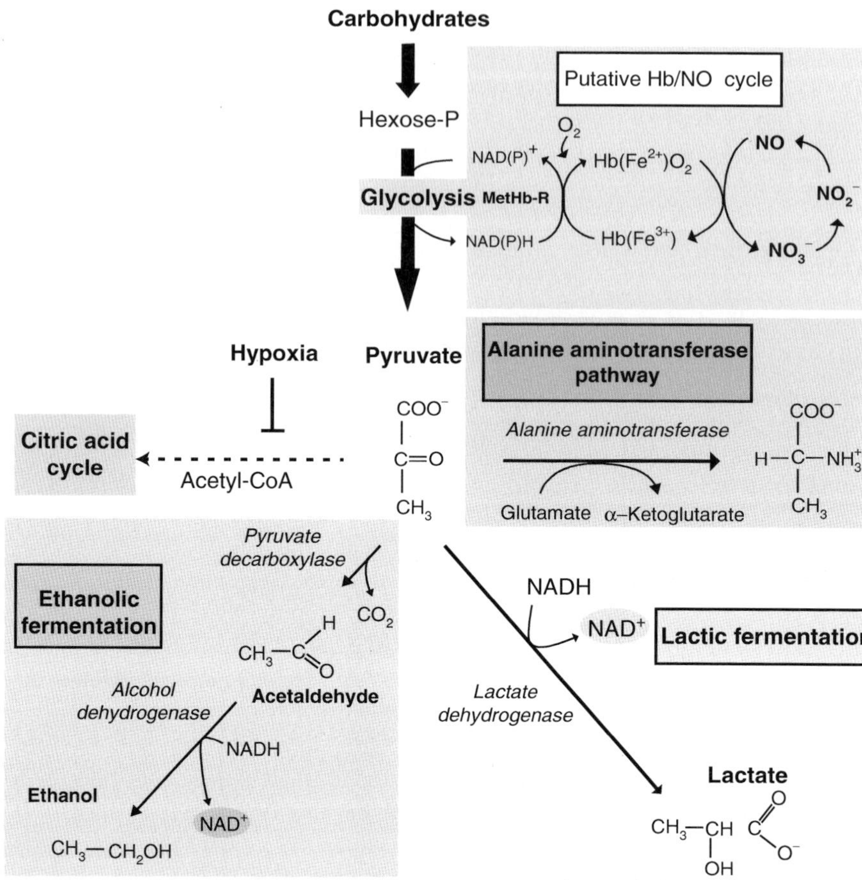


**Fig. 14.2** Schematic diagram of major plant metabolic pathways under flooding stress (hypoxia). Hypoxia leading to decreased mitochondrial respiration with increased glycolytic flux and fermentation pathways. Nitrate as intermediate electron acceptor and participate in oxidation of NAD(P)H

uptake of inorganic nutrients (Buwalda et al. 1988a; Trought and Drew 1980b), the regulation of cytoplasmic pH and membrane potentials (Greenway and Gibbs 2003) and the efflux of internal cell constituents such as $K^+$, $Cl^-$, organic and amino acids and 'basic and acidic metabolites' (reviewed by Buwalda et al. 1988b). Thirdly, it can decrease stomatal conductance and/or leaf water potentials (Bradford and Hsiao 1982; Else et al. 2001; Huang et al. 1995; Jackson and Hall 1987; Kriedemann and Sands 1984; van der Moezel et al. 1989a). Schematic diagram of major plant metabolic pathways under flooding stress (hypoxia) has been indicated in Fig. 14.3. Hypoxia leads to decreased mitochondrial respiration with increased glycolytic flux and fermentation pathways. Nitrate acts as intermediate electron acceptor and participates in oxidation of NAD(P)H.

External symptoms of injury include variations, viz. downward bending of leaf petioles, stem swelling (particularly in small plants), chlorosis, oedema, red or purple pigmentation in leaves (of pear and some other plants), browning of leaf margins, reduction or cessation of growth (more pronounced in roots than stems), twig dieback, death of roots, wilting, leaf drop and death of the entire plant. Seedlings develop symptoms more quickly than do large plants. Plants with roots injured by waterlogged soil may subsequently suffer drought stress or death when, after the soil drains, the root system is unable to meet transpirational demands of the top. Plants

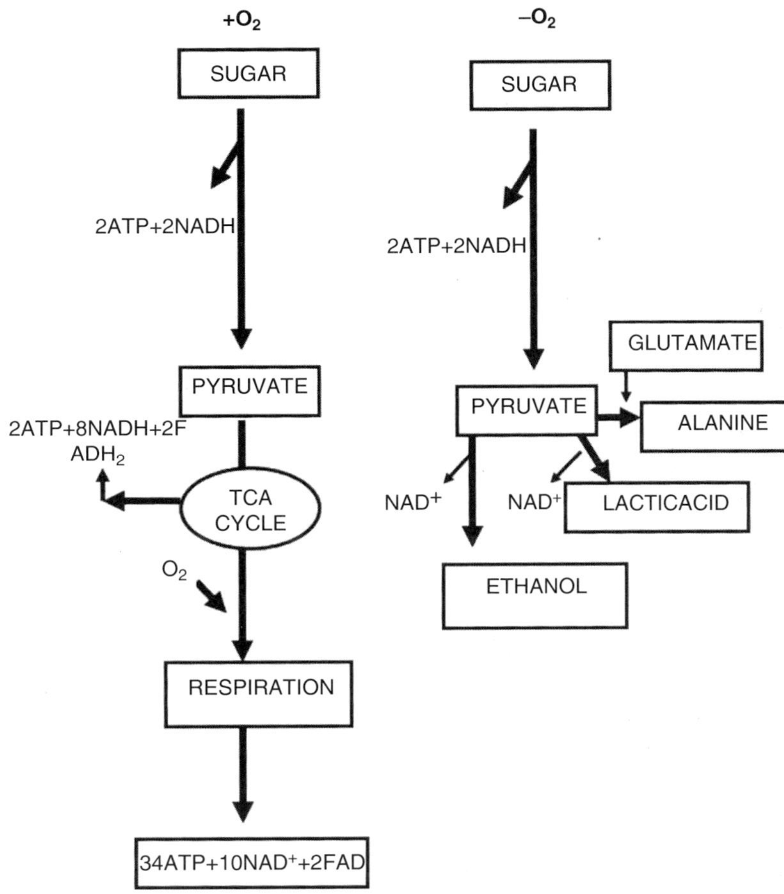


**Fig. 14.3** Low-oxygen conditions lead to metabolic shift in plants

stressed or injured by waterlogging also become abnormally susceptible to certain fungal pathogens. Most often *Phytophthora* species cause root rot in periodically waterlogged soils.

Other physiological changes in plants as a result of root asphyxiation include an increase in the production of the ethylene precursor, ACC (1-aminocyclopropane-1-carboxylic acid) (Bradford and Yang 1980); a reduction in cytokinin and gibberellin synthesis in the roots (Kramer and Boyer 1995); and an increase in abscisic acid concentration in leaves which has been implicated in causing stomatal closure (Kramer and Boyer 1995; Else et al. 1995). Ultimately, hypoxia or anoxia of the soil can result in root tissue damage, inhibition of the vegetative and reproductive growth, changes in plant anatomy, premature senescence and plant mortality (Drew 1997; Kozlowski 1997).

Considerably low amounts of oxygen in the root zone hampers root respiration resulting in a limited supply of energy required for nutrient uptake and transport (Boru et al. 2003a). This nutrient deficiency ultimately disturbs a range of physiological processes such as reduced root, stem and leaf growth; chlorosis and necrosis resulting in premature leaf senescence, stomatal closure, photosynthesis and respiration; and increased susceptibility to diseases in plants subjected to waterlogged conditions (Drew and Sisworo 1979; Liao and Lin 2001; Bange et al. 2004). Despite adversely affecting a multitude of physiological and biochemical processes, waterlogging is also known to considerably inhibit the uptake of some important nutrients such as N, P, K, $Ca^{2+}$, $Mg^{2+}$ and $Fe^{2+}$ by crop plants (Gutierrez Boem et al. 1996). This can occur as a result of an alteration in nutrient availability in the soil environment either by removal, oxidation or leaching to deep soil profiles (Orchard and So 1985; Akhtar and Memon 2009). Available literature indicates that the concentration of N, P, K and $Ca^{2+}$ may decrease while that of $Fe^{2+}$, $Na^+$ and $Cl^-$ increase in the leaves and stems of waterlogged plants (Meek et al. 1980; Hocking et al. 1985, 1987; Reicosky et al. 1985; Hodgson 1990). Under waterlogged conditions, $Fe^{3+}$ and $Mn^{3+}$ may rapidly reduce to $Fe^{2+}$ and $Mn^{2+}$, respectively, as reported by several researchers (Nathanson et al. 1984; Marschner 1986; Mortvedt et al. 1991). This results in increased availability of these nutrients in the rooting environment of crop plants. It has been reported that the roots of some plants (such as rice) adapted to waterlogging stress can avoid the uptake of $Fe^{2+}$ and $Mn^{2+}$ ions by releasing oxygen into the rhizosphere for the oxidation of $Fe^{2+}$ and $Mn^{2+}$ (Mengel et al. 2001). However, waterlogging-sensitive plants like cotton, wheat and barley are not able to oxidise $Fe^{2+}$ and $Mn^{2+}$, and consequently $Fe^{2+}$ and $Mn^{2+}$ toxicity may occur under waterlogged conditions (Drew 1988).

Synthesis and translocation of growth regulators (gibberellins and cytokinins) in roots slows, and concentrations of auxins and ethylene in stems increase. Mycorrhizal fungi, which associate with plant roots symbiotically, are also adversely affected, further suppressing plant uptake of mineral nutrients, especially phosphorus. Internal water deficit in some plants increases until they die, but many kinds of plants regain the normal degree of hydration, while their stomata remain closed during flooding. Stomata of some tolerant plants reopen as the plant adapts to flooding.

The development of hypoxia in soils presents a range of challenges for plants. Roots normally require oxygen for optimal production of adenosine triphosphate (ATP) from sugars. Under aerobic conditions, glucose in roots is oxidised to 6 mol of $CO_2$ and 6 mol of $H_2O$, and up to 38 mol of ATP are produced. However, under waterlogged (anaerobic) conditions, glucose is oxidised to produce 2 mol of ethanol and 2 mol of $CO_2$ with a yield of only 2 mol of ATP. Thus, the transfer of roots from aerobic to anaerobic conditions can decrease ATP production by about 95%. This has important consequences for the metabolism of plants growing in waterlogged soils as ATP is the fuel for nearly all cellular processes.

Waterlogging and submergence lead to reduced gas exchange between the plant tissue and the atmosphere (Armstrong 1979). Oxygen diffuses about 10,000 times more slowly in water than in air, so flooding leads to lack of sufficient $O_2$ for normal root respiration (Drew 1997; Kramer and Boyer 1995).

As a result of soil hypoxia or anoxia, a change from aerobic to anaerobic respiration takes place in the root. This causes a reduction in ATP production resulting in reduced energy for normal plant metabolic processes. During anaerobic root respiration, potentially toxic metabolites such as ethanol, lactic acid, acetaldehyde and cyanogenic compounds can accumulate in the plant. Finally, cytosolic acidosis can occur in the cells, due to lactic acid accumulation in the cytoplasm, which results in cell death (Kozlowski 1997; Liao and Lin 2001; Jackson 2002).

Waterlogging can cause stomatal closure under both nonsaline (Bradford and Hsiao 1982; Else et al. 2001; Jackson and Hall 1987) and saline conditions (Huang et al. 1995b; Kriedemann and Sands 1984; van der Moezel et al. 1989a). Waterlogging (hypoxia) affects membrane selectivity and transport of nutrients (Buwalda et al. 1988a; Drew 1983; Morard and Silvestre 1996) and the retention of ions and small molecular weight metabolites by roots (Buwalda et al. 1988b). Two kinds of damage to membrane barriers or processes can be postulated: (1) substantial loss of membrane integrity such that transpiration results in the uptake and transport of ions by mass flow. This hypothesis was suggested to explain the effects of hypoxia in decreasing the concentrations of phosphate, potassium, calcium and magnesium in the xylem of wheat (Trought and Drew 1980c); and (2) more specific effects on ion influx and efflux mediated by deficits of energy (ATP).

The dramatic decrease of gas diffusion in water compared with diffusion in air is a major problem for terrestrial crop plants and limits the entry of $CO_2$ for photosynthesis and $O_2$ for respiration. The main problems during submergence are shortage of oxygen due to slow diffusion rates of gases in water and unfavourable conditions of light and carbon dioxide supply. Collectively, these factors lead to loss of biomass and eventually death of the submerged plants. These $O_2$ restrictive conditions dramatically affect plant growth, development and survival. It has been suggested that physiological and molecular studies of the mechanisms of anoxia tolerance in wild plants that still possess long-term anoxia tolerance are more likely to provide evidence of physiological mechanisms than studies of crop plants (Crawford and Brändle 1996).

## Mechanism of Waterlogging Tolerance in Plants

In many aquatic and amphibious species, the debilitating effects of flooding stress are overcome by an oxygen-dependent, ethylene-mediated stimulation of underwater shoot elongation that encourages renewed contact with the aerial environment. With the exception, of rice species this strategy offers few opportunities for molecular dissection of signal sensing and transduction processes involved in the submergence escape. Greater activities of glycolytic and fermentative enzymes, increased availability of soluble sugars and involvement of antioxidant defence mechanism against post-stress oxidative damages are the main metabolic mechanisms for waterlogging tolerance in wheat.

Rice, the second food crop in the world, can tolerate some submergence as paddy rice or deep-water rice. It is well adapted to flooding of the roots because of its ability to transport oxygen efficiently from the aerial parts of the plant to the roots. Other crops such as canola and barley are very sensitive to waterlogging and can experience significant yield losses. For all these crop plants, it would be important to improve waterlogging and flooding tolerance. In order to do this, it will be critical to understand the physiology of flooding tolerance/sensitivity and to identify genes important in mounting a response (Dennis et al. 2000).

The survival of plants is closely related to carbohydrates, the substrate for respiration. There are two major aspects of the correlation of plants' flooding adaptability and carbohydrates. First, there are morphological and physiological responses of terrestrial plant species that enable the positive effects of carbohydrates on underwater plant performance: plants usually elongate or reduce underwater elongation and maintain a higher level of root carbohydrates that facilitates survival. Second, plants change the expression of hormone, enzyme and gene, adjusting carbohydrate metabolism to flooding.

Through adaptive evolutionary processes, many plant species can survive after a long- or short-term submergence. There are two major approaches to analyse the plant response and adaptation to submergence stress; one is to study the decrease of plant oxygen concentration, and the other is to investigate the increase of plant ethylene. However, for plants growing in intermittently waterlogged soils, it might be best to adopt a combination of strategies, maintaining membrane semipermeability and decreasing stomatal conductance after waterlogging while at the same time developing new adventitious roots containing aerenchyma if possible.

Under submergence stress, plants employ the following adaptation strategies (Shuduan et al. 2010):

1. Morphological adaptation via stem elongation and forming adventitious roots and aeration tissues
2. Change in metabolic pathways and energy production through anaerobic metabolism
3. Regulation of physiological activities by changing the hormone levels of ethylene, gibberellin and abscisic acid or variation in morphology and anatomy

4. Elimination of poisonous active oxygenic free radical by antioxidase system under anaerobic condition

Responses to avoid the adverse effects of submergence include underwater photosynthesis, aerenchyma formation and enhanced shoot elongation. The molecular regulatory networks involved in these responses, including the putative signals to sense submergence, are discussed, and suggestions are made on how to unravel the mechanistic basis of the induced expression of various adaptations that alleviate $O_2$ shortage underwater. Molecular biology and bioinformatics techniques can be used to study the mechanisms of plant adaptation to submergence at gene level. Recent progress has been made in the isolation and functional analyses of genes controlling yield and tolerance to abiotic stresses. In addition, promising new methods are being developed for identifying additional genes and variants of interest and putting these to practical use in crop improvement.

## Morphological, Anatomical and Metabolic Adaptations

Some morphological adaptations for flooding stress include development of air-space tissue (aerenchyma) within tissues and ventilation roots (pneumatophores), morphological changes like formation of hypertrophied lenticels, the initiation of adventitious roots and/or the development of aerenchyma. Many plants respond to waterlogging by forming aerenchyma in adventitious roots (Drew 1983; Vartapetian and Jackson 1997). However, aerenchyma is only capable of delivering $O_2$ into anaerobic environments over short distances. Armstrong (1979) has developed a series of relationships between root porosity, oxygen consumption and the maximum length of an internal aerobic pathway. These relationships suggest that at porosities of 15% and the rates of oxygen use typical of the roots of cereals, there would be sufficient $O_2$ transport within roots for them to reach lengths of ~10–20 cm (Armstrong 1979). Experimental evidence confirms that aerenchyma does not form pre-established roots unless they are relatively short. For wheat, this length is ~10–20 cm (Thomson et al. 1990).

Our knowledge of the basic adaptive mechanisms of plants to soil waterlogging has benefited from large scale genomic and proteomic approaches; however, the diversity of the adaptive responses involved underlines the difficulty when studying this stress. This update reviews our current comprehension of the metabolic, physiological and morphological responses and adaptations of plants to soil waterlogging.

## Morphological and Anatomical Adaptations

### Root Growth

A common adaptation of plants to waterlogging is the survival and growth of seminal roots and production of numerous adventitious roots with aerenchyma (Belford 1981; Trought and Drew 1982; Drew 1983; Smirnoff and Crawford 1983; Justin and Armstrong 1987; Barrett-Lennard et al. 1988; Thomson et al. 1992; Huang et al. 1994a). Huang et al. (1994a) reported a drastic decrease (42–50%) in length of the longest seminal root and also in total length of seminal roots by 14-day hypoxia for waterlogging intolerant genotypes (Bayles, BR34, Coker-9766 and FL302). The above hypoxic stress had no significant effect on the growth of seminal roots for tolerant genotypes (Gore and Savannah). Total root dry mass was reduced for all genotypes except for Savannah (Huang et al. 1994a).

The root growth in waterlogging in tolerant genotypes is drastically suppressed by stress. However, the tolerant genotypes have the ability to continue their root growth under the stress in some extent. However, the waterlogging tolerance of a plant is determined not only by its capability to undergo morphological adaptations (Fig. 14.4) but also by the ability to recover from transient waterlogging or hypoxia of the root system (Krizek 1982; Huang et al. 1994a, 1997). The growth of many species that are tolerant to hypoxic conditions can also be reduced when the roots are waterlogged, but unlike sensitive ones, tolerant species rapidly resume their

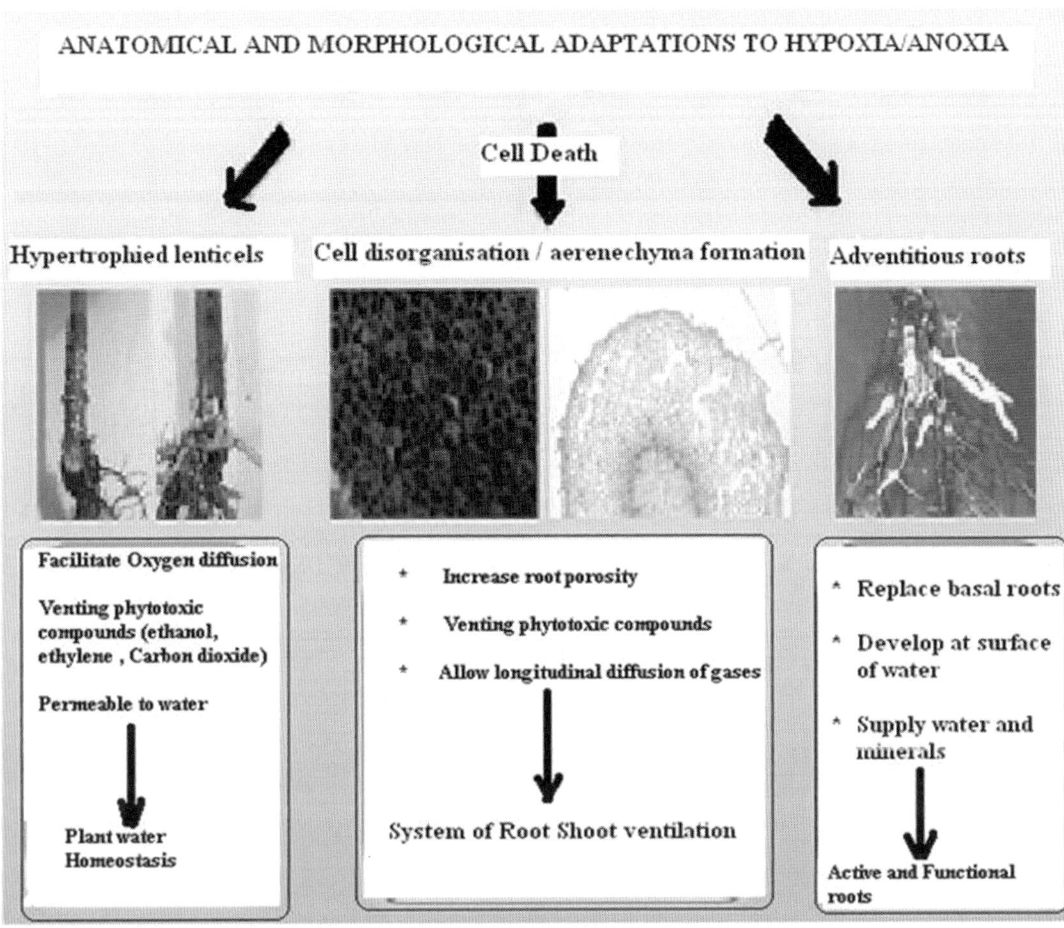


**Fig. 14.4** Submergence stress-induced anatomical morphological changes in plants

growth a short period after the resumption of aeration in roots (Crawford 1982).

### Aerenchyma Formation and Increased Root Porosity

Formation of aerenchyma has been observed in the roots of wheat when grown under low $O_2$ concentrations (Benjamin and Greenway 1979; Trought and Drew 1980c; Belford 1981; Erdmann and Wiedenroth 1986; Barrett-Lennard et al. 1988; Thomson et al. 1990, 1992; Drew 1991; Huang et al. 1994a, b; Watkin et al. 1998; McDonald et al. 2001a, b; Haque et al. 2010). Aerenchyma formation increases the porosity of roots above the usual levels contributed by intercellular spaces (Colmer 2003). The aerenchyma is usually formed within 5–7 days of the onset of hypoxia in wheat (Thomson et al. 1990). Increased root porosity or anatomical investigation may be the evidence of aerenchyma formation (Fig. 14.4).

Aerenchyma provides a low-resistance internal pathway for the movement of $O_2$ from the shoots to the roots (Armstrong 1979; Armstrong and Webb 1985; Drew et al. 1985) and allows the roots to respire aerobically and to maintain growth under hypoxic conditions. Moreover, a part of oxygen transported to plant root tips through the aerenchyma leaks out into the surrounding soil and results in a small zone of oxygenated soil around the roots providing an aerobic environment for microorganisms that can prevent the influx of potentially toxic soil components (Visser et al. 1997; Armstrong and

Armstrong 1988; Colmer 2003) such as nitrites and sulphides of Fe, Cu and Mn. Therefore, aerenchyma formation is thought to be one of the most important morphological adaptations for the tolerance to hypoxic or anoxic stress. The aerenchyma in stems and roots can be distinguished into lysigenous and schizogenous aerenchyma on the basis of the process of formation (Jackson and Armstrong 1999; Evans 2003; Visser and Voesenek 2004).

Lysigenous aerenchyma is created through cell disintegration (death) in the primary cortex of adventitious roots (Drew et al. 1979, 1981; Justin and Armstrong 1991; Huang et al. 1997; Haque et al. 2010), whereas the schizogenous aerenchyma is formed by the separation of cells from each other, often accompanied by cell divisions and normal expansion (Jackson and Armstrong 1999; Colmer et al. 2004).

Under oxygen deficient condition, ethylene production is accelerated which in turn stimulates aerenchyma formation in adventitious roots and induces the growth of the roots (Drew et al. 1979; Jackson 1989). The immediate precursor of ethylene is 1-aminocyclopropane-1-carboxylic acid (ACC), which is synthesised to a large extent in roots (Bradford and Yang 1980). The activity of ACC synthase is stimulated in roots under flooding conditions (Cohen and Kende 1987). However, the conversion of ACC to ethylene requires oxygen, and the conversion reaction is blocked in an anaerobic root cell. The ACC is therefore translocated from the anaerobic root cells towards the more aerobic portions of the root or to the shoot. The lower portions of the stems are usually the site of highest ACC accumulation, and in the presence of oxygen, ethylene is released (Sairam et al. 2008).

The increase in root porosity of tolerant genotypes in response to waterlogging stress could represent adaptation to anaerobic or hypoxic conditions. High porosity in root tissue increases the possibility of $O_2$ diffusion from shoots to roots (Haldemann and Brändle 1983). The poorly developed adventitious root system and relatively low root porosity of hexaploid wheat are thought to contribute to its sensitivity to waterlogging (Thomson et al. 1992). Boru et al. (2003b) reported 12–20% (v/v) root porosity for tolerant wheat genotypes (Ducula, Prl/Sara and Vee/Myna) and 6–8% for sensitive genotypes (Seri-82 and Kite/Glen) under hypoxia.

### Barriers to Radial Oxygen Loss (ROL)

Oxygen in aerenchymatous roots may be consumed by respiration or be lost to the rhizosphere via radial diffusion from the root. The flux of oxygen from roots to rhizosphere is termed as radial oxygen loss (ROL) which usually oxygenates the rhizosphere of the plants growing in waterlogged soils (Armstrong 1979). However, ROL decreases the amount of $O_2$ supply to the apex of roots that solely depends on aerenchymatous $O_2$ and, therefore, would decrease the root growth in hypoxic or anoxic environment (Armstrong 1979; Jackson and Drew 1984).

The roots of many wetland plants contain a complete or partial barrier to ROL in their epidermis, exodermis or subepidermal layers (Armstrong 1971; Jackson and Drew 1984; Jackson and Armstrong 1999), whereas in non-wetland plants usually are lacking or having partial barriers resulting considerable loss in aerenchymatous $O_2$ in root through ROL. Wheat plants can form aerenchymatous adventitious root in response to waterlogging, which contains a partial barrier to ROL and can consume only 20% of the total $O_2$ entering a root through aerenchyma (Thomson et al. 1992). It is suggested that the loss in internal $O_2$ contributes to the poor growth of adventitious roots and intolerance of wheat to waterlogged soil (Thomson et al. 1992). In contrary, waterlogging-tolerant rice not only has a larger volume of aerenchyma but also has a strong barrier to ROL in basal regions of its adventitious roots and therefore deeper root penetration into waterlogged soil (Armstrong 1971; Thomson et al. 1992; Colmer et al. 1998). However, some wheat genotypes can increase suberin or lignin on epidermis or exodermis of root which may act as barriers to ROL and results in increased tolerance to waterlogging (Arikado 1959; Jackson and Drew 1984; Watkin et al. 1998; McDonald et al. 2001b).

## Metabolic Adaptation

The plant tissue under hypoxia or anoxia suffers from energy crisis (Gibbs and Greenway 2003) due to reduced root respiration in both waterlogging-tolerant and nontolerant plants (Marshall et al. 1973; Lambers 1976; Drew 1983, 1990). The tolerant plant species cope with the energy crisis through metabolic adaptation to oxygen deficiency. The metabolic adaptations to oxygen deficiency includes anaerobic respiration, maintenance of carbohydrate supply for anaerobic respiration, avoidance of cytoplasmic acidification and development of antioxidative defence system (Davies 1980; Armstrong et al. 1994; Drew 1997; Setter et al. 1997).

### Anaerobic Respiration

Plant cells produce energy in presence of oxygen through aerobic respiration which includes glycolysis, TCA or Krebs cycle and oxidative phosphorylation (Fig. 14.5). Under anoxic condition, Krebs cycle and oxidative phosphorylation are blocked, and cells inevitably undergo anaerobic respiration to fulfil the demand for energy (Davies 1980). Generation of energy under anaerobic condition is largely achieved through glycolysis. For the continued operation of glycolytic pathway, the regeneration of $NAD^+$, a cofactor from NADH, is essential (Drew 1997). Large quantities of pyruvate generated in glycolysis as an end product must be converted to alternative products to recycle NADH to $NAD^+$ (Fig. 14.5). Ethanolic fermentation or lactate fermentation is the most important process by which NADH can be recycled to $NAD^+$ during oxygen deficiency (Kennedy et al. 1992; Perata and Alpi 1993; Ricard et al. 1994).

The efficacy of energy production by glycolysis and fermentation is much lower than that of aerobic respiration (Fig. 14.5). Moreover, the end products of glycolytic and fermentative pathway, such as ethanol, lactic acid and carbon dioxide, pose an additional hazard to the cell. It is well reported that the maintenance of an active glycolysis and an induction of fermentative metabolism are adaptive mechanisms for plant tolerance to anoxia (Kennedy et al. 1992; Ricard et al. 1994; Drew 1997; Sairam et al. 2008). Various environmental factors affect submergence tolerance of rice (Fig. 14.6). In waterlogged environment, anoxia is always preceded by hypoxia (Settler and Waters 2003) and hypoxia is considered as hypoxic pretreatment (HPT) before exposing the plants to anoxia (Waters et al. 1991b). Hypoxia accelerates the induction of glycolytic and fermentative enzymes, for example, aldolase and enolase (Bouny and Saglio 1996; Germain et al. 1997). This induction can improve or at least sustain the glycolytic rate in anoxic plants contributing higher tolerance to anoxia.

Anoxic cells may undergo lactic fermentation rather ethanolic fermentation onset of anoxia, though ethanol rather than lactate is the less deteriorating end product of fermentation (Davies 1980). An accumulation of lactate promotes acidification of the cytoplasm (Roberts et al. 1984) of anoxia-sensitive plants, such as maize, wheat and barley (Menegus et al. 1989, 1991). However, the enhanced lactate transport out of the roots into the surrounding medium may help to avoid cytoplasmic acidification (Xia and Saglio 1992). Moreover, lowered cytoplasmic pH leads to the activation of PDC and inhibition of LDH (Davies 1980) resulting in a shift from lactate fermentation to ethanolic fermentation.

### Increased Availability of Soluble Sugars

Due to shifting of energy metabolism from aerobic to anaerobic mode under hypoxia or anoxia, the energy requirements of the tissue are greatly restricted as very few ATPs are generated per molecule of glucose. A high level of anaerobic metabolism in hypoxic or anoxic roots is therefore very important to supply the energy for the survival of plants (Jackson and Drew 1984). Thus, maintaining adequate levels of readily metabolisable (fermentable) sugars in hypoxic or anoxic roots is one of the adaptive mechanisms to waterlogging or oxygen-deficient environment (Setter et al. 1987; Xia and Saglio 1992; Sairam et al. 2009).

The amount of root sugar reserve and activity of sucrose-hydrolysing enzymes are important determinants for waterlogging tolerance of crop

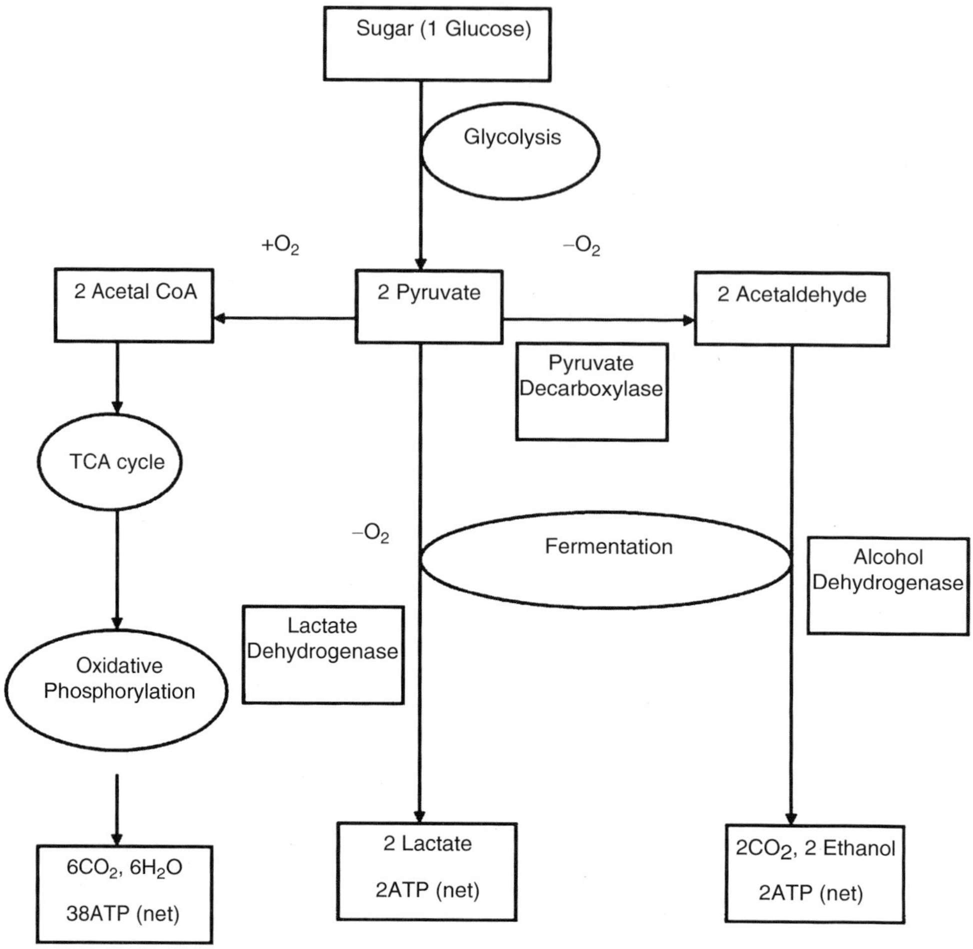


**Fig. 14.5** Shifting of aerobic to anaerobic respiration under anoxic condition

plants (Sairam et al. 2009). Zeng et al. (1999) reported that of the two enzymes involved in sucrose hydrolysis, the activity of invertase is down-regulated, while that of sucrose synthase (SS) is up-regulated in hypoxic maize seedlings. Therefore, the availability of sufficient sugar reserves in the roots with the increased activity of SS to provide reducing sugars for anaerobic respiration is one of the important mechanisms of waterlogging tolerance. The concentration of soluble carbohydrate in roots and shoots of wheat is increased when the crop is subjected to long-term oxygen deficit (Barrett-Lennard et al. 1982; Albrecht et al. 1993).

The accumulation of sugars has been attributed to the fact that growth is inhibited in hypoxically treated roots, while photosynthetic reactions are still active in the less challenged leaves (Mustroph and Albrecht 2003). Wheat genotypes tolerant to waterlogging accumulate more sugar in their roots in response to hypoxia compared to sensitive genotypes (Huang and Johnson 1995). A schematic diagram showing morphological and metabolic adaptive traits for waterlogging tolerance is shown in Fig. 14.7.

## Antioxidant Activities

To counter the hazardous effects of oxygen radicals, all aerobic organisms evolve a complex antioxidative defence system consisting of both antioxidants like ascorbate (AsA), glutathione (GSH), phenolic compounds, etc., and

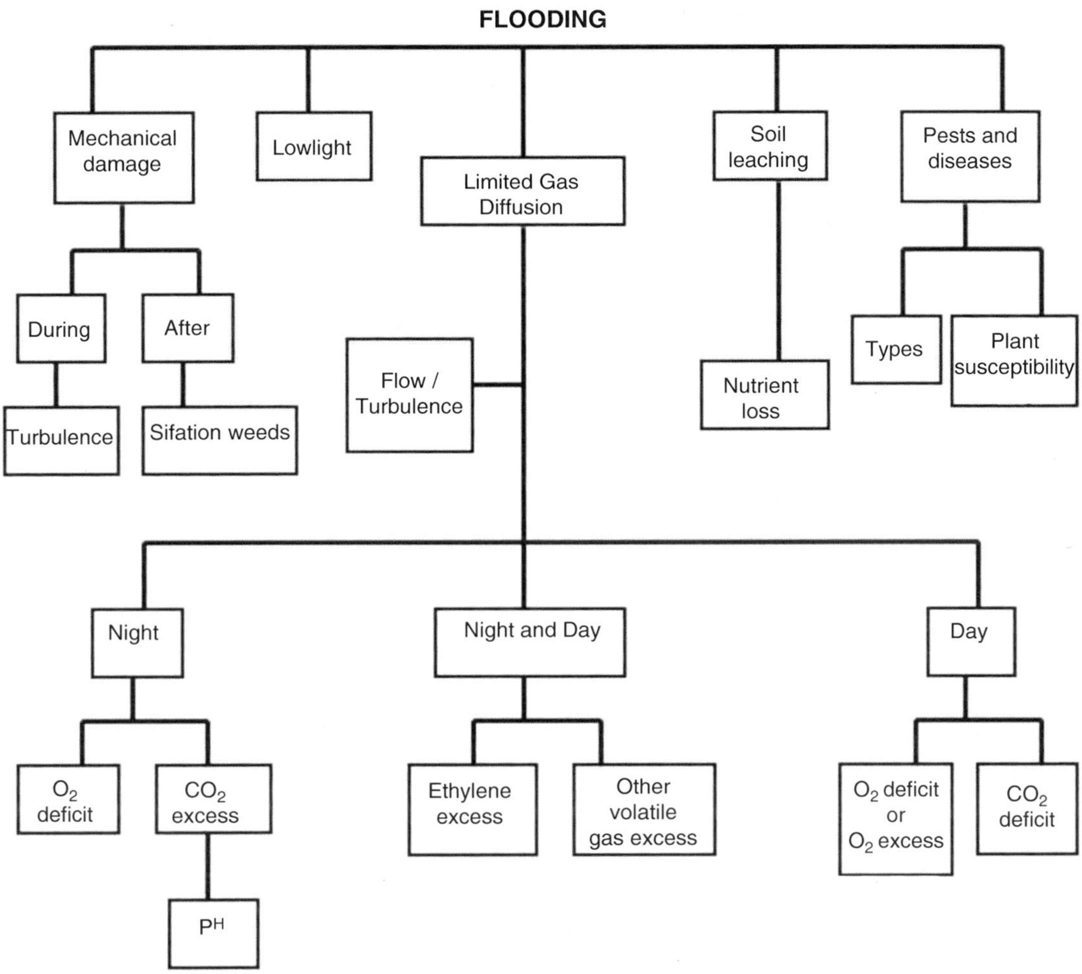


**Fig. 14.6** Submergence tolerance of rive affected by various environmental factors

antioxidative enzymes such as superoxide dismutase, catalase, peroxidases, glutathione reductase and ascorbate peroxidase (Zhang and Kirkham 1994; Foyer et al. 1997; Garnczarska 2005). A higher level of antioxidants and an increase in the activity of antioxidative enzymes are assumed to be adaptive mechanisms in overcoming certain stress situations (Foyer et al. 1995; Mishra et al. 1995). The tolerance to the stress may be improved by increased antioxidant capacity. Many recent attempts to improve stress tolerance in plants have been made by introducing and expressing genes encoding enzymes involved in the antioxidative defence system (Gupta et al. 1993; Foyer et al. 1995) (Fig. 14.8). The roots of young wheat plants are able to cope with the deleterious effects of oxygen radical generation induced by re-aeration after anoxia by means of their antioxidative defence system including increased capacity to scavenge radicals and elevated activities of enzymes of the AsA-GSH cycle to enable the restoration of the essential, highly reduced state of the antioxidants, AsA and GSH (Albrecht and Wiedenroth 1994; Biemelt et al. 1998).

An increase in the activity of antioxidant enzymes, viz. superoxide dismutase (SOD), ascorbate peroxidase (APX), glutathione reductase (GR) and catalase (CAT), and a network of low molecular mass antioxidants (ascorbate,

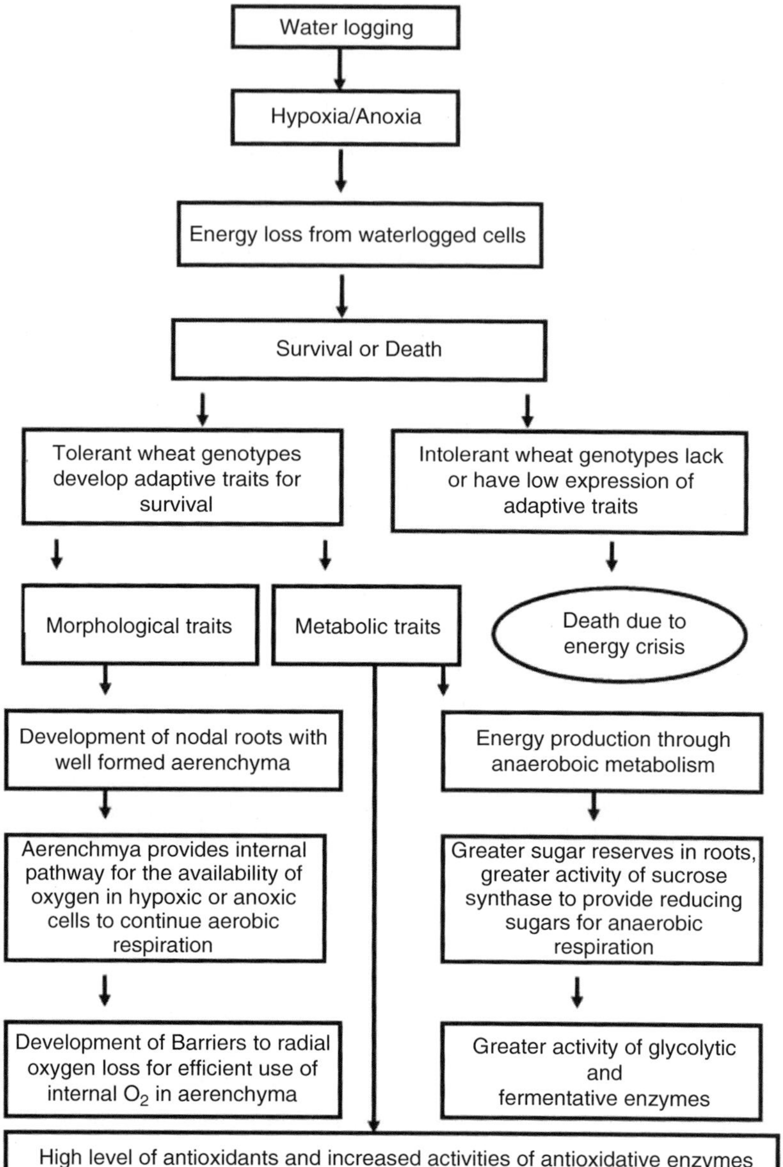


**Fig. 14.7** Schematic diagram showing morphological and metabolic adaptive traits for water logging tolerance

glutathione, phenolic compounds and α-tocopherol) in response to submergence stress have been reported by several workers (Elstner 1986; Bowler et al. 1992; Sairam et al. 2000, 2001, 2002). A 14-fold increase in SOD activity under waterlogged condition has been reported in Iris pseudacorus by Monk et al. (1989). An increase in total SOD activity has been reported in wheat roots under anoxia, and the degree of increase was positively correlated with duration of anoxia (Van Toai and Bolles 1991). Induction of enzymes involved in the ascorbate-glutathione cycle (APX, MDHAR, DHAR and GR) has been shown for anaerobically germinated rice seedlings and roots of wheat (*Triticum aestivum*) seedlings (Ushimaru et al. 1997; Albrecht and Wiedenroth 1994).

## Molecular Strategies

Studies on the effects of flooding at the molecular level may lead to improvements that will

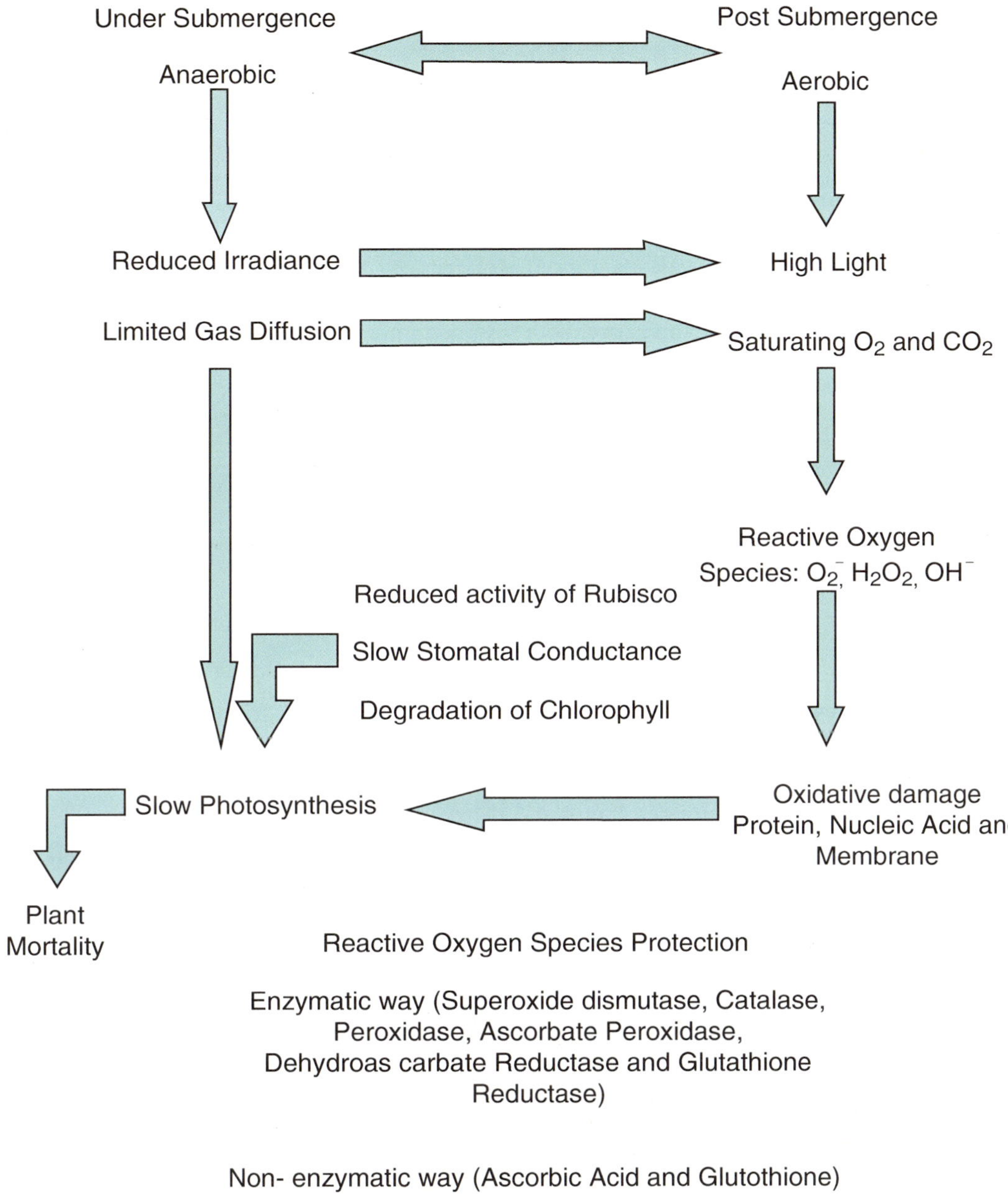


**Fig. 14.8** Enzymatic alterations under submergence stress

result in flood-tolerant crop plants. Flooding causes substantial stress for terrestrial plants, particularly if the floodwater completely submerges the shoot. Some important molecular strategies which are taken into account during construction of a tolerant crop variety are, viz. flooding and ethylene production, ethylene and aerenchyma formation, ethylene accumulation and adventitious root formation, shift in energy metabolism and anaerobiosis-induced proteins (ANP), hypoxia and non-symbiotic haemoglobins, haemoglobin and nitric oxide interaction, waterlogging, ROS production and antioxidant activity, hypoxic signalling and gene regulation (Sairam et al. 2008).

## Physiological and Biochemical Basis

Growth and development of the vast majority of vascular plant species is impeded by soil flooding and particularly by complete submergence, both of which can result in death. However, numerous wetland species are highly productive in flood-prone areas. This is achieved by means of a combination of life-history traits (Blom 1999) and certain key physiological adaptations and acclimations such as physical 'escape' from a submerged environment (Voesenek et al. 2003), avoidance of oxygen deficiency through effective internal aeration (Jackson and Armstrong 1999), anoxia tolerance (Gibbs and Greenway 2003) and a capacity to prevent, or repair, oxidative damage during re-aeration (Blokhina et al. 2003).

Waterlogging is a serious problem, which affects crop growth and yield in low-lying rainfed areas. The main cause of damage under waterlogging is oxygen deprivation, which affects nutrient and water uptake, so the plants show wilting even when surrounded by excess of water. Lack of oxygen shifts the energy metabolism from aerobic mode to anaerobic mode. Plants adapted to waterlogged conditions have mechanisms to cope with this stress such as aerenchyma formation, increased availability of soluble sugars, greater activity of glycolytic pathway and fermentation enzymes and involvement of antioxidant defence mechanism to cope with the post hypoxia/anoxia oxidative stress. Gaseous plant hormone ethylene plays an important role in modifying plant response to oxygen deficiency. It has been reported to induce genes of enzymes associated with aerenchyma formation, glycolysis and fermentation pathway. Besides, non-symbiotic haemoglobins and nitric oxide have also been suggested as an alternative to fermentation for maintaining lower redox potential (low NADH/NAD ratio) and thereby playing an important role in anaerobic stress tolerance and signalling.

Induction of phosphatase (P-ase) activities in the present study is likely to be caused by the low level of Pi under waterlogging treatment. These results suggest a dependence of the enzyme level on Pi availability as a signal for induction of P-ase activities in sorghum. Similar reports on the increase in P-ase activities in inverse reproportion to the low level of Pi have been demonstrated in numerous species and plant parts, viz. wheat leaves and roots (Barrett-Lennard et al. 1982; Mclachlan and Demarco 1982), maize leaves (Elliot and Lauchli 1986), sorghum roots (Furlani et al. 1984) and common beans roots (Helal 1990). The expression of higher P-ase activities in both tissues (shoots and roots) is suggestion of its global role in enhancing Pi availability and possibly recycling of organic Pi compounds. Under short-term waterlogging stress, P-ases play very important role to sustain the adverse environmental conditions in correlation to low phosphorus levels.

### Development of Stress-Tolerant Varieties Through Genetic Engineering

The tolerant genotypes of crop species can adapt to transient waterlogging by developing mechanisms related to morphology and metabolism to cope with the stress. In rice farming, flooding regimes are manipulated (e.g. paddy rice) or are accommodated by genotype selection (e.g. deep-water rice) to secure much of the world's production of this staple crop (Grist 1986). There have also been recent advances towards developing cultivars for lowland areas prone to short-duration flash flooding (Siangliw et al. 2003; Toojinda et al. 2003). The fact that anaerobic proteins are involved in a wide variety of cellular processes will certainly complicate genetic engineering approaches for flooding tolerance in plant.

The roots of comparatively tolerant genotypes contain greater sugar content (total, reducing and nonreducing sugar) than in susceptible genotypes of pigeon pea. Moreover, waterlogging induces to increase the content of reducing sugar through increased activity of SS in tolerant genotypes.

The tolerant genotypes show increased expression of mRNA for SS, while susceptible genotypes show very little expression under waterlogged condition (Sairam et al. 2009).

This carbohydrate accumulation might support fermentation of HPT roots over a long period of anoxic stress and could enhance tolerance against oxygen deficiency leading to a higher tolerance to anoxia. Moreover, exogenous supply of glucose prolongs the retention of root elongation potential under anoxic condition (Waters et al. 1991a). The ratio of the root-to-shoot sugar increases for waterlogging-tolerant wheat genotypes under hypoxia (Huang and Johnson 1995). The relatively large amount of sugars transported to root facilitates the energy supply for root respiration and ion uptake (Huang 1997).

It will be interesting to investigate the participation of an ROS-sensing mechanism involving PM-NADPH oxidase in different plant species. Again the paradoxical ROS may prove to be second messenger in the response mechanism. Further, it will be interesting to determine whether observed increases in NO evolution under flooding condition from roots or soils can contribute as a positive message in root-to-shoot communication.

Analyses of near-isogenic genotypes that differ in the adaptive response to oxygen deprivation are likely to yield critical information on regulatory mechanisms. Alterations in cytosolic pH and calcium may also have a role in the signalling processes. The importance of changes in adenylate charge, redox status and carbohydrate levels must also be considered. Many questions remain to be answered about the response of individual cells. What could be the basis of differential response between stress-tolerant and intolerant organs and species? Do these differ in cellular signalling and response mechanisms? Again we need to understand what signalling transduction pathways are activated or inhibited and how do multiple and interacting pathways control adaptive responses. The involvement of growth regulators such as ethylene, auxin, gibberellins and ABA in hypoxic regulation is also an interesting possibility. Protection of the energetic needs of meristematic cells, promotion or avoidance of programmed cell death in response to submergence through genetic engineering are yet to be further studied. How do cells in roots and aerial organs communicate over a long distance when there is an oxygen crisis in the roots? Understanding the cell-to-cell and long-distance signalling mechanisms that determine the organ and whole-plant response to oxygen deprivation, viz. regulation of leaf and internode elongation, petiole curvature, aerenchyma formation and adventitious root growth, is another inviting area for research.

Studies are required to isolate and characterise the genes involved in tolerance to anaerobic stress and to determine the molecular mechanisms and analysis of genes that confer increased flooding and anaerobic tolerance in crop plants. A simple post-translational modification of sucrose synthase by the addition/removal of phosphate can lead to potent changes in the tolerance of seedlings to anoxia (Subbaiah and Sachs 2003). Discovery of genes and proteins likely to be involved in structural modifications (aerenchyma formation and root tip death) indicate further that these mechanisms are multipronged and multicomponent.

Though the molecular basis of the adaptation to transient low-oxygen conditions has not been completely characterised, progress has been made towards identifying genes and gene products induced during low-oxygen conditions. Promoter elements and transcription factors involved in the regulation of anaerobically induced genes have been characterised. Transgenic plants may clarify the physiological role of the fermentation pathways and their contribution to flooding tolerance (Dennis et al. 2000).

## Genetic Diversity Strategies to Survive Flooding

Genetic diversity in the plant response to flooding includes alterations in architecture, metabolism and elongation growth associated with a low-$O_2$ escape strategy and an antithetical quiescence scheme that allows endurance of prolonged submergence. Not all species in flood-prone environments are flood tolerant. Some species avoid flooding by completing their life cycle between two subsequent flood events, whereas flooding periods are survived by dormant life stages (e.g. *Chenopodium rubrum* thrives infrequently flooded environments by timing its growth between floods and producing seeds that survive flooding (Van der Sman et al. 1992)). Established plants also use avoidance strategies through the development of anatomical and morphological traits. This amelioration response, here called the low-oxygen escape syndrome (LOES), facilitates the survival of submerged organs. Upon complete submergence, several species from flood-prone environments have the capacity to stimulate the elongation rate of petioles, stems or leaves. This fast elongation can restore contact between leaves and air but can also result in plant death if energy reserves are depleted before emergence.

Most plants are highly sensitive to anoxia during submergence. An important aspect of the adaptation to oxygen limitation includes metabolic changes such as avoidance of self-poisoning and cytoplasmic acidosis and maintenance of adequate supplies of energy and sugar. During anoxia, ATP and NAD+ are generated not in the Krebs cycle and the respiratory chain but via glycolysis and fermentation. A number of enzymes of the anaerobic pathways such as alcohol dehydrogenase and pyruvate decarboxylase induced during anoxia have been cloned and characterised (Bucher and Kuhlemeier 1993; Umeda and Uchimiya 1994). Using rice seedlings, Umeda and Uchimiya (1994) observed the coordinated expression of genes whose products are involved in glycolysis and alcohol fermentation under submergence stress. They analysed the mRNA level in submergence-tolerant rice FR13A and submergence-sensitive IR42 and showed these genes including glucose phosphate isomerase, phosphofructokinase, glyceraldehyde phosphate dehydrogenase and enolase may change in FR13A. Several anaerobic stress-inducing promoters have also been analysed. Kyozuka et al. (1994) demonstrated that maize Adhl promoter was strongly induced (up to 81-fold) in roots of seedling after 24 h of anaerobic treatment. For gene expression to obtain anaerobiosis-tolerant plants, a desirable promoter should not be active under aerobic condition.

Flooding is frequently accompanied with a reduction of cellular $O_2$ content that is particularly severe when photosynthesis is limited or absent. This necessitates the production of ATP and regeneration of $NAD^+$ through anaerobic respiration. The examination of gene regulation and function in model systems provides insight into low-$O_2$-sensing mechanisms and metabolic adjustments associated with controlled use of carbohydrate and ATP. At the developmental level, plants can escape the low-$O_2$ stress caused by flooding through multifaceted alterations in cellular and organ structure that promote access to and diffusion of $O_2$. These processes are driven by phytohormones, including ethylene, gibberellin, and abscisic acid. This exploration of natural variation in strategies that improve $O_2$ and carbohydrate status during flooding provides valuable resources for the improvement of crop endurance of an environmental adversity that is enhanced by global warming. Studies on the effects of flooding at the molecular level may lead to improvements that will result in flood-tolerant crop plants. Flooding leads to a deprivation of oxygen in plants. Many plants respond to oxygen deprivation with adaptations that allow short-term survival. Some plants show excellent tolerance to anoxia.

## Waterlogging and Crop Production

Many crops are sensitive to waterlogging and complete submergence. Just a few days of flooding can damage plants and will result in significant agricultural losses. It is therefore highly relevant to understand the traits that improve flooding tolerance and the genes and proteins underlying these traits. Owing to the global nature of flooding and the serious threat that floods will occur more often in the near future, it is to be expected that more knowledge on flooding tolerance will facilitate strongly the development of flood-tolerant crop varieties that can grow and yield on marginal, flood-prone land. The paper of Malik et al. (pp. 499–508) shows that hybridisation of wheat with *Hordeum marinum*, a waterlogging-tolerant wild relative, improves waterlogging tolerance of wheat. These amphiploids, in contrast to wheat, have more dry mass, a higher porosity in adventitious roots and develop a barrier to prevent radial $O_2$ loss of roots upon waterlogging.

The growing understanding of the molecular basis and genetic diversity in submergence and flooding acclimations provides opportunities to breed and engineer crops tolerant of these conditions that would benefit the world's farmers. The evaluation of diversity exposes plasticity in metabolic and developmental acclimations that enable distinct strategies that increase fitness in a flooded environment. Natural variation in acclimation schemes provides opportunities for development of crops with combinations of submergence tolerance traits that are optimal at specific developmental stages and under particular flooding regimes, which vary substantially worldwide. The first example of this is the use of marker-assisted breeding to introduce the submergence tolerance conferring Sub1 genotype to selected rice cultivars (Xu et al. 2006), which may appreciably benefit rice production in flood-prone lands in the Third World. The further exploration of the molecular basis of genetic diversity in flooding tolerances is critical given the global climate change scenarios that predict heavy precipitation in regions of our planet.

## Conclusion

Hypoxia or anoxia, consequences of waterlogging, results in energy crisis in waterlogged cells which triggers the tolerant genotypes to develop different traits associated with waterlogging tolerance at least to survive under the stress. The study of regulatory mechanisms and signalling events responsible for triggering responses to hypoxia or anoxia in wheat plants is a prospective area of research. A number of questions remain unanswered about the response of individual cells, viz. the basis for the differential response between waterlogging-tolerant and intolerant genotypes (do these genotypes differ in the responses in respect to cellular signalling and response mechanisms), signalling transduction pathways involved to control adaptive responses, communication of roots and shoot cells over a long distance during energy crisis, understanding the cell-to-cell and long-distance, signalling mechanisms which determine the organ and whole-plant response to hypoxia or anoxia, aerenchyma formation and adventitious root growth is another interesting area for research. Examination of regulatory mechanisms and signalling events responsible for triggering responses to oxygen-deficient conditions in plants is also an interesting area of research. So far we only know a part of the unfolding story, with many more questions still unanswered. Answering these questions will be of relevance to agriculture and will provide knowledge of the fundamental nature of anaerobic life.

Advances in genome biology, genetic resources and high-throughput technologies provide excellent resources for the exploration of oxygen-sensing mechanisms in plant cells. It is imperative to identify sensors and dissect the signalling pathways that occur at the cellular, tissue, organ and whole-plant level. So far we only know a little part of the unfolded story, with many more phenomena still unknown. Exploring these phenomena will be of relevance to waterlogging tolerance of crops and will provide knowledge of the fundamental nature of the crops under anaerobiosis.

## References

Akhtar MS, Memon M (2009) Biomass and nutrient uptake by rice and wheat: a three-way interaction of potassium, ammonium and soil type. Pak J Bot 41 (6):2965–2974

Albrecht G, Wiedenroth EM (1994) Protection against activated oxygen following re-aeration of hypoxically pre-treated wheat roots. The response of the glutathione system. J Exp Bot 45:449–455

Albrecht G, Kammerer S, Praanik W, Wiedenroth EM (1993) Fructan content of wheat seedling (*Triticum aestivum* L.) under hypoxia and following re-aeration. New Phytol 123:471–476

Arikado H (1959) Supplementary studies on the development of the ventilating systems in various plants growing on low land and on upland. Bull Fac Agric Mie Univ 20:1–24

Armstrong W (1971) Radial oxygen losses from intact rice roots as affected by distance from the apex, respiration and waterlogging. Physiol Plant 25:192–197

Armstrong W (1979) Aeration in higher plants. Adv Bot Res 7:225–332

Armstrong J, Armstrong W (1988) Phragmites australis: a preliminary study of soil oxidizing sites and internal gas transport pathways. New Phytol 108:373–382

Armstrong W, Drew M (2002) Root growth and metabolism under oxygen deficiency. In: Waisel EA, Kafkafi Y (eds) Plant roots: the hidden half. Marcel Dekker, New York, pp 729–761

Armstrong W, Webb T (1985) A critical oxygen pressure for root extension in rice. J Exp Bot 36:1573–1582

Armstrong W, Brändle R, Jackson MB (1994) Mechanisms of flood tolerance in plants. Acta Bot Neerl 43:307–358

Bailey-Serres J, Voesenek LACJ (2008) Flooding stress: acclimations and genetic diversity. Ann Rev Plant Biol 59:313–339. doi:10.1146/annurev.arplant.59.032607.092752

Bange MP, Milroy SP, Thongbai P (2004) Growth and yield of cotton in response to water logging. Field Crops Res 88:129–142. doi:10.1016/j.fcr.2003.12.002

Barrett-Lennard EG (1986) Effects of waterlogging on the growth and NaCl uptake by vascular plants under saline conditions. Reclam Reveg Res 5:245–261

Barrett-Lennard ED, Robson AD, Greenway H (1982) Effect of phosphorus deficiency and water deficit on phosphatase activities from wheat leaves. J Exp Bot 33:682–693

Barrett-Lennard EG, Leighton P, Buwalda F, Gibbs J, Armstrong W, Thomson CJ, Greenway H (1988) Effects of growing wheat in hypoxic nutrient solutions and of subsequent transfer to aerated solutions I. Growth and carbohydrate status of shoots and roots. Aust J Plant Physiol 15:585–598

Belford RK (1981) Response of winter wheat to prolonged water logging under outdoor conditions. J Agric Sci Camb 97:557–568

Benjamin LR, Greenway H (1979) Effects of a range of $O_2$ concentrations on the porosity of barley roots and on their sugar and protein concentrations. Ann Bot 43:383–391

Biemelt S, Keetman U, Albrecht G (1998) Re-aeration following hypoxia or anoxia leads to activation of the antioxidative defense system in roots of wheat seedlings. Plant Physiol 116:651–658

Biemelt S, Keetman U, Mock HP, Grimm B (2000) Expression and activity of isoenzymes of superoxide dismutase in wheat roots in response to hypoxia and anoxia. Plant Cell Environ 23:135–144

Blokhina OB, Fagerstedt KV, Chirkova TV (1999) Relationships between lipid peroxidation and anoxia tolerance in a range of species during post-anoxic reaeration. Physiol Plant 105:625–632

Blokhina O, Virolainen E, Fagerstedt KV (2003) Antioxidants, oxidative damage and oxygen deprivation stress: a review. Ann Bot 91:179–194. doi:10.1093/aob/mcf118

Blom CWPM (1999) Adaptations to flooding stress: from plant community to molecule. Plant Biol 1:261–273

Blom CWPM, Voesenek LACJ (1996) Flooding: the survivals strategies of plants. Trends Ecol Evol 11:290–295

Boivin P, Favre F, Hammecker C, Maeght JL, Delarivière J, Poussin JC, Wopereis MCS (2002) Processes driving soil solution chemistry in a flooded rice-cropped vertisol: analysis of long-time monitoring data. Geoderma 110:87–107

Boru G, Vantoai T, Alves J, Hua D, Knee M (2003a) Responses of soybean to oxygen deficiency and elevated root-zone carbon dioxide concentration. Ann Bot 91:447–453. doi:10.1093/aob/mcg040

Boru G, van Ginkel M, Trethowan RM, Boersma L, Kronstad WE (2003b) Oxygen use from solution by wheat genotypes differing in tolerance to waterlogging. Euphytica 132:151–158

Botha AM, Botha FC (1991) Effect of anoxia on the expression and molecular form of the pyrophosphate-dependent phosphofructokinase. Plant Cell Physiol 32:1299–1302

Bouny JM, Saglio PH (1996) Glycolytic flux and hexokinase activities in anoxic maize root tips acclimated by hypoxic pretreatment. Plant Physiol 111:187–194

Bowler C, Montagu MV, Inze D (1992) Superoxide dismutase and stress tolerance. Annu Rev Plant Physiol Plant Mol Biol 43:83–116

Bradford KJ, Hsiao TC (1982) Stomatal behaviour and water relations of waterlogged tomato plants. Plant Physiol 70:1508–1513

Bradford KJND, Yang SF (1980) Xylem transport of 1-aminocyclopropane-1-carboxylic acid, an ethylene precursor, in waterlogged tomato plants. Plant Physiol 65:322–326

Bucher M, Kuhlemeier C (1993) Long-term anoxia tolerance. Plant Physiol 103:441–448

Buwalda F, Barrett-Lennard EG, Greenway H, Davies BA (1988a) Effects of growing wheat in hypoxic nutrient solutions and of subsequent transfer to aerated solutions. II. Concentrations and uptake of nutrients and sodium in shoots and roots. Aust J Plant Physiol 15:599–612

Buwalda F, Thomson CJ, Steigner W, Barrett-Lennard EG, Gibbs J, Greenway H (1988b) Hypoxia induces membrane depolarization and potassium loss from wheat roots but does not increase their permeability to sorbitol. J Exp Bot 39:1169–1183

Carystinos GD, MacDonald HR, Monroy AF, Dhinsda RS, Poole RJ (1995) Vacuolar H+-translocating pyrophosphatase is induced by anoxia or chilling in seedlings of rice. Plant Physiol 108:641–649

Chirkova TV, Novitskaya LO, Blokhina OB (1998) Lipid peroxidation and antioxidant systems under anoxia in plants differing in their tolerance to oxygen deficiency. Russ J Plant Physiol 45:55–62

Cohen E, Kende H (1987) In vivo 1-aminocyclopropane-1-carboxylate synthase activity in internodes of deep water rice: Enhancement by submergence and low oxygen levels. Plant Physiol 84:282–286

Colmer TD (2003) Long-distance transport of gases in plants: a perspective on internal aeration and radial oxygen loss from roots. Plant Cell Environ 26:17–36

Colmer TD, Gibberd MR, Wiengweera A, Tinh TK (1998) The barrier to radial oxygen loss from roots of rice (*Oryza sativa* L.) is induced by growth in stagnant solution. J Exp Bot 49:1431–1436

Colmer TD, Peeters AJM, Wagemaker CAM, Vriezen WH, Ammerlaan A, Voesenek LACJ (2004) Expression of α-expansin genes during root acclimations to O2 deficiency in Rumex palustris. Plant Mol Biol 56:423–437

Conaty WC, Tan DKY, Constable GA, Sutton BG, Field DJ, Mamum EA (2008) Genetic variation for waterlogging tolerance in cotton. J Cotton Sci 12:53–61

Crawford RMM (1982) Physiological responses to flooding. In: Lange OL, Nobel PS, Osmond CB, Ziegler H (eds) Encyclopedia of plant physiology, new series, vol 12B, Physiological plant ecology II. Springer, Berlin, pp 453–477

Crawford RMM (1992) Oxygen availability as an ecological limit to plant distribution. Adv Ecol Res 23:93–185

Crawford RMM, Brändle R (1996) Oxygen deprivation stress in a changing environment. J Exp Bot 47:145–159. doi:10.1093/jxb/47.2.145

Dat J, Capelli N, Folzer H, Bourgeade P, Badot PM (2004) Sensing and signaling during plant flooding. Plant Physiol Biochem 42:273–282

Davies DD (1980) Anaerobic metabolism and the production of organic acids. In: Davies DD (ed) The biochemistry of plants, vol 2. Academic, New York, pp 581–611

Dennis ES, Dolferus R, Ellis M, Rahman M, Wu Y, Hoeren FU, Grover A, Ismond KP, Good AG, Peacock WJ (2000) Molecular strategies for improving waterlogging tolerance in plants. J Exp Bot 51:89–97

Drew MC (1983) Plant injury and adaptation to oxygen deficiency in the root environment: a review. Plant Soil 75:179–199

Drew MC (1988) Effects of flooding and oxygen deficiency on plant mineral nutrition. Adv Plant Nutr 3:115–159

Drew MC (1990) Sensing soil oxygen. Plant Cell Environ 13:681–693

Drew MC (1991) Oxygen deficiency in the root environment and plant mineral nutrition. In: Jackson MB, Davies DD, Lambers H (eds) Plant life under oxygen deprivation. Academic, The Hague, pp 303–316

Drew M (1997) Oxygen deficiency and root metabolism: injury and acclimation under hypoxia and anoxia. Ann Rev Plant Physiol Plant Mol Biol 48:223–250

Drew MC, Sisworo EJ (1979) The development of waterlogging damage in young barley plants in relation to plant nutrient status and changes in soil properties. New Phytol 82:301–314. doi:10.1111/j.1469-8137.1979.tb02656.x

Drew MC, Jackson MB, Giffard S (1979) Ethylene-promoted adventitious rooting and development of cortical air spaces (aerenchyma) in roots may be adaptive responses to flooding in *Zea mays* L. Planta 147:83–88

Drew MC, Jackson MB, Giffard SC, Campbell R (1981) Inhibition by silver ions of gas space (aerenchyma) formation in adventitious roots of Zea mays L. subjected to exogenous ethylene or to oxygen deficiency. Planta 153:217–224

Drew MC, Saglio PH, Pradet A (1985) Larger adenylate energy charge and ATP/ADP ratios in aerenchymatous roots of *Zea mays* in anaerobic media as a consequence of improved internal oxygen transport. Planta 165:51–58

Elliot GC, Lauchli A (1986) Evaluation of an acid phosphatase assay for detection of phosphorus deficiency in leaves of maize (*Zea mays* L.). J Plant Nutr 9:1469–1477

Else M, Hall K, Arnold G, Davies W, Jackson M (1995) Export of abscisic acid, 1-aminocyclopropane-1-carboxilyc acid, phosphate, and nitrate from roots to shoots of flooded tomato plants. Plant Physiol 107:377–384

Else MA, Coupland D, Dutton L, Jackson MB (2001) Decreased root hydraulic conductivity reduces leaf water potential, initiates stomatal closure and slows leaf expansion in flooded plants of castor oil (*Ricinus communis*) despite diminished delivery of ABA from the roots to shoots in xylem sap. Physiol Plant 111:46–54

Elstner EF (1986) Metabolism of activated oxygen species. In: Davies DD (ed) The biochemistry of plants, vol 11, Biochemistry of metabolism. Academic, San Diego, pp 253–315

Erdmann B, Wiedenroth EM (1986) Changes in the root system of wheat seedlings following root anaerobiosis. II. Morphology and anatomy of evolution forms. Ann Bot 58:607–616

Evans DE (2003) Aerenchyma formation. New Phytol 161:35–49

FAO (2002) Agriculture. Available online at: http://www.fao.org/waicent/FAOINFO/AGRICULT/ag1/ag11/gaez//nav.html. Accessed 18 Mar 2009

Foyer CH, Souriau N, Perret S, Lelandais M, Kunert KJ, Pruvost C, Jounanin L (1995) Overexpression of glutathione reductase but not glutathione synthetase leads to increases in antioxidant capacity and resistance to photoinhibition in poplar trees. Plant Physiol 109:1047–1057

Foyer CH, Looez-Delgado H, Dat JF, Scott IM (1997) Hydrogen peroxide and glutathione-associated mechanisms of acclamatory stress tolerance and signaling. Physiol Plant 100:241–254

Furlani AMC, Clark RB, Maranville JW, Ross WM (1984) Root phosphatase activity of sorghum genotypes grown with organic and inorganic sources of phosphorus. J Plant Nutr 7:1583–1595

Garnczarska M (2005) Response of the ascorbate—glutathione cycle to reaeration following hypoxia in lupine roots. Plant Physiol Biochem 43:583–590

Geigenberger P (2003) Response of plant metabolism to too little oxygen. Curr Opin Plant Biol 6:247–256

Germain V, Ricard B, Raymond P, Saglio PH (1997) The role of sugars, hexokinase and sucrose synthase in the determination of hypoxically induced tolerance to anoxia in tomato roots. Plant Physiol 114:167–175

Gibberd MR, Cocks PS (1997) Effect of waterlogging and soil pH on the micro-distribution of naturalized annual legumes. Aust J Agr Res 48:223–229

Gibberd MR, Gray JD, Cocks PS, Colmer TD (2001) Waterlogging tolerance among a diverse range of Trifolium accessions is related to root porosity, lateral root formation and 'aerotropic rooting'. Ann Bot 88:579–589

Gibbs J, Greenway H (2003) Mechanisms of anoxia tolerance in plants. I. Growth, survival and anaerobic catabolism. Funct Plant Biol 30:1–47

Greenway H, Gibbs J (2003) Mechanisms of anoxia tolerance in plants II. Energy requirements for maintenance and energy distribution to essential processes. Funct Plant Biol 30(10):999–1036

Grist DH (1986) Rice, 6th edn. Longman Group Ltd, New York

Gupta A, Webb RP, Holaday AS, Allen RD (1993) Overexpression of superoxide dismutase protects plants from oxidative stress: induction of ascorbate peroxidase in superoxide dismutase overexpressing plants. Plant Physiol 103:1067–1073

Gutierrez Boem FH, Lavado RS, Porcelli CA (1996) Note on the effects of winter and spring waterlogging on growth, chemical composition and yield of rapeseed. Field Crops Res 47:175–179. doi:10.1016/ 0378-4290 (96)00025-1

Haldemann C, Brändle R (1983) Avoidance of oxygen deficit stress and release of oxygen by stalked rhizomes of *Schoenoplectus lacustris*. Physiol Veg 21:109–113

Haque ME, Abe F, Kawaguchi K (2010) Formation and extension of lysigenous aerenchyma in seminal root cortex of spring wheat (*Triticum aestivum* cv. Bobwhite line SH 9826) seedlings under different strength of waterlogging. Plant Root 4:31–39

Helal HM (1990) Varietal differences in root phosphatase activity as related to the utilization of organic phosphates. Plant Soil 123:161–163

Hocking PJ, Reicosky DC, Meyer WS (1985) Nitrogen status of cotton subjected to two short term periods of waterlogging of varying severity using a sloping plot water table facility. Plant Soil 87:375–391. doi:10.1007/BF02181905

Hocking PJ, Reicosky DC, Meyer WS (1987) Effects of intermittent waterlogging on the mineral nutrition of cotton. Plant Soil 101:211–221. doi:10.1007/ BF02370647

Hodgson AS (1990) Micronutrients: are they important under waterlogging? In: Swallow D (ed) Fifth Australian cotton conference. ACGRA, Broadbeach, pp 165–170

Huang B (1997) Mechanisms of plant resistance to waterlogging. In: Basra AS, Basra RK (eds) Mechanisms of environmental stress resistance in plants. Harwood Academic Publishers, Amsterdam, pp 59–82

Huang B, Johnson JW (1995) Root respiration and carbohydrate status of two wheat genotypes in response to hypoxia. Ann Bot 75:427–432

Huang B, Johnson JW, Nesmith DS, Bridges DC (1994a) Root and shoot growth of wheat genotypes in response to hypoxia and subsequent resumption of aeration. Crop Sci 34:1538–1544

Huang B, Johnson JW, Nesmith S, Bridges DC (1994b) Growth, physiological and anatomical responses of two wheat genotypes to waterlogging and nutrient supply. J Exp Bot 45:193–202

Huang B, Ne Smith DS, Bridges DC, Johnson JW (1995a) Responses of squash to salinity, waterlogging, and subsequent drainage: I. Gas exchange, water relations, and nitrogen status. J Plant Nutr 18:127–140

Huang B, NeSmith DS, Bridges DC, Johnson JW (1995b) Responses of squash to salinity, waterlogging, and subsequent drainage: II. Root and shoot growth. J Plant Nutr 18:141–152

Huang B, Johnson JW, Box JE, Nesmith S (1997) Root characteristics and physiological activities of wheat in response to hypoxia and ethylene. Crop Sci 37:812–818

Huck MG (1970) Variation in taproot elongation rate as influenced by composition of the soil air. Agron J 62:815–818

Igamberdiev AU, Baron K, Manac'h-Little N, Stoimenova M, Hill RD (2005) The haemoglobin/nitric oxide cycle: involvement in flooding stress and effects on hormone signalling. Ann Bot 96:557–564

Jackson MB (1982) Ethylene as a growth promoting hormone under flooded conditions. In: Wareing PF (ed) Plant growth substances. Academic, London, pp 291–301

Jackson MB (1989) Regulation of aerenchyma formation in roots and shoots by oxygen and ethylene. In: Osborn DJ, Jackson MB (eds) Cell separation in plants. Springer, Berlin, pp 263–274

Jackson M (2002) Long-distance signaling from roots to shoots assessed: the flooding story. J Exp Bot 53:175–181

Jackson MB (2004) The impact of flooding stress on plants and crops. http://www.plantstress.com/Articles/index.asp. A website addressing plant environmental stress issues in agriculture, plant physiology, breeding, genetics and biotechnology

Jackson MB, Armstrong W (1999) Formation of aerenchyma and the processes of plant ventilation in relation to soil flooding and submergence. Plant Biol 1:274–287

Jackson MB, Colmer TD (2005) Response and adaptation by plants to flood- ing stress. Ann Bot 96:501–505

Jackson MB, Drew MC (1984) Effects of flooding on growth and metabolism of herbaceous plants. In: Kozlowski TT (ed) Flooding and plant growth. Academic, New York, pp 47–128

Jackson MB, Hall KC (1987) Early stomatal closure in water logged pea plants is mediated by abscisic acid in the absence of foliar water deficits. Plant Cell Environ 10:121–130

Jackson MB, Ram PC (2003) Physiological and molecular basis of susceptibility and tolerance of rice plants to complete submergence. Ann Bot 91:227–241. doi:10.1093/aob/mcf242

Justin SHFW, Armstrong W (1987) The anatomical characteristics of roots and plant response to soil flooding. New Phytol 106:465–495

Justin SHFW, Armstrong W (1991) Evidence for the involvement of ethene in aerenchyma formation in adventitious roots of rice (*Oryza sativa*). New Phytol 118:49–62

Kalashnikov YE, Balakhnina TI, Zakrzhevsky DA (1994) Effect of soil hypoxia on activation of oxygen and the system of protection from oxidative destruction in roots and leaves of *Hordeum vulgare*. Russ J Plant Physiol 41:583–588

Kennedy RA, Rumpho ME, Fox TC (1992) Anaerobic metabolism in plants. Plant Physiol 100:1–6

Kirk GJD, Solivas JL, Alberto MC (2003) Effects of flooding and redox conditions on solute diffusion in soil. Eur J Soil Sci 54:617–624

Kozlowski TT (1984) Plant responses to flooding of soil. Bioscience 34(3):162–167. doi:10.2307/1309751

Kozlowski TT (1997) Response of woody plants to flooding and salinity. Tree Physiol Monogr 1:1–29

Kramer P, Boyer J (1995) Water relations of plants and soils. Academic, San Diego

Kriedemann PE, Sands R (1984) Salt resistance and adaptation to root-zone hypoxia in sunflower. Aust J Plant Physiol 11:287–301

Krizek DT (1982) Plant response to atmospheric stress caused by waterlogging. In: Christiansen MN, Lewis CF (eds) Breeding plants for less favorable environments. Wiley, New York, pp 293–335

Kyozuka J, Olive MR, Peacock WJ, Dennis ES, Shimamoto K (1994) Promoter elements required for the developmental expression of the maize Adhl gene in transgenic rice. Plant Cell 6:799–810

Lambers H (1976) Respiration and NADH oxidation of the root of flood- intolerant Senecio species as affected by anaerobiosis. Plant Physiol 37:117–122

Liao CT, Lin CH (2001) Physiological adaptation of crop plants to flooding stress. Proc Natl Sci Counc ROC (B) 25:148–157

Lu Y, Watanabe A, Kimura M (2004) Contribution of plant photosynthates to dissolved organic carbon in a flooded rice soil. Biogeochemistry 71:1–15

Marschner H (1986) Mineral nutrition of higher plants. Academic, London

Marshall DR, Broue P, Pryor AJ (1973) Adaptive significance of alcohol dehydrogenase isozymes in maize. Nature 244:16–17

McDonald MP, Galwey NW, Colmer TD (2001a) Waterlogging tolerance in the tribe Triticeae: the adventitious roots of *Critesion marinum* have a relatively high porosity and a barrier to radial oxygen loss. Plant Cell Environ 24:585–596

McDonald MP, Galwey NW, Ellneskog-Staam P, Colmer TD (2001b) Evaluation of *Lophopyrum elongatum* as a source of genetic diversity to increase waterlogging tolerance of hexaploid wheat (*Triticum aestivum*). New Phytol 151:369–380

McKee WH, McKevlin MR (1993) Geochemical processes and nutrient uptake by plants in hydric soils. Environ Toxicol Chem 12:2197–2207

Mclachlan KD, Demarco DG (1982) Acid phosphatase activity of intact roots and phosphorus nutrition in plants: III. Its relation to phosphorus garnering by wheat and a comparison with leaf activity as a measure of phosphorus status. Aust J Agric Res 33:1–11

Meek BD, Owen-Bartlett EC, Stolzy LH, Labanauskas CK (1980) Cotton yield and nutrient uptake in relation to water table depth. Soil Sci Soc Am J 44:301–305. doi:10.2136/sssaj1980.03615995004400020020x

Menegus F, Cattaruzza L, Chersi A, Fronza G (1989) Differences in the anaerobic lactate- succinate production and in the changes of cell sap pH for plants with high and low resistance to anoxia. Plant Physiol 90:29–32

Menegus F, Cattaruzza L, Mattana M, Beffagna N, Ragg E (1991) Response to anoxia in rice and wheat seedlings. Changes in the pH of intracellular compartments, glucose-6- phosphate level, and metabolic rate. Plant Physiol 95:760–767

Mengel K, Kirkby EA, Kosegarten H (2001) Principles of plant nutrition. Kluwer Academic Publishers, Dordrecht/Nethelands, pp 849. ISBN 9781402000089

Mertens E (1991) Pyrophosphate-dependent phosphofructokinase, an anaerobic glycolytic enzyme? FEBS Lett 285:1–5

Mishra PN, Fatma T, Singhal GS (1995) Development of antioxidative defence system of wheat seedlings in response to high light. Physiol Plant 95:72–82

Mittler R, Vanderauwera S, Gollery M, Breusegem FV (2004) Reactive oxygen gene network of plants. Trends Plant Sci 9:490–498

Monk LS, Fagerstedt KV, Crawford RMM (1987) Superoxide dismutase as an anaerobic polypeptide. A key factor in recovery from oxygen deprivation in Iris pseudacorus? Plant Physiol 85:1016–1020

Monk LS, Fagerstedt KV, Crawford RMM (1989) Oxygen toxicity and superoxide dismutase as an antioxidant in physiological stress. Physiol Plant 76 (3):456–459

Morard E, Silvestre J (1996) Plant injury due to oxygen deficiency in the root environment of soilless culture: a review. Plant Soil 184:243–254

Mortvedt JJ, Cox FR, Shuman LM, Weich RM (eds) (1991) Micronutrients in agriculture. Soil Science Society of America, Madison

Mustroph A, Albrecht G (2003) Tolerance of crop plants to oxygen deficiency stress: fermentative activity and photosynthetic capacity of entire seedlings under hypoxia and anoxia. Physiol Plant 117:508–520

Narayanan S, Ruma D, Gitika B, Sharma SK, Pauline T, Ram MS, Ilavazhagan G, Sawhney RC, Kumarand D, Banerjee PK (2005) Antioxidant activities of sea buckthorn (*Hippophae rhamnoides*) during hypoxia induced oxidative stress in glial cells. Mol Cell Biochem 278:9–14

Nathanson K, Lawn RJ, de Jarun PLM, Byth DE (1984) Growth, nodulation and nitrogen accumulation by soybean in saturated soil culture. Field Crops Res 8:73–92. doi:10.1016/0378-4290(84)90053-4

Orchard PW, So HB (1985) The response of sorghum and sunflower to short-term waterlogging .2. Changes in the soil environment under waterlogged conditions. Plant Soil 88:407–419

Pang JY, Zhou MX, Mendham N, Shabala S (2004) Growth and physiological responses of six barley genotypes to waterlogging and subsequent recovery. Aust J Agr Res 55:895–906. doi:10.1071/AR03097

Perata P, Alpi A (1993) Plant response to anaerobiosis. Plant Sci 93:1–7

Pezeshki SR (2001) Wetland plant responses to soil flooding. Environ Exp Bot 46:299–312

Pezeshki SR, DeLaune RD (1998) Responses of seedlings of selected woody species to soil oxidation-reduction conditions. Environ Exp Bot 40:123–133

Ponnamperuma FN (1984) Effects of flooding on soils. In: Kozlowski TT (ed) Flooding and plant growth. Academic, Orlando, pp 9–45

Probert ME, Keating BA (2000) What soil constraints should be included in crop and forest models? Agric Ecosyst Environ 82:273–281

Reicosky DC, Meyer WS, Schaefer NL, Sides RD (1985) Cotton response to short-term waterlogging imposed with a water-table gradient facility. Agric Water Manag 10:127–143. doi:10.1016/0378-3774. (85) 90002-2

Ricard B, Couee I, Raymond P, Saglio PH, Saint-Ges V, Pradet A (1994) Plant metabolism under hypoxia and anoxia. Plant Physiol Biochem 32:1–10

Roberts JKM, Callis J, Wemmer D, Walbot V, Jardetzky O (1984) Mechanism of cytoplasmic pH regulation in hypoxic maize root tips and its role in survival under hypoxia. Proc Natl Acad Sci U S A 81:3379–3383

Sairam RK, Srivastava GC, Saxena DC (2000) Increased antioxidant activity under elevated temperatures: a mechanism of heat stress tolerance in wheat genotypes. Biol Plant 43:245–251

Sairam RK, Chandrasekhar V, Srivastava GC (2001) Comparison of hexaploid and tetraploid wheat cultivars in their response to water stress. Biol Plant 44:89–94

Sairam RK, Rao KV, Srivastava GC (2002) Differential response of wheat genotypes to long term salinity stress in relation to oxidative stress, antioxidant activity and osmolyte concentration. Plant Sci 163:1037–1046

Sairam RK, Kumutha D, Ezhilmathi K, Deshmukh PS, Srivastava GC (2008) Physiology and biochemistry of waterlogging tolerance in plants. Biol Plant 52:401–412

Sairam RK, Kumutha D, Chinnusamy V, Meena RC (2009) Waterlogging- induced increase in sugar mobilization, fermentation, and related gene expression in the roots of mung bean (Vigna radiata). J Plant Physiol 166:602–616

Scott P (2000) Resurrection plants and the secrets of eternal leaf. Ann Bot 85:159–166

Setter TL, Kupkanchanakul T, Pakinnaka L, Aguru Y, Greenway H (1987) Mineral nutrients in floodwater and floating rice growing at water depths up to two metres. Plant Soil 104:147–150. doi:10.1007/BF02370637

Setter TL, Ellis M, Laureles EV, Ella ES, Senadhira D, Mishra SB, Sarkarung S, Datta S (1997) Physiology and genetics of submergence tolerance in rice. Ann Bot 79(suppl A):67–77

Settler TL, Waters I (2003) Reviews of prospects for germplasm improvement for waterlogging tolerance in wheat, barley and oats. Plant Soil 253:1–34

Shuduan T, Zhu MY, Zhang KR, Zhang QF (2010) Physiological responses of bermudagrass (*Cynodon dactylon*) to submergence. Acta Physiologiae Plant 32 (1):133–140. doi:10.1007/s11738-009-0388-y

Siangliw M, Toojinda T, Tragoonrung S, Vanavichit A (2003) Thai jasmine rice carrying QTLch9 (SubQTL) is submergence tolerant. Ann Bot 91:255–261

Smirnoff N, Crawford RMM (1983) Variation in the structure and response to flooding of root aerenchyma in some wetland plants. Ann Bot 51:237–249

Smith KA, Russell RS (1969) Occurrence of ethylene, and its significance, in anaerobic soil. Nature 222:769–771

Subbaiah CC, Sachs MM (2003) Molecular and cellular adaptations of maize to flooding stress. Ann Bot 90:119–127

Thomson CJ, Armstrong W, Waters I, Greenway H (1990) Aerenchyma formation and associated oxygen movement in seminal and nodal roots of wheat. Plant Cell Environ 13:395–403

Thomson CJ, Colmer TD, Watkin ELJ, Greenway H (1992) Tolerance of wheat (*Triticum aestivum* cv. Gamenya and Kite) and triticale (*Triticosecale* cv. Muir) to waterlogging. New Phytol 120:335–344

Toojinda T, Siangliw M, Tragoonrung S, Vanavichit A (2003) Molecular genetics of submergence tolerance in rice: QTL analysis of key traits. Ann Bot 91:243–253

Trought MCT, Drew MC (1980a) The development of waterlogging damage in wheat seedlings (*Triticum aestivum* L.) I. Shoot and root growth in relation to changes in concentration of dissolved gases and solutes in the soil solutions. Plant Soil 54:77–94

Trought MCT, Drew MC (1980b) The development of waterlogging damage in wheat seedlings (Triticum aestivum L.) II. Accumulation and redistribution of nutrients by the shoot. Plant Soil 56(2):187–199

Trought MCT, Drew MC (1980c) The development of waterlogging damage in young wheat plants in anaerobic solution cultures. J Exp Bot 31:1573–1585

Trought MCT, Drew MC (1982) Effects of waterlogging on young wheat plants (*Triticum aestivum* L.) and on soil solutes at different temperatures. Plant Soil 69:311–326

Umeda M, Uchimiya H (1994) Differential transcript levels of genes associated with glycolysis and alcohol fermentation in rice plants (*Oryza sativa* L.) under submergence stress. Plant Physiol 106:1015–1022

Ushimaru T, Maki Y, Sano S, Koshiba K, Asada K, Tsuji H (1997) Induction of enzymes involved in the ascorbate-dependent antioxidative system, namely ascorbate peroxidase, mono dehydroascorbate reductase and dehydroascorbate reductase, after exposure to air of rice (*Oryza sativa*) seedlings germinated under water. Plant Cell Physiol 38:541–549

Van der Sman AJM, Blom CWPM, van de Steeg HM (1992) Phenology and seed production in *Chenopodium rubrum*, *Rumex maritimus* and *Rumex palustris* as related to photoperiod in river forelands. Can J Bot 70:392–400

Van Toai TT, Bolles CS (1991) Postanoxic injury in soybean (Glycine max) seedlings. Plant Physiol 97:588–592

Vartapetian BB, Jackson M (1997) Plant adaptations to anaerobic stress. Ann Bot 79:3–20

Visser EJW, Voesenek LACJ (2004) Acclimation to soil flooding – sensing and signal – transduction. Plant Soil 254:197–214

Visser EJW, Nabben RHM, Blom CWPM, Voesenek LACJ (1997) Elongation by primary lateral roots and adventitious roots during conditions of hypoxia and high ethylene concentrations. Plant Cell Environ 20:647–653

Visser EJW, Voesenek LACJ, Vartapetian BB, Jackson MB (2003) Flooding and plant growth. Ann Bot 91:107–109. doi:10.1093/aob/mcg014

Voesenek LACJ, Jackson MB, Toebes AHW, Vriezen WH, Colmer TD (2003) Desubmergence-induced ethylene production in Rumex palustris: regulation and ecophysiological significance. Plant J 33:341–352

Waters I, Kuiper PJC, Watkin E, Greenway H (1991a) Effects of anoxia on wheat seedlings I. Interaction between anoxia and other environmental factors. J Exp Bot 42:1427–1435

Waters I, Morrell S, Greenway H, Colmer TD (1991b) Effects of anoxia on wheat seedlings II. Effects of $O_2$ supply prior to anoxia on tolerance to anoxia, alcoholic fermentation and sugars. J Exp Bot 42:1437–1447

Watkin ELJ, Thomson CJ, Greenway H (1998) Root development in two wheat cultivars and one triticale cultivar grown in stagnant agar and aerated nutrient solution. Ann Bot 81:349–354

Webb T, Armstrong W (1983) The effects of anoxia and carbohydrates on the growth and viability of rice, pea and pumpkin roots. J Exp Bot 34:579–603

Wesseling J, van Wijk WR (1957) In: Luthin JN (ed) Drainage in agricultural lands. American Society for Agronomy, Madison, pp 461–504

Xia JH, Saglio PH (1992) Lactic acid efflux as a mechanism of hypoxic acclimation of maize root tips to anoxia. Plant Physiol 100:40–46

Xu K, Xu X, Fukao T, Canlas P, Maghirang-Rodriguez R, Heuer S, Ismail AM, Bailey-Serres J, Ronald PC, Mackill DJ (2006) Sub1 A is an ethylene-response-factor-like gene that confers submergence tolerance to rice. Nature 442:705–708

Yan B, Dai Q, Liu X, Huang S, Wang Z (1996) Flooding induced membrane damage, lipid oxidation and activated oxygen generation in corn leaves. Plant Soil 179:261–268

Zeng Y, Avigne WT, Koch KE (1999) Rapid repression of maize invertase by low oxygen. Invertase / sucrose synthase balance, sugar signaling potential and seedling survival. Plant Physiol 121:599–608

Zhang J, Kirkham MB (1994) Drought- stress-induced changes in activities of superoxide dismutase, catalase, and peroxidase in wheat species. Plant Cell Physiol 35:785–791

Zitomer RS, Lowry CV (1992) Regulation of gene expression by oxygen in *Saccharomyces cerevisiae*. Microbiol Rev 56:1–11

# Stress Tolerance in Plants: A Proteomics Approach

15

Gyana Ranjan Rout and Sunil Kumar Senapati

**Abstract**

Both abiotic and biotic stresses adversely affect the plant growth and productivity including crop plants. The development of stress-tolerant plants will be greatly advantageous for modern agriculture in areas that are prone to such stresses. In recent times, several advances have been made towards identifying potential stress-related genes which are capable of increasing the tolerance of plants to both abiotic and biotic stresses. This interaction between biotic and abiotic stresses is controlled by hormone signaling pathways that may induce or antagonize one another, in particular that of abscisic acid. Specificity in multiple stress responses is further controlled by a range of biochemical and molecular mechanisms that act together in a complex regulatory network. Transcription factors, kinase cascades, and reactive oxygen species are key components of this cross talk as are heat shock factors and small RNAs. This review emphasis on elucidating the proteins associated with abiotic and biotic stresses in plants.

## Introduction

Plants need suitable physiological condition as well as essential nutrients to grow in the natural environment. As a consequence of their sessile growth habit, plants have adapted dynamic responses to these stresses at the physiological, biochemical, and molecular levels, thus enabling them to survive under different environmental conditions. Most of the crop plants grow in environments that are suboptimal, which prevents the plants from attaining their full genetic potential for growth and reproduction (Bray et al. 2000; Rockstrom and Falkenmark 2000). Abiotic stress factors such as heat, cold, drought, salinity, and nutrient stress have a huge impact on world agriculture, and it has been suggested that they reduce average yields by >50% for most major crop plants (Wang et al. 2003). Further to this, plants must defend themselves from attack by a

G.R. Rout (✉)
Department of Agricultural Biotechnology, Orissa University of Agriculture and Technology, Bhubaneswar 751003, Odisha, India
e-mail: grrout@rediffmail.com; grrout2012@gmail.com

S.K. Senapati
Department of Biotechnology, Guru Ghasidas Viswa Vidhalaya, Bilaspur, Chhattisgarh, India
e-mail: sunilsenapati007@gmail.com

G.R. Rout and A.B. Das (eds.), *Molecular Stress Physiology of Plants*,
DOI 10.1007/978-81-322-0807-5_15, © Springer India 2013

vast range of pests and pathogens, including fungi, bacteria, viruses, nematodes, and herbivorous insects (Hammond-Kosack and Jones 2000). Each stress elicits a complex cellular and molecular response system implemented by the plant in order to prevent damage and ensure survival, but often at the detriment of growth and yield (Herms and Mattson 1992). Industrial activity combined with a low conscience of the consequences of environmental pollution during a long period created a worldwide problem of soil, air, and water contamination with various pollutants. Heavy metals are among the most widespread soil contaminants. Abiotic stresses usually cause protein dysfunction. The yield and quality of cereals are severely affected by heat stress in many countries (Treglia et al. 1999). Heat stress affects the grain yield and quality. Drought, soil salinity, and heavy metal significantly affect plant growth, development, and productivity, thus posing a severe threat to agriculture throughout the world. Among abiotic stresses, osmotic stress is one of the most severe, caused by drought, high salinity, and cold stresses in nature. Water is the most wide-ranging difficulty among abiotic stresses for production of crop in the world environment. Strategy is to obtain plants with higher performance under water stress conditions by identifying and modifying the molecular mechanisms that take place when the water availability becomes limiting. Plants, as sessile organisms, rely on proteomic plasticity to remodel themselves during periods of developmental change and to respond to biotic and abiotic stresses. In the last decade, methodological improvements have allowed comparative proteomic investigations of plants under stress which have allowed us to analyze biochemical pathways and the complex response of plants to environmental stimuli (Qureshi et al. 2007). More comprehensive approaches that include quantitative and qualitative analyses of gene expression products are necessary at the transcriptome, proteome, and metabolome levels. Abiotic stresses resulted the cellular dehydration, such as freezing and salt and water stress, often lead to similar changes in plant gene expression and metabolism (Cook et al. 2004; Kreps et al. 2002). The phytohormone abscisic acid (ABA) is produced under abiotic stress such as drought and high salinity. ABA is a key mediator in controlling plant response to abiotic stress by regulating stomatal closure and by triggering the activation of many stress-related genes, thereby increasing the tolerance of plants to the stresses. Many abiotic stress-responsive genes have been identified in plants, including rice and *Arabidopsis*, by using molecular techniques (Fowler and Thomashow 2002; Rabbani et al. 2003; Yamaguchi-Shinozaki and Shinozaki 2006; Nakashima et al. 2009). Stress-induced genes not only function to protect cells from abiotic stress through the production of important enzymes and metabolic proteins (functional proteins) but they also regulate signal transduction and gene expression in the stress response (regulatory proteins). Functional proteins contain hydrophilic proteins including dehydrins and "late embryogenesis abundant" (LEA) proteins and also enzymes that are required for the synthesis of osmoprotectants such as proline and sugars. Regulatory proteins that are activated in response to abiotic stresses, including transcription factors (TFs) such as DREBs (dehydration-responsive element-binding proteins), AREBs (ABA-responsive element-binding proteins), and NAC proteins, have been identified in *Arabidopsis* and rice (Yamaguchi-Shinozaki and Shinozaki 2006; Nakashima et al. 2009, 2012). Proteomics also makes an essential bridge between the transcriptome and metabolome (Wang et al. 2004; Gray and Heath 2005), complementing genomics research. Only by grouping all this information together is it possible to achieve a comprehensive and exhaustive analysis of the mechanism of plant defense against abiotic and biotic stresses. Upon several stress responses, protein, protein–protein interaction, and posttranslation modification have been also identified (Salekdeh et al. 2002b). Plants have evolved to live in environments where they are often exposed to different stress factors in combination. Being sessile, they have developed specific mechanisms that allow them to detect precise environmental changes and respond to complex stress conditions, minimizing damage while conserving valuable resources for growth and reproduction. Plants activate a specific and unique stress response when subjected to a combination of multiple stresses (Rizhsky et al. 2002). Keeping in the view of the above fact, the present review highlighted the abiotic and biotic stresses in plants in proteomic approach.

## Abiotic Stress in Plants

Abiotic stresses limit the productivity and growth potential of plants. Phytohormone like abscisic acid (ABA) is a key mediator in controlling plant response to abiotic stress by triggering the activation of many stress-related genes, thereby increasing the tolerance of plants to the stresses. Stress-induced genes not only function to protect cells from abiotic stress through the production of important enzymes and metabolic proteins (functional proteins) but they also regulate signal transduction and gene expression in the stress response (regulatory proteins). Functional proteins are required for the synthesis of osmoprotectants such as proline and sugars. Regulatory proteins that are activated in response to abiotic stresses, including transcription factors (TFs) such as DREBs (dehydration-responsive element-binding proteins), AREBs (ABA-responsive element-binding proteins), and NAC proteins, have been identified in *Arabidopsis* and rice (Yamaguchi-Shinozaki and Shinozaki 2006; Nakashima et al. 2009). Nakashima et al. (2012) reported that NAC proteins are plant-specific transcription factors and more than 100 NAC genes have been identified in *Arabidopsis* and rice to date. Tran et al. (2010) reported that NAC transcriptional factors which constitute one of the largest families of plant specific and help to enhance tolerance against various abiotic stresses. The cDNA encoding a NAC protein was first reported as the responsive to dehydration 26 (RD26) gene in *Arabidopsis* (Yamaguchi-Shinozaki et al. 1992). Many NAC proteins, including *Arabidopsis* CUC2, have important functions in plant development. Some NAC genes are upregulated during wounding and bacterial infection (Collinge and Boller 2001; Mysore et al. 2002; Hegedus et al. 2003), whereas others mediate viral resistance (Xie et al. 1999). NAC proteins were thought to be transcriptional activators as the *Arabidopsis* ATAF1/2 proteins can activate the CaMV 35S promoter in yeast cells. The *Arabidopsis* AtNAM (NARS2) protein was confirmed by Duval et al. (2002) to function as a transcriptional activator in a yeast system. Kikuchi et al. (2000) reported eight NAC genes in rice (OsNAC1 to OsNAC8) which encode proteins with a single NAC domain. Each OsNAC gene has a unique tissue-specific expression pattern, suggesting this family regulates the development of rice. Ooka et al. (2003) performed a comprehensive analysis of NAC family genes in rice and *Arabidopsis*. They identified 75 predicted NAC proteins in full-length cDNA datasets of rice and 105 predicted genes in the *Arabidopsis* genome. Recently, Nuruzzaman et al. (2010) conducted a genome-wide analysis of the NAC transcription factor family in rice and *Arabidopsis* by investigating 151 nonredundant NAC genes in rice and 117 in *Arabidopsis*. Morishita et al. (2009) also reported that ANAC078 in the NAC group is responsive to a combination of high light and heat stress. They also demonstrated that ANAC078 regulates flavonoid biosynthesis, leading to the accumulation of anthocyanins under high-light conditions. Hu et al. (2006) reported that the overexpression of the stress-responsive SNAC1 gene increased drought and salt tolerance in rice. SNAC1 enhanced drought resistance in transgenic rice plants at the reproductive stage during growth in fields under severe drought stress without affecting yields (Hu et al. 2006). Nogueira et al. (2005) reported that the SsNAC23 gene, which is homologous to the rice OsNAC5, is associated with cold, herbivory, and water stress in sugarcane. Tran et al. (2009) identified 31 unigenes containing complete open reading frames encoding GmNAC proteins in soybean. Analysis of the C-terminal regulatory domain using a yeast one-hybrid system indicated that among the 31 GmNAC proteins, 28 have transcriptional activation activity. Among them, nine GmNAC genes are induced by dehydration stress with differential induction levels in both shoots and roots. Sperotto et al. (2009) reported that OsNAC5 expression is upregulated by natural (aging) and induced senescence processes. They suggested that OsNAC5 is a novel senescence-associated ABA-dependent NAC transcription factor. The stress-responsive NAC proteins, including OsNAC5, might also function during the process of senescence as well as stress tolerance mediated through ABA. Yang et al. (2011) identified a NAC transcription factor VND INTERACTING2 (VNI2) that mediates signaling cross talk between salt stress response and the process of leaf aging. VNI2 regulates the cold-regulated (COR) and responsive to dehydration (RD) genes by binding directly to their promoters. Overexpression of COR or RD led to prolonged leaf longevity, as observed in the VNI2-overexpressing transgenic plants. Mostafa Kamal et al. (2010a, b) identified GTP-binding proteins, which are involved in signal transduction mechanism in plant systems, and these proteins regulate a flow of kinases that play a vital role in environmental stress signal transduction, and also high temperature seems to upregulate the synthesis of GTP-binding proteins resulting in increased kinase activity (Grover et al. 2001). Some heat-upregulated proteins showed the similarities to elongation factors (EF) and eucaryotic translation initiation factors (eIFs). Heat shock involves changes in the expression patterns of the eIFs, the EF 1-beta, EF 1-alpha, EF Tu, and eIF (4A, 4B, 4E, 4E-1, 4E-2, 5A-1, 5A-2, SUII), in wheat leaves (Gallie et al. 1998). Calcium is a universal molecule in both animals and plants, and the transient increase in $Ca^{2+}$ level during heat stress is well documented in plants. Heat shock triggers cytosolic $Ca^{2+}$ bursts, which is transducted by Ca2+ binding proteins(CBP) such as calmodin (CaM), calcineurin (CBL), and annexin and then upregulates the expression of heat shock proteins (HSPs) (Liu et al. 2003). Tubulin proteins are coupled to GTP-binding proteins, which play a role in heat resistance in plant (Segal and Feldman 1996). Mostafa Kamal et al. (2010a, b) also identified serine carboxypeptidase, glucose-1-phosphate, glucose-6-phosphate, and S-adenosyl-methionine synthetase proteins. Different abscisic acid-responsive proteins, LEA protein such as chaperonin, cys peroxiredoxin, ethylene response, and elongation factor TU are responsible for drought stress (Mostafa Kamal et al. 2010a, b). They also observed that cyclin-dependent kinase like zinc finger, transcription factor like MYB, lipid transfer proteins and WRKY are effective for drought stress. Cyclophilin, aquaporin, and chitinase play an important role in cold (upregulated) stress of wheat as reported by Houde et al. (2006). The effects of dehydration, cold temperature treatment, and osmotic and salt stress on the expression of an abscisic acid-responsive protein kinase mRNA (PKABA1) were determined in wheat seedlings (Holappa and Simmons 1995). Mitogen has three protein kinases, MAPK (mitogen-activated protein kinase) and a ribosomal kinase homologue, increased markedly and simultaneously when plants were treated with low temperature (Zhang et al. 2006). GLPs (glucagon-like peptide) function primarily as superoxide dismutase (SOD) to protect plants from the effects of oxidative stress (Khuri et al. 2001). ABA is a central regulator of many plant responses to environmental stresses, particularly osmotic stresses (Chinnusamy et al. 2008; Cramer 2010; Hubbard et al. 2010; Kim et al. 2010). The essential components of ABA signaling include receptors (PYR/PYL/RCAR), protein phosphatases (PP2C), and protein kinases (SnRK2/OST1) (Ma et al. 2009; Park et al. 2009). The PYR/PYL/RCAR proteins were identified as soluble ABA receptors by two independent groups as reported by Park et al. (2009). Leung and Giraudat (1998) identified 2C-type protein phosphatases (PP2C) including ABI1 and ABI2 from the ABA-insensitive *Arabidopsis* mutants abi1-1 and abi2-1, and they act as global negative regulators of ABA signaling. In yeast, the well-documented central regulators of protein synthesis and energy are SnRK1 (Snf1/MAPK) (mitogen-activated protein kinase), TOR1, and GCN2 (Zaborske et al. 2010; Staschke et al. 2010; Smeekens et al. 2010). These proteins are largely controlled by the phosphorylation of enzymes; all three are protein kinases acting as key hubs in the coordination of metabolism during stressful conditions (Hey et al. 2010). In plants, target of rapamycin (TOR) activity is inhibited by osmotic stress and ABA (Deprost et al. 2007), and GCN2 activity is stimulated by UV light, amino acid starvation, ethylene, and cold stress (Lageix et al. 2008).

## Water Stress

Drought is one of the major limiting factors of plant production worldwide. Plant adaptation to drought is the result of many different physiological and molecular mechanisms. Several studies have shown that physiological adaptations to water stress were associated with drought-induced proteins (Bray 1997). Proteomics has proved to be a powerful tool for the identification of proteins and mechanisms involved in drought response and tolerance (Riccardi et al. 1998; Kawasaki et al. 2000; Salekdeh et al. 2002a, b; Hajheidari et al. 2005). Out of 78 responsive proteins, 16 were identified, including proteins involved in the water stress response, the basic metabolic pathway, and lignification. Studies on mild drought stress in rice leaves showed that drought-induced changes in about 42 proteins were reversed completely within 10 days of rewatering. Molecular analyses of wild species such as drought-tolerant grasses would provide a better insight into genes and mechanisms by which plants may adapt to prolonged drought. Tolerant plants are able to maintain tissue water content, or survive a reduction in tissue water content, or recover more completely after rewatering (Cabuslay et al. 1999). The level of proline increased up to 20-fold in response to severe drought conditions as reported earlier (Pedrol et al. 2000; Clifford et al. 1998). Proline is considered as a compatible solute (Samaras et al. 1995) and an osmoprotectant (Serrano and Gaxiola 1994; Okuma et al. 2000). Proline may confer a protective effect by protecting or inducing stress-protective proteins (Khedr et al. 2003). They reported that severe salt stress caused an inhibition of the antioxidative enzymes catalase and peroxidase in *Pancratium maritimum* L., but the activity of these enzymes was also maintained significantly higher in the presence of proline. Therefore, plant cells require two different mechanisms which will enable the detoxification of excess ROS and fine modulation of ROS for signaling purposes. SOD acts as a first line of defense converting superoxide to the less toxic hydrogen peroxide molecules. The abundance of the cytosolic Cu-Zn SOD of *Elymus elongatum* increased up to twofold in response to drought. Cytosolic Cu-Zn SOD was shown to be upregulated in rice (Salekdeh et al. 2002b) and sugar beet (Hajheidari et al. 2005). The upregulation of this enzyme in plants grown under progressive stress reveals its important role in response to drought. The detoxification of $H_2O_2$ is accomplished with ascorbate peroxidase, glutathione peroxidase, catalase, and 2-Cys peroxiredoxin. Ascorbate peroxidase reduces $H_2O_2$ to water, with the concomitant generation of monodehydroascorbate (MDHA). MDHA is a radical with a short lifetime that disproportionates to ascorbate and dehydroascorbate (DHA). DHA is reduced to ascorbate by the action of DHA reductase, using glutathione as the reducing substrate. The most upregulated protein (4.8-fold) was identified as dehydroascorbate reductase suggesting its strong role in the detoxification of $H_2O_2$. Hydrogen peroxide ($H_2O_2$) is a key regulatory molecule in the response to stresses (Mittler 2002), and its ability to selectively induce a subset of defense genes like glutathione S-transferases and glutathione peroxidases without directly inducing other defense genes. The upregulation of both glutathione S-transferases and glutathione peroxidase may represent such co-expression mechanisms. Wehmeyer et al. (1996) suggested that small heat shock proteins (sHSPs) are among the several factors required for desiccation tolerance. Hajheidari et al. (2005) reported the upregulation of two sHSPs under drought stress in sugar beet grown in the field. These proteins showed sequence homology with abscisic acid- and stress-inducible protein (ASR). The upregulation of this gene has been also reported in response to drought stress in maize (Riccardi et al. 1998) and salt stress in a salt-tolerant rice genotype (Salekdeh et al. 2002a). Iusem et al. (1993) have reported that characterization of cDNA encodes ASR protein whose expression was activated by leaf water deficit and fruit ripening. Silhavy et al. (1995) identified a nuclear-targeting sequence motif in this gene isolated from Solanum chacoense Bitter which is extremely resistant to viruses, insects, and drought.

## Phytohormone Stress

A variety of plant hormones, including salicylic acid (SA), jasmonate (JA), ethylene, and abscisic acid, have been implicated in mediating responses to a wide range of biotic and abiotic stresses (Diaz et al. 2002; Thomma et al. 1998; Audenaert et al. 2002). The roles of these hormones are dependent upon the particular host–pathogen interaction (Knoester et al. 1998). On the basis of the interactions of hormones in which resistant responses to biotrophs require SA, whereas responses to necrotrophs require JA and ethylene (Feys and Parker 2000), in some instances, these hormones are involved in determining the level of host basal resistance (Delaney et al. 1994). In case of tomato, ethylene, JA, and SA all independently contribute to its resistance to *Botrytis cinerea* (Diaz et al. 2002). The host plant actively regulates the *Xanthomonas campestris* pv. vesicatoria-induced disease response via the sequential action of at least three hormones (JA, ethylene, and SA), which promote expansive cell death of its own tissue (O'Donnell et al. 2003). Further, the effect of phytohormones is also regulated by other factors. For example, the MAPK kinase, EDR1, negatively regulates SA-inducible defenses (Frye et al. 2001), whereas MAPK 4 appears to differentially regulate SA and JA signals (Petersen et al. 2000). These findings also suggest that MAPK modulates cross talk between different plant defense pathways (Hammond-Kosacky and Parkerz 2003).

## Osmotic Stress

Among abiotic stresses, osmotic stress is one of the most severe, caused by drought, high salinity and cold stresses in nature. Plants respond to osmotic stress at the morphological, anatomical, cellular, and molecular levels. To cope with osmotic-related stresses, plants have developed various responses such as production of osmolites for osmotic adjustment, synthesis of $Na^+/H^+$ antiporters for ion sequestration, and many other mechanisms (Bohnert et al. 1995). AtHKT1, which is a $Na^+$ transporter, mediates osmolality balance between xylem vessels and xylem parenchyma cells (Snarpi et al. 2005). The operation of these responses usually requires three steps: osmotic stress recognition, signal transduction, and production of components for the physiological response (Tamura et al. 2003). In plants, a hybrid-type histidine kinase (AtTHK1) functions as an osmosensor and transmits the stress signal to a downstream mitogen-activated protein kinase (MAPK) cascade (Urao et al. 1999). Lu and Neumann (1999) reported that when rice seedlings were exposed to osmotic stress modulated by polyethylene glycol 6000, growth in emerging first leaves of the intact plant was inhibited. Early inhibition of leaf growth was not related to changes in root size, osmotic potential gradients, or cell wall-yielding characteristics in the leaf expansion zone of stressed seedlings. Deak and Malamy (2005) demonstrated that osmotic stress represses the formation of autonomous lateral roots from lateral root primordia of *Arabidopsis*, while lateral root initiation was not greatly affected. Abscisic acid (ABA) and a newly identified gene, LRD2, are involved in osmotic repression of lateral root formation. Further examination revealed that both ABA and LRD2 control root system architecture even in the absence of osmotic stress. The molecular mechanisms that mediated responses to environmental cues could also be regulators of intrinsic developmental programs in the root. Zonia and Mnnik (2004) investigated that tobacco pollen tube cell volume changes in response to osmotic perturbation by activation of the phospholipid signaling pathway. Several intermediates in the phospholipid signaling pathway were detected during pollen tube growth. Hypo-osmotic stress induced a rapid increase in phosphatidic acid and a decrease in phosphatidylinositol phosphate. The fact that these signaling molecules are present during normal growth and the mechanism for osmotic response involved components of the biomechanical networks driving pollen tube cell elongation. In osmotically stressed wheat coleoptiles, reduced rates of phenylalanine ammonia-lyase and tyrosine ammonia-lyase activities suppress phenylalanine biosynthesis, resulting in a reduced level of wall-bound ferulic acid. This decrease in wall-bound ferulic acid may lead to reduced levels of diferulic acid, an important contributor to maintaining cell wall extensibility (Wakabayashi et al. 1997). The osmotic potential of soil alters the depth of root systems, the rate of root elongation, and the number of lateral roots (van der Weel et al. 2000); however, there are only a few reports about phenotypic changes in rice under osmotic stress (Lu and Neumann 1999).

Zang and Komatsu (2007) reported that 15 proteins are changed in the basal part of rice leaf sheaths by mannitol treatment. The basal section of leaf sheath may undergo more subtle changes in protein expression than the distal end, where significant morphological changes are evident. They reported that 327 proteins were detected by digital image analysis. Among these, 15 proteins significantly responded to osmotic stress by up- or downregulation: 12 proteins increased in amount and 3 proteins decreased. Ten of the identified proteins are recognized as important components for stress response such as two proteasome degradation system-related proteins, five endoplasmic reticulum (ER)-related proteins, two proteins related to detoxification, and uroporphyrinogen decarboxylase, which is a cell death-related protein (Mock and Grimm 1997). Two proteins matched annotated sequences for putative proteins of unknown function and the three remaining proteins responding to osmotic stress were not similar to other proteins in existing databases.

The glyoxalase system is ubiquitous in nature and consists of two enzymes: glyoxalase I and glyoxalase II, which act coordinately to convert 2-oxoaldehydes into 2-hydroxyacids using reduced glutathione as a cofactor. The glyoxalase pathway involving glyoxalase I and glyoxalase II enzymes is required for glutathione-based detoxification of methylglyoxal. Methylglyoxal is a primary physiological substrate for glyoxalase I. Besides detoxification of methylglyoxal, the glyoxalase system could also play a role in providing tolerance under stress by recycling glutathione that would be "trapped" spontaneously by methylglyoxal to form hemithioacetal, thereby maintaining glutathione homeostasis (Creighyton et al. 1988). Transgenic plants overexpressing glyoxalase I showed significant tolerance to methylglyoxal and high salt (Veena and Sopory 1999; Singla-Pareek et al. 2003).

## Heavy Metal Stress

Heavy metals (HMs) are among the most widespread soil contaminants which damage the physiological and metabolic processes. Certain heavy metals such as Cd, Hg, and Pb affect cell systems due to the increasing exposure of living organisms to these metals in the environment (DalCorso et al. 2008). The physiological and molecular basis of plant interactions with the environment has attracted considerable interest in recent years. Being sessile organisms, plants are constantly exposed during their life cycle to adverse environmental conditions that negatively affect growth, development, or productivity. The presence of toxic compounds, such as heavy metals (HMs), is one important factor that can cause damage to plants by altering major plant physiological and metabolic processes (Hossain et al. 2010; Villiers et al. 2011). Different metals and metalloids which are toxic to plants such as copper (Cu), iron (Fe), manganese (Mn), zinc (Zn), nickel (Ni), cobalt (Co), cadmium (Cd), and arsenic (As). Importantly, few HMs and transition metals such as sodium (Na), potassium (K), calcium (Ca), magnesium (Mg), Fe, Cu, Zn, Co, or Ni are essential micronutrients that are critically involved in the functional activities of large numbers of proteins involved in sustaining growth and development. However, at excess concentrations, these metal ions can become detrimental to living organisms, including plants. Although HMs are natural constituents of soils and occur naturally in the environment, contamination of soils by toxic metals and metalloids is of major concern worldwide (Villiers et al. 2011). Depending on their oxidation states, HMs can be highly reactive, resulting in toxicity of plant cells in many ways. At the cellular and molecular level, HM toxicity results in alterations of different plant physiological processes, including inactivation and denaturation of enzymes, proteins, blocking of functional groups of metabolically important molecules, displacement/substitution of essential metal ions from biomolecules and functional cellular units, conformational modifications and disruption of membrane integrity (Sharma and Dubey 2007), which is finally attributed to altered plant metabolism, inhibition of photosynthesis, respiration, and alerted activities of several key enzymes (Sharma and Dietz 2009; Hossain et al. 2010, 2012). In addition, HMs are known to disturb redox homeostasis by stimulating the formation of free radicals and reactive oxygen species (ROS) such as singlet oxygen ($1O^2$), superoxide radicals ($O^{2\bullet-}$), hydrogen peroxide ($H_2O_2$), and hydroxyl radicals (•OH) (Dubey 2011; Anjum et al. CR00152402012). Recently, methylglyoxal (MG), a cytotoxic compound, was also found to increase in response various stresses including HMs (Yadav et al. 2005a, b). An increase in MG level in plant cells further intensifies the production of ROS by interfering with different plant physiological and metabolic processes such as inactivation of the antioxidant defense system (Hoque et al. 2010; Hossain et al. 2011a, b) and interfering with vital plant physiological processes such as photosynthesis (Saito et al. 2011). This increase in ROS and MG exposes cells to oxidative stress leading to lipid peroxidation, biological macromolecule deterioration, membrane dismantling, ion leakage, and DNA strand cleavage, and finally death of plants (Navari-Izzo 1998; Romero-Puertas et al. 2002; Barconi et al. 2011).

Heavy metals (HMs) are intracellularly chelated through the synthesis of amino acids, organic acids, GSH, or HM-binding ligands such as metallothioneins (MTs), phytochelatins (PCs), compartmentation fcc within vacuoles, and upregulation of the antioxidant defense and glyoxalase systems to counter the deleterious effects caused by ROS and methylglyoxal (Cobbett 2000; Yadav 2010; Hossain et al. 2012). A large number of recent studies in plants involving sensitive, tolerant, mutant, transgenic, and hyperaccumulator-adopting strategies in the fields of physiology, genomics, proteomics, and metabolomics suggest that GSH by itself and its related metabolizing enzymes, proteins, and peptides play a pivotal role in HM tolerance by controlling different plant physiological processes, including ROS and MG detoxification, heavy metal uptake, translocation, chelation, and detoxification. Yang et al. (2007) reported that a total of 17 Al-responsive proteins were identified, with 12 of those being upregulated and 5 downregulated. Among the upregulated proteins are copper/zinc superoxide dismutase (Cu-Zn SOD), GST, and *S*-adenosylmethionine synthetase 2, which are the consistently known Al-induced enzymes previously detected at the transcriptional level in other plants. More importantly, a number of other identified proteins including cysteine synthase (CS), 1-aminocyclopropane-1-carboxylate oxidase, G protein b subunit-like protein, abscisic acid- and stress-induced protein, putative Avr9/Cf-9 rapidly elicited protein 141, and a 33 kDa secretory protein are novel Al-induced proteins. Most of these proteins are functionally associated with signaling transduction, antioxidation, and detoxification (Hossain et al. 2012).

## Mode of Action of Heavy Metals in Plant Cells

The toxicity of HMs is manifested in many ways when plant cells accumulate them at high levels. HMs are divided into two groups: redox active (Fe, Cu, Cr, Co) and redox inactive (Cd, Zn, Ni, Al, etc.). The redox active HMs are directly involved in the redox reaction in cells and result in the formation of $O_2^{\bullet-}$ and subsequently in $H_2O_2$ and OH production via the Haber-Weiss and Fenton reactions (Dietz et al. 1999; Schutzendubel and Polle 2002). Exposure of plants to redox inactive HMs also results in oxidative stress through indirect mechanisms such as interaction with the antioxidant defense system, disruption of the electron transport chain, or induction of lipid peroxidation. The latter can be due to an HM-induced increase in lipoxygenase (LOX) activity. Another important mechanism of HM toxicity is the ability of HMs to bind strongly to oxygen, nitrogen, and sulfur atoms

(Nieboer and Richardson 1980). This binding affinity is related to free enthalpy of the formation of the product of the HM and ligand with low solubility of these products. Because of these features, HMs can inactivate enzymes by binding to cysteine residues. Many enzymes need cofactors to work properly for both HM ions (such as $Fe^{2+}$, $Mg^{2+}$, $Cu^{2+}$, $Ca^{2+}$) and organic molecules (such as heme, biotin, FAD, NAD, or coenzyme A). The displacement of one HM ion by another leads to the inhibition or loss of enzyme activities. Divalent cations such as $Co^{2+}$, $Ni^{2+}$, and $Zn^{2+}$ displace $Mg^{2+}$ in ribulose-1,5-bisphosphate-carboxylase/oxygenase (RuBisCO) and result in a loss of activity (Wildner and Henkel 1979; Van Assche and Clijsters 1986). Displacement of $Ca^{2+}$ by $Cd^{2+}$ in calmodulin, an important protein in cellular signaling, led to the inhibition of calmodulin-dependent phosphodiesterase activity in radish (Rivetta et al. 1997). Additionally, HMs cause membrane damage through various mechanisms, including the oxidation of and cross-linking with protein thiols, inhibition of key membrane protein such as $H^{+}$-ATPase, or causing changes in the composition and fluidity of membrane lipids (Meharg 1993). Accumulation of MG, a cytotoxic compound, was found to increase in response to HM stress in plants due to impairment of the glyoxalase system that finally elicits oxidative stress by reducing the GSH content (Hossain and Fujita 2010; Hossain et al. 2011a, b, 2012).

Metallothioneins (MTs) are low-molecular-weight (4–8 kDa), Cys-rich, HM-binding, gene-encoded polypeptides that can bind heavy metals via the thiol groups of their Cys residues (Hamer 1986). Although the precise physiological function of MTs has not yet been fully elucidated, proposed roles include participation in maintaining the homeostasis of essential transition HMs, sequestration of toxic HMs, and protection against intracellular oxidative damage (Gasic and Korban 2006). Plant MTs are extremely diverse and have been subdivided into three classes based on the arrangement of the Cys residue (Zhou et al. 2006). The Cys-Cys, Cys-X-Cys, and Cys-X-X-Cys motifs (in which X denotes any amino acid) are characteristic and invariant for MTs. The organization or distribution of cysteine residues confers different MT isoforms and their ability to bind and sequester different HM ions for detoxification and homeostasis. The biosynthesis of MTs is regulated at the transcriptional level and is induced by several factors, including hormones, cytotoxic agents, and HMs, such as Cd, Zn, Hg, Cu, Au, Ag, Co, Ni, and Bi (Kagi 1991). Gene expression studies were performed to quantify mRNA levels in different tissues, at different developmental stages, and under stress conditions such as HM exposure. MT genes appear to be differentially regulated in a tissue-specific manner and in relation to the developmental stage and also in response to a number of stimuli, including heavy metals (Castiglione et al. 2007). Ahn et al. (2012) showed that three *Brassica rapa* MT genes (*BrMT1, BrMT2,* and *BrMT3*) are differentially regulated under various HM stresses.

## Cold Stress

Plants are frequently facing different environmental variations. Among these constraints, low temperature is one of the most crucial factors that impair the distribution, growth, and productivity of crops or wild plant species. The mechanisms underlying cold acclimation and tolerance have been intensively studied over the past years, and numerous studies on plant response to cold stress have been reported (Ruelland et al. 2009). Cold weather can result in pollen sterility in plants at the reproductive stage of development. Several studies of plant transcriptomes during cold stress have been performed in rice (Chen et al. 2002a, b; Fowler and Thomashow 2002; Kawamura and Uemura 2003; Amme et al. 2006; Goulas et al. 2006; Cui et al. 2005; Yan et al. 2006; Hashimoto and Komatsu 2007). Hashimoto and Komatsu (2007) reported that some proteins are involved in cold stress, and these are the chilling stress-responsive proteins. In general, the level of mRNA does not always correlate well with the level of protein mainly due to posttranscriptional regulation (Yan et al. 2006). As a global study of the proteins comprising the proteome, proteomics can be playing an increasingly important role in addressing plant response to environmental changes. The proteomes of various plants in response to different environmental factors including drought, salt, heat, and heavy metal have been investigated (Ouerghi et al. 2000; Hajduch et al. 2001; Salekdeh et al. 2002a; Majoul et al. 2003). The differentially expressed proteins have been identified and well-documented stress-responsive proteins and some novel cold-responsive proteins. Plants differ in their cold response and cold-tolerant species may develop efficient strategies to adapt to chilling environment. Bressan et al. (2001) reported that the large-scale transcriptional studies were hindered by the absence of genome sequence information.

## ROS-Scavenging Mechanisms in Cold Stress

Cold stress may disturb cellular redox homeostasis and promotes the production of reactive oxygen species (ROS) and reactive aldehydes (as 4-hydroxy-nonenal and methylglyoxal). Plants develop ROS-scavenging mechanisms to cope with the oxidative stress. Because chloroplast is one of the major sources of ROS, four proteins involved in chloroplast redox homeostasis, including 2-cys peroxiredoxin, chloroplast (BAS1), peroxiredoxin-2E, ferritin-1, and alanine-2-oxoglutarate aminotransferase 1 (GGT1) (Gao et al. 2009). BAS1 is involved in the detoxification of alkyl hydroperoxides and its gene expression is regulated under the control of the cellular redox state (Baier and Dietz 1997). Peroxiredoxin-2E reduces hydrogen peroxide and alkyl hydroperoxides with reducing equivalents provided through the thioredoxin or glutaredoxin system and was a component of chloroplast thioredoxin system (Bréhélin et al. 2003). Ferritin-1, an iron binding protein, is to protect plants from oxidative damage induced by manifold stresses (Deák et al. 1999). GGT1, an enzyme involved in photorespiration and photorespiration, has been suggested to be important for maintaining electron flow to prevent photoinhibition under stress conditions (Wingler et al. 2000). Methionine residues of proteins are a major target for oxidation by ROS and peptide methionine sulfoxide reductase plays protective roles in the cellular response to oxidative stress (Kwon et al. 2007). Aldo-keto reductases can detoxify lipid peroxidation products and glycolysis-derived reactive aldehydes that contribute significantly to cellular damages caused by abiotic stresses and improve scavenging capacity of the plant.

Cold stress may impose a great influence on the protein synthesis apparatus in plant. A chloroplast EF-G protein appeared in gel after 2, 5, and 24 days of cold treatment, but there was no spot at the corresponding position in gel of the control group. Because chloroplast EF-G is a housekeeping gene, there must be an EF-G protein spot elsewhere in gel of control group (Gao et al. 2009). It is postulated that posttranslational modification or alternative splicing may occur on EF-G in *T. halophila* under cold stress to maintain the normal operation of protein synthesis. Although, EF-G was previously reported to play a role in chloroplast biogenesis and development in *Arabidopsis*. PAB1 was reported to be involved in ubiquitin-dependent protein catabolic process. This may indicate that a high protein turnover rate is needed for plants to eliminate the misfolded polypeptides under cold environment (Amme et al. 2006; Goulas et al. 2006) and also in rice plant (Cui et al. 2005; Yan et al. 2006; Hashimoto and Komatsu 2007). They suggested that the cold stress responses in *Thellungiella* are similar, in general, to other plant species. Because *Thellungiella* is a cold-tolerant relative of *Arabidopsis*, comparison of the cold-induced rosette leaf proteome changes in *Thellungiella*, and *Arabidopsis* was performed to advance the understanding of the higher cold tolerance of *Thellungiella*. The identification of cold-responsive proteins in *Thellungiella* provides not only new insights into cold stress responses but also a good starting point for further investigation of their functions. However, considering the inability of 2-DE-based proteomic study to resolve membrane proteins and low abundant proteins, integrated transcriptomic, proteomic, and metabolomic approaches should be adopted to gain further insight into plant response to cold stress.

## Biotic Stress in Plants

Plants responding to biotic stresses produce several protective compounds and proteins such as pathogenesis-related (PR) proteins, directly disease-related proteins, and other proteins. Biotic is one of the serious stresses affecting plant growth and productivity. A clear perceptive of the molecular mechanisms involved in plants response to biotic stress is of fundamental importance to plant science (Mostafa Kamal et al. 2010a, b). The study of PR proteins is also important for crop production due to the fact that many plant-derived pathogen-related proteins have been identified as members of PR protein families 2, 3, 4, 5, 8, 10, and 14 (Hoffmann-Sommergruber 2002). Among these are pathogenesis-related proteins coded by the host plant that accumulate in response to pathogen infection or other signals related to plant defense responses. Several PR proteins have been characterized at the molecular level and have shown to have antifungal activity in vitro (Datta and Muthukrishnan 1999) and show enzymatic activity such as β-1,3-glucanase and chitinase (PR2 and PR3, respectively), both involved in the degradation of microbial cell wall structural polysaccharides (Legrand et al. 1987) and PR4 and PR9, characterized by ribonuclease and peroxidase activity, respectively (Caporale et al. 2004). Proteomic approaches have been successfully used in discovering the resistance mechanisms in maize against kernel rot caused by *Fusarium* or *Aspergillus* (Chen et al. 2004; Chen and Chen 2002; Sonia et al. 2004). Mostafa Kamal (2010a,b) reported that methyl jasmonate (MeJA) and its free acid jasmonic acid (JA) collectively referred to as jasmonates are important cellular regulators involved in diverse developmental process, such as seed germination, root growth, fertility, seed ripening, and senescence. In addition, jasmonate activate plant defense mechanisms in response to insect-driven wounding and various pathogens (Creelman and Mullet 1997). Carmona et al. (1993) reported that different purothionins are active against plant pathogens both in vitro and in vivo. Gamma-1-purothionin showed a higher structural analogy with scorpion toxins and against insect defensins which also present the cystine-stabilized alpha helical (CSH) motif (Bruix et al. 1993) and gamma-2-purothionin inhibits protein translation in cell-free systems resulting in the exhibited plant toxins for pathogen (Colilla et al. 1990). Mostafa Kamal et al. (2010a) indicated that the antifungal activity has been associated with two immunochemically distinct proteins; the proteins are homologous with thaumatin- and pathogenesis-related proteins of the PR5 family. These proteins have intensely sweet properties of thaumatin; multiple unrelated defense functions against insect and fungal pests can now be associated with the family of thaumatin-homologous proteins (Hejgaard et al. 1991). Antimicrobial proteins (MBP-1) inhibit spore germination or hyphal elongation of several plant pathogenic fungi, including two seed pathogens of maize (*Fusarium moniliforme* Sheld. and *Fusarium graminearum* (*Gibberella zeae* (Schw.) Petsch)) and several bacteria, including a bacterial pathogen of maize (*Clavibacter michiganense* ssp. nebraskense) (Duvick et al. 1992). Wheat calcium-dependent protein kinase (CDPK) genes were found to respond to various biotic and abiotic stimuli, including cold, hydrogen peroxide ($H_2O_2$), salt, drought, powdery mildew (*Blumeria graminis* tritici, BGT), as well as phytohormones abscisic acid (ABA) and gibberellic acid (GA) (Li et al. 2008). Chitinases are important components of plant defense in response to attack by pathogens as *F. graminearum* (Li et al. 2001). Dilbirligi and Gill (2003) reported that many RGA sequences in wheat are identified as disease resistance gene. Ascorbate peroxidase, peroxidase, and glutathione (GSH)-dependent dehydroascorbate reductase accumulate early in grain fill. SGT1, a component of R-gene-triggered disease resistance, and serpin, a serine protease inhibitor, are also present and may protect the developing grain against various pathogens (Wong et al. 2004). The pathogen resistance proteins such as serpin; chitinase, which hydrolyzes the structural carbohydrate of fungal cell walls; barwin/PR-4 protein, which is induced by fungal pathogens and binds chitin; and xylanase inhibitor protein, which inhibits a fungal enzyme that degrades plant cell walls (Hurkman and Tanaka 2007). Zhang et al. (2008) reported that a total of 375 functional ESTs have been identified by SSH from a cDNA library of resistant soybean varieties associated with *Phytophthora* root rot. They also observed that system acquired resistance (SAR) induced by *Phytophthora* in soybean primarily depended on salicylic acid-mediated signaling pathways, which was different from resistance mechanisms in *Arabidopsis*. Nevertheless, the functions of differentially expressed proteins induced by *Phytophthora* are almost the same in soybean and *Arabidopsis*, including material metabolism enzymes and various regulatory factors; this suggests that the molecular basis of system acquired resistance against *Phytophthora* is much the same in soybean and *Arabidopsis* (Sun et al. 2008). Plant disease resistance genes usually have conserved motifs such as the nucleotide-binding site (NBS). Based on the NBS domain, two putative resistance genes (*KR1* and *KR4*) were isolated from soybean. Sixteen identified proteins were potentially involved in protein degradation, defense signal transfer, oxidative stress, cell wall reinforcement, and energy and metabolism regulation (Yang, et al. 2010). In addition, differential expression was found by SSH in cDNA libraries from soybean varieties resistant to soybean cyst nematode (Cheng et al. 2007; Lu and Fang 2003).

## Interaction with Insect and Pest with Stress

*Trichoderma* spp. have been known as biocontrol agents for the control of plant diseases (Harman et al. 2004). In many cases, the beneficial fungi may induce systemic resistance that is mediated by alterations in plant gene expression (Alfano et al. 2007; Shoresh et al. 2010). They also reported that the plant in association with *Trichoderma* strains enhanced plant growth, but the effects, as with other plant growth-promoting microbes (Gamalero et al. 2009), are greater when plants are under suboptimal conditions or biotic, abiotic, or physiological stresses (Yildirim et al. 2006). Several recent reports indicate that the fungi enhances tolerance to abiotic stresses during plant growth, in part due to improved root growth, improvement in water-holding capacity of plants (Harman 2000; Bae et al. 2009), or enhancement in nutrient uptake; whereas, in the absence of stress, plant growth may or may not be enhanced. Although molecular studies indicate greater expression of gene families involved in plant protection against abiotic stresses or oxidative damage in *Trichoderma* spp.-treated plants, no experimental evidence has been presented correlating enhanced tolerance of plants colonized with biocontrol fungi to these changes in molecular level. These fungi are frequently applied as seed treatments, where they may improve plant stands and induce long-term improvements in plant quality (Harman 2000, 2006).

## Effects of Abiotic Stress on Plants: A Systems Biology

Recent advances in biotechnology have dramatically changed our capabilities for gene discovery and functional genomics. For the first time, we can now obtain a holistic "snapshot" of a cell with transcript, protein, and metabolite profiling (Umezawa 2011). Such a "systems biology" approach allows for a deeper understanding of physiologically complex processes and cellular function (Kitano 2002). Understanding the function of genes is a major challenge of the post-genomic era. While many of the functions of individual parts are unknown, their function can sometimes be inferred through association with other known parts, providing a better understanding of the biological system as a whole. High-throughput -omics technologies are facilitating the identification of new genes and gene function. In addition, network reconstructions at the genome scale are key to quantifying and characterizing the genotype to phenotype relationships (Feist and Palsson 2008).

Fundamentally, plants require energy (light), water, carbon, and mineral nutrients for growth. Abiotic stress reduces growth and yield below the optimum levels. Plant responses to abiotic stresses are dynamic and complex system (Skirycz and Inze 2010; Cramer 2010); they are both elastic (reversible) and plastic (irreversible).

The plant responses to stress are dependent on the tissue or organ affected by the stress. Transcriptional responses to stress are tissue or cell specific in roots and are quite different depending on the stress involved (Dinneny et al. 2008). In addition, the level and duration of stress (acute vs. chronic) can have a significant effect on the complexity of the response (Tattersall et al. 2007). Water deficit inhibits plant growth by reducing water uptake into the expanding cells and alters enzymatically the rheological properties of the cell wall, for example, by the activity of ROS (reactive oxygen species) on cell wall enzymes (Skirycz and Inze 2010). In addition, water deficit alters the cell wall nonenzymatically, for example, by the interaction of pectate and calcium (Boyer 2009). Furthermore, water conductance to the expanding cells is affected by aquaporin activity and xylem embolism (Parent et al. 2009; Boursiac et al. 2008). With long-term stress, photosynthesis declines due to stomatal limitations for $CO_2$ uptake and increased photoinhibition from difficulties in dissipating excess light energy (Pinheiro and Chaves 2011). One of the earliest metabolic responses to abiotic stresses and the inhibition of growth is the inhibition of protein synthesis (Good and Zaplachinski 1994; Vincent et al. 2007) and an increase in protein folding and processing (Liu and Howell 2010). Energy metabolism is affected as the stress becomes more severe (e.g., sugars, lipids, and photosynthesis) (Cramer et al. 2007; Kilian et al. 2007). Thus, there are gradual and complex changes in metabolism in response to stress.

## Regulatory Mechanism

The plant molecular responses to abiotic stresses involve interactions and cross-link with many molecular pathways (Takahashi et al. 2004). Reaction oxygen species (ROS) signaling in response to abiotic stresses and its interactions with hormones has been thoroughly reviewed (Mittler et al. 2010). ROS and RNS (reactive nitrogen species) form a coordinated network that regulates many plant responses to the environment. There are a large number of studies on the oxidative effects of ROS on plant responses to abiotic stress, but only a few studies documenting the nitrosative effects of RNS (Molassiotis and Fotopoulos 2011). Hormones are also important regulators of plant responses to abiotic stress. The two most important hormones are abscisic acid (ABA) and ethylene. ABA is a central regulator of many plant responses to environmental stresses, particularly osmotic stresses (Cramer 2010; Kim et al. 2010; Chinnusamy et al. 2008). There are slower responses to ABA involving transcriptional responses that regulate growth, germination, and protective mechanisms. Recently, the essential components of ABA signaling have been identified, and their mode of action was elaborated (Umezawa et al. 2010). The current model of ABA signaling includes three core components, receptors (PYR/PYL/RCAR), protein phosphatases (PP2C), and protein kinases (SnRK2/OST1). The PYR/PYL/RCAR proteins were identified as soluble ABA receptors by two independent groups (Ma et al. 2009; Park et al. 2009). The 2C-type protein phosphatases (PP2C) including ABI1 and ABI2 were first identified from the ABA-insensitive *Arabidopsis* mutants abi1-1 and abi2-1, and they act as global negative regulators of ABA signaling (Leung and Giraudat 1998). Yamaguchi-Shinozaki and Shinozaki (2006) reported that the transcriptional regulation of dehydration and salinity stresses have revealed both ABA-dependent and ABA-independent pathways. Cellular dehydration under water-limited conditions induces an increase in endogenous ABA levels that trigger downstream target genes encoding signaling factors, transcription factors, metabolic enzymes, and others. In the post-genomic era, comprehensive analyses using three systematic approaches have increased our understanding of the complex molecular regulatory networks associated with stress adaptation and tolerance. The first one is "transcriptomics" for the analysis of coding and noncoding RNAs and their expression profiles. The second one is "metabolomics" that is a powerful tool to analyze a large number of metabolites. The third one is "proteomics" in which protein and protein modification profiles offer an unprecedented understanding of regulatory networks. Protein complexes involved in signaling have been analyzed by a proteomics approach (Kaufmann et al. 2011). Hirai et al. (2004, 2007) identified MYB transcription factors regulating glucosinolate biosynthesis in *Arabidopsis* in response to S and N deficiency using an integrated transcriptomics and metabolomics approach. Genes and metabolites in glucosinolate metabolism were found to be coordinately regulated (Hirai et al. 2004). Co-expression analysis was used to identify two MYB transcription factors that positively regulate glucosinolate metabolism (Hirai et al. 2007). Mao et al. (2009) performed a gene co-expression network analysis of 1,094 microarrays of *Arabidopsis* using a nontargeted approach. They identified 382 modules in this network. The top three modules with the most nodes were photosynthesis, response to oxidative stress, and protein synthesis. Many of the modules also involved responses to environmental stresses. They constructed a cold-induced gene network from a subset of microarrays. Weston et al. (2008) used weighted co-expression analysis to define six modules for *Arabidopsis* responses to abiotic stress. Two hubs in the common response module were an ankyrin-repeat protein and genes involved in Ca signaling. They created a compendium of genomic signatures and linked them to their co-expression analysis.

## Proteomics Approach in Biotic Stress

High temperature and drought, which often occur together during the growing season and likely contribute to poor kernel development, have been reported to increase growth of the fungus and toxin production (Payne 1998). Jones et al. (1981) reported that irrigating corn fields to reduce drought stress also reduced fungal infection and aflatoxin contamination. Irrigation not only relieved drought stress but also reduced soil temperature. Lower soil temperature was found to reduce aflatoxin contamination in peanut (Hill et al. 1983). Increased aflatoxin contamination was observed in drought-treated peanuts with increased soil temperatures (Cole et al. 1985). Dorner et al. (1989) also reported that a higher soil temperature favors *Aspergillus flavus* growth and aflatoxin production. A study on the effect of drought on peanut resistance to *A. flavus* by Wotton and Strange (1987) found that fungal colonization was inversely related to water supply as was aflatoxin production. Payne et al. (1986) also concluded that water stress appears to be a major factor affecting aflatoxin contamination. Studies of aflatoxin and fumonisin contamination of corn grown under high or moderate heat stress (Abbas et al. 2002; Chen et al. 2004) demonstrate that heat stress also plays an important role in the susceptibility of corn to both aflatoxin and fumonisin contamination. Tubajika and Damann (2001) compared ear rot and aflatoxin production between nine drought-tolerant and two aflatoxin-resistant corn lines under field and laboratory conditions. They found that drought-tolerant lines all had significantly lower levels of ear rot and aflatoxin contamination compared to the aflatoxin-resistant controls, when grown under drought conditions (Tubajika and Damann 2001). A proteomics approach was recently employed to identify proteins whose level of expression associated with kernel resistance against *A. flavus* infection and aflatoxin production (Chen et al. 2002a, b). They found over a dozen proteins, either unique or fivefold upregulated in resistant lines, have been identified and sequenced (Chen et al. 2002a, b, 2004). These proteins can be grouped into three categories based on their peptide sequence homology, that is, storage proteins, such as globulin 1, globulin 2, and late embryogenesis abundant proteins (LEA3, LEA14); stress-related proteins, such as an aldose reductase (ALD), a peroxiredoxin antioxidant (PER1), a cold-regulated protein, a water stress-inducible protein, an anionic peroxidase, a glyoxalase I protein (GLX I), and several small heat shock proteins (HSP); and antifungal proteins, which include a trypsin inhibitor and a pathogenesis-related protein. The majority of protein identified was stress-related proteins and highly hydrophilic storage proteins. These data suggest that kernel resistance may require not only the presence of high levels of antifungal proteins but also that of high levels of stress-related proteins and highly hydrophilic storage proteins. Storage proteins have been reported to play an important role in stress tolerance. Chen et al. (2002a, b) reported that the members of the LEA genes family have been associated with plant responses to many different stresses including drought, salt, cold, heat, and wounding (Thomann et al. 1992). Transgenic expression of an LEA protein from barley demonstrated increased tolerance to water and salt stress in rice (Xu et al. 1996). Some stress-related proteins have been reported to not only confer stress tolerance but also enhance disease resistance. The expression of heat shock proteins, especially the small HSPs under stress, has been widely studied (Vierling 1991) and shown to possess molecular chaperone activity (Jacob et al. 1993). Aside from heat stress, HSPs are also induced by other stresses such as cold, drought, or salinity (Sabehat et al. 1998). The role of glyoxalase in stress tolerance is also highlighted in a recent study using transgenic tobacco plants overexpressing a *Brassica juncea* glyoxalase I (Veena et al. 1999). Park et al. (2001) and Shin et al. (2002) also found that transgenic expression of the tobacco stress-inducible gene 1 (Tsi1) induced expression of several pathogenesis-related genes under normal conditions, resulting in improved tolerance to salt and pathogens. The mechanisms plants use to adapt to abiotic and biotic stresses have been widely studied in a number of plants.

## Production of ROS in Plants

Current research effort has focused on the isolation of stress-responsive genes and their regulation as a means to understand the molecular events underlying the adaptation process. An increasing body of evidence suggests that a subset of plant responses to biotic and abiotic stress is shared, such as the generation of reactive oxygen species (ROS), the activation of mitogen-activated protein kinases (MAPKs), and hormone modulations (Mittler 2002) (Fig. 15.1). Organelles such as chloroplasts, mitochondria, and microbodies are a major source of ROS production in plant cells. Together with an extensive battery of oxidases, the plant cell is well armed for bountiful yet flexible ROS production. In chloroplasts, the primary sources of ROS production are the Mehler reaction and the antenna pigments (Asada and Takahashi 1987). Production of ROS by these sources is enhanced in plants by conditions limiting $CO_2$ fixation, such as drought, salt, and temperature stress, as well as by the combination of these conditions with high-light stress. In C3 plants, limiting $CO_2$ conditions can also activate the photorespiratory pathway (Foyer 2002). As part of this pathway, $H_2O_2$ is generated in peroxisomes by the enzymatic activity of glycolate oxidase. Production of $H_2O_2$ in microbodies can also occur during lipid catabolism as a side product of fatty acid oxidation. In mitochondria, overreduction of the electron transport chain is the main source of $O^{2-}$ production under specific stress conditions (Møller 2001). ROS in plant cells include the detoxifying reactions catalyzed by cytochromes in both the cytoplasm and the endoplasmic reticulum. They are thought to play a key role in ROS signaling and contain a multimeric flavocytochrome that forms an electron transport chain capable of reducing $O_2$ to $O^{2-}$. Chemical inhibitors of mammalian NADPH oxidase (such as diphenyleneiodonium) have been shown to block or impair ROS production during biotic or abiotic stresses in plants (Allan and Fluhr 1997). In addition to NADPH oxidases, pH-dependent cell wall peroxidases, germin-like oxalate oxidases, and amine oxidases have been proposed to generate ROS at the apoplast (Bolwell and Wojtaszek 1997; Walters 2003). Although much attention has been given to NADPH oxidases and their possible role in cell signaling, other ROS-producing mechanisms in the mitochondria, apoplast, and peroxisome are likely to play a role in ROS signaling in response to different stimuli or developmental signals.

## Modulation of ROS Signaling

ROS production is recognized as a common event in plant response to biotic and abiotic stresses (Mittler et al. 2010; Lamb and Dixon

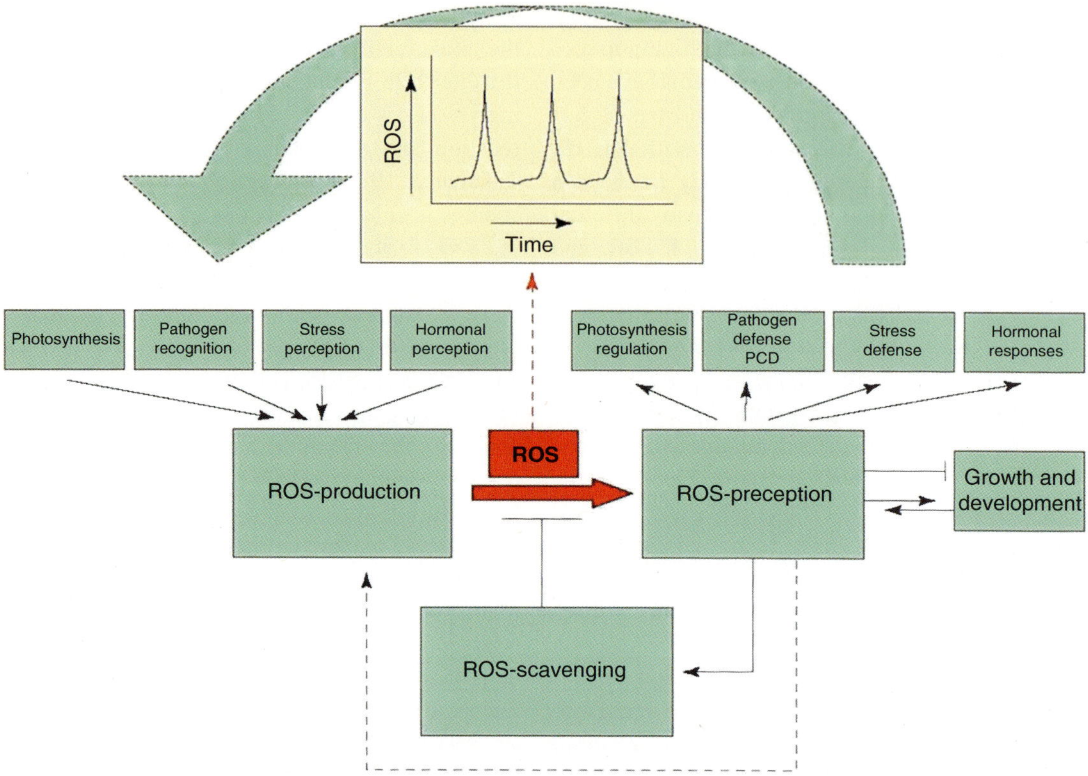


**Fig. 15.1** Modulation of reactive oxygen species (ROS) signaling by the reactive oxygen gene network of plants (Published by Mittler et al. 2004)

1997). The hypersensitive response (localized plant cell death at the infection site) to halt pathogen invasion during an incompatible host–pathogen interaction has also been reported to involve the production of ROS. Mittler et al. (2004) reported that calcium signaling is predominantly controlled in plants by storage and release; ROS signaling is controlled by production and scavenging. Different developmental or environmental signals feed into the ROS signaling network and perturb ROS homeostasis in a compartment-specific or even cell-specific manner. Perturbed ROS levels are perceived by different proteins, enzymes, or receptors and modulate different developmental, metabolic, and defense pathways. ROS can be generated by various enzymatic activities, of which the best studied are NADPH oxidases, and removed by an array of ROS-scavenging enzymes. The intensity, duration, and localization of the different ROS signals are determined by interplay between the ROS-producing and ROS-scavenging pathways of the cell. This process requires a tight mode of regulation and might involve amplification and/or feedback inhibition loops. In addition to regulating the intensity and duration of the different ROS signals, the ROS-scavenging pathways are also responsible for maintaining a low steady-state baseline of ROS on which the different signals can be registered. The reactive oxygen gene network therefore modulates the steady-state level of ROS in the different cellular compartments for signaling purposes as well as for protection against oxidative damage. It is possible that the use of ROS as versatile signaling molecules originated from their proposed use to sense stress. Most forms of biotic or abiotic stress disrupt the metabolic balance of cells, resulting in enhanced production of ROS. Simple organisms, such as bacteria or yeast, sense the enhanced production of ROS using redox-sensitive transcription factors and other molecular sensors, activate different ROS defense pathways, and regulate their metabolic pathways to lower the production rate of ROS (Costa and Moradas-Ferreira 2001; Georgiou 2002). The signal transduction pathway was demonstrated by Mittler et al. (2004) (Fig. 15.2).

## Stress and Transcription Factors

Studies of transcriptional activation of some stress-responsive genes have also led to the identification of cis-acting elements ABRE (ABA-responsive element) and DRE (dehydration-responsive element)/CRT (C-repeat) that function in ABA-dependent and ABA-independent gene expression in response to stress, respectively (Seki et al. 2003). Transcription factors belonging to the ethylene-responsive element-binding factor family that bind to DRE/CRT were also isolated (Liu et al. 1998). The genes encoding these transcription factors are induced early and transiently in response to cold, and these transcription factors, in turn, activate the expression of target genes. Liu et al. (1998) reported that the transcription factors (DREB2A and DREB2B) are also induced by dehydration and promote the expression of various genes involved in drought stress tolerance. The expression of a new DNA-binding protein DBF1 that specifically interact with the DRE2 cis-element of a corn rab17 gene promoter is induced by ABA, dehydration, and high salinity (Kizis and Pages 2002). Another example of a transcription factor is calcium-dependent protein kinases (CDPKs). CDPKs are implicated as important sensors of $Ca^{2+}$ flux in plants in response to stress (Ludwig et al. 2004). CDPKs are encoded by multigene families, and expression levels of these genes are spatially and temporally controlled throughout development. In addition, subsets of CDPKs are involved in signal transduction during stress including cold, salt, and drought or pathogen infection. A new transcription factor, BOS1 (*Botrytis susceptible 1*), was found to be required for both biotic and abiotic stress responses in *Arabidopsis* as reported by Mengiste et al. (2003). This complex network of interactions allows plants to respond in a highly specific fashion to the exact combination of environmental stresses encountered.

## Identification of the Differentially Expressed Proteins

Many cold stress-responsive proteins have been identified (Gao et al. 2009). These proteins inclu-deputative 2,3-bisphosphoglycerate-independent phosphoglycerate mutase, glycine-rich RNA-binding protein 7 (GRP7), ferritin-1, EF-G, phosphoglycerate kinase, ATP synthase CF1 beta chain, UDP-glucose pyrophosphorylase, cysteine proteinase inhibitor, RuBisCO small chain, carbonic anhydrase, and SAL1 phosphatase. New proteins such as isopropylmalate synthase, cytochrome b6–f complex iron–sulfur subunit, and SNF2 domain-containing protein/helicase domain-containing protein/zinc finger (C3HC4 type RING finger) family protein have been identified in response to cold stress. These novel cold-responsive proteins may play important roles in cold tolerance of *Thellungiella halophila*. The stress related proteins are involved in RNA metabolism and energy pathway, protein synthesis, folding and degradation as well as cell wall and cytoskeleton, metabolisms of nitrogen and sulfur, and signal transduction (Emanuelsson et al. 2000). The cold stress-responsive proteins linked to RNA metabolism, defense response, protein synthesis, and folding and degradation were all upregulated markedly during cold treatment. It suggests that these cellular processes were enhanced by exposure to cold stress. The identified protein involved in photosynthesis and energy pathway displayed diverse change patterns. *T. halophila* is a valuable model system for investigating the mechanisms involved in plant cold stress tolerance. Proteomic analysis of cold stress response in *T. halophila* can help to identify key regulators of cold tolerance in plants. Regulation of chloroplast functions under cold stress. The expression regulation of chloroplast proteins is of central importance in cold adaptation (Foyer et al. 1997), and a substantial portion (46%, 23/50) of the identified proteins were predicted to be localized in chloroplast. Eighteen of these have

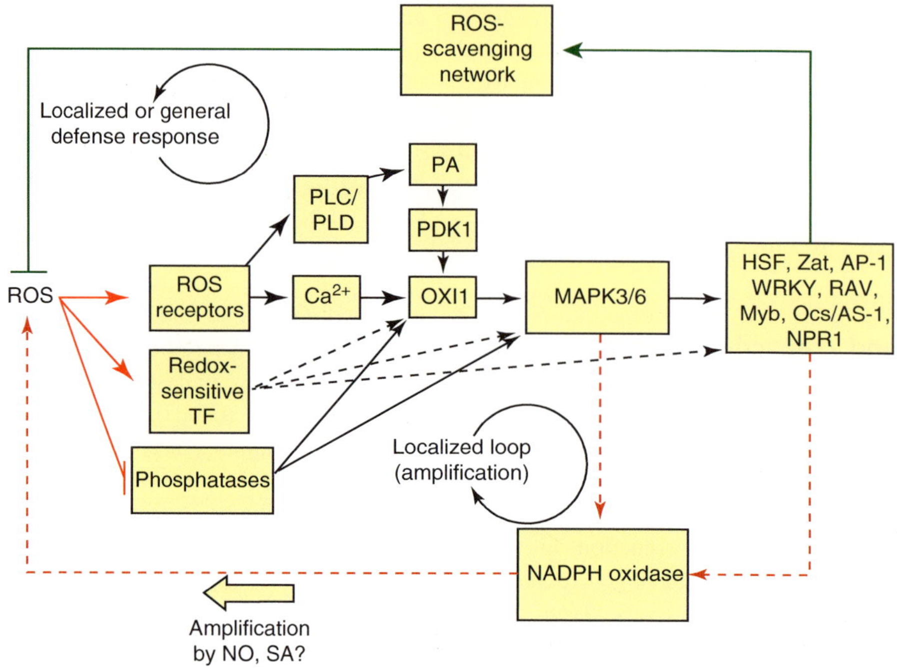


**Fig. 15.2** Model of the reactive oxygen species (ROS) signal transduction pathway. ROS can be detected by at least three mechanisms (ROS receptors, redox-sensitive transcription factors, and phosphatases). *HSF* heat shock factor, *PDK* phosphoinositide-dependent kinase, *TF* transcription factor (Published by Mittler et al. 2004)

been identified to reside in the chloroplast by previous proteomic studies according to the Plant Proteome Database (PPDB, http://ppdb.tc.cornell.edu/) (Sun et al. 2009). These chloroplast-located cold stress-responsive proteins are associated with various aspects of chloroplast, including Calvin cycle and electron transport (discussed below), chloroplast RNA processing (putative RNA-binding protein cp29), and chloroplast protein synthesis and folding (EF-G and heat-shock protein 70), as well as chloroplast redox homeostasis (BAS1, Peroxiredoxin-2E and Ferritin- 1). At the same time, a chloroplast-localized carbonic anhydrase, which facilitates $CO_2$ move across the chloroplast envelope, was found to decrease in abundance after 5 and 24 days of cold treatment. The changes in Calvin cycle enzymes observed in this study might be associated with decrease in photosynthetic $CO_2$ assimilation, possibly resulting in decreased growth rate in the long term. Enhanced RNA metabolism, including RNA processing, transporting from nucleus to cytoplasm, and mRNA secondary structure stability may be impaired under cold stress, especially for the cold stress-induced defense-related transcripts (Zhu et al. 2007). Several RNA-binding proteins and helicase-like proteins were all found to be upregulated by cold stress in the present study, including GRP7, glycine-rich RNA-binding protein GRP1A (GRP1A), putative RNA-binding protein cp29, and SNF2 domain-containing protein/helicase domain-containing protein/zinc finger (C3HC4-type RING finger) family proteins indicating enhanced RNA metabolism may play an important role in cold tolerance of *T. halophila*. Among these cold-responsive nucleic acid-binding proteins, GRPs are suggested to play an important role in posttranscriptional regulation of gene expression in plants under various stress conditions (Mousavi and Hotta 2005). GRP7 has been demonstrated to play a role in the export of mRNAs from the nucleus to the cytoplasm under cold stress conditions in *Arabidopsis* (Kim et al. 2008). RNA-binding protein cp29 is a subunit of the photosystem II and its phosphorylation was reported to relate with the cold tolerance in maize (Mauro et al. 1997). A RNA-binding protein cp29 was proved to be induced by cold stress in *Arabidopsis* (Amme et al. 2006). The increase in two chloroplast RNA-binding proteins may help to enhance or maintain the chloroplast RNA synthesis under cold conditions. In addition, GRPs and chloroplast RNA-binding proteins are suggested to be part of the plant innate immunity system and increase in their abundance may promote the expression of immunity-related mRNA (Fu et al. 2007).

## Conclusion

In conclusion, the proteomic analysis is a very useful tool for providing complex information about differences in the plant proteome during abiotic and biotic stresses. It is almost necessary to clarify the differential function of the individual stress-responsive NAC genes for the control of abiotic stress tolerance and the other biological processes including biotic stress tolerance and growth regulation in order to fully utilize the potential of NAC transcription factors. Large-scale transcriptome analyses coupled with microarray, proteomic, and metabolomic analysis of plants perturbed at the levels of individual or multiple components of the ROS network will be essential for future studies. In the future, a combination of reverse genetics, genomics, and proteomic approaches in various developmental stages and stress conditions will provide us with critical information to elucidate the functional differences of the stress-responsive NAC factors and their relationship in transcriptional control. It is a challenge for plant scientists/environmentalists/biotechnologists in the twenty-first century to develop stable multiple stress tolerance traits in agronomically important crop plants, thus improving yields particularly in areas with adverse environmental conditions and contributing to food security.

## References


Abbas HK, Williams WP, Windham GL, Pringle HC, Xie W, Shier WT (2002) Aflatoxin and fumonisin contamination of commercial corn (*Zea mays*) hybrids in Mississippi. J Agric Food Chem 50:5246–5254

Ahn YO, Kim SH, Lee J, Kim HR, Lee HS, Kwak SS (2012) Three *Brassica rapa* metallothionein genes are differentially regulated under various stress conditions. Mol Biol Rep 39:2059–2067

Alfano G, Ivey MLL, Cakir C, Bos JIB, Miller SA, Madden LV, Kamoun S, Hoitink HAJ (2007) Systemic modulation of gene expression in tomato by Trichoderma hamatum 382. Phytopathology 97:429–437

Allan AC, Fluhr R (1997) Two distinct sources of elicited reactive oxygen species in tobacco epidermal cells. Plant Cell 9:1559–1572

Amme S, Matros A, Schlesier B, Mock HP (2006) Proteome analysis of cold stress response in *Arabidopsis thaliana* using DIGE technology. J Exp Bot 57:1537–1546

Anjum NA, Ahmad I, Mohmood I (2012) Modulation of glutathione and its related enzymes in plants responses to toxic metals and metalloids – a review. Environ Exp Bot 75:307–324

Asada K, Takahashi M (1987) Production and scavenging of active oxygen in photosynthesis. In: Kyle DJ et al (eds) Photoinhibition, vol 9, Topics in photosynthesis. Elsevier, Amsterdam/New York, pp 227–287

Audenaert K, De Meyer GB, Hofte MM (2002) Abscisic acid determines basal susceptibility of tomato to *Botrytis cinerea* and suppresses salicylic acid-dependent signaling mechanisms. Plant Physiol 128:491–501

Bae H, Sicher RC, Kim MS, Kim S-H, Strem MD, Melnick RL, Bailey BA (2009) The beneficial endophyte Trichoderma hamatum isolate DIS 219b promotes growth and delays the onset of the drought response in Theobroma cacao. J Exp Bot 60:3279–3295

Baier M, Dietz K-J (1997) The plant 2-Cys peroxiredoxin BAS1 is a nuclear-encoded chloroplast protein: its expressional regulation, phylogenetic origin, and implications for its specific physiological function in plants. Plant J 60:282–314

Barconi D, Bernardini G, Santucci A (2011) Linking protein oxidation to environmental pollutants: redox proteome approaches. J Proteomics 74:2324–2337

Bohnert HJ, Nelson DE, Jensen RG (1995) Adaptations to environmental stresses. Plant Cell 7:1099–1111

Bolwell GP, Wojtaszek P (1997) Mechanisms for the generation of reactive oxygen species in plant defence – broad perspective. Physiol Mol Plant Pathol 51:347–366

Boursiac Y, Boudet J, Postaire O, Luu DT, Tournaire-Roux C, Maurel C (2008) Stimulus-induced down-regulation of root water transport involves reactive oxygen species-activated cell signalling and plasma membrane intrinsic protein internalization. Plant J 56 (2):207–218

Boyer JS (2009) Evans review: cell wall biosynthesis and the molecular mechanism of plant enlargement. Funct Plant Biol 36(5):383–394

Bray EA (1997) Plant responses to water deficit. Trends Plant Sci 2:48–54

Bray EA, Bailey-Serres J, Weretilnyk E (2000) Responses to abiotic stresses. In: Gruissem W, Buchannan B, Jones R (eds) Biochemistry and molecular biology of plants, vol 23. American Society of Plant Physiologists, Rockville, pp 1158–1203

Bréhélin C, Meyer EH, de Souris JP, Bonnard G, Meyer Y (2003) Resemblance and dissemblance of *Arabidopsis* type II peroxiredoxins: similar sequences for divergent gene expression, protein localization, and activity. Plant Physiol 132:2045–2057

Bressan RA, Zhang C, Zhang H, Hasegawa PM, Bohnert HJ, Zhu JK (2001) Learning from the *Arabidopsis* experience: the next gene search paradigm. Plant Physiol 127:1354–1360

Bruix M, Jimenez MA, Santoro J, Gonzalez C, Colilla FJ, Mendez E, Rico M (1993) Solution structure of gamma 1-H and gamma 1-P thionins from barley and wheat endosperm determined by 1H-NMR: a structural motif common to toxic arthropod proteins. Biochemistry 32:715–724

Cabuslay G, Ito O, Alejar A (1999) Genotypic differences in physiological responses to water deficit in rice. In: O'Toole J, Ito O, Hardy B (eds) Genetic improvement of rice for water-limited environments. International Rice Research Institute, Los Baños, pp 99–116

Carmona MJ, Molina A, Fernandez JA, Lopez-Fando JJ, Garcia-Olmedo F (1993) Expression of the a-thionin gene from barley in tobacco confers enhanced resistance to bacterial pathogens. Plant J 3:457–462

Caporale C, Di Berardino I, Leonardi L, Bertini L, Cascone A, Buonocore V, Caruso C (2004) Wheat pathogenesis-related proteins of class 4 have ribonuclease activity. FEBS Lett 575:71–76

Castiglione S, Franchin C, Fossati T, Lingua G, Torrigiani P, Biondi S (2007) High zinc concentrations reduce rooting capacity and alter metallothionein gene expression in white poplar (Populus alba L. cv. Villafranca). Chemosphere 67:1117–1126

Chen CH, Chen ZX (2002) Potentiation of developmentally regulated plant defense response by AtWRKY18, a pathogen induced *Arabidopsis* transcription factor. Plant Physiol 129:706–716

Chen W, Provart NJ, Glazebrook J, Katagiri F, Chang HS, Eulgem T, Mauch F, Luan S, Zou G, Whitham SA, Budworth PR, Tao Y, Xie Z, Chen X, Lam S, Kreps JA, Harper JF, Si-Ammour A, Mauch-Mani B, Heinlein M, Kobayashi K, Hohn T, Dangl JL, Wang X, Zhu T (2002a) Expression profile matrix of *Arabidopsis* transcription factor genes suggests their putative functions in response to environmental stresses. Plant Cell 14:559–574

Chen ZY, Brown RL, Damann KE, Cleveland TE (2002b) Identification of unique or elevated levels of kernel proteins in aflatoxin-resistant maize genotypes through proteome analysis. Phytopathology 92:1084–1094

Chen ZY, Brown RL, Cleveland TE (2004) Evidence for an association in corn between stress tolerance and resistance to *Aspergillus* flavus infection and aflatoxin contamination. Afr J Biotechnol 3:693–699

Cheng JP, Chen B, Xu YM, Zhou XA, Lu CM, Wei WH (2007) Construction of suppression subtractive cDNA library and analysis of differentially expressed genes during maturing period seed of high and low oil content soybean variety. Chin J Oil Crop Sci 29:365–371

Chinnusamy V, Gong Z, Zhu JK (2008) Abscisic acid-mediated epigenetic processes in plant development and stress responses. J Integr Plant Biol 50 (10):1187–1195

Clifford SC, Arndt S, Corlett JE, Joshi S, Sankhla N, Popp M, Jones HG (1998) The role of solute accumulation, osmotic adjustment and changes in cell wall elasticity in drought tolerance in Ziziphus mauritiana (Lamk.). J Exp Bot 49:967–977

Cobbett CS (2000) Phytochelatins and their roles in heavy metal detoxification. Plant Physiol 123:825–832

Cole RJ, Sanders TH, Hill RA, Blankenship PD (1985) Mean geocarposphere temperatures that induce preharvest aflatoxin contamination of peanuts under drought stress. Mycopathologia 91:41–46

Colilla FJ, Rocher A, Mendez E (1990) Gamma-purothionins: amino acid sequence of two polypeptides of a new family of thionins from wheat endosperm. FEBS Lett 270:191–194

Collinge M, Boller T (2001) Differential induction of two potato genes, Stprx2 and StNAC, in response to infection by Phytophthora infestans and to wounding. Plant Mol Biol 46:521–529

Cook D, Fowler S, Fiehn O, Thomashow MF (2004) A prominent role for the CBF cold response pathway in configuring the low-temperature metabolome of *Arabidopsis*. Proc Natl Acad Sci U S A 101:15243–15248

Costa V, Moradas-Ferreira P (2001) Oxidative stress and signal transduction in Saccharomyces cerevisiae: insights into ageing, apoptosis and diseases. Mol Aspects Med 22:217–246

Cramer GR (2010) Abiotic stress & plant responses from the whole vine to the genes. Aust J Grape Wine Res 16:86–93

Cramer GR, Ergul A, Grimplet J, Tillett RL, Tattersall EA, Bohlman MC, Vincent D, Sonderegger J, Evans J, Osborne C, Quilici D, Schlauch KA, Schooley DA, Cushman JC (2007) Water and salinity stress in grapevines: early and late changes in transcript and metabolite profiles. Funct Integr Genomics 7(2):111–134

Creelman RA, Mullet JE (1997) Biosynthesis and action of jasmonates in plants. Ann Rev Plant Physiol Plant Mol Biol 48:355–381

Creighyton DJ, Magliorini M, Pourmotabbed T, Guha MK (1988) Optimization of efficiency in the glyoxalase pathway. Biochemistry 27:7376–7384

Cui S, Huang F, Wang J, Ma X, Cheng Y, Liu J (2005) A proteomic analysis of cold stress responses in rice seedlings. Proteomics 5:3162–3172

DalCorso G, Farinati S, Maistri S, Furini F (2008) How plants cope with cadmium: staking all on metabolism and gene expression. J Integr Plant Biol 50(10):1268–1280

Datta SK, Muthukrishnan S (1999) Pathogenesis-related proteins in plants. CRC Press, Boca Raton

Deak KI, Malamy J (2005) Osmotic regulation of root system architecture. Plant J 43:17–28

Deák M, Horváth GV, Davletova S, Török K, Sass L, Vass I, Barna B, Király Z, Dudits D (1999) Plants ectopically expressing the iron binding protein, ferritin, are tolerant to oxidative damage and pathogens. Nat Biotechnol 17:192–196

Delaney TP, Uknes S, Vernooij B, Friedrich L, Weymann K, Negrotto D, Gaffney T, Gut-Rella M, Kessmann H, Ward E (1994) A central role of salicylic acid in plant disease resistance. Science 266:1247–1250

Deprost D, Yao L, Sormani R, Moreau M, Leterreux G, Nicolai M, Bedu M, Robaglia C, Meye C (2007) The Arabidopsis TOR kinase links plant growth, yield, stress resistance and mRNA translation. EMBO Rep 8(9):864–870

Diaz J, ten Have A, van Kan JA (2002) The role of ethylene and wound signaling in resistance of tomato to *Botrytis cinerea*. Plant Physiol 129:1341–1351

Dietz KJ, Bair M, Kramer U (1999) Free radical and reactive oxygen species as mediators of heavy metal toxicity in plants. In: Prasad MNV, Hagemeyer J (eds) Heavy metal stress in plants from molecules to ecosystems. Springer, Berlin, pp 73–79

Dilbirligi M, Gill KS (2003) Identification and analysis of expressed resistance gene sequences in wheat. Plant Mol Biol 53:771–787

Dinneny JR, Long TA, Wang JY, Jung JW, Mace D, Pointer S, Barron C, Brady SM, Schiefelbein J, Benfey PN (2008) Cell identity mediates the response of *Arabidopsis* roots to abiotic stress. Science 320 (5878):942–945

Dorner JW, Cole RJ, Sanders TH, Blankenship PD (1989) Interrelationship of kernel water activity, soil temperature, maturity, and phytoalexin production in preharvest aflatoxin contamination of drought-stressed peanuts. Mycopathologia 105:117–128

Dubey RS (2011) Metal toxicity, oxidative stress and antioxidative defense system in plants. In: Gupta SD (ed) Reactive oxygen species and antioxidants in higher plants. CRC Press, Boca Raton, USA, pp 177–203

Duval M, Hsieh TF, Kim SY, Thomas TL (2002) Molecular characterization of AtNAM: a member of the Arabidopsis NAC domain superfamily, Plant Mol. Biol 50:237–248

Duvick JP, Rood T, Rao AG, Marshak DR (1992) Purification and characterization of a novel antimicrobial peptide from maize (*Zea mays* L.) kernels. J Biol Chem 267:18814–18820

Emanuelsson O, Nielsen H, Brunak S, von Heijne G (2000) Predicting subcellular localization of proteins based on their N-terminal amino acid sequence. J Mol Biol 300:1005–1016

Feist AM, Palsson BO (2008) The growing scope of applications of genome-scale metabolic reconstructions using *Escherichia coli*. Nat Biotechnol 26 (6):659–667

Feys BJ, Parker JE (2000) Interplay of signaling pathways in plant disease resistance. Trends Genet 16: 449–455

Fowler S, Thomashow MF (2002) Arabidopsis transcriptome profiling indicates that multiple regulatory pathways are activated during cold acclimation in addition to the CBF cold response pathway. Plant Cell 14:1675–1690

Foyer CH (2002) The contribution of photosynthetic oxygen metabolism to oxidative stress in plants. In: Inzé D, Van Montagu M (eds) Oxidative stress in plants. Taylor and Francis, New York, pp 33–68

Foyer CH, Lelandais M, Kunert KJ (1997) Photooxidative stress in plants. Physiol Plant 92:696–717

Frye CA, Tang D, Innes RW (2001) Negative regulation of defense responses in plants by a conserved MAPKK kinase. Proc Natl Acad Sci U S A 98:373–378

Fu ZQ, Guo M, Jeong BR, Tian F, Elthon TE, Cerny RL, Staiger D, Alfano JR (2007) A type III eVector ADP-ribosylates RNA-binding proteins and quells plant immunity. Nature 447:284–288

Gallie DR, Le H, Caldwell C, Browning KS (1998) Analysis of translation elongation factors from wheat during development and following heat shock. Biochem Biophys Res Commun 245:295–300

Gamalero G, Erta G, Glick BR (2009) The use of microorganisms to facilitate the growth of plants in saline soils. Springer, Heidelberg

Gao F, Zhou Y, Zhu W, Li X, Fan L, Zhang G (2009) Proteomic analysis of coldstress-responsive proteins in Thellungiella rosette leaves. Planta 230:1033–1046

Gasic K, Korban S (2006) Heavy metal stress. In: Rao KVM, Raghavendra AS, Reddy KJ (eds) Physiology and molecular biology of stress tolerance in plants. Springer, Dordrecht, pp 219–254

Georgiou G (2002) How to flip the (redox) switch. Cell 111:607–610

Good AG, Zaplachinski ST (1994) The effects of drought stress on free amino acid accumulation and protein synthesis in *Brassica napus*. Physiol Plant 90(1):9–14

Goulas E, Schubert M, Kieselbach T, Kleczkowski LA, Gardeström P, Schröder W, Hurry V (2006) The chloroplast lumen and stromal proteomes of *Arabidopsis thaliana* show differential sensitivity to short- and long-term exposure to low temperature. Plant J 47:720–734

Gray GR, Heath D (2005) A global reorganization of the metabolome in *Arabidopsis* during cold acclimation is revealed by metabolome fingerprinting. Plant Physiol 124:236–248

Grover A, Kapoor A, Lakshmi OS, Agarwal S (2001) Understanding molecular alphabets of the plant abiotic stress responses. Plant Mol Biol 80:206–216

Hajduch M, Rakwal R, Agrawal GK, Yonekura M, Pretova A (2001) High resolution two-dimensional electrophoresis separation of proteins from metal-stressed rice (Oryza sativa L.) leaves: drastic reductions/fragmentation of ribulose-1, 5-bisphosphate carboxylase oxygenase and induction of stress-related proteins. Electrophoresis 22:2824–2831

Hajheidari M, Abdollahian-Noghabi M, Askari H, Hedari M, Sadeghian SY, Ober ES, Hosseini Salekdeh GH (2005) Proteome analysis of sugar beet leaves under drought stress. Proteomics 5:950–960

Hamer DH (1986) Metallothionein. Annu Rev Biochem 55:913–951

Hammond-Kosack KE, Jones JDG (2000) Response to plant pathogens. In: Buchannan B, Gruissem W, Jones R (eds) Biochemistry and molecular biology of plants. American Society of Plant Physiologists, Rockville, pp 1102–1157

Hammond-Kosacky KE, Parkerz JE (2003) Deciphering plant– pathogen communication: fresh perspectives for molecular resistance breeding. Curr Opin Biotechnol 14:177–193

Harman GE (2000) Myths and dogmas of biocontrol: changes in perceptions derived from research on Trichoderma harzianum T-22. Plant Dis 84:377–393

Harman GE (2006) Overview of mechanisms and uses of *Trichoderma* spp. Phytopathology 96:190–194

Harman GE, Howell CR, Viterbo A, Chet I, Lorito M (2004) Trichoderma species—opportunistic, a virulent plant symbionts. Nat Rev Microbiol 2:43–56

Hashimoto M, Komatsu S (2007) Proteomic analysis of rice seedlings during cold stress. Proteomics 2007 (7):1293–1302

Hegedus D, Yu M, Baldwin D, Gruber M, Sharpe A, Parkin I, Whitwill S, Lydiate D (2003) Molecular characterization of *Brassica napus* NAC domain transcritional activators induced in response to biotic and abiotic stress. Plant Mol Biol 53:383–397

Hejgaard J, Jacobsen S, Svendsen I (1991) Two antifungal thaumatin like proteins from barley grain. FEBS Lett 291:27–131

Herms DA, Mattson WJ (1992) The dilemma of plants—to grow or defend. Q Rev Biol 67:283–335

Hey SJ, Byrne E, Halford NG (2010) The interface between metabolic and stress signaling. Ann Bot 105 (2):197–203

Hill RA, Blankenship PD, Cole RJ, Sanders TH (1983) Effects of soil moisture and temperature on preharvest invasion of peanuts by the *Aspergillus flavus* group and subsequent aflatoxin development. Appl Environ Microbiol 45:628–633

Hirai MY, Yano M, Goodenowe DB, Kanaya S, Kimura T, Awazuhara M, Arita M, Fujiwara T, Saito K (2004) Integration of transcriptomics and metabolomics for understanding of global responses to nutritional stresses in Arabidopsis thaliana. Proc Natl Acad Sci USA 101:10205–10210

Hirai MY, Sugiyama K, Sawada Y, Tohge T, Obayashi T, Suzuki A, Araki R, Sakurai N, Suzuki H, Aoki K, Goda H, Nishizawa OI, Shibata D, Saito K (2007) Omics-based identification of Arabidopsis Myb transcription factors regulating aliphatic glucosinolate biosynthesis. Proc Natl Acad Sci USA 104:6478–6483

Hoffmann-Sommergruber K (2002) Pathogenesis-related (PR)-proteins identified as allergens. Biochem Soc Trans 30:930–935

Holappa LD, Simmons MKW (1995) The wheat abscisic acid-responsive protein kinase mRNA, PKABA1, is up-regulated by dehydration, cold temperature, and osmotic stress. Plant Physiol 108:1203–1210

Hoque MA, Uraji M, Akhter Banu MN, Mori IC, Nakamura Y, Murata Y (2010) The effects of methylglyoxal on glutathione S-transferase from *Nicotiana tabacum*. Biosci Biotechnol Biochem 74(10):2124–2126

Hossain MA, Fujita M (2010) Evidence for a role of exogenous glycinebetaine and proline in antioxidant defense and methylglyoxal detoxification systems in mung bean seedlings under salt stress. Physiol Mol Biol Plant 16:19–29

Hossain MA, Hossain MZ, Fujita M (2009) Stress-induced changes of methylglyoxal level and glyoxalase I activity in pumpkin seedlings and cDNA cloning of glyoxalase I gene. Aust J Crop Sci 3(2):53–64

Hossain MA, Hasanuzzaman M, Fujita M (2010) Up-regulation of antioxidant and glyoxalase systems by exogenous glycinebetaine and proline in mung bean confertolerance to cadmium stress. Physiol Mol Biol Plant 16(3):259–272

Hossain MA, Hasanuzzaman M, Fujita M (2011a) Coordinate induction of antioxidant defense and glyoxalase system by exogenous proline and glycinebetaine is correlated with salt tolerance in mung bean. Front Agric China 5(1):1–14

Hossain MA, da Silva JAT, Fujita M (2011b) Glyoxalase system and reactive oxygen species detoxification system in plant abiotic stress response and tolerance: an intimate relationship. In: Shanker AK, Venkateswarlu B (eds) Abiotic stress in plants-mechanisms and adaptations. INTECH-Open Access Publisher, Rijeka, pp 235–266

Hossain MA, Hossain MD, Rohman MM, da Silva JAT, Fujita M (2012) Onion major compounds (flavonoids, organosulfurs) and highly expressed glutathione-related enzymes: possible physiological interaction, gene cloning and abiotic stress response. In: Aguirre CB, Jaramillo LM (eds) Onion consumption and health. Nova, New York

Houde M, Belcaid M, Ouellet F, Danyluk J, Monroy AF, Dryanova A, Gulick P, Bergeron A, Laroche A, Links MG, MacCarthy L, Crosby WL, Sarhan F (2006) Wheat EST resources for functional genomics of abiotic stress. BMC Genomics 7:149–166

Hu H, Dai M, Yao J, Xiao B, Li X, Zhang Q, Xiong L (2006) Over expressing a NAM, ATAF, and CUC (NAC) transcription factor enhances drought resistance and salt tolerance in rice. Proc Natl Acad Sci U S A 103:12987–12992

Hubbard KE, Nishimura N, Hitomi K, Getzoff ED, Schroeder JI (2010) Early abscisic acid signal transduction mechanisms: newly discovered components and newly emerging questions. Genes 24(16):1695–1708

Hurkman WJ, Tanaka CK (2007) Extraction of wheat endosperm proteins for proteome analysis. J Chromatogr 849:344–350

Iusem ND, Bartholomew DM, Hitz WD, Scolnik PA (1993) Tomato (Lycopersicon esculentum) transcript induced by water deficit and ripening. Plant Physiol 102:1353–1354

Jacob U, Gaestel M, Engel K, Buchner J (1993) Small heat shock proteins are molecular chaperones. J Biol Chem 268:1517–1520

Jones RK, Duncan HE, Hamilton PB (1981) Planting date, harvest date, and irrigation effects on infection and aflatoxin production by *Aspergillus flavus* in field corn. Phytopathology 71:810–816

Kagi JHR (1991) Overview of metallothionein. Methods Enzymol 205:613–626

Kaufmann K, Smaczniak C, De VS, Angenent GC, Karlova R (2011) Proteomics insights into plant signaling and development. Proteomics 11 (4):744–755

Kawamura Y, Uemura M (2003) Mass spectrometric approach for identifying putative plasma membrane proteins of *Arabidopsis* leaves associated with cold acclimation. Plant J 36:141–154

Kawasaki S, Miyake C, Kohci T, Fujii S, Uchida M, Yokota A (2000) Response of wild watermelon to drought stress: accumulation of an ArgE homologue and citrulline in leaves during water deficits. Plant Cell Physiol 41:864–873

Khedr AHA, Abbas MA, Abdel Wahid AA, Quick WP, Abogadallah GM (2003) Proline induces the expression of salt stress- responsive proteins and may improve the adaptation of *Pancratium maritimum* L. to salt-stress. J Exp Bot 54:2553–2562

Khuri S, Bakker FT, Dunwell JM (2001) Phylogeny, function, and evolution of the cupins, a structurally conserved, functionally diverse superfamily of proteins. Mol Biol Evol 18:593–605

Kikuchi K, Ueguchi-Tanaka M, Yoshida KT, Nagato Y, Matsusoka M, Hirano HY (2000) Molecular analysis of the NAC gene family in rice. Mol Gen Genet 54:1047–1051

Kilian J, Whitehead D, Horak J, Wanke D, Weinl S, Batistic O, D'Angelo C, Bornberg-Bauer E, Kudla J, Harter K (2007) The AtGenExpress global stress expression data set: protocols, evaluation and model data analysis of UV-B light, drought and cold stress responses. Plant J 50(2):347–363

Kim JS, Jung HJ, Lee HJ, Kim KA, Goh CH, Woo Y, Oh SH, Han YS, Kang H (2008) Glycine rich RNA-binding protein 7a Vects abiotic stress responses by regulating stomata opening and closing in *Arabidopsis thaliana*. Plant J 55:455–466

Kim TH, Bohmer M, Hu H, Nishimura N, Schroeder JI (2010) Guard cell signal transduction network: advances in understanding abscisic acid, CO2, and Ca2+ signaling. Annu Rev Plant Biol 61:561–591

Kitano H (2002) Systems biology: a brief overview. Science 295(5560):1662–1664

Kizis D, Pages M (2002) Maize DRE-binding proteins DBF1 and DBF2 are involved in rab17 regulation through the drought-responsive element in an ABA-dependent pathway. Plant J 30:679–689

Knoester M, Van Loon LC, Den V, Heuvel J, Hennig J, Bol JF, Linthorst HJM (1998) Ethylene-insensitive tobacco lacks nonhost resistance against soil-borne fungi. Proc Natl Acad Sci U S A 95:1933–1937

Kreps JA, Wu Y, Chang HS, Zhu T, Wang X, Harper JF (2002) Transcriptome changes for *Arabidopsis* in response to salt, osmotic, and cold stress. Plant Physiol 130:2129–2141

Kwon SJ, Kwon SI, Bae MS, Cho EJ, Park OK (2007) Role of the methionine sulfoxide reductase msrB3 in cold acclimation in *Arabidopsis* plant and cell physiology. Plant Cell Physiol 48:1713–1723

Lageix S, Lanet E, Pouch-Pelissier MN, Espagnol MC, Robaglia C, Deragon JM, Pelissier T (2008) Arabidopsis eIF2alpha kinase GCN2 is essential for growth in stress conditions and is activated by wounding. BMC Plant Biol 8:134

Lamb C, Dixon RA (1997) The oxidative burst in plant disease resistance. Annu Rev Plant Physiol Plant Mol Biol 48:251–275

Legrand M, Kauffmann S, Geoffroy P, Fritig B (1987) Biological function of pathogenesis related proteins: four PR proteins of tobacco are chitinases. Proc Natl Acad Sci U S A 84:6750–6754

Leung J, Giraudat J (1998) Abscisic acid signal transduction. Annu Rev Plant Mol Biol 49:199–222

Li W, Faris JD, Muthukrishnan S, Liu D, Chen P, Gill BS (2001) Isolation and characterization of cDNA clones of acidic chitinases and beta-1, 3-Glucanases from wheat spike induced by Fusarium graminearum. Theor Appl Genet 102:353–362

Li AL, Zhu YF, Tan XM, Wang X, Wei B, Guo HZ, Zhang ZL, Chen XB, Zhao GY, Kong XY, Jia JZ, Mao L (2008) Evolutionary and functional study of the CDPK gene family in wheat (*Triticum aestivum* L.). Plant Mol Biol 66:429–443

Liu JX, Howell SH (2010) Endoplasmic reticulum protein quality control and its relationship to environmental stress responses in plants. Plant Cell 22(9):2930–2942

Liu Q, Kasuga M, Sakuma Y, Abe H, Miura S, Yamaguchi-Shinozaki K, Shinozaki K (1998) Two transcription factors, DREB1 and DREB2, with an EREBP/AP2 DNA binding domain separate two cellular signal transduction pathways in drought- and low temperature responsive gene expression, respectively, in Arabidopsis. Plant Cell 10:1391–1406

Liu HT, Li B, Shang ZL, Li XZ, Mu RL, Sun DY, Zhou RG (2003) Calmodulin is involved in heat shock signal transduction in wheat. Plant Physiol 132:1186–1195

Lu B, Fang XJ (2003) Express analysis of cDNA induced by infecting roots of soybean cyst nematode race 4 in soybean ZZD2315. Mol Plant Breed 1: 193–200

Lu Z, Neumann PM (1999) Water stress inhibits hydraulic conductance and leaf growth in rice seedling but the transport of water via mercury-sensitive water channels in the root. Plant Physiol 120:143–152

Ludwig AA, Romeis T, Jones JD (2004) CDPK-mediated signaling pathways: specificity and cross-talk. J Exp Bot 55:181–188

Ma Y, Szostkiewicz I, Korte A, Moes D, Yang Y, Christmann A, Gril E (2009) Regulators of PP2C phosphatase activity function as abscisic acid sensors. Science 324:1064–1068

Majoul T, Bancel E, Triboï E, Ben Hamida J, Branlard G (2003) Proteomic analysis of the effect of heat stress on hexaploid wheat grain: characterization of heat-responsive proteins from total endosperm. Proteomics 3:175–183

Mao L, Van HJL, Dash S, Dickerson JA (2009) Arabidopsis gene co-expression network and its functional modules. BMC Bioinformatics 10:346

Mauro S, Dainese P, Lannoye R, Bassi R (1997) Cold-resistant and cold-sensitive maize lines differ in the phosphorylation of the photosystem II subunit, CP29. Plant Physiol 115:171–180

Meharg AA (1993) The role of plasmalemma in metal tolerance in angiosperm. Physiol Plant 88 (1):191–198

Mengiste T, Chen X, Salmeron J, Dietrich R (2003) The BOTRYTIS SUSCEPTIBLE1 gene encodes an R2R3MYB transcription factor protein that is required for biotic and abiotic stress responses in *Arabidopsis*. Plant Cell 15:2551–2565

Mittler R (2002) Oxidative stress, antioxidants and stress tolerance. Trends Plant Sci 7:405–410

Mittler R, Vanderauwera S, Gollery M, Van Breusegem F (2004) Reactive oxygen gene network of plants. Trends Plant Sci 9:490–498

Mittler R, Vanderauwera S, Suzuki N, Miller G, Tognetti VB, Vandepoele K, Gollery M, Shulaev V, Van BF (2010) ROS signaling: the new wave? Trends Plant Sci 16(6):300–309

Mock HP, Grimm B (1997) Reduction of uroporphyrinogen decarboxylase by antisense RNA expression affects activities of other enzymes involved in tetrapyrrole biosynthesis and leads to light dependent necrosis. Plant Physiol 113:1101–1112

Molassiotis A, Fotopoulos V (2011) Oxidative and nitrosative signaling in plants: two branches in the same tree? Plant Signal Behav 6(2):210–214

Møller IM (2001) Plant mitochondria and oxidative stress: electron transport, NADPH turnover, and metabolism of reactive oxygen species. Annu Rev Plant Physiol Plant Mol Biol 52:561–591

Morishita T, Kojima Y, Maruta T, Nishizawa-Yokoi A, Yabuta Y, Shigeoka S (2009) Arabidopsis NAC transcription factor, ANAC078, regulates flavonoid biosynthesis under high-light. Plant Cell Physiol 50:2210–2222

Mostafa Kamal AH, Kim KH, Shin KH, Kim DE, Oh MW, Choi JS, Hirano H, Heo HY, Woo SH (2010a) Proteomics-based dissection of biotic stress responsive proteins in bread wheat (*Triticum aestivum* L.). Afr J Biotechnol 9:7239–7255

Mostafa Kamal A, Kim AH, Shin KH, Choi JS, Baik NK, Tsujimoto H, Heo HY, Park CS, Wool SH (2010b) Abiotic stress responsive proteins of wheat grain determined using proteomics technique. Aust J Crop Sci 4(3):196–208

Mousavi A, Hotta Y (2005) Glycine-rich proteins: a class of novel proteins. Appl Biochem Biotechnol 120:169–174

Mysore KS, Crasta OR, Tuori RP, Folkerts O, Swirsky PB, Martin GB (2002) Comprehensive transcript profiling of Pto- and Prf-mediated host defense responses to infection by Pseudomonas syringae pv. tomato. Plant J 32:299–315

Nakashima K, Ito Y, Yamaguchi-Shinozaki K (2009) Transcriptional regulatory networks in response to abiotic stresses in Arabidopsis and grasses. Plant Physiol 149:88–95

Nakashima K, Takasaki H, Mizoi J, Shinozaki K, Yamaguchi-Shinozaki K (2012) NAC transcription factors in plant abiotic stress responses. Biochim Biophys Acta 1819:97–103

Navari-Izzo F (1998) Thylakoid-bound and stromal antioxidative enzymes in wheat treated with excess copper. Physiol Plant 104:630–638

Nieboer E, Richardson DHS (1980) The replacement of the nondescript term heavy Metals' by a biologically and chemically significant classification of metal ions. Environ Pollut Ser B Chem Phys 1(1):3–26

Nogueira FTS, Schlögl PS, Camargo SR, Fernandez JH, De Rosa VE Jr, Pompermayer P, Arruda P (2005) SsNAC23, a member of the NAC domain protein family, is associated with cold, herbivory and water stress in sugarcane. Plant Sci 169:93–106

Nuruzzaman M, Manimekalai R, Sharoni AM, Satoh K, Kondoh H, Ooka H, Kikuchi S (2010) Genome-wide analysis of NAC transcription factor family in rice. Gene 465:30–44

O'Donnell PJ, Schmelz E, Block A, Miersch O, Wasternack C, Jones JB, Klee HJ (2003) Multiple hormones act sequentially to mediate a susceptible tomato pathogen defense response. Plant Physiol 133:1181–1189

Okuma E, Soeda K, Tada M, Murata Y (2000) Exogenous proline mitigates the inhibition of growth of Nicotiana tabacum cultured cells under saline conditions. Soil Sci Plant Nutr 46:257–263

Ooka H, Satoh K, Doi K, Nagata T, Otomo Y, Murakami K, Matsubara K, Osato N, Kawai J, Carninci P, Hayashizaki Y, Suzuki K, Kojima K, Takahara Y, Yamamoto K, Kikuchi S (2003) Comprehensive analysis of NAC family genes in *Oryza sativa* and *Arabidopsis thaliana*. DNA Res 10:239–247

Ouerghi Z, Rémy R, Ouelhazi L, Ayadi A, Brulfert J (2000) Two dimensional electrophoresis of soluble leaf proteins, isolated from two wheat species (Triticum durum and Triticum aestivum) differing in sensitivity towards NaCl. Electrophoresis 21:248–2491

Parent B, Hachez C, Redondo E, Simonneau T, Chaumont F, Tardieu F (2009) Drought and abscisic acid effects on aquaporin content translate into changes in hydraulic conductivity and leaf growth rate: a trans-scale approach. Plant Physiol 149(4):2000–2012

Park JM, Park CJ, Lee SB, Ham BK, Shin R, Paek KH (2001) Over expression of the tobacco Tsi1 gene encoding an EREBP/AP2- type transcription factor enhances resistance against pathogen attack and osmotic stress in tobacco. Plant Cell 13:1035–1046

Park SY, Fung P, Nishimura N, Jensen DR, Fujii H, Zhao Y, Lumba S, Santiago J, Rodrigues A, Chow TF, Alfred SE, Bonetta D, Finkelstein R, Provart NJ, Desveaux D, Rodriguez PL, McCourt P, Zhu JK, Schroeder JI, Volkman BF, Cutler SR (2009) Abscisic acid inhibits type 2C protein phosphatases via the PYR/PYL family of START proteins. Science 324:1068–1071

Payne GA (1998) Process of contamination by aflatoxin producing fungi and their impact on crops. In: Sinha KK, Bhatnagar D (eds) Mycotoxins in agriculture and food safety. Marcel Dekker, New York, pp 279–306

Payne GA, Cassel DK, Adkins CR (1986) Reduction of aflatoxin contamination in corn by irrigation and tillage. Phytopathology 76:679–684

Pedrol N, Ramos P, Reigosa MJ (2000) Phenotypic plasticity and acclimation to water deficits in velvet-grass: a long-term greenhouse experiment. Changes in leaf morphology, photosynthesis and stress-induced metabolites. J Plant Physiol 157:383–393

Petersen M, Brodersen P, Naested H, Andreasson E, Lindhart U, Johansen B, Nielsen HB, Lacy M, Austin MJ, Parker JE (2000) Arabidopsis MAP kinase 4 negatively regulates systemic acquired resistance. Cell 103:1111–1120

Pinheiro C, Chaves MM (2011) Photosynthesis and drought: can we make metabolic connections from available data? J Exp Bot 62(3):869–882

Qureshi MI, Qadir S, Zolla L (2007) Proteomics-based dissection of stress-responsive in plants. Plant Physiol 164:1239–1260

Rabbani MA, Maruyama K, Abe H, Khan MA, Katsura K, Ito Y, Yoshiwara K, Seki M, Shinozaki K, Yamaguchi-Shinozaki K (2003) Monitoring expression profiles of rice genes under cold, drought, and high-salinity stresses and abscisic acid application using cDNA microarray and RNA gel-blot analyses. Plant Physiol 133:1755–1767

Riccardi F, Gazeau P, de Vienne D, Zivy M (1998) Protein changes in response to progressive water deficit in maize: quantitative variations and identification. Plant Physiol 117:1253–1263

Rivetta A, Negrini N, Cocucci M (1997) Involvement of $Ca^{2+}$-calmodulin in $Cd^{2+}$ toxicity during the early phases of radish (*Raphanus sativus* L.) seed germination. Plant Cell Environ 20:600–608

Rizhsky L, Liang H, Mittler R (2002) The combined effect of drought stress and heat shock on gene expression in tobacco. Plant Physiol 130:1143–1151

Rockstrom J, Falkenmark M (2000) Semiarid crop production from a hydrological perspective: gap between potential and actual yields. Crit Rev Plant Sci 19:319–346

Romero-Puertas MC, Palma JM, Gomez M, Del R, Sandalio LM (2002) Cadmium causes the oxidative modification of proteins in pea plants. Plant Cell Environ 25:677–686

Ruelland E, Vaultier MN, Zachowski A, Hurry V (2009) Cold signalling and cold acclimation in plants. Adv Bot Res 49:35–150

Sabehat A, Lurie S, Weiss D (1998) Expression of small heat-shock proteins at low temperatures. A possible role in protecting against chilling injuries. Plant Physiol 117:651–658

Saito R, Yamamoto H, Makino A, Sugimoto T, Miyake C (2011) Methylglyoxal functions as Hill oxidant and stimulates the photoreduction of $O_2$ at photosystem I: a symptom of plant diabetes. Plant Cell Environ 34 (9):1454–1464

Salekdeh GH, Siopongco HJ, Wade LJ, Ghareyazie B, Bennett J (2002a) A proteomics approach to analysing drought and salt-responsiveness in rice. Field Crop Res 76:199–219

Salekdeh GH, Siopongco HJ, Wade LJ, Ghareyazie B, Bennett J (2002b) Proteomics analysis of rice leaves during drought stress and recovery. Proteomics 2:1131–1145

Samaras Y, Bressan RA, Csonka LN, Gracia-Rios MG, Paino D'Urzo M, Rhodes D (1995) Proline accumulation during drought and salinity. In: Smirnoff N (ed) Environment and plant metabolism: flexibility and accumulation. Oxford publishers, UK, pp 161–187

Schutzendubel A, Polle A (2002) Plant responses to abiotic stresses: heavy metal-induced oxidative stress and protection by mycorrhization. J Exp Bot 53: 1351–1365

Segal G, Feldman M (1996) Tubb1, a beta-tubulin cDNA from common wheat. EMBL/GenBank/DDBJ databases

Seki M, Kamei A, Yamaguchi-Shinozaki K, Shinozaki K (2003) Molecular responses to drought, salinity and frost: common and different paths for plant protection. Curr Opin Biotechnology 14:194–199

Serrano R, Gaxiola R (1994) Microbial models and stress tolerance in plants. Crit Rev Plant Sci 13:121–138

Sharma SS, Dietz KJ (2009) The relationship between metal toxicity and cellular redox imbalance. Trends Plant Sci 14(1):43–50

Sharma P, Dubey RS (2007) Involvement of oxidative stress and role of antioxidative defense system in growing rice seedlings exposed to toxic concentrations of aluminum. Plant Cell Rep 26(11): 2027–2038

Shin R, Park JM, An JM, Paek KH (2002) Ectopic expression of Tsi1 in transgenic hot pepper plants enhances host resistance to viral, bacterial, and oomycete pathogens. Mol Plant Microbe Interact 15:983–989

Shoresh M, Mastouri F, Harman G (2010) Induced systemic resistance and plant responses to fungal biocontrol agents. Annu Rev Phytopathol 48:21–43

Silhavy D, Hutvagner G, Barta E, Banfalvi Z (1995) Isolation and characterization of a water-stress-inducible cDNA clone from Solanum chacoense. Plant Mol Biol 27:587–595

Singla-Pareek SL, Reddy MK, Sopory SK (2003) Genetic engineering of the glyoxalase pathway in tobacco leads to enhanced salinity tolerance. Proc Natl Acad Sci U S A 100:14672–14677

Skirycz A, Inze D (2010) More from less: plant growth under limited water. Curr Opin Biotechnol 21 (2):197–203

Smeekens S, Ma J, Hanson J, Rolland F (2010) Sugar signals and molecular networks controlling plant growth. Curr Opin Plant Biol 13(3):274–279

Snarpi SM, Horie T, Motoda J, Kubo M, Yang H, Yoda K, Hrie R, Chan WY, Leung HY, Hattori K, Konomi M, Osumi M, Yamagami M, Schroeder JI, Uozumi N (2005) Enhanced salt tolerance mediated by AtHKT1 transporter-induced Na unloading from xylem vessels to xylem parenchyma cells. Plant J 44:928–938

Sonia C, Montserrat C, Maria C, Joaquin A, Blanca SS (2004) The defense response of germinating maize embryos against fungal infection: a proteomics approach. Proteomics 4:383–396

Sperotto RA, Ricachenevsky FK, Duarte GL, Boff T, Lopes KL, Sperb ER, Grusak MA, Fett JP (2009) Identification of up-regulated genes in flag leaves during rice grain filling and characterization of OsNAC5, a new ABA-dependent transcription factor. Planta 230:985–1002

Staschke KA, Dey S, Zaborske JM, Palam LR, McClintick JN, Pan T, Edenberg HJ, Wek RC (2010) Integration of general amino acid control and target of rapamycin (TOR) regulatory pathways in nitrogen assimilation in yeast. J Biol Chem 285 (22):16893–16911

Sun GZ, Zhu ZD, Wu XF, Cui YL, Wang XM (2008) Comparative proteomics of Arabidopsis induced by salicylic acid, wound and Phytophthora sojae. Scientia Agri Sin 4:1030–1039

Sun Q, Zybailov B, Majeran W, Friso G, Olinares PD, van Wijk KJ (2009) PPDB, the plant proteomics database at Cornell. Nucleic Acids Res 37:969–974

Takahashi S, Seki M, Ishida J, Satou M, Sakurai T, Narusaka M, Kamiya A, Nakajima M, Enju A, Akiyama K, Yamaguchi-Shinozaki K, Shinozaki K (2004) Monitoring the expression profiles of genes induced by hyper-osmotic, high salinity, and oxidative stress and abscisic acid treatment in Arabidopsis cell culture using a full-length cDNA microarray. Plant Mol Biol 56(1):29–55

Tamura T, Hara K, Yamaguchi Y, Koizumi N, Sano H (2003) Osmotic stress tolerance of transgenic tobacco expressing a gene encoding a membrane-located receptor-like protein from tobacco plants. Plant Physiol 131:454–462

Tattersall EA, Grimplet J, Deluc L, Wheatley MD, Vincent D, Osborne C, Ergul A, Lomen E, Blank RR, Schlauch KA, Cushman JC, Cramer GR (2007) Transcript abundance profiles reveal larger and more complex responses of grapevine to chilling compared to osmotic and salinity stress. Funct Integr Genomics 7 (4):317–333

Thomann EB, Sollinger J, White C, Rivin CJ (1992) Accumulation of group 3 late embryogenesis abundant proteins in *Zea mays* embryos. Plant Physiol 99:607–614

Thomma B, Eggermont K, Penninckx I, Mauch-Mani B, Vogelsang R, Cammue BPA, Broekaert WF (1998) Separate jasmonate dependent and salicylate-dependent defense response pathways in Arabidopsis are essential for resistance to distinct microbial pathogens. Proc Natl Acad Sci U S A 95:15107–15111

Tran LS, Quach TN, Guttikonda SK, Aldrich DL, Kumar R, Neelakandan A, Valliyodan B, Nguyen HT (2009) Molecular characterization of stress-inducible GmNAC genes in soybean. Mol Genet Genomics 281:647–664

Tran LSP, Nishiyama R, Yamaguchi-Shinozaki K, Shinozaki K (2010) Potential utilization of NAC transcription factors to enhance abiotic stress tolerance in plants by biotechnological approach. GM Crops 1(1):32–39

Treglia A, Spano G, Rampino P, Giangrande E, Nocco G, Mita G, Di Fonzo N, Peppotta C (1999) Identification by *in vitro* translation and Northern blot analysis of heat shock mRNAs isolated from wheat seeds exposed to different temperatures during ripening. Cereal Sci 30:33–38

Tubajika KM, Damann KE (2001) Sources of resistance to aflatoxin production in maize. J Agric Food Chem 49:2652–2656

Umezawa T (2011) Systems biology approaches to abscisic acid signaling. J Plant Res 124(4):539–548

Umezawa T, Nakashima K, Miyakawa T, Kuromori T, Tanokura M, Shinozaki K, Yamaguchi-Shinozaki K (2010) Molecular basis of the core regulatory network in abiotic responses: sensing, signaling and transport. Plant Cell Physiol 51:1821–1839

Urao T, Yakubov B, Satoh R, Yamaguchi-Shinozaki K, Seki M, Hirayama T, Shinozaki K (1999) A transmembrane hybrid-type histidine kinase in Arabidopsis functions as an osmosensor. Plant Cell 11:1743–1754

Van Assche F, Clijsters H (1986) Inhibition of photosynthesis in *Phaseolus vulgaris* by treatment with toxic concentration of zinc: effect on ribulose-1,5-bisphosphate carboxylase/ oxygenase. J Plant Physiol 125:355–360

van der Weel CM, Spollen WF, Sharo RE, Baskin TI (2000) Growth of *Arabidopsis thaliana* seedlings under water deficit studied by control of water potential in nutrient-agar media. J Exp Bot 51:1555–1562

Veena RVS, Sopory SK (1999) Glyoxalase 1 from *Brassica juncea*: molecular cloning, regulation and its over-expression confer tolerance in transgenic tobacco under stress. Plant J 17:385–395

Vierling E (1991) The roles of heat shock proteins in plants. Annu Rev Plant Physiol Plant Mol Biol 42:579–620

Villiers F, Ducruix C, Hugouvieux V (2011) Investigating the plant response to cadmium exposure by proteomic and metabolomic approaches. Proteomics 11 (9):1650–1663

Vincent D, Ergul A, Bohlman MC, Tattersall EA, Tillett RL, Wheatley MD, Woolsey R, Quilici DR, Joets J, Schlauch K, Schooley DA, Cushman JC, Cramer GR (2007) Proteomic analysis reveals differences between *Vitis vinifera* L. cv. Chardonnay and cv. Cabernet Sauvignon and their responses to water deficit and salinity. J Exp Bot 58(7):1873–1892

Wakabayashi K, Hoson T, Kamisaka S (1997) Osmotic stress suppresses cell wall stiffening and the increase in cell wall-bound ferulic and diferulic acids in wheat coleoptiles. Plant Physiol 113:967–973

Walters DR (2003) Polyamines and plant disease. Phytochemistry 64:97–107

Wang WX, Vinocur B, Altman A (2003) Plant responses to drought, salinity and extreme temperatures: towards genetic engineering for stress tolerance. Planta 218:1–14

Wang W, Vinocur B, Shoseyov O, Altman A (2004) Role of plant heat-shock proteins and molecular chaperones in the abiotic stress response. Trends Plant Sci 9:244–252

Wehmeyer N, Hernandez LD, Finkelstein RR, Vierling E (1996) Synthesis of small heat- shock proteins is part of the developmental program of late seed maturation. Plant Physiol 112:747–757

Weston DJ, Gunter LE, Rogers A, Wullschleger SD (2008) Connecting genes, co- expression modules, and molecular signatures to environmental stress phenotypes in plants. BMC Syst Biol 2:16

Wildner GF, Henkel J (1979) The effect of divalent metal ions on the activity of $Mg^{2+}$ depleted ribulose-1,5-bisphosphate oxygenase. Planta 146:223–228

Wingler A, Lea PJ, Quick WP, Leegood RC (2000) Photorespiration: metabolic pathways and their role in stress protection. Philos Trans R Soc Lond B Biol Sci 355:1517–1529

Wong JH, Cai N, Tanaka CK, Vensel WH, Hurkman WJ, Buchanan BB (2004) Thioredoxin reduction alters the solubility of proteins of wheat starch endosperm: an early event in cereal germination. Plant Cell Physiol 45:407–415

Wotton HR, Strange RN (1987) Increased susceptibility and reduced phytoalexin accumulation in drought-stressed peanut kernels challenged with *Aspergillus flavus*. Appl Environ Microbiol 53:270–273

Xie Q, Sanz-Burgos AP, Guo H, García JA, Gutiérrez C (1999) GRAB proteins, novel members of the NAC domain family, isolated by their interaction with a geminivirus protein. Plant Mol Biol 39:647–656

Xu D, Duan X, Wang B, Hong B, Ho THD, Wu R (1996) Expression of a late embryogenesis abundant protein gene HVA1, from barley confers tolerance to water deficit and salt stress in transgenic rice. Plant Physiol 110:249–257

Yadav SK (2010) Heavy metals toxicity in plants: an overview on the role of glutathione and phytochelatins in heavy metal stress tolerance of plants. S Afr J Bot 76:167–179

Yadav SK, Singla-Pareek SL, Ray M, Reddy MK, Sopory SK (2005a) Methylglyoxal levels in plants under salinity stress are dependent on glyoxalase I and glutathione. Biochem Biophys Res Commun 337:61–67

Yadav SK, Singla-Pareek SL, Reddy MK, Sopory SK (2005b) Transgenic tobacco plants over expressing glyoxalase enzymes resist an increase in methylglyoxal and maintain higher reduced glutathione levels under salinity stress. FEBS Lett 579:6265–6271

Yamaguchi-Shinozaki K, Shinozaki K (2006) Transcriptional regulatory networks in cellular responses and tolerance to dehydration and cold stresses. Annu Rev Plant Biol 57:781–803

Yamaguchi-Shinozaki K, Koizumi M, Urao S, Shinozaki K (1992) Molecular cloning and characterization of 9 cDNAs for genes that are responsive to desiccation in *Arabidopsis* thaliana: sequence analysis of one cDNA clone that encodes a putative transmembrane channel protein. Plant Cell Physiol 33:217–224

Yan SP, Zhang QY, Tang ZC, Su WA, Sun WN (2006) Comparative proteomic analysis provides new insights into chilling stress responses in rice. Mol Cell Proteomics 5:484–496

Yang Q, Wang Y, Zhang J, Shi W, Qian C, Peng X (2007) Identification of aluminum-responsive proteins in rice roots by a proteomic approach: cysteine synthase as a key player in Al response. Proteomics 7:737–749

Yang H, Huang Y, Zhi H, Yu D (2010) Proteomics-based analysis of novel genes involved in response toward soybean mosaic virus infection. Mol Biol Rep 38:511–521

Yang SD, Seo PJ, Yoon HK, Park CM (2011) The Arabidopsis NAC transcription factor VNI2 integrates abscisic acid signals into leaf senescence via the COR/RD genes. Plant Cell 23:2155–2168

Yildirim E, Taylor AG, Spittler TD (2006) Ameliorative effects of biological treatments on growth of squash plants under salt stress. Sci Hortic (Amst) 111:1–6

Zaborske JM, Wu X, Wek RC, Pan T (2010) Selective control of amino acid metabolism by the GCN2 eIF2 kinase pathway in Saccharomyces cerevisiae. BMC Biochem 11:29

Zang X, Komatsu S (2007) A proteomics approach for identifying osmotic-stress-related proteins in rice. Phytochemistry 68:426–437

Zhang T, Liu Y, Yang T, Zhang L, Xu L, An L (2006) Diverse signals converge at MAPK cascades in plants. Plant Physiol Biochem 44:274–283

Zhang C, Meng Q, Gai J, Yu D (2008) Cloning and functional characterization of an O-acetylserine (thiol) lyase-encoding gene in wild soybean (*Glycine soja*). Mol Biol Rep 35:527–534

Zhou G, Xu Y, Li J, Yang L, Liu JY (2006) Molecular analyses of the metallothionein gene family in rice (*Oryza sativa* L.). J Biochem Mol Biol 39:595–606

Zhu J, Dong CH, Zhu JK (2007) Interplay between cold-responsive gene regulation, metabolism and RNA processing during plant cold acclimation. Curr Opin Plant Biol 10:290–295

Zonia L, Mnnik T (2004) Osmotically induced cells swelling versus cell shrinking elicits specific changes in phospholipid signals in tobacco pollen tubes. Plant Physiol 134:813–823

# Marker-Assisted Breeding for Stress Resistance in Crop Plants

# 16

Jogeswar Panigrahi, Ramya Ranjan Mishra, Alok Ranjan Sahu, Sobha Chandra Rath, and Chitta Ranjan Kole

**Abstract**

Molecular markers have extensively been used for tagging and mapping of genes and QTLs conferring resistance to biotic and abiotic stresses. These tools have also been used for screening of germplasms, fingerprinting, and marker-assisted breeding in crop systems. This chapter presents an overview on the basic concepts of molecular mapping and marker-assisted breeding and its most widely used applications in crop improvement programs, viz., marker-assisted backcross breeding, gene introgression, gene pyramiding, and marker-assisted selection at an early generation, with emphasis on stress-related traits and examples from several crops. We have also discussed some quantitative aspects of marker-based introgression, backcross breeding, and gene pyramiding programs. We have also added a note on breeding by design and genomic selection as tools for future breeding endeavors aiming at introgression of stress resistance into high-yielding cultivars. Harnessing the full potential of marker-aided breeding for improvement of stress resistance in crop systems will require a multidisciplinary approach and integrated knowledge of the molecular and physiological processes influencing the stress-related traits. Hence, marker-aided breeding for stress resistance in the post-genomic era poses great challenge for molecular breeders to realize the target objectives.

## Introduction

Crop plants are constantly challenged by pathogens, insects, and the harsh environment; hence, one of the major objectives of plant breeding has always been adaptation of crop plants. After the rediscovery of the Mendel's laws of inheritance in 1900 and the "chromosomal theory of inheritance" by Walter Sutton (1903), the strategies of breeding through directed crosses and selection

J. Panigrahi (✉) • R.R. Mishra • A.R. Sahu • S.C. Rath
Plant Biotechnology Laboratory, School of Life Sciences, Sambalpur University, Jyoti Vihar, Sambalpur 768019, Odisha, India
e-mail: drjpanigrahi@gmail.com; ramya_jp@rediffmail.com; alok.btgene@gmail.com; prath77@gmail.com

C.R. Kole
Institute of Nutraceutical Research, Clemson University, Clemson, SC 29634, USA
e-mail: ckole@clemson.edu

G.R. Rout and A.B. Das (eds.), *Molecular Stress Physiology of Plants*,
DOI 10.1007/978-81-322-0807-5_16, © Springer India 2013

for the desired recombinants were developed, which led to an enormous and steady increase in crop productivity and quality. This progress of conventional breeding is based entirely on the availability of genetic variation and its utilization for selection of host resistance against the pathogens/insects and host tolerance to harsh environment. Even though this process has proven to be very effective and ecofriendly means of reducing losses due to stresses, the effect of environment in many cases masks the effect of genotype. Therefore, the phenotype-based selection does not reflect the true genetic potential of a crop genotype. Hence, it has been realized that a selection procedure indirectly at the genotype level would augment the efficiency of breeding efforts (Dekkers and Hospital 2002).

Development of DNA (or molecular) markers in the 1980s laid the foundation for the utility of indirect selection strategies in plant breeding. Although DNA markers have manifold uses in crop breeding, the most promising one for cultivar development is marker-assisted breeding (MAB). The use of DNA markers that are tightly linked to the target loci as a substitute for or to assist phenotypic screening during the processes of breeding for the development of cultivars conferring resistance to biotic and abiotic stresses and other attributes required by the plant breeders for the genetic augmentation of crop species is considered as MAB.

## Fundamentals of Stress Resistance in Crop Plants

Stress encountered by crop plants is the external conditions that adversely affect the growth, development, and ultimately productivity. These external stresses can be biotic, imposed by pathogenic organisms (bacteria, fungi, virus, invertebrates, and weeds), or abiotic, arising from an excess or deficiency in the physical and chemical environmental parameters (water logging, drought, extreme temperature, soil salinity, soil alkalinity, soil mineral content, phytotoxic compounds, etc.). Being immobile in nature, crop plants continuously protect themselves against these stresses either with altered and induced gene expression and cellular metabolism or with inherent physical and physiological defense mechanisms. The former one is considered as sensitive stress tolerance, whereas the later one is referred as stress resistance. This stress resistance is mostly unique to individual plant species/breeding lines and might have developed during its coevolution with stress-causing agents through changes in the genetic makeup or through activation of dormant or preexisting resistance genes.

## Biotic vs. Abiotic Stress Resistance in Crop Plants

Biotic stress resistance is usually recognized as being either monogenically or polygenically controlled. In most cases, monogenic resistance is race-specific and fits the "gene-for-gene" hypothesis (Flor 1971). In some cases, biotic stress resistance is under the control of polygenes involving quantitative variation at multiple loci. This might be due to complex interactions between genes of host and genes of pathogen under varying environments. In contrast, abiotic stress resistance or tolerance, most often than not, is always under the control of multiple alleles, influenced by genotype-environment interaction. Thus, abiotic stress resistance is mostly quantitative with its continuous variation, moderate heritability, and environmental sensitivity and has been considered as a polygenic character (Geiger and Heun 1989). Therefore, complex plant resistance to insects, pathogens, and abiotic stresses requires an understanding on the dynamics of genetic variation for resistance traits within natural population. This dynamics depends largely on three major factors: (1) the extent to which a resistance trait is genetically controlled and heritable, (2) the probability with which chance effects can change the distribution of the resistance trait (genetic drift), and (3) the nature of natural selection that may act on the resistance trait (Simms and Rausher 1992). Hence, mapping simple inherited trait loci (SITL), quantitative trait loci (QTL), and candidate gene loci should resolve the dynamics of the above-said factors for better planning of resistance breeding in crop species.

## Source of Stress Resistance

Enriched germplasm of any crop species including its primary, secondary, and tertiary gene pools is the key source of alleles conferring stress resistance. Thus, any crop improvement program including molecular breeding approaches depends on the conservation of plant genetic resources, in particular the allied genetic diversity. Although the germplasm of secondary and tertiary gene pool represents the most important resource of resistance genes in an array of crop species (see for review Kole 2011), they have not yet been utilized well as compared to the genes from the primary gene pool for the breeding programs, due to constraints of cross-incompatibility, particularly among the too-distant genotypes, and the negative association of resistance with various undesirable traits.

There are many notable instances where extremely useful traits have originated from the wild relatives, both at the interspecific and intergeneric levels. This is exemplified with some major crops including rice, wheat, chickpea, tomato, pigeon pea, mung bean, and lettuce (Table 16.1).

## Genetics of Stress Resistance

Stress resistance is a type of genetic control, in which the host itself plays the role of an antagonist either individually or with interaction to the environment. To explain the genetic basis of stress resistance, two pairs of terms – monogenic and polygenic, and vertical (qualitative) and horizontal (quantitative) – have been used more often in genetics. According to Van der Plank (1963), vertical resistance is effective against one or a few physiological races of a given pathogen, while horizontal resistance is resistance to many races. In vertical resistance, there is a differential genetic response between the host plant varieties and the races of the same plant pathogenic organism. There is evidence that only one pair of alleles representing a gene mostly governs vertical resistance. In horizontal resistance, there is no manifestation of differential response; it is generally polygenic, that is, controlled by many genes, which are not specific for disease resistance, but they occur in healthy plants, regulating normal processes that combined express resistance (Van der Plank 1968).

Literature reveals that vertical resistance is mostly mono- or oligogenic and horizontal resistance is polygenic. Although there are many examples where this correlation is true, this is not a generalized concept (Bergamin Filho et al. 1995). In sorghum, resistance to *Periconia circinata* is monogenic and horizontal, and in rye, resistance to *Puccinia hordei* is polygenic but presents a differential interaction with the pathogen races (Parlevliet 1977). Monogenic vertical and polygenic horizontal resistance can occur in the same genotype. According to Parlevliet (1989), the selection of horizontal resistance in the presence of monogenic vertical resistance can produce an undesirable effect, resulting in high frequencies of genes with vertical resistance. Horizontal resistance was considered partial, due to mechanisms that partially hinder pathogen growth in the host tissues or harsh environment. This resistance may be an expression of near immunity or of a resistance reaction similar to that of vertical resistance (Eskes 1980).

The effects of horizontal resistance are quantitative and multitude. In plants, horizontal resistance with polygenic inheritance showed greater durability than vertical resistance with monogenic inheritance (Van der Plank 1968; Parlevliet 1977). Theoretical considerations indicate that polygenic resistance systems are more able to tolerate genetic changes in the pathogen and/or environment than monogenic systems. Thus, a polygenic resistance will be more durable and stable than a monogenic resistance, across the environment (Bergamin Filho et al. 1995), and the polygenic resistance is also considered as insurance policy to monogenic resistance (Castro et al. 2003).

**Table 16.1** Examples of agronomically important traits originated from wild relatives in various crops

| Crop | Desirable agronomic traits | Source | References |
|---|---|---|---|
| Rice | Blast resistance | *Oryza rhizomatis, O. longiglumis* | Das et al. (2012) and Jena (2010) |
| | Resistance to bacterial leaf blight | *O. rufipogon, O. nivara, O. longistaminata, O. minuta* | Jena (2010) |
| | Resistance to brown plant hopper | *O. punctata, O. officinalis, O. minuta, O. latifolia* | Jena (2010) |
| | Resistance to green leaf hoper | *O. eichingeri, O. officinalis* | Jena (2010) |
| | Salt tolerance | *O. coarctata* | Jena (2010) |
| | Adaptation to aerobic soil | *O. granulata* | Jena (2010) |
| | Cold tolerance at the seedling stage | *O. rufipogon* | Koseki et al. (2010) |
| Wheat | Leaf rust, stem rust, powdery mildew | *Aegilops sharonensis* | Olivera et al. (2008) |
| Tomato | Tomato yellow leaf curl virus resistance | *Solanum pimpinellifolium, S. peruvianum, S. chilense, S. habrochaites, S. cheesmaniae* | Chen et al. (2011a) |
| | Salt tolerant | *S. pimpinellifolium* | Sun et al. (2010) |
| | Crown and root rot<br>Bacterial canker<br>Early blight | *S. habrochaites* | Quesada-Ocampo and Hausbeck (2010) |
| | Salt tolerant | *S. pimpinellifolium* | Sun et al. (2010) |
| | | *Lycopersicon pennellii* | Foolad (2007) |
| | Crown and root rot<br>Bacterial canker<br>Early blight | *S. habrochaites* | Quesada-Ocampo and Hausbeck (2010) |
| | Fusarium wilt | *L. pimpinellifolium* | Foolad (2007) |
| | Powdery mildew | *L. chilense, L. hirsutum* | Foolad (2007) |
| | Cold tolerance | *L. pimpinellifolium, L. hirsutum* | Foolad (2007) |
| | Drought tolerance | *L. pimpinellifolium, L. pennellii* | Foolad (2007) |
| Chickpea | Ascochyta blight | *Cicer echinospermum, C. pinnatifidum, C. bijugum, C. judaicum, C. montbretii* | Singh et al. (2008) |
| | Fusarium wilt | *C. bijugum, C. judaicum, C. pinnatifidum, C. reticulatum, C. echinospermum, C. cuneatum* | Singh et al. (2008) |
| | Bruchids | *C. echinospermum, C. bijugum, C. judaicum* | Singh et al. (2008) |
| | Cyst nematode | *C. pinnatifidum, C. Bijugum, C. reticulatum* | Singh et al. (2008) |
| | Freezing tolerance | *C. bijugum, C. reticulatum, C. echinospermum, C. pinnatifidum* | Singh et al. (2008) |
| Pigeon pea | Resistance to Phytophthora blight | *Cajanus platycarpus* | Mallikarjuna et al. (2011) |
| | Sterility mosaic disease | *C. albicans, C. platycarpus, C. cajanifolius, C. lineatus, C. scarabaeoides, C. sericeus* | Kumar et al. (2005) |
| | Resistance to pod borer, pod fly, early flowering, photoperiod insensitivity | *C. scarabaeoides, C. platycarpus* | Saxena et al. (1990), Mallikarjuna et al. (2011) |

(continued)

**Table 16.1** (continued)

| Crop | Desirable agronomic traits | Source | References |
|---|---|---|---|
| | Soil salinity tolerance | *C. scarabaeoides*, *C. albicans* | Rao et al. (1981), Subbarao et al. (1991) |
| Mung bean | Resistance to bruchid | *Vigna radiata var. sublobata* | Tomoka et al. (1992) |
| | | *V. nepalensis* | Somata et al. (2008) |
| | Resistance to drought | *Vigna radiata var. sublobata* | Sholihin (2002) |
| Lettuce | Resistance to aphid | *Lactuca virosa* | Jansen (1996) |

## Concept of Marker-Assisted Breeding (MAB): Exploiting DNA Marker-Based Linkage Maps Through Conventional Crop Breeding

Developments in plant molecular genetics have provided new tools for mining and tagging novel alleles for genetic improvement of crops. Among these tools, DNA markers are potential one and can enhance the efficiency of crop breeding programs through their use in MAB. In MAB, the selection of target traits can be achieved indirectly using molecular markers that are tightly linked to the underlying genes or that have been developed from the actual gene sequences (Xu and Crouch 2008). More specifically, plant breeding will benefit from the use of DNA markers through:

1. More effective identification, quantification, and characterization of genetic variation in all available germplasms including primary, secondary, and tertiary gene pools (Tanksley et al. 1989; Tanksley and McCouch 1997; Gur and Zamir 2004)
2. Tagging, cloning, and introgressing SITL and/or quantitative trait loci (QTL) useful for enhancing the agronomic performance (Dudley 1993; Gibson and Somerville 1993; Paterson 1998; Gur and Zamir 2004; Peters et al. 2003; Kole and Gupta 2004; Pena 2004; Holland 2004; Salvi and Tuberosa 2005)
3. Augmenting (differentiating, selecting, pyramiding, and integrating) genetic variation in breeding populations (Stuber 1992; Xu 1997; Collard et al. 2005; Francia et al. 2005; Varshney et al. 2005; Wang et al. 2007; Collard and Mackill 2008)

Although molecular markers have manifold utility in crop improvement programs, this chapter focuses on the use of molecular markers in MAB to improve the efficiency and scope of cultivar development for specific traits conferring resistance to biotic and abiotic stresses.

## Prerequisites of MAB

MAB aims directly at those genes, which control the trait to be improved, but this assumes that genotypes at these target genes can be recognized easily, unambiguously, and at an appropriate time without impeding the breeding program. The prerequisites for this MAB are:

- Development of suitable molecular markers and their characterization
- Establishment of marker to gene(s) linkage through molecular maps
- Establishment of marker-trait association and validation of molecular markers across the population

### Development of Suitable Molecular Markers and Their Characterization

Several types of DNA markers including restriction fragment length polymorphism (RFLP), random amplified polymorphic DNA (RAPD), amplified fragment length polymorphism (AFLP), inter-simple sequence repeat (ISSR), microsatellites or simple sequence repeat (SSR), expressed sequence tag (EST), cleaved amplified polymorphic sequence (CAPS), diversity arrays technology (DArT), and single nucleotide polymorphism (SNP) have been used for map construction in several crops (Doveri et al. 2008). Each marker system has its own advantages and disadvantages, and the various factors to be considered in selecting one or more of these marker systems have been described (Semagn et al. 2006; Panigrahi 2011). There are five main considerations for the use of DNA markers in marker-assisted breeding: level of polymorphism, reliability and reproducibility, quantity and quality of DNA required, technical procedure for marker assay (genotyping), and cost of marker development (Mackill and Ni 2000; Mohler and Singrun 2004; Collard and Mackill 2008).

*Level of polymorphism*: Ideally, the marker should be highly polymorphic in breeding material (i.e., it should discriminate between different genotypes), especially in core breeding material.

*Reliability and reproducibility*: Markers should be tightly linked to target loci, preferably less than 5-cM genetic distance. The use of flanking markers in alternate phases (coupling and repulsion) or intragenic markers will greatly increase the reliability of the markers to predict phenotype. The marker should be reproducible across the laboratories.

*DNA quantity and quality*: Some marker techniques require large amounts and high quality of DNA, which may sometimes be difficult to obtain in practice, and this adds to the cost of the procedures.

*Technical procedure for marker assay*: The level of simplicity and the time required for the technique are critical considerations. High-throughput simple and quick methods are highly desirable.

*Cost*: The marker assay must be cost-effective in order for MAS to be feasible.

The most widely used markers in major crops including cereals and legumes are simple SSRs or microsatellites (Gupta et al. 1999; Gupta and Varshney 2000; Kole and Gupta 2004; Li et al. 2008; Kumar et al. 2011), whereas in oilseed, Brassicas are RFLPs (Panigrahi et al. 2009). Both SSR and RFLP are highly reproducible, codominant in inheritance, relatively simple and cheap to use, and generally highly polymorphic. The only disadvantages of SSRs are that they typically require polyacrylamide gel electrophoresis and generally give information only about a single locus per assay, although multiplexing of several markers is possible. These problems have been overcome in many cases by selecting SSR markers that have large enough size differences for detection in agarose gels, as well as multiplexing several markers in a single reaction (Kalia et al. 2011). Sequence tagged site (STS), sequence characterized amplified region (SCAR), or single nucleotide polymorphism (SNP) markers that are derived from specific DNA markers (e.g., RFLPs, RAPDs) that are linked to a gene or quantitative trait loci (QTL) are also extremely useful for MAS (Shan et al. 1999; Sanchez et al. 2000; Sharp et al. 2001; Collard and Mackill 2008; Kumar et al. 2011).

## Establishment of Marker-Gene(s)' Linkage Through Molecular Maps

A linkage map is the graphical representation of the chromosomes showing the position of loci contrasting between two different parents (Paterson 1996). Linkage maps indicate the position and relative genetic distances between markers along chromosomes, which is analogous to landmarks along a highway. The molecular marker systems along with genetic linkage analysis led to the development of high-density DNA marker maps (i.e., with many markers of known location, interspersed at short intervals throughout the genome) in several crop species, which is the prerequisite for various applications of MAB. Among them, the most important one is to identify chromosomal locations containing genes and QTLs associated with stress.

The construction of molecular maps involves the classical strategy of linkage mapping aided by polymorphic molecular markers as traits, and this comprises primarily the following steps: (a) selection of genetically diverse parents aiming at optimum DNA polymorphism, (b) raising of segregating population(s) for assessment on recombination events, (c) selection of molecular marker system for genotyping of mapping population, (d) screening of parents for polymorphism and genotyping the mapping population using polymorphic DNA markers, (e) phenotyping of the mapping population aiming to study the inheritance of desired traits, and (f) linkage analysis and construction of linkage maps.

To establish the framework of linkage groups representing a genome, several software programs can be used. These include Linkage I (Suiter et al. 1983), MapMaker v2.0 (Lander et al. 1987), MapMaker/EXP 3.0 (Lincoln et al. 1992a), CRI-MAP (Weaver et al. 1992), GMendel (Liu and Knapp 1992), JoinMap (Starn 1993), MapManager (Manly 1995), and MapDisto (Lorieux 2012). Choice of a program, however, may depend on the kind of computer available (Macintosh/IBM), criteria of assembling marker loci and maps, and targeted use of the genetic map besides several other factors. After construction of the genetic map, it is worthwhile to align the marker data as per the order in the linkage groups to verify any scoring error and completeness of molecular map.

### Completeness of the Molecular Map

Depiction of linkage groups of the same number as the haploid number of chromosomes and absence of too many unlinked markers is the primary criteria to assume that a genetic map is complete (Kole and Gupta 2004). An organism with a small genome and less number of chromosomes would obviously span a shorter distance and require lesser number of markers to cover its chromosomes, as compared to an organism with a large genome with many gametic chromosomes. Average recombination fraction between marker loci is another criterion to judge coverage. In general, recombination fraction of less than 5% with lower range of variation is thought to be optimum. Addition of new markers particularly in case of recombinant inbred and doubled haploid population-based maps to any of the linkage groups in the primary genetic map also provides a positive indication of optimum coverage. Tendency of a marker to show linkage to the markers of another linkage group is obviously a negative indication. More confirmation of completeness can be obtained by using repetitive DNA elements from telomeres as probes and determining the strength of linkage to the terminal markers of the linkage group in the primary genetic linkage map (Ganal et al. 1992; Burr et al. 1992).

### Interpretation of a Molecular Map

A molecular map provides useful firsthand information about the nature of genome organization and reorganization (Kole and Gupta 2004). Clustering of marker loci might indicate that they are located near the centromere region. Mapping of a group of markers with preferential skewness in a genomic location is an evidence for its involvement with biological fitness (Kole et al. 1997), and this particular location might contain loci of some deleterious alleles from either of the parents. Detection of duplicated loci in clusters might indicate genome reorganization due to chromosomal aberrations. When such duplications are too frequent, particularly for loci detected with cDNA probes or cloned genes, it might suggest paleoploidy (Kole et al. 1997).

### Establishment of Marker to Trait Association and Validation of Molecular Markers

Establishment of marker to trait associations is one of the prerequisites for MAB. Demonstrated linkages between target traits/genes and molecular markers are traditionally based on genetic mapping experiments, and this is important to confirm that these associations are consistent across the mapping populations and breeding lines. For efficient MAB, marker(s) should either co-segregate or be closely linked with the target trait, with a distance of 2 cM or less (Xu 2003). Markers associated with major genes or QTL in one population may be used directly for MAB and MAS in other breeding materials. For genes with relatively minor contributions, the cross-population comparison of genes, alleles, and gene effects is required because of multilocus and multiallelic features that characterize most quantitative traits. To find close marker-trait associations, a two-step process could be involved (especially for quantitative traits). The first step is based on a primary mapping population derived from genetically diverse parents, often with complicated genetic backgrounds. The second step is based on near-isogenic lines (NILs) that share a common genetic background and differ only at the target locus. There are many factors including genetic background and allele distribution around the target locus that is related to the detection of marker-trait associations determines the efficiency of MAB (Xu 2003).

The flow diagram shows identification, validation, and implementation of marker-trait association in MAB programs at four stages:

## MAB of Simple Inherited Trait Loci (SITLs) Conferring Stress Resistance

MAB for simple inherited trait loci (SITLs) including monogenic traits and oligogenic traits, which follows simple Mendelian inheritance, is quite straight forward approach. It involves:

i. Identification of suitable DNA marker that either co-segregates with stress resistance gene/locus or a pair of flanking markers

**Identification, Validation, and Implementation of Marker-Trait Association in MAB Programs**

*Stage I: Identification of marker-trait association for stress resistance*

Parent-1 × Parent-2

(Both the parents should be genetically diversed and contrasting for stress response)

↓

Development of segregating mapping population ($F_2$, RILs, DHLs, BC populations)

↓

Phenotyping and genotyping of segregating population

↓

Establishment of genetic linkage map and identification of markers linked to stress resistance locus/loci

*Stage II: Validation of target locus in different genetic background*

Parent-3 × Parent-2 (must be differing for the stress response from parent-3)

↓

Development of segregating progeny ($F_2$, RILs, DHLs, BC populations)

↓

Screening with identified marker(s) to identify individuals with allele from parent-2

↓

Phenotype segregating population with parent-2 allele at marker locus

↓

Estimate validity of marker in predicting stress resistant or tolerant phenotype

*Stage III: Test the usefulness of marker in characterization core germplasm*

Separate the germplasm into lines that will serve as donor or recipient of the stress resistance

↓

Is the marker able to differentiate between donor and recipient germplasm?

Yes ←→ No

Yes ↓ Marker suitable for use

No ↓ Identify alternative marker loci closely flanked the trait locus

↓

Screen the marker against the germplasm

↓

Identify markers able to detect polymorphism between donor and recipient and segregating lines *at par*

*Stage IV: Implementation of the marker to the breeding program*

Prepare simple genotyping protocol (marker assay) for the marker detection and its allele size

↓

Provide the marker data along with allelic information to the molecular breeders for the marker-assisted breeding (MAB) for stress resistance

around the locus conferring resistance and validation by testing its applicability and reliability in predicting the stress resistance in the related populations

ii. Deployment of identified marker(s) in various breeding programs, such as marker-assisted backcross breeding (MABC), marker-assisted gene pyramiding (MAGP), and marker-assisted selection (MAS) at early generations

## Molecular Mapping of Major Genes Conferring Stress Resistance

Molecular mapping of loci controlling biotic stress resistance, most often under the control of oligogenes, is considered as gene mapping or SITL mapping. The strategy of gene mapping is quite simple. The phenotypic expression of the biotic stress resistance is recorded on the individuals of a mapping population; they are converted to genotypes depending upon which parent they resemble, followed by linkage analysis with the marker data already available. In some cases, the phenotypic expression may be recorded using metric data or scores. For mapping of the trait loci, these quantified phenotypic data are converted to discrete forms resembling the parents, followed by routine linkage analysis.

The alternative strategy is using the genetic stocks, commonly referred to as near-isogenic lines (NILs) that are available from conventional breeding programs for various major genes conferring stress resistance. The construction of NILs for a specific gene is carried out by repeated backcrossing of a donor parent (DP) to a recurrent parent (RP). Backcrossing accompanied by selection for the gene of interest and recovery of the recurrent parent is continued until the newly developed line is nearly isogenic with the RP, except for the chromosome segment that has the target gene. The basic idea of the NIL gene-mapping technique is to identify molecular markers located in the linkage block surrounding the introgressed stress resistance gene. Linkage between a molecular marker and the target gene can be assessed by determining the marker genotypes of the RP, its NIL derivative, and their corresponding DP. Putative evidence of linkage is obtained whenever the NIL has the same marker genotype as its DP, but a different marker genotype than its RP. Since only a few genotypes (RP, NIL, and DP) have to be assayed, the NIL gene-mapping technique provides a proper tool for narrowing down a large set of randomly chosen markers to a subset of a few markers, some of which are very tightly linked to the target gene introgressed into NIL. With this subset of molecular markers, however, a traditional segregation analysis would have to be performed in order to find a tightly linked marker to the target gene.

The third strategy employed in major gene mapping is called bulked segregant analysis (BSA, Michelmore et al. 1991), which is popularly known as molecular tagging of genes. BSA strategy employs mostly $F_2$ segregating population for tagging and mapping the major gene. In this method, the homozygotes of the contrasting phenotypes (resistance and susceptible to stress) are from the $F_2$ based on the $F_3$ segregation analysis. The DNA from homozygotes of each phenotype is pooled to form a pool similar to resistant parent (which is referred as resistant bulk) and pool similar to susceptible parent (which is referred as susceptible bulk). The genotypes to be assayed for markers include resistant parent, resistant bulk, susceptible parent, and susceptible bulk. Putative markers associated with alternative phenotypes are identified, that is, known as gene tagging, and subject to traditional linkage analysis to identify the tightly linked marker to the target gene – molecular mapping of gene.

Literature reveals mapping of an array of such loci conferring biotic stress resistance in several crop species using all the above strategies. A list of mapped SITLs conferring stress resistance in different crop systems is furnished in Table 16.2.

## MAB for Major Genes Conferring Stress Resistance

Tight linkage of a marker to a target gene conferring stress resistance can be exploited for indirect selection of monogenic and oligogenic traits in a breeding program. Application of this method presupposes that the initial population is polymorphic for the marker and the target gene, and both are in extreme linkage disequilibrium. Instead of testing the phenotypes of the target gene itself, selection in segregating generations is based on determination of marker genotype. Only those individuals that carry the desired marker allele(s) are selected (Melchinger 1990). However, use of two linked markers bracketing the target gene should improve the reliability of MAB (Collard and Mackill 2008). These flanked molecular markers can be effectively employed by backcross breeding, gene pyramiding, gene introgression, and early generation selection of superior cultivars aiming at complete stress resistance and higher stress tolerance.

### Marker-Assisted Backcross Breeding (MABC) for Major Genes Conferring Stress Resistance

Backcross breeding has been a widely used strategy for almost a century to incorporate major gene(s) conferring resistance to biotic and abiotic stresses into an adapted or elite variety, which is possessed with large number of desirable agronomic traits but is susceptible to the target stress (Allard 1999). In repeated crossings, the hybrids ($BC^{1\text{-}n}$) are backcrossed with the recurrent parent until most of the genes stemming from the donor parent are eliminated except stress resistance (Becker 1993). This conventional backcross breeding programs are planned on the assumption that the proportion of the recurrent parent genome is recovered at the rate of $\{1-(1/2)^{n+1}\}$ for each "n" generation of backcrossing. Thus, after six backcrosses, we expect to recover 99% of the recurrent parent. This approach was first described

**Table 16.2** A list of simple inherited trait loci (SITL) conferring stress resistance mapped using various kinds of molecular markers in some major crops

| Crop | Trait | Gene | Marker type | Marker and distance in cM | LG/ Chr | References |
|---|---|---|---|---|---|---|
| Rice | Leaf blast resistance | *Pi-1* | RFLP | *NPB181(3.5)* | 11 | Yu (1991) |
| | | *Pi-2(t)* | RAPD | *RG64(2.1)* | 6 | Hittalmani et al. (1995a) |
| | | *Pi-4(t)* | RFLP | *RG869(15.3)* | 12 | Yu et al. (1991) |
| | | *Pi-ta* | RFLP | *RZ397(3.3)* | 12 | Yu et al. (1991) |
| | | *Pi-5(t)* | RFLP | *RG498 (5.0), RG788 (10.0)* | 4 | Wang et al. (1994) |
| | | *Pi-6(t)* | RFLP | *RG869(20)* | 12 | Yu (1991) |
| | | *Pi-7(t)* | RFLP | *RG103(5.1)* | 11 | Wang et al. (1994) |
| | | *Pi-10(t)* | RAPD | *RRF6, RRH18* | 5 | Naqvi et al. (1995) |
| | | *Pib* | RFLP | *RZ123* | 2 | Miyamoto et al. (1996) |
| | | *Pi-9* | RFLP | RG16, pBA 14, pBA8 | 6 | Liu et al. (2002) |
| | | *Pi-38* | SSLP | RM 206, RM21 | 11 | Gowda et al. (2006) |
| | | | AFLP | AF-1, AF-2, AF-3 | | |
| | | *Pi-1 (t)* | SSR | RM224 | NA | Prasad et al. (2009) |
| | Bacterial blight resistance | *Xa-1* | RFLP | *Npb235 (3.3)* | 4 | Yoshimura et al. (1992) |
| | | *Xa-2* | RFLP | *Npb 235(3.4), Npb197(9.4)* | 4 | Yoshimura et al. (1992) |
| | | *Xa-3* | RFLP | *Npb181(2.3), Npb 78(3.5)* | 11 | Yoshimura et al. (1992) |
| | | *Xa-4* | RFLP | *Npb181(1.7), Npb 78(1.7)* | 11 | Yoshimura et al. (1992) |
| | | *Xa-1* | RAPD | | | Yoshimura et al. (1995b) |
| | | *Xa-5* | RFLP | *RG556 (0-1)* | 5 | Mc Couch et al. (1992) |
| | | *Xa-10* | RAPD | *OPO07(5.3)* | 11 | Yoshimura et al. (1995a) |
| | | *Xa-13* | RFLP | *RZ390 (0.0), RG 136(3.8)* | 8 | Yoshimura et al. (1995a) |
| | | *Xa-21* | RAPD | *Pta818(0.0), Pta 248(1.0)* | 11 | Ronald et al. (1992) |
| | | *Xa-7* | RGA | RNM20576, MY4 | 6 | Zhang et al. (2009b) |
| | | *xa34*(t) | SSR | RM493 (4.29) and RM446 (3.05) | 1 | Chen et al. (2011b) |
| | Gall midge resistance | *Gm-2* | RFLP | RG329(1.3), RG476 (3.4) | 4 | Mohan et al. (1994) |
| | | *Gm4-t* | RAPD | E20570 (0.4) from R1813 and 1.4 from S1633b | 8 | Nair et al. (1996) Mohan et al. (1997) |
| | | *Gm-7* | AFLP/ SCAR | SA 598 | 4 | Sardesai et al. (2002) |
| | | | SSR | RM22550 (0.9) and RM547 (1.9) | 8 | Nanda et al. (2010) |
| | | *Gm11t* | SSR | RM28574 (4.4) and RM28706 (3.8 ) | 12 | Himabindu et al. (2010) |
| | Brown plant hopper | *bph4* | SSR | RM589 (1.1) and RM586 (0.6) | 6 | Jairin et al. (2010) |
| | | *BPH12* | SSR | RM16459 and RM1305 (1.9) | 4 | Qiu et al. (2012) |
| | | *bph20(t)* | SSR | RM435 (2.7) and RM540 (2.5) | 6 | Yang et al. (2012) |
| | | | | BYL7 and BYL8 (2.5) | | |
| | | *bph21(t)* | SSR | RM222 (7.9) and RM244 (4.0) | 10 | Yang et al. (2012) |
| | Rice stripe virus | *RSV1* | SSR, SRAP | RM457 (4.5) and SR1 (2.9) | 11 | Zhao et al. (2010) |

(continued)

**Table 16.2** (continued)

| Crop | Trait | Gene | Marker type | Marker and distance in cM | LG/ Chr | References |
|---|---|---|---|---|---|---|
| Wheat | Powdery mildew resistance | *Pm* | RFLP | Xwg516 | 2B | Rong et al. (1998) |
| | | *Pml* | RFLP | WHS178(3) | NA | Jahoor (1998) |
| | | *Pm2* | RFLP | Xbcd1871(3.5) | 5D | Ma et al. (1994) |
| | | *Pm2* | RFLP | Xfba393 | 5DS | Nelson et al. (1995) |
| | | *Pm3b* | RFLP | Xbcd1434 (1.3) | 1AS | Ma et al. (1994) |
| | | *Pm4a* | RFLP | Xbcd1231-2A (1.5) | 2AL | Ma et al. (1994) |
| | | *Pm4a* | RFLP | Xcdo678-2A (1.6) | 2AL | Ma et al. (1994) |
| | | *Pml2* | RFLP | Xpsr10,Xpsr106,Nor2, Xpsr141,Xpsr113 | 6DS | Jia et al. (1996) |
| | | *Pml2* | RFLP | Xpsr142, Xpsr149, Xpsr2, Xpsr605 | 6DL | Jia et al. (1996) |
| | | *Pm34* | SSR | Xbarc1775D (5.4), Xbarc1445D (2.6) | 5D | Miranda et al. (2006) |
| | | *PmY39* | SSR | Xgwm257, Xgwm296, Xgwm319 (0) | 2B | Zhu et al. (2006) |
| | | *Pm35* | SSR | Xcfd (10.3), Xgdm (8.6), Xcfd (11.9) | 5DL | Miranda et al. (2006) |
| | | *PmTb7A.2* | STS | MAG2185 and MAG1759 | 7A | Chhuneja et al. (2012) |
| | | *PmTb7A.1* | DArT and SSR | wPt4553 and Xcfa2019 (4.3) | | |
| | | *Pm-M53* | AFLP | Apm109 (1.0) and Apm161 (3.0) | 5D | Li et al. (2011) |
| | | *PmHNK54* | SSR | Xbarc5 (5.0) and Xgwm312 (6.0) | 2AL | Xu et al. (2011b) |
| | | *Pm4d* | STS | Xgwm526 (3.4) and Xbarc122 (1.0) | 2AL | Schmolke et al. (2012) |
| | | *PmAS846* | EST | BJ261635 and CJ840011 | 5BL | Xue et al. (2012) |
| | | *Pm45* | SSR | Xcfd80, Xmag6139 (0.7) and Xmag6140 (2.7) | 6DS | Ma et al. (2011) |
| | Cyst nematode resistance | *Crel* | RFLP | Xglk605 (7.3), Xglk588 (8.4) | 2BL. | Williams et al. (1994) |
| | | *Ccn-D1* | RAPD | OPE20 (tight) | 2DS | Eastwood et al. (1994) |
| | High-temperature adult-plant stripe rust resistance | *HTAP* | RGAP SSR | Xbarc182 and Xwgp5258 (2.3) | 7BL | Ren et al. (2012) |
| | High-temperature resistance to stripe rust | *Yrxy2* | SSR | Two markers flanked at (4.0) and (6.4) | 2A | Zhou et al. (2011) |
| | | *Yrxy1* | RGAP | M8 (2.3) and M9 (3.5) | 7A | |
| | | | SSR | Xbarc49 (15.8) and Xwmc422 (26.1) | | |
| | Streak mosaic virus resistance | *Wsml* | STS | Wg232 (tight) | 4 L | Talbert et al. (1996) |
| | | *WSSMV* | RFLP | Xbcd1095, Xcdo373 | 2DL | Khan et al. (2000) |
| | Hessian fly resistance | *H23* | RFLP | XksuH4 (6.9), XksuG48a (15.6) | 6D | Ma et al. (1993) |
| | | *H24* | RFLP | Xcnl BCD 451 (5.9), Xcnl CDO 482 (5.9), Xksu G48b (12.9) | 3DL | Ma et al. (1993) |

(continued)

**Table 16.2** (continued)

| Crop | Trait | Gene | Marker type | Marker and distance in cM | LG/ Chr | References |
|---|---|---|---|---|---|---|
| | Stem rust resistance | *Sr2* | RFLP | Xbcd1871 (3.5) | 5B | Nelson et al. (1995) |
| | | *Sr40* | SSR | Xwmc 344 (0.7) | 2B | Wu et al. (2009) |
| | | *Sr13* | SSR | barc37 (4.0) | 6A | Admassu et al. (2011) |
| | Leaf rust resistance | *Lr1* | RFLP | PSR580(13.8), PSR567 | 5DL | Feuillet et al. (1995) |
| | | *Lr3* | RFLP | Xmwg 798 (0) | 6BL | Sacco et al. (1998) |
| | | *Lr13* | RFLP | Xbcd 1709(<7.9), Xpsrs 912 (9.1) | 2B | Seyfarth et al. (1998) |
| | | *Lr32* | RFLP | Xbcd 1278 (3.6), Xcdo 395 (6.9) | 3DS | Autrique et al. (1995) |
| | | *Lr34* | SSR | Xgwm1220, Xgwm295 | 7D | Spielmeyer et al. (2005) |
| | | *Lr52 (W)* | SSR | Xgwm 443 (16.5) | 5B | Hiebert et al. (2005) |
| | | *Lr34* | EST-SSR | SWM-10 | 7DS | Bosolini et al. (2006) |
| | | *Lr22a* | SSR | GWM-296 | 2DS | Hiebert et al. (2007) |
| | | *Lr1* | RGA | WR-03 (0.07) | 5DL | Qiu et al. (2007) |
| | | *Lr41* | SSR | Xbarc 124 (<1.0) | 2DS | Sun et al. (2009) |
| | Stripe rust resistance | *Yr15* | RFLP<br>STMS | Nor1 B (11)<br>WMS533(4.5) | 1B | Sun et al. (1999) and Fahima et al. (1997) |
| | | *yrH 52* | RFLP | Nor1 (1.4) | 1B | Peng et al. (1999) |
| | | | STMS | NA (0.02-0.35) | | |
| | | *Yr36* | STS | Xbarc 101 (2.0) | 6B | Uauy et al. (2005) |
| | | *YrCH42* | SSR | Xgwm498(1.6), Xbarc-187 (2.3) | 1B | Li et al. (2006a) |
| | | *YrZH84* | SSR | Xcfa2040-7B(1.4), Xbarc32-7B (4.8) | 7BL | Li et al. (2006b) |
| | | *Yr34* | RGA | Awn Inhibitor gene A-1 (12.2) | 5AL | Barian et al. (2006) |
| | | *Yr26* | SSR/ STS | Xwe173 (1.4)/ Xbarc181 (6.7) | 1B | Wang et al. (2008) |
| | | *YrZak* | SSR | Xwgp 102, Xgwm 501 | 2B | Sui et al. (2009) |
| | Loose smut resistance | *T 10* | RAPD | UBC 353 (14) | 2BL | Procunier et al. (1997) |
| | | | RFLP | Xcrc4.2 (14), Xcrc 153.2 (10) | 2BL | Procunier et al. (1997) |
| | Septoria nodorum blotch resistance | | RAPD | UBC521 (15), RC37 (13) | 3A | Cao et al. (1998) |
| | Scab resistance | *FHB* | AFLP | XeageMetal (16.2) ksuH16 (12.7) Xbcd1331(7.3) | 3BS, 2AL, 6BS | Anderson et al. (1998) |
| | | | RFLP | Xcdo 1387 (7.4) | 4AL | Anderson et al. (1998) |
| | | | | Xcdo 524 (5.9) | 6BS | |
| | Yellow mosaic virus resistance | *YmYF* | SSR | Xwmc41 (8.1), Xgwm349 (11.6) | 2D | Liu et al. (2005) |
| Barley | Stem rust resistance | *rpg 4* | RAPD | | 7 M | Borokova et al. (1995) |
| | Powdery mildew resistance | *MiLa* | RFLP | WG645 (Co-segregating) | 2 | Saghai-Maroof et al. (1994) |

(continued)

**Table 16.2** (continued)

| Crop | Trait | Gene | Marker type | Marker and distance in cM | LG/ Chr | References |
|---|---|---|---|---|---|---|
| | Aluminum tolerance | *Al* | RFLP | Xwg464 (21.6) | 4H | Echart et al. (2006) |
| | Crown rust resistance | *rpc-1* | RAPD | OPO-08$_{700}$ (2.5) | 3H | Agrama et al. (2004) |
| | Leaf rust resistance | *Rph-5* | RFLP<br>AFLP | VT-1(0.3)<br>C970 (0.5) | 3H | Mamadov et al. (2003) |
| | | *Rph-7* | RFLP | Hv3Lrk (3.20) | 3HS | Brunner et al. (2000) |
| Chickpea | Resistance to Fusarium wilt | *Foc1* | SSR | H3A12 (3.9) and TA110 (2.1) | | Gowda et al. (2009) |
| | | *Foc2* | SSR | TA96 (0.2) and H3A12 (2.7) | | |
| | | *Foc3* | SSR | H1B06y (0.2) and TA194 (0.7) | | |
| Mung bean | Bruchid resistance | *Bruchid* | RFLP | PA882 (3.6) | 8 | Young et al. (1992) |
| | | *Br-2* | RAPD, STS | OPC-06 (11.0)<br>STSbr2 (5.8) | 8 | Sun et al. (2008) |
| Pigeon pea | Sterility mosaic disease resistance | *SMD* | AFLP | E-CAA/M-GTG$_{150}$ (5.7) | NA | Ganapathy et al. (2009) |
| | | | | E-CAA/M-GTG$_{60}$ (4.8) | | |
| | | | | E-CAG/M-GCC$_{150}$ (5.2) | | |
| | | | | E-CAG/M-GCC$_{120}$ (20.1) | | |
| | Fusarium wilt resistance | – | RAPD | OPM03$_{704}$ and OPAC11$_{500}$ | NA | Kotresh et al. (2006) |
| Soybean | SMV resistance | *Rsv* | RFLP/ SSR | PA 186 (1.5), pK644a(2.1), SM176(0.5) | NA | Yu et al. (1994) |
| Tomato | Fusarium wilt resistance | *L2* | RFLP | TG105 (0.0 ±4.8) | 11 | Sarfatti et al. (1989) |
| | TMV resistance | *Tm2* | RFLP | TG3 (0.3), CD3 (0.9) | 9 | Young and Tanksley (1989a) |
| | Powdery mildew resistance | *Lv* | RAPD | OP57,OP58, OP218 | 12 | Chunwongse et al. (1994) |
| | | *Lv* | RFLP | CT211, CT 219 (5.5) | 12 | Chunwongse et al. (1994) |
| | | *Ol-2* | RAPD | OPU3$_{1500}$ | 4 | de Giovanni et al. (2004) |
| | | *Ol-1, Lv* | AFLP | – | 6,12 | Bai et al. (2003) |
| | Salt tolerance | | Isozyme | Est3, Prx7, Pgdh2, pgt1 | 1, 3, 12 | Foolad and Jones (1993) |
| Tomato yellow leaf curl virus resistance | *Ty-2* | RFLP | Tg36 | 11 | | Hanson et al. (2006) |
| | *Ty-3* | SCAR | | cLET31-P16, T1079, P169c | 6 | Ji and Scott (2006) |
| | Alternaria stem canker | *ASc* | RFLP | YAC (11.2) | 3 | Mesbah et al. (1999) and vander Blezen et al. (1995) |
| | Bacterial spot | *Xv4* | RFLP | TG599, TG134 | 3 | Astuamonge et al. (2000) |
| | | *Bs4* | AFLP | P11M6, P11M1 | 5 | Ballvora et al. (2001) |
| Mustard | White rust resistance | *Ac21* | RAPD | WR2 (7), WR1a/3 (1.4) | NA | Prabhu et al. (1998) |
| | | *Ac2 (t)* | RAPD | OPB06$_{1000}$ (5.5) | NA | Mukherjee et al. (2001) |
| Rapeseed | Turnip mosaic virus resistance | *TuRBO1* | RFLP | pO120b (3.6) | 6 | Walsh et al. (1999) |
| | Black leg resistance | *LEM1* | RFLP | wg2a3b (1), wg5a1a (1.9) | 6 | Figdore et al. (1993) |
| | | *LepR3* | SSR | SR12281a, SN2428Rb(2.9) | N10 | Yu et al. (2008) |
| | White rust resistance | *ACA1* | RFLP | tg6a12 (12.9) | 9 | Ferreira et al. (1995b) |
| Indian rape | White rust resistance | *Aca1* | RFLP | wg6c1a (5.0), ec2b3a (5.5) | 4 | Kole et al. (2002a, b) and Kole et al. (1996) |

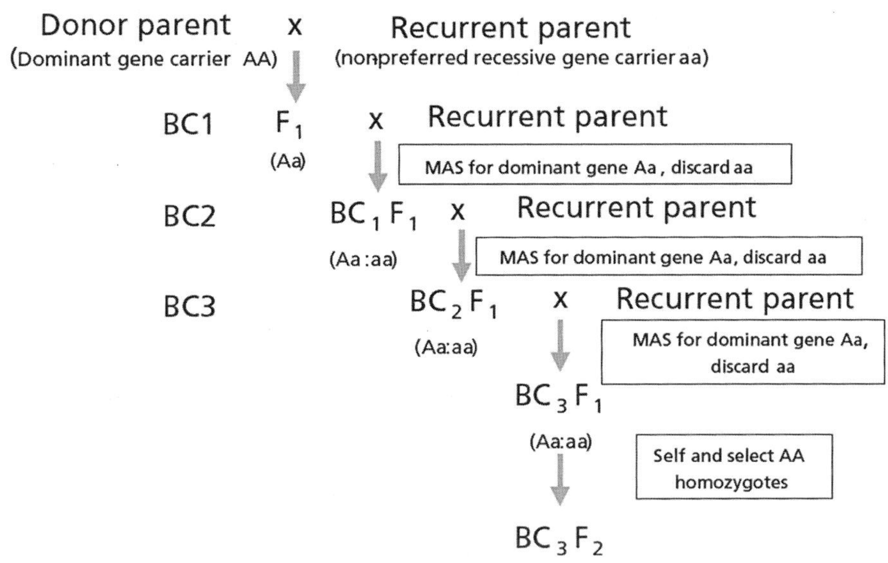


**Fig. 16.1** Flow diagram showing marker-assisted backcrossing approach to transfer a single dominant allele conferring stress resistance

in 1922 and was practiced between the 1930s and 1960s in several crops (Stoskopf et al. 1993). DNA markers are now increasingly used in backcross breeding to increase the efficiency of selection at three levels, viz., foreground selection (i.e., to trace the presence of target allele or gene), background selection (i.e., selection of reconstructed recurrent parent genome in backcross program), and recombinant selection (i.e., identification of BC progeny with the target gene and recombination events between the target locus and linked and/or flanked markers to minimize linkage drag) (Holland 2004; Collard and Mackill 2008).

Marker-assisted foreground selection is useful for stress-associated traits that have tedious and time-consuming phenotypic screening procedures. By this approach, the reproductive traits can also be selected at the sapling stage, and this allows for the identification of best plants at an early stage for backcrossing as well as phenotyping. Furthermore, both dominant and recessive alleles conferring stress resistance can also be selected and introgressed by this approach (Figs. 16.1 and 16.2), which is time consuming and difficult to do using conventional breeding.

Marker-assisted background selection was initially proposed by Young and Tanksley (1989b), and the term was coined by Hospital and Charcosset (1997). The background selection refers to the use of closely linked flanking markers for selection of recombinant and unlinked background markers to select the genetic makeup of recurrent parent (RP). Background markers are unlinked to the target trait on all other chromosomes, and these markers can be used to select recurrent genome against the donor genome. Hence, this approach is quite useful, for accelerated reconstitution of recurrent genome along with the desired donor fragment. With conventional backcrossing, it takes a minimum of six BC generations to reconstitute the RP, which may be possessed with several donor chromosome fragments unlinked to the target gene. Using marker-assisted background selection approach, the reconstituted RP genome can be achieved in two to four BC generations (Visscher et al. 1996; Hospital and Charcosset 1997; Frisch et al. 1999a, b) even by eliminating undesired donor fragment, thus saving two to four BC generations. Thus, use of marker-assisted background selection accelerates the development of NIL at par with RP containing desired stress resistance and has been referred to as "complete line conversion" (Ribaut et al. 2002).

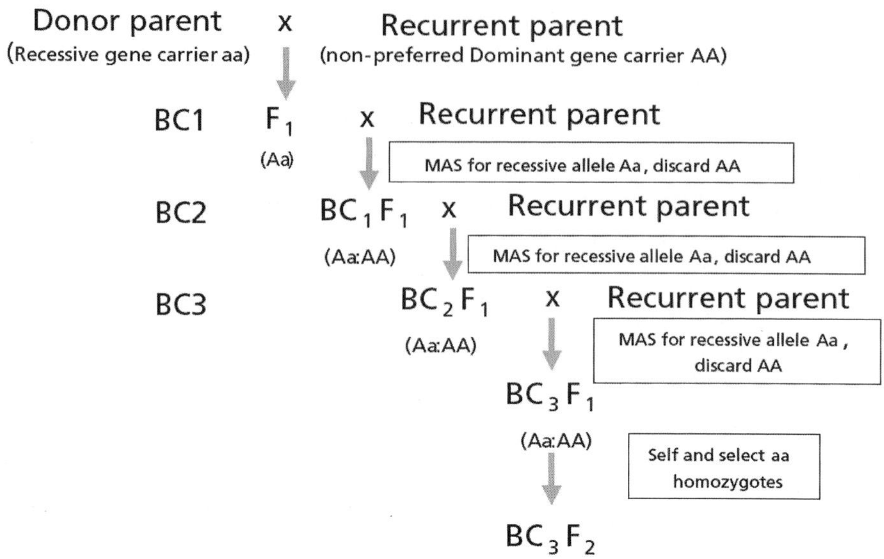


**Fig. 16.2** Flow diagram showing marker-assisted backcrossing approach to transfer a single recessive allele associated with stress resistance

## Foreground and Background Selection Through MAB: A Case Study

Marker-assisted backcross breeding strategy has proven valuable in accelerating the backcross programs to incorporate two major blast-resistant genes *Pi-2(t)* and *Pi-1* from pyramided donor parent into the background of IR64, an elite cultivar widely grown in Asia. Simultaneously recurrent parent genome was recovered with the help of both RAPD and RFLP markers to have the introgressed line identical with that of IR64 (Hittalmani et al., personal communication). During each generation of $F_1$, $BC_1F_1$, and $BC_2F_2$, the progenies were screened for tracking the major resistant genes *Pi-2(t)* and *Pi-1* using tightly linked markers (Fig. 16.3). The broad-based *Pi-2(t)* major blast-resistant gene, present in chromosome 6, was confirmed by using the previously established codominant RG64 SAP marker (Hittalmani et al. 1995a). Similarly, the tightly linked RM-144/c481 (2 cM) rice microsatellite marker confirmed the introgression of complimentary resistance *Pi-1* gene. This resistant gene is present on chromosome number 11 and is linked to c481 clone, which is a PCR-based STS marker (Mew et al. 1994; Kurata et al. 1994). Simultaneously, progenies from $F_1$, $BC_1F_1$, and $BC_2F_1$, having *Pi-2(t)* and *Pil* major resistance genes, were screened for the fast recovery of recurrent parent genome (R.PG) using both RAPD and SSR primers. Through background screening, it was possible to recover 96% of IR64 background by $BC_2F_1$ generation (Fig. 16.3).

Marker-assisted recombinant selection reduces the size of the donor segment of the genome containing the target locus. This approach is important because the decrease of donor fragment size reduces the deleterious effect of many undesirable genes, linked to the target gene, that negatively affect crop performance – this is referred to as "minimization of linkage drag" (Hospital 2005). Using conventional breeding methods, the donor segment can remain very large even with many BC generations (Young and Tanksley 1989a; Ribaut and Hoisington 1998; Salina et al. 2003) because the conventional breeding is phenotype-biased selection. By using flanking markers around a stress-resistant gene (e.g., less than 5 cM on either side), linkage drag can be minimized (Collard and Mackill 2008). Further, these markers enable the breeders to detect the double recombination events occurring on both sides of stress resistance locus, which are extremely rare, and this recombinant selection is usually performed using at least two BC generations (Frisch et al. 1999b).

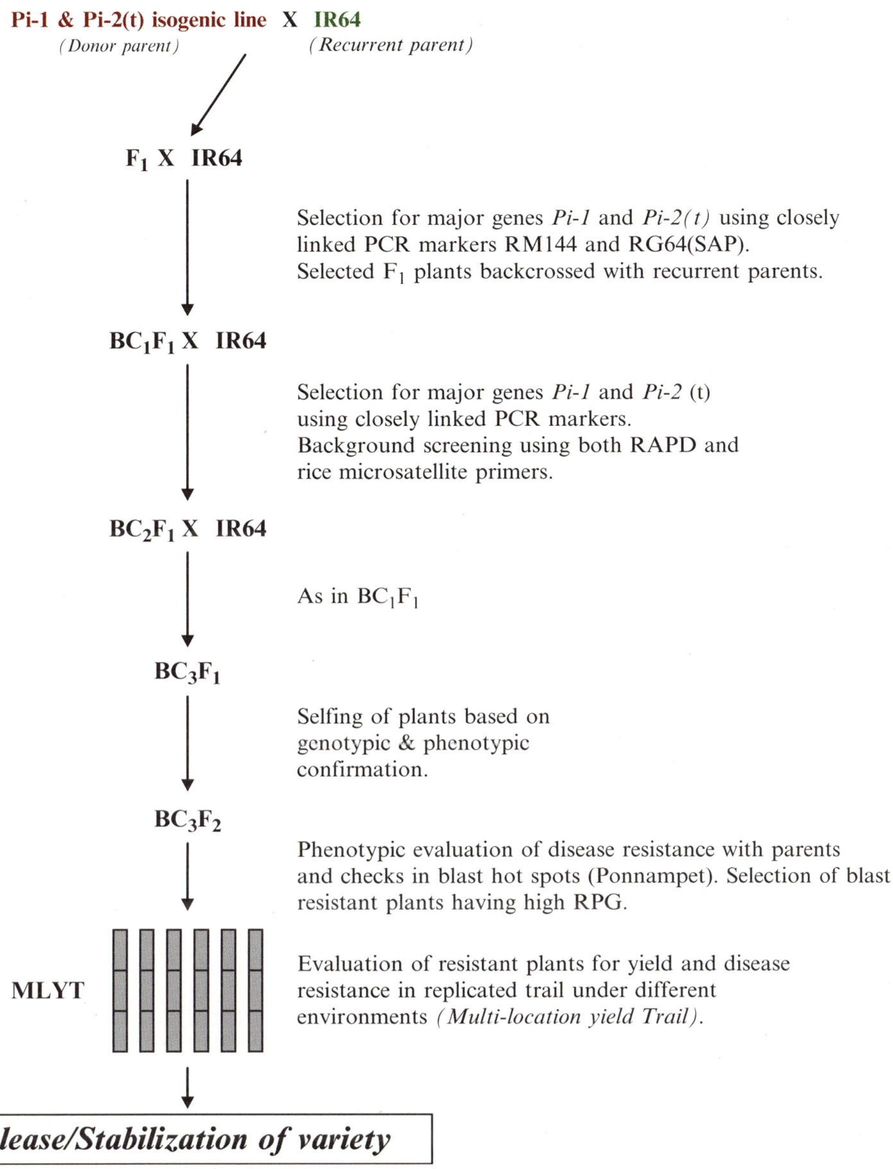


**Fig. 16.3** Schematic sketch of marker-assisted backcross breeding for introgression of two major blast resistance genes, *Pi-1* and *Pi-2(t)*, in rice

## Removal of Linkage Drag Using MAB: A Case Study

Keygene, an agrobased company, was involved in a marker-assisted breeding approach for the development of a novel lettuce variety resistant to the aphid, *Nasonovia ribisnigri*, during mid-nineties (Jansen 1996). The resistance gene to this aphid was present in its wild relative, *Lactuca virosa*, which could be introgressed from a by repeated backcrossing. Although the backcrossing program developed aphid-resistant genotype, the new product was of extremely poor quality, bearing yellow leaves and a greatly reduced head. Marker analysis revealed that this reduced quality was caused by a negative trait closely linked to aphid resistance, so-called linkage drag. In this case, the linkage drag was recessive, only appeared in the

**Table 16.3** Example of marker-assisted backcross breeding of SITLs conferring stress resistance for the development of cultivars and advance breeding lines in several crop systems

| Crop | Agronomic traits | Markers used | End product | References |
|---|---|---|---|---|
| Rice | Bacterial blight resistance (*Xa21*) | RFLP, PCR-based markers | Minghui-63 (*Xa21*): A hybrid restorer line | Chen et al. (2000) |
| | | AFLP | IV-6078 (*Xa21*): A hybrid restorer line | Chen et al. (2001) |
| | Bacterial blight resistance (*Xa5*) | STS | Angke-A cultivar | Bustamam et al. (2002) |
| | Bacterial blight resistance (*Xa7*) | STS | Conde-A cultivar | Bustamam et al. (2002) |
| Wheat | Leaf rust resistance | RFLP, CAPS | Isogenic lines of three spring cultivars – Express, Kern, and UC 1037 | Helguera et al. 2005 |
| | Leaf rust resistance (*Lr 21*) | SSR | Advanced breeding lines | Somers et al. (2005) |
| | Leaf rust resistance (*Lr 47*) | PCR-based markers | *Yecora rojoLr47*-cultivar | Chicaiza et al. (2006) |
| | Cereal cyst nematode | RFLP | Adv. breeding lines (DH population based) | Ogbonnaya et al. (2001) |
| | Stripe rust resistance (*Yr-36*) | PCR-based markers | Yecora rojo *yr36-GplB1*, *UC1113 yr36-GplB1*- two cultivars | Chicaiza et al. (2006) |
| | Stem rust resistance (*sr-38*) | PCR-based markers | Anza *Lr37*/*Yr17*/ *Sr38* | Chicaiza et al. (2006) |
| Barley | Resistance to boron toxicity | SSR | Advanced breeding lines | Emebiri et al. (2009) |
| | Resistance to *BaYMV I-III* | RFLP | Mokkei 01530- A Cultivar | Okada et al. (2003) |
| | Resistance to spot net blotch | SSR | Line *WI3586-1747* | Eglinton et al. (2006) |
| Maize | Resistance Southwestern corn borer (SWCB) | RFLP | Advanced breeding lines | Willcox et al. (2002) |
| Tomato | Resistance to tomato spotted wilt virus | CAPS | Advanced breeding lines | Langella et al. (2004) |

homozygous state, thereby increasing the difficulty in phenotype-based selection of recombination events. With the identification of DNA markers flanking the aphid-resistant locus, individuals having recombination events in the vicinity of the gene were preselected for further analysis. More than 1,000 $F_2$ plants were screened this way, leading to the selection of some 100 individuals bearing a recombination or even double recombination in the vicinity of the aphid resistance gene. Only those individuals needed to be phenotyped for both the resistance and the absence of the negative characteristics at $F_3$ level. This approach eventually led to the selection of an individual bearing recombination events very close to each side of the gene, thereby removing the recessive linkage drag located on both sides of the resistance gene, in addition to being tightly linked. By removing the linkage drag, the mildly sweet, crunchy iceberg lettuce with aphid resistance (Cultivar-*Fortunas Rz*) was developed by the seed breeding company Rijk Zwaan, Berlin.

Some examples of MABC for traits conferring stress resistance in rice, wheat, barley, maize, tomato, and soybean are furnished in Table 16.3.

## Marker-Assisted Gene Pyramiding (MAGP) for Major Genes Conferring Stress Resistance

Pyramiding stress resistance genes to develop durable resistant variety is another strategy of MAB. Breakdown of disease or pest resistance conferred by single major gene is a common phenomenon in several crops, due to the

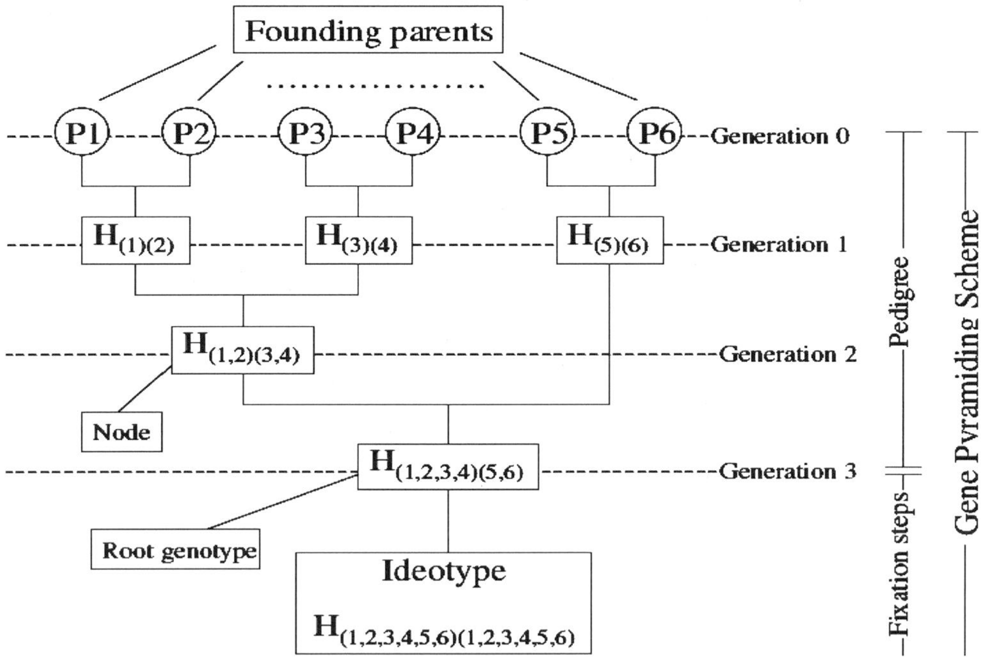


**Fig. 16.4** Gene pyramiding strategy in plants (Redrawn from Servin et al. 2004)

emergence of new pathogenic races and/or biotypes after its wide cultivation (Kiyosawa 1982; Ou 1985). Therefore, breeding for durable resistance would be more effective if two or more major and minor resistance genes are combined into a single genetic background. This will lessen the chances of biotic resistance breakdown by continuing evolution of new races of pathogens and biotypes of pests. Pyramiding of two or more major and minor biotic stress resistance genes through conventional breeding method is possible, but it is not easy to identify the plants containing more than one gene simultaneously due to the different genetic behavior of genes, the mode of gene action *inter se*, the appearance of same phenotype and destructive bioassay, etc. Since DNA marker assays are nondestructive and markers for multiple specific genes can be tested simultaneously using a single DNA sample without phenotyping (Hittalmani et al. 1995b, 2000; Yoshimura et al. 1995a; Huang et al. 1997; Werner et al. 2005; Saghai Maroof et al. 2008), the MAGP substantially enhances the selection process in crop breeding.

A schematic approach for accumulating six genes into a single genotype was demonstrated (Fig. 16.4) by Servin et al. (2004). Accordingly, the gene pyramiding scheme consists of two steps, viz., pedigree step and fixation step. The *pedigree step* aims at accumulating of all target genes in a single genotype called the root genotype, and the *fixation step* aims at fixing all the target genes into a homozygous state, that is, to derive the ideal genotype (ideotype). Although the pedigree step may be common, several different procedures can be used to undergo fixation in gene pyramiding.

The marker-assisted gene pyramiding involves the following steps in general:

- Fine-map the donor genes conferring stress resistance and develop tightly linked flanking markers for each resistant gene.
- The plant with known tagged gene (X-1) is crossed with the second plant carrying another target gene (X-2) and similarly with the other genes.
- The $F_1$ plants of the above crosses are screened for the presence of both the resistance genes (X-1 and X-2) by assaying for the presence both pair of flanking markers.
- $F_2$ individual plants are screened for the presence of all combination of genes using

**Table 16.4** MAGP of SITLs conferring resistance to biotic and abiotic stress in some major crops

| Crop | Agronomic trait | Pyramided gene(s) | References |
|---|---|---|---|
| Rice | Blast resistance | *Pi1*, *Piz5*, and *Pita* | Hittalmani et al. (2000) |
| | Bacterial blight resistance | *Xa4*, *Xa5*, *Xa13*, and *Xa 21* | Huang et al. (1997) |
| | | *Xa7*and *Xa 21* | Zhang et al.(2006) |
| | | *Xa5*, *Xa13*, and *Xa 21* | Kottapalli et al. (2010) |
| | | *Xa3* and *Xa 26* | Li et al. (2012) |
| | Bacterial blight, yellow stem borer and sheath blight | *Xa21*, *cry1Ab*, *RC7 chitinase*, and *cry1Ac* | Datta et al. (2002) |
| | Bacterial blight, yellow stem borer | *Xa21*, *cry1Ab*, and *cry1Ac* | Jiang et al. (2004) |
| | Brown plant hopper | *Bph1* and *Bph2* | Sharma et al. (2004) |
| | Gall midge resistance | *Gm1* and *Gm 4* | Kumaravadivel et al. (2006) |
| | | *Gm2* and *Gm6(t)* | Kattiyar et al. (2001) |
| Wheat | Powdery mildew | *Pm2* and *Pm4a* | Liu et al. (2000) |
| | Leaf rust resistance | *Lr41*, *Lr42* and *Lr43* | Cox et al. ( 1994) |
| | Green bug resistance | *Gb2*, *Gb3* and *Gb6* | Porter et al. (2000) |
| | Cyst nematode resistance | *Cre X* and *Cre Y* | Barloy et al. (2007) |
| Barley | Yellow mosaic virus resistance | *rym1* and *rym5* | Okada et al. (2004) |
| | | *rym4*, *rym9*, *rym11* | Werner et al. (2005) |
| Pearl millet | Downy mildew resistance | *P7-3* and *P7-4* | Hash et al. (2006) |
| Soybean | Soybean mosaic virus resistance | *Rsv1*, *Rsv3* and *Rsv4* | Shi et al. (2009) and Saghaimaroof et al. (2008) |
| Grapevine | Powdery mildew and downy mildew resistance | *Run1* and *Rpv1* | Eibach et al. (2007) |

markers, and plants carrying all gene combination are forwarded to the next generation.

- Homozygous plants carrying all the desired combination of target genes identified and further tested for their traits.

MAGP has been well demonstrated in several crops including rice, wheat, pearl millet, barley, and soybean (Table 16.4). The example cited in the Table 16.4 evidenced that the combination of multiple resistance genes (effective against specific races of a pathogen) can provide durable (broad spectrum) resistance. The ability of a pathogen to overcome two or more resistance genes by mutation is considered much lower compared with the invading of stress resistance controlled by a single gene (Kloppers and Pretorius 1997; Shanti et al. 2001; Singh et al. 2001). In this section, we have only cited the case of gene pyramiding to improve qualitative stress-related traits such as disease and pest resistance. In section "Marker-Assisted Pyramiding of QTLs Conferring Stress Resistance", we will discuss the marker-assisted pyramiding of QTLs conferring resistance or tolerance to biotic and abiotic stress.

## Marker-Assisted Gene Introgression from Breeding Lines and Wild Relatives

Evaluation of the exotic and adapted germplasm using DNA marker is an integral part of MAB, and it played important role in the acquisition, storage, and use of plant genetic resources in breeding endeavors (Bretting and Widrlechner 1995; Xu et al. 2003). The wild allies of the domesticated crops, including its progenitors and the species in secondary and tertiary gene pools, provide plant breeders with a handy source of desirable genes (Tanksley 1983). The use of wild germplasm and putative progenitors in crop improvement gained importance in the 1970s and 1980s (Hajjar and Hodgkin 2007; Dwivedi et al. 2008) and is continuing in several orphan crops till date (Panigrahi et al. 2007; Varshney et al. 2010; Mishra et al. 2012). The advancements in genomics have considerably increased the use of genes from exotic germplasms, in particular

**Table 16.5** Wild allies of domesticated crop species used for the release of stress resistance cultivars in the last 20 years

| Crop | No. of pest or disease resistance trait introgressed | No. of abiotic stress resistance trait introgressed | Total no. of stress-related traits[a] introgressed |
|---|---|---|---|
| Rice | 7 | 3 | 10 |
| Wheat | 11 | – | 11 |
| Maize | 1 | – | 1 |
| Barley | – | 1 | 1 |
| Potato | 6 | – | 6 |
| Tomato | 10 | 2 | 12 |
| Lettuce | 3 | – | 3 |

Partly adapted from Hajjar and Hodgkin (2007)
[a]Total numbers of individual traits obtained from wild species are indicated in the last column for each crop

from the secondary gene pools, being the source agroeconomic genes. Hajjar and Hodgkin (2007) reported on the use of more than 60 wild species to introgress desired genes conferring stress resistance and other agronomically important traits into 13 agricultural crops during mid-1980s to 2005. The most common use of wild relatives is as a source of pest and disease resistance, although other characteristics including abiotic stress tolerance, yield attributes, improved quality, and cytoplasmic male sterility and fertility restoration also have been improved using crop wild relatives in individual cases. The number of traits introgressed into elite cultivars by MAB is furnished in Table 16.5 (adapted from Hajjar and Hodgkin 2007).

The process by which beneficial traits from core collection of germplasm are transferred to the elite cultivars of a crop is called introgression. Mostly single major genes conferring stress resistance, identified in exotic or unadapted germplasm, are transferred into adapted breeding material by backcrossing (Simmonds 1993). The use of these exotic germplasm in MAB can aid in transferring genes with minimal linkage drag, thus making the introgression of genes is one of the preferential application of modern breeding. Introgression libraries (IL) allow for the identification of favorable alleles in exotic germplasm, which can be exploited for improving elite breeding material. ILs are obtained by a cross between the unadapted germplasm and an elite recurrent parent, which is followed by several generations of recurrent backcrossing. A set of polymorphic markers that can distinguish between parental alleles is used to trace chromosome segments through the crosses. The backcrossing is followed by at least one generation of selfing, which leads to plants homozygous at targeted introgressed segments (Zamir 2001). This systematic introgression of individual, short, marker-defined chromosome segments into the elite background results in a library of ILs. Each line carrying different parts of the donor genome and the library can be used to screen for favorable alleles obtained from the exotic unadapted germplasm. Introgression lines were first developed in tomato in 1994 and have since then been adopted for several crops, including barley, maize, wheat, and rye (Falke et al. 2009).

## Marker-Assisted Selection (MAS) of Superior Genotypes Having Major Genes Conferring Stress Resistance at Early Generations

Molecular markers can be employed at any stage of a plant-breeding program. Hence, MAS has great advantage in early generation selections by eliminating undesirable gene combinations. Subsequently, the breeders focus on a lesser number of high-priority lines of desirable allelic or gene combination. MAS-based early generation selection not only selects suitable gene combinations but also ensures a high probability of retaining superior breeding lines (Eathington et al. 1997). An important prerequisite for successful early generation selection with MAS is large populations and low heritability of the target traits, as under individual selection, the relative efficiency of MAS is greatest for characters with low heritability (Lande and Thompson 1990). Results from Kuchel et al. (2007) and Bonnett et al. (2005) showed that maximum gain can be achieved at lowest cost in marker-assisted wheat breeding when molecular markers, closely linked to target genes, are utilized to enrich target loci within segregating populations in early generations. When the linkage between the marker and the selected trait loci conferring stress resistance is not very close, the probability of recombination between the marker and trait loci will increase, thereby increasing the number of gene combination for early generation MAS. This leads to more cost of genotyping a larger number of plants. To avoid this disadvantage, Ribaut and Betran (1999) developed a strategy involving MAS at an early generation called single large-scale MAS (SLS–MAS). This approach used flanking markers (less than 5 cM, on both sides of a target locus) in a single MAS step. Alternatively, using codominant DNA markers, it is possible to fix specific alleles in their homozygous state as early as the $F_2$ generation. However, this may require large population sizes; thus, in practical terms, a small number of loci may be fixed at each generation (Koebner and Summers 2003). Another strategy is to "enrich" rather than fix alleles by selecting homozygotes and heterozygotes for a target locus within a population in order to reduce the size of the breeding populations required (Bonnett et al. 2005).

## MAB of Polygenic Traits Conferring Stress Resistance

Most of the traits conferring abiotic stress tolerance and pest/disease resistance are under the control of many genes, and each gene has either major or minor contribution on the expression of the traits. Hence, these traits are referred as polygenic or quantitative traits. The genetic loci associated with these complex traits are referred as quantitative trait loci (QTL). QTL can be of three types, viz., (a) major QTL, (b) major plus minor QTL, and (c) minor QTL. Usually the major QTLs exhibit Mendelian inheritance, whereas other two types deviate from Mendelian nature of inheritance, which makes the situation difficult to trace them. There is always chance of losing a QTL having minor effect during phenotypic selection. This can be overcome by utilization of molecular markers (Peleman and vander Voort 2003a; Anuradha and Balasubramani 2011). Employment of various molecular markers along with statistical analysis substantially improved the genetic studies of quantitative traits, even those controlling abiotic and biotic stress resistance in crop plants.

## Molecular Mapping of Polygenic Traits Conferring Stress Resistance

Linkage between a genetic marker and quantitative trait was first demonstrated by Sax (1923) through associating the seed size (a quantitative trait) with seed color (a morphological marker) in *Phaseolus vulgaris*. Later on, Rasmusson (1933) and Everson and Schaller (1955) reported linkages between single genetic markers and quantitative traits. However, associating specific QTL with genetic marker(s) and locating them on linkage groups was possible with the construction of complete molecular genetic maps that allowed systematic searches of an entire genome for the QTLs influencing a trait (Thoday 1961; Paterson et al. 1988). Four widely used methods for detecting QTLs are single-marker analysis, simple interval mapping, composite interval mapping, and multiple interval mapping (Edwards et al. 1987; Lander and Botstein 1989; Zeng 1994).

The concept of interval mapping (Lander and Botstein 1989) to detect such QTLs in marker intervals using molecular maps has became a popular strategy of mapping polygenes. Presently a number of excellent computer programs are available for QTL mapping, but most widely used computer program is MapMaker/QTL 1.1 (Lincoln et al. 1992b). In this program, the likelihood positions are determined by scanning the whole genome as described by the complete map, interval-wise, by fixing generally a likelihood of odds (LOD) for the presence of a QTL at the default precision level of LOD 2. The program provides several useful genetic information besides the putative chromosomal locations of the gene clusters. It provides the degree of likelihood of presence of a QTL, contribution of parental alleles to the QTL, the extent of variation explained by a QTL, and the nature of gene action depending on the mapping population (additive, dominant or recessive, and epistatic). The program allows for detection of QTLs with minor effects by rescanning of the genome by fixing the major ones, verification of linked QTLs whether spurious or true, identification of the confidential interval of a QTL, and formulation of multilocus model (Zeng 1994) with several QTLs that provide information on the ultimate picture of the genetics of the character. One can use the linked markers to evaluate the nature of gene action between QTLs such as epistasis. It is also possible to learn about the nature of variation of a character and to transform the data to normal distribution, besides obtaining the values of mean, skewness, kurtosis, variance, etc. of the character. A list of important QTLs associated with biotic and abiotic stresses identified in some major crops is furnished in Table 16.6.

Keeping track of all genes involved in these complex quantitative traits controlling stress resistance during a breeding program is a great challenge. Peleman and vander Voort (2003a) suggested the simplifications of these complex analyses to dissect the contributions of each QTLs of the genome on the trait values. Such simplifications can be obtained at several levels:

(a) *Simplification of the phenotype:* Division of a complex stress resistance phenotype into its separate genetic components. For example, drought tolerance is determined by a vast array of component characters, such as stomatal frequency, stomatal conductance, canopy temperature, and chlorophyll fluorescence. Mapping the genes involved in these separate components provides a better understanding of drought tolerance and a higher chance of success.

(b) *Simplification of the mapping:* Separating the effect of each QTL by generating near-isogenic lines (NILs), using the technique of introgression line libraries (Eshed and Zamir 1995) and reverse QTL mapping (RQM; Wye et al. 2000; Peleman et al. 2005), enables more precise measurement of effect of the QTL and thereby the precise/fine-mapping of the QTL. Fine-mapping of a QTL and high resolution maping around the target QTL are essential steps for the exploitation of QTL in marker-assisted breeding (Kole et al. 1996, 1997 and 2001). These approaches significantly aid in unraveling the complexity of quantitative traits conferring biotic and abiotic stress resistance.

However, the above strategies of QTL analysis for MAB of polygenic traits has not improved much in last couple of decades (Bernado 2008; Xu and Crouch 2008). Hence, the current approach of MAS for these quantitative traits is focused on association mapping, which is based upon linkage disequilibrium between alleles at marker loci and QTL controlling desired traits (Bachlava et al. 2009). This genomic selection process used natural population where linkage disequilibria are due to random genetic drift and several round hybridizations between genetically different lines. The detailed understanding of association mapping approaches has been discussed recently in two reviews (Zhu et al. 2008; Rafalski 2010). The basic steps involved in association mapping are well illustrated by Kumar et al. (2011) and the same is depicted in the flow diagram in Fig. 16.5 (redrawn from Kumar et al. 2011).

## Marker-Assisted Breeding for Polygenic Traits Conferring Stress Resistance

Marker-assisted breeding for major and minor genes conferring abiotic and biotic stress resistance requires the development of suitable type and number of markers flanking the trait loci and simultaneous manipulation of several QTLs in order to have a significant impact on quantitative stress resistance. For this reason, repetitions of field tests are required to characterize the effects of QTLs and genetic interactions (additive, dominant, and epistasis) and to evaluate their stability

**Table 16.6** A list of QTLs detected for important agronomic traits related to biotic and abiotic stress in some major crops

| Crop | Character | QTL | LG | References |
|---|---|---|---|---|
| Rice | Osmotic adjustment, dehydration tolerance | 15 | 81, 3, 5, 7, 8 | Lilley et al. (1996) |
| | Root trait-related drought resistance | 28 | 2,3,4,6,7,11 | Ali (1999) |
| | Drought avoidance | 36 | 1,2,3,4,5,6,7,9,10,11,12 | Courtois et al. (2000) |
| | Root morphology and drought avoidance | 15 | – | Champoux et al. (1995) |
| | Drought tolerance | DTQTL | 1,3,4,8,9 | Kanjoo et al. (2011) |
| | Heat tolerance at flowering stage | qHTSF1.1, qHTSF4.1 | 1,4 | Ye et al. (2012) |
| | Submergence tolerance | 4 QTLs | 1,2,9,12 | Septiningsih et al. (2012) |
| | Cold tolerance | qCTB7 | 7 | Zhou et al. (2010) |
| | Salt tolerance | Saltol | 1 | Nejad et al. (2008) |
| | Salt tolerance | ST QTL | 1,3,4,7,8,9,10,12 | Kanjoo et al. (2011) |
| | Salt tolerance | Saltol, 2 QTLs | 1,8,10 | Islam et al. (2011) |
| | Green rice leaf hopper resistance | qGRH9 | 9 | Fujita et al. (2010) |
| | RYMV resistance | 12 | 1,2,3,4,7,8, 9,12 | Albar et al.(1998) |
| | Stripe virus resistance | qSTV11 $^{HAB}$ -1, qSTV11 $^{HAB}$ -2 | 11 | Zhang et al. (2012) |
| | Stripe virus disease | qStv4,qStv9,qStv12 | 4,9,12 | Wang et al. (2011) |
| | Stripe virus resistance | qSTV1.1, qSTV1.2, qSTV11 | 1,11 | Zhang et al. (2010) |
| | Stripe disease resistance | qSTV7, qSTV11$^{KAS}$ | 7,11 | Zhang et al. (2011) |
| | Sheath blight resistance | 6 | 2,3,8,9,12 | Li et al. (1995) |
| | Sheath blight resistance | qSBR11-1 | 11 | Channamallikarjuna et al. (2010) |
| | Sheath blight resistance | qShB1, qShB2, qShB3 and qShB5 | 1,2,3,5 | Xu et al. (2011a) |
| | Small brown plant hopper | Qsbph11e, Qsbph11f, Qsbph11g | 11 | Duan et al. (2010) |
| | Small brown plant hopper | qSBPH1, qSBPH2, qSBPH11 | 1,2,11 | Zhang et al. (2010) |
| Wheat | Resistance to stripe rust | YrR61, QYr.uga-6AS | 2A | Hao et al. (2011) |
| | Resistance to stripe rust | 8 QTLs | 2B,3B,1A, 4A, 4B, 5B | Chen et al. (2012) |
| | Resistance to stripe rust | Yrq1 | 2DS | Cao et al. (2012) |
| | Fusarium head blight | 2 QTLs | 5AS, 5AL | Chu et al. (2011) |
| | Wheat yellow mosaic bymovirus resistance | QYm.njau-3B.1, QYm.njau-5A.1 QYm.njau-7B.1 | 3BS, 5AL, 7BS | Zhu et al. (2012) |
| | Resistance to Fusarium infection | Fhb5 | - | Xue et al. (2011) |
| | Temperature-sensitive resistance to stripe rust | QYr-tem-5B.1, QYr-tem-5B.2 | 5B | Feng et al. (2011) |
| | Resistance to soil-borne cereal mosaic virus | QSbm.ubo-2BS | 2BS | Maccaferri et al. (2011) |

(continued)

**Table 16.6** (continued)

| Crop | Character | QTL | LG | References |
|---|---|---|---|---|
| Maize | European corn borer resistance | 7 | – | Schon et al. (1993) |
| | Plant resistance | 11 | 1,2, 3, 4, 5, 6, 7, 8, 10 | Khairallah et al. (1998) |
| | SE corn borer resistance | 7 | 3, 5, 6, 8, 9 | Khairallah et al. (1998) |
| Mung bean | Powdery mildew resistance | 3 | 3, 7,8 | Young et al. (1993) |
| | Bruchid resistance | 7 | 1, 2 | Somata et al. (2008) |
| Pigeon pea | Sterility mosaic disease | qSMD1, qSMD2, qSMD3, qSMD4, qSMD5, qSMD6 | 1,2,3,9 | Gnanesh et al. (2011) |
| Chickpea | Resistance to Fusarium wilt | $QTL_{Foc02}$, $QTL_{Foc5}$ | 2 | Cobos et al. (2009) |
| | Botrytis grey mould resistance | QTL1, QTL2, QTL3 | 6A, 3 | Anuradha et al. (2011) |
| | Ascochyta blight resistance | QTL (1-5) | 2,3,4,6,8 | Anbessa et al. (2009) |
| | *Drought-related traits* Grain yield plant$^{-1}$ | 2 QTLs | 1,3 | Rehman (2009) |
| | Harvest index | 5 QTLs | 1,3,4,8 | Rehman (2009) |
| | Drought tolerance score | 7 QTLs | 1,2,3,4,6,7,8 | Rehman (2009) |
| | Days to flowering | 4 QTLs | 1,3,4,6 | Rehman (2009) |
| | Days to maturity | 4 QTLs | 1,3,7 | Rehman (2009) |
| | Plant height | 4 QTLs | 1,3,6,8 | Rehman (2009) |
| | Stomatal conductance | 1 QTL | 7 | Rehman (2009) |
| | Canopy temperature differential | 3 QTLs | 1,3,6 | Rehman (2009) |
| Indian rape | Freezing tolerance | 6 | 2, 4, 5, 7, 9, 10 | Teutonico et al. (1995) |
| | Freezing tolerance | 3 | 1, 9 | Kole et al. (2002b) |
| | Winter survival | 5 | 1,2,3,7 | Kole et al. (2002b) |
| | White rust resistance | 2 | 4, 2 | Kole et al. (2002a) |
| Rape seed | Vernalization requiring flowering time | 3 | 9, 12, 16 | Ferreira et al. (1995a) |

across environments (Ribaut and Hoisington 1998; Zheng et al. 2000). Otherwise, MAB will be a biased selection process. The basic steps involved in MAB of polygenic traits are:

- Marker conversion: Establishment of simple and reliable marker genotyping method.
- QTL confirmation: Identification and QTL along with its confidence limit using molecular map.
- Fine-mapping of QTL by using substitution mapping.
- QTL validation: Verification of the identified QTL and its contribution on the desired trait under different genetic backgrounds. Preferably this can be done either by advanced backcross-QTL (AB-QTL) analysis or QTL-NIL analysis.
- Marker validation: Testing the level of polymorphism of the most tightly linked markers (within 5–10 cM) flanking the target locus and also assessing the reliability of markers to predict the phenotype.

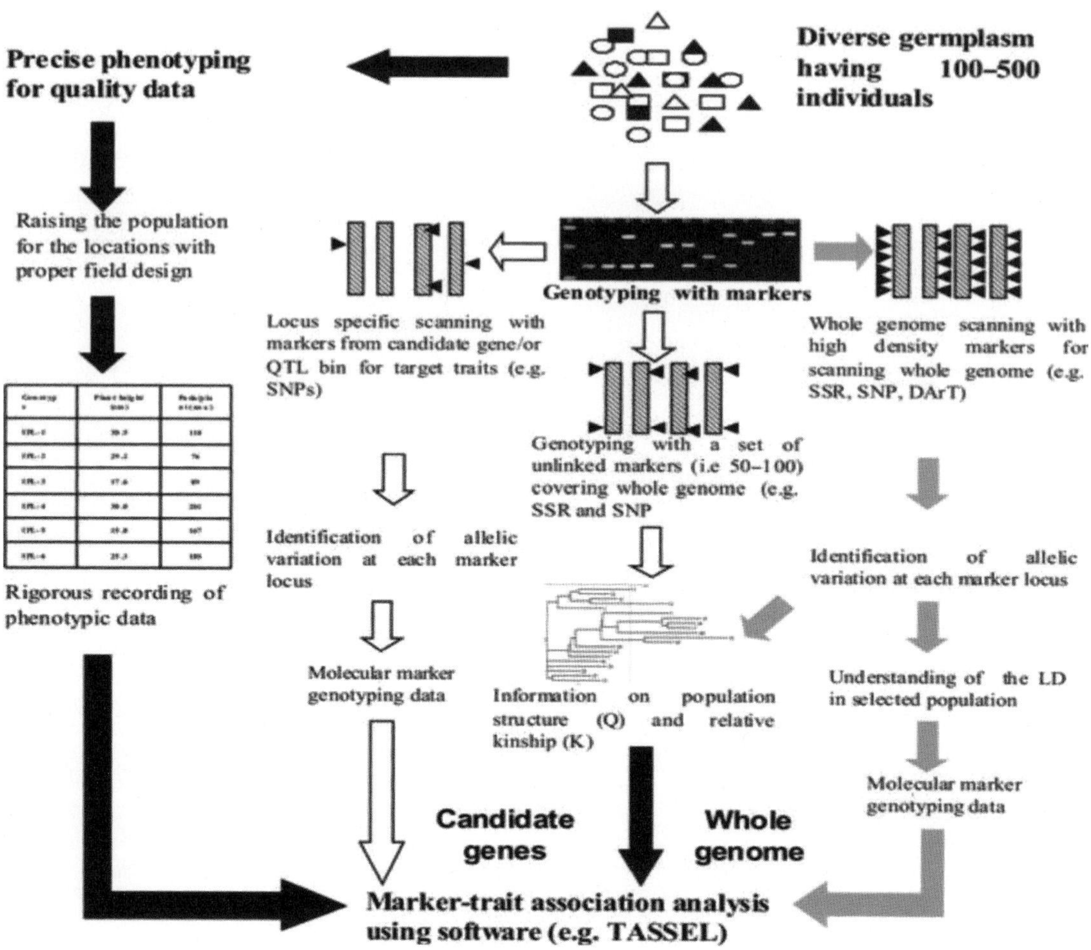


**Fig. 16.5** Schematic representation of MAB approaches for polygenic traits (Redrawn from Kumar et al. 2011)

As discussed in the earlier section, MAB for polygenic traits also followed the common strategies such as backcross breeding and introgression of QTLs and pyramiding of QTLs.

## Marker-Assisted Backcross Breeding (MABC) for QTLs Conferring Stress Resistance

Just like backcross breeding for a major gene, MABC approaches have also been used to introgress several QTLs conferring resistance to biotic and abiotic stresses into adapted or elite variety in several crop plants including rice, maize, and tomato with reasonable success. This approach can also be used at three levels, viz., foreground selection, recombinant selection, and background selection. Since several genes are involved in the expression of a quantitative trait, these genes have smaller individual effects on the phenotype, and the effect of individual genes is not easily monitored during backcross breeding due to several factors (see for review Serraj et al. 2011). However, the development of improved field experimental design, robust mathematical models, and statistical approaches enhanced the QTL mapping and introgressive breeding in several crops. Two powerful approaches, viz., AB-QTL analysis and association mapping, were used for QTL mapping and its backcross breeding endeavors in several crops (Table 16.7).

**Table 16.7** Example of marker assisted back cross breeding of QTLs conferring stress resistance for the development of cultivars and advance breeding lines in several crop systems

| Crop | Agronomic traits | Markers used | End product | References |
|---|---|---|---|---|
| Rice | Rice yellow mottle virus resistance | RFLP, SSR | Advanced breeding line | Ahmadi et al. (2001) |
| | Drought tolerance | RFLP, SSR | Near isogenic lines | Steele et al. (2006) |
| | Submergence tolerance (*SUB1*) | SSR | Swarna-SUB1: A cultivar | Neeraja et al. (2007) |
| | | SSR | OM 1490/IR 64 -Sub1based advanced breeding lines (10 no.) | Lang et al. (2011) |
| | | SSR | ASS996-*SUB*1: A cultivar | Cuc et al. (2012) |
| Wheat | Fusarium head blight resistance | SSR | Breeding lines | Miedaner et al. (2006) |
| | | SSR | Advanced breeding lines | Sommers et al. (2005) |
| | | SSR | Advanced breeding lines | Wilde et al. (2008) |
| Barley | Resistance to stripe rust | RFLP, RAPD, AFLP | Tango-A cultivar | Hayes et al. (2003) |
| Maize | Resistance south-western corn borer (SWCB) | RFLP | Advanced breeding lines | Willcox et al. (2002) |
| Tomato | Resistance to black mold | RFLP, CAPS | Experimental breeding lines | Robert et al. (2001) |
| | Resistance to early blight | RFLP | Advanced breeding lines | Foolad et al. (2002) |

## MABC of QTLs Conferring Submergence Tolerance in Rice: A Case Study

Rice landraces tolerant of up to 2 weeks of complete submergence were collected from farmers' fields in the 1950s. Success in fine-mapping of *SUBMERGENCE1* (*SUB1*), a major QTL from the submergence tolerant landrace FR13A, on chromosome 9 of rice genome (Xu and Mackill 1996) has enabled marker-assisted breeding of high-yielding rice Swarna-Sub1 (Neeraja et al. 2007) that is capable of enduring transient complete submergence in a short span of 2 years. Now this QTL has been introgressed into several cultivars of rice growing across the world.

## Marker-Assisted Pyramiding of QTLs Conferring Stress Resistance

Introgression of quantitative resistance controlled by QTLs provides another promising strategy to develop durable biotic and abiotic stress resistance. Castro et al. (2003) depicted quantitative resistance as an insurance policy in case of the breakdown of qualitative resistance. QTL pyramiding is an efficient strategy for improvement of biotic stress resistance in crops. This strategy is based on the combination of desirable QTLs through conventional crossing using molecular markers. Once desirable QTLs are detected, a strategy for QTL pyramiding employs the use of NILs each harboring only one target QTL. The NIL-QTLs are produced by backcrossing and marker selection. A parent line with a positive QTL is backcrossed with a recurrent parent lacking the QTL. Subsequently, a line that carries only a positive QTL region from the mother line in the recurrent parent genome background is selected by molecular markers. The NILs can be used to accurately evaluate the effect of each QTL individually. Once QTLs with important effects are identified in this manner, the appropriate NIL-QTLs are crossed to pyramid two or more QTLs in the same background. A notable example of the combination of quantitative resistance was the pyramiding of a single stripe rust gene and two

**Table 16.8** Marker assisted pyramiding of QTLs conferring resistance to biotic and abiotic stress in some major crop systems

| Crop | Agronomic trait | Pyramided QTL(s) | References |
|---|---|---|---|
| Rice | Sheath blight resistance | *qSB7, qSB9* and *qSB 11* | Yin et al. (2008) |
| | Leaf folder resistance | *qRLF3, qRLF4* and *qRLF-8* | Rao et al. (2010) |
| Wheat | Crown rot resistance | *QCr.usq-1D.1* | Bovill et al. (2010) |
| | | *QCr.usq-2B.2* and | |
| | | *QCr.usq-4B.2* | |
| Barley | Stripe rust resistance | *QTl4,QTL-7* and *Rspx* | Castro et al. (2003) |
| | | *QTl5* and *Rspx* | |
| | Yellow dwarf virus resistance | *Ryd2* and *Ryd3* | Riedel et al. (2011) |
| Lettuce | Downy mildew resistance | *Rbq 4, Rbq 5, Rbq 6, Rbq 7* and *Rbq 11* | Zhang et al. (2009a) |

QTLs (Castro et al. 2003). In theory, MAB could also be used to pyramid QTLs from multiple parents, and the marker-aided pyramiding could also facilitate the combination of QTLs for abiotic stress tolerance, especially QTLs effective at different growth stages. Marker-aided QTL pyramiding might also be used to combine QTLs especially where one QTL masks the presence of other QTLs (Sanchez et al. 2000; Zhu et al. 2006). This was experimentally validated for two interacting resistance QTLs for rice yellow mottle virus (Ahmadi et al. 2001) and downy mildew resistance in lettuce (Zhang et al. 2009a). The details of some more successful marker-aided QTL pyramiding have been furnished in Table 16.8.

## Recent Approaches in MAB: Breeding by Design and Genomic Selection

Breeding by design is a novel *in silico* concept that aims to control all allelic variation for all the genes/QTLs of agronomic importance. Peleman and vander Voort (2003b) proposed that by understanding the genetic basis of all agronomically important characters and the allelic variation at those loci, the breeder would be able to design superior genotypes *in silico*, which demonstrates that DNA markers not only improve selection processes but can aid in creating novel genotypes bearing superior agroeconomic traits on farmers' demand as well as scientific requirement. This strategy involves three steps, viz.:

(a) Mapping of loci involved underlying all agronomically relevant traits preferably through IL libraries, which are extremely powerful tool not only for mapping but also in reducing the genetic complexity of quantitative traits by separating them into a set of monogenic loci. An IL library consists of a series of lines harboring a single homozygous donor segment introgressed into a uniform cultivated background, which also referred as near-isogenic introgression lines (NIILs) by Tuinstra et al. (1997).
(b) Assessment of the allelic variation through intensive chromosome haplotyping and extensive phenotyping of all agronomic traits.
(c) Designing superior genotypes comprising a combination of favorable alleles at all loci through accurate selection of recombination events using flanking markers to collate different favorable alleles next to each other.

This designed breeding approach involves an integrative complementary application of technological tools and the materials currently available to develop superior varieties. The success of this approach would essentially depend on the availability of extremely saturated marker maps and precision of phenotyping.

Another strategy of MAB is "genomic selection (GS)," which is based upon the simultaneous estimation of effects on the phenotype of all available loci, haplotypes, and markers without a previous selection of markers with effects on the phenotype (Jannink et al. 2010). GS requires availability of phenotypic and genotypic data and the establishment of a genetic model based upon linkage disequilibrium, so that the differences in the phenotype are explained by the markers analyzed (Lorenz et al. 2011).

All three approaches were extensively used in present-day molecular breeding and the results will be realized soon.

## Conclusion

MAB for vertical (qualitative) and horizontal (quantitative) resistance to biotic and abiotic stress is a holistic approach. Till date, MAB was applied in small scale for crop breeding, but there will be a greater level of adoption within the next decade and beyond. Factors that should lead to a greater adoption of MAB in crops include:

- Establishment of facilities for marker genotyping and staff training within many crop breeding institutes in different countries
- Currently available (and constantly increasing) data on genes and QTLs controlling traits and the identification of linked and flanked markers
- Development of effective and durable strategies for using markers in breeding
- Establishment and curation of public databases for QTLs, genes, and marker data
- Available resource for generating new markers from DNA sequence data arising from genome-sequencing projects and research in functional genomics

It is also critical that future endeavors in MAB are based upon lessons that have been learned from past successes and (especially) failures in using MAB. Further optimization of marker genotyping methods in terms of cost-effectiveness and a greater level of integration between molecular and conventional breeding (especially in designing efficient and cost-effective strategies) represent the main challenges for the greater adoption and impact of MAB on crop breeding in the near future.

## References


Admassu B, Perovic D, Friedt W, Ordon F (2011) Genetic mapping of the stem rust (*Puccinia graminis* f. sp. Tritici Eriks. & E. Henn ) resistance gene *Sr13* in wheat (*Triticum aestivum* L.). Theor Appl Genet 122:643–648

Agrama HA, Dahleen L, Wentz M, Jin Y, Steffenson B (2004) Molecular mapping of the crown rust resistance gene *Rpcl* in barley. Phytopathology 94:858–861

Ahmadi N, Albar L, Pressoir G, Pinel A, Fargette D, Ghesquiere A (2001) Genetic basis and mapping of the resistance to rice yellow mottle virus. III. Analysis of QTL efficiency in introgressed progenies confirmed the hypothesis of complementary epistasis between two resistance QTLs. Theor Appl Genet 103:1084–1092

Albar L, Lorieux M et al (1998) Genetic basis and mapping of the resistance to rice yellow mottle virus. I. QTLs identification and relationship between resistance and plant morphology. Theor Appl Genet 97:1145–1154

Ali (1999) Mapping quantitative trait loci for root traits related to drought resistance in rice (*Oryza sativa* L.) using AFLP markers. PhD thesis, Texas Tech University, Texas

Allard RW (1999) Principles of plant breeding, 2nd edn. Wiley, New York

Anbessa Y, Taran B, Warkentin TD, Tullu A, Vandenberg A (2009) Genetic analyses and conservation of QTL for ascochyta blight resistance in chickpea (*Cicer arietinum* L.). Theor Appl Genet 119:757–765

Anderson IA, Waldron BL, Moreno-Sevilla B, Stack RW, Frohberg RC (1998) Detection of *Fusarium* head blight resistance QTL in wheat using AFLPs and RFLPs. In: Slinkard AE (ed) Proceedings of the 9th international wheat genetics symposium, vol 1. University of Saskatchewan, Saskatoon, pp 135–137

Anuradha J, Balasubramani G (2011) Recent molecular advances to combat abiotic stress tolerance in crop plants. Biotechnol Mol Biol Rev 6:31–58

Anuradha C, Gaur PM, Pande S, Gali KK, Ganesh M, Kumar J, Varshney RK (2011) Mapping QTL for resistance to botrytis. Euphytica 182:1–9

Astua Monge G, Minsavage GV, Stall RE, Vallejos CE, Davis MJ, Jones JB (2000) *Xv4-vrxv4*: a new gene for gene interaction identified between *Xanthomonas campestris* pv. *vesicatoria* race $T_3$ and the wild tomato relative *Lycopersicon pennellii*. Mol Plant Microbe Interact 13:1346–1355

Autrique E, Singh RP, Tanksley SD, Sorells ME (1995) Molecular markers for four leaf rust resistance genes introgressed into wheat from wild relatives. Genome 38:75–83

Bachlava E, Dewey RE, Burton JW, Cardinal AJ (2009) Mapping candidate genes for oleate biosynthesis and their association with unsaturated fatty acid seed content in soybean. Mol Breed 23:337–347

Bai Y, Huang C, van der Hulst R, Meijer-Dekens F, Bonnema G, Lindhout P (2003) QTLs for tomato powdery mildew resistance (*Oidium lycopersici*) in *Lycopersicon parviflorum* G1.1601 co-localize with two qualitative powdery mildew resistance genes. Mol Plant Microbe Interact 16:169–176

Ballvora A, Pierre M, van den Ackerveken G et al (2001) Genetic mapping and functional analysis of the tomato *Bs4* locus governing recognition of the *Xanthomonas campestris* pv. *vesicatoria AvrBs4* protein. Mol Plant Microbe Interact 14:629–638

Barian HS, Parry N, Barclay IR et al (2006) Identification and characterization of stripe rust resistance gene *Yr34* in common wheat. Theor Appl Genet 112: 1143–1148

Barloy D, Lemoine J, Abelard P, Tanguy AM, Rand R, Jahier J (2007) Marker assisted pyramiding of two cereal cyst nematode resistance genes from *Aegilops variabilis* in wheat. Mol Breed 20:31–40

Becker H (1993) Pflanzenzu¨ chtung. Verlag Eugen Ulmer, Stuttgart

Bergamin Filho A, Kimati H, Amorim L (1995) Manual de fitopatologia: principios e conceitos, 3rd edn. Ceres, Sao Paulo

Bernado R (2008) Molecular markers and selection for complex traits in plants: learning from the last 20 years. Crop Sci 48:1649–1664

Bonnett DG, Rebetzke GJ, Spielmeyer W (2005) Strategies for efficient implementation of molecular markers in wheat breeding. Mol Breed 15:75–85

Borovkova LG, Steffenson BJ, Jin Y et al (1995) Identification of molecular markers linked to the stem rust resistance gene *rpg4* in barley. Phytopathology 85:181–185

Bossolini E, Krattinger SG, Keller B (2006) Development of simple sequence repeat markers specific for the *Lr34* resistance region of wheat using sequence information from rice and *Aegilops tauschii*. Theor Appl Genet 113:1049–1062

Bovill WD, Horne M, Herde D, Davis M, Widermuth GB, Sutherland MW (2010) Pyramiding QTL increases seedling resistance to crown rot (*Fusarium pseudograminearum*) of wheat (*Triticum aestivum*). Theor Appl Genet 121:127–136

Bretting PK, Widrlechner MP (1995) Genetic markers and plant genetic resource management. Plant Breed Rev 13:11–86

Brunner S, Keller B, Feuillet C (2000) Molecular mapping of *Rph-7.g* leaf rust resistance gene in barley (*Hordeum vulgare*). Theor Appl Genet 101:783–788

Burr B, Burr FA, Matz EC, Romero-Severson J (1992) Pinning down loose ends: mapping telomeres, and factors affecting their length. Plant Cell 4:953–960

Bustamam M, Tabien RE, Suwarno A, Abalos MC, Kadir TS, Ona I, Bernardo M, Veracruz CM, Leung H (2002) Asian rice biotechnology network: improving popular cultivars through marker assisted backcrossing by the NARES. In: Proceedings of international rice congress, Beijing

Cao W, Huges GR, Ma H, Dong Z (1998) Development of DNA based markers for resistance to *Septoria nodorum* blotch in durum wheat. In: Slinkard AE (ed) Proceedings of the 9th international wheat genetics symposium, vol 3. Univ Saskatchewan, Saskatoon, pp 95–97

Cao X, Zhou J, Gong X, Zhao G, Jia J, Qi X (2012) Identification and validation of a major quantitative trait locus for slow rusting resistance to stripe rust in wheat. J Integr Plant Biol. doi:10.1111/j.1744-7909.2012.01111.x

Castro AJ, Capettini F, Corey AE et al (2003) Mapping and pyramiding of qualitative and quantitative resistance to stripe rust in barley. Theor Appl Genet 107:922–930

Champoux MC, Wang GL, Sarkarung S, Mackill DJ, O'Toole JE, Huang N, Mc Couch SR (1995) Locating genes associated with root morphology and drought avoidance in rice via linkage to molecular markers. Theor Appl Genet 90:969–981

Channamallikarjuna V, Sonah H, Prasad M, Rao GJN, Chand S, Upreti HC, Singh NK, Sharma TR (2010) Identification of major quantitative trait loci *qSBR11-1* for sheath blight resistance in rice. Mol Breed 25:155–166

Chen S, Lin XH, Xu CG, Zhang Q (2000) Improvement of bacterial blight resistance of 'Minghui 63', an elite restorer line of hybrid rice, by molecular marker assisted selection. Crop Sci 40:239–244

Chen S, Xu CG, Lin XH, Zhang Q (2001) Improving bacterial blight resistance of '6078', an elite restorer line of hybrid rice, by molecular marker-assisted selection. Plant Breed 120:133–137

Chen H, Fu R, Li X (2011a) Tomato yellow leaf virus (TYLCV): the structure, ecotypes and the resistance germplasm resources in tomato. Afr J Biotechnol 10:13361–13367

Chen S, Liu X, Zeng L, Ouyang D, Yang J, Zhu X (2011b) Genetic analysis and molecular mapping of a novel recessive gene *xa34(t)* for resistance against *Xanthomonas oryzae* pv. *oryzae*. Theor Appl Genet 122:1331–1338

Chen J, Chu C, Souza EJ, Guttieri MJ, Chen X, Xu S, Hole D, Zemetra R (2012) Genome wide identification of QTL conferring high temperature adult plant (HTAP) resistance to stripe rust (*Puccinia striiformis* f. sp. *tritici*) in wheat. Mol Breed 29:791–800

Chhuneja P, Kumar K, Stirnweis D, Hurni S, Keller B, Dhaliwal HS, Singh K (2012) Identification and mapping of two powdery mildew resistance genes in *Triticum boeoticum* L. Theor Appl Genet 124:1051–1058

Chicaiza O, Khan IA, Zhang X, Brevis JC, Jackson L, Chen X, Dubcovsky J (2006) Registration of five wheat isogenic lines for leaf rust and stripe rust resistance genes. Crop Sci 46:485–487

Chu C, Niu Z, Zhong S, Chao S, Friesen TL, Halley S, Elias EM, Dong Y, Faris JD, Xu SS (2011) Identification and molecular mapping of two QTLs with major effects for resistance to *Fusarium* head blight in wheat. Theor Appl Genet 123:1107–1119

Chunwongse J, Bunn TB, Crossman C, Jiang J, Tanksley SD (1994) Chromosomal localization and molecular marker tagging of the powdery mildew resistance gene *(Lv)* in tomato. Theor Appl Genet 89:76–79

Cobos MJ, Winter P, Kharrat M, Cubero JI, Gil J, Millan T, Rubio J (2009) Genetic analysis of agronomic traits in a wide cross of chickpea. Field Crops Res 111:130–136

Collard BCY, Mackill MJ (2008) Marker assisted selection: an approach for precision plant breeding in the twenty first century. Philos Trans R Soc B 363:557–572

Collard BCY, Jahufer MZZ, Brouwer JB, Pang ECR (2005) An introduction to markers, quantitative trait loci (QTL) mapping, and marker-assisted selection for crop improvement: the basic concepts. Euphytica 142:169–196

Courtois B, Mclaren G, Sinha PK, Prasad K, Yadav R, Shen L (2000) Mapping QTLs associated with drought avoidance in upland rice. Mol Breed 6:55–66

Cox TS, Raupp WJ, Gill BS (1994) Leaf rust-resistance genes- *Lr41*, *Lr42* and *Lr43* transferred from *Triticum tauschii* to common wheat. Crop Sci 34:339–343

Cuc LM, Huyen LTN, Hien PTM, Hang VTT, Dam NQ, Mui PT, Quangi VD, Ismail AM, Ham LH (2012) Application of marker assisted backcrossing to introgress the submergence tolerance QTL *SUB*1 into the Vietnam elite rice variety-AS996. Am J Plant Sci 3:528–536

Das A, Soubam D, Singh PK, Thakur S, Singh NK, Sharma TR (2012) A novel blast resistance gene, *Pi54rh* cloned from wild species of rice. Oryza rhizomatis confers broad spectrum resistance to Magnaporthe oryzae. Funct Integr Genomics. doi:10.1007/s10142-012-0284-1

Datta K, Baisakh N, Thet KM, Tu J, Datta S (2002) Pyramiding transgenes for multiple resistance in rice against bacterial blight, yellow stem borer and sheath blight. Theor Appl Genet 106:1–8

de Giovanni C, Dell'Orco P, Bruno A, Ciccarese F, Lotti C, Ricciardi L (2004) Identification of PCR-based markers (RAPD, AFLP) linked to a novel powdery mildew resistance gene (*ol-2*) in tomato. Plant Sci 166:41–48

Dekkers JCM, Hospital F (2002) The use of molecular genetics in the improvement of agricultural populations. Nat Rev Genet 3:22–32

Doveri S, Lee D, Maheswaran M, Powell W (2008) Molecular markers: history, features and applications. In: Kole C, Abbott AG (eds) Principles and practices of plant genomics. Science publisher USA, Enfield

Duan CX, Su N, Cheng ZJ, Lei CL, Wang JL, Zhai HQ, Wan JM (2010) QTL analysis for the resistance to small brown planthopper (*Laodelphax striatellus* Fallen) in rice using backcross inbred lines. Plant Breed 129:63–67

Dudley JW (1993) Molecular markers in plant improvement: manipulation of genes affecting quantitative traits. Crop Sci 33:660–668

Dwivedi SL, Stalker HT, Blair MW, Bertoli DJ, Nielen S, Ortiz R (2008) Enhancing crop gene pools with beneficial traits using wild relatives. Plant Breed Rev 30:179–230

Eastwood RF, Lagudah ES, Appels R (1994) A direct search for DNA sequences tightly linked to cereal cyst nematode resistance genes in *Triticum tauschii.* Genome 37:311–319

Eathington SR, Dudley JW, Rufener GK (1997) Usefulness of marker-QTL associations in early generation selection. Crop Sci 37:1686–1693

Echart CL, Neto JFB, Cavali SS (2006) Aluminium tolerance in Barley: molecular mapping analysis. R Bras Agrociencia Pelotas 12:15–20

Edwards MD, Stuber CW, Wendel JF (1987) Molecular-marker facilitated investigations of quantitative-trait loci in maize. I. Numbers, genomic distribution and types of gene action. Genetics 116:113–125

Eglinton JK, Coventry SJ, Chalmers K (2006) Breeding outcomes from molecular genetics.-Breeding for success: diversity in action. In: Proceedings of the 13th Australasian plant breeding conference, 1 Christchurch

Eibach R, Zyprian E, Welter L, Topfer R (2007) The use of molecular markers for pyramiding resistance genes in grapevine breeding. Vitis 46:120–124

Emebiri L, Michael P, Moody D (2009) Enhanced tolerance to boron toxicity in two-rowed barley by marker-assisted introgression of favourable alleles derived from Sahara 3771. Plant Soil 314:77–85

Eshed Y, Zamir D (1995) An introgression line population of *Lycopersicon pennellii* in the cultivated tomato enables the identification and fine mapping of yield associated QTL. Genetics 141:1147–1162

Eskes A (1980) Bases Geneticas da resistencia horizontal a pathogenos em plantas. Ciencia e Cultura 32:1464–1472

Everson E, Schaller CW (1955) The genetics of yield differences associated with awn barbing in the barley hybrid ('Lion' x 'Atlas 10') x 'Atlas'. Agron J 47:276–280

Fahima T, Chague V, Sun G, Korol A, Ronin Y, Roder M, Grama A, Neva E (1997) Identification and potential use of PCR markers flanking the *Triticum dicoccoides-derived* stripe rust resistance gene *Yr15* in wheat. In: Proceedings of wheat international 5th congress plant molecular biology, Singapore

Falke K, Susic Z, Wilde P, Wortmann H, Mohring J, Piepho HP, Geiger H, Miedaner T (2009) Testcross performance of rye introgression lines developed by marker assisted backcrossing using an Iranian accession as donor. Theor Appl Genet 118:1225–1238

Feng V, Zuo LL, Zhang ZY, Lin RM, Cao YY, Xu SC (2011) Quantitative trait loci for temperature sensitive resistance to *Puccinia striiformis* f. sp. *tritici* in wheat cultivar Flinor. Euphytica 178:321–329

Ferreira ME, Satagopan JM, Yandell BS, Williams PH, Osborn TC (1995a) Mapping loci controlling vernalization requirement and flowering time in *Brassica napus*. Theor Appl Genet 90:727–732

Ferreira ME, Williams PH, Osborn TC (1995b) Mapping of a locus controlling resistance to *Albugo candida* in *Brassica napus* using molecular markers. Phytopathology 85:218–220

Feuillet C, Messmer M, Schachermayr G, Keller B (1995) Genetic and physical characterization of the *Lr1* leaf rust locus in wheat *(Triticum aestivum* L.). Mol Gen Genet 248:553–562

Figdore SS, Ferreira ME, Slocum MK, Williams PH (1993) Association of RFLP markers with trait loci affecting club root resistance and morphological characters in *Brassica oleracea* L. Euphytica 63:33–44

Flor HH (1971) Current status of the gene for gene concept. Annu Rev Phytopathol 9:275–296

Foolad MR (2007) Genome mapping and molecular breeding of tomato. Int J Plant Genomics. doi:10.1155/2007/64358

Foolad MR, Jones RA (1993) Mapping salt-tolerance genes in tomato *(Lycopersicon esculentum)* using trait-based marker analysis. Theor Appl Genet 87:184–192

Foolad M, Zhang L, Khan A, Nino-Liu D, Lin G (2002) Identification of QTLs for early blight (*Alternaria solani*) resistance in tomato using backcross populations of a *Lycopersicon esculentum* x *L. hirsutum* cross. Theor Appl Genet 104:945–958

Francia E, Tacconi G, Crosatti C, Barabaschi D, Bulgarelli D, Dall'Aglio E, Vale G (2005) Marker assisted selection in crop plants. Plant Cell Tissue Org Cult 82:317–342

Frisch M, Bohn M, Melchinger AE (1999a) Comparison of selection strategies for marker assisted backcrossing of a gene. Crop Sci 39:1295–1301

Frisch M, Bohn M, Melchinger AE (1999b) Minimum sample size and optimal positioning of flanking markers in marker-assisted backcrossing for transfer of a target gene. Crop Sci 39:967–975

Fujita D, Doi K, Yoshimura A, Yasui H (2010) A major QTL for resistance to green rice leafhopper (*Nephotettix cincticeps* Uhler) derived from African rice (*Oryza glaberrima* Steud.). Breed Sci 60: 336–341

Ganal MW, Broun P, Tanksley SD (1992) Genetic mapping of tandemly repeated telomeric DNA sequences in tomato *Lycopersicon esculentum*. Genomics 14:444–448

Ganapathy KN, Byre GM, Venkatesha SC et al (2009) Identification of AFLP markers linked sterility mosaic disease in pigeonpea *Cajanus cajan* (L.) Millsp. Int J Integr Biol 7:145–149

Geiger HH, Heun M (1989) Genetics of quantitative resistance to fungal diseases. Annu Rev Phytopathol 27:317–341

Gibson S, Somerville C (1993) Isolating plant genes. Trends Biotechnol 11:306–313

Gnanesh BN, Bohra A, Sharma M, Byregowda M, Pandey S, Wesley V, Saxena RK, Saxena KB, Kavi Kishor PB, Varshney RK (2011) Genetic mapping and quantitative trait locus analysis of resistance to sterility mosaic disease in pigeonpea [*Cajanus cajan* (L.) Millsp.]. Field Crops Res 123:53–61

Gowda M, Roy Barman S, Chatto BB (2006) Molecular mapping of a novel blast resistance gene *Pi38* in rice using SSLP & AFLP markers. Plant Breed 125:596–599

Gowda SJM, Radhika P, Kadoo NY, Mhase LB, Gupta VS (2009) Molecular mapping of wilt resistance genes in chickpea. Mol Breed 24:177–183

Gupta PK, Varshney RK (2000) The development and use of microsatellite markers for genetic analysis and plant breeding with emphasis on bread wheat. Euphytica 113:163–185

Gupta PK, Varshney RK, Sharma PC, Ramesh B (1999) Molecular markers and their application in wheat breeding: a review. Plant Breed 118:369–390

Gur A, Zamir D (2004) Unused natural variation can lift yield barriers in plant breeding. PLoS Biol 2:245

Hajjar R, Hodgkin T (2007) The use of wild relatives in crop improvement: a survey of developments over the last 20 years. Euphytica 156:1–13

Hanson PM, Green SK, Kuo G (2006) *Ty-2*, a gene on chromosome 11 conditioning gemini virus resistance in tomato. Rep Tomato Genet Coop 56:17–18

Hao Y, Chen Z, Wang Y, Bland D, Buck J, Guedira GB, Johnson J (2011) Characterization of a major QTL for adult plant resistance to stripe rust in US soft red winter wheat. Theor Appl Genet 123:1401–1411

Hash CT, Thakur RP, Rao VP, Bhasker Raj AG (2006) Evidence for enhanced resistance to diverse isolates of pearl through gene pyramiding millet downy mildew. Int Sorghum Millets Newsl 47:134–138

Hayes PM, Corey AE, Mundt C, Toojinda T, Vivar H (2003) Registration of 'Tango' barley. Crop Sci 43:729–731

Helguera M, Vanzetti L, Soria M, Khan IA, Kolmer J, Dubcovsky J (2005) PCR markers for *Triticum speltoides* leaf rust resistance gene *Lr51* and their use to develop isogenic hard red spring wheat lines. Crop Sci 45:728–734

Hiebert C, Thomas J, McCallum B (2005) Locating the broad-spectrum wheat leaf rust resistance gene *Lr52* (*LrW*) to chromosome 5B by a new cytogenetic method. Theor Appl Genet 110:1453–1457

Hiebert CW, Thomas JB, Somers DJ, McCallum BD, Fox SL (2007) Microsatellite mapping of adult-plant leaf rust resistance gene *Lr22a* in wheat. Theor Appl Genet 115:877–884

Himabindu K, Suneetha K, Sama VSAK, Bentur JS (2010) A new rice gall midge resistance gene in the breeding line CR57-MR1523, mapping with flanking markers and development of NILs. Euphytica 174:179–187

Hittalmani S, Mew T, Foolad MR, Huang N (1995a) Identification of blast resistance gene *Pi-2(t)* in a segregating population of rice. Theor Appl Genet 91:9–14

Hittalmani S, Mew T, Huang N, Khush GS (1995b) Disease resistance breeding for blast and bacterial leaf blight in rice. In: Fragile lives in fragile ecosystems. Proceedings of international rice research conference, IRRI, Manila

Hittalmani S, Parco A, Mew TV, Zeigler RS, Huang N (2000) Fine mapping and DNA marker assisted pyramiding of the three major genes for blast resistance in rice. Theor Appl Genet 100:1121–1128

Holland JB (2004) Implementation of molecular markers for quantitative traits in breeding programs- Challenges and opportunities. In: Fischer T (ed) New directions for a diverse planet. Proceedings of the 4th international crop science congress, The Regional Institute Ltd., Gosford, NSW, Australia (www.cropscience.org.au)

Hospital F (2005) Selection in backcross programmes. Philos Trans R Soc B 360:1503–1511

Hospital F, Charcosset A (1997) Marker assisted introgression of quantitative trait loci. Genetics 147:1469–1485

Huang N, Angeles ER, Domingo J, Magpantay G, Singh S, Bennett J, Khush GS (1997) Pyramiding of bacterial blight resistance genes via DNA marker aided selection in rice. Theor Appl Genet 95:313–320

Islam MR, Salam MA, Hassan L, Collard BCY, Singh RK, Gregorio GB (2011) QTL mapping for salinity tolerance at seedling stage in rice. Emir J Food Agric 23:137–146

Jahoor A (1998) Marker assisted breeding in cereals, specially with respect to synteny among loci for mildew resistance. In: Gupta PK (ed) Genetics and biotechnology for crop improvement. Rastogi Publications, Meerut

Jairin J, Sansen K, Wongboon W, Kothcharerk J (2010) Detection of a brown planthopper resistance gene *bph4* at the same chromosomal position of *Bph3* using two different genetic backgrounds of rice. Breed Sci 60:71–75

Jannink JL, Lorenz AJ, Iwata H (2010) Genomic selection in plant breeding: from theory to practice. Brief Funct Genomics 9:166–177

Jansen JP (1996) Aphid resistance in composites. International application published under the patent cooperation treaty (PCT) No. WO 97/46080

Jena KK (2010) The species of the genus *Oryza* and transfer of useful genes from wild species into cultivated rice, O. sativa. Breed Sci 60:518–523

Ji YF, Scott JW (2006) *Ty-3*, a begomo virus resistance locus linked to *Ty-1* on chromosome 6 of tomato. Rep Tomato Genet Coop 56:22–25

Jia J, Devos KM, Chao S, Miller TE, Reader SM, Gale MD (1996) RFLP based maps of the homoeologous group-6 chromosomes of wheat and their application in the tagging of *Pm12*, a powdery mildew resistance gene transferred from *Aegilops speltoides* to wheat. Theor Appl Genet 92:559–565

Jiang GH, Xu CG, Tu JM, Li XH, He YQ, Zhang QF (2004) Pyramiding of insect and disease resistance genes into an elite *indica*, cytoplasm male sterile restorer line of rice, 'Minghui 63'. Plant Breed 123:112–116

Kalia RK, Rai MK, Kalia S, Singh R, Dhawan AK (2011) Microsatellite markers: an overview of the recent progress in plants. Euphytica 177:309–334

Kanjoo V, Jearakongman S, Punyawaew K, Siangliw JL, Siangliw M, Vanavichit A, Toojinda T (2011) Co-location of quantitative trait loci for drought and salinity tolerance in rice. Thai J Genet 4:126–138

Katiyar S, Verulkar S, Chandel G, Zhang Y, Huang B, Bennet J (2001) Genetic analysis and pyramiding of two gall midge resistance genes (*Gm 2* and *Gm 6 t*) in rice (*Oryzae sativa* L.). Euphytica 122:327–334

Khairallah MM, Bohn M, Jiang C (1998) Molecular mapping of QTL for southwestern corn borer resistance, plant height and flowering in tropical maize. Plant Breed 117:309–318

Khan AA, Bergstrom GC, Nelson JC, Sorrells ME (2000) Identification of RFLP markers for resistance to wheat spindle streak mosaic bymovirus (WSSMV) disease. Genome 43:477–482

Kiyosawa S (1982) Genetics and epidemiological modelling of breakdown of plant disease resistance. Annu Rev Phytopathol 20:93–117

Kloppers FJ, Pretorius ZA (1997) Effects of combinations amongst genes *Lr13*, *Lr34* and *Lr37* on components of resistance in wheat to leaf rust. Plant Pathol 46:737–750

Koebner RMD, Summers RW (2003) 21st century wheat breeding: plot selection or plate detection? Trends Biotechnol 21:59–63

Kole C (2011) Wild crop relatives: genomic and breeding resources. Springer, Heidelberg/Dordrecht/London/New York

Kole C, Gupta PK (2004) Genome mapping and map based cloning. In: Jain HK, Kharkwal MC (eds) Plant breeding: Mendelian to molecular approaches. Narosa Publishing House, New Delhi

Kole C, Kole P, Vogelzang R, Osborn TC (1997) Genetic linkage map of a *Brassica rapa* recombinant inbred population. J Hered 88:553–557

Kole C, Thormann CE, Karlsson BH, Palta JP, Gaffney P, Yandell B, Osborn TC (2002a) Comparative mapping of loci controlling winter survival and related traits in oilseed *Brassica rapa* and *B. napus*. Mol Breed 9:201–210

Kole C, Williams PH, Rimmer SR, Osborn TC (2002b) Linkage mapping of genes controlling resistance to white rust (*Albugo candida*) in *Brassica rapa* (syn. *campestris*) and comparative mapping to *B. napus* and *Arabidopsis thaliana*. Genome 45:22–27

Kole C, Teutonico R, Mengistu A, Williams PH, Osborn TC (1996) Molecular mapping of a locus controlling resistance to *Albugo candida* in *Brassica rapa*. Phytopathology 86:367–369

Kole C, Quijada P, Michaels AD, Amasino RM, Osborn TC (2001) Evidence for homology of flowering-time genes VFR2 from *Brassica rapa* and FLC from *Arabidopsis thaliana*. Theor Appl Genet 102:425–430

Koseki M, Kitazawa N, Yonebayashi S, Maehara Y, Wang ZX, Minobe Y (2010) Identification and fine mapping of a major quantitative trait locus originating from wild rice, controlling cold tolerance at the seedling stage. Mol Genet Genomics 284:45–54

Kotresh H, Fakrudin B, Punnuri SM, Rajkumar BK, Thudi M, Paramesh H, Lohithaswa H, Kuruvinashetti MS (2006) Identification of two RAPD markers genetically linked to a recessive allele of a *Fusarium* wilt resistance gene in pigeonpea (*Cajanus cajan* L. Millsp.). Euphytica 149:113–120

Kottapalli KR, Narasu ML, Jena KK (2010) Effective strategy for pyramiding three bacterial blight resistance genes into fine grain rice cultivar, Samba Mahsuri, using sequence tagged site markers. Biotechnol Lett 32:989–996

Kuchel H, Fox R, Rainheimer J, Mosionek L, Willey N, Bariana H, Jefferies S (2007) The successful application of a marker assisted wheat breeding strategy. Mol Breed 20:295–308

Kumar PL, Latha TKS, Kulkarni NK, Raghavendra N, Saxena KB, Waliyar F, Rangaswamy KT, Muniyappa V, Doriswamy S, Jones AT (2005) Broad-based resistance to pigeonpea sterility mosaic disease in wild relatives of pigeonpea (*Cajanus*: Phaseoleae). Ann Appl Biol 146:371–379

Kumar J, Choudhury AK, Solanki RK, Pratap A (2011) Towards MAS in pulses: a review. Plant Breed 130:297–313

Kumaravadivel N, Uma MD, Saravanan PA, Suresh H (2006) Molecular marker-assisted selection and pyramiding genes for gall midge resistance in rice suitable for Tamil Nadu Region. In: ABSTRACTS- 2nd international rice congress, IRRI, Manila

Kurata N, Nagamura Y, Yamamoto K, Harushima Y, Sue N, Wu J, Antonio BA, Shomura A, Shimizu T, Lin SY et al (1994) A 300 kilobase interval genetic map of rice including 883 expressed sequence. Nature Genet 8:365–372

Lande R, Thompson R (1990) Efficiency of marker assisted selection in the improvement of quantitative traits. Genetics 124:743–756

Lander ES, Botstein D (1989) Mapping Mendelian factors underlying quantitative traits using RFLP linkage maps. Genetics 121:185–199

Lander ES, Green P, Abrahamson J, Barlow A, Dally MJ, Lincoln SE, Newburg L (1987) MAPMAKER: an interactive computer package for constructing primary genetic linkage maps of experimental and natural populations. Genomics 1:174–181

Lang NT, Tao NV, Buu BC (2011) Marker assisted backcrossing (MAB) for rice submergence tolerance in Mekong Delta. Omonrice 18:11–21

Langella R, Ercolano M, Monti L, Frusciante L, Barone A (2004) Molecular marker assisted transfer of resistance to TSWV in Tomato elite lines. J Hort Sci Biotechnol 79:806–810

Li ZB, Pinson SRM, Marchetti MA, Stansel JW, Park WD (1995) Characterization of quantitative trait loci (QTLs) in cultivated rice to field resistance to sheath blight (*Rhizoctonia solani*). Theor Appl Genet 91:382–388

Li GQ, Li ZF, Yang WY et al (2006a) Molecular mapping of stripe rust resistance gene *YrCH*42 in Chinese wheat cultivar Chuanmai 42 and its allelism with *Yr*24 and *Yr*26. Theor Appl Genet 112:1434–1440

Li ZF, Zheng TC, He ZH et al (2006b) Molecular tagging of stripe rust resistance gene *YrZH*84 in Chinese wheat line Zhou 8425B. Theor Appl Genet 112:1098–1103

Li TH, Li YX, Li ZC, Zhang HL, Qi YW, Wang T (2008) Simple sequence repeat analysis of genetic diversity in primary core collection of peach (*Prunus persica*). J Integr Plant Biol 50:102–110

Li T, Zhang Z, Hu Y, Duan X, Xin Z (2011) Identification and molecular mapping of a resistance gene to powdery mildew from the synthetic wheat line M53. J Appl Genet 52:137–143

Li HJ, Li XH, Xiao JH, Wing RA, Wang SP (2012) Ortholog alleles at *Xa3*/*Xa26* locus confer conserved race-specific resistance against *Xanthomonas oryzae* in rice. Mol Plant 5:281–290

Lilley JM, Ludlow MM, McCouch SR, O'Toole JC (1996) Locating QTL for osmotic adjustment and dehydration tolerance in rice. J Exp Bot 47:1427–1436

Lincoln S, Dally M, Lander ES (1992b) Mapping genes controlling quantitative traits with MAPMAKER/ QTL 1.1. Whitehead Institute Technical Reports (2nd edn), Cambridge, MA

Lincoln SE, Dally MJ, Lander ES (1992a) Constructing genetic linkage maps with MAPMAKER/ EXP 3.0. Whitehead Institute Technical Reports (3rd edn), Cambridge, MA

Liu BH, Knapp SJ (1992) G-Mendel:- a program for Mendelian segregation and linkage analysis of individual or multiple progeny populations using log-likelihood ratios. J Hered 81:407

Liu J, Liu D, Tao W, Li W, Wang S, Chen P, Cheng S, Gao D (2000) Molecular marker-facilitated pyramiding of different genes for powdery mildew resistance in wheat. Plant Breed 119:21–24

Liu G, Lu G, Zeng L, Wang GL (2002) Two broad spectrum blast resistance genes, *Pi9(t)* and *Pi2(t)* are physically linked on rice chromosome 6. Mol Genet Genomics 267:472–480

Liu W, Nie H, Wang S et al (2005) Mapping a resistance gene in wheat cultivar Yangfu 9311 to yellow mosaic virus, using microsatellite markers. Theor Appl Genet 111:651–657

Lorenz AJ, Chao S, Asoro FG, Heffner EL, Hayashi T, Iwata H, Smith KP, Sorrells ME, Jannink JL (2011) Genomic selection in plant breeding: knowledge and prospects. Adv Agron 110:77–123

Lorieux M (2012) MapDisto: fast and efficient computation of genetic linkage maps. Mol Breed. doi:10.1007/s11032-012-9706-y

Ma ZQ, Gill BS, Sorrells ME, Tanksley SD (1993) RFLP markers linked to two Hessian fly resistance genes in wheat (*Triticum aestivum* L.) from *Triticum tauschii* (Coss;) Schal. Theor Appl Genet 85:750–754

Ma ZQ, Sorrells ME, Tanksley SD (1994) RFLP markers linked to powdery mildew resistance genes *Pm1*, *Pm2*, *Pm3* and *Pm4* in wheat. Genome 37:871–875

Ma H, Kong Z, Fu B, Li N, Zhang L, Jia H, Ma Z (2011) Identification and mapping of a new powdery mildew resistance gene on chromosome 6D of common wheat. Theor Appl Genet 123:1099–1106

Maccaferri M, Ratti C, Autonell RC, Vallega V, Demontis A, Stefanelli S, Tuberosa R, Sanguineti MC (2011)

Resistance to Soil-borne cereal mosaic virus in durum wheat is controlled by a major QTL on chromosome arm 2BS and minor loci. Theor Appl Genet 123:527–544

Mackill DJ, Ni J (2000) Molecular mapping and marker assisted selection for major gene traits in rice. In: Khush GS, Brar DS, Hardy B (eds) Proceedings of the 4th international rice genetics symposium. IRRI, Los Banos

Mallikarjuna N, Senapathy S, Jadhav DR, Saxena K, Sharma HC, Upadhyaya HD, Rathore A, Varshney R (2011) Progress in the utilization of *Cajanus platycarpus* (Benth.) Maesen in pigeonpea improvement. Plant Breed 130:507–514

Mamadov JA, Biyashev RM, Griffery CA, Saghaimaroof MA (2003) Molecular mapping of leaf rust resistance gene *Rph-5* in barley. Crop Sci 43:388–393

Manly KF (1995) New functions in MAP MANAGER, a macro computer program for genomic mapping. In: Proceedings of plant genome III, San Diego, CA

Mc Couch SR, Abenes ML, Angeles R, Khush GS, Tanksley SD (1992) Molecular tagging of a recessive gene *Xa-5* for resistance to bacterial blight of rice. Rice Genet Newsl 8:143–145

Melchinger AE (1990) Use of molecular markers in breeding for oligogenic disease resistance. Plant Breed 104:1–19

Mesbah LA, Kneppers TJA, Takken FLW, Laurent P, Hille J, Nijkamp HJJ (1999) Genetic and physical analysis of a YAC contig spanning the fungal disease resistance locus *Asc* of tomato (*Lycopersicon esculentum*). Mol Gen Genet 261:50–57

Mew TV, Parco A, Hittalmani S et al (1994) A 300 kb interval rice genetic map including 883 expressed sequences. Nat Genet 8:365–372

Michelmore RW, Paran I, Kesseli RV (1991) Identification of markers linked to disease resistance genes by bulked segregant analysis: a rapid method to detect markers in specific genomic regions by using segregating populations. Proc Natl Acad Sci USA 88:9828–9832

Miedaner T, Wilde F, Steiner B, Buerstmayr H, Korzun V, Ebmeyer E (2006) Stacking quantitative trait loci (QTL) for *Fusarium* head blight resistance from non-adapted sources in an European elite spring wheat background and assessing their effects on deoxynivalenol (DON) content and disease severity. Theor Appl Genet 112:562–569

Miranda LM, Murphy JP, Marshall D, Leath S (2006) *Pm34*: a new powdery mildew resistance gene transferred from *Aegilops tauschii* Coss. to common wheat (*Triticum aestivum* L.). Theor Appl Genet 113:1497–1504

Mishra RR, Sahu AR, Rath SC, Mishra SP, Panigrahi J (2012) Cyto-morphological and molecular characterization of *Cajanus cajan* x *C. scarabaeoides* $F_1$ Hybrid. Nucleus: doi:10.1007/s13237-012-0050-8

Miyamoto M, Ando I, Rybka K, Kodama O, Kawasaki S (1996) High resolution mapping of the *indica*-derived rice blast resistance genes. 1. *Pi-b*. Mol Plant Microbe Interact 9:6–13

Mohan M, Nair S, Bentur JS, Prasad Rao U, Bennet J (1994) RFLP and RAPD mapping of the rice *Gm-2* gene that confers resistance to biotype 1 of gall midge *(Orseolia oryzae)*. Theor Appl Genet 87:782–788

Mohan M, Sathyanarayanan PV, Kumar A, Srivastava MN, Nair S (1997) Molecular mapping of a resistance specific PCR based marker linked to a gall midge resistance gene (*Gm4t*) in rice. Theor Appl Genet 95:777–782

Mohler V, Singrun C (2004) General considerations: marker assisted selection. In: Lorz H, wenzel G (eds) Biotechnology in agriculture and forestry, vol 55, Molecular marker systems. Springer, Berlin

Mukherjee AK, Mohapatra T, Varshney A, Sharma R, Sharma RP (2001) Molecular mapping of a locus controlling resistance to *Albugo candida* in Indian mustard. Plant Breed 120:481–487

Myint KKM, Fujita D, Matsumura M, Sonoda T, Yoshimura A, Hideshi Yasui H (2012) Mapping and pyramiding of two major genes for resistance to the brown planthopper (*Nilaparvata lugens* [Sta°l]) in the rice cultivar ADR52. Theor Appl Genet 124:495–504

Nair S, Kumar A, Srivastava MN, Mohan M (1996) PCR-based DNA markers linked to a gall midge resistance gene, *Gm4t* has potential for marker aided selection in rice. Theor Appl Genet 92:660–665

Nanda A, Mohanty SK, Panda RS, Behera L, Prakash A, Sahu SC (2010) Flanking microsatellite markers for breeding varieties against Asian rice gallmidge. Trop Plant Biol 3:219–226

Naqvi NJ, Bonman JM, Mackill DJ, Nelson RJ, Chattoo BJ (1995) Identification of RAPD markers linked to a major blast resistance gene in rice. Mol Breed 1:341–348

Neeraja CN, Rodriguez RM, Pamplona A, Heuer S, Collard BCY, Septiningsih EM, Vergara G, Sanchez D, Xu K, Ismail AM, Mackill DJ (2007) A marker assisted backcross approach for developing submergence-tolerant rice cultivars. Theor Appl Genet 115:767–776

Nejad GM, Arzani A, Rezai AM, Singh RK, Gregorio GB (2008) Assessment of rice genotypes for salt tolerance using microsatellite markers associated with the *saltol* QTL. Afr J Biotechnol 7:730–736

Nelson JC, Sorrells ME, Van Deynze AE, Lu LH, Atkinson MD, Bernard M, Leroy P, Faris J, Anderson JA (1995) Molecular mapping of wheat major genes and rearrangements in homoeologous groups 4, 5 and 7. Genetics 141:721–731

Ogbonnaya FC, Subrahmanyam NC, Moullet O, De Majnik J, Eagles HA, Brown JS, Kollmorgen JF, Eastwood RF, Lagudah ES, Appels R (2001) Diagnostic DNA markers for cereal cyst nematode resistance in bread wheat. Aust J Agric Res 52:1367–1374

Okada Y, Kashiwazaki S, Kanatani R, Arai S, Ito K (2003) Effects of barley yellow mosaic disease resistant gene *rym1* on the infection by strains of Barley yellow mosaic virus *and* Barley mild mosaic virus. Theor Appl Genet 106:181–189

Okada Y, Kanatani R, Arai S, Ito K (2004) Interaction between Barley yellow mosaic disease resistance gene rym1 and rym5, in the response to *BaYMV* strains. Breed Sci 54:319–325

Olivera PD, Millet E, Anikster Y, Steffenson BJ (2008) Genetics of resistance to wheat leaf rust, stem rust, and powdery mildew in *Aegilops sharonensis*. Phytopathology 98:353–358

Panigrahi J (2011) Molecular mapping and map based cloning of genes in plants. In: Thangadurai D (ed) Plant biotechnology and transgenics. Bentham Science Publishers, Oak Park

Panigrahi J, Kumar DR, Mishra M, Mishra RP, Jena P (2007) Genomic relationships in the genus *Cajanus* as revealed by seed protein (albumin and globulin) polymorphisms. Plant Biotechnol Rep 1:109–116

Panigrahi J, Patnaik A, Kole P, Kole C (2009) Addition of RFLP markers to the genetic linkage map of *Brassica rapa* L. (syn. *campestris*) and detection of new QTLs for *Albugo candida* race-7( *AC-7*). Z. Naturoforsch 64C:882–890

Parlevliet JE (1977) Variation for partial resistance in a cultivar of rye, *Secale cereale*, to brown rust. Puccinia recondita sp. secalis. Cereal Rusts Bullet 5:13–16

Parlevliet JE (1989) Identification and evaluation of quantitative resistance. In: Leonard KJ, Fry FE (eds) Plant disease epidemiology: genetics, resistance and management, vol 2. McGraw-Hill Publishers, New York

Paterson AH (1996) Physical mapping and map-based cloning: bridging the gap between DNA markers and genes. In: Paterson AH (ed) Genome mapping in plants. Academic/RG Landes, New York/Austin

Paterson AH (1998) Molecular dissection of complex traits. CRC Press, Boca Raton

Paterson AH, Lander ES, Hewitt JD, Paterson S, Lincoln SE, Tanksley SD (1988) Resolution of quantitative traits into Mendelian factors by using a complete RFLP linkage map. Nature 335:721–726

Peleman JD, vander Voort JR (2003a) The challenges in marker assisted breeding. In: Van Hintum TJL, Lebeda A, Pink D, Schutt JW (eds) Eucarpia leafy vegetables 2003

Peleman JD, vander Voort JR (2003b) Breeding by design. Trends Plant Sci 8:330–334

Peleman JD, Wye C, Zethof J, Sorensen AP, Verbakel H et al (2005) Quantitative trait locus (QTL) isogenic recombinant analysis: a method for high-resolution mapping of QTL within a single population. Genetics 171:1341–1352

Pena L (2004) Transgenic plants: methods and protocols. Humana Press, Totowa

Peng JH, Fahima T, Roder MS, Li YC, Dahn A, Grama A, Ronin YJ, Korol AB, Nevo E (1999) Microsatellite tagging of the stripe rust resistance gene *YrH52* derived from wild emmer wheat *Triticum dicoccoides* and suggestive negative crossover interference on chromosome IB. Theor Appl Genet 98:862–872

Peters JL, Cnudde F, Gerats T (2003) Forward genetics and map based cloning approaches. Trends Plant Sci 8:484–491

Porter DR, Burd JD, Shufran KA, Webster JA (2000) Efficacy of pyramiding greenbug (Homoptera: Aphididae) resistance genes in wheat. J Econ Entomol 93(4):1315–1318

Prabhu KV, Somens DJ, Rakow G, Gugel RK (1998) Molecular markers linked to white rust resistance in mustard *Brassica juncea*. Theor Appl Genet 97:865–870

Prasad MS, Kanthi BA, Balachandran SM, Seshumadhav M, Mohan KM, Viraktmah BC (2009) Molecular mapping of rice blast resistance gene *Pi-1(t)* in the elite indica variety Samba Mahsuri. World J Microbiol Biotechnol 25:1765–1769

Procunier JD, Knox RE, Bernier AM, Gray MA, Howes NK (1997) DNA markers linked to a T10 loose smut resistance gene in wheat *(Triticum aestivum* L.). Genome 40:176–179

Qiu JW, Schurch AC, Yahiaoui N, Dong LL, Fan HJ, Zhang ZJ, Keller B, Ling HQ (2007) Physical mapping and identification of a candidate for the leaf rust resistance gene *Lr1* of wheat. Theor Appl Genet 115:159–168

Qiu Y, Guo J, Jing S, Zhu L, He G (2010) High-resolution mapping of the brown planthopper resistance gene *Bph6* in rice and characterizing its resistance in the 9311 and Nipponbare near isogenic backgrounds. Theor Appl Genet 12:1601–1611

Qiu Y, Guo J, Jing S, Zhu L, He G (2012) Development and characterization of *japonica* rice lines carrying the brown planthopper resistance genes *BPH12* and *BPH6*. Theor Appl Genet 124:485–494

Quesada-Ocampo M, Hausbeck MK (2010) Resistance in tomato and wild relatives to crown and root rot caused by (*Phytophthora capsici*. L.). Phytopathology 100:619–627

Rafalski JA (2010) Association genetics in crop improvement. Curr Opin Plant Biol 13:174–180

Rao IM, Venkataratnam N, Sheldrake AR, Rao IM (1981) Field screening of pigeonpea for tolerance to soil salinity. Int Chickpea Pigeonpea Newsl 1:23

Rao Y, Dong G, Zeng D, Hu J, Zeng L, Gao Z, Zhang G, Guo L, Qian Q (2010) Genetic analysis of leaf folder resistance in rice. J Genet Genomics 37:325–331

Rasmussoin M (1933) A contribution to the theory of quantitative character inheritance. Hereditas 18:245–261

Rehman AU (2009) Characterization and molecular mapping of drought tolerance in kabuli chickpea (*Cicer arietinum* L.). PhD thesis, University of Saskatchewan, Saskatoon, Canada

Ren RS, Wang MN, Chen XM, Zhang ZJ (2012) Characterization and molecular mapping of *Yr52* for high-temperature adult-plant resistance to stripe rust in spring wheat germplasm PI 183527. Theor Appl Genet. doi:10.1007/s00122-012-1877-8

Ribaut JM, Betran J (1999) Single large scale marker assisted selection (SLS-MAS). Mol Breed 5:531–541

Ribaut JM, Hoisington D (1998) Marker assisted selection: new tools and strategies. Trends Plant Sci 3:236–239

Ribaut JM, Jiang C, Hoisington D (2002) Simulation experiments on efficiencies of gene introgression by backcrossing. Crop Sci 42:557–565

Riedel C, Habekuss A, Schliephake E, Niks R, Broer I, Ordon F (2011) Pyramiding of *Ryd2* and *Ryd3* conferring tolerance to a German isolate of barley yellow dwarf virus-PAV (BYDV-PAV-ASL-1) leads to quantitative resistance against this isolate. Theor Appl Genet 123:69–76

Robert VJM, West MAL, Inai S, Caines A, Arntzen L, Smith JK, Clair DA (2001) Marker assisted introgression of black mold resistance QTL alleles from wild *Lycopersicon cheesmanii* to cultivated tomato (*L. esculentum*) and evaluation of QTL phenotypic effects. Mol Breed 8:217–233

Ronald PC, Albano B, Tabien R, Abenes L, Wu KK, McCouch SR, Tanksley SD (1992) Genetic and physical analysis of the rice bacterial blight disease resistance locus *Xa21*. Mol Gen Genet 236:113–120

Rong JK, Millet E, Manisterski J, Feldman M (1998) A powdery mildew resistance gene from wild emmer transferred into common wheat and tagged by molecular markers. In: Slinkard AE (ed) Proceedings of the 9th international wheat genetics symposium, vol 3. University of Saskatchewan, Saskatoon, pp 148–150

Sacco F, Suarez EY, Naranjo T (1998) Mapping of the leaf rust resistance gene *Lr3* on chromosome 6B of Sinvalocho MA wheat. Genome 41:686–690

Saghai Maroof MA, Zhang Q, Biyashev RM (1994) Molecular marker analyses of powdery mildew resistance in barley. Theor Appl Genet 88:733–740

Saghai Maroof MA, Jeong SC, Gunduz I, Tucker DM, Buss GR, Tolin SA (2008) Pyramiding of soybean mosaic virus resistance genes by marker assisted selection. Crop Sci 48:517–526

Salina E, Dobrovolskaya O, Efremova T, Leonova I, Roder MS (2003) Microsatellite monitoring of recombination around the *Vrn-B1* locus of wheat during early backcross breeding. Plant Breed 122:116–119

Salvi S, Tuberosa R (2005) To clone or not to clone plant QTLs: present and future challenges. Trends Plant Sci 10:297–304

Sanchez AC, Brar DS, Huang N, Li Z, Khush GS (2000) Sequence tagged site marker-assisted selection for three bacterial blight resistance genes in rice. Crop Sci 40:792–797

Sardesai N, Kumar A, Rajyshri KR, Nair S, Mohan M (2002) Identification and mapping of an AFLP marker linked to *Gm-7*, a gall midge resistance gene and its conversion to a SCAR marker for its utility in MAS in rice. Theor Appl Genet 105:691–698

Sarfatti M, Katan J, Fluhr R, Zamir D (1989) An RFLP marker in tomato linked to the *Fusarium oxysporum* resistance gene $I_2$. Theor Appl Genet 78:755–759

Sax K (1923) The association of size differences with seed coat pattern and pigmentation in *Phaseolus vulgaris*. Genetics 8:552–560

Saxena KB, Singh L, Reddy MV, Singh U, Leteef SS, Sharma SB et al (1990) Intra-species variation in *Atylosia scarabaeoides* (L.) Benth., a wild relative of pigeonpea (*Cajanus cajan* (L.) Millsp.). Euphytica 49:185–191

Schmolke M, Mohler V, Hartl L, Zeller FJ, Hsam SLK (2012) A new powdery mildew resistance allele at the *Pm4* wheat locus transferred from einkorn (*Triticum-monococcum*). Mol Breed 29:449–456

Schon CC, Lee M, Melchinger AE, Guthrie WD, Woodman WL (1993) Mapping and characterization of QTL affecting resistance against second generation European corn borer in maize with the aid of RFLPs. Heredity 70:648–659

Semagn K, Bjornstad A, Ndjiondjop MN (2006) An overview of molecular marker methods for plant. Afr J Biotechnol 25:2540–2569

Septiningsih EM, Sanchez DL, Singh N, Sendon PMD, Pamplona AM, Heuer S, Mackill DJ (2012) Identifying novel QTLs for submergence tolerance in rice cultivars IR72 and Madabaru. Theor Appl Genet 124:867–874

Serraj R, Mc Nally KL, Slamet-laedin I, Kohli A, Haefele SM, Atlin G, Kumar A (2011) Drought resistance improvement in rice: an integrated genetic and resource management strategy. Plant Prod Sci 14:1–14

Servin B, Martin OC, Mezard M, Hospital F (2004) Towards a theory of marker assisted gene pyramiding. Genetics 168:513–523

Seyfarth RC, Feuillet C, Keller B (1998) Development and characterization of molecular markers for the adult leaf rust resistance genes *Lr13* and *Lr35* in wheat. In: Slinkard AE (ed) Proceedings of the 9th international wheat genetics symposium, vol 3. University of Saskatchewan, Saskatoon, pp 154–155

Shan X, Blake TK, Talbert LE (1999) Conversion of AFLP markers to sequence-specific PCR markers in barley and wheat. Theor Appl Genet 98:1072–1078

Shanti ML, George MLC, Cruz CMV, Bernardo MA, Nelson RJ, Leung H, Reddy JN, Sridhar R (2001) Identification of resistance genes effective against rice bacterial blight pathogen in eastern India. Plant Dis 85:506–512

Sharma PN, Torii A, Takumi S, Mori N, Nakamura C (2004) Marker-assisted pyramiding of brown planthopper (*Nilaparvata lugens* Stål) resistance genes *Bph1* and *Bph2* on rice chromosome 12. Hereditas 140:61–69

Sharp PJ, Johnston S, Brown R et al (2001) Validation of molecular markers for wheat breeding. Aust J Agric Res 52:1357–1366

Shi A, Chen P, Li D, Zheng C, Zhang B, Hou A (2009) Pyramiding multiple genes for resistance to soybean mosaic virus in soybean using molecular markers. Mol Breed 23:113–124

Sholihin HDM (2002) Molecular mapping of drought *resistance* in mungbean (*Vigna radiate*): 2. QTL linked to drought resistance. J Bioteknol Pertanian 7:55–61

Simmonds NW (1993) Introgression and incorporation: strategies for the use of crop genetic resources. Biol Rev 68:539–562

Simms EL, Rausher MD (1992) Quantitative genetics. In: Fritz RS, Simms EL (eds) Ecology and evolution of plant resistance. University of Chicago Press, Chicago

Singh S, Sidhu JS, Huang N, Vikal Y, Li Z, Brar DS, Dhaliwal HS, Khush GS (2001) Pyramiding three bacterial blight resistance genes (*xa5, xa13* and *Xa21*) using marker-assisted selection into indica rice cultivar PR106. Theor Appl Genet 102:1011–1015

Singh R, Sharma P, Varshney RK, Sharma SK, Singh NK (2008) Chickpea improvement: role of wild species and genetic markers. Biotechnol Genet Eng Rev 25:267–314

Somers D, Thomas J, Depauw R, Fox S, Humphreys G, Fedak G (2005) Assembling complex genotypes to resist *Fusarium* in wheat (*Triticum aestivum* L.). Theor Appl Genet 111:1623–1631

Somta P, Kaga A, Tomooka N, Isemura T, Vaughan DA, Srinives P (2008) Mapping of quantitative trait loci for a new source of resistance to bruchids in the wild species *Vigna nepalensis* Tateishi & Maxted (*Vigna* subgenus *Ceratotropis*). Theor Appl Genet 117:621–628

Spielmeyer W, McIntosh RA, Kolmer J, Lagudah ES (2005) Powdery mildew resistance and *Lr34/Yr18* genes for durable resistance to leaf and stripe rust cosegregate at a locus on the short arm of chromosome 7D of wheat. Theor Appl Genet 111:731–735

Starn P (1993) Construction of integrated genetic linkage maps by means of a new computer package: Join Map. Plant J 3:739–744

Steele K, Price A, Shashidhar H, Witcombe J (2006) Marker assisted selection to introgress rice QTLs controlling root traits into an Indian upland rice variety. Theor Appl Genet 112:208–221

Stoskopf NC, Tomes DT, Christie BR (1993) Plant breeding: theory and practice. Westview Press Inc., San Francisco/Oxford

Stuber CW (1992) Biochemical and molecular markers in plant breeding. Plant Breed Rev 9:37–61

Subbarao GV, Johansen C, Jana MK, Rao JVDKK (1991) Comparative salinity responses among pigeonpea genotypes and their wild relatives. Crop Sci 31:415–418

Sui XX, Wang MN, Chen XM (2009) Molecular mapping of a stripe rust resistance gene in spring wheat cultivar *ZAK*. Phytopathology 99:1209–1215

Suiter KA, Wendel JF, Case JS (1983) Linkage-I: a Pascal computer program for the detection and analysis of genetic linkage. J Hered 74:203–204

Sun GL, Fahima T, Korol AB, Turpeinien T, Grama A, Ronin Y, Nevo E (1999) Identification of molecular markers linked to *Yr* 15 stripe resistance gene of wheat originated in wild emmer wheat, *Triticum dicoccoides*. Theor Appl Genet 95:622–628

Sun L, Chen X, Wang S, Wang L, Liu C, Mei L, Xu N (2008) Heredity analysis and gene mapping of bruchid resistance of a mung bean cultivar V2709. Agric Sci China 7:672–677

Sun X, Bai G, Carver BF (2009) Molecular markers for wheat leaf rust resistance gene *Lr41*. Mol Breed 23:311–321

Sun W, Xu X, Zhu H, Liu A, Liu L, Li J, Hua X (2010) Comparative transcriptomic profiling of a salt tolerant wild tomato species and a salt sensitive tomato cultivar. Plant Cell Physiol 51:997–1006

Sutton WS (1903) The chromosomes in heredity. Biol Bull 4:231–251

Talbert LE, Bruckner E, Smith LY, Sears R, Martin TJ (1996) Development of PCR markers linked to resistance to wheat streak mosaic virus in wheat. Theor Appl Genet 93:463–467

Tanksley SD (1983) Molecular markers in plant breeding. Plant Mol Biol Rep 1:3–8

Tanksley SD, McCouch SR (1997) Seed banks and molecular maps: unlocking genetic potential from the wild. Science 277:1063–1066

Tanksley SD, Young ND, Paterson AH, Bonierbale MW (1989) RFLP mapping in plant breeding: new tools for an old science. Biotechnology 7:257–263

Teutonico RA, Yandell BS, Satagopan JM, Ferreira ME, Palta JP, Osborn TC (1995) Genetic analysis and mapping of genes controlling freezing tolerance in oil seed *Brassica*. Mol Breed 1:329–339

Thoday JM (1961) Location of polygenes. Nature 191:368–370

Tomooka N, Lairungreang C, Nakeeraks P, Egawa Y, Thavarasook C (1992) Development of bruchid resistant mungbean line using wild mungbeam germplasm in Thailand. Plant Breed 109:60–66

Tuinstra MR, Ejeta G, Goldsbrough PB (1997) Heterogeneous inbred family analysis: an approach for developing near-isogenic lines that differ at quantitative trait loci. Theor Appl Genet 95:1005–1011

Uauy C, Brevis JC, Chen X et al (2005) High-temperature adult-plant (HTAP) stripe rust resistance gene *Yr36* from *Triticum turgidum* ssp. *dicoccoides* is closely linked to the grain protein content locus *Gpc-B1*. Theor Appl Genet 112:97–105

van der Biezen EA, Glagotskaya T, Overduin B, Nijkamp HJJ, Hille J (1995) Inheritance and genetic mapping of resistance to *Alternaria alternata* f. sp. *lycopersici* in *Lycopersicon pennellii*. Mol Gen Genet 247:453–461

vander Plank JE (1963) Plant diseases: epidemics and control. Academic, New York

vander Plank JE (1968) Disease resistance in plants. Academic, New York

Varshney RK, Graner A, Sorrells ME (2005) Genomics assisted breeding for crop improvement. Trends Plant Sci 10:621–630

Varshney RK, Penmetsa RV et al (2010) Pigeonpea genomics initiative (PGI): an international effort to improve crop productivity of pigeonpea (*Cajanus cajan* L.). Mol Breed 26:393–408

Visscher PM, Haley CS, Thompson R (1996) Marker assisted introgression in backcross breeding programs. Genetics 144:1923–1932

Walsh JA, Sharpe AG, Jenner CE, Lydiate DJ (1999) Characterization of resistance to turnip mosaic virus in oilseed rape *(Brassica napus)* and genetic mapping of *TURBO 1*. Theor Appl Genet 99:1149–1154

Wang GL, Mackill DJ, Bonman JM, McCouch SR, Champoux MC, Nelson RJ (1994) RFLP mapping of genes conferring complete and partial resistance to blast in a durably resistant rice cultivar. Genetics 136:1421–1434

Wang J, Chapman SC, Bonnett DG, Rebetzke GJ, Crouch J (2007) Applications of population genetic theory and simulation models to efficiently pyramid multiple genes via marker assisted selection. Crop Sci 47:582–588

Wang C, Zhang Y, Han D et al (2008) SSR and STS markers for wheat stripe rust resistance gene *Yr26*. Euphytica 159:359–366

Wang B, Jiang L, Zhang Y, Zhang W, Wang M, Cheng X, Liu X, Zhai H, Wan J (2011) QTL mapping for resistance to strip virus disease in rice. Plant Breed 130:321–327

Weaver R, Helms C, Mishra SK, Donis-Keller H (1992) Software for analysis and manipulation of genetic linkage data. Am J Hum Genet 50:1267–1274

Werner K, Friedt W, Ordon F (2005) Strategies for pyramiding resistance genes against the barley yellow mosaic virus complex (*BaMMV*, *BaYMV*, *BaYMV-2*). Mol Breed 16:45–55

Wilde F, Schon C, Korzun V, Ebmeyer E, Schmolke M, Harlt L, Miedaner T (2008) Marker-based introduction of three quantitative trait loci conferring resistance to *Fusarium* head blight into an independent elite winter wheat breeding population. Theor Appl Genet 117:29–35

Willcox MC, Khairallah MM, Bergvinson D, Crossa J, Deutsch JA, Edmeades GO, Gonzalez-De-Leon D, Jiang C, Jewell DC, Mihm JA, Williams WP, Hoisington D (2002) Selection for resistance to southwestern corn borer using marker assisted and conventional backcrossing. Crop Sci 42:1516–1528

Williams KJ, Fisher JM, Langridge P (1994) Identification of RFLP markers linked to the cereal cyst nematode resistant gene *(ere)* in wheat. Theor Appl Genet 89:927–930

Wu S, Pumphrey M, Bai G (2009) Molecular mapping of stem rust resistance gene *Sr40* in wheat. Crop Sci 49:1681–1686

Wye C, vander Voort RJ, Peleman J (2000) Reverse QTL mapping (RQM), an efficient approach for high resolution mapping of QTL. In: Plant and animal genome VII conference VII, San Diego

Xu Y (1997) Quantitative trait loci: separating, pyramiding, and cloning. Plant Breed Rev 15:85–139

Xu Y (2003) Developing marker assisted selection strategies for breeding hybrid rice. Plant Breed Rev 23:73–174

Xu Y, Crouch JH (2008) Marker assisted selection in plant breeding: from publication to practice. Crop Sci 48:391–407

Xu K, Mackill DJ (1996) A major locus for submergence tolerance mapped on rice chromosome 9. Mol Breed 2:219–224

Xu YB, Ishi T, Mc Couch SR (2003) Marker assisted evaluation of germplasm resources for plant breeding. In: Mew TW, Brar DS, Peng S, Dawe D, Hardy B (eds) Rice science-innovations and impact for livelihood. IRRI, Manila

Xu Q, Yuan X, Yu H, Wang Y, Tang S, Wei X (2011a) Mapping quantitative trait loci for sheath blight resistance in rice using double haploid population. Plant Breed 130:404–406

Xu W, Li C, Hu L, Wang H, Dong H, Zhang J, Zan X (2011b) Identification and molecular mapping of *PmHNK54*: a novel powdery mildew resistance gene in common wheat. Plant Breed 130:603–607

Xue S, Xu F, Tang M, Zhou Y, Li G, An X, Lin F, Xu H, Jia H, Zhang L et al (2011) Precise mapping *Fhb5*, a major QTL conditioning resistance to *Fusarium* infection in bread wheat (*Triticum aestivum* L.). Theor Appl Genet 123:1055–1063

Xue F, Ji W, Wang C, Zhang H, Yang B (2012) High density mapping and marker development for the powdery mildew resistance gene *PmAS846* derived from wild emmer wheat (*Triticum turgidum* var. *dicoccoides*). Theor Appl Genet 124:1549–1560

Yang L, Li RB, Li YR, Huang FK, Chen YZ, Huang SS, Huang LF, Liu C, Ma ZF, Huang DH et al (2012) Genetic mapping of *bph20(t)* and *bph21(t)* loci conferring brown planthopper resistance to *Nilaparvata lugens* Stal in rice (*Oryza sativa* L.). Euphytica 183:161–171

Ye C, Argayoso MA, Redona ED, Sierra SN, Laza MA, Dilla CJ, Mo Y, Thomson MJ, Chin J, Delavina CB, Diaz GQ, Hernandez JE (2012) Mapping QTL for heat tolerance at flowering stage in rice using SNP markers. Plant Breed 131:33–41

Yin Y, Zuo S, Wang H, Chen Z, Ma Y, Zhang Y, Gu S, Pan X (2008) Pyramiding effects of three quantitative trait loci for resistance to sheath blight using near isogenic lines of rice. Chin J Rice Sci 22:340–346

Yoshimura S, Yoshimura A, Saito A, Kishimoto N, Kawase M, Yano M, Nakagahra N, Ogawa T, Iwata N (1992) RFLP analysis of introgressed segments in three near isogenic lines of rice for bacterial blight resistance genes, *Xa-1, Xa-3* and *Xa-4*. Jpn J Genet 67:29–37

Yoshimura S, Yoshimura A, Iwata N, McCouch SR, Abenes ML, Baraoidan MR, Mew TW, Nelson RJ (1995a) Tagging and combining bacterial blight resistant genes in rice using RAPD and RFLP markers. Mol Breed 1:375–387

Yoshimura S, Yoshimura A, Nelson RJ, Mew TW, Iwata N (1995b) Tagging *Xa-1*, the bacterial blight resistance gene in rice, by using RAPD markers. Breed Sci 45:81–85

Young ND, Tanksley SD (1989a) RFLP analysis of the size of chromosomal segments retained around the *Tm-2a* locus of tomato during back cross breeding. Theor Appl Genet 77:353–359

Young ND, Tanksley SD (1989b) Restriction fragment length polymorphism maps and the concept of graphical genotypes. Theor Appl Genet 77:95–101

Young ND, Kumar L, Menancio-Hautea D, Danesh D, Talekar NS, Shanmugasundaraum S, Kim DH (1992) RFLP mapping of a major bruchid resistance gene in mungbean *(Vigna radiata* L. Wilczek). Theor Appl Genet 84:839–844

Young ND, Danesh D, Menancio-Hautea D, Kumar L (1993) Mapping oligogenic resistance to powdery mildew in mungbean with RFLPs. Theor Appl Genet 87:243–249

Yu ZH (1991) Molecular mapping of rice *(Oryza sativa* L.) genes via linkage to restriction fragment polymorphism (RFLP) markers. PhD thesis, Cornell University, Ithaca

Yu ZH, Mackill DJ, Bonman JM, Tanksley SD (1991) Tagging genes for blast resistance in rice via linkage to RFLP markers. Theor Appl Genet 81:471–476

Yu YG, Saghai-Maroof MA, Buss GR, Maughan PI, Tolin SA (1994) RFLP and microsatellite mapping of a gene for soybean mosaic virus resistance. Phytopathology 84:60–64

Yu F, Lydiate DJ, Rimmer SR (2008) Identification and mapping of third black leg resistance locus in *Brassica napus* derived from *B. rapa* subsp. *sylvestris*. Genome 51:64–72

Zamir D (2001) Improving plant breeding with exotic genetic libraries. Nat Rev Genet 2:983–989

Zeng ZB (1994) Precision mapping of quantitative trait loci. Genetics 136:1457–1468

Zhang J, Li X, Jiang G, Xu Y, He Y (2006) Pyramiding of *Xa7* and *Xa21* for the improvement of disease resistance to bacterial blight in hybrid rice. Plant Breed 125:600–605

Zhang NW, Pelgrom K, Niks RE, Visser RGF, Jeuken MJW (2009a) Three combined quantitative trait loci from non host *Lactuca saligna* are sufficient to provide complete resistance of lettuce against Bremia Lactucae. Mol Plant Microbe Interact 22:1160–1168

Zhang Y, Wang J, Pan J (2009b) Identification and molecular mapping of the rice bacterial blight resistance gene allelic to *Xa-7* from an elite restorer line Zhenhui 084. Eur J Plant Physiol 125:235–244

Zhang YX, Jiang L, Liu LL, Wang BX, Shen YY, Wang Q, Cheng XN, Wan JM (2010) Quantitative trait loci associated with resistance to rice stripe virus and small brown planthopper infestation in rice. Crop Sci 50:1854–1862

Zhang YX, Wang Q, Jiang L, Liu LL, Wang BX, Shen YY, Cheng XN, Wan JM (2011) Fine mapping of *qSTV11KAS*, a major QTL for rice stripe disease resistance. Theor Appl Genet 122:1591–1604

Zhang YX, Wang Q, Jiang L, Wang BX, Liu LL, Shen YY, Cheng XN, Jian-Min Wan JM (2012) Detection and fine mapping of two quantitative trait loci for partial resistance to stripe virus in rice (*Oryza sativa* L.). Mol Breed. doi:10.1007/s11032-012-9725-8

Zhao F, Cai Z, Hu T, Yao H, Wang L, Dong N, Wang B, Ru Z, Zhai W (2010) Genetic analysis and molecular mapping of a novel gene conferring resistance to rice stripe virus. Plant Mol Biol Rep 28:512–518

Zheng HG, Babu RC, Pathan MS, Ali L, Huang N, Courtois B, Nguyen HT (2000) Quantitative trait loci for root-penetration ability and root thickness in rice: comparison of genetic backgrounds. Genome 1:53–61

Zhou L, Zeng Y, Zheng W, Tang B, Yang S, Zhang H, Li J, Li Z (2010) Fine mapping a QTL *qCTB7* for cold tolerance at the booting stage on rice chromosome 7 using a near-isogenic line. Theor Appl Genet 121:895–905

Zhou XL, Wang WL, Wang LL, Hou DY, Jing JX, Wang Y, Xu ZQ, Yao Q, Yin JL, Ma DF (2011) Genetics and molecular mapping of genes for high-temperature resistance to stripe rust in wheat cultivar *Xiaoyan 54*. Theor Appl Genet 123:431–438

Zhu S, Walker DR, Boerma HR, Parrot WA (2006) Fine mapping of a major insect resistance QTL in soybean and its interaction with minor resistance QTLs. Crop Sci 46:1094–1099

Zhu C, Gore M, Buckler ES, Yu J (2008) Status and prospects of association mapping in plants. Plant Genome 1:5–20

Zhu X, Wang H, Guo J, Wu Z, Cao A, Bie T, Nie M, You FM, Cheng Z, Xiao J et al (2012) Mapping and validation of quantitative trait loci associated with wheat yellow mosaic bymovirus resistance in bread wheat. Theor Appl Genet 124:177–188

# DNA Methylation-Associated Epigenetic Changes in Stress Tolerance of Plants

17

Mahmoud W. Yaish

**Abstract**

Plants require optimum environmental conditions to grow, develop, and reproduce. Abiotic and biotic stresses have direct, negative effects on the biochemical and physiological processes which is associated with plant growth and development. These processes, under stress conditions, are significantly modified to increase a plant's tolerance and to allow it to reproduce in the shortest possible time leads to escape or to minimize its exposure to unfavorable environmental conditions. As a consequence of these changes on its life cycle, a significant reduction in plant yield is expected. Plants have evolved several strategies to cope with environmental stresses which include expression level alteration of some genes through the introduction of epigenetic modifications, such as DNA methylation. DNA methylation plays a key role in gene expression by enhancing RNA-directed DNA methylation (RdDM) of genes and by inducing some histone modifications. Plants sometimes inherit their tolerance to stresses through the transmission of methylated genes from the parents. They may also produce new alleles by favoring homologous recombination at less methylated loci. However, sometimes this type of inheritance is not stable. DNA methylation may be significantly affected by the environment and cannot be experimentally manipulated or maintained. Therefore, extra care should be taken when designing strategies intended on producing plants with novel traits based on variations in DNA methylation. This chapter dealt with a brief account on epigenetic changes due to DNA methylation, histone modifications, and small RNA interference to modify gene expression pattern throughout the growth and developmental stages of plants to adjust different biotic and abiotic plants responses. The chapter will discuss also the possible use of genetic modifications to induce epigenetic changes that may improve plant traits, especially a

M.W. Yaish (✉)
Department of Biology, College of Science, Sultan Qaboos University, P.O. Box 36, 123, Muscat, Oman
e-mail: myaish@squ.edu.om

G.R. Rout and A.B. Das (eds.), *Molecular Stress Physiology of Plants*,
DOI 10.1007/978-81-322-0807-5_17, © Springer India 2013

plant's ability to grow under abiotic and biotic stresses, and will try to answer fundamental questions on how DNA methylation, chromatin alteration, and small RNA molecules control gene expression.

## Introduction

Plants exposed to unpleasant environmental conditions such as high temperature, high salt, and drought try to modify their growth and developmental processes to minimize the amount of damage caused by the environmental stress. These modifications are typically transient and reversible, and their induction level depends on the exposure time to a particular environmental factor. These alterations initially occur at the epigenetic level in plants. Epigenetics is defined as the heritable alteration of gene expression without changing the basic DNA sequence (Bonasio et al. 2010). These alterations, such as DNA methylation, histone modifications, and small RNA interference, can play an individualized role or work in concert to modify gene expression pattern throughout the growth and developmental stages of plants and to protect a plant from the adverse environmental stimuli such as abiotic and biotic stresses.

The importance of epigenetic alterations that occur in plants under stressful conditions stems from the fact that these epigenetic modulations can be inherited through epigenetic memory (Boyko and Kovalchuk 2010). This process may lead to genetic alterations by producing new alleles that resulted from changes in homologous recombination frequency during meiosis (Engler et al. 1993; Mirouze et al. 2012), which leads to the production of new traits and significantly contributes to biological diversity.

The relationship between epigenetic changes and stress tolerance in plants has been previously discussed in several reviews (Boyko and Kovalchuk 2008; Chen et al. 2010b; Alvarez et al. 2010; Chinnusamy and Zhu 2009). New information has recently been published regarding the same issue. Therefore, this chapter aims to highlight recent advances regarding DNA methylation and their impact on the study of gene expression and phenotype. It will also discuss the relationship between DNA methylation and other epigenetic alterations, such as histone modifications and the production of microRNA molecules (miRNA). The chapter will discuss also the possible use of genetic modifications to induce epigenetic changes that may improve plant traits, especially a plant's ability to grow under abiotic and biotic stresses, and will try to answer fundamental questions on how DNA methylation, chromatin alteration, and small RNA molecules control gene expression. What are the molecular mechanisms that have been discovered so far and used to manipulate gene expression? And how a scientist can take advantage of this information to improve plant traits?

Understanding the molecular mechanisms behind stress-induced gene regulation will facilitate breeding programs that aim to improve plant traits and may minimize the need for excessive genetic modifications.

## DNA Methylation

DNA methylation is the covalent binding of a methyl group to the fifth-position carbon in a cytosine nucleotide ring of a DNA molecule. It often occurs where cytosine is linked by a phosphodiester bond (p) to guanine (CpG). In plants, DNA methylation has been found in three cytosine contexts, including methylation at CpG sites, CpHG sites (H representing A, C, or T), and CpHH sites.

Methylated cytosines can account for more than 30% of the nucleotides in plants (Gruenbaum et al. 1981a, b). Global DNA analysis and bisulfite DNA sequencing of the *Arabidopsis* genome revealed that about 20% of cytosines in the genome are methylated, and the methylation is located either in the promoter region or in the gene body where both sites can have an impact on gene expression (Zhang et al. 2006; Cokus et al. 2008). In wild-type plants, the methylation level of some genes dynamically changes throughout plant's growth and development in response to the exposure of an environmental stimulus. DNA methylation helps plants endure stress conditions by preventing unfavorable genetic rearrangement at a specific locus (Boyko et al. 2010a; Chinnusamy and Zhu 2009).

Despite some exceptions, it is widely accepted that the amount of DNA methylation within a promoter sequence is conversely related to gene expression level (Zilberman et al. 2007). Recently, genome-wide analysis revealed that methylation within the coding region occasionally has a positive effect on gene expression (Zhang et al. 2006). This information suggests the presence of a gene expression control mechanism in which the methyl groups of a structural gene and its promoter are redistributed in such a way that keeps the global DNA methylation level constant and the genome stable throughout plant growth and development.

Plant varieties with a reduced level of methylation are able to grow, develop, and compete with other varieties for the available environmental resources. For example, density-tolerant varieties of maize hybrids have a lower global DNA methylation level than density-sensitive ones. This may suggest that at high planting density, the density-tolerant plants are able to increase their gene expression and metabolism which, in turn, produces a higher yield (Guo et al. 2006; Tani et al. 2005).

DNA methylation is catalyzed by cytosine methyltransferases. Mutations within the coding region of genes involved in DNA methylation, such as *DECREASE IN DNA METHYLATION1* and *2* (*DDM1* and *DDM2*) (Kakutani et al. 1996; Vongs et al. 1993; Jackson et al. 2004), *DNA METHYLTRANSFERASE1* (*MET1*) (Vongs et al. 1993), and DNA demethylase *REPRESSOR OF SILENCING GENES1* (*ROS1*) (Agius et al. 2006), cause significant alterations in DNA methylation that lead to phenotypic changes of plants (Bartee and Bender 2001; Singer et al. 2001).

DNA methylation can also be altered by chemicals that covalently bind to the DNA methyltransferases active sites and inhibiting their catalytic activity (Santi et al. 1983). The cytidine analogue 5-azacytidine (5-azaC) (Santi et al. 1983) and zebularine (Cheng et al. 2003) are commonly used in plants as agents to inhibit DNA methylation. The use of these chemicals enhances hypomethylation and genome-wide transcriptional reactivation of silenced genes and leads to alterations in plant growth and development (Yaish et al. 2009; Borowska et al. 2011; Castilho et al. 1999).

Reducing global DNA methylation in plants does not necessarily reduce DNA methylation within the target gene. In contrast, a low level of global DNA methylation is often associated with hypermethylation at a specific gene. For example, overexpression of an antisense methyltransferase gene in *Arabidopsis*, which causes an overall decrease in genomic cytosines, causes hypermethylation of the SUPERMAN gene (Jacobsen and Meyerowitz 1997). Similarly, reducing the global DNA methylation in a cancer cell leads to hypermethylation of the tumor suppressor gene (Ehrlich 2002).

## DNA Methylation Controls Flowering in Plants Under Stress Conditions

Flowering is a reproductive stage that occurs prior to the appearance of fruits in seed-bearing plants. Therefore, it is a critical stage which determines the overall yield of plants. Flowering is highly sensitive to stress conditions. In fact, stressed plants tend to flower earlier than plants growing under optimum environmental conditions. Early flowering helps plants to complete their life cycle (from seed to seed) in less time and also reduces their exposure period to unfavorable conditions. Stress environmental factors that affect flowering have been discussed in another review (Wada and Takeno 2010).

DNA methylation controls the expression of some genes including those involved in flowering. Alterations in global DNA methylation can change the flowering time in plants. Mutations within the *METHYLTRANSFERASE1* gene (*MET1*) lead to late flowering in *Arabidopsis*. In addition, *met1* and *chromomethylase3* (*cmt3*) mutant lines cause embryonic malformation, abnormal cell division, seed viability, and improper auxin gradient (Xiao et al. 2006). DNA methylation controls the plant's requirement for cold temperatures during the vernalization process. For instance, treating wheat seeds with 5-azaC reduces the cold temperatures required by plants to induce flowering (Brock and Davidson 1994). In *Arabidopsis*, similar results were obtained when seeds were treated with 5-azaC (Burn et al. 1993; Finnegan et al. 1998).

Proteins harboring a methyl-CpG binding protein (MBD) are involved in various epigenetic modulation processes in plants (Berg et al. 2003; Springer and Kaeppler 2005). Molecular analysis of *methyl-CpG binding protein 9* (*atmbd9*) mutants in *Arabidopsis* revealed an increase in global DNA methylation that included Flowering Loci C (FLC), a transcription factor that represses flowering. These mutations led to early flowering and an increase in multiple axillary branching (Yaish et al. 2009; Peng et al. 2006). Treating early-flowering *atmbd9* mutant lines with 5-azaC leads to the recovery of wild-type flowering time and normal FLC expression level (Peng et al. 2006; Yaish et al. 2009). *Atmbd8* is able to control the flowering time in the *Arabidopsis* vernalization-responsive C24 ecotype. The mutant line *methyl-CpG binding protein 8-1* (*atmbd8-1*) showed a delay in flowering time under both long- and short-day photoperiods through an unknown mechanism as FLC expression was not affected in *atmbd8-1*, but the expression of *FLOWERING LOCUS T* (*FT*) and *SUPPRESSOR OF CBP1* (*SOC1*), which are major flowering promoters, was downregulated in the mutant (Stangeland et al. 2009).

The previous examples confirm that DNA methylation level has a significant effect on a plant's flowering time and its cold temperature requirement. Changes in DNA methylation level, whether due to a mutation within the methyltransferases and other associated genes, or due to 5-azaC treatment, have a global effect on the plant genome and the transcriptome. Therefore, changes in DNA methylation level are not targeting a specific gene but rather a set of genes. As a result, a change in flowering time, for example, is often accompanied by a pleiotropic phenotype in the plants. This notion limits the possible use of both genetic alterations of DNA methyltransferases and 5-azaC treatment to improve plant traits. In fact, an increase of global DNA methylation may be associated with other contradictory effects on specific genes.

## DNA Methylation in Response to Abiotic and Biotic Stresses

Plants exposed to external stress experience a series of physiological changes to avoid significant damage and also to complete their life cycle quickly, so they may provide seeds for the next generation and maintain the species. These changes are associated with alterations in gene expression which are controlled, to a certain extent, by the amount and the pattern of DNA methylation within the locus as reviewed in Yaish et al. (2011).

The DNA methylation process is widely accepted as a genome-protective mechanism against unfavorable factors which may alter a DNA sequence (Bender 1998). Methylation level within the genome is rapidly and dynamically affected by environmental changes. In *Mesembryanthemum crystallinum L.*, a facultative halophyte plant which is able to switch from C3 to crassulacean acid metabolism (CAM) photosynthesis system (Bloom 1979; Hofner et al. 1987; Vernon and Bohnert 1992), DNA methylation at the CCWGG (where W represents A or T) satellite sequences plays a key role in salt adaptation and ability to switch from C3 to CAM photosynthesis system in this plant. Although DNA methylation changes were not detected within the promoter of key photosynthesis genes, such as phosphoenolpyruvate carboxylase, the salt adaptation mechanism, which includes the switch to CAM photosynthesis system, was attributed to the observed change in DNA methylation status in the satellite sequence (Dyachenko et al. 2006). This process results in an alteration in chromatin structure and leads to global gene expression changes. It is likely that these changes occur in a set of genes associated with a process involved in the switch to CAM upon salt stress.

In some rice genotypes, drought stress increases DNA methylation and only 70% of the total changes in DNA methylation reset to the normal level even after recovery in non-drought conditions (Wang et al. 2010). Reducing the level of DNA methylation often has a negative effect on the plant's ability to tolerate environmental stresses. For example, reduction of DNA methylation decreases the ability of *Arabidopsis* to cope with salt stress. *Met1-3* mutants are hypersensitive to salt stress due to a major loss of cytosine methylation in a putative small RNA target region that lowers the expression of the *Arabidopsis* sodium transporter gene (*AtHKT1*), which is essential for salt tolerance (Baek et al. 2011). In addition, methylation levels are useful for distinguishing salt-tolerant varieties. For example, 10 days after salt stress, the salt-tolerant wheat variety has a higher level of methylation than the salt-sensitive variety (Zhong and Wang 2007). Under drought conditions, hypermethylation is detected in the root tip DNA of peas (*Pisum sativum L.)* (Peng and Zhang 2009).

Cold temperature decreases the amount of DNA methyltransferase (Steward et al. 2000) and reduces DNA methylation by 10% in corn roots (Steward et al. 2002) and also in *Antirrhinum majus* (Hashida et al. 2003). Stresses produced by heavy metals such as cadmium, nickel, and chromium have varied effects on plants in terms of global DNA methylation. Peng and Zang (2009) have reviewed this issue thoroughly and found that DNA methylation status depends on the type of heavy metal and the plant species.

DNA methylation is also widely considered to be a protective mechanism against endonuclease digestion and undesired transposition (Bender 1998). In general, global genomic methylation increases, and resistance-related gene methylation decreases after viral infection (Boyko et al. 2007; Kovalchuk et al. 2003). An overall increase of methylation promotes genomic stability when a plant is attacked by a virus, whereas a decrease in resistance gene methylation level promotes genetic recombination and consequently the production of new genes that can help to resist the pathogen (Engler et al. 1993). In fact, a low level of DNA methylation alters the homologues recombination frequency during meiotic cell division and increases the chance of producing novel traits (Engler et al. 1993; Mirouze et al. 2012). An increase in homologues recombination frequency was obtained when plants are exposed to biotic stress, such as viral infections, and abiotic stress, such as high salt, heavy metals, and cold temperature (Bilichak et al. 2012; Boyko et al. 2010b). Homologues recombination may produce new heritable loci which increase plant tolerance. However, in most cases, the production of such a locus requires several homologues recombination events accompanied by constant exposure to the stress for several plant generations. In the long term, low methylation level will enhance the evolution of a resistance gene under selective pressure which may give rise to a permanent genetic change and a de novo resistance trait (Boyko and Kovalchuk 2011).

An increase in global DNA methylation will reduce global transcription and therefore slow the energy consumption of the cell which is required during pathogenic attack or other stress related to environmental challenges. On the other hand, a hypomethylation resistance gene will augment its expression and help the cell to face a temporary challenge.

## Histone Modifications and DNA Methylation

Histone modification plays an important role in epigenetic regulation. Histones are alkaline proteins associated with nuclear DNA and are the main protein component of a chromatin. Their N-terminal residues undergo post translational modifications, specifically acetylation, methylation, phosphorylation, ubiquitination, glycosylation, carbonylation, ADP-ribosylation, sumoylation, and biotination. Modifications in histone cause alternations in the nucleosomal structure and change DNA expression activity. Early works by Allfrey et al. (1964) described the role of histone acetylation and methylation in controlling eukaryotic gene expression. Histone modifications are controlled by environmental factors including abiotic and biotic stresses. Modifications within the histones alter the ability of plants to tolerate adverse environmental conditions such as high salt, drought, and bacterial infections. Tables 17.1 and 17.2 summarize a few examples where posttranslational modifications within histones are associated with environmental stresses and physiological disorders in plants, respectively.

MBD proteins harbor a binding domain specific to DNA methyl group and function as a platform for attaching other chromatin-remodeling proteins which are involved in regulating gene expression (Berg et al. 2003; Springer and Kaeppler 2005). For example, Arabidopsis METHYL-CpG BINDING PROTEIN 7 (AtMBD7) binds to arginine methyltransferase (PRMT11)(Scebba et al. 2007). The Arabidopsis METHYL-CpG BINDING PROTEIN 5–7 (AtMBD5 to AtMBD7) bind to the DDM1 protein in vitro (Zemach et al. 2008). Both proteins function as chromatin modifiers that control gene expression in plants. Loss-of-function studies showed that *Arabidopsis* with an *atmbd9* mutation displays a pleiotropic phenotype that leads to a decrease in histone acetylation and an increase in DNA methylation at the *FLC* locus (Peng et al. 2006; Yaish et al. 2009). As a result, *FLC* transcription declines and leads to an early-flowering phenotype.

DNA methylation dictates histone H3K4 methylation in human cells (Okitsu and Hsieh 2007). Similarly, in *Arabidopsis*, analysis of the *ddm1* mutant revealed that a decrease in DNA methylation is associated with a gain in H3K4me and a loss in H3K9me (Gendrel et al. 2002). In addition, a mutation within the *KRYPTONITE* gene, which encodes a member of the suppressor of variegation 3–9 (Su(var)3-9) family of histone methyltransferases, reduces H3K9me, eliminates DNA methylation, and decreases gene silencing (Jackson et al. 2004). Recently, DNA methylation analysis of plants growing under stress showed that DNA methylation is associated with H3K9me2 enrichment and H3K9ac depletion in the histones of salt-stressed progenies (Bilichak et al. 2012). The same study showed that DNA methylation and histone modification is associated with global gene repression and salt tolerance.

## Small RNA Is Associated with DNA Methylation and Environmental Stress Tolerance

MicroRNA (miRNA) molecules encoded by 20–24 nucleotides control gene expression in plants at the posttranscriptional level. Mutation analysis of the miRNA biogenesis machinery *DICER-LIKE 1–4* genes (*DCL1*, *DCL2*, *DCL3*, and *DCL4*) revealed an essential role for these genes in controlling gene expression in *Arabidopsis* (Laubinger et al. 2010) and epigenetic transgenerational memory of the progenies (Boyko and Kovalchuk 2010). In addition, miRNA is able to direct DNA methylation to a specific locus using the RNA-directed DNA methylation (RdDM) mechanism (Matzke et al. 2001, 2007; Pikaard 2006) by binding miRNA to the target genes. MiRNAs modulate growth and development in plants, including flowering (Schmid et al. 2003), and control the expression of genes in response to biotic (Ruiz-Ferrer and Voinnet 2009; Madlung and Comai 2004; Covarrubias and Reyes 2010) and abiotic stresses (Madlung and Comai 2004; Hirayama and Shinozaki 2010; Urano et al. 2010). DNA methylation is a prerequisite for gene silencing via small RNA-directed methylation. De novo methylation of *FWA*, a homeodomain floral transcription factor, is promoted by miRNA which initially targets the methylated parts of the locus and then stimulates farther methylation resulting instable gene silencing (Chan et al. 2006).

**Table 17.1** Epigenetic modifiers associated with abiotic and biotic stresses

| Species | Modifier | Type | Phenotype | Remark | Reference |
|---|---|---|---|---|---|
| *Arabidopsis thaliana* | Met1-3 | Cytosine methyltransferase | Mutants are hypersensitive to salt stress | Mutants loss cytosine DNA methylation | Baek et al. (2011) |
| *Arabidopsis thaliana* | AtHKT1 | Sodium transporter | Low expression causes hypersensitive to salt stress | Loss of methylation at the putative small RNA target region causes AtHKT1 repression | Baek et al. (2011) |
| *Arabidopsis thaliana* | AtNAP1-3 | Chaperone, NUCLEOSOME ASSEMBLY PROTEIN 1-3 | Mutants showed less tolerance to salt stress | Overexpression showed hyposensitivity to ABA | Liu et al. (2009) |
| *Arabidopsis thaliana* | AtSAP18 | Associated with histone deacetylation complex | Mutant is more sensitive to salt conditions and is impaired in chlorophyll synthesis | Transcriptional repressors and is associated with ethylene-responsive element binding factors (ERFs) under environmental stress | Song and Galbraith (2006) |
| *Arabidopsis thaliana* | HDA19 | Histone deacetylase | Mutants are hypersensitive to ABA and salt stress | Modulate seed germination and salt stress response | Chen and Wu (2010) |
| *Arabidopsis thaliana* | HDA6 | Histone deacetylase | Mutants are hypersensitive to ABA and salt stress | Modulates the expression of ABA and abiotic stress-responsive genes | Chen et al. (2010a) |
| *Arabidopsis thaliana* | HDA19 | Histone deacetylase | Overexpression increases resistant to *A. brassicicola* pathogen | Regulates genes involved in jasmonic acid (JA) and ethylene signaling | Zhou et al. (2005) |
| *Arabidopsis thaliana* | SKB1 | Kinase binding protein | Mutants are salt hypersensitive, delay in flowering, and showed growth retardation | Enhance histone 4 arginine3 (H4R3) symmetric dimethylation (H4R3sme2) | Zhang et al. (2011) |
| *Arabidopsis thaliana* | ADA2b | A component of the histone acetylation complex GCN5 | Mutants exhibit pleiotropic developmental defects and altered responses to low-temperature stress | Involved in maintaining H3 and H4 acetylation level | Kaldis et al. (2011) |
| *Arabidopsis thaliana* | SGF29A | A component of the histone acetylation complex GCN5 | Mutants are more tolerant to salt stress | Its function masked by ADA2b | Kaldis et al. (2011) |
| *Arabidopsis thaliana* | AtPP2C-6-6 | Phosphatase | Mutants upregulate the salt and stress inducible genes | Dephosphorylates the GCN5 and reduces its activity | Servet et al. (2008) |
| *Arabidopsis thaliana* | HOS15 | Associated with histone deacetylase | Mutant showed hypersensitive to freezing temperatures | Negative regulator of the histone 4 (H4) acetyltransferase | Zhu et al. (2008) |
| *Arabidopsis thaliana* | AtEML | GCN5-interacting protein | Controlled by cold and salt treatments | N.A. | Gao et al. (2007) |
| *Arabidopsis thaliana* | AtHD2C | Histone deacetylase | Overexpression reduces transpiration and | Downregulated by ABA | |

(continued)

**Table 17.1** (continued)

| Species | Modifier | Type | Phenotype | Remark | Reference |
|---|---|---|---|---|---|
| | | | enhances tolerance to salt and drought stresses, and the transgenic showed an ABA insensitive phenotype | | Sridha and Wu (2006) |
| *Arabidopsis thaliana* | DCL2 | Dicer | Mutants lose the transgenerational capacity of stress tolerance | Mutants showed impaired stress-induced DNA methylation and homologous recombination frequency | Boyko et al. (2010a) |
| *Brassica napus* | bnKCP1 | Putative kinase | Induced by cold stress | Interacts with HDA19 | Gao et al. (2003) |
| *Pisum sativum L. cv. Lincoln* | PsSNF5 | Chromatin-remodeling complexes SNF5-like | Induced by ABA, drought and low water content during the last stage of embryo development | N.A. | Rios et al. (2007) |

Molecular analysis of the *Arabidopsis* miRNA molecules and their targets revealed a conserved stress response pattern similar to other plant species (Sunkar and Zhu 2004). In rice, global gene expression analysis showed a fundamental role for miRNAs in controlling gene expression when plants are exposed to stress conditions such as cold temperatures, drought, high salt, and abscisic acid (ABA) treatment (Shen et al. 2010). MiRNAs isolated recently from *Arabidopsis* showed they have a critical role in ABA and salt tolerance in plants. For example, miR159 regulates the expression of *MYB101* and *MYB33* transcription factors by controlling their cleavage (Reyes and Chua 2007), and miR160 is another potential ABA regulatory miRNA molecule that is induced by ABA (Liu et al. 2007).

Defects in the miRNA synthesis machinery lead to phenotypic defects in plants. For example, mutations within the *ABA-HYPERSENSITIVE* (*ABH1*) and *CAP BINDING PROTEINS 20* (*CBP20*) genes (Papp et al. 2004), which encode cap-binding factors that are necessary for RNA maturation, lead to ABA hypersensitivity and enhance drought tolerance (Hugouvieux et al. 2001; Kwak et al. 2005). In addition, mutant lines for *STRESS RESPONSE SUPPRESSOR1* and *2* (*STRS1* and *2*) genes, which encode DEAD-box RNA helicases, display higher tolerance to drought than the wild-type (Kant et al. 2007).

## Use of Epiallelic Polymorphism in Plant Breeding

Changes in DNA methylation produce phenotypic variations in plants. For example, the extensive methylation of two direct repeats in the 5′-region of the imprinted *fwa* (homeodomain-containing transcription factor) gene caused the late-flowering phenotype (Soppe et al. 2000; Kinoshita et al. 2004). While the basic DNA sequences of both the wild-type and the *fwa* line are the same, the *fwa* mutant line shows a lower level of ectopic gene expression in the vegetative tissue due to hypermethylation caused by a mutation in the methylation maintenance gene *MET1*(Kankel et al. 2003; Kinoshita et al. 2007). Likewise, hypermethylation of a putative *ANAPHASE-PROMOTING COMPLEX 13* (*APC 13*) gene in the *ddm1* mutant causes a reduction in apical dominancy and produces an abnormal floral architecture (Kakutani 1997; Saze et al. 2008; Kakutani et al. 1996). In the tomato, hypermethylation of an SBP-box (*SQUAMOSA* promoter binding protein-like) gene at the colorless non-ripening (*Cnr*)

**Table 17.2** Epigenetic modifiers associated with physiological impairs

| Species | Modifier | Type | Phenotype | Remark | Reference |
|---|---|---|---|---|---|
| *Arabidopsis thaliana* | ELO1, ELO2, and ELO3 | Elongator complex, co-localized with histone acetyltransferases | Mutants showed malformation of leaves and roots | Alters the expression of auxin, ethylene, and JA-related genes | Nelissen et al. (2010) |
| *Arabidopsis thaliana* | MET1 and CMT3 | CpG methyltransferase and CpNpG and CpNpN chromomethylase | Mutants showed improper embryo development, cell division, and auxin gradient | DNA methylation regulates embryogenesis and seed viability | Xiao et al. (2006) |
| *Arabidopsis thaliana* | PRZ1 or AtADA2b | Chromatin-remodeling component | Mutants showed block the auxin effects on morphogenesis | Essential for proper histone acetylation at auxin-controlled loci | Anzola et al. (2010) |
| *Arabidopsis thaliana* | GCN5 (general control non-repressed protein5) | Histone acetyltransferase | N.A. | Control the production of miRNA at transcriptional and posttranscriptional levels | Kim et al. (2009). |
| *Arabidopsis thaliana* | TU8 | Heterochromatin protein 1 | Mutation causes defects in the induction of secondary metabolite biosynthesis | N.A. | Bennett et al. (2005) |
| *Hordeum vulgare L.* | HvFIE and HvE(Z) | Polycomb group, histone methyl transferase | Their expression varies based on the cultivars and seed size | Induced by ABA | Kapazoglou et al. (2010) |
| *Hordeum vulgare L.* | HvMYS, HvELP3, and HvGCN5 | Histone acetyltransferases | Their expression pattern varies between cultivars with varying seed size and weight | Induced by ABA | Papaefthimiou et al. (2010) |
| *Hordeum vulgare L.* | HvHDAC2-1 and HvHDAC2-2 | Histone deacetylase | Their expression varies between cultivars | Their expression is affected by JA, ABA, and salicylic acid (SA) | Demetriou et al. (2009) |

locus results in colorless fruits with substantial loss of cell-to-cell adhesion (Manning et al. 2006). Hypermethylation of the *CYCLOIDEA* gene, a class II TEOSINTE BRANCHED 1-CYCLOIDEA-PCF (TCP) transcriptional activator, confers irregular floral symmetry in *Linaria vulgaris* (Cubas et al. 1999). Similarly, hypermethylation of *FLC* induces early flowering in *atmbd9* (Yaish et al. 2009). The loss of methylation at the *Xa21G* promoter region in line 2 confers resistance to rice blight, while the fully methylated promoter is susceptible to the blight (Akimoto et al. 2007).

Genetic variation and phenotypic diversity are basic materials for the selection and improvement of breeding programs. Alterations in DNA methylation produce epialleles which may also yield phenotypic variations. Epiallele can be defined as any two or more genetically identical genes that are epigenetically distinct due to methylation. These epiallele-based phenotypic variants can be used in breeding programs to improve a plant's tolerance to stresses. However, DNA methylation within the genome can be stable, unstable, or stochastic. Stable DNA methylation takes place when a consistent and heritable variation in DNA methylation occurs at specific loci due to an environmental factor for several generations and persists even in the absence of the triggering environmental factor. However, unstable DNA methylation occurs when the variation disappears once the plants return to normal environmental conditions or when it is not transmittable to the next generation. In some cases, a DNA methylation variation takes place randomly and vanishes in the same way (Zhao et al. 2007). In this case, it is very difficult to monitor or use these stochastic types of variation to produce plants with a novel trait.

Epigenetic modifications lead to an adaptive evolutionary mechanism in plants. DNA methylation at a specific locus can be inherited through meiosis (Bender 2004) and shows gene expression diversity within the individuals of the same plant species when grown under diverse environmental conditions.

Traits associated with methylated or unmethylated loci can be identified based on epigenetic mapping and the corresponding quantitative trait loci (QTLs) and then identified using positional cloning strategy. A high-resolution methylation map was constructed for *Arabidopsis* by first enriching methylated DNA using immunoprecipitation followed by microarray chip analysis (Cokus et al. 2008; Zhang et al. 2006). In canola (Long et al. 2011), the epigenetic map was constructed mainly using methylation-sensitive amplified polymorphism (MSAP) markers. Despite the importance of epigenetic maps in plant breeding, the use of such maps for improving plant traits is still uncommon.

## Conclusion and Future Prospective

DNA methylation controls gene expression through the binding of methyl group to DNA cytosines, the enhancement of further DNA methylation using RdDM, and the induction of histone modifications. DNA methylation can be affected by environmental cues or be inherited as epialleles and often associated with tolerance for abiotic and biotic stresses. Epialleles are potentially useful when used in plant tolerance improvement programs. However, molecular control of DNA methylation and the inheritance of epigenes are not currently manageable using available knowledge and technology. Future research should focus on better understanding of the effect of histone modifiers and miRNAs on DNA methylation and vice versa. Better understanding of the epigenetic mechanisms will facilitate future research aims to control the pattern of gene expression and the epigenetic inheritance of the stress tolerance phenotype in plants.

**Acknowledgements** This work was supported by a generous grant from the college of Science, Sultan Qaboos University IG/Sci/Biol/11/04.

## References

Agius F, Kapoor A, Zhu JK (2006) Role of the Arabidopsis DNA glycosylase/lyase ROS1 in active DNA demethylation. Proc Natl Acad Sci USA 103(31):11796–11801

Akimoto K, Katakami H, Kim HJ, Ogawa E, Sano CM, Wada Y, Sano H (2007) Epigenetic inheritance in rice plants. Ann Bot 100(2):205–217

Allfrey VG, Faulkner R, Mirsky AE (1964) Acetylation and methylation of histones and their possible role in the regulation of RNA synthesis. Proc Natl Acad Sci USA 51:786–794

Alvarez ME, Nota F, Cambiagno DA (2010) Epigenetic control of plant immunity. Mol Plant Pathol 11 (4):563–576

Anzola JM, Sieberer T, Ortbauer M, Butt H, Korbei B, Weinhofer I, Mullner AE, Luschnig C (2010) Putative Arabidopsis transcriptional adaptor protein (PROPORZ1) is required to modulate histone acetylation in response to auxin. Proc Natl Acad Sci USA 107 (22):10308–10313

Baek D, Jiang J, Chung JS, Wang B, Chen J, Xin Z, Shi H (2011) Regulated AtHKT1 gene expression by a distal enhancer element and DNA methylation in the promoter plays an important role in salt tolerance. Plant Cell Physiol 52(1):149–161

Bartee L, Bender J (2001) Two Arabidopsis methylation-deficiency mutations confer only partial effects on a methylated endogenous gene family. Nucleic Acids Res 29(10):2127–2134

Bender J (1998) Cytosine methylation of repeated sequences in eukaryotes: the role of DNA pairing. Trends Biochem Sci 23(7):252–256

Bender J (2004) DNA methylation and epigenetics. Annu Rev Plant Biol 55:41–68. doi:10.1146/annurev.arplant.55.031903.141641

Bennett RN, Wenke T, Freudenberg B, Mellon FA, Ludwig-Muller J (2005) The tu8 mutation of Arabidopsis thaliana encoding a heterochromatin protein 1 homolog causes defects in the induction of secondary metabolite biosynthesis. Plant Biol (Stuttg) 7 (4):348–357

Berg A, Meza TJ, Mahic M, Thorstensen T, Kristiansen K, Aalen RB (2003) Ten members of the Arabidopsis gene family encoding methyl-CpG-binding domain proteins are transcriptionally active and at least one, AtMBD11, is crucial for normal development. Nucleic Acids Res 31(18):5291–5304

Bilichak A, Ilnystkyy Y, Hollunder J, Kovalchuk I (2012) The progeny of Arabidopsis thaliana plants exposed to salt exhibit changes in DNA methylation, histone modifications and gene expression. PLoS One 7(1): doi:e3051510.1371/journal.pone.0030515

Bloom AJ (1979) Salt requirement for crassulacean acid metabolism in the annual succulent, Mesembryanthemum crystallinum. Plant Physiol 63(4):749–753

Bonasio R, Tu S, Reinberg D (2010) Molecular signals of epigenetic states. Science 330(6004):612–616

Borowska N, Idziak D, Hasterok R (2011) DNA methylation patterns of Brachypodium distachyon chromosomes and their alteration by 5-azacytidine treatment. Chromosome Res 19(8):955–967

Boyko A, Kovalchuk I (2008) Epigenetic control of plant stress response. Environ Mol Mutagen 49(1):61–72

Boyko A, Kovalchuk I (2010) Transgenerational response to stress in Arabidopsis thaliana. Plant Signal Behav 5 (8):995–998

Boyko A, Kovalchuk I (2011) Genetic and epigenetic effects of plant-pathogen interactions: an evolutionary perspective. Mol Plant 4(6):1014–1023. doi:10.1093/mp/ssr022

Boyko A, Kathiria P, Zemp FJ, Yao Y, Pogribny I, Kovalchuk I (2007) Transgenerational changes in the genome stability and methylation in pathogen-infected plants: (virus-induced plant genome instability). Nucleic Acids Res 35(5):1714–1725

Boyko A, Blevins T, Yao Y, Golubov A, Bilichak A, Ilnytskyy Y, Hollander J, Meins F Jr, Kovalchuk I (2010a) Transgenerational adaptation of Arabidopsis to stress requires DNA methylation and the function of Dicer-like proteins. PLoS One 5(3):e9514

Boyko A, Golubov A, Bilichak A, Kovalchuk I (2010b) Chlorine ions but not sodium ions alter genome stability of Arabidopsis thaliana. Plant Cell Physiol 51 (6):1066–1078. doi:10.1093/pcp/pcq048

Brock RD, Davidson JL (1994) 5-azacytidine and gamma rays partially substitute for cold treatment in vernalising winter wheat. Environ Exp Bot 31:195–199

Burn JE, Bagnall DJ, Metzger JD, Dennis ES, Peacock WJ (1993) DNA methylation, vernalization, and the initiation of flowering. Proc Natl Acad Sci USA 90 (1):287–291

Castilho A, Neves N, Rufini-Castiglione M, Viegas W, Heslop-Harrison JS (1999) 5-Methylcytosine distribution and genome organization in triticale before and after treatment with 5-azacytidine. J Cell Sci 112(Pt 23):4397–4404

Chan SW, Zhang X, Bernatavichute YV, Jacobsen SE (2006) Two-step recruitment of RNA-directed DNA methylation to tandem repeats. PLoS Biol 4(11):e363

Chen LT, Wu K (2010) Role of histone deacetylases HDA6 and HDA19 in ABA and abiotic stress response. Plant Signal Behav 5(10)

Chen LT, Luo M, Wang YY, Wu K (2010a) Involvement of Arabidopsis histone deacetylase HDA6 in ABA and salt stress response. J Exp Bot 61(12):3345–3353

Chen M, Lv S, Meng Y (2010b) Epigenetic performers in plants. Dev Growth Differ 52(6):555–566

Cheng JC, Matsen CB, Gonzales FA, Ye W, Greer S, Marquez VE, Jones PA, Selker EU (2003) Inhibition of DNA methylation and reactivation of silenced genes by zebularine. J Natl Cancer Inst 95 (5):399–409

Chinnusamy V, Zhu JK (2009) Epigenetic regulation of stress responses in plants. Curr Opin Plant Biol 12 (2):133–139

Cokus SJ, Feng S, Zhang X, Chen Z, Merriman B, Haudenschild CD, Pradhan S, Nelson SF, Pellegrini M, Jacobsen SE (2008) Shotgun bisulphite sequencing of the Arabidopsis genome reveals DNA methylation patterning. Nature 452(7184):215–219. doi:10.1038/nature06745

Covarrubias AA, Reyes JL (2010) Post-transcriptional gene regulation of salinity and drought responses by plant microRNAs. Plant Cell Environ 33(4):481–489

Cubas P, Vincent C, Coen E (1999) An epigenetic mutation responsible for natural variation in floral symmetry. Nature 401(6749):157–161

Demetriou K, Kapazoglou A, Tondelli A, Francia E, Stanca MA, Bladenopoulos K, Tsaftaris AS (2009) Epigenetic chromatin modifiers in barley: I. Cloning, mapping and expression analysis of the plant specific HD2 family of histone deacetylases from barley, during seed development and after hormonal treatment. Physiol Plant 136(3):358–368

Dyachenko OV, Zakharchenko NS, Shevchuk TV, Bohnert HJ, Cushman JC, Buryanov YI (2006) Effect of hypermethylation of CCWGG sequences in DNA of Mesembryanthemum crystallinum plants on their adaptation to salt stress. Biochemistry (Mosc) 71 (4):461–465

Ehrlich M (2002) DNA methylation in cancer: too much, but also too little. Oncogene 21(35):5400–5413. doi:10.1038/sj.onc.1205651

Engler P, Weng A, Storb U (1993) Influence of CpG methylation and target spacing on V(D)J recombination in a transgenic substrate. Mol Cell Biol 13(1):571–577

Finnegan EJ, Genger RK, Kovac K, Peacock WJ, Dennis ES (1998) DNA methylation and the promotion of flowering by vernalization. Proc Natl Acad Sci USA 95(10):5824–5829

Gao MJ, Schafer UA, Parkin IA, Hegedus DD, Lydiate DJ, Hannoufa A (2003) A novel protein from Brassica napus has a putative KID domain and responds to low temperature. Plant J 33(6):1073–1086

Gao MJ, Hegedus DD, Sharpe AG, Robinson SJ, Lydiate DJ, Hannoufa A (2007) Isolation and characterization of a GCN5-interacting protein from Arabidopsis thaliana. Planta 225(6):1367–1379

Gendrel AV, Lippman Z, Yordan C, Colot V, Martienssen RA (2002) Dependence of heterochromatic histone H3 methylation patterns on the Arabidopsis gene DDM1. Science 297(5588):1871–1873. doi:10.1126/science.1074950

Gruenbaum Y, Naveh-Many T, Cedar H, Razin A (1981a) Sequence specificity of methylation in higher plant DNA. Nature 292(5826):860–862

Gruenbaum Y, Stein R, Cedar H, Razin A (1981b) Methylation of CpG sequences in eukaryotic DNA. FEBS Lett 124(1):67–71

Guo M, Rupe MA, Yang X, Crasta O, Zinselmeier C, Smith OS, Bowen B (2006) Genome-wide transcript analysis of maize hybrids: allelic additive gene expression and yield heterosis. Theor Appl Genet 113(5):831–845. doi:10.1007/s00122-006-0335-x

Hashida SN, Kitamura K, Mikami T, Kishima Y (2003) Temperature shift coordinately changes the activity and the methylation state of transposon Tam3 in Antirrhinum majus. Plant Physiol 132(3):1207–1216

Hirayama T, Shinozaki K (2010) Research on plant abiotic stress responses in the post-genome era: past, present and future. Plant J 61(6):1041–1052

Hofner R, Vazquez-Moreno L, Winter K, Bohnert HJ, Schmitt JM (1987) Induction of crassulacean acid metabolism in Mesembryanthemum crystallinum by high salinity: mass increase and de novo synthesis of PEP-carboxylase. Plant Physiol 83(4):915–919

Hugouvieux V, Kwak JM, Schroeder JI (2001) An mRNA cap binding protein, ABH1, modulates early abscisic acid signal transduction in Arabidopsis. Cell 106 (4):477–487

Jackson JP, Johnson L, Jasencakova Z, Zhang X, PerezBurgos L, Singh PB, Cheng X, Schubert I, Jenuwein T, Jacobsen SE (2004) Dimethylation of histone H3 lysine 9 is a critical mark for DNA methylation and gene silencing in Arabidopsis thaliana. Chromosoma 112(6):308–315

Jacobsen SE, Meyerowitz EM (1997) Hypermethylated SUPERMAN epigenetic alleles in arabidopsis. Science 277(5329):1100–1103

Kakutani T (1997) Genetic characterization of late-flowering traits induced by DNA hypomethylation mutation in Arabidopsis thaliana. Plant J Cell Mol Biol 12(6):1447–1451

Kakutani T, Jeddeloh JA, Flowers SK, Munakata K, Richards EJ (1996) Developmental abnormalities and epimutations associated with DNA hypomethylation mutations. Proc Natl Acad Sci USA 93 (22):12406–12411

Kaldis A, Tsementzi D, Tanriverdi O, Vlachonasios KE (2011) Arabidopsis thaliana transcriptional co-activators ADA2b and SGF29a are implicated in salt stress responses. Planta 233(4):749–762

Kankel MW, Ramsey DE, Stokes TL, Flowers SK, Haag JR, Jeddeloh JA, Riddle NC, Verbsky ML, Richards EJ (2003) Arabidopsis MET1 cytosine methyltransferase mutants. Genetics 163(3):1109–1122

Kant P, Kant S, Gordon M, Shaked R, Barak S (2007) STRESS RESPONSE SUPPRESSOR1 and STRESS RESPONSE SUPPRESSOR2, two DEAD-box RNA helicases that attenuate Arabidopsis responses to multiple abiotic stresses. Plant Physiol 145 (3):814–830

Kapazoglou A, Tondelli A, Papaefthimiou D, Ampatzidou H, Francia E, Stanca MA, Bladenopoulos K, Tsaftaris AS (2010) Epigenetic chromatin modifiers in barley: IV. The study of barley polycomb group (PcG) genes during seed development and in response to external ABA. BMC Plant Biol 10:73

Kim W, Benhamed M, Servet C, Latrasse D, Zhang W, Delarue M, Zhou DX (2009) Histone acetyltransferase GCN5 interferes with the miRNA pathway in Arabidopsis. Cell Res 19(7):899–909

Kinoshita T, Miura A, Choi YH, Kinoshita Y, Cao XF, Jacobsen SE, Fischer RL, Kakutani T (2004) One-way control of FWA imprinting in Arabidopsis endosperm by DNA methylation. Science 303(5657):521–523. doi:10.1126/science.1089835

Kinoshita Y, Saze H, Kinoshita T, Miura A, Soppe WJ, Koornneef M, Kakutani T (2007) Control of FWA gene silencing in Arabidopsis thaliana by SINE-related direct repeats. Plant J 49(1):38–45

Kovalchuk I, Kovalchuk O, Kalck V, Boyko V, Filkowski J, Heinlein M, Hohn B (2003) Pathogen-induced systemic plant signal triggers DNA rearrangements. Nature 423(6941):760–762

Kwak KJ, Kim YO, Kang H (2005) Characterization of transgenic Arabidopsis plants overexpressing GR-RBP4 under high salinity, dehydration, or cold stress. J Exp Bot 56(421):3007–3016

Laubinger S, Zeller G, Henz SR, Buechel S, Sachsenberg T, Wang JW, Ratsch G, Weigel D (2010) Global effects of the small RNA biogenesis machinery on the Arabidopsis thaliana transcriptome. Proc Natl Acad Sci USA 107(41):17466–17473

Liu PP, Montgomery TA, Fahlgren N, Kasschau KD, Nonogaki H, Carrington JC (2007) Repression of AUXIN RESPONSE FACTOR10 by microRNA160 is critical for seed germination and post-germination stages. Plant J 52(1):133–146

Liu ZQ, Gao J, Dong AW, Shen WH (2009) A truncated Arabidopsis NUCLEOSOME ASSEMBLY PROTEIN 1, AtNAP1;3 T, alters plant growth responses to abscisic acid and salt in the Atnap1;3-2 mutant. Mol Plant 2(4):688–699

Long Y, Xia W, Li R, Wang J, Shao M, Feng J, King GJ, Meng J (2011) Epigenetic QTL mapping in Brassica napus. Genetics 189(3):1093–U1585. doi:10.1534/genetics.111.131615

Madlung A, Comai L (2004) The effect of stress on genome regulation and structure. Ann Bot 94(4):481–495

Manning K, Tor M, Poole M, Hong Y, Thompson AJ, King GJ, Giovannoni JJ, Seymour GB (2006) A naturally occurring epigenetic mutation in a gene encoding an SBP-box transcription factor inhibits tomato fruit ripening. Nat Genet 38(8):948–952. doi:10.1038/ng1841

Matzke MA, Matzke AJ, Pruss GJ, Vance VB (2001) RNA-based silencing strategies in plants. Curr Opin Genet Dev 11(2):221–227

Matzke M, Kanno T, Huettel B, Daxinger L, Matzke AJ (2007) Targets of RNA-directed DNA methylation. Curr Opin Plant Biol 10(5):512–519

Mirouze M, Lieberman-Lazarovich M, Aversano R, Bucher E, Nicolet J, Reinders J, Paszkowski J (2012) Loss of DNA methylation affects the recombination landscape in Arabidopsis. Proc Natl Acad Sci USA 109(15):5880–5885. doi:10.1073/pnas.1120841109

Nelissen H, De Groeve S, Fleury D, Neyt P, Bruno L, Bitonti MB, Vandenbussche F, Van der Straeten D, Yamaguchi T, Tsukaya H, Witters E, De Jaeger G, Houben A, Van Lijsebettens M (2010) Plant elongator regulates auxin-related genes during RNA polymerase II transcription elongation. Proc Natl Acad Sci USA 107(4):1678–1683

Okitsu CY, Hsieh C-L (2007) DNA methylation dictates histone H3K4 methylation. Mol Cell Biol 27 (7):2746–2757. doi:10.1128/mcb.02291-06

Papaefthimiou D, Likotrafiti E, Kapazoglou A, Bladenopoulos K, Tsaftaris A (2010) Epigenetic chromatin modifiers in barley: III. Isolation and characterization of the barley GNAT-MYST family of histone acetyltransferases and responses to exogenous ABA. Plant Physiol Biochem 48(2–3):98–107

Papp I, Mur LA, Dalmadi A, Dulai S, Koncz C (2004) A mutation in the Cap Binding Protein 20 gene confers drought tolerance to Arabidopsis. Plant Mol Biol 55 (5):679–686

Peng H, Zhang J (2009) Plant genomic DNA methylation in response to stresses: potential applications and challenges in plant breeding. Prog Nat Sci 19 (9):1037–1045. doi:10.1016/j.pnsc.2008.10.014

Peng M, Cui Y, Bi YM, Rothstein SJ (2006) AtMBD9: a protein with a methyl-CpG-binding domain regulates flowering time and shoot branching in Arabidopsis. Plant J Cell Mol Biol 46(2):282–296

Pikaard CS (2006) Cell biology of the Arabidopsis nuclear siRNA pathway for RNA-directed chromatin modification. Cold Spring Harb Symp Quant Biol 71:473–480

Reyes JL, Chua NH (2007) ABA induction of miR159 controls transcript levels of two MYB factors during Arabidopsis seed germination. Plant J 49(4):592–606

Rios G, Gagete AP, Castillo J, Berbel A, Franco L, Rodrigo MI (2007) Abscisic acid and desiccation-dependent expression of a novel putative SNF5-type chromatin-remodeling gene in Pisum sativum. Plant Physiol Biochem 45(6–7):427–435

Ruiz-Ferrer V, Voinnet O (2009) Roles of plant small RNAs in biotic stress responses. Annu Rev Plant Biol 60:485–510

Santi DV, Garrett CE, Barr PJ (1983) On the mechanism of inhibition of DNA-cytosine methyltransferases by cytosine analogs. Cell 33(1):9–10

Saze H, Shiraishi A, Miura A, Kakutani T (2008) Control of genic DNA methylation by a jmjC domain – containing protein in Arabidopsis thaliana. Science 319(5862):462–465. doi:10.1126/science.1150987

Scebba F, De Bastiani M, Bernacchia G, Andreucci A, Galli A, Pitto L (2007) PRMT11: a new Arabidopsis MBD7 protein partner with arginine methyltransferase activity. Plant J 52(2):210–222

Schmid M, Uhlenhaut NH, Godard F, Demar M, Bressan R, Weigel D, Lohmann JU (2003) Dissection of floral induction pathways using global expression analysis. Development 130(24):6001–6012

Servet C, Benhamed M, Latrasse D, Kim W, Delarue M, Zhou DX (2008) Characterization of a phosphatase 2C protein as an interacting partner of the histone acetyltransferase GCN5 in Arabidopsis. Biochim Biophys Acta 1779(6–7):376–382

Shen J, Xie K, Xiong L (2010) Global expression profiling of rice microRNAs by one-tube stem-loop reverse transcription quantitative PCR revealed important roles of microRNAs in abiotic stress responses. Mol Genet Genomics 284(6):477–488

Singer T, Yordan C, Martienssen RA (2001) Robertson's Mutator transposons in A. thaliana are regulated by the chromatin-remodeling gene Decrease in DNA Methylation (DDM1). Genes Dev 15(5):591–602

Song CP, Galbraith DW (2006) AtSAP18, an orthologue of human SAP18, is involved in the regulation of salt stress and mediates transcriptional repression in Arabidopsis. Plant Mol Biol 60(2):241–257

Soppe WJ, Jacobsen SE, Alonso-Blanco C, Jackson JP, Kakutani T, Koornneef M, Peeters AJ (2000) The late flowering phenotype of fwa mutants is caused by gain-of-function epigenetic alleles of a homeodomain gene. Mol Cell 6(4):791–802

Springer NM, Kaeppler SM (2005) Evolutionary divergence of monocot and dicot methyl-CpG-binding domain proteins. Plant Physiol 138(1):92–104

Sridha S, Wu K (2006) Identification of AtHD2C as a novel regulator of abscisic acid responses in Arabidopsis. Plant J 46(1):124–133

Stangeland B, Rosenhave EM, Winge P, Berg A, Amundsen SS, Karabeg M, Mandal A, Bones AM, Grini PE, Aalen RB (2009) AtMBD8 is involved in control of flowering time in the C24 ecotype of Arabidopsis thaliana. Physiol Plant 136(1):110–126

Steward N, Kusano T, Sano H (2000) Expression of ZmMET1, a gene encoding a DNA methyltransferase from maize, is associated not only with DNA replication in actively proliferating cells, but also with altered DNA methylation status in cold-stressed quiescent cells. Nucleic Acids Res 28(17):3250–3259

Steward N, Ito M, Yamaguchi Y, Koizumi N, Sano H (2002) Periodic DNA methylation in maize nucleosomes and demethylation by environmental stress. J Biol Chem 277(40):37741–37746

Sunkar R, Zhu JK (2004) Novel and stress-regulated microRNAs and other small RNAs from Arabidopsis. Plant Cell 16(8):2001–2019

Tani E, Polidoros AN, Nianiou-Obeidat I, Tsaftaris AS (2005) DNA methylation patterns are differently affected by planting density in maize inbreds and their hybrids. Maydica 50:19–23

Urano K, Kurihara Y, Seki M, Shinozaki K (2010) 'Omics' analyses of regulatory networks in plant abiotic stress responses. Curr Opin Plant Biol 13 (2):132–138

Vernon DM, Bohnert HJ (1992) A novel methyl transferase induced by osmotic stress in the facultative halophyte Mesembryanthemum crystallinum. EMBO J 11 (6):2077–2085

Vongs A, Kakutani T, Martienssen RA, Richards EJ (1993) Arabidopsis thaliana DNA methylation mutants. Science (New York, NY) 260(5116): 1926–1928

Wada KC, Takeno K (2010) Stress-induced flowering. Plant Signal Behav 5(8):944–947

Wang WS, Pan YJ, Zhao XQ, Dwivedi D, Zhu LH, Ali J, Fu BY, Li ZK (2010) Drought-induced site-specific DNA methylation and its association with drought tolerance in rice (Oryza sativa L.). J Exp Bot 62 (6):1951–1960

Xiao W, Custard KD, Brown RC, Lemmon BE, Harada JJ, Goldberg RB, Fischer RL (2006) DNA methylation is critical for Arabidopsis embryogenesis and seed viability. Plant Cell 18(4):805–814

Yaish MW, Peng M, Rothstein SJ (2009) AtMBD9 modulates Arabidopsis development through the dual epigenetic pathways of DNA methylation and histone acetylation. Plant J 59(1):123–135

Yaish MW, Colasanti J, Rothstein SJ (2011) The role of epigenetic processes in controlling flowering time in plants exposed to stress. J Exp Bot 62(11): 3727–3735

Zemach A, Gaspan O, Grafi G (2008) The three methyl-CpG-binding domains of AtMBD7 control its subnuclear localization and mobility. J Biol Chem 283 (13):8406–8411

Zhang X, Yazaki J, Sundaresan A, Cokus S, Chan SW, Chen H, Henderson IR, Shinn P, Pellegrini M, Jacobsen SE, Ecker JR (2006) Genome-wide high-resolution mapping and functional analysis of DNA methylation in arabidopsis. Cell 126(6):1189–1201. doi:10.1016/j.cell.2006.08.003

Zhang Z, Zhang S, Zhang Y, Wang X, Li D, Li Q, Yue M, Zhang YE, Xu Y, Xue Y, Chong K, Bao S (2011) Arabidopsis floral initiator SKB1 confers high salt tolerance by regulating transcription and pre-mRNA splicing through altering histone H4R3 and small nuclear ribonucleoprotein LSM4 methylation. Plant Cell 23(1):396–411. doi:10.1105/tpc.110.081356

Zhao X, Chai Y, Liu B (2007) Epigenetic inheritance and variation of DNA methylation level and pattern in maize intra-specific hybrids. Plant Sci 172 (5):930–938. doi:10.1016/j.plantsci.2007.01.002

Zhong L, Wang J-B (2007) The role of DNA hypermethylation in salt resistance of Triticum aestivum L. Wuhan Zhiwuxue Yanjiu 25(1):102–104

Zhou C, Zhang L, Duan J, Miki B, Wu K (2005) HISTONE DEACETYLASE19 is involved in jasmonic acid and ethylene signaling of pathogen response in Arabidopsis. Plant Cell 17(4):1196–1204

Zhu J, Jeong JC, Zhu Y, Sokolchik I, Miyazaki S, Zhu JK, Hasegawa PM, Bohnert HJ, Shi H, Yun DJ, Bressan RA (2008) Involvement of Arabidopsis HOS15 in histone deacetylation and cold tolerance. Proc Natl Acad Sci USA 105(12):4945–4950

Zilberman D, Gehring M, Tran RK, Ballinger T, Henikoff S (2007) Genome-wide analysis of Arabidopsis thaliana DNA methylation uncovers an interdependence between methylation and transcription. Nat Genet 39 (1):61–69

Printed by Printforce, the Netherlands